*Edited by Gustaaf Van Tendeloo,
Dirk Van Dyck, and Stephen J. Pennycook*

Handbook of Nanoscopy

Further Reading

Ohser, J. Schladitz, K.

3D Images of Materials Structures

Processing and Analysis

2009
Hardcover
ISBN: 978-3-527-31203-0

Codd, S. L., Seymour, J. D. (eds.)

Magnetic Resonance Microscopy

Spatially Resolved NMR Techniques and Applications

2009
Hardcover
ISBN: 978-3-527-32008-0

Maev, R. G.

Acoustic Microscopy

Fundamentals and Applications

2008
Hardcover
ISBN: 978-3-527-40744-6

Fukumura, H., Irie, M., Iwasawa, Y., Masuhara, H., Uosaki, K. (eds.)

Molecular Nano Dynamics

Vol. I: Spectroscopic Methods and Nanostructures/Vol. II: Active Surfaces, Single Crystals and Single Biocells

2009
Hardcover
ISBN: 978-3-527-32017-2

Roters, F., Eisenlohr, P. Bieler, T. R., Raabe, D.

Crystal Plasticity Finite Element Methods

in Materials Science and Engineering

2010
Hardcover
ISBN: 978-3-527-32447-7

Guo, J. (ed.)

X-Rays in Nanoscience

Spectroscopy, Spectromicroscopy, and Scattering Techniques

2010
Hardcover
ISBN: 978-3-527-32288-6

Tsukruk, V., Singamaneni, S.

Scanning Probe Interrogation of Soft Matter

2012
Hardcover
ISBN: 978-3-527-32743-0

Edited by Gustaaf Van Tendeloo, Dirk Van Dyck, and
Stephen J. Pennycook

Handbook of Nanoscopy

Volume 2

WILEY-VCH Verlag GmbH & Co. KGaA

The Editors

Prof. Gustaaf Van Tendeloo
Univ. of Antwerp (RUCA)
EMAT
Groenenborgerlaan 171
2020 Antwerpe
Belgium

Prof. Dirk Van Dyck
Univ. of Antwerp (RUCA)
EMAT
Groenenborgerlaan 171
2020 Antwerp
Belgium

Prof. Dr. Stephen J. Pennycook
Oak Ridge National Lab.
Condensed Matter Science Div.
Oak Ridge, TN 37831-6030
USA

All books published by **Wiley-VCH** are carefully produced. Nevertheless, authors, editors, and publisher do not warrant the information contained in these books, including this book, to be free of errors. Readers are advised to keep in mind that statements, data, illustrations, procedural details or other items may inadvertently be inaccurate.

Library of Congress Card No.: applied for

British Library Cataloguing-in-Publication Data
A catalogue record for this book is available from the British Library.

Bibliographic information published by the Deutsche Nationalbibliothek
The Deutsche Nationalbibliothek lists this publication in the Deutsche Nationalbibliografie; detailed bibliographic data are available on the Internet at <http://dnb.d-nb.de>.

© 2012 Wiley-VCH Verlag & Co. KGaA, Boschstr. 12, 69469 Weinheim, Germany

All rights reserved (including those of translation into other languages). No part of this book may be reproduced in any form – by photoprinting, microfilm, or any other means – nor transmitted or translated into a machine language without written permission from the publishers. Registered names, trademarks, etc. used in this book, even when not specifically marked as such, are not to be considered unprotected by law.

Cover Design Adam-Design, Weinheim
Typesetting Laserwords Private Limited, Chennai, India
Printing and Binding betz-druck GmbH, Darmstadt

Printed in the Federal Republic of Germany
Printed on acid-free paper

Print ISBN: 978-3-527-31706-6
ePDF ISBN: 978-3-527-64188-8
oBook ISBN: 978-3-527-64186-4
ePub ISBN: 978-3-527-64187-1
Mobi ISBN: 978-3-527-64189-5

Contents to Volume 1

Preface *XVII*
List of Contributors *XIX*

The Past, the Present, and the Future of Nanoscopy *1*
Gustaav Van Tendeloo and Dirk Van Dyck

Part I Methods *9*

1 Transmission Electron Microscopy *11*
   Marc De Graef

2 Atomic Resolution Electron Microscopy *45*
   Dirk Van Dyck

3 Ultrahigh-Resolution Transmission Electron Microscopy at Negative Spherical Aberration *81*
   Knut W. Urban, Juri Barthel, Lothar Houben, Chun-Lin Jia, Markus Lentzen, Andreas Thust, and Karsten Tillmann

4 Z-Contrast Imaging *109*
   Stephen J. Pennycook, Anrew R. Lupini, Albina Y. Borisevich, and Mark P. Oxley

5 Electron Holography *153*
   Hannes Lichte

6 Lorentz Microscopy and Electron Holography of Magnetic Materials *221*
   Rafal E. Dunin-Borkowski, Takeshi Kasama, Marco Beleggia, and Giulio Pozzi

7 Electron Tomography *253*
   Paul Anthony Midgley and Sara Bals

| 8 | **Statistical Parameter Estimation Theory – A Tool for Quantitative Electron Microscopy** 281
Sandra Van Aert |
|---|---|
| 9 | **Dynamic Transmission Electron Microscopy** 309
Nigel D. Browning, Geoffrey H. Campbell, James E. Evans, Thomas B. LaGrange, Katherine L. Jungjohann, Judy S. Kim, Daniel J. Masiel, and Bryan W. Reed |
| 10 | **Transmission Electron Microscopy as Nanolab** 345
Frans D. Tichelaar, Marijn A. van Huis, and Henny W. Zandbergen |
| 11 | **Atomic-Resolution Environmental Transmission Electron Microscopy** 375
Pratibha L. Gai and Edward D. Boyes |
| 12 | **Speckles in Images and Diffraction Patterns** 405
Michael M. J. Treacy |
| 13 | **Coherent Electron Diffractive Imaging** 437
J.M. Zuo and Weijie Huang |
| 14 | **Sample Preparation Techniques for Transmission Electron Microscopy** 473
Vasfi Burak Özdöl, Vesna Srot, and Peter A.van Aken |
| 15 | **Scanning Probe Microscopy – History, Background, and State of the Art** 499
Ralf Heiderhoff and Ludwig Josef Balk |
| 16 | **Scanning Probe Microscopy – Forces and Currents in the Nanoscale World** 539
Brian J. Rodriguez, Roger Proksch, Peter Maksymovych, and Sergei V. Kalinin |
| 17 | **Scanning Beam Methods** 615
David Joy |
| 18 | **Fundamentals of the Focused Ion Beam System** 645
Nan Yao |

Contents to Volume 2

Preface *XIX*
List of Contributors *XXI*

19 Low-Energy Electron Microscopy 673
Ernst Bauer
19.1 Introduction *673*
19.2 Theoretical Foundations *673*
19.3 Instrumentation *677*
19.4 Areas of Application *684*
19.4.1 Clean Surfaces *684*
19.4.2 Adsorption Layers *684*
19.4.3 Thin Films *689*
19.5 Discussion *693*
19.6 Concluding Remarks *694*
References *694*

20 Spin-Polarized Low-Energy Electron Microscopy 697
Ernst Bauer
20.1 Introduction *697*
20.2 Theoretical Foundations *697*
20.3 Instrumentation *700*
20.4 Areas of Application *701*
20.5 Discussion *705*
20.6 Concluding Remarks *706*
References *706*

21 Imaging Secondary Ion Mass Spectroscopy 709
Katie L. Moore, Markus Schröder, and Chris R. M. Grovenor
21.1 Fundamentals *709*
21.1.1 Physical Principles *710*
21.1.1.1 The Sputtering Process *710*
21.1.2 Technical Details *712*
21.1.2.1 Yield *712*
21.1.2.2 Resolution *714*
21.1.2.3 Sensitivity, Transmission, and Mass Resolution *715*
21.1.2.4 Steady State *716*
21.1.2.5 Quantification–Relative Sensitivity Factors *717*
21.1.2.6 Artifacts in SIMS *718*
21.1.2.7 Sample Charging *719*
21.2 SIMS Techniques *720*
21.2.1 Instrumentation *720*
21.2.1.1 Time of Flight SIMS *720*
21.2.1.2 Ion Microprobes *722*

21.2.1.3	Comparison of Techniques	725
21.2.2	Imaging and Data Analysis	727
21.3	Biological SIMS	731
21.3.1	Sample Preparation of Biological Materials for SIMS Analysis	735
21.3.1.1	Chemical Fixation	735
21.3.1.2	Cryofixation	736
21.4	Conclusions	740
	References	740

22	**Soft X-Ray Imaging and Spectromicroscopy**	**745**
	Adam P. Hitchcock	
22.1	Introduction	745
22.2	Experimental Techniques	749
22.2.1	Full-Field Transmission X-Ray Microscopy (TXM)	749
22.2.2	Full-Field Photoelectron Microscopy (X-PEEM)	751
22.2.3	Scanning Transmission X-Ray Microscopy (STXM)	752
22.2.4	Scanning Yield Techniques (SPEM, LEXRF-STXM, STM-SXM)	757
22.2.5	Sample Preparation Issues	762
22.2.6	Radiation Damage	764
22.3	Data Analysis Methods	764
22.3.1	Chemical Mapping – Fitting to Known Reference Spectra	764
22.3.2	Chemical Mapping – Unsupervised Statistical Analysis Methods	769
22.4	Selected Applications	771
22.4.1	Polymer Microstructure	771
22.4.2	Surfaces and Interfaces	773
22.4.3	3D Imaging (Tomography) and 3D Chemical Mapping (Spectrotomography)	774
22.4.4	*In situ* Techniques	777
22.5	Future Outlook and Summary	780
	Acknowledgments	782
	References	783

23	**Atom Probe Tomography: Principle and Applications**	**793**
	Frederic Danoix and François Vurpillot	
23.1	Introduction	793
23.2	Basic Principles	794
23.2.1	Field Production	794
23.2.2	Field Evaporation and Pulsed Field Evaporation	796
23.2.3	Image Formation	799
23.2.4	Pulsed Field Evaporation	802
23.2.4.1	Field Pulsing	803
23.2.4.2	Thermal Pulsing	804
23.3	Field Ion Microscopy	806
23.3.1	Introduction	806
23.3.2	Image Interpretation	807

23.3.3	Field Ion Microscopy in Materials Science *808*	
23.3.3.1	Point Defects *808*	
23.3.3.2	Linear Defects *809*	
23.3.3.3	Planar Defects *810*	
23.3.3.4	Phase Contrast *810*	
23.3.3.5	Three-Dimensional Field Ion Microscopy *811*	
23.4	Atom Probe Tomography *813*	
23.4.1	State of the Art *813*	
23.4.2	APT in Materials Science *815*	
23.4.2.1	Specimen Preparation *815*	
23.4.2.2	Phase Composition Measurements *816*	
23.4.2.3	Segregation *817*	
23.4.2.4	Early Stages of Phase Transformation *820*	
23.5	Conclusion *828*	
	References *828*	

24 Signal and Noise Maximum Likelihood Estimation in MRI *833*
Jan Sijbers

24.1	Probability Density Functions in MRI *833*	
24.1.1	Gaussian PDF *833*	
24.1.1.1	Moments of the Gaussian PDF *834*	
24.1.2	Rician PDF *834*	
24.1.2.1	Moments of the Rician PDF *835*	
24.1.3	Generalized Rician PDF *835*	
24.1.3.1	Moments of the Generalized Rician PDF *836*	
24.1.4	PDF of Phase Data *836*	
24.2	Signal Amplitude Estimation *837*	
24.2.1	Introduction *837*	
24.2.2	Signal Amplitude Estimation from Complex Data *838*	
24.2.2.1	Region of Constant Amplitude and Phase *838*	
24.2.2.2	Region of Constant Amplitude and Different Phases *839*	
24.2.3	Signal Amplitude Estimation from Magnitude Data *840*	
24.2.3.1	Region of Constant Amplitude and Known Noise Variance *840*	
24.2.3.2	Region of Constant Amplitude and Unknown Noise Variance *842*	
24.2.4	Discussion *843*	
24.3	Noise Variance Estimation *844*	
24.3.1	Introduction *844*	
24.3.2	Noise Variance Estimation from Complex Data *844*	
24.3.2.1	Region of Constant Amplitude and Phase *844*	
24.3.2.2	Region of Constant Amplitude and Different Phases *845*	
24.3.2.3	Background Region *845*	
24.3.3	Noise Variance Estimation from Magnitude Data *846*	
24.3.3.1	Background Region *846*	
24.3.3.2	Nonbackground Region *848*	
24.3.3.3	Double Acquisition Method *849*	

24.3.4	Discussion 850
24.4	Conclusions 850
	References 851

25 3-D Surface Reconstruction from Stereo Scanning Electron Microscopy Images 855
Shafik Huq, Andreas Koschan, and Mongi Abidi

25.1	Introduction 855
25.1.1	Geometric Calibration of SEM/LC-SEM 857
25.1.2	LC-SEM Stereo Image Acquisition 860
25.2	Matching Stereo Images 862
25.2.1	Matching with Energy-Minimizing Grid 862
25.2.1.1	Cost Function with Mean Field Annealing 863
25.2.1.2	Annealing with Inverse Temperature 864
25.2.1.3	Optimization Algorithm 865
25.2.1.4	Matching Results 865
25.2.2	Matching with Belief Propagation 866
25.2.3	Matching with Local Support Window 869
25.2.4	Matching with Graph Cuts 871
25.3	Conclusions 874
	Acknowledgments 875
	References 875
	Further Reading 876

Part II Applications 877

26 Nanoparticles 879
Miguel López-Haro, Juan José Delgado, Juan Carlos Hernández-Garrido, Juan de Dios López-Castro, César Mira, Susana Trasobares, Ana Belén Hungría, José Antonio Pérez-Omil, and José Juan Calvino

26.1	Introduction 879
26.2	Imaging Nanoparticles 883
26.2.1	Nanoparticle Size 884
26.2.2	Nanoparticle Bulk Structure 887
26.2.3	Nanoparticle Shape and Surface Structure 900
26.2.4	Nanoparticle Lattice Distortions and Interface Structure 911
26.3	Electron Tomography of Nanoparticles 920
26.4	Nanoanalytical Characterization of Nanoparticles 925
26.4.1	Optical Information on Individual Nanoparticles 929
26.4.2	Chemical Identification and Quantification of Individual Nanoparticles 930
26.4.3	Monitoring the Chemical Bonding and Oxidation State; Analysis of the EELS Fine Structure 936
26.4.4	Chemical Imaging at the Atomic Scale 939
26.5	*In situ* TEM Characterization of Nanoparticles 941

26.5.1	Environmental Cells for *In situ* Investigations of Gas–Solid Interaction *941*	
26.5.2	Synthesis of Nanostructured Materials *944*	
26.5.3	Nanoparticle Mobility *947*	
26.5.4	Nanoparticles under Working Conditions *948*	
	References *949*	

27 Nanowires and Nanotubes *961*
Yong Ding and Zhong Lin Wang

27.1	Introduction *961*
27.2	Structures of Nanowires and Nanotubes *961*
27.2.1	Determination of the Growth Directions of Nanowires *962*
27.2.2	The Three-Dimensional Structure of Nanowires *968*
27.2.3	Surface Reconstruction of Nanowires *969*
27.2.4	Chiral Indices of Carbon Nanotubes *971*
27.3	Defects in Nanowires *974*
27.3.1	Point Defects *974*
27.3.2	Dislocations *975*
27.3.3	Planar Defects: Twins *976*
27.3.4	Planar Defects: Stacking Faults *977*
27.3.5	Planar Defects: Inversion Domain Walls *981*
27.4	*In situ* Observation of the Growth Process of Nanowires and Nanotubes *983*
27.4.1	Si Nanowires Catalyzed by Au and Pd Nanoparticles *983*
27.4.2	*In situ* Observation of the Growth of Carbon Nanotubes *984*
27.5	*In situ* Mechanical Properties of Nanotubes and Nanowires *985*
27.5.1	Young's Modulus Measured by Quantifying Thermal Vibration Amplitude *985*
27.5.2	Bending Modulus by Electric-Field-Induced Mechanical Resonance in TEM *986*
27.6	*In situ* Electric Transport Property of Carbon Nanotubes *988*
27.7	*In situ* TEM Investigation of Electrochemical Properties of Nanowires *991*
27.8	Summary *991*
	References *991*

28 Carbon Nanoforms *995*
Carla Bittencourt and Gustaaf Van Tendeloo

28.1	Imaging Carbon Nanoforms Using Conventional Electron Microscopy *998*
28.1.1	Carbon Nanotubes *998*
28.1.2	Crop Circles *1005*
28.1.3	Carbon Beads *1007*
28.1.4	Carbon Nanoscrolls *1008*
28.1.5	Herringbone Carbon Nanofibers *1009*

28.1.6	Carbon Nanocones *1009*
28.1.7	Carbon Onionlike Particles *1010*
28.1.8	Carbon Spiroid *1011*
28.1.9	Nanodiamonds *1012*
28.1.10	Carbon Nanospheres *1013*
28.1.11	Fullerides *1014*
28.1.12	Carbon Nanoform Spatial Distribution in a Cell *1015*
28.1.13	*In situ* Evaluation of Tensile Loading of Carbon Nanotubes *1015*
28.1.14	Observation of Field Emission Sites in MWCNTs *1016*
28.2	Analysis of Carbon Nanoforms Using Aberration-Corrected Electron Microscopes *1016*
28.2.1	Nanodiamond *1018*
28.2.2	Carbon Peapods *1020*
28.2.3	Determination of the Chiral Angle of SWCNTs *1020*
28.2.4	Deformation of sp^2-Carbon Nanoforms during TEM Observation *1021*
28.2.5	Visualization of Chemical Reactions *1023*
28.2.6	Imaging at 30 and 60 kV *1025*
28.2.7	Graphene *1026*
28.2.8	Defects in Graphene *1028*
28.2.9	Graphene Edges *1028*
28.2.10	Electronic and Bonding Structure of Edge Atoms *1030*
28.2.11	Elemental Analysis of the Graphene Edges *1030*
28.2.12	Graphene Grain Structure *1032*
28.2.13	Engineering of Carbon Nanoforms: *In situ* TEM *1034*
28.2.13.1	Graphene *1034*
28.2.13.2	Fullerene *1035*
28.2.13.3	Standing Carbon Chain *1035*
28.3	Ultrafast Electron Microscopy *1037*
28.3.1	Four-Dimensional Ultrafast Electron Microscopy (4D-UEM) *1037*
28.3.2	Photon-Induced Near-Field Electron Microscopy (PINEM) *1039*
28.4	Scanning Tunneling Microscopy (STM) *1039*
28.5	Scanning Photocurrent Microscopy (SPCM) *1042*
28.6	X-Ray Electrostatic Force Microscopy (X-EFM) *1046*
28.7	Atomic Force Microscopy *1046*
28.7.1	Conductance Atomic Force Microscopy *1046*
28.7.2	Chemical Force Microscopy *1047*
28.7.3	Magnetic Force Microscopy *1047*
28.8	Scanning Near-Field Optical Microscope *1049*
28.9	Tip-Enhanced Raman and Confocal Microscopy *1049*
28.10	Tip-Enhanced Photoluminescence Microscopy *1052*
28.11	Fluorescence Quenching Microscopy *1053*
28.12	Fluorescence Microscopy *1054*
28.13	Single-Shot Extreme Ultraviolet Laser Imaging *1056*
28.14	Nanoscale Soft X-Ray Imaging *1056*

28.14.1	Near-Edge Absorption X-Ray Spectromicroscopy	*1056*
28.15	Scanning Photoelectron Microscopy	*1059*
	Acknowledgments	*1060*
	References	*1060*

29 Metals and Alloys *1071*
Dominique Schryvers

29.1	Formation of Nanoscale Deformation Twins by Shockley Partial Dislocation Passage	*1071*
29.2	Minimal Strain at Austenite–Martensite Interface in Ti-Ni-Pd	*1073*
29.3	Atomic Structure of Ni_4Ti_3 Precipitates in Ni-Ti	*1077*
29.4	Ni-Ti Matrix Deformation and Concentration Gradients in the Vicinity of Ni_4Ti_3 Precipitates	*1081*
29.5	Elastic Constant Measurements of Ni_4Ti_3 Precipitates	*1084*
29.6	New APB-Like Defect in Ti-Pd Martensite Determined by HRSTEM	*1085*
29.7	Strain Effects in Metallic Nanobeams	*1085*
29.8	Adiabatic Shear Bands in Ti6Al4V	*1088*
29.9	Electron Tomography	*1089*
29.10	The Ultimate Resolution	*1091*
	Acknowledgments	*1094*
	References	*1094*

30 *In situ* Transmission Electron Microscopy on Metals *1099*
J.Th.M. De Hosson

30.1	Introduction	*1099*
30.2	*In situ* TEM Experiments	*1100*
30.2.1	Stage Design	*1101*
30.2.2	Specimen Geometry and Specimen Preparation	*1102*
30.2.3	Load Rate Control versus Displacement Rate Control	*1104*
30.3	Grain Boundary Dislocation Dynamics Metals	*1110*
30.4	*In situ* TEM Tensile Experiments	*1117*
30.5	*In situ* TEM Compression Experiments	*1125*
30.6	Conclusions	*1146*
	Acknowledgments	*1147*
	References	*1148*

31 Semiconductors and Semiconducting Devices *1153*
Hugo Bender

31.1	Introduction	*1153*
31.2	Nanoscopic Applications on Silicon-Based Semiconductor Devices	*1156*
31.2.1	Metrology	*1156*
31.2.2	Epitaxy	*1156*
31.2.3	Strain	*1158*

31.2.4	Doping 1161
31.2.5	Gate Structures 1163
31.2.6	Silicides 1164
31.2.7	Metalization 1165
31.2.8	Three-Dimensional Structures 1168
31.2.9	Bonding and Packaging 1170
31.2.10	Failure Analysis and Debugging 1171
31.2.11	New Materials 1172
31.3	Conclusions 1172
	Acknowledgments 1172
	References 1173

32 Complex Oxide Materials   1179
Maria Varela, Timothy J. Pennycook, Jaume Gazquez, Albina Y. Borisevich, Sokrates T. Pantelides, and Stephen J. Pennycook

32.1	Introduction 1179
32.2	Aberration-Corrected Spectrum Imaging in the STEM 1180
32.3	Imaging of Oxygen Lattice Distortions in Perovskites and Oxide Thin Films and Interfaces 1183
32.4	Atomic-Resolution Effects in the Fine Structure – Further Insights into Oxide Interface Properties 1189
32.5	Applications of Ionic Conductors: Studies of Colossal Ionic Conductivity in Oxide Superlattices 1193
32.6	Applications of Cobaltites: Spin-State Mapping with Atomic Resolution 1198
32.7	Summary 1204
	Acknowledgments 1204
	References 1205

33 Application of Transmission Electron Microscopy in the Research of Inorganic Photovoltaic Materials   1213
Yanfa Yan

33.1	Introduction 1213
33.2	Experimental 1215
33.3	Atomic Structure and Electronic Properties of c-Si/a-Si:H Heterointerfaces 1215
33.4	Interfaces and Defects in CdTe Solar Cells 1220
33.5	Influences of Oxygen on Interdiffusion at CdS/CdTe Heterojunctions 1228
33.6	Microstructure Evolution of $Cu(In,Ga)Se_2$ Films from Cu Rich to In Rich 1231
33.7	Microstructure of Surface Layers in $Cu(In,Ga)Se_2$ Thin Films 1235
33.8	Chemical Fluctuation-Induced Nanodomains in $Cu(In,Ga)Se_2$ Films 1238
33.9	Conclusions and Future Directions 1242

Acknowledgment *1244*
References *1244*

34 Polymers *1247*
Joachim Loos
34.1 Foreword *1247*
34.2 A Brief Introduction on Printable Solar Cells *1248*
34.3 Morphology Requirements of Photoactive Layers in PSCs *1249*
34.4 Our Characterization Toolbox *1250*
34.5 How It All Started: First Morphology Studies *1251*
34.6 Contrast Creation in Purely Carbon-Based BHJ Photoactive Layers *1253*
34.6.1 Scanning Transmission Electron Microscopy *1253*
34.6.2 Energy-Filtered Transmission Electron Microscopy(EFTEM) *1255*
34.6.3 Conductive Atomic Force Microscopy(C-AFM) *1257*
34.7 Nanoscale Volume Information: Electron Tomography of PSCs *1259*
34.8 One Example of Electron Tomographic Investigation: P3HT/PCBM *1263*
34.9 Quantification of Volume Data *1266*
34.10 Outlook and Concluding Remarks *1268*
Acknowledgment *1269*
References *1269*

35 Ferroic and Multiferroic Materials *1273*
Ekhard Salje
35.1 Multiferroicity *1273*
35.2 Ferroic Domain Patterns and Their Microscopical Observation *1276*
35.3 The Internal Structure of Domain Walls *1280*
35.4 Domain Structures Related to Amorphization *1287*
35.5 Dynamical Properties of Domain Boundaries *1290*
35.6 Conclusion *1295*
References *1296*

36 Three-Dimensional Imaging of Biomaterials with Electron Tomography *1303*
Montserrat Bárcena, Roman I. Koning, and Abraham J. Koster
36.1 Introduction *1303*
36.1.1 Electron Microscopy of Biological Samples *1303*
36.1.2 Historical Remarks on Biological Electron Microscopy *1304*
36.2 Biological Tomographic Techniques *1306*
36.2.1 Basic Principle *1306*
36.2.2 Basic Practice *1307*
36.2.3 Sample Preparation *1309*
36.2.4 Thickness and Instrumentation *1313*

36.2.5	Thickness, Sampling, and Resolution	*1314*
36.2.6	Radiation Damage and Electron Dose	*1316*
36.2.7	Contrast Formation and Imaging Conditions	*1318*
36.3	Examples of Electron Tomography Biomaterials	*1318*
36.3.1	Whole Cells *1319*	
36.3.2	Interior of the Cell *1320*	
36.3.2.1	Viral Infection *1320*	
36.3.2.2	Cellular Morphology *1321*	
36.3.3	Viruses and Macromolecular Complexes *1323*	
36.3.3.1	Virus Structure *1323*	
36.3.3.2	Subtomogram Averaging *1324*	
36.4	Outlook *1325*	
36.4.1	Toward Higher Resolution *1326*	
36.4.2	Toward Larger Volumes *1327*	
	References *1327*	

37 Small Organic Molecules and Higher Homologs *1335*
Ute Kolb and Tatiana E. Gorelik

37.1	Introduction *1335*	
37.2	Optical Microscopy *1339*	
37.3	Scanning Electron Microscopy–SEM *1342*	
37.4	Atomic Force and Scanning Tunneling Microscopy (AFM and STM) *1343*	
37.5	Transmission Electron Microscopy (TEM) *1343*	
37.5.1	TEM: the Instrument *1344*	
37.5.2	TEM Sample Preparation *1346*	
37.5.2.1	Direct Deposition *1346*	
37.5.2.2	Vitrification *1348*	
37.5.2.3	Staining *1349*	
37.5.2.4	Surface Topology: Freeze Fracture and Etching *1350*	
37.5.3	Beam Sensitivity *1351*	
37.5.4	Morphological Questions *1354*	
37.5.4.1	Vesicles and Micelles *1354*	
37.5.4.2	Fibers *1357*	
37.5.4.3	Electron Tomography *1357*	
37.5.5	Structural Questions *1358*	
37.5.5.1	Polymer Chain Packing *1358*	
37.5.5.2	High-Resolution TEM *1360*	
37.5.5.3	Electron Diffraction *1361*	
37.5.6	Crystallographic Questions *1362*	
37.5.6.1	Lattice Cell Parameters Determination *1362*	
37.5.6.2	Structural Analysis Based on Electron Diffraction Data *1365*	
37.5.6.3	Combination of Electron Diffraction with Different Methods to Facilitate Structure Analysis *1365*	

37.5.6.4	Off-Zone Data Collection: Automated Diffraction Tomography (ADT) *1369*
37.6	Summary *1372*
	References *1373*

Index *1381*

Preface

Since the edition of the previous "Handbook of Microscopy" in 1997 the world of microscopy has gone through a significant transition.

In electron microscopy the introduction of aberration correctors has pushed the resolution down to the sub-Angstrom regime, detectors are able to detect single electrons, spectrometers are able to record spectra from single atoms. Moreover the object space is increased which allows to integrate these techniques in the same instrument under full computer support without compromising on the performance. Thus apart from the increased resolution from microscopy to nanoscopy, even towards picoscopy, the EM is gradually transforming from an imaging device into a true nanoscale laboratory that delivers reliable quantitative data on the nanoscale close to the physical and technical limits. In parallel, scanning probe methods have undergone a similar evolution towards increased functionality, flexibility and integration.

As a consequence the whole field of microscopy is gradually shifting from the instrument to the application, from describing to measuring and to understanding the structure/property relations, from nanoscopy to nanology.

But these instruments will need a different generation of nanoscopists who need not only to master the increased flexibility and multifunctionality of the instruments, but to choose and combine the experimental possibilities to fit the material problem to be investigated.

It is the purpose of this new edition of the "Handbook of Nanoscopy" to provide an ideal reference base of knowledge for the future user.

Volume 1 elaborates on the basic principles underlying the different nanoscopical methods with a critical analysis of the merits, drawbacks and future prospects. Volume 2 focuses on a broad category of materials from the viewpoint of how the different nanoscopical measurements can contribute to solving materials structures and problems.

The handbook is written in a very readable style at a level of a general audience. Whenever relevant for deepening the knowledge, proper references are given.

Gustaaf Van Tendeloo, Dirk Van Dyck, and Stephen J. Pennycook

List of Contributors

Mongi Abidi
The University of Tennessee
Min H. Kao
Department of Electrical
Engineering and Computer
Science
Imaging Robotics and Intelligent
Systems (IRIS) Lab
209 Ferris Hall
Knoxville
TN 37996-2100
USA

Ludwig Josef Balk
Bergische Universität Wuppertal
Fachbereich Elektronik
Informationstechnik
Medientechnik
Lehrstuhl für Elektronische
Bauelemente
Rainer-Gruenter-Str. 21
42119 Wuppertal
Germany

Sara Bals
University of Antwerp
Department of Physics
EMAT
Groenenborgerlaan 171
2020 Antwerp
Belgium

Montserrat Bárcena
Leiden University Medical Center
Department of Molecular Cell
Biology
Section Electron Microscopy
Einthovenweg 20
2333 ZC
The Netherlands

Juri Barthel
Forschungszentrum Jülich
GmbH
Peter Grünberg Institute and
Ernst Ruska Centre for
Microscopy and Spectroscopy
with Electrons
D-52425 Jülich
Germany

Ernst Bauer
Arizona State University
Department of Physics
Tempe
AZ 85287-1504
USA

Marco Beleggia
Technical University of Denmark
Center for Electron Nanoscopy
DK-2800 Kongens Lyngby
Denmark

Hugo Bender
Imec
Kapeldreef 75
Leuven 3001
Belgium

Carla Bittencourt
University of Antwerp
EMAT
Groenenborgerlaan 171
B-2020 Antwerp
Belgium

Albina Y. Borisevich
Oak Ridge National Laboratory
Materials Science and Technology
Division
Oak Ridge
TN 37831-6071
USA

Edward D. Boyes
The University of York
The York JEOL Nanocentre
Departments of Physics
Helix House
Heslington
York, YO10 5BR
UK

and

The University of York
The York JEOL Nanocentre
Department of Electronics
Helix House
Heslington
York, YO10 5BR
UK

Nigel D. Browning
Lawrence Livermore National
Laboratory
Condensed Matter and Materials
Division
Physical and Life Sciences
Directorate
7000 East Avenue
Livermore
CA 94550
USA

and

University of California-Davis
Department of Chemical
Engineering and Materials
Science
One Shields Ave
Davis, CA 95616
USA

and

University of California-Davis
Department of Molecular and
Cellular Biology
One Shields Ave
Davis, CA 95616
USA

and

Pacific Northwest National
Laboratory
902 Battelle Boulevard
Richland
WA 99352
USA

José Juan Calvino
Facultad de Ciencias de la
Universidad de Cádiz
Departamento de Ciencia de los
Materiales e Ingeniería
Metalúrgica y Química
Inorgánica
Campus Rio San Pedro
Puerto Real
11510-Cádiz
Spain

Geoffrey H. Campbell
Lawrence Livermore National
Laboratory
Condensed Matter and Materials
Division
Physical and Life Sciences
Directorate
7000 East Avenue
Livermore
CA 94550
USA

Frederic Danoix
Université de Rouen
Groupe de Physique des
Matériaux, UMR CNRS 6634
Site universitaire du Madrillet
Saint Etienne du Rouvray
76801
France

Juan José Delgado
Facultad de Ciencias de la
Universidad de Cádiz
Departamento de Ciencia de los
Materiales e Ingeniería
Metalúrgica y Química
Inorgánica
Campus Rio San Pedro
Puerto Real
11510-Cádiz
Spain

Marc De Graef
Carnegie Mellon University
Materials Science and
Engineering Department
5000 Forbes Avenue
Pittsburgh
PA 15213-3890
USA

J.Th.M. De Hosson
University of Groningen
Department of Applied Physics
Zernike Institute for Advanced
Materials and Materials
Innovation Institute
Nijenborgh 4
9747 AG Groningen
The Netherlands

Yong Ding
School of Materials Science and
Engineering
Georgia Institute of Technology
Atlanta
GA 30332-0245
USA

Rafal E. Dunin-Borkowski
Forschungszentrum Jülich
GmbH
Peter Grünberg Institute and
Ernst Ruska Centre for
Microscopy and Spectroscopy
with Electrons
D-52425 Jülich
Germany

and

Technical University of Denmark
Center for Electron Nanoscopy
DK-2800 Kongens
Lyngby
Denmark

James E. Evans
Lawrence Livermore National Laboratory
Condensed Matter and Materials Division
Physical and Life Sciences Directorate
7000 East Avenue
Livermore
CA 94550
USA

and

University of California-Davis
Department of Molecular and Cellular Biology
One Shields Ave
Davis
CA 95616
USA

and

Pacific Northwest National Laboratory
902 Battelle Boulevard
Richland
WA 99352
USA

Pratibha L. Gai
The University of York
The York JEOL Nanocentre
Department of Chemistry
Helix House
Heslington
York, YO10 5BR
UK

and

The University of York
The York JEOL Nanocentre
Department of Physics
Helix House
Heslington
York, YO10 5BR
UK

Jaume Gazquez
Oak Ridge National Laboratory
Materials Science and Technology Division
Oak Ridge
TN 37831-6071
USA

and

Universidad Complutense de Madrid
Departamento de Fisica Aplicada III Avda.
Complutense s/n
28040 Madrid
Spain

and

Instituto de Ciencia de Materiales de Barcelona-CSIC
Campus de la UAB
08193 Bellaterra
Spain

Tatiana E. Gorelik
Johannes Gutenberg-Universität
Mainz
Institut für Physikalische Chemie
Welderweg 11
55099 Mainz
Germany

Chris R. M. Grovenor
University of Oxford
Department of Materials
Parks Road
Oxford OX1 3PH
UK

Ralf Heiderhoff
Bergische Universität Wuppertal
Fachbereich Elektronik
Informationstechnik
Medientechnik
Lehrstuhl für Elektronische
Bauelemente
Rainer-Gruenter-Str. 21
42119 Wuppertal
Germany

Juan Carlos Hernández-Garrido
Facultad de Ciencias de la
Universidad de Cádiz
Departamento de Ciencia de los
Materiales e Ingeniería
Metalúrgica y Química
Inorgánica
Campus Rio San Pedro
Puerto Real
11510-Cádiz
Spain

Adam P. Hitchcock
McMaster University
Department of Chemistry and
Chemical Biology
Brockhouse Institute for
Materials Research
1280 Main Street West
Hamilton
ON L8S 4M1
Canada

Lothar Houben
Forschungszentrum Jülich
GmbH
Peter Grünberg Institute and
Ernst Ruska Centre for
Microscopy and Spectroscopy
with Electrons
D-52425 Jülich
Germany

Weijie Huang
University of Illinois at
Urbana-Champaign
Department of Materials Science
and Engineering and Materials
Research Laboratory
Urbana
IL 61801
USA

and

Carl Zeiss SMT Inc.
One Corporation Way
Peabody
MA 01960
USA

Ana Belén Hungría
Facultad de Ciencias de la
Universidad de Cádiz
Departamento de Ciencia de los
Materiales e Ingeniería
Metalúrgica y Química
Inorgánica
Campus Rio San Pedro
Puerto Real
11510-Cádiz
Spain

Shafik Huq
The University of Tennessee
Min H. Kao
Department of Electrical
Engineering and Computer
Science
Imaging, Robotics, and
Intelligent Systems (IRIS) Lab
209 Ferris Hall
Knoxville
TN 37996-2100
USA

Chun-Lin Jia
Forschungszentrum Jülich
GmbH
Peter Grünberg Institute and
Ernst Ruska Centre for
Microscopy and Spectroscopy
with Electrons
D-52425 Jülich
Germany

and

Xi'an Jiaotong University
International Centre for
Dielectrics Research (ICDR)
School of Electronic and
Information Engineering
28 Xianning West Road
Xi'an 710049
China

David Joy
University of Tennessee
Science and Engineering
Research Facility
Knoxville
TN 37996-2200
USA

and

Center for Nano Material Science
Oak Ridge National Laboratory
Oak Ridge
TN 37831
USA

Katherine L. Jungjohann
University of California-Davis
Department of Chemical
Engineering and Materials
Science
One Shields Ave
Davis, CA 95616
USA

Sergei V. Kalinin
Oak Ridge National Laboratory
Oak Ridge
TN 37922
USA

Takeshi Kasama
Technical University of Denmark
Center for Electron Nanoscopy
DK-2800 Kongens Lyngby
Denmark

Judy S. Kim
University of California-Davis
Department of Chemical
Engineering and Materials
Science
One Shields Ave
Davis
CA 95616
USA

Ute Kolb
Johannes Gutenberg-Universität
Mainz
Institut für Physikalische Chemie
Welderweg 11
55099 Mainz
Germany

Roman I. Koning
Leiden University Medical Center
Department of Molecular Cell
Biology
Section Electron Microscopy
Einthovenweg 20
2333 ZC
Leiden
The Netherlands

Andreas Koschan
The University of Tennessee
Min H. Kao
Department of Electrical
Engineering and Computer
Science
Imaging Robotics and Intelligent
Systems (IRIS) Lab
330 Ferris Hall
Knoxville
TN 37996-2100
USA

Abraham J. Koster
Leiden University Medical Center
Department of Molecular Cell
Biology
Section Electron Microscopy
Einthovenweg 20
2333 ZC
Leiden
The Netherlands

Thomas B. LaGrange
Lawrence Livermore National
Laboratory
Condensed Matter and Materials
Division
Physical and Life Sciences
Directorate
7000 East Avenue
Livermore
CA 94550
USA

Markus Lentzen
Forschungszentrum Jülich
GmbH
Peter Grünberg Institute and
Ernst Ruska Centre for
Microscopy and Spectroscopy
with Electrons
D-52425 Jülich
Germany

Hannes Lichte
Technische Universität Dresden
Triebenberg Laboratory
Institute for Structure Physics
01062 Dresden
Germany

Joachim Loos
University of Glasgow
School of Physics and Astronomy
Kelvin Building
Glasgow G12 8QQ
Scotland
UK

Juan de Dios López-Castro
Facultad de Ciencias de la
Universidad de Cádiz
Departamento de Ciencia de los
Materiales e Ingeniería
Metalúrgica y Química
Inorgánica
Campus Rio San Pedro
Puerto Real
11510-Cádiz
Spain

Andrew R. Lupini
Oak Ridge National Laboratory
Materials Science and Technology
Division
Oak Ridge
TN 37831-6071
USA

Peter Maksymovych
Oak Ridge National Laboratory
Oak Ridge
TN 37922
USA

Daniel J. Masiel
University of California-Davis
Department of Chemical
Engineering and Materials
Science
One Shields Ave
Davis
CA 95616
USA

Paul Anthony Midgley
University of Cambridge
Department of Material Science
and Metallurgy
Pembroke Street
Cambridge CB2 3QZ
UK

César Mira
Facultad de Ciencias de la
Universidad de Cádiz
Departamento de Ciencia de los
Materiales e Ingeniería
Metalúrgica y Química
Inorgánica
Campus Rio San Pedro
Puerto Real
11510-Cádiz
Spain

Katie L. Moore
University of Oxford
Department of Materials
Parks Road
Oxford OX1 3PH
UK

Mark P. Oxley
Oak Ridge National Laboratory
Materials Science and Technology
Division
Oak Ridge
TN 37831-6071
USA

and

Vanderbilt University
Department of Physics and
Astronomy
Nashville
TN 37235
USA

Vasfi Burak Özdöl
Stuttgart Center for Electron
Microscopy
Max Planck Institute for
Intelligent Systems
Heisenbergstr. 3
70569 Stuttgart
Germany

Sokrates T. Pantelides
Oak Ridge National Laboratory
Materials Science & Technology
Divsion
1 Bethel
Valley Road
Oak Ridge
TN 37831-6071
USA

and

Vanderbilt University
Department of Physics and
Astronomy
Nashville
TN 37235
USA

Stephen J. Pennycook
Oak Ridge National Laboratory
Materials Science and Technology
Division
Oak Ridge
TN 37831-6071
USA

and

Vanderbilt University
Department of Physics and
Astronomy
Nashville
TN 37235
USA

Timothy J. Pennycook
Oak Ridge National Laboratory
Materials Science & Technology
Divsion
1 Bethel Valley Road
Oak Ridge
TN 37831-6071
USA

and

Vanderbilt University
Department of Physics and
Astronomy
Nashville
TN 37235
USA

José Antonio Pérez-Omil
Facultad de Ciencias de la
Universidad de Cádiz
Departamento de Ciencia de los
Materiales e Ingeniería
Metalúrgica y Química
Inorgánica
Campus Rio San Pedro
Puerto Real
11510-Cádiz
Spain

Giulio Pozzi
Universita' di Bologna
Dipartimento di Fisica
V. le B. Pichat 6/2
40127 Bologna
Italy

Roger Proksch
Asylum Research
Santa Barbara
CA 93117
USA

Bryan W. Reed
Lawrence Livermore National Laboratory
Condensed Matter and Materials Division
Physical and Life Sciences Directorate
7000 East Avenue
Livermore
CA 94550
USA

Brian J. Rodriguez
University College Dublin
Conway Institute of Biomolecular and Biomedical Research and School of Physics
Belfield
Dublin 4
Ireland

Ekhard Salje
Cambrige University
Department Earth Science
Downing Street
Cambridge CB2 3EQ
UK

Markus Schröder
University of Oxford
Department of Materials
Parks Road
Oxford
OX1 3PH
UK

Dominique Schryvers
University of Antwerp
EMAT
Groenenborgerlaan 171
B-2020 Antwerp
Belgium

Jan Sijbers
University of Antwerp (CDE)
Vision Lab
Department of Physics
Universiteitsplein 1 (N.1.13)
B-2610 Wilrijk
Belgium

Vesna Srot
Stuttgart Center for Electron Microscopy
Max Planck Institute for Metals Research
Heisenbergstr. 3
70569 Stuttgart
Germany

Andreas Thust
Forschungszentrum Jülich GmbH
Peter Grünberg Institute and Ernst Ruska Centre for Microscopy and Spectroscopy with Electrons
D-52425 Jülich
Germany

Frans D. Tichelaar
Delft University of Technology
Applied Sciences
Kavli Institute of Nanoscience
Lorentzweg 1
NL-2628CJ Delft
The Netherlands

Karsten Tillmann
Forschungszentrum Jülich GmbH
Peter Grünberg Institute and Ernst Ruska Centre for Microscopy and Spectroscopy with Electrons
D-52425 Jülich
Germany

Susana Trasobares
Facultad de Ciencias de la
Universidad de Cádiz
Departamento de Ciencia de los
Materiales e Ingeniería
Metalúrgica y Química
Inorgánica
Campus Rio San Pedro
Puerto Real
11510-Cádiz
Spain

Michael M. J. Treacy
Arizona State University
Department of Physics
Bateman Building
B-147
Tyler Mall
Tempe
AZ 85287-1504
USA

Knut W. Urban
Forschungszentrum Jülich
GmbH
Peter Grünberg Institute and
Ernst Ruska Centre for
Microscopy and Spectroscopy
with Electrons
D-52425 Jülich
Germany

Sandra Van Aert
University of Antwerp
Electron Microscopy for Materials
Research (EMAT)
Groenenborgerlaan 171
2020 Antwerp
Belgium

Peter A. Van Aken
Stuttgart Center for Electron
Microscopy
Max Planck Institute for
Intelligent Systems
Heisenbergstr. 3
70569 Stuttgart
Germany

Dirk Van Dyck
University of Antwerp (UA)
EMAT
Groenenborgerlaan 171
2020 Antwerpen
Belgium

Marijn A. van Huis
Delft University of Technology
Applied Sciences
Kavli Institute of Nanoscience
Lorentzweg 1
NL-2628CJ Delft
The Netherlands

Gustaaf Van Tendeloo
University of Antwerp
EMAT
Groenenborgerlaan 171
B-2020 Antwerp
Belgium

Maria Varela
Materials Science & Technology
Division
Oak Ridge National Laboratory
1 Bethel Valley Road
Oak Ridge
TN 37831-6071
USA

François Vurpillot
Université de Rouen
Groupe de Physique des
Matériaux
UMR CNRS 6634
Site universitaire du Madrillet
Saint Etienne du Rouvray
76801
France

Zhong Lin Wang
School of Materials Science and
Engineering
Georgia Institute of Technology
Atlanta
GA 30332-0245
USA

Yanfa Yan
Department of Physics and
Astronomy
The University of Toledo
2801 Bancroft street
Toledo
Ohio 43606
USA

Nan Yao
Princeton University
Princeton Institute for the Science
and Technology of Materials
120 Bowen Hall
70 Prospect Avenue
Princeton
NJ 08540
USA

Henny W. Zandbergen
Delft University of Technology
Applied Sciences
Kavli Institute of Nanoscience
Lorentzweg 1
NL-2628CJ Delft
The Netherlands

J.M. Zuo
University of Illinois at
Urbana-Champaign
Department of Materials Science
and Engineering and Materials
Research Laboratory
Urbana
IL 61801
USA

19
Low-Energy Electron Microscopy
Ernst Bauer

19.1
Introduction

Low-energy electron microscopy (LEEM) is a surface-imaging method in which the surface is illuminated at normal incidence by slow electrons and the elastically reflected electrons are used to form an image of the surface. It is based on the high backscattering cross section that all materials have at low energies. In particular, in crystalline materials, the backscattered electrons are concentrated in one or more diffraction beams. One of them, usually the specular beam (the (00) beam), can be used to image the surface, analogous to the manner in which the image is generated in the transmission electron microscope (TEM) by the transmitted or diffracted beams. Likewise, other diffracted beams can be used to form images, called *dark-field images* in contrast to the bright-field images produced by the (00) beam.

LEEM allows real-time studies and is particularly useful for the study of processes on metal and semiconductor single-crystal surfaces but can also be applied to insulator surfaces, provided that charging is eliminated. It is easily combined with other surface-imaging methods such as mirror electron microscopy (MEM) or photoemission electron microscopy (PEEM). The lateral resolution of LEEM is limited to the 1–10 nm range by the aberrations of the cathode objective lens, whose cathode is the specimen. The depth resolution is in favorable cases in the sub-Angstrom range. The field of view can be as large as 100 μm, depending on the lens used. Many surface science and technology problems do not require a better resolution but a large field of view. Therefore, LEEM has a wide range of applications.

19.2
Theoretical Foundations

LEEM can be best understood by looking at the low-energy electron diffraction (LEED) pattern that is formed at the back focal plane of the cathode lens. Even

Handbook of Nanoscopy, First Edition. Edited by Gustaaf Van Tendeloo, Dirk Van Dyck, and Stephen J. Pennycook.
© 2012 Wiley-VCH Verlag GmbH & Co. KGaA. Published 2012 by Wiley-VCH Verlag GmbH & Co. KGaA.

in the absence of lens aberrations, the LEED pattern is not the Fourier transform of the potential distribution in the surface layer of the specimen with which the incident electron wave interacts. This is due to the strong electron–specimen interaction, which makes the first Born approximation invalid. As a consequence, the kinematical diffraction theory is insufficient for the calculation of the intensity of the diffracted beams and therefore also of the LEEM image intensity. Rather, the dynamical theory has to be used as described in various textbooks [1–4].

However, presently no quantitative information is drawn from intensity differences in LEEM images. Only the magnitude of the difference which determines the contrast is important. As an example, Figure 19.1 shows the reflection coefficient $R(E)$ of the specularly reflected beam ((00) beam) from a W(110) and a W(100) surface as a function of energy E [5, 6]. The two $R(E)$ curves differ drastically in their energy dependence, so that it is always possible to find strong contrast between (110)- and (100)-oriented grains, for example, at about 2 or 14 eV. On the other hand, no contrast may be seen if the energy is chosen wrongly, for example, at about 6 or 11 eV. The figure also shows that reflectivities as high as 60% may be obtained. Such high values are connected with bandgaps or regions with a low density of states (steep bands) in the band structure of the crystal along the propagation direction of the wave. For energies in bandgaps, no wave propagation is possible, and the wave is strongly damped and totally reflected: the crystal is a reactive medium. Ideally, the reflectivity should be 100%, but inelastic

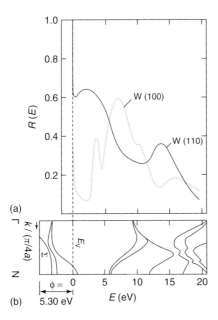

Figure 19.1 Specular reflection coefficient of the W(110) and W(100) surfaces as a function of electron energy (a) and band structure of W along the ΓN ([110]) direction (b) [6].

scattering, which is predominantly in the forward direction, and phonon and defect scattering decrease it to the observed value. The missing electrons appear in the specimen current [6]. Similar considerations apply to energy ranges with a low density of states, but with increasing energy, part of the intensity is lost into other diffraction channels. For example, at 13.6 eV where $R(E)$ of the (00) beam has again a maximum on the W(110) surface, (10) diffraction beams in addition to the specular beam are excited as easily seen by an Ewald construction (taking the inner potential $V_0 = 17.4$ eV into account). In addition, elastic backscattering decreases and inelastic scattering increases with energy and reduces R even more.

Diffraction contrast is not only strong between grains with different orientations but also between regions on a single-crystal surface that differ in surface structure, such as between reconstructed and unreconstructed surface regions or clean and adsorbate-covered regions, because these regions also differ in their $I_{00}(E)$ dependence. Examples that will be discussed later are the contrast between (1×1) and (7×7) structures on Si(111) or between clean Mo(110) and copper monolayer islands on this surface. A frequent situation in surface science is the presence of an ordered overlayer in two or more azimuthal orientations. They are indistinguishable at normal incidence because the diffraction conditions are identical. Contrast can be obtained by changing the diffraction conditions, that is, by tilting the beam or by imaging with one of the nonspecular overlayer beams. The resulting contrast has been termed tilted bright field and dark-field contrast, respectively, in contrast to the bright-field contrast when imaging with the (00) beam.

The contrast discussed up to now depends on the periodic arrangement of the atoms in the surface region of the specimen. One of the main goals of surface microscopy is, however, to image deviations from this periodic arrangement, that is, surface defects. The most important defects are monatomic steps, because many surface properties and processes depend strongly on them. LEED is uniquely suited for the determination of the average step distribution which determines the profiles of the diffraction spots [7, 8]. The profiles are analyzed in terms of the kinematic theory because the backscattered intensity may be separated into a dynamic and a kinematic term, $|F^2|$ and $|G^2|$, respectively. The structure factor $|F^2|$ contains the information on the size and shape of the two-dimensional unit cell and the distribution of the atoms in it, the lattice factor $|G^2|$ on the size and shape of the periodic arrangement of the unit cells. A step terminates this arrangement by introducing a vertical phase shift that changes $|G^2|$. In bright-field LEEM the step contrast can be easily understood in terms of this phase shift. A step of height d produces a path difference $2d$ between the waves reflected from the adjoining terraces. When $(2n + 1)\lambda/2 = 2d$, the waves are out of phase and interfere destructively on defocusing or as a consequence of the limited phase contrast transfer function of the objective lens. By determining the λ values – that is, the energies $E = 1.5/\lambda^2$ (in eV), where λ is in nanometers – at which the contrast is strongest, the step height can be determined with high accuracy. LEEM thus allows the determination of the local step distribution. An example of the step contrast is shown in Figure 19.2.

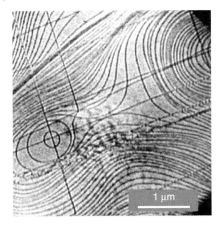

Figure 19.2 LEEM image of the step distribution on a Mo(110) surface. The rounded steps were formed by sublimation during the preceding high-temperature treatment while the straight steps are the results of glide during cooling. Electron energy 14 eV. (Adapted from Ref. [9].)

Another contrast that is also a consequence of path differences between different regions on the specimen is the quantum size effect contrast, which appears in thin films with parallel top and bottom boundaries. It is the electron–optical counterpart to reflection-reducing or reflection-enhancing layers in optics: the reflectivity has a maximum when $2d = n\lambda_f$ and a minimum when $2d = (2n+1)\lambda_f/2$ (λ_f is the wavelength in the film). Because of multiple reflection of the electron wave at the boundaries, the maxima and minima occur at well-defined thicknesses so that the thickness and thickness differences can be determined with high accuracy. Finally, it has to be kept in mind that the specimen is the cathode in a cathode lens. This means that the specimen is in a high field, and any deviation from planarity will cause a field distortion that will deflect the electrons. These deflections produce a topographic contrast, in particular at very low energy, which is sometimes helpful. However, if the field distortions of the surface features overlap, image interpretation becomes difficult if not impossible.

An important aspect of any surface microscopy is the information depth. In LEEM, as in any other surface probe that uses slow electrons, the information depth is, in general, determined mainly by inelastic scattering but also by thermal diffuse scattering and elastic backscattering. Elastic backscattering is important mainly at very low energies at which frequently only the specular beam can be excited, as discussed above, and a bandgap exists. The information depth is then of the order of the wavelength of the electron, which is 3–4 Å at zero energy for realistic mean inner potentials due to the exponential damping in total reflection. Thermal diffuse scattering, in particular, multiphonon scattering, which causes scattering predominantly far from the (00) beam, becomes important at high temperatures.

Inelastic scattering due to collective and one-electron excitations dominates the information depth at most temperatures. The inelastic mean free path has a minimum value of 3–5 Å around 50 eV, which is predominantly caused by plasmon excitation. Toward lower energies below the plasmon excitation threshold, it increases rapidly in general, and toward higher energies, more slowly. The attenuation length in the LEEM energy range (0–200 eV) thus ranges from a few Angstroms to 10 Å. In materials in which the d band is only partially occupied, low-energy excitations with high cross section are possible, and the inelastic mean free path may decrease to less than 3 Å. On the other hand, in large bandgap materials, much larger values may occur. Depth information may also be obtained indirectly, for example, on buried interfaces via the strain field of the interfacial dislocations, which may give sufficient diffraction contrast at layer thicknesses of the order of 100 Å.

Additional information on electron–specimen interaction and on contrast formation may be found in the literature [10–12].

19.3
Instrumentation

A LEEM instrument differs in several aspects from conventional electron microscopes. The very nature of its purpose, the imaging of well-defined surfaces, requires ultrahigh vacuum (UHV) with a base pressure in the low 10^{-10} to high 10^{-11} mbar range. As in the case of the mirror microscope, illuminating and imaging must be separated if an image is to be produced. Otherwise, only diffraction is possible. Finally, the specimen must be at a high negative potential or, if the specimen should be at ground potential, the complete optical system must be at high positive potential. Otherwise, a LEEM instrument makes use of the components of standard electron microscopes, albeit using UHV technology. The first (ill-fated) system was built with glass technology using electrostatic lenses and 90° deflection. The first all-metal systems used magnetic lenses and 60° deflection [9, 13–15]. As an example, the first instrument, developed from one of the prototypes [13, 16] is shown in Figure 19.3, together with the schematic of the electron optics. Another design uses 90° deflection, which has the advantage that illumination and imaging system are in line [17]. The desire to have the specimen at ground potential has led to the development of electrostatic lens systems, the first of which makes use of double 45° deflections [18], while the second one uses very small deflections [19].

The resolution of all these instruments is limited by the chromatic and spherical aberration of the objective lens [20]. These can be eliminated by adding an electron mirror, which has aberrations of opposite sign, to the electron optics. Several instruments make use of this aberration correction possibility, the SMART instrument [21] and two commercial instruments, one originating from the 90° deflector system [22], the other, shown in Figure 19.4, from the 60° deflector system.

A LEEM instrument is also suitable for Auger electron emission microscopy (AEEM) and secondary electron emission microscopy (SEEM) when different

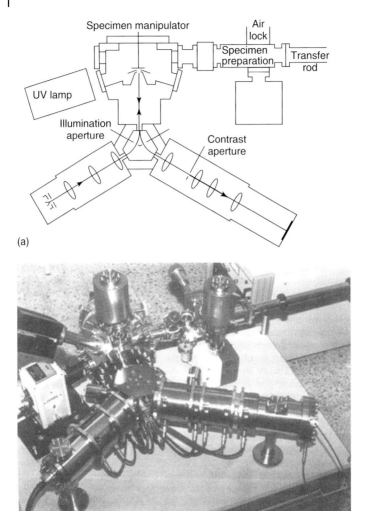

Figure 19.3 Schematic (a) and photo (b) of a LEEM instrument with a 60° beam separator. (Adapted from Ref. [20].)

electron energies in the illumination and in the imaging sections are used. The illumination system must then allow a wide range of illumination conditions and the imaging section must be able to accommodate an imaging energy filter. For efficient SEEM, primary beam energies up to about 500 eV are needed because the secondary electron yields of most materials peak at a few hundred electronvolts, while the secondary electron energy distribution peaks at a few electronvolts. In AEEM, primary electrons up to a few kiloelectronvolts are necessary in order to be able to detect all elements, while the characteristic energies of the Auger electrons range from a few tens of electronvolts to about 1 keV. Owing to the inconvenience

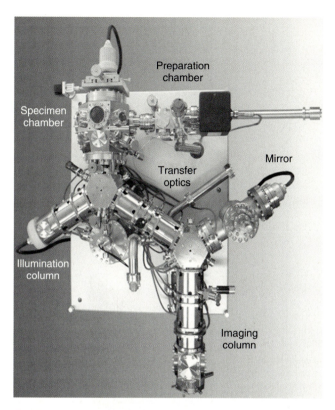

Figure 19.4 Aberration-corrected LEEM instrument with a 60° beam separator. (Courtesy Th. Schmidt.)

of the switching between LEEM and these imaging modes and because of the increasing availability of spectroscopic X-ray photoemission electron microscopy (XPEEM), AEEM and SEEM are barely used.

For LEEM, and in particular for LEED, the specimen should be illuminated with a parallel beam. This is achieved by demagnifying the crossover of the electron gun into the back focal plane of the objective lens with condenser lenses and a transfer lens. The electron source is either a LaB_6 or a W(310) field emitter. The beam aperture angle in the back focal plane, and therefore, the illuminated area on the specimen, is controlled by an aperture in the beam separator. A variety of beam separators are use, two of them for 60° deflection are shown in Figure 19.5: (a) the separator was used in the first instrument, the "close-packed prism array" and (b) the one used in later instruments [17, 23]. Various objective lenses are shown in Figure 19.5(c)–(f). In the magnetic triode lens (Figure 19.5f), the pole piece next to the specimen is electrically isolated so that its potential can be varied from ground potential to near-specimen potential. This allows studying surfaces at low fields [12]. The contrast aperture is not placed in the back focal plane of the objective lens where the LEED pattern is located as is usually done in standard

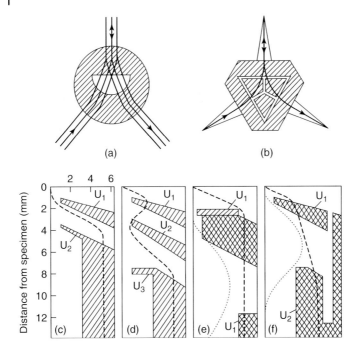

Figure 19.5 (a,b) 60° separators. (a) Separator used in the first experiments. (b) A double focusing separator that allows to use incident and emitted electrons with different energies. (c–f) Objective lenses. (c,d) electrostatic diode and triode, (e,f) magnetic diode and triode. Only the right half of the cross section is shown. The electrostatic potential and magnetic field on the optical axis are indicated by dashed and dotted lines, respectively. (Adapted from Ref. [20].)

electron microscopes for two reasons: (i) for dark-field imaging, that is, imaging with nonspecular beams, the aperture has to be shifted laterally and would intercept the incident beam for large shifts and (ii) the aperture would limit the intensity in the AEEM and SEEM illumination modes. Therefore, the contrast aperture is placed further downstream in an image of the LEED pattern (see Figure 19.3). The objective lens is focused in such manner that the image of the specimen is in the center of the separator, where it is, in principle, achromatic. For selected-area LEED, a field-limiting aperture can be placed in this position. However, in most applications, the illumination aperture is sufficient for this purpose. Therefore, the port for the manipulator of the field-limiting aperture in the 60° deflector instrument can be used for a fluorescent screen with which the primary beam may be inspected (cathode inspection, illumination system alignment). The contrast aperture in the image plane of the LEED pattern limits not only the angular aperture used in further image formation but acts also to a certain extent as a rough energy filter due to the chromatic aberration of the deflection field. This fact is actually used in the 90° instrument as a simple energy filter [24].

The lenses after the beam separator serve to image the LEED pattern into the plane of the contrast aperture, usually located in the following lens, to switch between LEED and LEEM and finally to project the LEED pattern or LEEM image onto the final detection system. In addition to the optical components mentioned up to now, the instrument contains a number of deflection systems for beam alignment and three stigmators, one in the objective lens, and one each in the illumination and imaging columns. The last two not only correct lens astigmatism but also the residual cylinder lens action of the beam separator.

Many LEEM instruments are now equipped with an energy filter, in particular in connection with XPEEM, AEEM, or SEEM [13, 21]. The energy filter uses a (simulated) hemispherical electrostatic sector field. Before the filter, the electrons are decelerated to about one tenth of the energy they have in the rest of the instrument except in the cathode lens. The final energy-filtered image is magnified by projective lenses onto a double-channel plate-fluorescent screen detector and recorded with a CCD camera. The channel plates, together with the CCD camera limit the resolution of the microscope at magnifications above about 10000 times [25]. Direct electronic detection eliminates this deficiency [26]. Of course, if the instrument is to be used only for LEEM, LEED, MEM, and emission electron microscopy with very slow electrons such as thermionic electron emission microscopy (TEEM) or near-ultraviolet PEEM, the energy filter and decelerating and accelerating lenses are not necessary, and the instrument can be much simpler, similar to the first instruments [9, 14]. Nevertheless, for LEED, the energy filter is useful for the elimination of the secondary electrons.

The objective lens is located in a large specimen chamber with ports pointing at the specimen position for mounting accessories for *in situ* experiments (gas sources, evaporators, etc.), for illumination with a high-pressure mercury short arc lamp for PEEM, or for connection to a synchrotron radiation beamline. The specimen is mounted on an eucentric $x-y-z$ manipulator on which it can be heated to 1600 °C. Specimen exchange is through an airlock via a specimen preparation chamber (see Figure 19.3). The microscopes are usually pumped by sputter ion and titanium sublimation pumps, and have a base pressure in the high 10^{-11} mbar range in the specimen chamber after a bakeout at 180 °C.

Provided that vibrations, current and high-voltage fluctuations, and alternating current fields are low enough and assuming no resolution limitation by the detection system, the resolution of LEEM is limited mainly by the lens aberrations of the cathode lens, the aberrations of the beam separator, and, in a spectroscopic instrument, the aberrations of the energy filter. With a well-designed beam separator and energy filter and a well-aligned system, the cathode lens determines the ultimate resolution, in particular, its acceleration field region. In an idealized situation, the lens can be divided into two sections: a homogeneous acceleration field section, which forms a virtual image behind the specimen, and a subsequent conventional electron lens, which forms a real image of the virtual image. The aberrations of the acceleration field are dominating. They produce aberration discs d_i in the image plane, which can be easily calculated analytically. The chromatic aberration, which is dominating at low energies, is proportional to the emission angle α, the spherical aberration, dominating at high energies, proportional to α^3. In addition, the angle-limiting aperture causes diffraction, which results in a diffraction disc of confusion d_d. The disc of confusion resulting from these three contributions can be estimated from $d = (d_c^2 + d_s^2 + d_d^2)^{1/2}$. For each combination of start energy E, energy width ΔE, and final energy E_o there is an optimum aperture for best resolution. An example of this method for the determination of the resolution is shown in Figure 19.6a [27]. The magnitude of the resolution is not significantly influenced by the lens effect of aperture at the end of the otherwise homogeneous field [28]. A more exact determination of the optimum resolution requires analytical calculations. Some results for the dependence of d on start energy are shown in Figure 19.6b for three cathode lens types used in LEEM instruments for an energy width ΔE of 0.5 eV [29]. With a field emission electron gun ($\Delta E \approx 0.25$ eV), a slightly better resolution can be achieved but $d \approx 5$ nm is about the limit without aberration correction. Aberration correction with an electron mirror eliminates the contributions of d_c and d_s [30] so that the resolution is determined by higher order aberrations, which become significant at larger α values so that larger apertures can be used [21, 22]. The resulting improvement is illustrated in Figure 19.6c [21].

Resolution and contrast are intimately connected. This is immediately evident in the wave-optical treatment of the imaging process [31–33] as illustrated in Figure 19.7 [31] by the intensity distribution across a step, which causes a phase difference of $\pi/2$ between the waves reflected from the adjoining terraces. A slight defocus increases the contrast but reduces the resolution. The resulting asymmetry distinguishes between upper and lower terraces. The wave-optical treatment, which has long been used in TEM [34] (see also Chapter 1) allows one to take into account all factors that are important for the image formation such as finite energy spread of the electron beam, finite size of the electron source, high voltage and lens current instabilities, and angle-limiting aperture, in the contrast transfer function $T(q)$, where q is the spatial frequency. Because of the large phase changes in LEEM $T(q)$ has to be used in a generalized form, which does not involve the weak phase approximation made in TEM. This function defines two q values, the point resolution limit q_P at the first zero of $T(q)$ and the information limit q_I, at which the damped oscillations of $T(q)$ approach zero. In practice q_P is used to define the resolution limit. Numerical calculations using this approach for an aberration-corrected and an uncorrected LEEM system give a resolution limit of about 0.9 and 3 nm, respectively, while geometric-optical calculations give 1.1 and 5 nm, respectively, for a $\pi/2$ phase object, $E = 10$ eV, and $\Delta E = 0.25$ eV[33]. The best resolutions reported in 2010 with aberration-corrected systems were around 2 nm [22, 35].

Another method to correct objective lens aberrations is focus variation, originally developed in TEM [36, 37] for instruments without aberration-correcting components. It allows phase retrieval from a series of images taken at different focus, provided that ΔE is sufficiently small so that spherical aberration is dominating. It has also shown some success in PEEM [38] but has not been applied yet to LEEM.

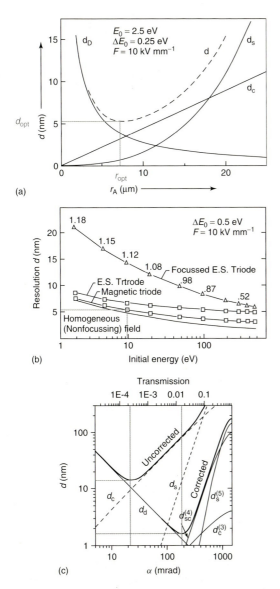

Figure 19.6 Theoretical resolution of cathode lenses. (a) Homogeneous field. Spherical aberration d_s, chromatic aberration d_c and diffraction disc d_d as a function of the aperture radius [27]. (b) Resolution of the electrostatic triode, electrostatic tetrode, and magnetic triode (Figure 19.5b) as a function of start energy. The numbers on the upper curve are the field strength at the specimen needed for a focused image [29]. (c) Aberrations of the magnetic diode lens of the SMART system and resolution without and with aberration correction as a function of angle at $E = 10$ eV and $\Delta E = 2$ eV. The resolution and transmission at the optimum aperture are indicated by horizontal and vertical lines, respectively. (Adapted from Refs. [21, 27, 29].)

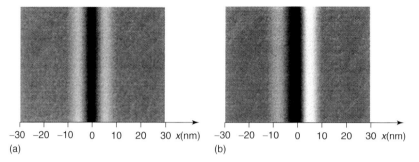

Figure 19.7 Theoretical intensity distribution at a surface step which causes a phase difference of $\pi/2$ between the waves reflected from the adjoining terraces. (a) in focus and (b) slightly defocused [31].

19.4
Areas of Application

19.4.1
Clean Surfaces

Because of its high depth resolution, LEEM is very well suited for the study of surface topography. The imaging of monatomic steps has already been mentioned (Figure 19.2), but etch pits, hillocks, or screw dislocations can be imaged as well. The real-time and high-temperature capabilities of LEEM also allow real-time studies of changes of the topography due to surface diffusion, electromigration, faceting, and other processes such as sublimation or surface phase transitions. Si surfaces are particularly suited for this purpose because of the availability of high-purity Si, which ensures negligible contamination by segregation. For this reason, a large amount of work has been done on Si(111) and (100) surfaces, which brought considerable insight into the thermodynamics and kinetics of surface processes as reviewed in [39, 40]. An example is the $Si(111) - (1 \times 1) \longleftrightarrow (7 \times 7)$ phase transition, which demonstrated, in the early phase of LEEM, the power of this method by revealing a wide variety of morphologies (Figure 19.8) [41] with little difference visible in laterally averaging techniques such as LEED. Similar studies have also been performed on metal and oxide surfaces. In particular, the latter have shown, amongst other interesting results, the importance of diffusion in the bulk for surfaces processes, for example, for TiO_2 (110) surfaces [42].

19.4.2
Adsorption Layers

Surfaces are frequently covered intentionally or unintentionally with gas atoms or molecules from the gaseous environment or via segregation of impurities in the bulk. These not only modify the surface composition but can also modify the

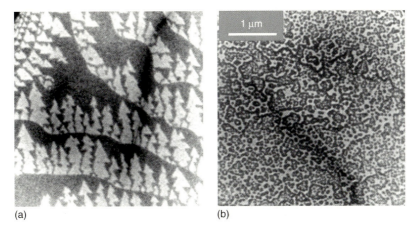

Figure 19.8 LEEM images of the Si(111) surface (a) heated briefly to 1300 K and cooled rapidly and (b) annealed at 1200 K for several hours and cooled slowly. LEED showed in both cases an excellent (7 × 7) pattern, only with slightly higher background in (b). The bright features have (7 × 7) structure, the dark (1 × 1) structure. (Adapted from Ref. [41].)

surface topography. When the interactions between the atoms are repulsive, they form superstructures, which are frequently characteristic for certain atoms and their coverage. This is illustrated by the LEEM images of Figure 19.9 [43], which shows various stages of the spiral chemical wave propagation in the reaction of NO and H_2 on Rh(110). The images are dark-field images taken with diffracted beams characteristic for different surface phases that form in this reaction. Many foreign atoms or molecules on surfaces have overall attractive lateral interactions even if the nearest-neighbor interactions are repulsive. Therefore, they form islands, which produce structural contrast. This is also the case when the adsorbate and substrate have the same lateral periodicity as in submonolayers of several metals on W(110) and Mo(110). The contrast between island and substrate is due to differences in backscattering cross section and phase. An example is the formation of striped phases of several metals such as Au and Pd on Mo(110) [44] and W[110] [45–47] surfaces in the submonolayer range at high temperatures. Figure 19.10 shows the striped phase of submonolayer Au on W(110).

Many adsorbates experience order–disorder transitions with increasing temperature before desorption occurs. Many of these transitions appear to be continuous in laterally averaging studies such as LEED because they average over the two phases involved in the transition. The (5 × 1) $\longleftrightarrow$ (1 × 1) transition at 0.4 monolayers of Au on Si(111) [48] is a good example. For these studies, the ability of imaging at high temperatures is of particular importance because it allows following the kinetics of the transition. If an adsorbate is deposited locally, then surface diffusion can be studied by following the propagation of the boundary between characteristic adsorbate structures that differ in coverage or structure. Similarly, reaction fronts between two different adsorbates can be used to follow chemical reactions on

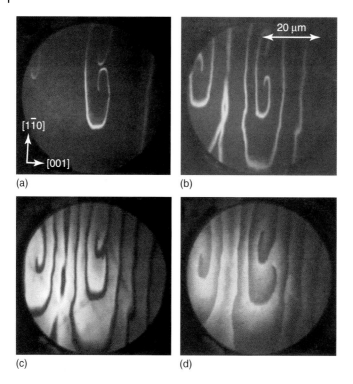

Figure 19.9 Dark-field LEEM images taken during the reaction between NO and H_2 on the Rh(110) surface. (a) and (b) are taken with diffraction beams characteristic for regions covered with N, (c) for a region covered with N and O, and (d) for a region covered only by O. (Adapted from Ref. [43].)

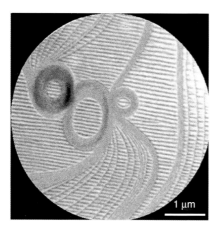

Figure 19.10 Video frame from the growth of the striped phase of Au on W(110) at 1100 K. (Courtesy T.O. Menteş.)

surfaces as already illustrated in Figure 19.9. In many cases, adsorbed atoms form a surface alloy with the substrate such as several metals on W(100) at 0.5 monolayer coverage [49] or Pb on Cu(111) at submonolayer coverage, which phase-separates with increasing coverage, forming self-assembled patterns [50]. LEEM allows also to determine alloying into deeper layers with high lateral resolution because the bright-field image in LEEM is determined by the intensity of the (00) beam. A local intensity analysis as function of the energy ($I(V)$) analysis) allows the extraction of the distribution of atoms normal to the surface if their backscattering cross sections differ sufficiently. An example is the surface alloying of Pd with the Cu(100) surface [51].

Adsorption of impurities from the interior of the crystal, that is, segregation, is also a good subject for LEEM. This is illustrated in Figure 19.11a [52] by the segregation of boron to the Si(100) surface, which does not change the periodicity of the surface but causes strong changes in the surface topography due to strain and anisotropic preferential adsorption on steps. Another example is the segregation of carbon to the (110) surface of molybdenum (Figure 19.11b). Here the surface reconstructs. The LEED pattern was interpreted in terms of epitaxial Mo_2C crystals [53] but scanning tunneling microscopy (STM) showed an ordered incorporation of carbon in the W lattice [54]. This is a good illustration of the limits of LEED and LEEM.

Another frequent surface phenomenon is the formation of two-dimensional compounds by interactions with reactive gases. While in many surface processes, steps play an important role, this does not seem to be the case in the high-temperature oxidation of W(100) at low pressure. Once domains of chemisorbed oxygen have formed – which involves a considerable amount of surface rearrangement – the

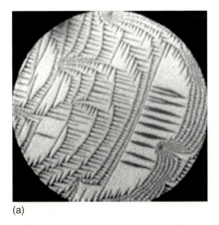
(a)

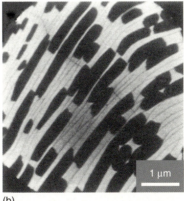
(b)

Figure 19.11 (a) Monatomic step structures formed upon the segregation of boron to the Si(100) surface; diffraction contrast due to the different orientation of the (1 × 2) structure on the two terraces. Electron energy 4.2 eV. (Adapted from Ref. [52]) (b) Segregation of carbon to the Mo(110) surface; diffraction contrast due to localized C segregation (dark regions). Electron energy 3 eV. (Adapted from Ref. [53].)

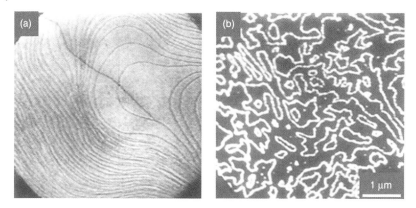

Figure 19.12 (a) Clean W(100) surface and (b) onset of two-dimensional oxide formation. Electron energy 7 and 16 eV, respectively. (Adapted from Ref. [55].)

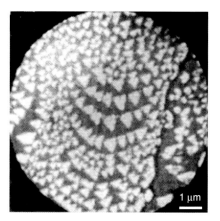

Figure 19.13 Video frame from the two-dimensional silicon nitride growth on Si(111) in 10^{-7} mbar ammonia at 1150 K. Electron energy 3 eV. (Adapted from Ref. [57].)

domain boundaries are the weak points that determine the further oxidation process. Figure 19.12 shows the initial step structure on a clean W(100) surface after a high-temperature flash and the transition from the chemisorbed layer to the two-dimensional oxide, which starts at the domain walls and spreads from them rapidly across the surface [55]. Another example is the oxidation of Ru(0001), which is of importance in catalysis [56]. Two-dimensional silicon nitride formation in the early stages of the nitridation of Si(111) in ammonia at high temperatures proceeds quite differently. A typical frame from a video taken during this process is shown in Figure 19.13. The nucleation clearly occurs in steps, and the (8 × 8) nitride structure then spreads across the terraces in a manner similar to the growth of the (7 × 7) structure on the clean Si(111) surface. Again, depending on nitridation conditions and the purity of the surface, a wide variety of morphologies and kinetics is observed. The combination with STM has proven very useful in understanding structure and growth of the two-dimensional nitride [57].

19.4.3
Thin Films

One of the major application areas of LEEM is the study of the early stages of the growth of thin films, in particular of how surface imperfections influence the growth and how reaction layers grow. Most metal–metal and metal–semiconductor film–substrate combinations alloy or react, at least at high temperatures, but there are also some less complicated processes, the simplest being homoepitaxy. One of the most important processes is the homoepitaxy of silicon on Si(100), which has been studied extensively with LEEM. At low supersaturation, at which no nucleation can take place on the terraces, new terraces nucleate only at imperfections, and spread over the existing terraces in a step-flow growth mode. The growth rate is highly anisotropic, reflecting the large difference between the step energies of steps parallel and perpendicular to the dimer rows on this (2×1)-reconstructed surface. This leads to the almond-shaped terraces seen in Figure 19.14a. At somewhat higher supersaturation, nucleation on the terraces occurs (Figure 19.14b), and the images rapidly become too complex for analysis [58].

The next simplest growth process is heteroepitaxy without alloying or reaction. The growth of Pb on Si(111) is an example. LEEM shows that Pb forms a monolayer initially and this is followed by the growth of three-dimensional (3D) crystals (Figure 19.15a–c). The tendency to 3D growth can be suppressed by depositing first an Au monolayer ("interfactant") (Figure 19.15d–f) [59]. Other typical examples are metals on W(110). Below 400–500 K these layers grow in a quasi-monolayer-by-monolayer mode, but at the upper temperature limit they grow at significantly different rates on different terraces. Some terraces have

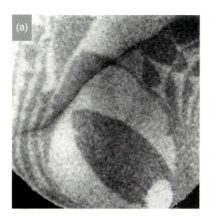
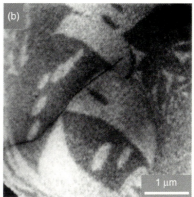

Figure 19.14 Video frames from the homoepitaxy of silicon on Si(100) at (a) very low supersaturation and (b) low supersaturation. Electron energy 5 eV. (Adapted from Ref. [58].)

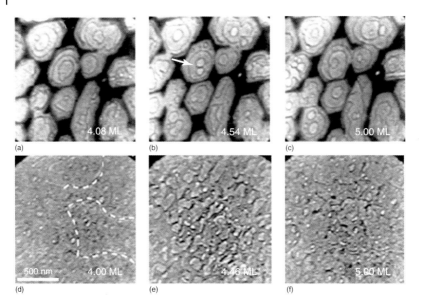

Figure 19.15 Video frames from the growth of Pb on Si(111) on the clean (7 × 7) surface (a)–(c) and on a Au monolayer with (6 × 6) structure (d)–(f). The arrow indicates two monolayer islands that have formed on a terrace (b) and which coalesce to form the next terrace (c). Electron energy 8 eV. (Adapted from Ref. [59].)

preferred nucleation sites from which the monolayers grow rapidly along the terraces, obviously incorporating atoms that diffuse across the steps from other terraces. This produces considerable thickness variations from terrace to terrace, which can be studied well by making use of the quantum size effect (Figure 19.16) [60]. An example can be found in the chapter 20 on spin-polarized low energy electron microscopy (SPLEEM) of this volume, in which the agglomeration of low temperature-deposited layers on annealing is also discussed. At high temperatures, only one or two adsorbed layers form on which flat three-dimensional crystals grow across the terraces [61]. Another case in which the quantum size effect has been useful in the study of the growth process is the growth of graphite on SiC(0001) from the first monolayer (graphene) to multilayers [62]. The growth of organic films can also be studied with LEEM, an example being pentacene films which are of interest for thin-film field effect transistors [63].

The most complex film growth process involves reaction with the substrate. The growth of Ge films on Si surfaces is a good example. After a wetting layer, three-dimensional crystals with a variety of thickness-, temperature-, and composition-dependent shapes are formed, a process that has been studied extensively with LEEM because if its technological importance [64]. For the same reason, the growth of metal films on Si is of interest. Many metals form compounds with Si, for example, Cu and Co. LEEM is very well suited for the study of the

19.4 Areas of Application | 691

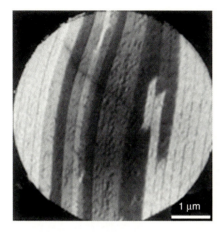

Figure 19.16 Terrace-limited monolayer-by-monolayer growth of copper on Mo(110) imaged via quantum size effect contrast. Electron energy 4 eV. (Adapted from Ref. [60].)

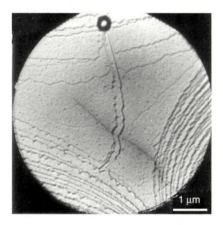

Figure 19.17 Cu silicide particle with its reaction trail on a Si(111) surface. Electron energy 4 eV. (Adapted from Ref. [65].)

growth of these films and their disintegration at high temperatures into individual compound crystals. Figure 19.17 shows an example. After completion of an initial two-dimensional reaction layer, three-dimensional Cu silicide crystals form with several orientations. One of them migrates across the surface leaving a reaction trail behind [65]. $CoSi_2$ layers have attracted considerable attention because of their possible applications in microelectronics. When these films are annealed at high temperature they break up into three-dimensional crystals. $CoSi_2$ sublimes more slowly than silicon and, therefore, large hillocks grow, with $CoSi_2$ crystals on top (Figure 19.18) [66]. This example shows that three-dimensional structures can also be imaged quite well, in spite of the field distortion caused by them, provided they

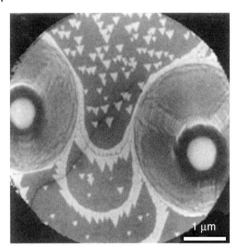

Figure 19.18 Hillocks on Si(111) with $CoSi_2$ crystals on top formed by faster sublimation of the substrate. Electron energy 10 eV. (Adapted from Ref. [66].)

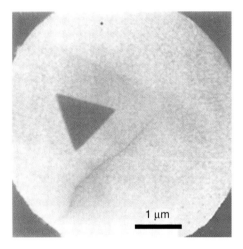

Figure 19.19 Au–Si eutectic particle in a Si(111) surface at about 900 K. (Adapted from Ref. [48].)

are well separated. Frequently particles appear dark at most or all electron energies, either because their surfaces are tilted so that they reflect only in narrow energy ranges along the optical axis ("facet reflections") or because they are not crystalline. An example is the dark triangle in Figure 19.19. It is a liquid Au–Si eutectic pyramid in a Si(111) surface which has well-defined crystallographic interfaces. It is dark because there is no well-defined specular beam due to the lack of crystallinity [48].

19.5
Discussion

When should LEEM be used for surface or thin-film studies, and when not? The situations when it should not be used are obvious: when a resolution of better than a few nanometers is needed, when the specimen is not UHV compatible because of high vapor or dissociation pressures, when it is very electron beam-sensitive due to electron-stimulated dissociation or desorption, or when it is very insulating or very rough. Another limitation is the gas pressure at the specimen. These situations are, however, not as LEEM-exclusive as they appear. Aberration-corrected instruments allow a skilled operator to reach 2 nm and less for suitable objects. Specimen cooling will reduce the vapor pressure problems and the use of electron energies below the thresholds for dissociation or desorption, which are easily accessible in LEEM, will eliminate specimen damage. Charging of insulating samples can be frequently eliminated by heating, illumination with UV light, or by simultaneous irradiation with electrons [67]. The only remedy for roughness is smoothing the surface by proper preparation procedures. As far as admissible pressures are concerned, LEEM images have been taken at pressures up to the 10^{-5} mbar range in the specimen chamber without noticeable image deterioration. This is possible because beam separator, illumination, and imaging columns are pumped separately from the specimen chamber, and the narrow constrictions between the chamber and the columns allow the maintenance of a large pressure gradient. It is mainly high-voltage stability which limits the maximum operating pressure. A major prerequisite is, of course, that the system is always clean, otherwise carbon contamination will make well-defined surface studies impossible.

The major competitor to LEEM in surface studies used to be high-energy reflection electron microscopy (REM) and standard scanning electron microscopy (SEM). Both types of instruments are available with surface science-compatible pressures, at least in the specimen region, and bakeout possibilities. Advantages of REM are higher resolution perpendicular to the direction of the electron beam and access to the specimen at nearly normal incidence. Although the disadvantage of the strong foreshortening in the beam direction caused by the grazing incidence of the electron beam can be partially eliminated [68], LEEM has completely displaced REM. The advantage of the second competitor of LEEM, SEM, is that it can be used for rough surfaces and that several spectroscopic methods can be easily combined with SEM. However, image acquisition in LEEM is much faster than in SEM because of the parallel versus sequential detection. Also the surface sensitivity is usually poorer in SEM than in LEEM and the contrast is more limited. The last two disadvantages can be reduced by going to very low energies, for example, by decelerating the electrons immediately before the specimen (scanning low-energy electron microscopy (SLEEM) [69]), but this introduces one of the main disadvantages of LEEM, the field distortions on rough samples. A great advantage of LEEM compared to the other two techniques is its easy combination with other imaging modes such as TEEM, PEEM, or SEEM and in particular with PEEM using synchrotron radiation for illumination (XPEEM) [70], which allows chemical and magnetic characterization irrespective of the structure of the specimen. Most synchrotron radiation sources are now equipped with instruments combining XPEEM with LEEM and energy filtering (spectroscopic photo emission and low energy electron microscope (SPELEEM)). A LEEM instrument can also be operated in the mirror mode in which the electrons are reflected immediately before the surface (MEM). MEM has been used to study surface potential variations, semiconductor doping profiles, and ferroelectric and ferromagnetic domains [71]. MEM is very useful for the study of surfaces consisting of very small crystals with random orientation, on noncrystalline surfaces, for example, before the removal of amorphous surface layers such as SiO_2 on crystalline silicon or the (chemo)mechanical polishing layer on single-crystal surfaces. MEM has therefore been developed as an inspection tool for technological surfaces [67]. LEED, which is easily obtained by removing the contrast aperture and changing the excitation of the intermediate lens, is indispensable for structural identification via the diffraction pattern.

Finally, a comparison with the various forms of scanning probe microscopies such as STM or atomic force microscopy (AFM) is necessary. STM and AFM have an intrinsically higher resolution but are much slower when imaging a large region due to their mechanical scanning image acquisition. When operated at high resolution, their field of view is small. Another, albeit minor, restriction is in deposition studies, because of the shadowing by the scanning tip. Experience has shown that STM and LEEM are more complementary than competing techniques.

19.6
Concluding Remarks

LEEM has developed over the last 25 years into a mayor surface-imaging technique. It is now frequently combined with XPEEM. Several commercial instruments are now available, two of them with aberration correction. This has lead to a rapid increase of the number of studies and publications. Only a few more or less randomly selected examples from the most active LEEM groups could be mentioned here, together with some of the early work. More information can be found in several reviews [10–12, 20, 40].

References

1. Pendry, J.B. (1974) *Low Energy Electron Diffraction*, Academic Press, New York.
2. Van Hove, M.A. and Tong, S.Y. (1979) *Surface Crystallography by LEED*, Springer, Berlin.
3. Clarke, L.J. (1985) *Surface Crystallography*, John Wiley & Sons, Ltd, Chichester.
4. Van Hove, M.A., Weinberg, W.H., and Chen, C.-M. (1986) *Low-Energy Electron Diffraction*, Springer, Berlin.
5. Herlt, H.-J., Feder, R., Meister, G., and Bauer, E. (1981) *Solid State Commun.*, 38, 973–976.
6. Herlt, H.-J. (1982) *Elastische Rückstreung sehr langsamer Elektronen an reinen und an gasbedeckten*

Wolfram-Einkristalloberflächen, PhD thesis, TU Clausthal.
7. (a) Henzler, M. (1982) *Appl. Surf. Sci.*, **11/12**, 450–469; (b) Henzler, M. (1984) *Appl. Phys. A*, **34**, 205–214.
8. Yang, H.-N., Wang, G.-C., and Lu, T.-M. (1993) *Diffraction from Rough Surfaces and Dynamic Growth Fronts*, World Scientific, Singapore.
9. Telieps, W. and Bauer, E. (1985) *Ultramicroscopy*, **17**, 57–66.
10. Bauer, E. and Telieps W. (1988) in *Surface and Interface Characterization by Electron Optical Methods* (eds. A. Howie and U. Valdre), Plenum Press, New York, pp. 195–233.
11. Bauer, E. (1990) in *Chemistry and Physics of Solid Surfaces VIII* (eds R. Vanselow and R. Howe), Springer, Berlin, pp. 267–287.
12. Bauer, E. (2007) in *Science of Microscopy* (eds. P. Hawkes and J. Spence), Kluwer/Springer Academic Publishers, pp. 606–656.
13. Veneklasen, L.H. (1991) *Ultramicroscopy*, **36**, 76–90.
14. Tromp, R.M. and Reuter, M. (1991) *Ultramicroscopy*, **36**, 91–106.
15. Bauer, E. (1994) *Surf. Sci.*, **299/300**, 102–115.
16. Veneklasen, L.H. (1992) *Rev. Sci. Instrum.*, **63**, 5513–5532.
17. Tromp, R.M., Mankos, M., Reuter, M.C., Ellis, A.W., and Copel, M. (1998) *Surf. Rev. Lett.*, **5**, 1189–1197.
18. Grzelakowski, K. and Bauer, E. (1996) *Rev. Sci. Instrum.*, **67**, 742–747.
19. Adamec, P., Bauer, E., and Lencova, B. (1998) *Rev. Sci. Instrum.*, **69**, 3583–3587.
20. Bauer, E.G. (1994) *Rep. Prog. Phys.*, **57**, 895–938.
21. Schmidt, Th. *et al.* (2002) *Surf. Rev. Lett.* **9**, 223–232.
22. Tromp, R.M., Hannon, J.B., Ellis, A.W., Wan, W., Berghaus, A., and Schaff, O. (2010) *Ultramicroscopy*, **110**, 852–861.
23. Kolarik, V., Mankos, M., and Veneklasen, L. (1991) *Optik*, **87**, 1–12.
24. Tromp, R.M., Hannon, J.B., Fujikawa, Y., Berghaus, A., and Schaff, O. (2009) *J. Phys.: Condens. Matter*, **21**, 314007-1–314007-13.
25. Shimizu, H., Yasue, T., and Koshikawa, T. (2007) 6th International Symposium on Atomic Level Characterizations of New Materials and Devices, pp. 203–206.
26. Van Gastel, R., Sikharulidze, I., Schramm, S., Abrahams, J.P., Poelsema, B., Tromp, R.M., and van der Molen, S.J. (2009) *Ultramicroscopy*, **110**, 33–35.
27. Bauer, E. (1985) *Ultramicroscopy*, **17**, 51–56.
28. Lenz, M. and Muellerová, I. (1992) *Ultramicroscopy*, **41**, 411–417.
29. Chmelik, J., Veneklasen, L., and Marx, G. (1989) *Optik*, **83**, 155–160.
30. Rose, H. and Preikszas, D. (1992) *Optik*, **92**, 31–44.
31. Müller, Th. (1995) *Bildentstehung in LEEM*, MS thesis, TU Clausthal.
32. Pang, A.B., Müller, Th., Altman, M.S., and Bauer, E. (2009) *J. Phys. Cond. Matter*, **21**, 314006-1–314006-10.
33. Schramm, S., Pang, A.B., Altman M.S., and Tromp, R. (2011) *Ultramicroscopy*, doi.10.1016/j.ultramic.2011.11.005
34. Frank, J. (1973) *Optik*, **38**, 519–536.
35. Schmidt, Th. *et al.* (2010) *Ultramicroscopy*, **110**, 1358–1361.
36. Ikuta, T. (1989) *J. Electron Microsc.*, **38**, 415–422.
37. Coene, W., Janssen, G., Op de Beeck, M., and van Dyck, D. (1992) *Phys. Rev. Lett.*, **69**, 3743–3746.
38. Koshikawa, T., Shimizu, H., Amakawa, R., Ikuta, T., Yasue, T., and Bauer, E. (2005) *J. Phys.: Condens. Matter*, **17**, S1371–S1380.
39. Hannon, J.B. and Tromp, R.M. (2000) *Annu. Rev. Mater. Res.*, **33**, 263–288.
40. Altman, M.S. (2010) *J. Phys.: Condens. Matter*, **22**, 084017-1–084017-14.
41. Telieps, W. (1987) *Appl. Phys. A*, **44**, 55–61.
42. McCarty, K.F. and Bartelt, N.C. (2003) *Surf. Sci.*, **527**, L203–L212; **540**, 157–171; **543**, 185–206.
43. Schmidt, Th., Schaak, A., Günther, S., Ressel, B., Bauer, E., and Imbihl, R. (2000) *Surf. Sci.*, **318**, 549–554.
44. Mundschau, M., Bauer, E., Telieps, W., and Swiech, W. (1989) *Surf. Sci.*, **213**, 381–392.
45. Duden, Th. (1996) *Entwicklung und Anwendung der Polarisationsmanipulation in*

der Niederenergie-Elektronenmikroskopie, PhD thesis, TU Clausthal.

46. de la Figuera, J., Léonard, F., Bartelt, N.C., Stumpf, R., and McCarty, K.F. (2008) *Phys. Rev. Lett.*, **100**, 186102-1–186102-4.
47. Menteş, T.O., Locatelli, A., Aballe, L., and Bauer, E. (2008) *Phys. Rev. Lett.*, **101**, 085701-1–085701-4.
48. Swiech, W., Bauer, E., and Mundschau, M. (1991) *Surf. Sci.*, **253**, 283–296.
49. Man, K.L., Feng, Y.J., and Altman, M.S. (2006) *Phys. Rev. B*, **74**, 085420-1–085420-11.
50. van Gastel, R., Plass, R., Bartelt, N.C., and Kellogg, G.L. (2003) *Phys. Rev. Lett.*, **91**, 055503-1–055503-4.
51. Sun, J., Hannon, J.B., Kellogg, G.L., and Pohl, K. (2007) *Phys. Rev. B*, **76**, 205414-1–205414-100.
52. Jones, D.E., Pelz, J.P., Hong, Y., Bauer, E., and Tsong, I.S.T. (1996) *Phys. Rev. Lett.*, **77**, 330–333.
53. Mundschau, M., Bauer, E., and Swiech, W. (1988) *Catal. Lett.*, **1**, 405–411.
54. Bode, M., Pascal, R., and Wiesendanger, R. (1995) *Surf. Sci.*, **344**, 185–191.
55. Altman, M.S. and Bauer, E. (1996) *Surf. Sci.*, **347**, 265–279.
56. Flege, J.I., Hrbek, J., and Sutter, P. (2008) *Phys. Rev. B*, **78**, 165407-1–165407-5.
57. Bauer, E., Wei, Y., Müller, T., Pavlovska, A., and Tsong, I.S.T. (1995) *Phys. Rev. B*, **51**, 17891–17901.
58. Swiech, W. and Bauer, E. (1991) *Surf. Sci.*, **255**, 219–228.
59. Schmidt Th. and Bauer, E. (2000) *Phys. Rev. B*, **62**, 15815–15825.
60. Mundschau, M., Bauer, E., and Swiech, W. (1989) *J. Appl. Phys.*, **65**, 581–584.
61. McCarty, K.F. *et al.* (2009) *New J. Phys.*, **11**, 043001-1–043001-37.
62. Hibino, H., Kageshima, H., Maeda, F., Nagase, M., Kobayashi, Y., and Yamaguchi, H. (2008) *Phys. Rev. B*, **77**, 075413-1–075413-7.
63. Al-Mahboob, A., Sadowski, J.T., Fujikawa, Y., Nakajima, K., and Sakurai, T. (2008) *Phys. Rev. B*, **77**, 035426-1–035426-6.
64. Ross, F.M., Tromp, R.M., and Reuter, M.C. (1999) *Science*, **286**, 1931–1934.
65. Mundschau, M., Bauer, E., and Swiech, W. (1989) *J. Appl. Phys.*, **65**, 4747–4752.
66. Bauer, E., Mundschau, M., Swiech, W., and Telieps, W. (1991) *J. Vac. Sci. Technol.*, **19**, 1007–1013.
67. Mankos, M., Adler, D., Veneklasen, L., and Bauer, E. (2008) *Phys. Procedia*, **1**, 485–504.
68. Müller, P. and Métois, J.J. (2005) *Surf. Sci.*, **599**, 187–195.
69. Frank, L., Müllerová, I., Faulina, K., and Bauer, E. (1999) *Scanning*, **21**, 1–13.
70. Locatelli, A. and Bauer, E. (2008) *J. Phys.: Condens. Matter.*, **20**, 82202–82024.
71. Luk'yanov, A.E., Spivak, G.V., and Gvozdover, R.S. (1974) *Sov. Phys.-Usp.*, **16**, 529–552.

20
Spin-Polarized Low-Energy Electron Microscopy
Ernst Bauer

20.1
Introduction

Spin-polarized low-energy electron microscopy (SPLEEM) is a method of imaging the magnetic microstructure of surfaces and thin films with slow specularly reflected electrons. It is based on the fact that electron scattering is spin dependent via the spin–spin and spin–orbit interactions between the incident electron and the specimen. In specular reflection, only the spin–spin interaction occurs. If the specimen has regions with preferred spin alignment at length scales above the resolution limit of low-energy electron microscopy (LEEM), then these regions can be imaged via the contribution of the spin–spin interaction to the total scattering potential. Thus, magnetic and structural information is obtained simultaneously. SPLEEM is easily combined with low-energy electron diffraction (LEED), mirror electron microscopy (MEM) and the various types of emission microscopy (photo-electron, secondary electron, or thermionic electron emission microscopy). Lateral and depth resolution, information depth, and field of view are comparable to that of LEEM (see Chapter 19). Therefore, SPLEEM is an excellent method for the study of the correlation between magnetic structure, microstructure, and crystal structure.

20.2
Theoretical Foundations

Polarized electrons [1] have been used for some time in the study of the structure and magnetism of surfaces by spin-polarized low-energy electron diffraction (SPLEED) [2–5]. SPLEED is a laterally averaging method. In SPLEEM, the sample is used as a cathode in a cathode lens electron microscope and the diffracted electrons are used for imaging the surface in the same manner as in a standard LEEM instrument. Thus, LEEM and SPLEEM are laterally resolving methods. The difference between SPLEEM and LEEM results from the fact that the incident beam is spin polarized in SPLEEM. The exchange interaction between the spin-polarized beam electrons and the spin-polarized electrons in a ferromagnet causes contrast in addition to

Handbook of Nanoscopy, First Edition. Edited by Gustaaf Van Tendeloo, Dirk Van Dyck, and Stephen J. Pennycook.
© 2012 Wiley-VCH Verlag GmbH & Co. KGaA. Published 2012 by Wiley-VCH Verlag GmbH & Co. KGaA.

the structural contrast, which depends on the polarization $\boldsymbol{P}$ of the incident beam and the magnetization $\boldsymbol{M}$ resulting from the preferred alignment of the spins in the ferromagnet. This contrast is proportional to $\boldsymbol{P} \cdot \boldsymbol{M} = I_M$, so that the reflected intensity may be written as $I = I_0 + I_M$. When $\boldsymbol{P}$ is reversed, I_M changes sign while the polarization-independent contribution I_0 is unaffected. The difference of the intensities $I^{\pm} = I_0 + I_M^{\pm}$ of images taken with $\pm\boldsymbol{P}$, usually normalized with the sum of $I_0 = I^+ + I^-$ and with the degree of polarization $P = |\boldsymbol{P}| \leq 1$, $(I_M^+ - I_M^-)/(I^+ + I^-)P = 2I_M^+/I_0 P = A$, is called the *exchange asymmetry*, and gives an image of the $\boldsymbol{M}$ distribution in the sample. The direction of $\boldsymbol{M}$ can be easily determined by maximizing A by rotating $\boldsymbol{P}$ parallel/antiparallel to $\boldsymbol{M}$. Extraction of the magnitude of $\boldsymbol{M}$ requires an A analysis in terms of a dynamical SPLEED theory or an empirical calibration and is sometimes complicated because of the spin-dependent quantum size effects (QSEs) discussed below.

SPLEEM is particularly useful for the study of crystalline ferromagnetic materials. These have a spin-dependent band structure, with majority-spin and minority-spin bands usually separated by a few tenths of an electron volt (eV) to about 1 eV. Figure 20.1 shows such an exchange-split band structure (J. Noffke, personal communication). The [0001] direction is a frequently encountered orientation in cobalt layers. Below the two energy bands is a large energy gap. Electrons with energies in such a gap cannot propagate in the crystal and are totally reflected (see Chapter 19). This is true for majority-spin electrons up to 1 eV and for minority-spin electrons up to 2 eV. Between these two energies there is an increasing excess of minority-spin electrons in the reflected intensity because majority-spin electrons can penetrate into the crystal. Thus, A and, therefore, magnetic contrast are large.

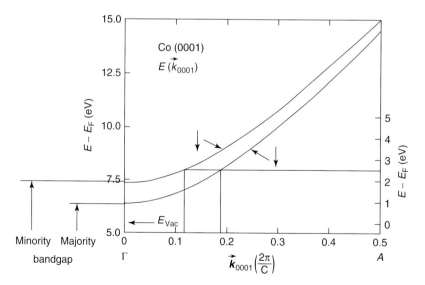

Figure 20.1 Band structure of cobalt above the vacuum level in the [0001] direction. The Fermi energy E_F is 5.3 eV below the vacuum level E_V.

Above 2 eV, *A* rapidly decreases because both types of electrons can now penetrate into the crystal.

One of the currently most important fields in magnetic materials is the study of ultrathin magnetic layers. These layers frequently show pronounced QSE oscillations in *A*. They can be understood by inspecting Figure 20.1: to every energy *E* there are two *k* values, one for majority-spin electrons, the other for minority-spin electrons. Although *k* is not a good quantum number in very thin films, it is still defined well enough to allow thickness-dependent standing waves in the layer, which occur at different thicknesses for fixed *E* or at different *E* for fixed thickness for majority-spin and minority-spin electrons. This causes spin-dependent oscillations and a significant enhancement of *A*, as seen in Figure 20.2. Such standing waves in the layer are formed not only in energy regions in which the substrate has a band gap, but generally whenever there is poor matching of the wave functions at the interface. Numerous studies of this spin-dependent QSE effect have been made since its first observation, for example, to determine the band structure above the vacuum level [6], as reviewed in [7]. The QSE effect in *A* makes determination of the magnitude of *M* difficult, but is very helpful for contrast enhancement.

The spin-averaged information depth of SPLEEM is the same as in LEEM. Then, the inverse mean free path $1/\lambda$ of electrons with energies of a few electronvolts above the vacuum level in d metals is approximately proportional to the total number of unoccupied d states [9]. Thus, there is no universal behavior, but

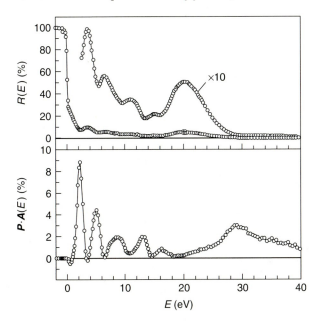

Figure 20.2 Specular intensity and exchange asymmetry of a six monolayer thick [0001]-oriented cobalt layer on a W(110) surface as a function of energy. (Adapted from Ref. [8].)

rather a pronounced material dependence. For nickel $\lambda \approx 1.1$ nm, for cobalt $\lambda \approx 0.8$ nm, and for iron $\lambda \approx 0.6$ nm. Below a kinetic energy of about 5 eV, λ becomes increasingly spin dependent, with $\lambda_\uparrow/\lambda_\downarrow$ reaching a value of about 4 at 0 eV for Fe, for example [10]. In addition to the limitation by inelastic scattering, the information depth is strongly limited in band gaps, for example, on Co(0001) below 1 eV.

20.3
Instrumentation

A SPLEEM instrument is very similar to a LEEM instrument (see Chapter 19) but differs in the illumination system and in the cathode objective lens. The stray field of magnetic lenses at the specimen position is usually small enough so that it does not influence magnetic specimens noticeably. Nevertheless, an electrostatic tetrode objective lens, which has a resolution comparable to that of a magnetic lens (see Chapter 19), is preferred. The magnetic sector is unavoidable, but changes the polarization direction of the incident beam only slightly because of the low field and short path in it. The illumination system incorporates the polarization

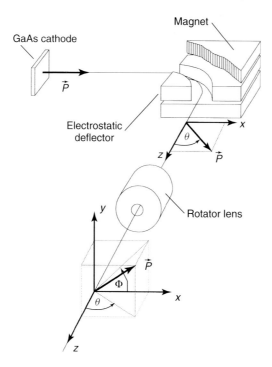

Figure 20.3 Polarization manipulator. The combined electric and magnetic 90° sector allows to rotate **P** in the x-z plane (θ), the rotator lens around the z-axis. (Adapted from Ref. [12].)

manipulator, which is needed to adjust the direction of P. In the first instruments [11, 12], it consists of a crossed $E \times B$ field 90° deflector and a magnetic spin rotator lens, as schematically shown in Figure 20.3 [12], while a newer instrument uses a Wien filter [13].

The spin-polarized electrons are produced by photoelectron emission from III–V semiconductor cathodes by left or right circular polarized light from a diode laser. The simplest cathode is a GaAs (100) single crystal surface [14], which is covered with a Cs–O layer in order to approach negative electron affinity. Electrons are excited from the spin-orbit-split valence band of GaAs into the conduction band, from which they can escape without having to overcome a barrier. The maximum theoretical polarization obtainable from this cathode is 50% but in practice 20–25% is usual. Much higher polarizations (up to 90%) can be obtained from complex III–V semiconductor multilayer cathodes, in particular when illuminated from the back [15]. Contrary to the LaB_6 emitter used in LEEM, which has thousands of operating hours, the activation of these photocathodes has a useful life of the order of hours to weeks, depending on the vacuum, and activation is still more an art than a science.

The resolution in SPLEEM is comparable to that in LEEM, but the signal-to-noise ratio is poorer because of the need for image subtraction if purely magnetic images are desired. In as much as $I_M \ll I_0$, longer exposures (in the 1 s range) are needed for a good signal-to-noise ratio when using GaAs cathodes. With multilayer cathodes, image acquisition times in the 0.01 s range can still give good magnetic contrast [13]. The field of view is up to about 50 μm, similar to that in LEEM.

20.4
Areas of Application

The magnetic domain structure on the surface of bulk magnetic materials is generally too large grained for SPLEEM studies. Exceptions are, for example, closure domains on surfaces such as on the Co(0001) surface or domain walls [16]. The major field of SPLEEM is, therefore, the study of thin films and superlattices whose magnetic properties depend strongly on film thickness and structure, in particular on the interface structure.

One of the questions is to what extent the substrate influences the magnetic properties. Cobalt layers on W(110) illustrate this influence. On clean W(110), cobalt grows with the closest-packed hcp plane parallel to the substrate ((0001) orientation), on the W_2C-covered surface it grows in the fcc structure with (100) orientation [17]. In both cases, the magnetization is in-plane and has a pronounced uniaxial anisotropy with the easy axis in the hcp layer parallel to W[110] [17]. At room temperature, magnetic contrast starts at three monolayers, although the magnetic moment is still small at this thickness. The magnetic domains are initially small, but rapidly coalesce with increasing film thickness and form large domains as seen in Figure 20.4 [18]. Steps on the substrate surface have little influence on the size and shape of the domains. The Neel-type domain walls between the domains can be imaged with P perpendicular to M in the domains (Figure 20.4d [18]). In contrast

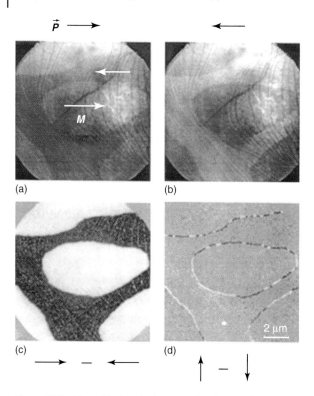

Figure 20.4 Magnetic domain images of a six monolayer thick cobalt layer on a W(110) surface. (a,b) images taken with the indicated **P** directions, (c,d) difference images obtained with the indicated **P** directions. The W[1–10] direction is horizontal. Electron energy 2 eV. (Adapted from Ref. [18].)

to cobalt on W(110), which is in-plane magnetized from the very beginning, cobalt on Au(111) has initially out-of-plane magnetization, which switches to in-plane at about 4.4 monolayers [19]. Such spin reorientation transitions occur also in other thin ferromagnetic films on various substrates, both out-of-plane to in-plane and in-plane to in-plane as reviewed in [7] and in the general references mentioned at the end. These transitions are due to the change of the various magnetic anisotropy energy contributions, such as dipolar, interface, and magnetoelastic anisotropy energy to the total energy with increasing thickness or temperature. Monolayer films are usually paramagnetic at room temperature but in some cases the Curie temperature T_C is above room temperature so that the phase transition can be studied conveniently. The Fe monolayer on two monolayers of Au on W(110) is an example [20]. Starting from 6 K below T_C (Figure 20.5a), paramagnetic regions develop on narrower terraces and spread with increasing temperature to wider terraces (Figure 20.5b, 1.65 K below T_C) until the complete film is completely paramagnetic. This coexistence of ferro- and paramagnetic states is caused by finite size effects and illustrates the influence of steps on the magnetic properties on ultrathin films.

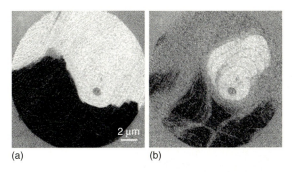

(a)   (b)

Figure 20.5 Phase transition from the ferromagnetic to the paramagnetic state of a Fe monolayer on two Au monolayers on a W(110) surface with various atomic terrace width regions. Electron energy 3.5 eV. (Adapted from Ref. [20].)

Ferromagnetic nanostructures, which can be obtained, for example, by film growth at elevated temperatures or by annealing of continuous films, can also be studied well with SPLEEM, provided that they are single crystalline and their top surface is perpendicular to the electron beam. Figure 20.6 [21] shows some examples of Fe nanostructures obtained by annealing films with different thicknesses at different temperatures. Depending on thickness and temperature, either single domain wires or closure domain structures are formed.

Nonmagnetic overlayers on magnetic layers can have a strong influence on their magnetic structure. An example is gold on Co(0001) layers (M.S. Altman, H. Pinkvos, E. Bauer, unpublished). Figure 20.7a is the SPLEEM image of a 10 monolayer thick cobalt layer deposited on W(110) at 400 K. One monolayer of gold deposited onto this cobalt layer at room temperature has no influence on the domain structure, but when two gold monolayers are deposited, most of the cobalt layer switches to out-of-plane magnetization – so that $M \perp P$ and $A = 0$ – but on some terraces the original in-plane magnetization remains (Figure 20.7b), even if additional gold is deposited. This figure illustrates very well how the topography of the substrate can be propagated through a deposited layer and influences its magnetic properties. In contrast to gold, copper overlayers have no influence on the domain structure in cobalt layers up to the largest overlayer thickness studied (14 layers). Only pronounced QSE asymmetry oscillations are seen as a function of copper layer thickness [22].

One of the most interesting subjects in thin film magnetism is the magnetic coupling between magnetic layers through nonmagnetic layers. It oscillates frequently between ferromagnetic and antiferromagnetic or the so-called biquadratic coupling. SPLEEM is particularly well suited for the study of this phenomenon because of the small sampling depth, which allows correlation of the magnetization of the top layer with that of the bottom layer. An example is shown in Figure 20.8 [23] for sandwiches with five (a–c) and six (d–f) Au monolayers between the Co layers. At five Au monolayers, the coupling is antiferromagnetic; at six Au monolayers, it is biquadratic with M in the top layer rotated $45°$ with respect to the bottom layer M.

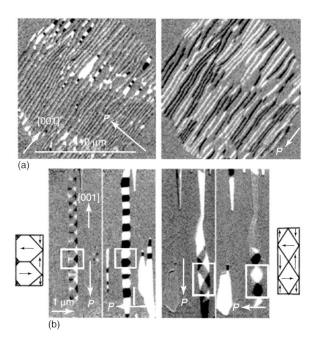

Figure 20.6 Iron nanostructures on a W(110) surface. (a) Ten monolayer thick film annealed at 650 K and (b) 3.5 monolayer thick film annealed at 720 K. The domain structure in the frames marked with thin lines is illustrated on the sides. (Adapted from Ref. [21].)

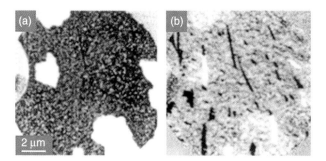

Figure 20.7 Influence of a gold overlayer on the magnetic domain structure of a 10 monolayer thick cobalt layer on W(110). Images taken with **P** in-plane of (a) uncovered Co layer and (b) of the same area covered with two Au monolayers. In (a) **M** is in-plane, in (b) out-of-plane except in on several narrow terraces. Electron energy 1.5 eV (M.S Altman, H. Pinkvos, E. Bauer, unpublished).

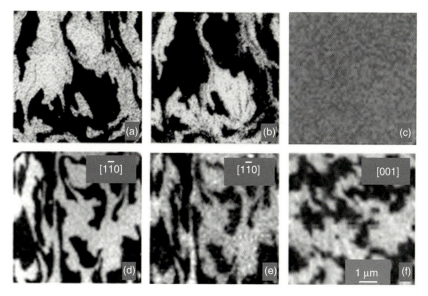

Figure 20.8 Magnetic interlayer coupling between cobalt layers on W(110) through gold layers. (a–c) Antiferromagnetic coupling through a five monolayer thick Au film, (d–f) biquadratic coupling through a six monolayer thick Au film. The Co layers are seven monolayers thick. (a,d) are images of the bottom Co layer covered with Au, (b,c,e,f) of the top Co layer. In all images **P** is in-plane, in (a,b,d,e) **P** is parallel to the easy axis, in (c,f) **P** is perpendicular to it. The combination of the **M** components in (e,f) shows many small domains with **M** rotated 45°. Electron energy 1.5 eV. (Adapted from Ref. [23].)

20.5 Discussion

When should SPLEEM be used for the study of the magnetic microstructure of materials, and when not? Section 20.4 has mentioned a number of interesting applications, but there are many more such as magnetic switching phenomena in pulsed fields, or magnetization processes in fields perpendicular to the surface. Fields parallel to the surface can only be applied in a pulsed manner because of the beam deflection that they cause, while fields normal to the surface require only refocusing. The areas for which SPLEEM is not well suited are essentially the same as in LEEM (see Chapter 19): rough surfaces, high vapor pressure materials, and so on.

There are many other magnetic imaging methods. Most of them do not image magnetization but the internal or external magnetic field distribution caused by the magnetization distribution such as Lorentz microscopy (see Chapter 6), electron holography (see Chapter 5), or magnetic force microscopy. These techniques are to a large extent complementary to SPLEEM, as is MEM [24], which can be easily combined with LEEM. The most important competitor to SPLEEM is scanning electron microscopy with polarization analysis (SEMPA), which images the magnetization distribution in the specimen via polarization analysis of the secondary electrons. The advantages of SEMPA over SPLEEM are an easy combination with electron spectroscopy, unimportance of the crystallinity of the specimen for high brightness, and the absence of a high electric field at the specimen surface. The most important advantages of SPLEEM over SEMPA are rapid image acquisition and easy combination with LEED and various emission microscopy methods.

20.6
Concluding Remarks

After 20 years of development and applications, SPLEEM is now a mature technique with a commercial instrument available. Up to now, only two noncommercial instruments have produced all the data reported here, in [7] and in the general references [25–27].

References

1. Kessler, J. (1985) *Polarized Electrons*, 2nd edn, Springer, Berlin.
2. Pierce, D.T. and Celotta, R.J. (1981) *Adv. Electron. Electron Phys.*, **56**, 219–289.
3. (a) Feder, R. (1981) *J. Phys. C*, **14**, 2049–2091; (b) Feder, R. (1983) *Phys. Scr.*, **T4**, 47–51.
4. Feder, R. (ed.) (1985) *Polarized Electrons in Surface Physics*, World Scientific, Singapore.
5. Kirschner, J. (1985) *Polarized Electrons at Surfaces*, Springer, Berlin.
6. Zdyb, R. and Bauer, E. (2002) *Phys. Rev. Lett.*, **88**, 166403-1–166403-4.
7. Rougemaille, N. and Schmid, A.K. (2010) *Eur. Phys. J.-Appl. Phys.*, **50**, 20101-1–20101-18.
8. Wurm, K. (1994) Spin-polarizierte LEEM-Untersuchungen an dünnen Kobalt-Epitaxieschichten auf W(110). Master Thesis, TU Clausthal.
9. Siegmann, H.C. (1994) *Surf. Sci.*, **307–309**, 1076–1086.
10. Hong, J. and Mills, D.L. (2000) *Phys. Rev. B*, **62**, 5589–5600.
11. Grzelakowski, K., Duden, T., Bauer, E., Poppa, H., and Chiang, S. (1994) *IEEE Trans. Mag.*, **30**, 4500–4502.
12. Duden, T. and Bauer, E. (1995) *Rev. Sci. Instrum.*, **66**, 2861–2864.
13. Suzuki, M. *et al.* (2010) *Appl. Phys. Express*, **3**, 026601-1–026601-3.
14. Pierce, D.T. (1995) in *Experimental Methods in the Physical Sciences*, vol. **29A** (eds F.B. Dunning and R.G. Hulet), Academic Press, San Diego, CA, pp. 1–38.
15. Jin, X. *et al.* (2008) *Appl. Phys. Express*, **1**, 045002-1–045002-3.
16. Altman, M.S., Pinkvos, H., Hurst, J., Poppa, H., Marx, G., and Bauer, E. (1991) *MRS Symp. Proc.*, **232**, 125–132.
17. Pinkvos, H., Poppa, H., Bauer, E., and Hurst, J. (1992) *Ultramicroscopy*, **47**, 339–345.
18. Bauer, E., Duden, T., Pinkvos, H., Poppa, H., and Wurm, K. (1996) *J. Magn. Magn. Mater.*, **156**, 1–6.
19. Duden, T. and Bauer, E. (1997) *MRS Symp. Proc.*, **475**, 283–288.
20. Zdyb, R. and Bauer, E. (2008) *Phys. Rev. Lett.*, **100**, 155704-1–155704-4.
21. Zdyb, R., Pavlovska, A., Jalochowski, M., and Bauer, E. (2006) *Surf. Sci.*, **600**, 1586–1591.
22. Poppa, H., Pinkvos, H., Wurm, K., and Bauer, E. (1993) *MRS Symp. Proc.*, **313**, 219–226.
23. Duden, T. and Bauer, E. (1999) *Phys. Rev. B*, **59**, 474–479.
24. Luk'yanov, A.E., Spivak, G.V., and Gvozdover, R.S. (1974) *Sov. Phys.-Usp.*, **16**, 529–552.
25. Bauer, E. (2005) in *Magnetic Microscopy of Nanostructures* (eds H. Hopster and

H.P. Oepen), Springer, Berlin, pp. 111–136.

26. Bauer, E. (2005) in *Modern Techniques for Characterizing Magnetic Materials* (ed. Y. Zhu), Kluwer Academic Publication, Boston, pp. 361–379.

27. Bauer, E. (2007) in *The Handbook of Magnetism and Advanced Magnetic Materials*, vol. 3 (eds H. Kronmueller and S. Parkin), John Wiley & Sons, Ltd, Chichester, pp. 1470–1487.

21
Imaging Secondary Ion Mass Spectroscopy

Katie L. Moore, Markus Schröder, and Chris R. M. Grovenor

Secondary ion mass spectrometry (SIMS) is the analysis of charged particles emitted by bombardment of a surface with energetic primary ions. Analysis of the mass-to-charge ratio of these particles enables detailed analysis of the surface chemistry. SIMS is a major technique for investigating a wide range of materials because of its combination of very high chemical sensitivity, excellent mass resolution, surface sensitivity, and lateral and depth resolution not matched by any other technique. SIMS is being used for the analysis of an increasingly wide range of materials from semiconductors and metals to polymers and biological materials.

21.1
Fundamentals

Secondary ion mass spectroscopy (SIMS) is the process by which the surface to be analyzed is bombarded by an energetic primary beam of ions, causing particles to be emitted from the surface. These emitted (secondary) particles will be a combination of electrons, neutral species, atoms, molecules, or atomic and cluster ions [1–3]. It is the secondary ions that are detected and analyzed to give information about the chemical composition of the sample. The electrons can also be detected and can give an indication of the surface topography and sample structure. SIMS is able to detect all elements (and isotopes) from hydrogen to uranium with parts per billion sensitivity for some elements as well as being able to detect and distinguish between molecular ions [2]. High lateral resolution can be obtained (down to 50 nm for some instruments) and it can be applied to any type of material that is stable under vacuum. It is this unique combination of characteristics that makes SIMS a very powerful technique in the physical and life sciences.

Handbook of Nanoscopy, First Edition. Edited by Gustaaf Van Tendeloo, Dirk Van Dyck, and Stephen J. Pennycook.
© 2012 Wiley-VCH Verlag GmbH & Co. KGaA. Published 2012 by Wiley-VCH Verlag GmbH & Co. KGaA.

21.1.1
Physical Principles

21.1.1.1 The Sputtering Process

A schematic of the sputtering process is shown in Figure 21.1. The bombardment of a surface by energetic ions or atoms to cause erosion is known as *sputtering* [1]. When a high-energy particle hits the surface, its energy is transferred to the sample by a series of inelastic collisions. In order to achieve desorption of particles from the sample surface, the momentum transfer perpendicular to the surface must exceed the binding energy of the particles (0.5–30 eV depending on the surface work function). Sometimes, particles are desorbed from the surface by the direct impact of the primary ion but these "direct-knock-on" processes are rare. Normally, the primary particle penetrates up to 20 nm into the sample and loses its energy by a series of collisions (the linear collision cascade model [4]) and only momentum transfers by secondary recoils return to the surface. The diameter of the affected area is defined as the area disturbed not only by the primary ion impact but also the resurfacing collision cascade, and can reach diameters of 5–10 nm [5]. Only ions from the immediate surface are sputtered from the sample [6]; hence, SIMS is a very surface-selective technique. The collision process also results in mixing of the atoms in the near-surface region of the solid to a depth of ~10 nm [3] and roughening of the surface. Benninghoven gave a definition for the so-called "static limit" for SIMS analysis [7]. If the probability of an incident ion hitting an area

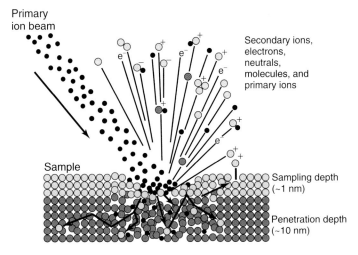

Figure 21.1 A schematic diagram showing the principles of SIMS analysis. Note that the primary ion beam is not drawn to scale and should be several hundred times larger. The collision processes occurring when energetic particles hit a surface and the accompanying surface mixing are also illustrated. (Adapted from Evans Analytical Group SIMS Theory Tutorial http://www.eaglabs.com/training/tutorials/sims_theory_tutorial/index.php (04 October 2010).)

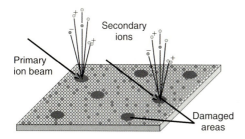

Figure 21.2 Schematic diagram showing the damaged areas after ion impacts during static SIMS analysis. (Adapted from CAMECA *http://www.cameca.com/instruments-for-research/sims-static.aspx* (04 October 2010).)

twice during one measurement is below 1%, one can assume static conditions and every impact yields signals from undisturbed surface areas (Figure 21.2). Static SIMS uses a primary ion beam with a very low current density to remain within the static limit, and this is the key principle of time of flight SIMS (ToF-SIMS).

Dynamic SIMS uses a much higher primary ion current density (above 10^{12} ions cm^{-2}) [8], far exceeding the static limit, resulting in fast erosion of the surface and formation of a crater. Static SIMS can provide molecular information, whereas dynamic SIMS provides mostly elemental and isotopic information because the rate of damage to the sample surface is much higher, resulting in the breaking of most chemical bonds. Near-surface analysis can still be undertaken, but the faster erosion rate allows bulk analysis of the sample, including chemical depth profiling and two- and three-dimensional imaging.

If the surface is covered with molecular species, we have to distinguish, in the static limit, between two main regions surrounding the point of impact. The secondary ion flux originating from the area closest to the impact point is dominated by molecular fragments and atomic species. Only intact molecular ions originate from the surrounding area. The energy transfer to the surface beyond this outer area is too low to result in desorption of secondary ions, as shown in Figure 21.3.

The model of the linear collision cascade is limited to atomic primary ions and fails to describe the processes taking place during the impact of cluster- and polyatomic primary ions. Cluster- and polyatomic primary ions tend to fragment in close proximity to the sample surface and each resulting fragment carries a fraction of the total kinetic energy, resulting in a large number of low-energy collision cascades originating from the same point of impact. This energy is then deposited fairly close to the surface, resulting in a reduced fragmentation of molecular species, but desorbing a large amount of material from the area in the immediate impact vicinity. Several models exist to describe the processes that result in the formation of charged secondary particles; information on the more reliable models can be found in Refs. [10–14].

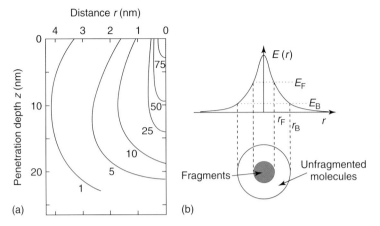

Figure 21.3 (a) Energy deposition by perpendicular impact of a 10 keV Xe^+ primary ion on an organic surface as a function of distance to impact point (r) and depth (z). Numbers indicate deposited energy density (eV/nm^3). (Redrawn from Whitlow, H.J., Hautala, M., and Sundqvist, B.U.R. (1987) Collision cascade parameters for slow particles impinging on biomolecule targets. *Int. J. Mass Spectrom. Ion Process.* **78**, 329–340, Ref. [9].) (b) Statistically averaged energy distribution ($E(r)$) at a surface as a function of distance r to impact point. (Redrawn from Benninghoven, A. (1979) Molecular secondary ion emission, in *Secondary Ion Mass Spectrometry, SIMS-II: Proceedings of the Second International Conference on Secondary Ion Mass Spectrometry (SIMS II)*, Stanford University, Stanford, California, USA, 1979, Springer Series in Chemical Physics, Vol. xiii, Springer-Verlag, Berlin, p. 298, Ref. [10].)

21.1.2
Technical Details

In order to extract quantitative chemical information from SIMS analysis, it is important to understand how the interaction of the primary ion beam with the specific sample type affects the yield of secondary ions.

21.1.2.1 Yield

Sputter yield is defined as the number of particles removed per primary particle impact, which normally lies between 0.1 and 10 [1]. The primary ion beam energy, the mass of the incident ions, and the angle of incidence all affect the sputter yield, and the yield is also dependent on the physical state of the sample. The crystallinity of the sample surface can also have a direct effect on the sputter yield as certain crystallographic planes sputter faster than others [1]. Surface topography also has a large effect on the average sputtering yield, and this is particularly evident when the roughness is of a scale similar to the dimensions of the damage cascade, which results in much higher yields in comparison to a flat surface [15]. Holes or pores in the sample can reduce the sputter yield by trapping sputtered material.

In addition, not all elements sputter at the same rate; therefore, in multielement samples, sputtering results in the preferential loss of one component and the surface becomes enriched in the element with the lower elemental yield [1, 16–18].

The very important influence of yield exerted by the material being analyzed, the so-called matrix effect, will be described below.

Ionization Efficiency The SIMS ionization efficiency is known as the secondary ion yield, defined as the fraction of sputtered atoms that become ionized [19]. During the sputtering process, very few of the species emitted are, in fact, spontaneously ionized; for most materials, the fraction of charged secondary particles is approximately 1%, so the majority of emitted species are uncharged and cannot be directly analyzed.

Secondary ion yields depend on many factors, the most influential being properties of the element itself; the ionization potential for positive ions, and the electron affinity for negative ions. The choice of primary beam can also affect the ion efficiency. An oxygen beam increases the positive ion yield, whereas a cesium ion beam increases the negative ion yield. Oxygen bombardment creates oxygen–metal bonds in an oxygen-rich zone. When these bonds break during ionization, the oxygen becomes preferentially negatively charged because of its high electron affinity, and its high ionization potential prevents it from becoming positively charged. This enhances the probability of the local formation of M^+ ions. Cesium implantation, on the other hand, decreases the surface work function, which means that more secondary electrons are excited over the surface potential barrier; hence, there are more electrons available leading to increased negative ion formation [19]. For elements with a high ionization yield, using a cesium beam may increase the average number of secondary ions emitted per primary ion to 20% [19]. The optimum choice of primary ion species is dependent on the position in the periodic table of the element to be analyzed, as shown in Figure 21.4 for cesium and oxygen primary beams.

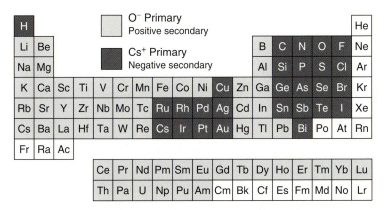

Figure 21.4 The ion yield is dependent on the choice of the primary ion beam. Generally, the ion yield for elements on the left-hand side of the periodic table is maximized with an oxygen beam and elements on the right with a cesium beam. (Redrawn from Evans Analytical Group SIMS Theory Tutorial (04 October 2010) http://www.eaglabs.com/training/tutorials/sims_theory_tutorial/beameff.php using data from Ref. [1].)

With the emergence of cluster- and polyatomic primary ion sources, information on the distribution of complete molecules is now easily accessible, so the choice of primary ion is not only determined by the effect on ionization of one particular atomic species but also on the overall analytical question. While ionization efficiency is one consideration that needs to be taken into account when choosing the primary ion beam, the focusing ability for imaging is also important. The best imaging sources use field ionization of liquid metals, with gallium liquid metal ion guns (LMIGs) being the most popular closely followed by the newer generation of gold and bismuth LMIGs.

21.1.2.2 Resolution
Resolution in SIMS has three different meanings, which are all dependent on several factors. These are discussed in the following sections.

Lateral Resolution The ultimate spatial resolution of a SIMS experiment is determined by the size of the collision cascade initiated by bombardment of a single ion into the 1 nm sampling depth from which sputtered ions can escape. Monte Carlo simulation of a collision cascade from a single primary ion indicates that the ultimate resolution is about 2 nm at this 1 nm sampling depth [2]. Of course, to obtain a measurable secondary ion signal, a primary ion beam containing very many ions must be used. In practice, the spatial resolution is determined by the primary beam diameter. With a 50 nm beam, the resolution cannot be better than 50 nm, and the pixel size on any image should be matched as closely as possible to the beam size to minimize any further degradation of resolution. The best lateral resolution can be achieved with gallium liquid metal ion sources [2] as they can be focused on smaller diameters than other sources, although there is still a trade-off between current and diameter [20].

Mass Resolution Mass resolution is just as important as the lateral resolution, as it is this which determines the selectivity of the instrument between ions of very similar masses.

Mass resolution, R, is defined in Eq. (21.1), where M_1 is the mass of the ion of interest and M_2 is the mass of an adjacent interference peak [21].

$$R = \frac{M_1}{M_1 - M_2} = \frac{M}{\Delta M} \tag{21.1}$$

Some elements and molecular fragments have nominally the same mass, for example, $^{79}Br^-$ and $^{31}P^{16}O_3^-$ and $^{40}Ca^{16}O^-$ and $^{56}Fe^-$. Fortunately, the small deviation in the atomic masses from integer values, the so-called mass deficits, means that with a high-resolution mass spectrometer most ions can be separated. However, even with many technological advances in SIMS instrumentation, one of the main problems associated with SIMS analysis of biological materials is the difficulty in distinguishing with confidence between elemental and molecular species with nominally the same mass. Great care must be taken to ensure that the correct species is identified or the results are meaningless. At lower (<15 amu)

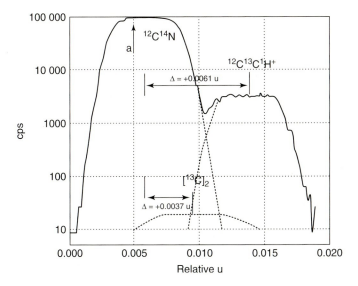

Figure 21.5 A high-resolution mass spectra around mass 26 from a coral sample spiked with ^{13}C. The arrow marked "a" indicates where the detector should be positioned to avoid any interference from the $^{13}C_2^-$ peak. (Reproduced with permission from "Subcellular imaging of isotopically labeled carbon compounds in a biological sample by ion microprobe (NanoSIMS)" by Clode, P.L., Stern, R.A., and Marshall, A.T. (2007) Subcellular imaging of isotopically labeled carbon compounds in a biological sample by ion microprobe (NanoSIMS). *Microsc. Res. Tech.*, **70** (3), 220–229, Ref. [22]. Copyright 2007, with permission from John Wiley & Sons.)

and higher (>100 amu) masses, there are relatively few mass interferences but almost any combination of carbon, hydrogen, oxygen, phosphorus, and nitrogen is possible as molecular fragments from biological samples. Figure 21.5 shows the spectrum around mass 26 from a coral sample spiked with ^{13}C. The $^{12}C^{14}N^-$ and $^{12}C^{13}C^1H^-$ peaks are clearly resolved, but the $^{13}C_2^-$ peak requires a mass resolution of ~7200 to separate it fully from the $^{12}C^{14}N^-$ peak [22]. To avoid any possible interference from the small $^{13}C_2^-$ peak, the detector should be positioned at the arrow marked "a."

Depth Resolution The final measure of resolution is important when creating a depth profile through a material. The depth resolution is determined by the thickness of the sampling depth from which the secondary ions originate in a single measurement [23], which is generally about 1 nm [19]. This resolution is degraded by sputtering effects such as atomic mixing, which implants surface atoms some 10 nm into the material, and ion-induced surface topography, which usually becomes more severe with depth [24].

21.1.2.3 Sensitivity, Transmission, and Mass Resolution

In ion microscopy, the three key experimental parameters in an analysis are *sensitivity, transmission,* and *mass resolution,* and the microscopist would generally like, and to some extent needs, all of these to be maximized. Unfortunately, these three parameters are related and maximizing one will inevitably result in compromising another. The sensitivity of an SIMS instrument is determined by the ionization efficiency of the element to be analyzed and the optical transmission of the ion optics. The first has been seen to be related to ionization potential or electron affinity as well as chemical environment and the choice of primary beam [6]. The transmission of the system is given as the ratio of the number of ions formed at the sample surface to the number of ions actually detected, that is, it is a measure of the number of ions lost along the path to the detectors.

To achieve high mass resolution in a magnetic sector SIMS instrument, slits are used to cut off the edges of the secondary beam of ions to produce narrow peaks, as shown in Figure 21.5. This obviously lowers the transmission of the microscope and hence the sensitivity is reduced. Identification of peaks is usually performed under high-resolution conditions to avoid mass interferences and the slits are then opened for analysis, reducing the mass resolution but increasing the sensitivity. In a SIMS instrument that uses lenses and apertures to shape the primary beam, there is another constraint. To achieve a high lateral resolution, the primary beam needs to be made as small as possible, which in turn reduces the current in the beam and the sensitivity of the microscope, as fewer secondary ions are generated.

In a ToF-SIMS instrument, the transmission is given by the limited acceptance angle of the extraction electrode and the apertures following the extraction ion optics. The angle of acceptance also has an impact on the mass resolution, which in a ToF-SIMS instrument is largely determined by the primary ion pulse length that gives the starting point of the time measurement. Most primary ion sources used for analysis purposes are therefore equipped with buncher optics to shorten the primary ion pulse down to 1–2 ns. The limited angular acceptance of the mass analyzer prevents transmission of secondary ions with angled trajectories, which would take a longer flight path through the mass analyzer and compromise the mass resolution. The transmission is fixed for a given instrument, whereas the mass resolution can be compromised by a change in the pulse length of the primary ion source. A lengthening of the pulse might be necessary to increase the signal intensity and thereby the sensitivity for a specific ion or to make focusing the primary ion beam easier to achieve better lateral resolution.

21.1.2.4 Steady State

In dynamic SIMS, the secondary ion signals from a sample under primary ion bombardment vary in a complex and independent way at the start of the analysis before reaching a steady state. This initial region is known as the transient [25]. Any analysis performed in this transient region will not be quantitative, and comparisons between different samples or analysis on different days become impossible. Figure 21.6 shows this effect in a biological specimen, and we can see that in this case a Cs^+ dose of 10^{17} atoms/cm^3 is needed to achieve steady state [6]. In some cases, it may not be possible to achieve this steady state at all, for example, if the sample is very thin or so sensitive to the ion beam that it may be

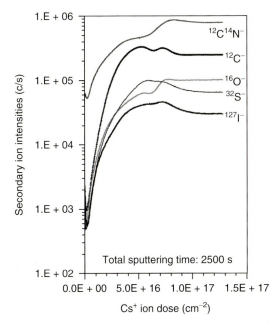

Figure 21.6 Complex ion signal variation from a thyroid follicle as a function of the amount of Cs$^+$ implanted before reaching steady state. (Reprinted from Guerquin-Kern, J.L., Wu, T.D., Quintana, C., and Croisy, A. (2005) Progress in analytical imaging of the cell by dynamic secondary ion mass spectrometry (SIMS microscopy). *Biochim. Biophys. Acta*, **1724** (3), 228–238, Ref. [6], Copyright (2005), with permission from Elsevier.)

completely sputtered away before this criterion has been reached. Alternatively, the structure and chemistry of the sample may change significantly with depth and all the close-to-surface information would be lost before quantitative analysis can be performed. In static SIMS, the dose is far too low to reach steady state so the analysis if often considered to be semiquantitative. As SIMS is a destructive method, it is not possible to improve the accuracy of the measurement simply by analyzing the same area again.

21.1.2.5 Quantification – Relative Sensitivity Factors

One of the major problems with SIMS is that quantification of the sample composition is very difficult because each element and molecular species has a different ion yield. With careful calibration, each ion of interest can have a measured relative sensitivity factor (RSF), and this can be used to give an indication of its suitability for SIMS analysis.

RSF tables have been established for most elements in silicon [2, 19] driven by the semiconductor industry, but RSF tables are now starting to be made for other matrices, including GaN [26]. However, there is a real need to establish these tables for other materials because without this, quantification is almost impossible. There has been, for instance, very little work on RSF values for elements in biological matrices [27]. The Si tables are still useful in planning experiments because variations of six orders of magnitude have been measured for different elements. For instance, under O^- bombardment the RSF of Na^+ ions in Si is 100 times larger than for Si^+ ions, but the value for Br^+ is $\times 1000$ lower [2, 3]. Without an accurate knowledge of these RSF values, it is obvious that no reliable information on the relative concentration of these two elements in silicon could be obtained. It is useful in SIMS analysis of other materials to use these RSF values in silicon to give at least a rough indication of the likely ion yield.

21.1.2.6 Artifacts in SIMS

As with any technique, there are some important artifacts that have to be borne in mind while conducting SIMS analysis. Some of the effects that can affect secondary ion signals are phase boundaries, implantation, topography, crater edges, holes, halo effects, diffusion, surface and atomic mixing, the matrix effect, preferential sputtering, induced roughening, redeposition of sputtered material, and ion channeling [28]. Some of these effects are controlled by the design of the instrumentation and some are intrinsic to the sputtering process. Some can be minimized or eliminated by careful specimen preparation or instrument and experimental design, but others always need to be considered. This is quite a lengthy list of artifacts and considering the problems associated with quantification and interpretation one might wonder why SIMS is so popular as an analysis technique. Nevertheless, the combination of sensitivity, mass resolution, and lateral resolution outweighs these problems in many situations. Some of these artifacts are discussed in the following sections.

Matrix Effects In general, we expect the ion yield and secondary ion intensity to be closely related to the concentration of the element in the sample, but the chemistry and crystallography of the surrounding matrix can also have a pronounced effect. This is known as the *matrix effect* and is the general term used to describe differences in sensitivity for a given element in samples of different composition, structure, or crystallographic alignment [2]. These differences result from changes in the ionization efficiency and/or the sputtering yield, and the effect is generally further complicated with the use of reactive primary ions. The matrix effect makes it more difficult to measure the concentration of the bulk element than a trace element [29], and quantification at internal interfaces is very challenging [2]. A well-known example of this is given by Wilson *et al.* [2]. The yield of ions of many elements changes across a Si/SiO_2 interface because of the yield enhancement caused by the presence of oxygen. Some differences are shown in Table 21.1:

Detecting cluster ions containing the primary beam element such as MCs^+ (where M is the atom of interest and Cs^+ the bombarding species) considerably reduces the variations in intensities due to differences in the sputtering yield, but does not avoid matrix effects altogether [2, 16, 30]. This method allows measurement of the matrix composition, but it is often necessary to make a standard sample of known concentration which is similar to the material to be analyzed to improve the quantification [29]. Biological materials are particularly prone to this problem

Table 21.1 Yield changes at a Si/SiO$_2$ interface under O$_2$ primary ion bombardment.

Element	Relative yield enhancement in SiO$_2$
B	6
P	60–70
As	20
F	2

because there tends to be a large variation in the density in different parts of biological materials, which makes the matrix effect very complicated [31]. For example, the cell membrane is structurally and chemically quite different to the nucleus.

Surface Topography It is very important that the sample surface is flat in order to get reliable SIMS information, as any topography can either enhance or suppress the secondary ion emission. Like secondary electron emission in a scanning electron microscope, the yield of ions at a step edge is often higher, and holes can trap sputtered species leading to a lower yield [28]. Topography, if present initially, cannot be removed by sputtering [2]. Flat samples also allow optimum transmission to be maintained through the optical path [6].

Topography also has a serious effect on depth profiling experiments as the primary ion beam, in particular reactive ions, causes rippling of the crater bottom [24]. As a general rule, the surface topography always roughens with increasing ion dose. As the surface does not remain flat during the profile, the depth resolution is degraded because not all of the ions come from the same position in the sample [3].

21.1.2.7 Sample Charging

SIMS uses a primary beam of charged particles, and for insulating samples this results in a buildup of charge on the surface. In dynamic SIMS, a positive primary ion beam can be used to produce negative secondary ions, and the main effect of positive surface charging is the inhibition or total suppression of negative ions [1]. To minimize charging, insulating specimens are usually coated with a few nanometers of a metal, such as gold or platinum [2], and it is necessary to sputter this away before analysis can be started. For some very insulating samples, a metallic coating is insufficient, or in the case of TOF-SIMS, where only the top monolayer investigated, it would prevent analysis of the surface of interest. In these cases, an electron flood gun may be used. The bombardment of the surface by low-energy electrons is usually sufficient to compensate for the charge buildup [32].

21.2
SIMS Techniques

21.2.1
Instrumentation

21.2.1.1 Time of Flight SIMS

As an example of a modern ToF-SIMS, the IonTOF ToF-SIMS V is the latest in a long line of TOF-based mass spectrometers. New liquid metal ion sources using Bi^+ and cluster ions such as Bi_3^+ as the primary ion have helped to establish this instrument as a chemical imaging tool at the submicron scale for atomic and especially for molecular surface species.

The ToF-SIMS V's main components are, in addition to the mass spectrometer and the different optional primary ion sources, a five-axes-manipulator with a cooling stage option, an electron flood gun for charge compensation and a detector for ion-induced secondary electrons. The samples can be observed via two independent optical camera systems, as shown in Figure 21.7. As with all SIMS instruments, the ToF-SIMS V is operated under high-vacuum conditions. Sample transfer from ambient conditions into the vacuum of the main chamber is via an airlock with optional cooling for samples that would otherwise sublimate or evaporate into the vacuum, contaminating the instrument, or would simply be lost before analysis. Some instruments are equipped with docking stations for transfer devices, making it possible to transfer samples from one high-vacuum instrument to the ToF-SIMS and vice versa without exposing the sample to ambient conditions. This is particularly important for ToF-SIMS analysis where only the top surface is analyzed and any surface contamination must be avoided.

The two most important components in a TOF-SIMS instrument are the primary ion source and the TOF mass spectrometer itself. For high-resolution imaging an LMIG is used. The advantage of LMIG sources, in general, over the more commonly used gas ionization sources is the higher brightness and smaller area of emission [34], making it possible to focus the primary beam down to <50 nm. Bismuth is used in modern LMIG sources as it forms stable positive clusters during the field ionization process [35]. The atomic ion and different cluster ions can be separated in the primary ion column. Varying the primary ion species, in particular switching from monoatomic to cluster primary ions, enhances the emission of intact molecular species [36].

In order to image a sample surface, two modes of operation are possible. For small areas, the primary ion beam is rastered over the surface either randomly, to minimize charging problems from single impacts, or in a linear fashion, and a complete secondary ion mass spectrum is acquired for each pixel. The position and signal intensity for a specific secondary ion of interest can then be summed up over several scans making up the secondary ion image, with the total primary ion current density being kept well below the static limit. Secondary ion species of interest can be selected in post measurement analysis from the collected raw data, making it possible to sum up related species to enhance signal intensity and

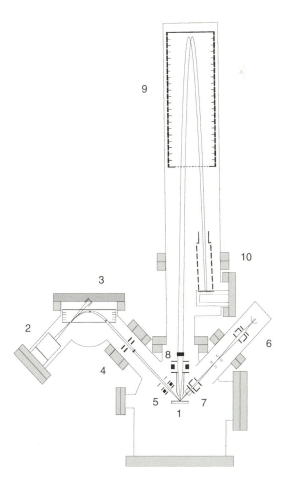

Figure 21.7 Schematic diagram of the ToF-SIMS V showing the main components: (1) sample stage (manipulator), (2–5) electron impact sputter gun ((2) ion formation chamber, (3) 90° deflector and pulser, (4) buncher, (5) target lens), (6–7) LMIG ((6) ion formation, (7) target lens), (8–10) mass spectrometer ((8) extraction optics, (9) ion reflector, (10) post acceleration and detector). (Adapted from Schröder, M. (2006) Flugzeit-sekundärionenmassenspektrometrie an thiol self assembly monolagen auf gold, PhD thesis. Physikalisches Institut, Westfälische Wilhelms-Universität, Münster, Ref. [33].)

thereby contrast. The second mode of imaging is used to obtain a view exceeding the scanning range of the deflection plates of the primary ion source. For this so-called macro-rastering, the primary ion beam stays in the same position and the sample stage is moved. The limitation for the image size is only the mechanical movement range of the sample stage, and the lateral resolution in this case is limited by the mechanical accuracy of the sample stage rather than the beam size. As an illustration of the recent development of very large scale mapping, the two images in Figure 21.8 are ToF-SIMS maps for characteristic molecular fragments

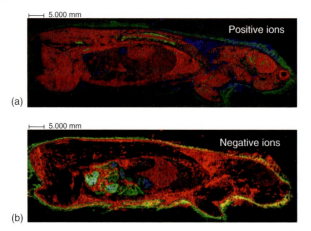

Figure 21.8 ToF-SIMS images, presented as three-color overlays, of a section through a whole mouse. Field of view 28 × 84 mm² acquired with a Bi_3^+ primary ion source. (a) Positive secondary ions: red phosphatidylcholine fragment (m/z 224); green cholesterol (m/z 369 and 385); and blue diacylglycerol (m/z 577). (b) Negative secondary ions: red sum of stearic (m/z 255) and oleic (m/z 281) fatty acid carboxylates; green cholesterol sulfate (m/z 465); and blue taurocholic acid carboxylate (m/z 514). (With kind permission from Springer Science + Business Media: Brunelle, A. and Laprevote, O. (2009) Lipid imaging with cluster time-of-flight secondary ion mass spectrometry. *Anal. Bioanal. Chem.*, **393** (1), 31–35, Alain Brunelle, Figure 4.)

of a whole mouse body [37]. These images take a long time to acquire, up to 12 h, and the spatial precision is limited to about 100 μm.

The ToF-SIMS V is equipped with a TOF mass spectrometer with a 10 keV post-acceleration stage in front of the detector and with 20 keV post-acceleration as an optional upgrade for better performance in the analysis of high mass fragments from organic materials. A combination of a microchannel plate and a photomultiplier works as a single ion counting detector. High mass resolution is achieved by short primary ion pulses, a long flight path, and the introduction of an ion mirror (reflectron-type TOF mass spectrometer). The ion mirror makes it possible to achieve first-order energy focusing, which further increases the mass resolution [38]. In order to increase the transmission, modern TOF mass spectrometers also use gridless ion reflectors.

21.2.1.2 Ion Microprobes

The CAMECA NanoSIMS 50 is an example of a modern dynamic SIMS instrument optimized for high lateral resolution analysis while maintaining high mass resolution ($M/\Delta M = 5000$) and high sensitivity (parts per billion for some elements) [39]. For instance, a mass resolution of 3500 is achievable while still maintaining 100% transmission [40].

In order to achieve these operating characteristics, the NanoSIMS has several unique features such as coaxial optics and a normal incidence primary ion beam. Working in the ion microprobe mode, the primary ion beam is rastered across the

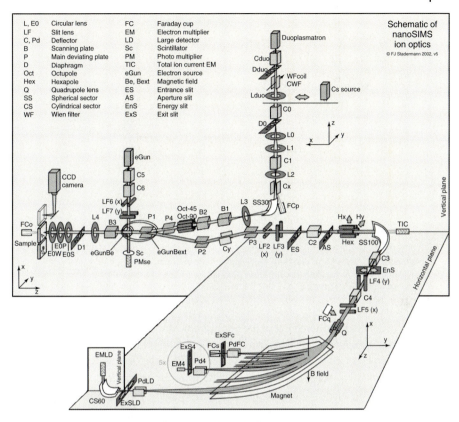

Figure 21.9 Schematic diagram of the NanoSIMS 50 showing the ion optics. (NanoSIMS schematic courtesy of F.J. Stadermann.)

surface and secondary ions are detected as a function of the beam position on the sample, resulting in high lateral resolution and collection efficiency. Figure 21.9 shows schematically the ion optics for the NanoSIMS 50 [41].

The use of reactive primary ions, in this case cesium or oxygen, increases the secondary ion yield of negative and positive ions respectively. If a liquid metal gallium source is used (to achieve a resolution of <50 nm) the elemental signals are 100–1000 times lower [40] which would make trace element analysis almost impossible.

The CAMECA NanoSIMS 50 utilizes a novel probe forming system which solves many of the problems that were previously encountered in the design of dynamic SIMS instruments. The main conflict in the design of conventional ion optics is that the objective lens of the primary ion column must be as close as possible to the sample to produce a small and intense beam, but the extraction optics must also be placed very close to the sample to collect as many secondary ions as possible. The primary and secondary optics both have an irreducible physical size so in

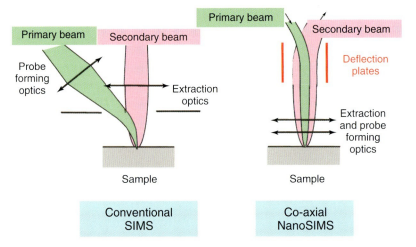

Figure 21.10 Schematic diagram showing the conventional and NanoSIMS probe forming systems. (Reproduced with permission from CAMECA.)

conventional optics neither is in the optimum position. The NanoSIMS solves this problem by simultaneously focusing the primary ion beam and collecting most of the secondary ions through colinear optics as illustrated in Figure 21.10 [39]. The coaxial optics has several advantages including a shorter working distance, smaller probe size, higher collection efficiency (improved transmission), and minimization of shadowing effects because of the normal incidence on the sample [40].

The main disadvantages with this technique are that the primary and secondary ions must be of opposite polarity and equal energy. This constraint means that minimization of matrix effects using the MCs$^+$ technique, for instance, is not possible, and oxygen flooding cannot be used [40]. In common with most optical designs, the smaller the primary ion beam diaphragm (to improve lateral resolution), the lower the sensitivity, transmission, and mass resolution.

The NanoSIMS uses a mass analyzer in a Mauttach-Herzog [3] configuration with a 90° spherical electrostatic sector and an asymmetric magnet, shown schematically in Figure 21.11. This type of mass spectrometer has reasonable transmission, unrivalled mass resolution ($M/\Delta M > 10^4$) high energy acceptance, and preserves the lateral resolution, which allows high mass resolution analysis with a small spot size [1].

After mass selection, the secondary ions are detected using electron multipliers, which can accurately measure low-intensity ion signals. Another advantage of the NanoSIMS design over previous magnetic sector SIMS instruments is the ability to collect several ion signals at the same time, five ions for the NanoSIMS 50 and seven for the NanoSIMS 50L [39]. This multicollection capability allows a more accurate reconstruction of the sample chemistry as each ion image comes from exactly the same volume, which is not the case for images from mass spectrometers that have

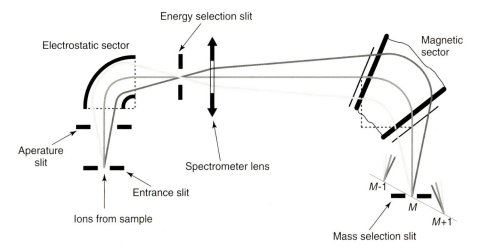

Figure 21.11 Schematic diagram of a double focusing mass spectrometer showing the electrostatic and magnetic sectors. (Redrawn from Migeon, H.N. (2008) SIMS Instrumentation: Spectrometers. NANOBEAMS PhD School Week 2, Ref. [16].)

to acquire images sequentially. This ensures reliable isotopic measurements and image registration.

21.2.1.3 Comparison of Techniques

The two instruments discussed above share a wide range of applications but there are fundamental differences between the state-of-the-art static and dynamic SIMS instruments in terms of their design aims. The NanoSIMS 50 has been designed mainly for imaging, with a lateral resolution of about 50 nm *and* relatively high mass resolution. The instrument is limited to the analysis of atomic secondary ions and small molecular fragments because of primary dose densities well beyond the static limit. The fixed number of detectors is another limitation. The ToF-SIMS V is designed to be a more versatile mass spectrometer. The full mass range is covered for each mass spectrum, and both atomic and large molecular information is accessible up to about 10 000 Da. Since the primary and secondary ion columns are separate features, it is possible to switch primary ion species easily to accommodate different analytical needs. For high lateral resolution applications, the ToF-SIMS V struggles to reach 50 nm lateral resolution while maintaining an adequate mass resolution, so that clear separation of isobaric ions is difficult in many practical situations.

A recent collaboration under the EU 6th Framework project "Nanobeams" compared the imaging capabilities of the latest ToF-SIMS instrumentation and the NanoSIMS. An immiscible blend of polystyrene (PS) and poly(methylmethacrylate) (PMMA) was used as a model system to study the nanoscale morphology of self-assembled polymer films, but with the PMMA being fully deuterated to allow for isotopic analysis. An IonTOF ToF-SIMS V instrument equipped with a Bi liquid metal ion source was used, and in order to minimize fragmentation of the larger characteristic ions from each of the polymers Bi_3^+ and Bi_5^+ cluster ions were selected. NanoSIMS imaging with a Cs^+ primary ion beam allowed for higher resolution imaging, and the combination of these two techniques provides complementary information. A ToF-SIMS instrument can acquire a full mass spectrum at each pixel and therefore provide much more information about the chemical nature of the sample, whereas a dynamic SIMS instrument can give clearer images with higher resolution. Typical images from these two instruments from the same samples are shown in Figure 21.12.

SIMS falls somewhere in between the best imaging and elemental detection techniques. Although the lateral resolution of SIMS is not as good as energy-dispersive X-ray spectroscopy and electron energy loss spectroscopy in a transmission electron microscope [3, 42], these techniques can neither distinguish between isotopes of the same element, prohibiting the use of isotopic spiking experiments [6], nor detect molecular species. In contrast, SIMS does not have the sensitivity of some chemical analysis techniques, but these generally derive their sensitivity from sampling a very large volume, which means there is no opportunity to localize elements since spot

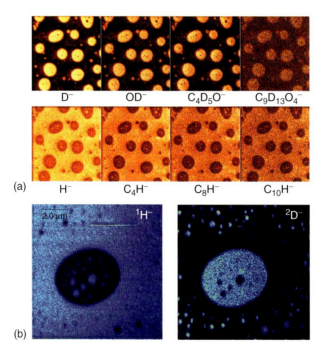

Figure 21.12 Comparison of the images and information obtainable from (a) ToF-SIMS and (b) NanoSIMS on the same samples. The ion images in (a) show the characteristic fragments from d-PMMA (top row) and from the PS (bottom row), image size 31 μm. The 10 μm² NanoSIMS images in (b) show many small d-PMMA regions in the PS matrix, which are hardly visible in the ToF-SIMS images.

sizes can be as large as 5–15 mm [43]. Synchrotron-based X-ray techniques, such as X-ray Fluorescence, are approaching the lateral resolution of SIMS analysis, but the spatial resolution is significantly worse because of the large interaction volume of the X-rays. However, synchrotron mapping can be over a very large scale, requires minimal sample preparation as the samples can be analyzed in air [44], and the sensitivity is better than that of SIMS, approaching the ppb range [45]. Although other techniques have individual characteristics that are better than SIMS, it is the combination of characteristics listed previously that makes SIMS an attractive technique.

21.2.2
Imaging and Data Analysis

The data acquired by the two different types of SIMS instruments is fundamentally different. Whereas a ToF-SIMS instrument acquires a complete mass spectrum for each pixel, a dynamic SIMS instrument is limited to the acquisition of a few selected ions. The ToF-SIMS user thus does the whole data analysis after acquisition, and the major challenge is to establish a reliable mass calibration in order to correctly identify specific ion signals. The next step is to find with confidence ion signals that relate to the specific analytical problem. A comparison with the vast libraries of mass spectra is sometimes necessary or at least helpful. From the set of raw data, it is then possible to reconstruct images for each potentially relevant line from the mass spectrum.

The NanoSIMS 50 operator has to do the ion selection beforehand, and in some cases has to design the whole experiment with the restriction to five secondary ions in mind. The raw data obtained is one image per secondary ion per scan, although it is possible to acquire several images in succession, which can be combined to form a 3D representation of the chemistry of the sample.

The next level of image data analysis is similar for both techniques. All mathematical operations are possible. Adding images pixel by pixel of, for example, molecular fragments from the same parent molecule is a widely used means to increase intensity and thereby contrast. Background noise can be reduced or even eliminated either by subtracting an image of the background signal or subtracting an average threshold value from the intensity for each pixel. For normalizing purposes, it is possible to divide images pixel by pixel. It is common to use signal intensity changes as a function of the position on the specimen to illustrate changes in sample chemistry. These line scans, often carried out for different secondary ions along the same line, can give a detailed insight into the relative positions of structures for different secondary ions. An example of this is shown in Figure 21.13 from an oxidized stainless steel sample. The line scan clearly shows that the boron is accumulated at the Cr-rich oxide/magnetite interface on surfaces that have been exposed to the water chemistry characteristic of a pressurized water nuclear reactor [46].

These data analysis tools do not go beyond the simple analysis of signal intensities. More sophisticated analysis approaches normally need correlation between the larger number of mass lines that the ToF-SIMS can provide. These techniques

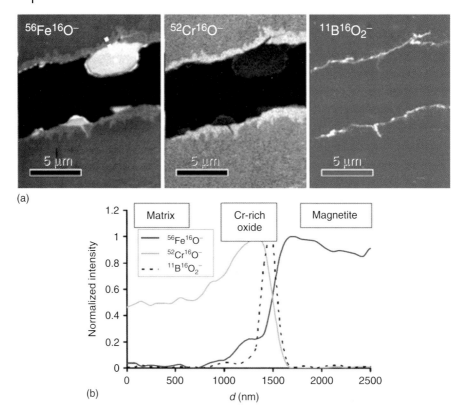

Figure 21.13 (a) NanoSIMS images showing oxidation on the surface of a stainless steel sample. The dotted white line indicates the position of the line scan shown in (b), which shows that boron is located at the interface of the Cr-rich oxide and the magnetite ($^{56}56^{16}O^-$). (Reprinted from Lozano-Perez, S., Schroder, M., Yamada, T., Terachi, T., English, C.A., and Grovenor, C.R.M. (2008) Using NanoSIMS to map trace elements in stainless steels from nuclear reactors. *Appl. Surf. Sci.*, **255** (4), 1541–1543, Ref. [46]. Copyright (2008), with permission from Elsevier.)

include principle component analysis (PCA) [47] and maximum autocorrelation factor (MAF) analysis [48, 49].

A larger challenge from the analytical as well as the technical side is true 3D imaging of complex samples. When using an ion-based technique to erode an inhomogeneous sample surface in order to expose the next layer for image acquisition, we need to ensure that the sample is sputtered evenly. Different materials sputter at different rates, which can lead to errors in assigning an accurate depth scale. Ion-beam-based sputtering also has the problem that there is a potential degradation of the sample as a result of accumulated damage from the primary ion beam [36]. Organic and biological samples are highly susceptible to ion-induced fragmentation and successive signal degradation [36]. Since the surface roughness has to be taken into account during SIMS reconstruction of a series of 2D images, every image has to be interpreted as the projection of a rough surface structure. Only in combination with other techniques such as atomic force microscopy (AFM) or optical profilometers, which provide the necessary topography information before and after the actual SIMS measurement, is it possible to attempt a reconstruction of the true three-dimensional chemical structure of a sample.

For the case when the structures of interest sit on top of a flat and chemically distinguishable sample surface, it is possible to redefine a baseline of the depth scale after the whole structure is removed. An example of this approach is shown in Figure 21.14 [50, 51]. In this experiment, several layers of normal rat kidney cells were grown on cover slips and chemically fixed. The cells form a rough surface structure, schematically shown in the top image of Figure 21.14a. After sputtering through the whole cellular layer and acquiring images during (or in this case in between) sputtering cycles, it is possible to deconvolute the images and

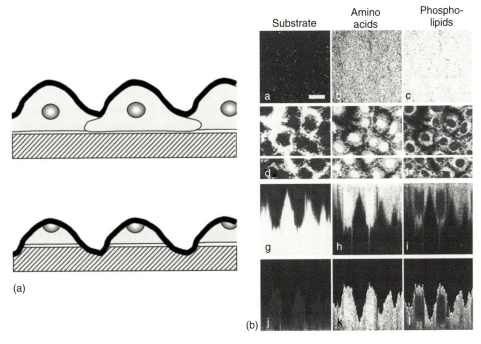

Figure 21.14 (A) Schematic diagram of the surface structure of rat kidney cells grown on a cover slip, with the dark black line indicating the uppermost monolayer before sputtering (top) and part way through sputtering (bottom) [50], and (B) (a–c) ToF-SIMS secondary ion images of the top surface of the sample before sputtering, (d–f) within the cell after 45 slices, (g–i) cross sections through the stack along the white line shown in (d–f), (j–l) cross sections after rescaling the data correcting the z-axis. The substrate signal is Na^+, the amino acid and phospholipid signals are summed from several ions. Scale bar = 20 µm. (Breitenstein, D., Rommel, C.E., Mollers, R., Wegener, J., and Hagenhoff, B. (2007) The chemical composition of animal cells and their intracellular compartments reconstructed from 3D mass spectrometry. *Angew. Chem. Int. Ed.*, **46** (28), 5332–5335, Ref. [50]. Copyright Wiley-VCH Verlag GmbH & Co. KGaA. Reproduced with permission.)

establish a "real" depth scale. Na$^+$ was used as the ion of interest for the glass slide and several ion signals were summed up for the amino acids (middle column, Figure 21.14b) and phospholipids (right column). After several sputtering cycles, the original homogeneous surface reveals distinct round features, correlating with cellular compartments (Figure 21.14b) (d–f). The white line across Figure 21.14b), images (d–f) indicates the position of a section in the z-direction, derived from the acquired raw data. The corresponding ion images show a clear contrast between the Na$^+$ signal originating from the glass slide and the cellular signals (Figure 21.14b) images (g–i). In the next step, the onset of the Na$^+$ signal defines the baseline of the corrected depth scale. The adjusted ion images, (j–l), show a more realistic picture of the surface structure and depth distribution of the cellular features.

Non-ion-beam-related techniques, for example, sequential microtomy, do not suffer from these problems, but are limited in terms of depth resolution and lateral drift because of the difficulty in realigning the sample after movement of the sample from the stage to the microtome and back into the instrument. An example of this sequential microtomy and ToF-SIMS analysis is shown in Figure 21.15 [52].

In this example, foraminifera of the type *ammonia tepida* were transferred into a small saltwater droplet on a sample holder, plunge frozen in liquid propane and further cooled to liquid nitrogen temperatures [52]. This kind of preparation keeps the organism preserved as close to the living state as possible. The sample was introduced into a high-vacuum microtome chamber attached to a ToF-SIMS instrument for analysis. In order to gain 3D information for the whole organism, which is almost 400 μm in diameter, the sample was sliced with the microtome until the top part of the specimen was uncovered. The sample was subsequently transferred under high vacuum-conditions into the analysis chamber of the instrument where

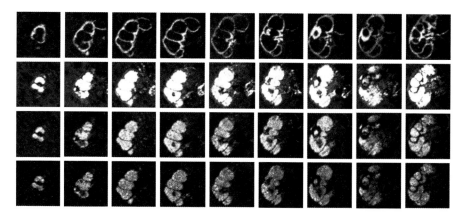

Figure 21.15 Positive ToF-SIMS images of a foraminifera cryosectioned 18 times (25 keV Bi$_3^+$, 400 × 400 μm^2). Ca$^+$ (top line), K$^+$ (second line), C$_3$H$_8$N$^+$ (third line), C$_5$H$_{15}$PNO$_4^+$ (bottom line). (Reproduced with permission from Möller, J. (2008) Analysen kryopräparierter nicht-dehydrierter probensysteme mit hilfe eines neu entwickelten ToF-SIMS-instruments mit integrierter Hochvakuumkryoschnittapparatur, in PhD thesis in Physikalisches Institut, Westfälische Wilhelms-Universität, Münster, Ref. [52].)

image acquisition was carried out using a 25 keV Bi_3^+ primary ion beam and a 400 × 400 µm² raster [52]. After each measurement, the sample was returned to the microtome chamber in order to remove a further 15–25 µm. This series of microtome slices and consecutive image acquisition from the newly uncovered area can be repeated through the entire sample thickness without any distortion of the surface due to preferential sputtering. In Figure 21.15, every second image of a series of 18 consecutive images is shown for Ca^+, characteristic for the calcite shell, K^+, which is enriched in the living parts of the cellular organism and two organic compounds ($C_3H_8N^+$ and $C_5H_{15}PNO_4^+$), both originating from the cellular body of the foraminifera.

21.3
Biological SIMS

SIMS has its roots in the semiconductor industry and its development was driven by the need to accurately measure low concentrations of dopants in semiconductors [2]. However, it has become widely used for element localization in geochemistry, metallurgy, and electronics [53]. The SIMS community has for a long time been dominated by chemists and physicists who have mainly focused on improving the instrumentation with little interaction with the biological community [54]. SIMS is essentially a physical science technique, so why is it increasingly being used to study biological specimens? The main reasons are its high sensitivity, detection of all elements and isotopes, molecular imaging combined with a reasonable lateral resolution, and the potential for 3D imaging [55, 56].

The earliest experiments using SIMS to image biological specimens were performed by Galle [57] in 1970, who studied sodium and potassium ions in renal tissue and human blood. The first SIMS instruments were, however, not suitable for biological SIMS. The primary ion beam was large and intense, with a diameter of 25–250 µm, an energy of 10 keV, and a beam current of 10–100 nA [6]. These conditions are highly destructive for most biological samples, which tended to be completely vaporized after one analysis. Trace element analysis, which requires long acquisition times, was almost impossible. Other problems included poor mass resolution ($M/\Delta M \sim 300$) and a lateral resolution of around 1 µm [6]. Both of these values were insufficient for biological analysis. For example, it was not possible to separate $^{12}C^{14}N^-$ ($m = 26.045$) and $^{12}C_2^1H_2^-$ ions ($m = 26.024$) or $^{31}P^{16}O_3^-$ and $^{89}Br^-$ ions, or to obtain much intercellular detail in images.

The ability of SIMS to map molecular ions was quickly found to be particularly useful for biological materials. Perhaps the most useful ion is CN^-, which is a characteristic fragment of amino acids and therefore allows mapping of protein distributions [28]. It is also widely used to show the general morphology of biological samples [58] since membranes also have a high yield of CN^-, as shown in later examples.

SIMS analysis of biological materials generally falls into two classes. The first is elemental mapping, which exploits the sensitivity of SIMS combined with its high mass resolution. Trace elements are important in many biological processes and are important in the human diet. In order to understand how these trace elements are sequestered and localized, it is necessary to be able to map them with subcellular resolution. While there are a number of instruments that are capable of detecting parts per million (ppm) and even parts per billion (ppb) concentrations, they usually have limited or no spatial resolution. An example of the value of combining high-sensitivity analysis with high-resolution imaging is shown by Moore *et al.*, who used the NanoSIMS to map ppm concentrations of arsenic in rice grain [59]. Arsenic contamination of ground water is a severe problem in countries like Bangladesh, and this has resulted in rice grain containing elevated levels of arsenic as the rice is grown in paddy fields irrigated with this contaminated water. Understanding where the arsenic is localized in the grain can help to plan strategies to reduce the human uptake. Figure 21.16 shows that the arsenic is localized in a region just below the bran layer, so in order to reduce the arsenic concentration the rice grain needs to be milled to remove more than just the outer bran layer.

As rice grains dry out naturally on the plant, there was no need for complex sample preparation in this experiment as there was minimal residual water left in the grain. Samples were simply mounted in resin, sectioned with a razor blade, and coated with Pt to prevent charging [59].

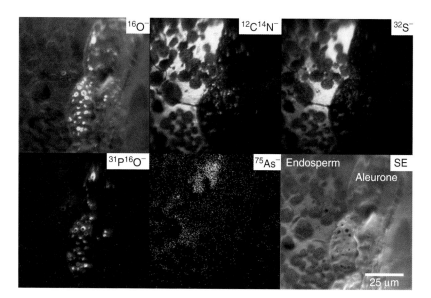

Figure 21.16 NanoSIMS ion maps of the edge of a rice grain showing two cells of the outer bran layer (marked aleurone) and part of the starchy endosperm, which extends across the majority of the rice grain. Maps for $^{16}O^-$, $^{12}C^{14}N^-$, $^{32}S^-$, $^{31}P^{16}O^-$, $^{75}As^-$, and the secondary electron (SE) signal were acquired with the Cs^+ primary ion beam. The arsenic is localized in the white rice part of the grain rather than in the bran layer. (Reproduced with permission from Moore, K.L., Schroder, M., Lombi, E., Zhao, F.J., McGrath, S.P., Hawkesford, M.J., Shewry, P.R., and Grovenor, C.R.M. (2010) NanoSIMS analysis of arsenic and selenium in cereal grain. *New Phytol.*, **185** (2), 434–445, Ref. [59]. Copyright 2010, with permission from John Wiley & Sons.)

The second main use of SIMS is in ratio or spiking measurements, where stable isotopes [22, 60] or "exotic" atoms [6, 61, 62] are incorporated as labels into cells or tissue samples as a means to understand transport and sequestration mechanisms using the high mass resolution and spatial resolution of SIMS. Labels such as ^{13}C or ^{15}N are treated by cells almost exactly as the more naturally occurring isotopes [22], but are easily distinguished by SIMS (Figure 21.5). This technique is more commonly used in dynamic SIMS instruments because there are fewer mass interferences and, unlike ToF-SIMS instruments, they are not able to map for the larger fragments that could provide direct information about the position of specific molecules. There are several elements that have isotopes with a low natural abundance that can be used for spiking experiments, many of which are relevant to biological materials, including ^{2}H, ^{13}C, ^{15}N, ^{18}O, ^{34}S, ^{41}K, and ^{44}Ca.

In Figure 21.17 ^{15}N has been used to monitor protein renewal in the mouse cochlea using ^{15}N-leucine and to demonstrate quantitative analysis of isotopes in subcellular compartments. The mouse cochlea is a highly organized tissue with several different cell types. Located in a subcellular structure of this tissue is the stereocilium, the mechanosensing organelle of hair cells. As nitrogen is not easily ionized in SIMS analysis, it must be measured as the CN$^-$ ion [60]. Initial experiments on untreated control samples proved that the NanoSIMS could be used to measure a ^{15}N/^{14}N ratio of 0.368% in the cochlea very similar to the natural isotope ratio of 0.3673% even though the structures are only 250 nm across [62]. Figure 21.17 shows a NanoSIMS image of the sterocilia, labeled Sb1, from a mouse that was fed on a ^{15}N-leucine diet for nine days. The ^{12}C$^-$, ^{13}C$^-$, ^{12}C^{14}N$^-$, and ^{12}C^{15}N$^-$ images were acquired simultaneously, and the ^{12}C^{14}N$^-$/^{12}C^{15}N$^-$ and ^{12}C$^-$/^{13}C$^-$ ratio images were obtained by dividing the images pixel by pixel after analysis. There is strong contrast in the ^{12}C^{14}N$^-$/^{12}C^{15}N$^-$ ratio image because ^{15}N has been incorporated into the proteins, but the ^{12}C$^-$/^{13}C$^-$ ratio image has no contrast because no additional ^{13}C was added. A hue saturation intensity (HSI) image can be used to illustrate the ^{15}N incorporation into the tissue, which can then be expressed as a percentage of protein renewal, with values ranging from 0% (blue) to 60% (pink) as shown in Figure 21.17g.

In the second example, ^{13}C ions have been mapped to understand the uptake and metabolism of carbon in coral. This animal was chosen because it is thought that symbiotic algae photosynthetically fix inorganic carbon from seawater and subsequently translocate this to the coral host. In order to understand coral reef formation and growth and storage of these fixed carbon compounds, it is important to understand the role of the individual cell types [22]. Colonies of coral were incubated in seawater artificially enhanced with NaH^{13}CO$_3$ before fixing, decalcification, embedding, and sectioning. The decalcification step ensures that any H^{13}CO$_3$ that has not been translocated is converted into ^{13}CO$_2$ and lost from the sample. The ^{13}C$^-$ signal was readily detected with the NanoSIMS and the subcellular resolution enabled the ^{13}C distribution to be clearly visualized within the algal symbionts, as can be seen in Figure 21.18. The authors demonstrate that it is necessary to have sufficient mass resolution to be able to distinguish ^{13}C from the isobaric interference of ^{12}C^{1}H$^-$ ($M/\Delta M > 1660$), and that the position of the detector needs to be

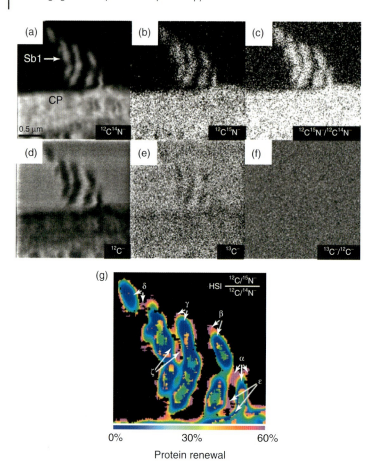

Figure 21.17 NanoSIMS images of stereocilia from a mouse cochlea fed with ^{15}N-Leucine for nine days. (a) $^{12}C^{14}N^-$, (b) $^{12}C^{15}N^-$, (c) $^{12}C^{14}N^-/^{12}C^{15}N^-$ ratio image, (d) $^{12}C^-$, (e) $^{13}C^-$, (f) $^{12}C^-/^{13}C^-$ ratio image, and (g) HSI image of the $^{12}C^{15}N/^{12}C^{14}N$ ratio derived from (a) and (b). The colors correspond to the excess ^{15}N derived from the measured $^{12}C^{15}N^-/^{12}C^{14}N^-$ isotope ratio, expressed as a percentage of the ^{15}N excess in the feed, which is a measure of protein renewal. The HSI image is 3 μm × 3 μm. (Reproduced from Lechene, C., Hillion, F., McMahon, G., Benson, D., Kleinfeld, A., Kampf, J.P., Distel, D., Luyten, Y., Bonventre, J., Hentschel, D., Park, K., Ito, S., Schwartz, M., Benichou, G., and Slodzian, G. (2006) High-resolution quantitative imaging of mammalian and bacterial cells using stable isotope mass spectrometry. *J. Biol.*, **5** (6), 20, Ref. [62]. Open Access article.)

carefully chosen to avoid the low mass tail of the $^{12}C^1H^-$ as shown in Figure 21.18B. Taking a ratio of the ^{13}C image with the ^{12}C image, as shown in Figure 21.18A(d), minimizes the effects of topography and matrix effects which could alter the ion yield of the C^- ions. The $^{13}C/^{12}C$ image in Figure 21.18 shows regions of significant ^{13}C accumulation in a ring-like structure in the algal symbiont.

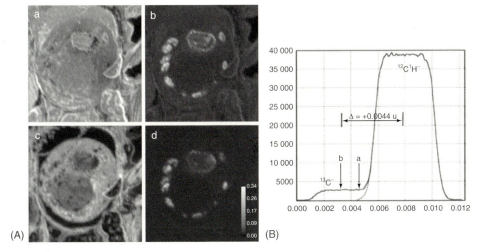

Figure 21.18 (A) High-resolution NanoSIMS images of a single symbiotic alga in a ^{13}C-labeled coral Galaxea fascicularis: (a) ^{12}C$^-$; (b) ^{13}C$^-$; (c) ^{12}C^{14}N$^-$; and (d) ^{13}C$^-$/^{12}C$^-$. FOV = 10 μm. Scale on (d) indicates ^{13}C$^-$/^{12}C$^-$ ratio values with a minimum of 0 and a maximum of 0.34. (B) Mass scan around mass 13 showing the interference of the ^{12}C^{1}H$^-$ peak on the ^{13}C$^-$ peak, positioning of the detector at point b avoids interference from the low mass tail of the ^{12}C^{1}H$^-$ peak, which would occur at point a. (Reproduced with permission from Clode, P.L., Stern, R.A., and Marshall, A.T. (2007) Subcellular imaging of isotopically labeled carbon compounds in a biological sample by ion microprobe (NanoSIMS). *Microsc. Res. Tech.*, **70** (3), 220–229, Ref. [22]. Copyright 2007, with permission from John Wiley & Sons.)

21.3.1
Sample Preparation of Biological Materials for SIMS Analysis

SIMS analysis needs to be performed under ultra high-vacuum conditions ($<10^{-7}$ Pa) to prevent scattering of the primary and secondary ions by residual gas molecules and analysis of contaminants adsorbed onto the surface [1, 21]. Biological materials inherently consist of a large fraction of water (about 80% [63, 64]) and this must be removed before analysis can be undertaken. This process is the most difficult and causes the most problems in the study of samples from the life sciences [65]. Guerquin *et al.* [6] state that "An ideal sample preparation should preserve both the structural and chemical integrity of the living cell. However, this requirement must be adapted to the instrument resolution and appropriate to the specificity of the analysis under investigation."

There is considerable literature on these sample preparation issues and all the authors stress the care that must be taken to achieve a final sample with chemistry representative of the living sample.

21.3.1.1 Chemical Fixation
This type of relatively simple preparation is suitable for situations where the analyte of interest is tightly bound to a cellular structure, for example, to a membrane,

organelle, or DNA. This process involves immersing the tissue into a fixative such as 2.5% glutaraldehyde in 0.1 M cacodylate buffer, dehydration in a series of increasing ethanol solutions, and finally resin embedding [66]. Although this method is suitable for structural analysis by electron microscopy, and is one of the simplest methods of preparing biological specimens, it causes complete redistribution of diffusible ions such as Na^+ and other small molecules, and is therefore an inappropriate method if the *in vivo* distribution of these ions is to be studied with SIMS [64, 67, 68].

An example of chemical fixation is shown in the following example. The most common method currently used to locate markers and labels in cells is to use an optical technique such as fluorescence microscopy. This technique, although powerful and very widely used, suffers from several problems [69]. It requires the use of bulky dyes that can modify the behavior of the molecule of interest in the cell [6], is diffraction-limited to 200–700 nm, and the molecule of interest may not be naturally fluorescent. Imaging SIMS is increasingly finding use in the study of the metabolisms of cells as well as in research on targeted drug delivery and the localization of metals, as the technique and instruments improve. By using an element or isotope that is not naturally found in biological samples, small molecules can be labeled without affecting the shape, binding potential, or lipophilicity, and the chemical localization of many molecules or non-fluorescent compounds can be directly explored. Lau *et al.* [69] recently presented results comparing on the same cell the distribution of BrdU in HeLa cells mapped with conventional fluorescent immunochemistry and NanoSIMS analysis of the $^{79}Br^-$ and $^{80}Br^-$ ions. A typical result is shown in Figure 21.19 and the distribution of BrdU from the two techniques is identical. These authors subsequently demonstrated the imaging of the distribution of a Mo-containing drug, ATN-224, a therapeutic copper chelator for which there is no fluorescent marker. In this case, NanoSIMS imaging not only provided a means of locating the drug in the cell, but when combined with organelle-specific fluorescent staining the drug could be located to individual lysosomes [69].

21.3.1.2 Cryofixation

Cryofixation is the main alternative to chemical fixation [28, 70]. This route attempts to immobilize all the molecules and ions in their *in vivo* state by freezing the sample so fast that ice forms in the amorphous state and takes up the same volume as the liquid water it replaces [71, 72]. Growth of ice crystals must be avoided as they are normally acicular and will damage the cells [73]. Having made this choice of sample preparation, there are various routes that can be taken [67, 74–78].

If the SIMS instrument has a cryostage, then it is possible to analyze the sample in the hydrated state. This has the advantage that complex sample preparation methods do not need to be used and the redistribution of diffusible ions should be avoided; however, the ion yields from a frozen sample can be difficult to quantify, water is preferentially removed along the z-axis, and matrix effects can be severe [79–81]. Figure 21.20 shows cryo-ToF-SIMS images of the metaxylem from a cross section of a bean stem, indicating how this technique can be used to monitor the

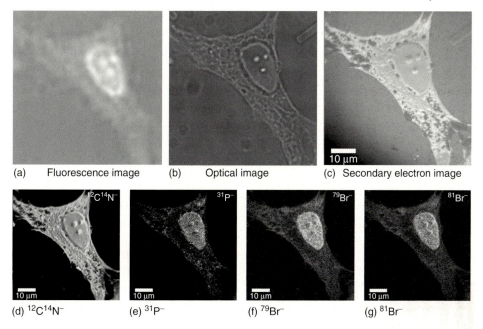

Figure 21.19 Images of the same HeLa cell taken with (a) fluorescent, (b) optical, and (c–g) NanoSIMS microscopes. The position of the nucleoli is identical in all the images. (Reproduced with permission from Lau, K.H., Christlieb, M., Schröder, M., Sheldon, H., Harris, A.L., and Grovenor, C.R.M. (2010) Development of a new bimodal imaging methodology; a combination of fluorescence microscopy and high-resolution secondary ion mass spectrometry. *J. Microsc.*, **240** (1), 21–31, Ref. [69]. Copyright 2010, with permission from John Wiley & Sons.)

uptake of nutrients into plant specimens [82]. This image was taken with a spatial resolution of ~1μm, allowing the distribution of the elements of interest to be mapped at a subcellular level. For instance, it is possible to see the dendritic ice crystals within each of the cells, showing that the sample is not entirely vitrified. The water distribution in the sample is shown in Figure 21.20a by $^1H_3^{16}O^+$, which is the most abundant water-related ion species in this case [82]. Figure 21.20e clearly shows that the tracer for K transport is predominantly localized in the xylem vessels with some spreading to neighboring tissues. In contrast, the Na is localized in the xylem parenchyma, the cells surrounding the xylem. This is consistent with what is understood about nutrient transport properties in these cells, and shows the potential of cryo-ToF-SIMS in detecting multiple tracers at high resolution and sensitivity for studying these mechanisms in very great detail.

For microscopes that do not have a cryostage, the best preparation route is to fix and dehydrate the sample at a low temperature to avoid redistribution artifacts, and then infiltrate with resin. Embedded samples can then be sectioned using a microtome and deposited on a silicon wafer for analysis. As an example of the preservation of the structure of delicate (highly vacuolated) plant samples

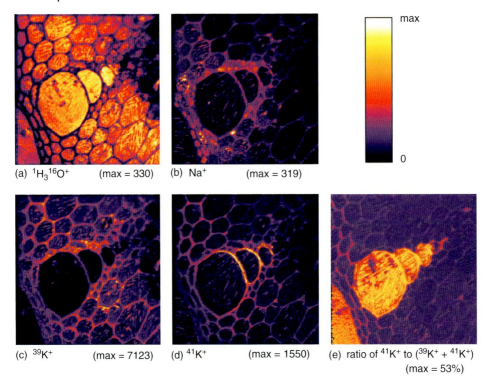

Figure 21.20 Cryo-TOF-SIMS images of the surface of a bean stem cross section showing the metaxylem. The images show the maps of $^1H_3{}^{16}O^+$, $^{23}Na^+$, $^{39}K^+$, and $^{41}K^+$. (e) Shows the ratio of $^{41}K^+$ to ($^{39}K^+ + {}^{41}K^+$) in percent. The field of view is 195 μm × 195 μm [82].

by high-pressure freezing (HPF), Figure 21.21 shows elemental maps from the nickel hyperaccumulator *Alyssum lesbiacum* at the subcellular scale [83]. Ultrarapid cryofixation is generally accepted to be the preferred method for preserving the most challenging biological material in the *in vivo* state [56, 71, 83, 84], and so HPF followed by freeze substitution with acetone and resin embedding was chosen for this sample. Samples were sectioned with a microtome to a thickness of 0.5 μm and deposited onto Si wafers. The images in Figure 21.21 show a cross section of a leaf with a stomata complex, epidermal cells, and palisade mesophyll cells. The left-hand set of images (Figure 21.21a) were taken with the Cs^+ beam and the right (Figure 21.21b) with the O^- beam. The Cs^+ beam gives a much higher resolution, better than 80 nm in the CN^- image, allowing features such as the cell cytoplasm and starch grains within a single chloroplast in one of the two stomatal guard cells to be clearly resolved [83]. The nickel distribution (dry shoot Ni content of 7.87 ± 1.89 g kg^{-1}) is uniform across the central vacuole of the epidermal cells. Analysis of the electropositive ions, however, shows that the *in vivo* distribution

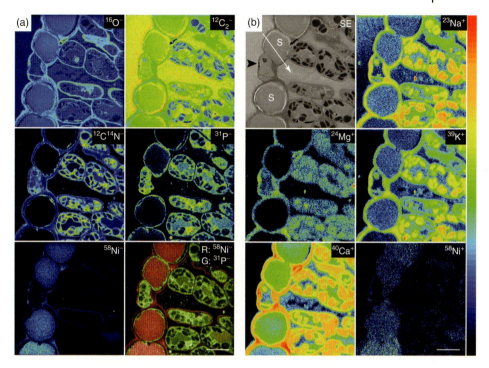

Figure 21.21 NanoSIMS analysis of a cross section of a nickel rich Alyssum Lesbiacum leaf prepared by high-pressure freezing and freeze substitution. (a) NanoSIMS maps from a peripheral region of the leaf cross section including a stomatal complex obtained using the Cs$^+$ primary ion beam showing ^{16}O$^-$, ^{12}C$_2^-$, ^{12}C^{14}N$^-$, ^{31}P$^-$, and ^{58}Ni$^-$ ion maps and a color overlay for Ni and P. (b) NanoSIMS maps obtained using the O$^-$ primary ion beam for the distribution of ^{23}Na$^+$, ^{24}Mg$^+$, ^{39}K$^+$, ^{40}Ca$^+$, and ^{58}Ni$^+$ signals from the same area shown in (a). The secondary electron (SE) image was acquired with the Cs$^+$ beam and shows the morphology of the sample. Scale bar= 10 μm. (Reproduced with permission from TECHNICAL ADVANCE: Smart, K.E., Smith, J.A.C., Kilburn, M.R., Martin, B.G.H., Hawes, C., and Grovenor, C.R.M. (2010) High-resolution elemental localization in vacuolate plant cells by nanoscale secondary ion mass spectrometry. *Plant J.*, **63** (5), 870–879, Ref. [83]. Copyright 2010, with permission from John Wiley & Sons.)

of the highly diffusible ions such as ^{23}Na$^+$, ^{24}Mg$^+$, ^{39}K$^+$, and ^{40}Ca$^+$ have been altered significantly during the sample preparation. It is now well established that the Na/K and K/Ca ratios give an important calibration of the quality of the sample preparation and less rigorous sample preparation will result in the complete redistribution of these elements [65, 83, 85]. For example, the K/Ca ratio in the epidermal cell vacuoles is much lower than would be expected for living cells [83], so even with the use of advanced sample preparation techniques that achieve excellent morphological preservation it is hard to avoid redistribution of the fastest diffusing ions.

21.4
Conclusions

Both static and dynamic SIMS instrumentation has been improved in the past 15 years to the state where high-resolution imaging of elemental and molecular species can be reliably carried out even on dilute species. The range of sample types that can be analyzed has also been expanded, so that imaging SIMS is now a general purpose analytical technique covering the whole range of physical science and biological materials.

References

1. Vickerman, J.C.E., Brown, A.E., and Reed, N.M.E. (1989) *Secondary Ion Mass Spectrometry Principles and Applications*, Clarendon Press.
2. Wilson, R.G., Stevie, F.A., and Magee, C.W. (1989) *Secondary Ion Mass Spectrometry: A Practical Handbook for Depth Profiling and Bulk Impurity Analysis*, vol. 1, John Wiley & Sons, Inc., New York, (various pagings).
3. Benninghoven, A., Reudenauer, F.G., and Werner, H.W. (1987) *Secondary Ion Mass Spectrometry: Basic Concepts, Instrumental Aspects, Applications, and Trends (Chemical Analysis)*, John Wiley & Sons, Inc., New York.
4. Sigmund, P. (1969) Theory of sputtering. I. Sputtering yield of amorphous and polycrystalline targets. *Phys. Rev.*, **184**(2), 383.
5. CAMECA http://www.cameca.com/instruments-for-research/sims-static.aspx (accessed 04 October 2010).
6. Guerquin-Kern, J.L., Wu, T.D., Quintana, C., and Croisy, A. (2005) Progress in analytical imaging of the cell by dynamic secondary ion mass spectrometry (SIMS microscopy). *Biochim. Biophys. Acta*, **1724** (3), 228–238.
7. Benninghoven, A. (1971) Beobachtung von oberflachenreaktionen mit der statischen methode der sekundarionen-massenspektroskopie. I die methode. *Surf. Sci.*, **28** (2), 541–562.
8. CAMECA http://www.cameca.com/instruments-for-research/sims-dynamic.aspx (accessed 04 October 2010).
9. Whitlow, H.J., Hautala, M., and Sundqvist, B.U.R. (1987) Collision cascade parameters for slow particles impinging on biomolecule targets. *Int. J. Mass Spectrom. Ion Process.*, **78**, 329–340.
10. Benninghoven, A. (1979) in *Secondary Ion Mass Spectrometry, SIMS-II: Proceedings of the Second International Conference on Secondary Ion Mass Spectrometry (SIMS II), Stanford University, Stanford, California, USA, 1979*, Springer Series in Chemical Physics, Vol. xiii, Springer-Verlag, Berlin, p. 298.
11. Benninghoven, A. (1983) Some aspects of secondary ion mass-spectrometry of organic-compounds. *Int. J. Mass Spectrom. Ion Processes*, **53**, 85–99.
12. Yu, M.L. (1978) Work-function dependence of negative-ion production during sputtering. *Phys. Rev. Lett.*, **40** (9), 574–577.
13. Yu, M.L. and Lang, N.D. (1983) Direct evidence of electron-tunneling in the ionization of sputtered atoms. *Phys. Rev. Lett.*, **50** (2), 127–130.
14. Vickerman, J.C., and Briggs, D., SurfaceSpectra Limited (2001) *ToF-SIMS: Surface Analysis by Mass Spectrometry*, vol. **ix**, IM; SurfaceSpectra, Chichester, MA, 789 p.
15. Littmark, U. and Hofer, W.O. (1978) Influence of surface-structures on sputtering – angular-distribution and yield from faceted surfaces. *J. Mater. Sci.*, **13** (12), 2577–2586.
16. Philipp, P. (2008) Applications of SIMS: Introduction – Depth Profiling I - II, NANOBEAMS PhD School Week 2.
17. Werner, H.W. and Warmoltz, N. (1976) The influence of selective sputtering on

surface composition. *Surf. Sci.*, **57** (2), 706–714.
18. Albers, T., Neumann, M., Lipinsky, D., Wiedmann, L., and Benninghoven, A. (1994) Combined depth profile analysis with SNMS, SIMS and XPS - preferential sputtering and oxygen-transport in binary metal-oxide multilayer systems. *Surf. Interface Anal.*, **22** (1–12), 9–13.
19. Evans, C. http://www.eaglabs.com/training/tutorials/sims_theory_tutorial/index.php (accessed 04 October 2010).
20. Levi-Setti, R., Hallegot, P., Girod, C., Chabala, J.M., Li, J., Sodonis, A., and Wolbach, W. (1991) Critical issues in the application of a gallium probe to high-resolution secondary ion imaging. *Surf. Sci.*, **246** (1–3), 94–106.
21. Downard, K., Royal Society of Chemistry (Great Britain) (2004) *Mass Spectrometry: A Foundation Course*, vol. **xvi**, Royal Society of Chemistry, Cambridge, p. 210.
22. Clode, P.L., Stern, R.A., and Marshall, A.T. (2007) Subcellular imaging of isotopically labeled carbon compounds in a biological sample by ion microprobe (NanoSIMS). *Microsc. Res. Tech.*, **70** (3), 220–229.
23. Liebl, H. (1975) Ion probe microanalysis. *J. Phys. E*, **8** (10), 797–808.
24. van der Heide, P.A.W., Lim, M.S., Perry, S.S., and Bennett, J. (2003) A systematic study of the surface roughening and sputter rate variations occurring during SIMS ultrashallow depth profile analysis of Si with Cs+. *Nucl. Instrum. Methods. Phys. Res. B*, **201** (2), 413–425.
25. Wittmaack, K. (1996) Sputtering yield changes, surface movement and apparent profile shifts in SIMS depth analyses of silicon using oxygen primary ions. *Surf. Interface Anal.*, **24** (6), 389–398.
26. Evans, C. http://www.eaglabs.com/files/posters/PS703.pdf (accessed 04 October 2010).
27. Ramseyer, G.O. and Morrison, G.H. (1983) Relative sensitivity factors of elements in quantitative secondary ion mass-spectrometric analysis of biological reference materials. *Anal. Chem.*, **55** (12), 1963–1970.
28. Levi-Setti, R.R. (1988) Structural and microanalytical imaging of biological materials by scanning microscopy with heavy-ion probes. *Annu. Rev. Biophys. Biophys. Chem.*, **17**, 325–347.
29. Vickerman, J.C. and Gilmore, I.S. (2009) *Surface Analysis: The Principal Techniques*, vol. **xix**, 2nd edn, John Wiley & Sons, Ltd, Chichester, 666 p.
30. Gnaser, H. (1996) Initial stages of cesium incorporation on keV-Cs+-irradiated surfaces: positive-ion emission and work-function changes. *Phys. Rev. B*, **54** (23), 17141–17146.
31. Burns, M.S. (1988) Biological microanalysis by secondary ion mass-spectrometry – status and prospects. *Ultramicroscopy*, **24** (2–3), 269–281.
32. Migeon, H.N., Schuhmacher, M., and Slodzian, G. (1990) Analysis of insulating specimens with the Cameca Ims4f. *Surf. Interface Anal.*, **16** (1–12), 9–13.
33. Schröder, M. (2006) Flugzeit-Sekundärionenmassenspektrometrie an Thiol-self-assembly-Mmonolagen auf Gold, PhD thesis. Physikalisches Institut, Westfälische Wilhelms-Universität, Münster.
34. Krohn, V.E. and Ringo, G.R. (1975) Ion-source of high brightness using liquid-metal. *Appl. Phys. Lett.*, **27** (9), 479–481.
35. Kollmer, F. (2004) Cluster primary ion bombardment of organic materials. *Appl. Surf. Sci.*, **231–232**, 153–158.
36. Wucher, A. (2006) Molecular secondary ion formation under cluster bombardment: a fundamental review. *Appl. Surf. Sci.*, **252** (19), 6482–6489.
37. Brunelle, A. and Laprevote, O. (2009) Lipid imaging with cluster time-of-flight secondary ion mass spectrometry. *Anal. Bioanal. Chem.*, **393** (1), 31–35.
38. Alikhanov, S.G. (1957) A new impulse technique for ion mass measurements. *Sov. Phys. JETP-USSR*, **4** (3), 452–453.
39. Slodzian, G., Daigne, B., Girard, F., Boust, F., and Hillion, F. (1992) Scanning secondary ion analytical microscopy with parallel detection. *Biol. Cell*, **74** (1), 43–50.
40. CAMECA http://www.cameca.fr/doc_en_pdf/ns50_instrumentation_booklet.pdf (accessed June 2007), Instrumentation booklet.

41. Stadermann, F.J. http://www.nrims.hms.harvard.edu/slides/Stadermann.pdf (accessed 04 October 2010).
42. Newbury, D.E. (1986) *Advanced Scanninge electron Microscopy and X-ray Microanalysis*, vol. xii, Plenum, New York, London, 454 p.
43. Evans, C. http://www.eaglabs.com/techniques/analytical_techniques/gdms.php (accessed 04 October 2010).
44. Lombi, E., Scheckel, K.G., and Kempson, I.M. (2011) In situ analysis of metal(loid)s in plants: state of the art and artefacts. *Environ. Exp. Bot.*, **72** (1), 3–17.
45. Lombi, E., Scheckel, K.G., Pallon, J., Carey, A.M., Zhu, Y.G., and Meharg, A.A. (2009) Speciation and distribution of arsenic and localization of nutrients in rice grains. *New Phytol.*, **184** (1), 193–201.
46. Lozano-Perez, S., Schroder, M., Yamada, T., Terachi, T., English, C.A., and Grovenor, C.R.M. (2008) Using NanoSIMS to map trace elements in stainless steels from nuclear reactors. *Appl. Surf. Sci.*, **255** (4), 1541–1543.
47. Jolliffe, I.T. (2002) *Principal Component Analysis*, vol. xxix, 2nd edn, Springer, New York, London, 487 p.
48. Larsen, R. (2002) Decomposition using maximum autocorrelation factors. *J. Chemom.*, **16** (8–10), 427–435.
49. Tyler, B.J. (2006) Multivariate statistical image processing for molecular specific imaging in organic and bio-systems. *Appl. Surf. Sci.*, **252** (19), 6875–6882.
50. Breitenstein, D., Rommel, C.E., Mollers, R., Wegener, J., and Hagenhoff, B. (2007) The chemical composition of animal cells and their intracellular compartments reconstructed from 3D mass spectrometry. *Angew. Chem. Int. Ed.*, **46** (28), 5332–5335.
51. Breitenstein, D., Rommel, C.E., Stolwijk, J., Wegener, J., and Hagenhoff, B. (2008) The chemical composition of animal cells reconstructed from 2D and 3D ToF-SIMS analysis. *Appl. Surf. Sci.*, **255** (4), 1249–1256.
52. Möller, J. (2008) Analysen kryopräparierter nicht-dehydrierter Probensysteme mit Hilfe eines neu entwickelten ToF-SIMS-Instruments mit integrierter Hochvakuumkryoschnittapparatur, PhD thesis. Physikalisches Institut, Westfälische Wilhelms-Universität, Münster.
53. Fragu, P., Briancon, C., Fourre, C., Clerc, J., Casiraghi, O., Jeusset, J., Omri, F., and Halpern, S. (1992) Sims microscopy in the biomedical field. *Biol. Cell*, **74** (1), 5–18.
54. Heeren, R.M.A., McDonnell, L.A., Amstalden, E., Luxembourg, S.L., Altelaar, A.F.M., and Piersma, S.R. (2006) Why don't biologists use SIMS? A critical evaluation of imaging MS. *Appl. Surf. Sci.*, **252** (19), 6827–6835.
55. Burns, M.S. (1982) Applications of secondary ion mass-spectrometry (SIMS) in biological-research – a review. *J. Microsc. (Oxford)*, **127**, 237–258.
56. Chandra, S., Smith, D.R., and Morrison, G.H. (2000) Subcellular imaging by dynamic SIMS ion microscopy. *Anal. Chem.*, **72** (3), 104a–114a.
57. Galle, P. (1970) Sur une nouvelle méthode d'analyse cellulaire utilisant le phenomene d'emission ioniqe secondaire. *Ann. Phys., Biol. Med.*, **42**, 83.
58. Grignon, N., Halpern, S., Gojon, A., and Fragu, P. (1992) N-14 and N-15 imaging by SIMS microscopy in soybean leaves. *Biol. Cell*, **74** (1), 143–146.
59. Moore, K.L., Schroder, M., Lombi, E., Zhao, F.J., McGrath, S.P., Hawkesford, M.J., Shewry, P.R., and Grovenor, C.R.M. (2010) NanoSIMS analysis of arsenic and selenium in cereal grain. *New Phytol.*, **185** (2), 434–445.
60. Clode, P.L., Kilburn, M.R., Jones, D.L., Stockdale, E.A., Cliff, J.B., Herrmann, A.M., and Murphy, D.V. (2009) In situ mapping of nutrient uptake in the rhizosphere using nanoscale secondary ion mass spectrometry. *Plant Physiol.*, **151** (4), 1751–1757.
61. Behrens, S., Losekann, T., Pett-Ridge, J., Weber, P.K., Ng, W.O., Stevenson, B.S., Hutcheon, I.D., Relman, D.A., and Spormann, A.M. (2008) Linking microbial phylogeny to metabolic activity at the single-cell level by using enhanced

61. element labeling-catalyzed reporter deposition fluorescence in situ hybridization (EL-FISH) and NanoSIMS. *Appl. Environ. Microbiol.*, **74** (10), 3143–3150.
62. Lechene, C., Hillion, F., McMahon, G., Benson, D., Kleinfeld, A., Kampf, J.P., Distel, D., Luyten, Y., Bonventre, J., Hentschel, D., Park, K., Ito, S., Schwartz, M., Benichou, G., and Slodzian, G. (2006) High-resolution quantitative imaging of mammalian and bacterial cells using stable isotope mass spectrometry. *J. Biol.*, **5** (6), 20.
63. Cooke, R. and Kuntz, I.D. (1974) Properties of water in biological-systems. *Ann. Rev. Biophys. Bioeng.*, **3**, 95–126.
64. Mentre, P. (1992) Preservation of the diffusible cations for SIMS microscopy. 1. A problem related to the state of water in the cell. *Biol. Cell*, **74** (1), 19–30.
65. Chandra, S. and Morrison, G.H. (1995) Imaging ion and molecular-transport at subcellular resolution by secondary-ion mass-spectrometry. *Int. J. Mass Spectrom. Ion Process.*, **143**, 161–176.
66. Hayat, M.A. (2000) *Principles and Techniques of Electron Microscopy: Biological Applications*, vol. xix, 4th edn, Cambridge University Press, Cambridge, 543 p.
67. Chandra, S. and Morrison, G.H. (1992) Sample preparation of animal-tissues and cell-cultures for secondary ion mass-spectrometry (SIMS) microscopy. *Biol. Cell*, **74** (1), 31–42.
68. Stika, K.M., Bielat, K.L., and Morrison, G.H. (1980) Diffusible ion localization by ion microscopy – a comparison of chemically prepared and fast-frozen, freeze-dried, unfixed liver sections. *J. Microsc. (Oxford)*, **118**, 409–420.
69. Lau, K.H., Christlieb, M., Schröder, M., Sheldon, H., Harris, A.L., and Grovenor, C.R.M. (2010) Development of a new bimodal imaging methodology; a combination of fluorescence microscopy and high resolution secondary ion mass spectrometry. *J. Microsc.*, **240** (1), 21–31.
70. Dahl, R. and Staehelin, L.A. (1989) High-pressure freezing for the preservation of biological structure – theory and practice. *J. Electron Microsc. Tech.*, **13** (3), 165–174.
71. Steinbrecht, R.A. and Zierold, K. (1987) *Cryotechniques in Biological Eelectron Microscopy*, vol. xvii, Springer-Verlag, Berlin, London, 297 p.
72. Studer, D., Graber, W., Al-Amoudi, A., and Eggli, P. (2001) A new approach for cryofixation by high-pressure freezing. *J. Microsc. (Oxford)*, **203**, 285–294.
73. McDonald, K.L. and Auer, M. (2006) High-pressure freezing, cellular tomography, and structural cell biology. *Biotechniques*, **41** (2), 137–143.
74. Hajibagheri, M.A. and Flowers, T.J. (1993) Use of freeze-substitution and molecular distillation drying in the preparation of Dunaliella-Parva for ion localization studies by X-ray-microanalysis. *Microsc. Res. Tech.*, **24** (5), 395–399.
75. Giddings, T.H. and Staehelin, L.A. (1980) Ribosome binding-sites visualized on freeze-fractured membranes of the rough endoplasmic-reticulum. *J. Cell Biol.*, **85** (1), 147–152.
76. Livesey, S.A. and Linner, J.G. (1987) Cryofixation taking on a new look. *Nature*, **327** (6119), 255–256.
77. Kellenberger, E. (1991) The potential of cryofixation and freeze substitution - observations and theoretical considerations. *J. Microsc. (Oxford)*, **161**, 183–203.
78. Huang, C.X., Canny, M.J., Oates, K., and Mccully, M.E. (1994) Planing frozen-hydrated plant specimens for SEM observation and EDX microanalysis. *Microsc. Res. Tech.*, **28** (1), 67–74.
79. Chandra, S. (2008) Challenges of biological sample preparation for SIMS imaging of elements and molecules at subcellular resolution. *Appl. Surf. Sci.*, **255** (4), 1273–1284.
80. Chandra, S., Bernius, M.T., and Morrison, G.H. (1986) in *Proceedings of the Fifth International Conference on Secondary Ion Mass Spectrometry: SIMS V* (eds A. Benninghoven, R.J. Colton, and D.Simons), Springer-Verlag, Berlin, p. 429.
81. Derue, C., Gibouin, D., Lefebvre, F., Studer, D., Thellier, M., and Ripoll, C. (2006) Relative sensitivity factors of inorganic cations in frozen-hydrated

standards in secondary ion MS analysis. *Anal. Chem.*, **78** (8), 2471–2477.

82. Metzner, R., Schneider, H.U., Breuer, U., and Schroeder, W.H. (2008) Imaging nutrient distributions in plant tissue using time-of-flight secondary ion mass spectrometry and scanning electron microscopy. *Plant Physiol.*, **147** (4), 1774–1787.

83. Smart, K.E., Smith, J.A.C., Kilburn, M.R., Martin, B.G.H., Hawes, C., and Grovenor, C.R.M. (2010) High-resolution elemental localization in vacuolate plant cells by nanoscale secondary ion mass spectrometry. *Plant J.*, **63** (5), 870–879.

84. Sparks, J.P., Chandra, S., Derry, L.A., Parthasarathy, M.V., Daugherty, C.S., and Griffin, R. (2011) Subcellular localization of silicon and germanium in grass root and leaf tissues by SIMS: evidence for differential and active transport. *Biogeochemistry*, **104**, 237–249.

85. Derue, C., Gibouin, D., Demarty, M., Verdus, M.C., Lefebvre, F., Thellier, M., and Ripoll, C. (2006) Dynamic-SIMS imaging and quantification of inorganic ions in frozen-hydrated plant samples. *Microsc. Res. Tech.*, **69** (1), 53–63.

22
Soft X-Ray Imaging and Spectromicroscopy
Adam P. Hitchcock

22.1
Introduction

Over the past two decades synchrotron-based soft X-ray imaging techniques have developed into powerful tools for characterization of many different types of samples and phenomena. The suite of such techniques now rivals analytical electron and scanning probe microscopies in terms of breadth of methodologies and applications. It is important to recognize that soft X-ray imaging is a complement rather than a competitor to electron and scanning probe microscopies, typically with unique or superior analytical properties but almost always, lower spatial resolution. As the numbers of soft X-ray microscopy beamlines is steadily growing, access to these facilities is increasingly available. It is now very feasible to plan research strategies that integrate soft X-ray synchrotron imaging with a wide range of laboratory-based nanoscale microscopies, including scanning probe (Chapters 2, 15, 16), transmission electron (Chapters 1, 3-11, 14), and secondary electron (Chapters 17, 25) microscopies, as well as hard X-ray microscopies [1].

This chapter summarizes the most common soft X-ray microscopy techniques and gives insights into the instrumental issues and near future developments, along with selected examples that illustrate the power of each approach. Essentially all techniques described use synchrotron radiation since this is the most powerful and flexible source of soft X-rays currently available. This chapter will not give any general introduction to synchrotron radiation light sources, insertion devices, and beamlines, as these are well described elsewhere [2, 3]. Where specific issues of the X-ray source or beamline are relevant, such as degree of coherence, details are provided. While there are significant efforts to develop competitive laboratory-based soft X-ray microscopes based on higher harmonic lasers [4], laser plasma sources [5], or electron-to-X-ray conversion schemes [6], these are not competitive at present in terms of spatial resolution, intensity, or analytical performance, especially in the arena of nanoscopy.

Handbook of Nanoscopy, First Edition. Edited by Gustaaf Van Tendeloo, Dirk Van Dyck, and Stephen J. Pennycook.
© 2012 Wiley-VCH Verlag GmbH & Co. KGaA. Published 2012 by Wiley-VCH Verlag GmbH & Co. KGaA.

The field of soft X-ray microscopy has been reviewed regularly [7–17] during its development in the past 20–30 years. The more comprehensive reviews include an early overview of all types with emphasis on biological applications [7]; photoelectron imaging [10, 11]; a comprehensive review of methods using energy resolved photoelectrons, along with surface and materials applications [12]; a detailed exposition of the principles and applications of zone plate (ZP)-based microscopes [15]; a review of polymer applications [16]; and a recent focused review of biomaterials applications of X-ray photoemission electron microscopy (X-PEEM) [17]. The textbook, *Soft X-rays and Extreme Ultraviolet Radiation, Principles and Applications*, by David Attwood [3], is essential reading in this field. I will refer to specific reviews or articles for more detailed presentations on specific topics as they are presented. As a complement to this review, I maintain and regularly update a web accessible bibliography (*http://unicorn.mcmaster.ca/xrm-biblio/xrm_bib.html*) of publications in soft X-ray microscopy, which was originally published as a supplement to a review of polymer applications [16].

Soft X-ray imaging methods can be divided into three general types: (i) methods using focused X-rays, with the focusing achieved by mirrors – typically Kirkpatrick-Baez (KB) – or ZPs, including scanning transmission X-ray microscopy (STXM), which is based on the detection of the transmitted X-rays, and scanning photoelectron microscopy (SPEM), which is based on kinetic-energy-resolved detection of the photoelectron; (ii) full-field methods, which include transmission X-ray microscopy (TXM) in which ZPs are used both to control illumination and to project the distribution of the transmitted X-rays to an appropriate area detector and X-PEEM in which the distribution of ejected electrons (both primary and secondary) is magnified using electrostatic and/or magnetic electron magnification lenses and imaged with a suitable X-ray sensitive camera; (iii) coherent diffraction imaging (CDI) methods, in which the far-field coherent scattering signal is measured and inverted into a real space image via a variety of methods. Table 22.1 is a listing of the soft X-ray microscopes available around the world, to the best of my knowledge. Figure 22.1 [18] shows the schematic layout of the common methods of types (i) and (ii). CDI methods are very much at the frontier of soft X-ray imaging and will become increasingly important as fourth-generation light sources come on line and mature. However, this chapter focuses on direct imaging methods and does not cover CDI methods. Readers interested in CDI should consult other articles [19] and reviews [20].

Figure 22.2 [21] shows a conceptual plot of how the various branches of X-ray microscopy and related coherent diffraction (also called *"lensless"*) imaging techniques perform with respect to photon energy and spatial resolution. As this handbook focuses on deep submicron resolution (<100 nm) and this chapter deals with soft X-ray techniques, this review deals with techniques that perform within the dotted rectangle in the lower left part of this plot.

Table 22.1 Synchrotron soft X-ray microscopes.

Type	Facility	Name	City	Country	Source	E-range (eV)	Status
TXM	Alba	Mistral	Barcelona	Spain	BM	270–2600	Construction
TXM	ALS	XM1	Berkeley	USA	BM	250–900	Operating
TXM	ALS	XM2 (NCXT)	Berkeley	USA	BM	250–6000	Operating
TXM	Astrid	XRM	Aarhus	Denmark	BM	500	Operating
TXM	BESSY	U41-TXM	Berlin	Germany	Und-L	250–600	Operating
TXM	Diamond	B24 cryo-TXM	Chilton	UK	BM	250–2500	Construction
TXM	NSRL	TXM	Hefei	China	BM	500	Operating
TXM	Ritsumeikan	BL12	Kyoto	Japan	BM	500	Operating
X-PEEM	Alba	Circe	Barcelona	Spain		100–2000	Construction
X-PEEM	ALS	PEEM-2	Berkeley	USA	BM	175–1500	Operating
X-PEEM	ALS	PEEM-3	Berkeley	USA	EPU	150–2000	Operating
X-PEEM	BESSY	UE49 SMART	Berlin	Germany	EPU	100–1800	Operating
X-PEEM	BESSY	UE49 SPEEM	Berlin	Germany	EPU	100–1800	Commissioning
X-PEEM	CLS	CaPeRS	Saskatoon	Canada	EPU	130–2500	Operating
X-PEEM	Diamond	I06	Chilton	UK	EPU	80–2100	Operating
X-PEEM	Elettra	BL 1.2 L	Trieste	Italy	EPU	50–1000	Operating
X-PEEM	MAX-lab	1311	Lund	Sweden	Und-L	100–1400	Operating
X-PEEM	NSRRC	BL05B2	Hinschu	Taiwan	EPU	60–1400	Operating
X-PEEM	Photon factory	BL 18A	Tsukuba	Japan	BM	50–150	Operating
X-PEEM	SLS	SIM	Villigen	Switzerland	EPU	90–1200	Operating
X-PEEM	Soleil	Tempo	Saint-Aubin	France	EPU	50–1500	Operating
X-PEEM	Soleil	Hermes	Saint-Aubin	France	EPU	250–1500	Commissioning
X-PEEM	Spring-8	BL25SU	Hyogo	Japan	EPU	220–2000	Operating
X-PEEM	Spring-8	BL17SU	Hyogo	Japan	EPU	300–1800	Operating
X-PEEM	SRC	Sphinx	Stoughton	USA	Und-L	70–2000	Operating

(*continued overleaf*)

Table 22.1 (Continued)

Type	Facility	Name	City	Country	Source	E-range (eV)	Status
STXM	ALS	5.3.2.2	Berkeley	USA	BM	250–750	Operating
STXM	ALS	5.3.2.1	Berkeley	USA	BM	250–2500	Commissioning
STXM	ALS	11.0.2	Berkeley	USA	EPU	100–2000	Ooperating
STXM	BESSY	old-STXM	Berlin	Germany	BM	250–600	Decommissioned
STXM	BESSY	MAXYMUS	Berlin	Germany	EPU	250–1500	Operating
STXM	CLS	10ID1	Saskatoon	Canada	EPU	130–2500	Operating
STXM	Diamond	I08	Chilton	UK	EPU	250–2500	Construction
(S)TXM	Elettra	Twin-mic	Trieste	Italy	Und-L	250–2000	Operating
STXM	UVSOR	BL4U	Okazaki	Japan	Und-L	50–800	Construction
STXM	NSLS	X1A (2)	Upton	USA	Und-L	250–1000	Decommissioned
STXM	PLS	nanoscopy	Pohang	Korea	EPU	100–2000	Construction
STXM	SLS	PolLux	Villigen	Switzerland	BM	250–750	Operating
STXM	SLS	NanoXAS	Villigen	Switzerland	BM	250–750	Commissioning
STXM	Soleil	Hermes	Saint-Aubin	France	EPU	250–1500	Construction
STXM	SSRF	SXS	Shanghai	China	EPU	200–2000	Operating
STXM	SSRL	13-1	Stanford	USA	EPU	250–1000	Operating
SPEM	ALS	BL7.0	Berkeley	USA	Und-L	90–1300	Decommissioned
SPEM	ALS	Maestro	Berkeley	USA	EPU	90–1300	Commissioning
SPEM	Elettra	BL 2.2 L	Trieste	Italy	Und-L	200–1400	Operating
SPEM	Elettra	BL 3.2 L	Trieste	Italy	Und-L	27, 95	Construction
SPEM	MAX-lab	BL 31	Lund	Sweden	Und-L	15–150	Operating
SPEM	NSRRC	BL09A1	Hinschu	Taiwan	Und	60–1500	Operating
SPEM	PAL	8A1	Pohang	Korea	Und	20–2000	Operating
SPEM	Soleil	Antares	Saint-Aubin	France	EPU	50–1500	Operating

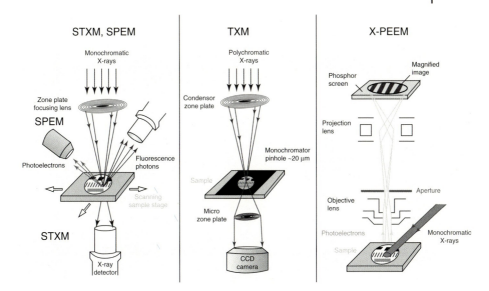

Figure 22.1 Schematic layout of the four common methods of soft X-ray microscopy, which include two focused probe methods, STXM and SPEM, and two full-field imaging methods, TXM and X-PEEM. In each case, the first of the two methods involves transmission of X-rays through the sample and thus is primarily a "bulk" sensitive technique, while the second method involves electron detection and thus is primarily a surface-sensitive technique. (From Ref. [18], used with permission from the author, Joachim Stöhr.)

22.2
Experimental Techniques

22.2.1
Full-Field Transmission X-Ray Microscopy (TXM)

This was the original synchrotron-based, soft X-ray imaging technique, developed in the 1960s by Schmal and Rudolf at Göttingen University [22, 23]. The optical layout is analogous to that of a conventional visible light microscope, with a condenser lens used to increase the light density at the sample and an imaging lens to magnify the transmitted light and transfer it to a detector. As indicated in Figure 22.3, a large ZP is used as both a condenser and a crude monochromator to concentrate a narrow band width of the incident X-rays to a few tens of microns, and a micro ZP is used to reimage the X-rays that are transmitted through the sample. Magnification is controlled by the position of the micro ZP and the recording device and by the properties of the imaging micro ZP. There has been a tremendous improvement in the quality of ZPs used for both full-field and scanning microscopes over the past 30 years, with several laboratories producing state-of-the-art ZPs capable of 10 nm imaging [24, 25] and commercial suppliers able to provide ZPs with 25 nm performance. With careful tuning and high-contrast samples, it is possible to

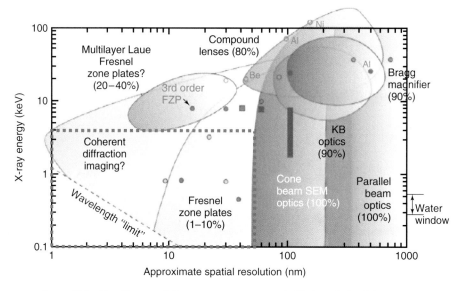

Figure 22.2 Classification of X-ray microscopy techniques based on focusing technologies, achievable spatial resolution, and photon energy range. Points indicate the current capabilities, whereas the shaded zones indicate the range of spatial resolution and photon energies where the technique is applicable as well as the expected potential for spatial resolution. Calculated resolutions are shown with open symbols, measured ones with filled symbols of the appropriate color, circles correspond to synchrotrons, and squares to laboratory sources. See the original article [21] for the sources of the specific performance data. The dotted rectangle indicates the regime relevant to this chapter. (From Ref. [21], used with permission from Materials Today via RightsLink license 2517731089158.)

increase the spatial resolution by using higher diffraction orders of the ZP [26]. Exposure times are typically a fraction of a second up to a few tens of seconds. Although TXMs have the tremendous advantage of parallel detection, there are limitations to the field of view, which is typically between 10 and 30 μm. Mechanical scanning of the sample and tessellation of carefully matched images is used to study wider fields of view.

In general, TXMs produce some of the highest spatial resolution images [25], are excellent tools for magnetic dynamics studies [28–30], and are the preferred method for tomography (see below) but are only rarely used for spectromicroscopy [33–34]. Most TXM instruments are located on bend magnets without any dispersion device other than the condenser ZP. The early applications of full-field microscopes were in water-window biological imaging [35]. More recently, applications have emerged in static magnetism (using off-axis partly circularly polarized X-rays) [27], time-resolved magnetic dynamics [28–30], and a wide range of materials such as cement [31–32]. Until recently, although most bend-magnet sources provide a much wider spectral range of X-rays, almost all TXM research has been carried out in the water window (300–520 eV) in order to study fully hydrated samples. In addition, although TXM condensers act as a crude monochromator, it takes

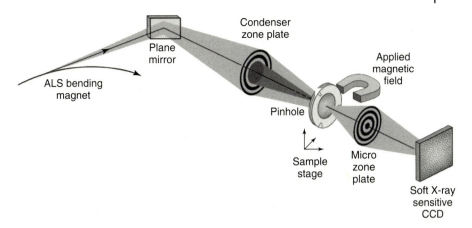

Figure 22.3 Layout of a bend-magnet-based full-field transmission X-ray microscope (TXM, in this case, XM-1 at the Advanced Light Source) [27]. This relatively simple approach provides very high flux and some energy tunability by moving the condenser zone plate. The sample is typically in air (with a short path length to windows enclosing the upstream and down stream optics), which allows relatively straightforward adaptation to modification of the sample environment, as illustrated in this figure, by the applied magnetic field. (From Ref. [27], used with permission from Materials Today via RightsLink license 2534370654740.)

considerable effort to reoptimize the optical setup for different photon energies, and thus most systems have little or no capability to scan the photon energy for spectroscopic applications. A major exception is the undulator-based full-field TXM at the U41 undulator beamline at the Berlin synchrotron, BESSY2 [33]. Soft X-ray undulator radiation has a high degree of coherence, yet the optical principles of TXM require incoherent illumination to avoid artifacts associated with coherent diffraction [36]. The BESSY2 TXM uses a novel rotating capillary optic as the condenser [37] to "bust coherence," thereby allowing use in a TXM of highly monochromated and tunable, but mostly incoherent, light. An alternate approach of rotating a section of condenser ZP has been implemented at the other undulator-based TXM, Twinmic at Elettra [38]. However, to my knowledge, this is not yet set up for spectroscopic full-field studies. Perhaps the most specialized TXM is that of the National Centre for X-ray Tomography (NCXT) at the Advanced Light Source (ALS), which is dedicated to high-throughput water-window tomography. Further discussion of this instrument and its capabilities is given in Section 22.4.3.

22.2.2
Full-Field Photoelectron Microscopy (X-PEEM)

Photoemission electron microscopy (PEEM) is the only soft X-ray imaging technique where there is a larger nonsynchrotron than synchrotron community. Laser or Hg lamp UV illumination is used in laboratory implementations, and topography and work function are the dominant contrast mechanisms. When mounted on a high-performance beam line at the synchrotron, PEEM instruments can exploit a much wider range of contrast mechanisms, including spectroscopy (elemental and chemical), X-ray linear dichroism (XLD) to study geometric alignment, X-ray magnetic circular dichroism (XMCD) to study ferromagnetism, and X-ray magnetic linear dichroism (XMLD) to study antiferromagnetism. While there are numerous implementations of the commercial instruments at synchrotrons, the two most advanced are custom-built instruments: the spectromicroscope for all relevant techniques (SMART) facility at BESSY2 [39–41] and the PEEM-3 facility at the ALS [42–44]. SMART has implemented aberration compensation electron optics to reduce both spherical and chromatic aberrations [41], with the goal to achieve spatial resolutions below 5 nm with X-ray illumination. PEEM-3 is in the process of implementing a similar system. Figure 22.4 presents a schematic of the electron optical layout of the PEEM-3 instrument at the ALS. Operational since 2008, as of fall 2010, the instrument is in PEEM-2.5 configuration, without the aberration compensation section in the dotted rectangle. Although these two instruments are still under development, the expectation is for very large improvements in efficiency for a spatial resolution in the 50–100 nm range (that of current synchrotron-based X-PEEMs) and to ultimately achieve sub-5 nm spatial resolution with a small percentage of overall transmission (which is a transmission similar to that of current X-PEEM instruments) (Figure 22.4b). These expectations are firmly based on experiences with the performance of analytical aberration-compensated electron microscopes [45, 46]. The SMART system at BESSY2 has achieved better than 4 nm in the low-energy electron microscopy (LEEM) mode of operation [42], which is a truly remarkable performance. However, to my knowledge, the improvements in the spatial resolution and transmission in the X-PEEM mode have not been as good as originally expected. In parallel to the developments at BESSY2 and the ALS, Tromp *et al.* [47] have developed an alternative aberration compensation concept, which has also achieved remarkable performance and is being made available commercially as a combined LEEM/PEEM instrument. I expect significant improvements in X-PEEM performance in the next few years, as these aberration-compensated instruments mature. It should be noted that the ultimate spatial resolution can be obtained only with optimal samples, ones that are totally flat, conducting, and with a uniform potential at the surface.

22.2.3
Scanning Transmission X-Ray Microscopy (STXM)

In a STXM, images are obtained serially by mechanically raster scanning the sample though the focal point of a ZP X-ray lens or (in a few cases) by scanning the ZP and order sorting aperture (OSA) synchronously while the sample is stationary. Figure 22.5a is a schematic of the focusing geometry of a ZP optic. The OSA is required to block the zeroth order (undiffracted) light, which is typically 5–20 times more intense than the first-order focused light. Figure 22.5b,c provides schematics of the polymer STXM [48] at the ALS on beamline 5.3.2.2 [49]. This was the first successful implementation of interferometric control of the (X,Y)

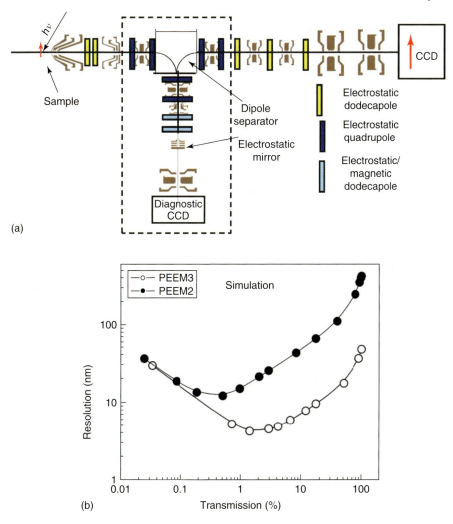

Figure 22.4 (a) Electron optical layout of the PEEM-3 instrument at the ALS [44]. Operational since 2008; as of spring 2011, the instrument is in PEEM-2.5 configuration, without the aberration compensation section in the dotted rectangle. (b) Comparison of predicted transmission of the optics as a function of spatial resolution, with and without the aberration compensation. (Andreas Scholl, private communication, used with permission.)

spatial relationship of the ZP and sample.[1] Similar approaches have now been implemented in other commercial and custom-built soft X-ray STXMs. The quality of the ZP is paramount to the intrinsic resolution capabilities of the STXM (as

1) The first STXM at the NSLS X-ray ring used a nondifferential interferometry system to monitor the (x,y) position of the sample stage but not the correlation of the sample position with the ZP [50].

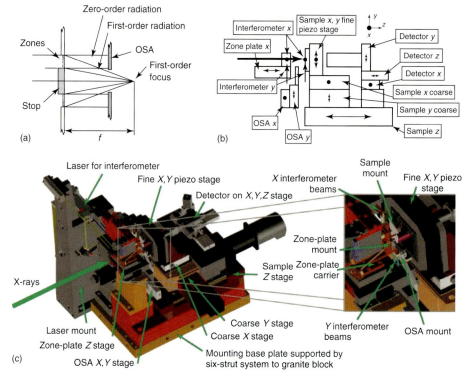

Figure 22.5 (a) Schematic of the focusing in a scanning transmission X-ray microscope (STXM). (b) Schematic of stage motions in the STXMs at ALS, CLS, and SLS. (c) Schematic of the polymer STXM [48] at the Advanced Light Source, beamline 5.3.2.2 [49]. This was the first successful implementation of interferometric control of the (X,Y) spatial relationship of the zone plate and sample. Similar approaches have been implemented in many other commercial and custom-built soft X-ray STXMs – see Table 22.1. (From Ref. [48], used with permission from IUCr – e-mail from Peter Strickland, 29 Sep, 2010.)

well as SPEM and TXM) and is limited by fabrication technologies that have been slowly but continuously improving over the past decades.

Since ZP focusing is so critical to soft X-ray microscopy, it is worth considering the dependence of the focusing on the ZP properties [3, 15]. A Fresnel ZP is a circular diffraction grating, ideally consisting of concentric metal rings alternating with circular slots mounted on a thin X-ray transparent support. The ZP provides a first-order diffracted beam when the path difference between neighboring open zones is such that the signals from all the open zones reinforce at the focus. To second order (which is a valid approximation at low numerical aperture (NA)), the focusing condition is given by [15]

$$r_n^2 = n\lambda f \tag{22.1}$$

where r_n is the radial position of the nth zone, λ is the X-ray wavelength, and f is the focal length. ZPs are strongly chromatic devices, with the focal length increasing linearly with photon energy. To get a good focus, the ZP must be illuminated with monochromatic light. The condition to achieve diffraction-limited resolution is given approximately by [51].

$$\frac{\Delta\lambda}{\lambda} \leq \frac{1}{n_z} \qquad (22.2)$$

where n_z is the number of zones. Thus, for a ZP with 800 zones (typical of ZPs used in modern STXMs) a resolving power better than 1000 is required. Typically beamlines for STXM are designed to provide a minimum of 3000 resolving power, and the undulator-based STXM beamlines achieve resolving powers of 10 000 or higher, such that there is no problem achieving diffraction-limited performance. However, energy resolution can limit the achievable spatial resolution if the diffraction-limiting apertures (which are usually combined with the exit slit function of the monochromator) are set too wide, in which case resolution degradation will occur. The NA of a ZP is given by [15].

$$NA = \frac{r_n}{f} = \frac{\lambda}{2\Delta r_n} \qquad (22.3)$$

where Δr_n is the outer zone width. The diffraction-limited resolution of a ZP according to the Rayleigh criteria [36] is given by

$$\delta_{\text{Rayleigh}} = \frac{0.61\lambda}{NA} = 1.22\Delta r_n \qquad (22.4)$$

and the focal length in μm is given by

$$f = \frac{D\Delta r_n}{\lambda} = \frac{D\Delta r_n E}{1239.8} \qquad (22.5)$$

where D is the diameter of the ZP in μm, Δr_n is the outer zone width in nm, and E is the photon energy in eV. While the above equations describe the performance of ZPs in first order, ZPs also produce odd higher orders of focusing, with a focal length of f/m, where m is the order of focusing. Thus, when a ZP is operated using its third-order focused spot, the spatial resolution is theoretically three times better than at first order, and with care, this approach can be used to improve spatial resolution [26]. However, the energy resolution needed to achieve diffraction-limited resolution for third-order operation is much higher ($\times 3$), and the intensity is much lower, than for first order, such that it is very challenging to find and optimize the third-order focused light from a ZP. Many other issues are involved in fully optimizing the properties and fabrication of high-quality ZPs. These are discussed in great detail in the comprehensive treatment by Howells, Jacobsen, and Warwick [15].

While the ZP optic places fundamental limits on the spatial resolution performance of STXM microscopes (as well as TXM and SPEM), there are many other factors that must be optimized to achieve the focusing potential. STXMs typically involve many high-precision motorized stages for positioning and scanning. These need to be carefully selected to provide the required precision, linearity, and step size resolution. A critical aspect is stabilizing the (x,y) focal position as photon energy is changed. Since the focal length varies linearly with photon energy (Eq.(4.5)), a typical near-edge X-ray absorption spectroscopy (NEXAFS) spectral scan of 40 eV width requires a change in the z-position of the ZP or sample of several hundred microns, which must be made with a lateral shift (rollout) much less than the spatial resolution (say, 10 nm). Since the best high-precision mechanical stages (which are needed to give the required range of motion) have rollout figures of 100 nm at best, it is not possible to preserve positioning on the sample by relying only on the properties of the mechanical stage(s) used to set the focal length. Indeed, before the development of interferometric control [48] of STXM mechanisms, drift in the image position was typically a few microns over the course of a 40 eV NEXAFS scan. While this can be largely corrected by software alignment procedures developed for image sequence or stack mode of data acquisition [52], the rollout issue makes point or linescan spectroscopy difficult or impossible, and the software algorithms for alignment have significant limitations, especially when the stack has reversals in image contrast with photon energy. In contrast, when properly tuned, the interferometric controlled STXMs achieve positional stability of better than 50 nm over very large photon energy ranges – typically the full photon range available on the beamline [48]. This active interferometric control approach also provides some compensation for environmental vibrations, typically up to 50 or 100 Hz.

Many different types of detectors have been developed for use with soft X-ray STXMs. These include gas proportional counters [53], phosphor-photomultiplier systems [48], Si photodiodes, avalanche photodiodes, diode arrays [54, 55], and imaging systems [56]. The last two types of detectors are able to provide dark field and phase contrast signals [56], in addition to the bright field signal. High detector efficiency is important to minimize the dose (and thus radiation damage) needed for a given statistical precision. The single photon counting systems typically achieve an efficiency of 30–50%, depending on the photon energy.

An advantage of the photon-in, photon-out character of STXM (and TXM) is that it is possible to operate these microscopes with the X-rays traversing a finite gas path. Thus, most of the TXMs operate with the sample in air, while STXMs typically have the sample and all of the microscope mechanism in a tank that can be evacuated and back filled with He, which is needed to prevent overheating of motors. Aside from the cryo-STXM at the National Synchrotron Light Source (NSLS I) ([57], presently decommissioned), the new MAXYMUS microscope at BESSY2, [58] , and the STXM at beamline 5.3.2.1 at the ALS [59] which are constructed with full ultrahigh vacuum (UHV) technologies, STXMs have been constructed using low-vacuum technologies and lubricated mechanical stages that use volatile organic lubricants. A consequence is that there is a significant problem of carbon contamination of samples when the focused, high-dose-density X-ray beam is left on the same spot on the sample for a time, giving a dose in the giga Gray (GGy) range (which can be a few seconds or even shorter). In order to get around the problem with carbon contamination of samples when using long exposures, as well as to implement some significant advances in the mechanical and controls technologies, a new STXM has recently been developed on a dedicated bend-magnet beamline at the ALS [59]. In addition to use of UHV technologies to ensure a clean, low-organic environment, it has been designed to scan the ZP as well as the sample, thereby allowing heavy, complex sample mountings, such as that for cryo-spectro-tomography, to be used. In addition, the beam line features an extended photon energy range, from 250 to 3000. The instrument is undergoing commissioning and will be available for general use by the end of 2012.

22.2.4
Scanning Yield Techniques (SPEM, LEXRF-STXM, STM-SXM)

This section describes systems in which a detection channel different from the transmitted X-ray beam is used in a focused probe X-ray microscope. This is typically done to provide sensitivity to a different aspect of the sample, such as its surface, which can be measured with total electron yield (TEY) detection in STXM or by SPEM; electrical properties, measured by electron beam induced current (EBIC) in STXM; optical luminescence properties [60]; or low-energy X-ray fluorescence (LEXRF) measured in STXM [61–65]. An emerging and very promising technique to achieve extremely high (perhaps subnanometer) spatial resolution is to operate a scanning probe microscope (SPM) – typically scanning tunneling microscope (STM) – with zone plate focussed X-ray illumination of the sample–tip junction. Modulation of the X-ray intensity induces a modulation of the STM signal to the extent that X-ray photoionization contributes to the detected signal. Encouraging preliminary results have been obtained at the Photon Factory [66], and the nanoXAS beamline at the Swiss Light Source [68] is dedicated to the development and exploitation of STM–STXM.

When electrons are the detected particle, there is enhanced surface sensitivity. This can be achieved in a STXM simply by measuring the sample current or, with single electron sensitivity, by using a channeltron or channelplate if the vacuum is good enough ($<5 \times 10^{-5}$ torr). While not as sensitive, sample current detection does allow measurements in very poor vacuum (10^{-2} torr), typical of present-generation STXMs. In order to detect electrons with a specific kinetic energy, a time-of-flight (TOF) or dispersive electron spectrometer is required, which necessitates a good high-vacuum or UHV environment in order to minimize inelastic scattering of the photoelectrons before dispersion and detection. Such SPEM [12] are implemented at several synchrotrons (NSRC, Taiwan; PLS, Korea; Elettra, Italy), with the Electra ESCA microscope system being perhaps the best developed and most productive to date. Figure 22.6a shows the basic concept of SPEM [12]. It uses a typical ZP-OSA probe-forming system as also used in STXM. However, in contrast to STXM, where the sample is usually orthogonal to the beam direction, in SPEM the sample is placed at an angle to the incoming X-rays in order to optimize the solid angle seen by the electron energy analyzer. Since the yield of energy-selected photoelectrons is much smaller than the intensity of the transmitted light or the TEY, it is important to have as efficient an electron spectrometer system as possible, which typically means using some type of multichannel detection in the electron analyzer (Figure 22.6b). Because of the low electron yields, the ZP systems used in SPEM are usually selected to have lower spatial resolution and higher

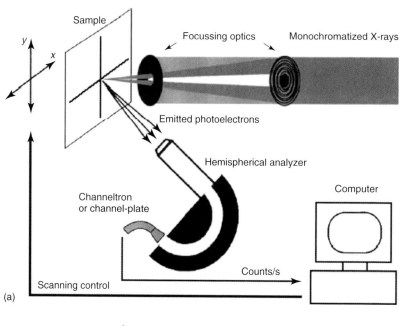

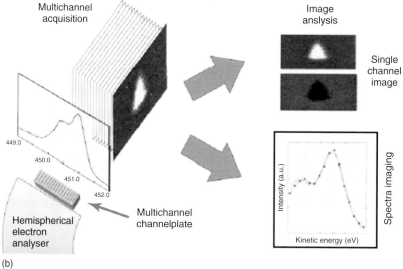

Figure 22.6 (a) Schematic of a scanning photoelectron microscope (SPEM). (b) Schematic of parallel detection in the electron spectrometer to gain efficiency in SPEM. (From Ref. [12], used with permission from Prog. Surf. Sci via Rights Link license 2518440745535.)

efficiency than those used in STXM, typically working at a few 100 nm spatial resolution [12]. Very recently, angular resolved photoemission (ARPES (angle resolved photoelectron spectroscopy)) has been measured in SPEM to study the band structure of materials with submicron spatial resolution (E.E. Rotenberg, private communication). There are beamlines and endstations under development at Elettra, ALS, and Soleil, which will provide sub-100 nm ARPES capability. The new Antares nanoARPES beamline at Soleil has already measured band structures on areas smaller than 100 nm [67].

X-ray fluorescence (XRF) detection in STXM offers a means to greatly improve the detection limits for trace elements in STXM. While XRF has been a major tool in the tender X-ray (2–10 keV) [69] and hard X-ray (10–100 keV) [1] spectral regions, it is only recently that XRF detection has been implemented in soft X-ray STXM. Recently, Kaulich *et al.* [63, 64, 70] pioneered this in the novel Twinmic combined TXM/STXM instrument at Elettra [38]. Figure 22.7 shows a schematic of the

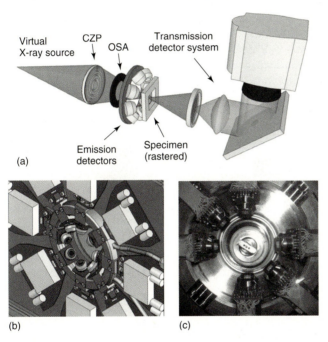

Figure 22.7 (a) Overall schematic of the Twinmic combined TXM/STXM microscope [72] at Elettra. The presample X-ray optics in the layout shown are those for full-field TXM. For STXM, the postspecimen imaging ZP (not shown) is removed and the condenser ZP (CZP) is replaced with a focusing ZP. Special features of Twinmic include a high sensitivity camera as the STXM detector, which allows phase and dark field imaging as well as conventional bright field imaging [56], and an array of silicon drift detectors (SDD) [64] which are used for detecting low levels of elements via low-energy X-ray fluorescence. (b) Schematic and (c) photograph of the six-SDD array currently installed in Twnimic. Each is attached to a water-cooled Cu support and has an advanced pulse processing amplifier directly behind the pnSensor SDD. (Burkhard Kaulich, private communication, used with permission.)

Twinmic set-up, along with a view of the individual silicon drift detectors (SDDs). Currently at Twinmic there is an array of six such detectors which, combined, provide a solid angle of detection of 0.2 sr [64]. Recently, SDDs have been added to the STXMs at the ALS and the Canadian Light Source (CLS). In each case, although the implementation uses only a single SDD, a very effective XRF detection system has been achieved because the detector is only a few millimeters away from the sample as opposed to the 28 mm distance in Twinmic. It is envisaged that there will be significant applications of XRF-STXM in the area of environmental science, where it can be important to detect trace metals for toxicology or other issues. As an example, the ALS system has been used to measure the oxidation state and map the distribution of arsenic in biofilms of Fe metabolizing bacteria (*Acidovorax sp.* strain BoFeN1), which produce copious amounts of various Fe minerals. These species are known to be tolerant to very high levels of toxic As, an important environmental problem in Bangladesh and certain parts of the United States, and *Acidovorax* has been identified in As-contaminated soils. It is not clear whether the accommodation to high As levels is because the As adsorbs to the extracellular Fe minerals or whether the bacteria are actively involved in accommodating to the As by some type of detoxification or biomineralization process. XRF-STXM studies [65, 71] have been able to contribute to this field by mapping the As in these biofilms, which were grown anaerobically for one week with 1 mM As^V in the culture medium. The results from this study (Figure 22.8) very nicely demonstrates that XRF, and especially absorption spectroscopy using XRF-yield, greatly extends the detection limits of STXM. Figure 22.8a compares the X-ray absorption spectrum (XAS) (blue) recorded from transmitted X-rays at the position indicated by the circle on the As map in Figure 22.8c to that measured from the photon energy dependence of the yield of XRF between 1300 and 1400 eV where the Mg $K\alpha$ and As $L\alpha$ X-ray emission lines are located (see Figure 22.8b) An advantage of XRF-STXM is that the energy dispersive detector provides maps of other elements, as well as the total absorption, which can provide useful complementary information to the element-specific maps derived from the XRF-yield NEXAFS signal. In this case, the combination of the XRF and XRF-NEXAFS provided maps of the Mg, Fe, and As elemental distributions [65, 71], while the O 1s and absorption signal below the O 1s edge gave maps of the biological material.

A novel development with the potential to dramatically improve the spatial resolution beyond the limitations of current and conceivable future ZP technologies is to combine a focused X-ray probe with a scanning probe implemented as a spatial constraining detector. Eguchi *et al.* [66] were the first to explore this concept but were only partly successful since they had only limited focusing in the Photon Factory end station that was modified to demonstrate XAS detection via an STM tip. Recently, the Swiss Light Source has constructed and is in the process of commissioning the NanoXAS beamline and microscope, a novel combination of a bend-magnet soft X-ray STXM and an SPM [73, 74]. Figure 22.9 shows a schematic of the overall NanoXAS concept as well as a cartoon of photoelectron trajectories under the influence of the applied field from the coaxial SPM tip, which is located at the focus of the ZP. Figure 22.9c shows a STXM image of a

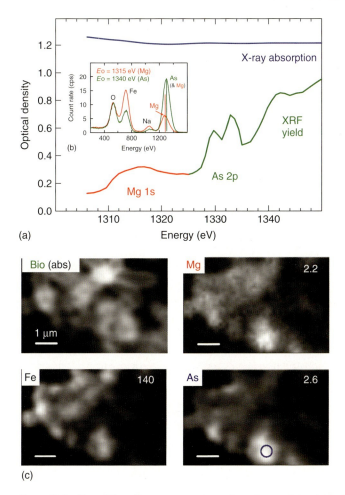

Figure 22.8 Use of X-ray fluorescence to improve sensitivity in studies of minority species such as As in association with *Acidovorax sp.* strain BoFeN1 [65, 71]. (a) X-ray absorption spectrum (blue) recorded with the X-ray beam at the position indicated by the circle on the As map in (c), compared to the NEXAFS spectrum of the same spot measured from the photon energy dependence of the yield of X-ray fluorescence (XRF-yield) between 1300 and 1350 eV. (b) X-ray fluorescence spectrum from the same spot, recorded at two incident photon energies (1315 and 1340 eV). (c) Maps of total absorption and the Mg, Fe, and As elemental distributions derived from XRF. The gray scale indicates counts per pixel for a 180 ms integration time [65, 71].

polystyrene/polymethyl-methacrylate (PS/PMMA) blend recorded at 285 eV, where the PS selectively absorbs, while Figure 22.9d presents topography and phase images recorded in the same region with the *in situ* SPM [68]. The ultimate goal of this instrument is enhancement of spatial resolution by using time modulation of the incident photon intensity and detecting the scanning probe signal in phase at the modulation frequency. At present, the SPM channel provides surface-sensitive

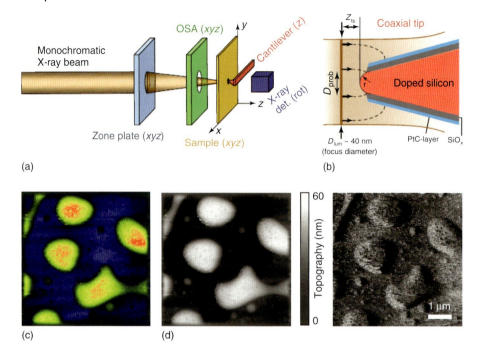

Figure 22.9 Operating principles and images recorded with NanoXAS, a novel soft X-ray STXM combined with an SPM at the Swiss Light Source [73]. (a) Schematic of NanoXAS. (b) Schematic view of the coaxial SPM tip located at the focus of the zone plate indicating the trajectories of the photoelectron under the influence of the applied field. (c) STXM image of a polystyrene-polymethylmethacrylate (PS/PMMA) blend recorded at 285 eV, where the PS selectively absorbs. (d) Topography and phase images recorded in the same region with the *in situ* SPM. The ultimate goal of this instrument is enhancement of spatial resolution by using incident photon intensity modulation and scanning probe pickup at the modulation frequency. At present, the SPM channel provides surface-sensitive imaging and some chemical sensitivity in phase mode, which is complementary to the bulk perspective of STXM. ((a,b) Ref. [73] – used with permission; (c,d) from Christoph Quitmann, private communication, used with permission.)

imaging and some chemical sensitivity in phase mode, which is complementary to the bulk perspective of STXM. Simulations suggest that a resolution well below 5 nm will be achievable [73]. However, at present, there are significant challenges in optimization of the device so as to enhance the modulated signal and suppress background from the parts of the focused X-ray beam that are outside of the pickup region of the SPM tip.

22.2.5
Sample Preparation Issues

For transmission techniques (TXM, STXM), the sample must be partially transparent at the X-ray energies of interest. The optimal thickness is such that the optical density (OD) at the most intensely absorbing energy is between 1 and 2 OD units. It is best to adapt sample preparation to achieve this if at all possible, rather than struggle with the challenges of absorption saturation in the case of samples that are too thick or that have insufficient contrast in the case of samples that are too thin. Signals below 0.1 OD (down to perhaps 0.01 OD) can be studied if data of sufficient statistical quality is measured. As most samples have an upper dose beyond which the material is too modified for the results to be analytically useful, it is not always possible to use longer times/higher doses or defocusing to build statistics in the case of weakly absorbing samples. The performance for strongly absorbing (nearly opaque) samples in most soft X-ray microscopes is such that there is significant spectral distortion for absorbance levels above 2–3 OD. The upper absorption limit before spectral distortion occurs varies a lot among different instruments since it depends on higher order and stray light content of the incident light as well as detector efficiency and backgrounds. The impact of these factors in turn depends on the details of the NEXAFS spectrum, with spectra having very sharp peaks such as the Ca 2p edge [75] being much more sensitive to absorption saturation artifacts than edges with less sharp features.

Samples for transmission soft X-ray microscopies can be prepared using all of the techniques that are used in transmission electron microscopy (TEM), including ultramicrotomy, cryo-ultramicrotomy, spin-coating, solvent casting, ion beam thinning, and focused ion beam (FIB) milling [76]. Although expensive relative to other methods, FIB is becoming increasingly common, as it allows precise extraction of a sample region of a particular morphology or at a specific spatial location. Modern FIB systems are able to handle materials with dramatically differing densities and toughness, including biological specimens and hard–soft composites that are very difficult to microtome without specimen artifacts. In addition, problems with ion implantation that plagued early FIB samples have been resolved with the addition of low-energy ion polishing techniques as the final stage of the FIB procedure. It is possible to measure single cell prokaryotic biological systems with TXM or STXM without sectioning, which is a considerable advantage over electron microscopic methods, in which biological sample preparation can be chemically complex and may introduce artifacts. Plunge-freeze vitrification is used to preserve microstructure without ice crystal formation during TXM tomography measurements.

For surface-sensitive techniques (SPEM, X-PEEM), UHV compatibility, surface flatness, and conductivity are required. The X-PEEM technique has the most stringent requirements, as there is a very strong electric field at the surface (10–20 kV/mm), which makes X-PEEM extremely sensitive to field emission that is readily induced (and occasionally fatal to the sample) if there are any particles or abrupt changes in topography. Metallographic polishing and metal coating procedures can be used to prepare even very insulating samples such as rocks or teeth for X-PEEM studies [77].

22.2.6
Radiation Damage

Radiation damage is inevitable whenever ionizing radiation is used for microscopy. In order to avoid artifacts in imaging and particularly when making more extensive analytical measurements, it is essential to be aware of the doses at which the sample experiences damage effects that will affect the results of the measurement. If the main goal is to study morphology, it is possible to extend the dose that can be used without artifacts by freezing the sample. This does not change the rate of chemical transformation [78, 79], but it does stop mass loss. Whenever NEXAFS spectra are being used analytically, it is important to determine the dose at which critical spectral features of the most radiation-sensitive component change their intensity because of radiation-induced structural changes. For some polymer materials, this dose can be extremely low – for example, PMMA, and similar esters have a critical dose – the dose to decrease (increase) a spectral feature by $1/e$ $(1 + 1/e)$ – of $\sim$60 MGy [80]. This dose is delivered in less than 100 ms of STXM operation under normal fluxes (10–100 MHz of photons in a 30 nm spot size), which means that damage-free measurements are very challenging. Typically, one must reduce the flux, dose density, or numbers of energy points sampled so that the dose delivered during a measurement of a given area is below 10% of the critical dose. In fact, with chemical amplification, it is possible to image damage in PMMA for exposures of less than 1 ms [81]. Despite these challenges, it is still much easier to perform X-ray absorption spectroscopy of highly radiation-sensitive materials in soft X-ray microscopes than using core level electron energy loss spectroscopy (EELS) in a TEM [82]. Although it depends on the spectral energy range that is required for meaningful analysis, quantitative comparisons [80, 82, 83] suggest an advantage of a factor of 100–1000 for soft X-ray absorption spectromicroscopy over electron energy loss spectromicroscopy at one core level edge, as measured by the G-factor (information/dose).

22.3
Data Analysis Methods

22.3.1
Chemical Mapping – Fitting to Known Reference Spectra

Figure 22.10 shows how quantitative chemical distributions can be derived from soft X-ray microscopy data, in this case a binary thin film blend of PS and PMMA, annealed on a SiO_x substrate, and subsequently transferred to microscopy grids for STXM characterization [84]. Figure 22.10a shows a set of as-recorded transmission images in the C 1s region to illustrate the types of changes in image contrast that occur in multicomponent samples as photon energy changes through the near-edge region. These changes reflect the spatial distribution and the mass–thickness properties of the chemical components. By fitting the spectral signal at each pixel

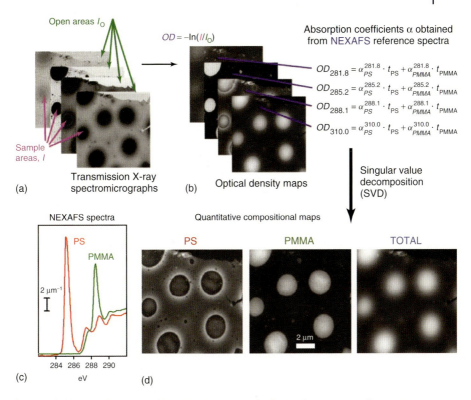

Figure 22.10 Principles of singular value decomposition (SVD) applied to a C 1s STXM image sequence (stack) from a PS-PMMA thin film [16, 84]. The transmission images (a) are converted to optical density (b) and a matrix transform that uses knowledge about the linear absorption coefficients (c) yields quantitative mass–thickness maps (d). (From Ref. [16], used with permission from Polymer via RightsLink license 2518500534762.)

to a linear combination of suitable reference signals (in this case, the C 1s spectra of PS and PMMA), it is possible to invert the set of measured transmission images to quantitative mass–thickness maps of the individual chemical components. If one can make reasonable assumptions about the density, then absolute thicknesses can be derived. The transmission data must first be converted to absorbance or OD using the Lambert-Beer law, $I = I_0 \exp(-OD)$. The set of OD images (Figure 22.10b) are then inverted to component maps. There are usually many more spectral points than chemical components (e.g., in this problem, there are only two chemical components but images were recorded at many more than two photon energies). In order to solve the inversion problem, one must have at least the same number of energy points in an image sequence as there are chemical components. More typically, there are many more – a reasonably good sampling of the NEXAFS spectral details can be achieved with 30–100 photon energies if non-uniform energy spacing is used. Quantitative component maps can be derived either by using conventional least squares procedures to fit a linear combination of suitable reference spectra (Figure 22.10c) to the spectrum at each pixel or by singular value decomposition (SVD) [85], which is an optimized matrix method to invert an overdetermined set of equations, as in this case. There is in general no exact solution to the inversion from measured data to component maps in the case of overdetermined systems. The optimum solution is a "close" result that in some sense satisfies all equations simultaneously. This closeness can be defined in a least squares sense and may be calculated by minimizing the residual error [86]. An illustration of the procedure and outcome is shown in Figure 22.10d. An interpretable model of the PS/PMMA thin film emerges from the SVD analysis. In this case, the analysis shows that the large thick PS droplets are surrounded by a sloping rim of PMMA in such a way that the total film thickness changes smoothly in height. During annealing, PS droplets form, which cannot sink into the PMMA matrix as the film is too thin. This thin film constraint and the balancing of interfacial energies forces the contact line between the PS and PMMA to be relatively high on the PS droplets, leading to a "rim" of PMMA that surrounds the PS droplets. Additional subtleties of the morphology in these thin films could be interpreted using the quantitative information gained from SVD [84].

An extension of the SVD procedure is often useful in cases where there may be some uncertainty about the absolute OD scales of the image sequence or one or more of the reference spectra. This can come about if only the shape and not the absolute intensity of the incident flux (I_0) is used in deriving OD scales. In such cases, scale errors in I_0 are equivalent to constant offsets in OD. Thus "stack fit" is a routine available in analysis of X-ray images and spectra (aXis2000) that automatically adds a constant term to the SVD analysis to compensate [87].

In transmission microscopies (TXM, STXM), the detected signal, after conversion to OD, is quantitatively related to the amount, and thus the intensity scales of the resulting component maps are in units of absolute thickness in nm, if the reference spectra are prepared on a linear absorbance scale (OD per nm of material, under its standard density). In situations such as X-PEEM or SPEM, where the measured intensities are only indirectly related to amounts, or if one does not convert reference spectra to an absolute intensity scale, one can still apply these methods, but the component maps then give information only about spatial distributions and relative amounts. In X-PEEM, if one can measure [80] or estimate the sampling depth for a given system and is comfortable assuming that the work function for

Figure 22.11 STXM analysis of a river biofilm exposed to 10 mg l^{-1} NiCl$_2$ before measurement [88]. (a) Optical density spectrum of the filamentous sheath in a Ni-exposed natural river biofilm (dashed line in the inset image) in the C 1s, O 1s, Mn 2p, Fe 2p, Ni 2p, Al 1s, and Si 1s edges, compared with the sum of the mass absorption coefficients for a best-fit composition. (b) Linear background subtracted spectra of the filamentous sheath at the C, O, Mn, Fe, Ni, and Si edges, along with the elemental edge jump signals (and uncertainties) used to deduce elemental amounts (for further details see [88]). (c) Component map of distribution of Ni (grayscale in nm), derived from two energy imaging at the Ni 2p edge. (d) Color-coded composite of the Mn (red), Fe (green), and Ni (blue) distributions superimposed on the gray scale map of the biology I(288.2 eV) − I(282 eV). (Figures 10 and 11 of Ref. [88], used with permission from Geobiology via RightsLink license 2518500066533.)

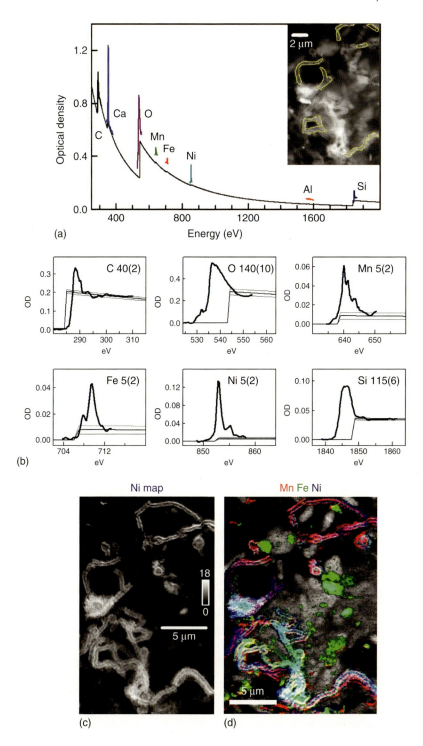

different regions of a surface is similar, then quantitative thickness analyses can be performed. An example of this is the quantitative determination of the distribution of proteins adsorbed on a phase-segregated blend surface (see Section 22.4.2).

Figure 22.11 shows a second example of quantitative mapping, in this case a complex environmental sample in which STXM was used to follow mineral and metal ion interactions, possibly mediated by bacterial biomineralization processes, in a natural river biofilm. The biofilm sample, taken from the South Saskatchewan river in Saskatoon, was intentionally spiked with 10 mg l^{-1} NiCl$_2$ during a 24 h period before the analysis to explore how different species accommodate toxic Ni^{2+} ions [88]. NEXAFS image sequences were measured for nine different elements (C, Ca, O, Mn, Fe, Ni, Na, Al, and Si). The inset image in Figure 22.11a shows a STXM image of a region of the biofilm where there is a variety of bacteria, both individual cells and groups of cells enclosed in a filamentous sheath. The metal ions were found preferentially in association with the filamentous structures, although there are also Fe-rich mineral grains in the field of view. The absorption spectrum at eight edges of the filamentous structure, taken from the identical area indicated in yellow outline in the inset in Figure 22.11a, is shown in Figure 22.11a, along with the elemental response expected for the average composition determined by SVD fits of the spectrum at each edge to appropriate reference spectra. Independent of the assumptions about chemical speciation (which can be challenging in a complex system like this), one can determine the elemental composition from the edge jump, as indicated in Figure 22.11b, where the edge jump for the determined composition, along with an error range, is plotted along with the average edge spectrum of the filaments. Figure 22.11c shows the quantitative Ni map, where the maximum thickness is only 18 nm, and the average thickness, mapped as NiO, (in an area determined to be Ni-rich), is less than 3 nm. Figure 22.11d is a color-coded composite of the Mn, Ni, and Fe signals. The Ni is found in association with Mn and Fe and is believed to accumulate through abiotic processes on Fe–Mn minerals created by biomineralization [88].

It is noteworthy that approaches similar to those outlined above – forward fitting by use of SVD or stack fit to derive relative or absolute quantitative maps from image sequences – can be used to analyze other types of image sequence signals, where a parameter other than incident photon energy is scanned. Examples include XRF maps (see Section 22.2.4) and linear dichroism maps (see Section 22.4.1) – the dependence of image contrast on the orientation of the E-vector of the soft X-rays relative to a sample at a given photon energy. Such sequences can be fit to the known functional response of linear dichroism, $I(\theta) = C + A\cos^2(\theta - \theta_{\text{ref}})$, where $I(\theta)$ is the intensity at a given E-vector angle, C is an angle-independent term, A is the amplitude of the angle-dependent signal, which has its maximum signal when $\theta = \theta_{\text{ref}}$, where θ_{ref} is the angle where the transition intensity is maximum (for 1s excitation, this corresponds to having the E-vector aligned with the transition moment vector, which, in simple systems, can be the direction of orbitals of the upper level to which the 1s electron is excited). This methodology is included in aXis2000 [87] and has been used to map distributions of β-sheet crystallites in dry and wet dragline spider silk [89, 90] and of sp^2 defects in carbon nanotubes [91–92, 93], in addition to the example given in Section 22.4.1.

22.3.2
Chemical Mapping – Unsupervised Statistical Analysis Methods

A concern with the types of forward fitting analyses described in the previous section is that they require prior knowledge of the chemical components present in the system under study. In many cases, this information is not available, or only some of the chemical species are known. Given the highly overdetermined nature of soft X-ray microscopic image sequences, it is of interest to consider multivariate statistical analysis (MSA) methods [94] that look for patterns in complex data of this type. MSA-based methods can be used for many aspects of the analysis: to identify the number of statistically valid chemical species that might be present; to derive meaningful reference spectra; and to convert the image sequence to quantitative chemical maps. The merit of this approach is that it provides a means to perform a "standard-less" analysis that effectively uses the highly redundant nature of the spectral domain information. Typically a STXM image sequence consists of $10^4 - 10^5$ spectra, for systems in which it is statistically meaningful to identify only five to seven components. Chris Jacobsen has provided the X-ray microscopy community with a useful tool for this type of analysis [95]. The procedure (PCA_GUI [96], which is embedded in aXis2000 [87]) first performs a principle component (PC) analysis, which identifies power-weighted eigenspectra and associated eigenimages. The user then selects an appropriate number of PCs to further process the data with a cluster analysis, which is a procedure to find a means of combining the PCs in such a way that they group together into clusters in "PC-space." These cluster spectra are typically much closer to true NEXAFS spectra of chemical components, although it is not always true that the cluster approach gives valid spectra and thus chemical identification. An added advantage of this approach is that a very significant noise reduction is achieved by restricting the analysis to include only the most significant PCs (typically 5–10, in a data set where the number of PCs (eigenspectra) is equal the number of pixels in each image). After rotating the PCs by identifying clustering in the PC-space, these reference spectra can be used in a final target analysis step to derive the chemical maps, which can be made quantitative or semiquantitative if the user supplies appropriate information about chemical composition and density. A very useful feature of the target analysis phase is that the spectra of chemical species known to be present can be introduced at this point, which often greatly improves the overall analysis.

Figure 22.12 presents results from analysis of a C 1s image sequence of a microtomed section of a core-shell microsphere composed of a divinyl benzene (DVB) core covered with a shell composed of a 50 : 50 DVB – ethylene glycol dimethylacrylate (EGDMA) mixture [97]. In this case, the PC analysis clearly indicates that only three statistically significant components were present. The reference spectra derived from the cluster analysis are displayed in Figure 22.12b, while three cuts through the PC-space (C1 vs C2, C1 vs C3, and C2 vs C3) (Figure 22.12b) show that

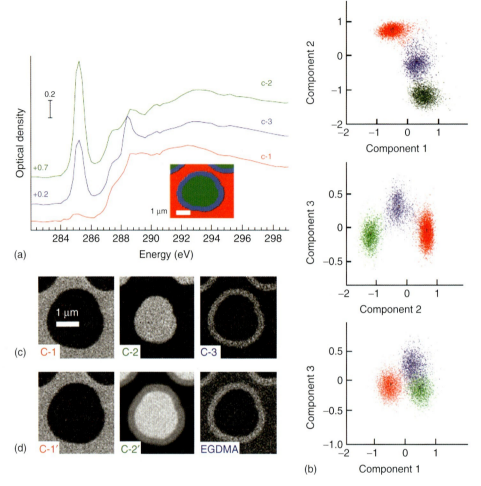

Figure 22.12 Results of a multivariate statistical analysis using a combination of principle component and target analysis applied to C 1s STXM results for a microtomed section of a core-shell microsphere [97]. (a) Rotated principle component spectra derived by PCA followed by a Euclidean cluster analysis [95]. Cluster 1 matches epoxy, and cluster 2 matches divinyl benzene (DVB). (b) Cluster plots of PC1 vs PC2; PC2 vs PC3; and PC1 vs PC3. (c) Component maps derived using the C-1, C-2, and C-3 target spectra. (d) Component maps derived by replacing C-3 target with the measured spectrum of ethylene glycol dimethylacrylate (EGDMA). Note the shell is known from the synthesis to be ~50 : 50 DVB:EGDMA. (From Ref. [16] used with permission from Polymer via RightsLink license 2518500534762.)

the cluster analysis (Euclidian metric) produced a very clean grouping of the data into three spectra and that the resulting spectra are very plausible NEXAFS spectra. The spectrum for component 1 is identical to that of epoxy resin, and that for component 2 is identical to that of DVB. Unfortunately, the spectrum for component 3 is identical to that of the shell, which is known from the material synthesis to be a composite of DVB and EGDMA. Thus, when these cluster spectra are used to fit the data, the result is simply a separation into the 3 spatial components (Figure 22.12c), without a clear indication that the shell is a mixture of two chemical species. When the spectrum of EGDMA is introduced in the target analysis step instead of component 3, the resulting maps (Figure 22.12d) are correct and provide the same answer as that from SVD. Both forward fitting analysis and MSA are valuable tools for the interpretation of soft X-ray spectromicroscopy data sets. It is usually wise to examine critical data sets with both approaches. A limitation of both approaches is that they always give an answer. It is essential that the analyst probes these answers to verify that they indeed represent the actual chemistry of the sample. Threshold masking techniques are provided in aXis2000, which allow extraction of the spectra associated with specific pixel subsets, defined on the basis of intensities in component maps derived by either forward fitting analysis or MSA. It is generally a good idea to extract such spatially selected spectra and investigate the quality of the analysis in spectral space. Counterparts to both SVD and stack fit stack analysis routines for fitting individual spectra are provided in aXis2000 for this purpose.

22.4
Selected Applications

22.4.1
Polymer Microstructure

Soft X-ray spectromicroscopy is optimally suited to study multicomponent polymer systems since (i) NEXAFS at the C 1s and other edges is well suited to differentiate even very chemically similar polymers [13, 98], (ii) many phase-segregated blends of technical and commercial importance have domain sizes easily resolved by current instruments, and (iii) electron beam analysis methods such as EELS in TEM have lower energy resolution (and thus are less able to differentiate species) and are significantly limited by the higher rate of radiation damage per unit analytical information [83]. Polymer microstructure analysis by soft X-ray microscopy is an application where there is significant industrial involvement, such as work on basic understanding of nanostructure of polyurethane foams [99, 100], superabsorbent gels [101], polymer reinforcements [102], and modification of polymer properties [103–105]. While the published work in this area tends to be fundamental, there are many cases where unpublished soft X-ray microscopic studies have played an important role in troubleshooting processing problems or assisting development of new or improved polymer processing.

A topical example of polymer analysis by STXM is in the area of organic solar cells. In contrast to semiconductor-based solar cells, the goal for organic solar cells is modest photoyields but ultralow cost fabrication [106]. Working devices must achieve a bicontinuous phase separation of donor and acceptor polymers with a mean size of the order of the exciton length (<10 nm) such that there is efficient separation and independent transport of the electron and hole generated by the

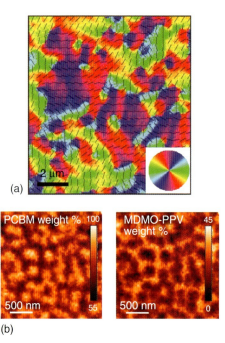

Figure 22.13 Example of linear dichroism mapping using STXM [109]. (a) In-plane molecular orientation map of an annealed thin film of poly(9,9 ƒ-dioctylfluorene-co-benzothiadiazole) (F8BT) derived from the linear dichroism at the C1s → π^* transition (285 eV) measured by sample rotation with a fixed horizontal E-vector. The π^* resonance is perpendicular to the polymer backbone, whose orientation in each domain is indicated via the black lines, the length of which indicate the estimated, fully extended length of the polymer backbone [109]. (b) Mapping the weight percentage composition of poly[2-methoxy-5-(3′,7′-dimethyloctyloxy)-1,4-phenylene vinylene] (MDMO-PPV) and (6,6)-phenyl-C_{61}-butyric acid (PCBM) in a 1 : 4 (w/w) blend (MDMO-PPV:PCBM) thin film, derived from images recorded at four energies (280, 284.4, 287.4, and 320 eV) [108]. (From B. Watts, private communication, permission granted.)

incident light. STXM is being applied to assist materials optimization in this area [107, 108]. Figure 22.13 presents two aspects of this work. The first demonstrates the ability of the linear dichroic signal in STXM microscopy to determine molecular orientation and domain structure in these types of materials. The second demonstrates imaging phase separation in a viable organic solar cell material. By recording images at a resonant photon energy with multiple sample orientations, the linear dichroism of an annealed thin film of poly(9,9-dioctylfluorene-co-benzothiadiazole) (F8BT) can be fit to a model to reveal the dominant molecular orientation within each pixel [109]. Figure 22.13a shows the in-plane orientation derived from the intensity of the C1s->$\pi^{*\sim}$(285 eV) – the color wheel in the inset indicates the correlation of color and orientation. The π^* resonance is perpendicular to the polymer backbone and thus the orientation of the polymer backbone can be determined. It is indicated via overplotted black lines, the lengths of which indicate the estimated fully extended length of the polymer backbone. Figure 22.13b shows maps of poly[2-methoxy-5-(3′,7′-dimethyloctyloxy)-1,4-phenylene vinylene] (MDMO-PPV) and (6,6)-phenyl-C_{61}-butyric acid (PCBM) in a 1 : 4 (w/w) blend (MDMO-PPV:PCBM) thin film. These maps were derived from images recorded at only four energies (280, 284.4, 287.4, and 320 eV), selected for their ability to differentiate these two species [108]. Such maps provide very useful feedback on the extent to which the desired interpenetrating continuous phase is actually achieved. The spatial resolution of the present STXM microscopes (20–30 nm) is still a little too coarse for this application, which really needs imaging of acceptor and donor domains, which are required to be within ∼10 nm of each other. Improvements to ZPs [25] and development of other soft X-ray imaging methods such as ptychography [110] as well as complementary chemically sensitive resonant scattering techniques [111, 112] are important to further progress in this area.

22.4.2
Surfaces and Interfaces

Magnetic materials are a critically important aspect of modern information storage and processing technologies. Soft X-ray microscopic and spectroscopic methods to study magnetism and magnetic dynamics are described in detail in the monograph by Stöhr and Siegmann [18]. The current generation of high-capacity hard disk storage systems relies heavily on research results obtained with synchrotron soft X-ray microscopies, spectroscopies, and dynamics measurements. Read–write heads exploiting giant magnetoresistance, out-of-plane recording media, and nanopatterned storage media all have benefited from soft X-ray microscopic and spectromicroscopic studies. As the performance continues to improve exponentially, increasingly small volumes of magnetic materials, with carefully controlled nanoscale materials chemistry are key. Nanoscale methods for their characterization are essential. Here, I give two examples, one based on polarization-dependent X-PEEM studies and the other using X-PEEM to investigate protein biomaterial interactions.

Interactions of ferromagnetic and antiferromagnetic domains are critical in the "pinning" of specific regions in complex magnetic sensors such as giant magnetoresistance heads, but the magnetic exchange interaction is quite complex and can be mediated or complicated by crystallographic effects. Figure 22.14 shows magnetically and structurally sensitive PEEM images recorded at specific Ni 2p and O 1s resonant peaks of the near-surface region of a polycrystalline sample of a NiO(001) surface [113]. Comparison of the contrast in identical regions reveals a complex interplay of crystal structure – orientation of the grains as deduced from the O 1s XLD – and antiferromagnetic magnetic order – as deduced from the Ni 2p linear X-ray linear magnetic dichroic (XLMD) signal. When a thin ferromagnetic Co layer is deposited on this surface, there is a perpendicular coupling between the Ni and Co moments. The chemical interactions at the Co-NiO interface drives a crystallographic transformation leading to reorientation of the Ni moments from the <112> to the <110> direction [113]. Detailed nanoscale dichroic studies show that the reorientation is driven by changes in magnetocrystalline anisotropy rather

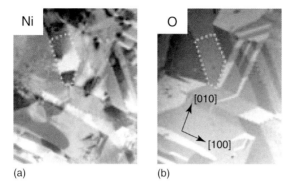

Figure 22.14 X-PEEM images (ALS PEEM-3) of the near-surface region of a Ni(001) single crystal recorded at (a) the Ni 2p$_{3/2}$ resonant peak showing X-ray linear magnetic dichroic contrast (XLMD) and (b) at the O 1s resonant peak showing X-ray linear dichroic contrast (XLD). Field of view is 8 μm. From the O 1s signal, the region outlined the dotted lines appears crystallographically to be a single domain, whereas the Ni 2p XMLD shows there is a complex magnetic domain structure within this and other individual crystal domains [113]. (From Hendrik Ohldag, private communication, permission granted.)

than exchange coupling. An additional example of magnetically sensitive soft X-ray microscopy in given in Section 4.4.4, where time-resolved XMCD-STXM is used to exemplify *in situ* conditioning of samples.

The second example of surface and interface applications of soft X-ray microscopy is in the area of soft materials, which are traditionally very difficult to study with electron-beam-based techniques. Figure 22.15 shows results from a combined C 1s and N 1s X-PEEM study of the adsorption of human serum albumin (HSA) on to a phase-segregated bend of PS and polylactic acid (PLA) [114]. This is part of a recently reviewed [17] set of X-PEEM studies of protein interactions with phase-segregated polymer blend surfaces, which are candidate biomaterials for use in medical application. The ∼50 nm thick, 40 : 60 PS:PLA blend film is spun coated on a Si wafer (0.7 wt% in toluene), annealed 6 h at 45 °C and exposed to a 0.005 mg ml^{-1} aqueous solution of HSA for 20 min. The colorized individual component maps are displayed across the top (Figure 22.15a–c), while Figure 22.15d is a rescaled color-coded composite of the derived component maps for HSA (blue) – from N 1s edge, PS (red), and PLA (green). This example nicely illustrates the advantage of using spectroscopic information from multiple core levels to improve sensitivity to specific components.

22.4.3
3D Imaging (Tomography) and 3D Chemical Mapping (Spectrotomography)

For the thin samples typical of soft X-ray transmission microscopy (TXM or STXM), 2D projection imaging is often sufficient. However, for thicker, low-density samples, such as biological or environmental samples, one needs information about the *z*-position of the absorbance in order to properly interpret the image and

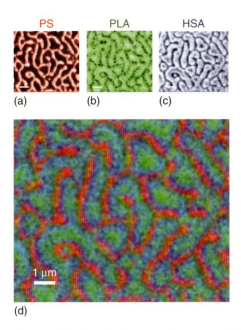

Figure 22.15 Analysis of protein (human serum albumin HSA) interactions with a polystyrene-polylactic acid (PS-PLA) polymer blend surface by C 1s and N 1s image sequences recorded in X-PEEM (ALS PEEM-2) [114]. The main image is a rescaled color-coded composite of the derived component maps for HSA (blue) – from N 1s edge, PS (red), and PLA (green). The ~50 nm thick, 40 : 60 PS:PLA blend film is spun coated on a Si wafer (0.7 wt% in toluene), annealed 6 h at 45 °C and exposed to a 0.005 mgml^{-1} aqueous solution of HSA for 20 min. (From Ref. [114], permission granted from ACS Biomacromolecules via RightsLink license 2518430179156.)

spectromicroscopic information. A simple way to achieve this is to record stereo pairs with the sample tilted between the two images by 10–30° [115]. However, for more complex cases, as in congested cells with many morphological elements distributed in 3D over a thickness of 1–3 μm, the better approach is to do a complete 3D visualization by serial section imaging [116] or angle scan tomography [117]. Full-field TXM is the most developed type of soft X-ray tomography. Samples can be loaded fully wet into pulled glass capillaries or mounted on whole or sections of TEM grids. The capillary approach is preferred not only because it allows full 180° rotation and thus avoids the "missing wedge" artifacts that exist with reduced angular range tomograms but also because it is a good way to retain fully wet samples in the case of room temperature studies and it is compatible with plunge-freeze vitrification techniques for cryotomography. The latter has become the accepted approach for both electron and X-ray tomography of biological samples since the plunge freezing prevents generation of large ice crystals that destroy the microstructure and the frozen samples do not suffer mass loss while recording the tomograms (although it is important to note that chemical damage from absorption of X-rays occurs at the same rate as at room temperature [79]).

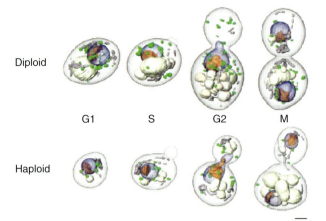

Figure 22.16 Three-dimensional images of S. cerevisiae yeast cells at different phases of the cell cycle [118], measured with the National Center for X-ray Tomography (NCXT) full-field cryo X-ray tomography instrument at the ALS. The top row shows diploid cells, whereas the bottom row shows haploid cells. Organelles are identified based on mass absorption at ~520 eV, and are color coded: nucleus, blue; nucleolus, orange; vacuoles, ivory; mitochondria, gray; lipid bodies, green. Scale bar is 1 μm. (From Ref. [118], used with permission from Yeast via RightsLink license XXXXX.)

At present, the most advanced facility for high-throughput soft X-ray tomography of biological samples is the NCXT at the ALS. Figure 22.16 shows results from a recent NCXT study [118] of the internal structure of *Saccharomyces cerevisiae* yeast cells at different phases of the cell cycle, derived from tomograms on eight different cells, measured at a single photon energy (~520 eV, in the water window). The top row shows diploid cells, and the bottom row shows haploid cells. The color coding of the internal structure is based on segmentation – differentiation on the basis of mass absorption at ~520 eV. The organelles are color coded: nucleus, blue; nucleolus, orange; vacuoles, ivory; mitochondria, gray; lipid bodies, green. As larger amounts of data are accumulated, it is very likely that this segmentation approach will produce very valuable insights into the internal structure of micron-scale biological organisms. Excellent work on studies of the 3D structure of biological specimens is also being carried out at the BESSY2 TXM facility, for example, the first study of mammalian tissue [119] and a recent study of the structure of *Vaccinia* virus [120].

If the full-field tomograms could be measured at a number of different photon energies, a more explicit connection between X-ray microscopic signal and chemical identification could be achieved. To date, the only full-field TXM system able to perform high-quality spectroscopy and spectrotomography is that at BESSY2 [34, 37]. Another approach to obtain 3D imaging is to carry out tomography in STXM. Although much slower than full-field TXM tomography, the superior spectroscopic capabilities of STXM provide an added dimension worth exploiting. Pioneering STXM tomographic experiments were done at NSLS [121], and the Stony Brook group developed and operated a UHV cryo-STXM for tomography at X-1A beamline at NSLS [122]. However, those earlier STXM tomographic studies used only a single photon energy, so they did not take advantage of the spectroscopic power of STXM. Recently, room temperature spectrotomography based on imaging over multiple photon energies and a range of sample orientations has been implemented in STXM. Measurements have been reported on wet samples in pulled glass capillaries [123] as well as dry samples on grid sections [124–126]. Although the glass capillary approach has the advantage of being able to measure the full 180° angular range, and thus does not suffer from "missing wedge" artifacts, to date it has not been able to prepare walls sufficiently thin to work at X-ray energies below 340 eV. In contrast, while the grid method typically can only be applied over ±70°, it has been possible to measure tomograms at energies as low as the S 2p region (150–180 eV) [126].

To illustrate the ability of STXM spectromicroscopy to provide quantitative 3D chemical mapping, Figure 22.17 shows results from a STXM spectrotomographic study of polyacrylate-filled PS latex microspheres in water in a pulled glass capillary [123]. Figure 22.17a is a single projection image at 530.0 eV, below the onset of the O 1s absorption. While this projection image shows the shell structure of the microspheres, the boundaries of the microspheres and the capillary are blurry because of curvature. After reconstruction of the tomogram from a set of 60 images (0–180°) at two photon energies (530.0, 532.0 eV), a much clearer view of the structure is obtained (Figure 22.17b). The difference in the OD at these two energies was used to quantify the amount of polyacrylate in each voxel, while the signal at 530 eV can be segmented to separately identify the glass wall, the water, and the PS shell. Figure 22.17c is a colorized rendering of the tomographic reconstruction in which the PS (red) and water (blue) are differentiated by OD at 530 eV; the glass wall has been removed, and the acrylate (green) is identified spectroscopically by the difference in OD at 532 and 530 eV. Figure 22.17d shows a quantitative projection of the tomogram of only the acrylate component in which the blue color scale corresponds to 2–8 nm of acrylate through the projected path while the green color scale corresponds to 8–15 nm in the projection. This type of information, in particular the ability to generate statistics on fill levels, rupture, and internal distributions over a number of different latex microspheres, is valuable feed back to the industrial partner in this study.

22.4.4
In situ Techniques

In recent years, a major thrust in the development of soft X-ray microscopies has been *in situ* studies, in which the properties of a sample are manipulated in some way while it is still in the microscope. When seeking information about structure–function relationships, *in situ* studies are highly desirable, as they can provide a very direct link between the nanostructure and how it can affect a particular process or how the nanostructure can be modified by a particular external stimulus. While there are examples in the literature of *in situ* experiments with all four types of soft X-ray microscopy this review has discussed, the TXM and STXM techniques are those with the most extensively developed *in situ* technologies because of the

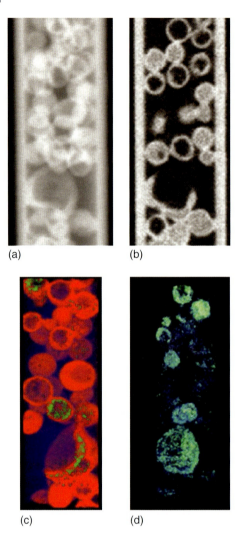

Figure 22.17 STXM tomography of polyacrylate-filled polystyrene microspheres in water in a pulled glass capillary [123]. (a) Projection image at 530.0 eV. (b) A 50 nm thick slice through the center of the tomographic reconstruction of 60 projections (0–180°/3°). (c) Colorized rendering of the reconstruction in which the PS (red) and water (blue) are differentiated by optical density at 530 eV, while the acrylate (green) is identified spectroscopically by the difference in OD at 532 and 530 eV. (d) Quantitative map of the acrylate component. The blue color scale corresponds to 2–8 nm of acrylate, and the green color scale corresponds to 8–15 nm in the projection. (From Ref. [123], used with permission from IUCr.)

relative ease with which the local sample environment can be adapted. The UHV requirements for SPEM and X-PEEM limit the flexibility to carry out experiments controlling the local chemical environment, and the strong fields at the sample in X-PEEM make it difficult to modify samples with applied electric or magnetic fields. In contrast, the photon-in/photon-out character of (S)TXM microscopes allow them to be readily adapted to make *in situ* modifications of the sample through controlled chemical, mechanical, thermal, or external fields. *In situ* experiments have been performed using electrochemistry [127], flow wet cells for chemical reaction [128], pulsed or AC current used for modifying magnetic properties [129], current flow in active semiconductor circuits [130], flow gas cells [131], variable humidity cells [90, 132], and variable temperatures [133].

Both TXM and STXM have been developed for magnetic dynamics experiments. While subnanosecond TXM experiments require few bunch operations to achieve subnanosecond time resolution, STXM techniques now exist to measure time-dependent magnetic properties of samples with 150 ps time resolution while the synchrotron is operating with full current in the multibunch mode. One very elegant recent example of this is the STXM study by Weigand *et al.* [129], which mapped the dynamics of switching of the out-of-plane magnetization (vortex core) caused by an applied AC current pulse (Figure 22.18). This experiment imaged and tracked with 250 ps time sampling and 10 nm positional precision the lateral position and polarity of the vortex core in a 500 nm permalloy pad after it was excited by different magnitudes of a GHz excitation pulse. The vortex core reversal was found to take place through coherent excitation by the leading and trailing edges of the pulse, which significantly lowered the field amplitude required for switching. The mechanism can be understood in terms of gyration around the vortex equilibrium positions, which are displaced by the applied field. The study showed that a much lower power is required to switch the polarity of the vortex

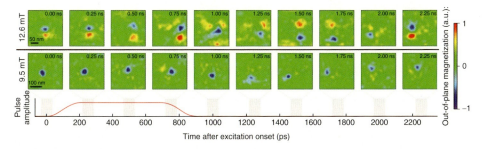

Figure 22.18 Use of time-resolved STXM to study the dynamics and mechanism of vortex core switching [129]. Each color-coded image shows successive 250 ps time steps of the out-of-plane magnetization of the center section of a 500 nm^2 permalloy element excited by in-plane magnetic pulses with a length of 700 ps. At 9.5 mT (below the switching threshold), the vortex core is displaced by the pulse and excited into gyrotropic gyration. At 12.6 mT (above the threshold), the dynamics of the two vortex core polarization states are superimposed, indicating vortex core switching. (From Ref. [129], http://prl.aps.org/abstract/PRL/v102/i7/e077201, copyright 2009 by the American Physical Society, used with permission.)

motion with an appropriately shaped pulsed field than with a DC field, thereby opening the possibility for practical magnetic storage or processing devices based on vortex polarity.

A second example of state-of-the-art *in situ* investigations is the work of the de Groot group, who have adapted STXM to study catalysts under realistic conditions of temperature and pressure [131, 134]. Figure 22.19a shows schematics of the catalyst cell that was built for both TEM and STXM use [135]. By using microfabrication techniques, heating is extremely localized and the whole structure is very compact and compatible with the spatial constrains of both STXM and TEM. While most of the Si_3N_4 support is thick, to be compatible with high pressure differentials (up to 2 bar), there are a series of 10 µm diameter holes between the heater traces where the Si_3N_4 has been further etched so as to be transparent to soft X-rays. There is provision for gas flow into the gap between the heater element and a cover window. Figure 22.19b,c presents localized Fe 2p NEXAFS spectra and color-coded component maps from a study of the evolution of an individual Fe-based Fischer–Tropsch synthesis catalyst particle during its activation phase (heating in atmospheric pressure of H_2 from ambient to 500 °C) [131]. Image sequences were recorded at the Fe 2p and O 1s edge as a function of temperature. Below 300 °C, the changes in the Fe catalyst were heterogeneous and depended on its interaction with the SiO_2 support. When the sample was heated further, the Fe environment became more homogeneous. By 450 °C, the catalyst consisted of Fe^0:Fe^{2+}:mixed Fe^{2+}/Fe^{3+} species in a ratio of about 1 : 1 : 1. Apparently, the enhanced mobility of Si species combined with the stabilization of Fe^{2+} leads to an almost homogeneous Fe distribution after reduction at this temperature. Fe^{2+} in close contact with SiO_2 is stabilized from further reduction to Fe^0.

22.5
Future Outlook and Summary

Soft X-ray microscopy has reached an intermediate phase of development where the major techniques – TXM, STXM, SPEM, and X-PEEM – are all well established with reliable and user-accessible instrumentation (see Table 22.1), which allows even inexperienced users to access its capabilities. The techniques have been applied to many different areas of science, medicine, and engineering, with a significant component of the work relying on materials characterization at the

Figure 22.19 (a) *In situ* microheater on a Si_3N_4 support with 10 µm wide X-ray transparent windows, which is being used for STXM studies of catalyst and other materials at temperatures as high as 700 °C, with multiatmosphere pressures of gases [135]. (b) Evolution with temperature under an atmosphere of H_2 of the distribution of Fe chemical states in a single grain of a Fischer–Tropsch catalyst [131] – red = α − Fe, green = magnetite (Fe_3O_4), and blue = Fe_2SiO_4. (c) Fe 2p spectra from the indicated regions, which correspond to relatively pure domains of each species. (Part (a) from Ref. [131], permission granted from Ultramicroscopy via RightsLink license 2534321437193; parts (b,c) from F. de Groot, private communication, permission granted.)

22.5 Future Outlook and Summary | 781

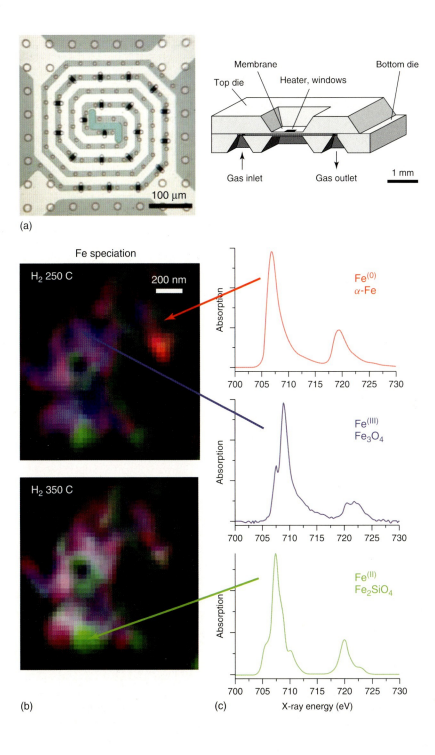

sub-50 nm spatial scale. There is ongoing development of both the microscopes and the ancillary instrumentation and methodology that is needed to apply them to specific problems. In the area of instrumentation, there will be significant breakthroughs in X-PEEM in the next few years, with compensation of both spherical and chromatic aberrations achieving sub-10 nm spatial resolution. The TXM and STXM performance continuously improves with advances in ZP fabrication technology. At the Tenth International Conference of X-ray Microscopy (Chicago, August 2010), there were a number of presentations in which both KB mirror focusing and ZP optics achieved sub-10 nm performance with high-contrast test samples. Translating that capability to routine operation with lower contrast, real-world samples are needed. Challenging applications such as measurement of the magnetic properties of nanoscale materials – for example, measuring the XMCD signals of individual 30 nm magnetite magnetosomes in magnetotactic bacteria, which has recently been achieved for the first time, using STXM [136] – are useful to bridge from artificial test structures to real-world problems.

A large part of the scientific impact of nanoscopy with soft X-ray microscopies will be the result of innovation in the area of *in situ* control of the sample environment. There are many groups actively working in this area, so new and more powerful applications are to be expected. Finally, there is the issue of access. The community of soft X-ray microscopy users is fully saturating the existing synchrotron-based microscopes. These instruments are typically the ones most in demand at every synchrotron, as determined by the competitive peer review process that governs access. There is clearly a need for building and operating additional beamlines and more soft X-ray microscopes. The most successful of the current instruments are those which are on dedicated beamlines whose optical properties have been optimized for the type of microscopy involved. This trend continues. Given the relatively large number of recently completed third-generation synchrotrons (CLS, Soleil, Diamond, Australian Synchrotron, Alba, Shanghai, and PETRA) as well as those funded and under construction (NSLS II, Taiwan, Saga-LS, SuperSOR, PLS upgrade), most of which do or will have soft X-ray microscopic instrumentation, I am optimistic that access to soft X-ray microscopies will be significantly easier in the near future. Finally, with the recent successes of the fourth-generation light sources at Flash and LCLS, there will be a new frontier of soft X-ray imaging at the femtosecond time scale, most likely dominated by single-shot coherent diffraction techniques. X-ray microscopy is already a significant contributor to the field of nanoscopy and will become increasingly important in the future.

Acknowledgments

I thank my group members and collaborators and many others who have provided material for this review: Peter Fischer, Frank de Groot, Burkhard Kaulich, Maya Kiskanova, Carolyn Larabell, Hendrik Ohldag, Christoph Quitmann, Andreas Scholl, Jo Stöhr, Ben Watts, and Phil Withers. Research by my group was carried out at the ALS and CLS and was supported by NSERC (Canada) and the Canada

Research Chair program. I especially thank Tolek Tyliszczak and David Kilcoyne for their contributions to developing and maintaining the STXM instruments at the ALS and Konstantin Kaznatcheev, Chithra Karunakaran, Jian Wang, Drew Bertwistle, and Yingshen Lu for developing and maintaining the CLS-SM beamline and STXM and PEEM instruments at the CLS. The ALS is supported by the Director, Office of Energy Research, Office of Basic Energy Sciences, Materials Sciences Division of the US Department of Energy, under Contract No. DE-AC03-76SF00098. The CLS is supported by NSERC, CIHR, NRC, and the University of Saskatchewan.

References

1. Bertsch, P.M. and Hunter, D.B. (2001) Applications of synchrotron-based x-ray microprobes. *Chem. Rev.*, **101**, 1809–1842.
2. Winick, H. (1995) *Synchrotron Radiation Sources: A Primer*, World Scientific, Singapore.
3. Attwood, D. (2000) *Soft X-rays and Extreme Ultraviolet Radiation, Principles and Applications*, Cambridge University Press.
4. Seres, J., Seres, E., Verhoef, A., Tempes, G., Streli, C., Wobrauschek, P., Yakovlev, V., Scrinzi, A., Spielmann, C., and Krausz, F. (2005) Source of coherent kiloelectronvolt X-rays. *Nature*, **433**, 596–598.
5. Bertilson, M., von Hofsten, O., Thieme, J., Lindblom, M., Holmberg, A., Takman, P., Vogt, U., and Hertz, H. (2009) First application experiments with the Stockholm compact soft x-ray microscope. Proceedings of the 9th International Conference on X-ray Microscopy, Zurich Switzerland, August 2008. *Journal of Physics: Conference Series*, **186**, 012025 - 1–012025-3.
6. Mainwaring, P. (2008) Application of the Gatan X-ray ultramicroscope to the investigation of material and biological samples. *Microsc. Today*, **16** (6) 14–17.
7. Kirz, J., Jacobsen, C., and Howells, M. (1995) Soft X-ray microscopes and their biological applications. *Q. Rev. Biophys.*, **28**, 33–130.
8. Ade, H. (1998) in *Experimental Methods in the Physical Sciences*, vol. **32** (eds J.A.R. Samson and D.L. Ederer), Academic Press, San Diego, CA, pp. 225–276.
9. Jacobsen, C. (1999) Soft X-ray microscopy. *Trends Cell Biol.*, **9**, 44–60.
10. Bauer, E. (2001) Photoelectron spectromicroscopy: present and future. *J. Electron. Spectrosc. Relat. Phenom.*, **114–116**, 975–987.
11. Bauer, E. (2001) Photoelectron microscopy. *J. Phys. Condens. Matter*, **13**, 11391–11404.
12. Guenther, S., Kaulich, B., Gregoratti, L., and Kiskinova, M. (2002) Photoelectron microscopy and applications in surface and material science. *Prog. Surf. Sci.*, **70**, 187–260.
13. Urquhart, S.G. and Ade, H. (2002) in *Chemical Applications of Synchrotron Radiation, Part I* (ed. T.K. Sham), World Scientific, Singapore, pp. 285–355.
14. Hitchcock, A.P., Stöver, H.D.H., Croll, L.M., and Childs, R.F. (2005) Chemical mapping of polymer microstructure using soft X-ray spectromicroscopy. *Aust. J. Chem.*, **58**, 423–432.
15. Howells, M., Jacobsen, C., and Warwick, T. (2007) in *Principles and Applications of Zone Plate X-Ray Microscopes in Science of Microscopy* (eds P. Hawkes and J. Spence), Springer, New York, pp. 835–926.
16. Ade, H. and Hitchcock, A.P. (2008) NEXAFS microscopy and resonant scattering: composition and orientation probed in real and reciprocal space. *Polymer*, **49**, 643–675.
17. Leung, B.O., Brash, J.L., and Hitchcock, A.P. (2010) Characterization of biomaterials by soft X-ray spectromicroscopy. *Materials*, **3**, 3911–3938.
18. Stöhr, J. and Siegmann, H.C. (2006) *Magnetism: From Fundamentals to*

19. Huang, X., Nelson, J., Kirz, J., Lima, E., Marchesini, S., Miao, H., Neiman, A.M., Shapiro, D., Steinbrener, J., Stewart, A., Turner, J.J., and Jacobsen, C.J. (2009) Soft X-ray diffraction microscopy of a frozen hydrated yeast cell. *Phys. Rev. Lett.*, **103**, 198101-1–198101-4.

20. Bergh, M., Huldt, G., Tîmneanu, N., Filipe, R., Maia, N.C., and Hajdu, J. (2008) Feasibility of imaging living cells at subnanometer resolutions by ultrafast X-ray diffraction. *Q. Rev. Biophys.*, **41**, 181–204.

21. Withers, P.J. (2007) X-ray nanotomography. *Mater. Today*, **10**, 26–34.

22. Schmahl, G. and Rudolph, D. (1969) Lichstarke Zoneplatten als abbildende Sytee für weiche Röntgenstrahlung (High power zone plates a image forming systems for soft X-rays). *Optik*, **29**, 577–585.

23. Schmahl, G., Rudolph, D., Niemann, B., and Christ, O. (1980) Zone-plate X-ray microscopy. *Q. Rev. Biophys.*, **13**, 297–315.

24. Chao, W.L., Harteneck, B.D., Liddle, J.A., Anderson, E.H., and Attwood, D.T. (2005) Soft X-ray microscopy at a spatial resolution better than 15 nm. *Nature*, **435**, 1210–1213.

25. Chao, W., Kim, J., Rekawa, S., Fischer, P., and Anderson, E.H. (2009) Demonstration of 12 nm resolution fresnel zone plate lens based soft X-ray microscopy. *Opt. Express*, **17**, 17669–17677.

26. Rehbein, S., Heim, S., Guttmann, P., Werner, S., and Schneider, G. (2009) Ultrahigh-resolution soft-x-ray microscopy with zone plates in high orders of diffraction. *Phys. Rev. Lett.*, **103**, 110801-1–110801-4.

27. Fischer, P., Kim, D.H., Chao, W.L., Liddle, J.A., Anderson, E.H., and Attwood, D.T. (2006) Soft X-ray microscopy of nanomagnetism. *Mater. Today*, **9**, 26–33.

28. Meier, G., Bolte, M., Eiselt, R., Krüger, B., Kim, D.-H., and Fischer, P. (2007) Direct imaging of stochastic domain-wall motion driven by nanosecond current pulses. *Phys. Rev. Lett.*, **98**, 187202-1–187202-4.

29. Fischer, P., Kim, D.-H., Mesler, B.L., Chao, W.L., and Anderson, E.H. (2007) Magnetic soft X-ray microscopy: imaging spin dynamics at the nanoscale. *J. Magn. Magn. Mater.*, **310**, 2689–2692.

30. Kasai, S., Fischer, P., Im, M.-Y., Yamada, K., Nakatani, Y., Kobayashi1, K., Kohnoand, H., and Ono, T. (2008) Probing the spin polarization of current by soft X-ray imaging of current-induced magnetic vortex dynamics. *Phys. Rev. Lett.*, **101**, 237203-1–237203-4.

31. Gartner, E.M., Kurtis, K.E., and Monteiro, P.J.M. (2000) Proposed mechanism of C-S-H growth tested by soft X-ray microscopy. *Cement Concrete Res.*, **30**, 817–822.

32. Silva, D.A., and Monteiro, P.J.M. (2005) Analysis of C-S-H hydration using soft X-rays transmission microscopy: effect of EVA copolymer.. *Cement Concrete Res.*, **35**, 2026–2032.

33. Heim, S., Guttmann, P., Rehbein, S., Werner, S., and Schneider, G. (2009) Energy-tunable full-field x-ray microscopy: cryo-tomography and nano-spectroscopy with the new BESSY TXM. Proceedings of the 9th International Conference on X-ray Microscopy, Zurich Switzerland August 2008. *J. Phys.: Conf. Ser.*, **186**, 012041-1–012041-3.

34. Guttmann, P., Bittencourt, C., Ke, X., Van Tendeloo, G., Umek, P., Arcon, D., Ewels, C.P., Rehbein, S., Heim, S., and Schneider, G. (2011) TXM-NEXAFS of TiO2-based nanostructures. 10th International Conference on X-Ray Microscopy, Chicago, August 12–16, 2010, *Am. Inst. Phys. Conf. Proc.* **1365**, 437–440.

35. Kihara, H., Yamamoto, A., Guttmann, P., and Schmahl, G. (1996) Observation of the internal membrane system of COS cells by X-ray microscopy. *J. Electron. Spectrosc. Relat. Phenom.*, **80**, 369–372.

36. Born, M. and Wolf, E. (2003) *Principles of Optics: Electromagnetic Theory of Propagation, Diffraction and Interference of Light*, Cambridge University Press.
37. Guttmann, P., Zeng, X., Feser, M., Heim, S., Yun, W., and Schneider, G. (2009) Ellipsoidal capillary as condenser for the BESSY full-field x-ray microscope. Proceedings of the 9th International Conference X-Ray Microscopy, Zurich Switzerland, August 2008. *J. Phy.: Conf. Ser.*, **186**, 012064 -1-012064-3.
38. Kaulich, B., Susini, J., David, C., Di Fabrizio, E., Morrison, G., Charalambous, P., Thieme, J., Wilhein, T., Kovac, J., Bacescu, D., Salome, M., Dhez, O., Weitkamp, T., Cabrini, S., Cojoc, D., Gianoncelli, A., Vogt, U., Podnar, M., Zangrando, M., Zacchigna, M., and Kiskinova, M. (2006) A European twin X-ray microscopy station commissioned at ELETTRA. Proceedings of the 8th International Conference on X-Ray Microscopy, IPAP Conference Proceedings Series 7 (eds S. Aoki, Kagoshima, Y. and Suzuki, Y.), pp. 22–25.
39. Wichtendahl, R., Fink, R., Kuhlenbeck, H., Preikszas, D., Rose, H., Spehr, R., Hartel, P., Engel, W., Schlögl, R., Freund, H.-J., Bradshaw, A.M., Lilienkamp, G., Schmidt, Th., Bauer, E., Brenner, G., and Umbach, E. (1998) SMART: an aberration-corrected XPEEMK/LEEM with energy filter. *Surf. Rev. Lett.*, **5**, 1249–1256.
40. Hartel, P., Preikszas, D., Spehr, R., Müller, H., and Rose, H. (2002) Mirror corrector for low-voltage electron microscopes. *Adv. Imaging Electron. Phys.*, **120**, 41–133.
41. Schmidt, Th., Marchetto, H., Lévesque, P.L., Groh, U., Maier, F., Preikszas, D., Hartel, P., Spehr, R., Lilienkamp, G., Engel, W., Fink, R., Bauer, E., Rose, H., Umbach, E., and Freund, H.-J. (2010) Double aberration correction in a low-energy electron microscope. *Ultramicroscopy*, **110**, 1358–1361.
42. Wan, W., Feng, J., Padmore, H.A., and Robin, D.S. (2004) Simulation of a mirror corrector for PEEM3. *Nucl. Instrum. Methods Phys. Res. A*, **519**, 222–229.
43. Feng, J., Forest, E., MacDowell, A.A., Marcus, M., Padmore, H., Raoux, S., Robin, D., Scholl, A., Schlueter, R., Schmid, P., Stöhr, J., Wan, W., Wei, D.H., and Wu, Y. (2005) An x-ray photoemission electron microscope using an electron mirror aberration corrector for the study of complex materials. *J. Phys C.: Condens. Matter*, **17**, S1339–S1350.
44. Feng, J. and Scholl, A. (2006) Photoemission electron microscopy (PEEM), in *Science of Microscopy* (eds P.W. Hawkes and J.C.H. Spence), Wiley-VCH Verlag GmbH, 657–695.
45. Haider, M., Uhlemann, S., Schwan, E., Rose, H., Kabius, B., and Urban, K. (1998) Electron microscopy image enhanced. *Nature*, **392**, 768–769.
46. Krivanek, O.L., Dellby, N., and Lupini, A.R. (1999) Towards sub-Å electron beams. *Ultramicroscopy*, **78**, 1–11.
47. Tromp, R., Hannon, J.B., Ellis, A.W., Wan, W., Berghaus, A., and Schaff, O. (2010) A new aberration-corrected, energy-filtered LEEM/PEEM instrument. I. Principles and design. *Ultramicroscopy*, **110**, 852–861.
48. Kilcoyne, A.L.D., Tylisczak, T., Steele, W.F., Fakra, S., Hitchcock, P., Franck, K., Anderson, E., Harteneck, B., Rightor, E.G., Mitchell, G.E., Hitchcock, A.P., Yang, L., Warwick, T., and Ade, H. (2003) Interferometer-controlled scanning transmission X-ray microscopes at the advanced light source. *J. Synchrotron Radiat.*, **10**, 125–136.
49. Warwick, T., Ade, H., Kilcoyne, A.L.D., Kritscher, M., Tylisczcak, T., Fakra, S., Hitchcock, A.P., Hitchcock, P., and Padmore, H.A. (2002) A new bend-magnet beamline for scanning transmission X-ray microscopy at the Advanced Light Source. *J. Synchrotron Radiat.*, **9**, 254–257.
50. Rarback, H., Shu, D., Feng, S.C., Ade, H., Kirz, J., McNulty, I., Kern, D.P., Chang, T.H.P., Vladimirsky, Y., Iskander, N., Attwood, D., McQuaid, K., and Rothman, S. (1988) Scanning X-ray microscope with 75-nm resolution. *Rev. Sci. Instrum.*, **59**, 52–59.

51. Thieme, J. (1988) Theoretical investigations of imaging properties of zone plates and zone plate systems using diffraction theory, in *Proceedings of the 2nd International Conference X-Ray Microscopy Stony Brook, USA, July 1987, X-ray Microscopy II* (eds D. Sayre, Howells, M. Kirz, J. and Rarback), H. Springer-Verlag, Berlin, 211–216.
52. Jacobsen, C., Wirick, S., Flynn, G., and Zimba, C. (2000) Soft X-ray spectroscopy from image sequences with sub-100 nm spatial resolution. *J. Microsc.*, **197**, 173–184.
53. Feser, M., Carlucci-Dayton, M., Jacobsen, C., Kirz, J., Neuhäusler, U., Smith, G., and Yu, B. (1998) Applications and instrumentation advances with the Stony Brook scanning transmission X-ray microscope. X-Ray Microfocusing: Applications and Techniques, San Diego California, July 22–23, 1998. *Proc. SPIE*, **3449**, 19–26.
54. Feser, M., Hornberger, B., Jacobsen, C., De Geronimo, G., Rehak, P., Holl, P., and Strüder, L. (2006) Integrating Silicon detector with segmentation for scanning transmission X-ray microscopy. *Nucl. Phys. Res. A*, **565**, 841–854.
55. Hornberger, B., Feser, M., and Jacobsen, C. (2007) Quantitative amplitude and phase contrast imaging in a scanning transmission X-ray microscope. *Ultramicroscopy*, **107**, 644–655.
56. Gianoncelli, A., Morrison, G.R., Kaulich, B., Bacescu, D., and Kovac, J. (2006) Scanning transmission x-ray microscopy with a configurable detector. *Appl. Phys. Lett.*, **89**, 251117-1–251117-3.
57. Maser, J., Osanna, A., Wang, Y., Jacobsen, C., Kirz, J., Spector, S., Winn, B., and Tennant, D. (2000) Soft X-ray Microscopy with a cryo scanning transmission X-ray microscope: I. Instrumentation, imaging and spectroscopy. *J. Microsc.*, **197**, 68–79.
58. Nolle, D., Weigand, M., Schütz, G., and Goering, E. (2011) High Contrast Magnetic and Nonmagnetic Sample Current Microscopy for Bulk and Transparent Samples Using Soft X-Rays. *Microsc. Microanal.*, **17**, 834–842.
59. Kilcoyne, A.L.D., Ade, H., Attwood, D. Hitchcock, A.P., McKean, P., Mitchell, G.E., Monteiro, P., Tyliszczak, T., and Warwick, T. (2010) A new Scanning Transmission X-ray Microscope at the ALS for operation up to 2500 eV. Proceedings of Synchrotron Radiation International, Melbourne, September 12–16, 2009. *AIP Conf. Proc.*, **1234**, 459–462.
60. Jacobsen, C., Lindaas, S., Williams, S., and Zhang, X. (1993) Scanning luminescence x-ray microscopy: imaging fluorescence dyes at suboptical resolution. *J. Microsc.*, **172**, 121–129.
61. Alberti, R., Longoni, A., Klatka, T., Guazzoni, C., Gianoncelli, A., Bacescu, D., and Kaulich, B. (2008) A low energy X-ray fluorescence spectrometer for elemental mapping X-scopy. IEEE Nuclear Science Symposium Conference, Record N14, pp. 31–33.
62. Alberti, R., Klatka, T., Longoni, A., Bacescu, D., Marcello, A., De Marco, A., Gianoncelli, A., and Kaulich, B. (2009) Development of a low-energy x-ray fluorescence system with sub-micrometer spatial resolution. *X-ray Spectrom.*, **38**, 205–210.
63. Gianoncelli, A., Kaulich, B., Alberti, R., Klatka, T., Longoni, A., De Marco, A., Marcello, A., and Kiskinova, M. (2009) Simultaneous soft X-ray transmission and emission microscopy. *Nucl. Instrum. Methods A*, **608**, 195–199.
64. Gianoncelli, A., Klatka, T. Alberti, R., Bacescu, D., De Marco, A., Marcello, A., Longoni, A., Kaulich, B., and Kiskinova, M. (2009) Development of a low-energy X-ray fluorescence system combined with X-ray microscopy. Proceedings of the 9th International Conference on X-ray Microscopy, Zurich Switzerland, August 2008. *J. Phys.: Conf. Ser.*, **186**, 012007-3.
65. Hitchcock, A.P., Tyliszczak, T., and Obst, M. (2012) Enhancing detection limits through X-ray fluorescence detection in soft X-ray scanning transmission X-ray microscopy. *Environmental Science and Technology*, in press.

66. Eguchi, T., Okuda, T., Matsushima, T., Kataoka, A., Harasawa, A., Akiyama, K., Kinoshita, T., Hasegawa, Y., Kawamori, M., Harujama, Y., and Matsui, S. (2006) Element specific imaging by scanning tunneling microscopy combined with synchrotron radiation light. *Appl. Phys. Lett.*, **89**, 2431191 - 1–2431191-3.

67. Ascensio, M.C., Imaging and nanoARPES: An innovative and powerful tool for the nanosciences, (2011) *Le Rayon de Soleil*, **21**, 4–6.

68. Schmid, I., Raabe, J., Fink, R.H., Wenzel, S., Hug, H., and Quitmann, C. (2011) Seeing and "feeling" polymer blends on a nano-scale. Proceedings of 10th International Conference on X-ray Microscopy, Chicago, August 14–18, 2010. *Am. Inst. Phys. Conf. Ser.*, **1365**, 449–452.

69. Susini, J., Somogyi, A., Barrett, R., Salomé, M., Bohic, S., Fayard, B., Eichert, D., Dhez, O., Bleuet, P., Martínez-Criado, G., and Tucoulou, R. (2004) The X-ray microscopy and micro-spectroscopy facility at the ESRF. Proceedings of International Symposium on Portable Synchrotron Light Sources and Advanced Applications, Shiga Japan, January 13–14. *AIP Conf. Proc.*, **716**, 18–21.

70. Kaulich, B., Gianoncelli, A., Beran A., Eichert, D., Kreft, I., Pongrac, P., Regvar, M., Vogel-Mikus, K., and Kiskinova, M. Low-energy (2009) X-ray fluorescence microscopy opening new opportunities for bio-related research. *J. R. Soc. Interface*, **6**, S641–S647.

71. Hitchcock, A.P., Tyliszczak, T., Obst, M., Swerhone, G.D.W., and Lawrence, J.R. (2010) Improving sensitivity in soft X-ray STXM using low energy X-ray fluorescence. *Microsc. Microanal.*, **16**, S - 2924–S-2925.

72. Kaulich, B., Bacescu, D., Cocco, D., Susini, J., David, C., DiFabrizio, E., Cabrini, S., Morrison, G., Thieme, J. Kiskinova, M. (2003) Twinmic: A European twin microscope station combining full-field imaging and scanning microscopy. Proceedings of 7th International Conference on X-Ray Microscopy ESRF, Grenoble, France, July 28 - August 2, 2002. *J Phys (France) IV*, **104**, 103–108.

73. Schmid, I., Raabe, J., Quitmann, C., Vranjkovic, S., Hug, H.J., Fink, R.H. (2009) NanoXAS, a novel concept for high resolution microscopy. Proceedings of the 9th Int. Conf. X-ray Microscopy, Aug 2008, Zurich, Switzerland. *J. Phys.: Conf. Ser.*, **186**, 012015 - 1–012015-3.

74. Schmid, I., Raabe, J., Sarafimov, B., Quitmann, C., Vranjkovic, S., Pellmont, Y., and Hug, H.J. (2010) Coaxial arrangement of a scanning probe and an X-ray microscope as a novel tool for nanoscience. *Ultramicroscopy*, **110**, 1267–1272.

75. Hanhan, S., Smith, A.M., Obst, M., and Hitchcock, A.P. (2009) Optimization of analysis of Ca 2p soft X-ray spectromicroscopy. *J. Electron. Spectrosc. Relat. Phenom.*, **173**, 44–49.

76. Sugiyama, M. and Sigesato, G. (2004) A review of focused ion beam technology and its applications in transmission electron microscopy. *J. Electron Microsc.*, **53**, 527–536.

77. De Stasio, G., Frazer, B.H., Gilbert, B., Richter, K.L., and Valley, J.W. (2003) Compensation of charging in X-PEEM: a successful test on mineral inclusions in 4.4 Ga old zircon. *Ultramicroscopy*, **98**, 57–62.

78. Zhang, X., Jacobsen, C., Lindaas, S., and Williams, S. (1995) Exposure strategies for PMMA from *in situ* XANES spectroscopy. *J. Vac. Sci. Technol. B*, **13**, 1477–1483.

79. Beetz, T. and Jacobsen, C. (2003) Soft X-ray radiation-damage studies in PMMA using a cryo-STXM. *J. Synchrotron Radiat.*, **10**, 280–283.

80. Wang, J., Morin, C., Hitchcock, A.P., Li, L., Zhang, X., Araki, T., Doran, A., and Scholl, A. (2009) Radiation damage in X-ray photoelectron emission microscopy: optimization for studies of radiation sensitive materials. *J. Electron. Spectrosc. Relat. Phenom.*, **170**, 25–36.

81. Leontowich, A.F.G. and Hitchcock, A.P. (2011) Zone plate focused soft X-ray lithography. *Appl. Phys. A: Mater. Sci. Process.*, **103** 1–11.

82. Wang, J., Botton, G.A., West, M.M., and Hitchcock, A.P. (2009) Quantitative evaluation of radiation damage to polyethylene terephthalate by soft X-rays and high energy electrons. *J. Phys. Chem. B*, **113**, 1869–1876.
83. Rightor, E.G., Hitchcock, A.P., Ade, H., Leapman, R.D., Urquhart, S.G., Smith, A.P., Mitchell, G.E., Fischer, D., Shin, H.J., and Warwick, T. (1997) Spectromicroscopy of poly(ethylene terephthalate): comparison of spectra and radiation damage rates in x-ray absorption and electron energy loss. *J. Phys. Chem. B*, **101**, 1950–1960.
84. Ade, H., Winesett, D.A., Smith, A.P., Qu, S., Ge, S., Sokolov, J., and Rafailovich, M. (1999) Phase segregation in polymer thin films: elucidations by X-ray and scanning force microscopy. *Europhys. Lett.*, **45**, 526–532.
85. Strang, G. (1988) *Linear Algebra and its Applications*, Harcourt Brace Jovanovich, San Diego, CA.
86. Press, W.H., Flannery, B.P., Teukolsky, S.A., and Vetterling, W.T. (1992) *Numerical Recipes in Fortran 77: The Art of Scientific Computing*, 2nd edn, Cambridge University Press.
87. Hitchcock, A.P. (2011) aXis2000 is free for noncommercial use. It is written in Interactive Data Language (IDL), http://unicorn.mcmaster.ca/aXis2000.html (accessed 15 December 2011).
88. Hitchcock, A.P., Dynes, J.J., Lawrence, J.R., Obst, M., Swerhone, G.D.W., Korber, D.R., and Leppard, G.G. (2009) Soft X-ray spectromicroscopy of nickel sorption in a natural river biofilm. *Geobiology*, **7**, 432–453.
89. Rousseau, M.E., Hernández Cruz, D., West, M.M., Hitchcock, A.P., and Pézolet, M. (2007) Nephilia clavipes spider dragline silk microstructure studied by scanning transmission X-ray microscopy. *J. Am. Chem. Soc.*, **129**, 3897–3905.
90. Lefevre, T., Pézolet, M., Hernández Cruz, D., West, M.M., Obst, M., Hitchcock, A.P., Karunakaran, C., and Kaznatcheev, K.V. (2009) Mapping molecular orientation in dry and wet dragline spider silk. Proceedings of the 9th International Conference X-ray Microscopy, August 2008, Zurich, Switzerland. *J. Phys.: Conf. Ser.*, **186**, 012089 - 1–012089-3.
91. Najafi, E., Hernández Cruz, D., Obst, M., Hitchcock, A.P., Douhard, B., Pireaux, J.-J., and Felten, A. (2008) Polarization dependence of the C 1s X-ray absorption spectra of individual multi-walled carbon nanotubes. *Small*, **4**, 2279–2285.
92. Felten, A., Gillon, X., Gulas, M., Pireaux, J.-J., Ke, X., Van Tendeloo, G., Bittencourt, C., Kilcoyne, A.L.D., Najafi, E., and Hitchcock, A.P. (2010) Measuring point defect density in individual carbon nanotubes using polarization-dependent X-ray microscopy. *ACS Nano*, **4**, 4431–4436.
93. Najafi, E., Wang, J., Hitchcock, A.P., Denommee, S., Ke, X., and Simard, B. (2010) Characterization of single-walled carbon nanotubes by scanning transmission X-ray spectromicroscopy: purification, order and dodecyl functionalization. *J. Am. Chem.*, **132**, 9020–9029.
94. Malinowski, E. (1991) *Factor Analysis in Chemistry*, 2nd ed, John Wiley & Sons, Inc., New York.
95. Lerotic, M., Jacobsen, C., Gillow, J.B., Francis, A.J., Wirick, S., Vogt, S., and Maser, J. (2005) Cluster analysis in soft X-ray spectromicroscopy: finding the patterns in complex specimens. *J. Electron. Spectrosc. Relat. Phenom.*, **144–147**, 1137–1143.
96. PCA_GUI and other useful IDL routines for analysis of soft X-ray microscopy data, http://xrm.phys.northwestern.edu (accessed 24 February 2011).
97. Koprinarov, I.N., Hitchcock, A.P., Li, W.H., Heng, Y.M., and Stöver, H.D.H. (2001) Quantitative compositional mapping of core-shell polymer microspheres by soft X-ray spectromicroscopy. *Macromolecules*, **34**, 4424–4429.
98. Urquhart, S.G. and Ade, H. (2002) Trends in the carbonyl core (C1s, O1s) → $\pi^*_{C=O}$ transition in the near edge X-ray absorption fine structure spectra

of organic molecules. *J. Phys. Chem. B*, **106**, 8531–8538.

99. Urquhart, S.G., Hitchcock, A.P., Smith, A.P., Ade, H.W., Lidy, W., Rightor, E.G., and Mitchell, G.E. (1999) NEXAFS spectromicroscopy of polymers: overview and quantitative analysis of polyurethane polymers. *J. Electron. Spectrosc. Relat. Phenom.*, **100**, 119–135.

100. Hitchcock, A.P., Koprinarov, I., Tyliszczak, T., Rightor, E.G., Mitchell, G.E., Dineen, M.T., Heyes, F., Lidy, W., Priester, R.D., Urquhart, S.G., Smith, A.P., and Ade, H. (2001) Optimization of scanning transmission x-ray microscopy for the identification and quantitation of reinforcing particles in polyurethanes. *Ultramicroscopy*, **88**, 33–49.

101. Mitchell, G.E., Wilson, L.R., Dineen, M.T., Urquhart, S.G., Hayes, F., Rightor, E.G., Hitchcock, A.P., and Ade, H. (2002) Quantitative characterization of microscopic variations in the cross-link density of gels. *Macromolecules*, **35**, 1336–1341.

102. Rightor, E.G., Urquhart, S.G., Hitchcock, A.P., Ade, H., Smith, A.P., Mitchell, G.E., Priester, R.D., Aneja, A., Appel, G., Wilkes, G., and Lidy, W.E. (2002) Identification and quantitation of urea precipitates in flexible polyurethane foam formulations by X-ray spectromicroscopy. *Macromolecules*, **35**, 5873–5882.

103. Croll, L.M., Stöver, H.D.H., and Hitchcock, A.P. (2005) Composite tectocapsules containing porous polymer microspheres as release gates. *Macromolecules*, **38**, 2903–2910.

104. Martin, Z., Jimenez, I., Gomez, M.A., Ade, H., and Kilcoyne, D.A. (2010) Interfacial interactions in PP/MMT/SEBS nanocomposites. *Macromolecules*, **43**, 448–453.

105. Martín, Z., Jiménez, I., Gómez-Fatou, M.A., West, M., and Hitchcock, A.P. (2011) Interfacial Interactions in Polypropylene-Organoclay-Elastomer nanocomposites: influence of polar modifications on the location of the clay. *Macromolecules*, **44**, 2179–2189.

106. Hoppe, H. and Sariciftci, N.S. (2004) Organic solar cells: an overview. *J. Mater. Res.*, **19**, 1924–1945.

107. McNeill, C.R., Watts, B., Thomsen, L., Belcher, W.J., Greenham, N.C., and Dastoor, P.C. (2006) Nanoscale quantitative chemical mapping of conjugated polymer blends. *Nano Lett.*, **6**, 1202–1206.

108. McNeill, C.R., Watts, B., Thomsen, L., Belcher, W.J., Kilcoyne, A.L.D., Greenham, N.C., and Dastoor, P.C. (2006) X-ray spectromicroscopy of polymer/fullerene composites: quantitative chemical mapping. *Small*, **2**, 1432–1435.

109. Watts, B., Schuettfort, T., and McNeill, C.R. (2011) Mapping of domain orientation and molecular order in polycrystalline semiconducting polymer films with soft x-ray microscopy. *Adv. Funct. Mater.* **21** 1122–1131.

110. Thibault, P., Dierolf, M., Menzel, A., Bunk, O., David, C., and Pfeiffer, F. (2008) High-resolution scanning X-ray diffraction microscopy. *Science*, **321**, 379–382.

111. Araki, T., Ade, H., Stubbs, J.M., Sundberg, D.C., Mitchell, G.E., Kortright, J.B., and Kilcoyne, A.L.D. (2006) Soft X-ray resonant scattering of structured polymer nanoparticle. *Appl. Phys. Lett.*, **89**, 1241061-1–1241061-3.

112. Mitchell, G.E., Landes, B.G., Lyons, J., Kern, B.J., Devon, M.J., Koprinarov, I., Gullikson, E.M., and Kortright, J.B. (2006) Molecular bond selective x-ray scattering for nanoscale analysis of soft matter. *Appl. Phys. Lett.*, **89**, 0441011-1–0441011-3.

113. Ohldag, H., van der Laan, G., and Arenholz, E. (2009) Correlation of crystallographic and magnetic domains at Co/NiO(001) interfaces. *Phys. Rev. B*, **79**, 052403-1–052403-4.

114. Leung, B.O., Wang, J., Brash, J.L., Hitchcock, A.P., Cornelius, R., Doran, A., and Scholl, A. (2009) An X-ray spectromicroscopy study of protein adsorption to a polystyrene-polylactide blend. *Biomacromolecules*, **10**, 1838–1845.

115. Gleber, S.-C., Sedlmair, J., Bertilson, M., von Hofsten, O., Heim, S., Guttmann, P., Hertz, H.M., Fischer, P.,

and Thieme, J. (2009) X-ray stereo microscopy for investigation of dynamics in soils. Proceedings of the 9th International Conference on X-ray Microscopy, Zurich Switzerland, August 2008. *J. Phys.: Conf. Ser.*, **186**, 012104 - 3.

116. Hitchcock, A.P., Araki, T., Ikeura-Sekiguchi, H., Iwata, N., and Tani, K. (2003) 3d chemical mapping of toners by serial section scanning transmission X-ray microscopy. Proceedings of the 7th International Conference on X-ray Microscopy, Grenoble France August. *J. Phys. IV France*, **104**, 509–512.

117. Le Gros, M.A., McDermott, G., and Larabell, C.A. (2005) X-ray tomography of whole cells. *Curr. Opin. Struct. Biol.*, **15**, 593–600.

118. Uchida, M., Sun, Y., McDermott, G., Knoechel, C., Le Gros, M.A., Parkinson, D., Drubin, D.G., and Larabell, C.A. (2011) Quantitative analysis of yeast internal architecture using soft X-ray tomography. *Yeast*, **28** 227–236.

119. Schneider, G., Guttmann, P., Heim, S., Rehbein, S., Mueller, F., Nagashima, K., Heymann, J.B., Müller, W.G., and McNally, J.G. (2010) Three-dimensional cellular ultrastructure resolved by X-ray microscopy. *Nat. Methods*, **7**, 985–987.

120. Carrascosa, J.L., Javier Chichón, F., Pereiro, E., Rodríguez, M.J., Fernández, J.J., Esteban, M., Heim, S., Guttmann, P., and Schneider, G. (2009) Cryo-x-ray tomography of Vaccinia Virus membranes and inner compartments. *J. Struct. Biol.*, **168**, 234–239.

121. Haddad, W.S., Trebes, J.E., Goodman, D.M., Lee, H.-R., McNulty, I., Anderson, E.H., and Zalensky, A.O. (1995) Ultrahigh-resolution soft x-ray tomography. X-ray Microbeam Technology and Applications, San Diego, CA. *Proc. SPIE*, **2516**, 102–107.

122. Wang, Y., Jacobsen, C., Maser, J., and Osanna, A. (2000) Soft x-ray microscopy with a cryo STXM: II. Tomography. *J. Microsc.*, **197**, 80–93.

123. Johansson, G.A., Tyliszczak, T., Mitchell, G.E., Keefe, M., and Hitchcock, A.P. (2007) Three dimensional chemical mapping by scanning transmission X-ray spectromicroscopy. *J. Synchrotron Radiat.*, **14**, 395–402.

124. Obst, M., Wang, J., and Hitchcock, A.P. (2009) 3-d chemical imaging with STXM tomography. Proceedings of the 9th International Conference X-ray Microscopy, August 2008, Zurich, Switzerland. *J. Phys.: Conf. Ser.*, **186**, 012045 - 1–012045-3.

125. Obst, M., Wang, J., and Hitchcock, A.P. (2009) Soft X-ray spectro-tomography study of cyanobacterial biomineral nucleation. *Geobiology*, **7**, 577–591.

126. Wang, J., Hitchcock, A.P., Karunakaran, C., Prange, A., Franz, B., Harkness, T., Lu, Y., Obst, M., and Hormes, J. (2011) 3D chemical and elemental imaging by STXM spectro-tomography. Proceedings of 10th International Conference on X-ray Microscopy. *Am. Inst. Phys. Conf. Ser.*, **1365** 215–218.

127. Guay, D., Stewart-Ornstein, J., Zhang, X., and Hitchcock, A.P. (2005) In situ spatial and time resolved studies of electrochemical reactions by scanning transmission X-ray microscopy. *Anal. Chem.*, **77**, 3479–3487.

128. Neuhäusler, U., Jacobsen, C., Schulze, D., Stott, D., and Abend, S. (2000) A specimen chamber for soft x-ray spectromicroscopy on aqueous and liquid samples. *J. Synchrotron Radiat.*, **7**, 110–112.

129. Weigand, M., Van Waeyenberge, B., Vansteenkiste, A., Curcic, M., Sackmann, V., Stoll, H., Tyliszczak, T., Kaznatcheev, K., Bertwistle, D., Woltersdorf, G., Back, C.H., and Schütz, G. (2009) Vortex core switching by coherent excitation with single in-plane magnetic field pulses. *Phys. Rev. Lett.*, **102**, 077201 - 1–077201-4.

130. Schneider, G., Rudolph, S., Meyer, A.M., Zschech, E., and Guttmann, P. (2005) X-ray microscopy: a powerful tool for electromigration studies in modern ICs. *Future Fab Int.*, **19**, 115–117.

131. de Smit, E., Swart, I., Creemer, J.F., Karunakaran, C., Bertwistle, D., Zandbergen, H.W., de Groot, F.M.F., and Weckhuysen, B.M. (2009)

Nanoscale chemical imaging of the reduction behavior of a single catalyst particle. *Angew. Chem. Int. Ed.*, **48**, 3632–3636.

132. Zhou, J., Wang, J., Fang, H., and Sham, T.-K. (2011) Investigating structural variation and water adsorption of SnO_2 coated carbon nanotube by nanoscale chemical imaging. *J. Mat. Chem.*, **21** 14622–14630.

133. Drake, I.J., Liu, T.C.N., Gilles, M.K., Tyliszczak, T., Kilcoyne, A.L.D., Shuh, D.K., Mathies, R.A., and Bell, A.T. (2004) An In-situ cell for characterization of solids by soft X-ray absorption. *Rev. Sci. Inst.*, **75**, 3242–3247.

134. de Smit, E., Swart, I., Creemer, J.F., Hoveling, G.H., Gilles, M.K., Tyliszczak, T., Kooyman, P.J., Zandbergen, H.W., Morin, C., Weckhuysen, B.M., and de Groot, F.M.F. (2008) Nanoscale chemical imaging of a working catalyst by scanning transmission X-ray microscopy. *Nature*, **456**, 222–225.

135. Creemer, J.F., Helveg, S., Hoveling, G.H., Ullmann, S., Molenbroek, A.M., Sarro, P.M., and Zandbergen, H.W. (2008) Atomic-scale electronmicroscopy at ambient pressure. *Ultramicroscopy*, **108**, 993–998.

136. Lam, K.P., Hitchcock, A.P., Obst, M., Lawrence, J.R., Swerhone, G.D.W., Leppard, G.G., Tyliszczak, T., Karunakaran, C., Wang, J., Kaznatcheev, K., Bazylinski, D., and Lins, U. (2010) X-ray magnetic circular dichroism of individual magnetosomes by Scanning Transmission X-ray Microscopy. *Chem. Geol.*, **270**, 110–116.

23
Atom Probe Tomography: Principle and Applications
Frederic Danoix and François Vurpillot

23.1
Introduction

Atom probe tomography (APT) is based on the pulsed field evaporation of surface atoms and the identification of field evaporated ions by time-of-flight mass spectrometry. The image formation can be described in a very scholarly manner by considering that the positions of atoms at the sample surface are derived from the impact coordinates of evaporated species onto the detector. To simplify, a simple point projection law is often involved (Figure 23.1). The in-depth investigation of the sample is provided by the layer-by-layer erosion of the specimen through the field evaporation mechanism. The high electric field (F) required (a few tens of volume per nanometer) to field evaporate surface atoms is derived from a high electric voltage (V_0) applied to the specimen prepared in the form of a sharply pointed needle (the tip radius (R) is close to 50 nm). The high magnification of the instrument (up to several millions), and thus its spatial resolution, is the result of the diverging projection that is produced during the flight of ions at the close vicinity of the tip surface. Time-of-flight mass spectrometry, which allows chemical identification, requires a pulsed field evaporation of ions.

APT is an evolution of the field ion microscope (FIM). FIM is based on the field ionization of rare gas atoms near the tip surface [1]. Ions originating from protruding surface atoms (high field) are received on a screen, making it possible to get a magnified image of the specimen surface resolved on the atomic scale.

Following this simple description, only a limited number of ingredients are necessary to understand and control the image formation for both instruments. The sample is defined generally as a very sharp needle, with a quasi-hemispherical end shape. This geometry is required to produce the mandatory high surface field. Indeed, the small curvature radius of the sample is responsible for the production of this field. Section 23.2.1 is devoted to the description of the field production. The controlled erosion of the tip is obtained through the process of field evaporation. A brief description of this fundamental physical process is presented in Section 23.2.2. The projection of ions, and therefore the image-to-tip back projection algorithm is

Handbook of Nanoscopy, First Edition. Edited by Gustaaf Van Tendeloo, Dirk Van Dyck, and Stephen J. Pennycook.
© 2012 Wiley-VCH Verlag GmbH & Co. KGaA. Published 2012 by Wiley-VCH Verlag GmbH & Co. KGaA.

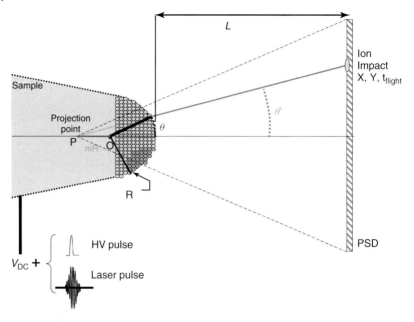

Figure 23.1 Basic principles of atom probe tomography. Note that the sample has a cylindrical symmetry.

determined completely by the trajectories of ions from the surface of the sample to the position sensitive detector (PSD). Section 23.2.3 focuses on these mechanisms.

Pulsed field evaporation, required for time-of-flight mass spectrometry, is ensured through two complementary mechanisms, that is, voltage pulsing or thermal pulsing. A review of these two processes is presented in Section 23.2.4.

After this survey of the fundamentals of FIM and APT, examples of application of these techniques in materials science are presented. Section 23.4 is devoted to FIM, and the information that can be derived from atomically resolved micrographs. In Section 23.5, a large overview of the application of APT is given, from classical metallurgy to the latest results obtained on phase transformations in semiconductors.

23.2
Basic Principles

23.2.1
Field Production

The mandatory colossal electric field F at the end of the apex is produced from the application of a voltage V_0 to the sample. The field is induced by the high curvature $1/R$ of the end apex. The field F is reduced from the value that can be calculated for a conductive sphere of radius R submitted to V_0 by a field factor β that takes into account not only the presence of the shank of the sample but also the global electrostatic geometry around the sample, as described in the following equation:

$$F = V/\beta R \tag{23.1}$$

Several investigators performed analytical or numerical calculations to extract the field factor β in atom probe. For instance, assuming hyperboloidal or paraboloidal approximations of the tip shape, the field factor is easily expressed.

With the distance between the electrodes h, Gomer recalls typical results of such calculations [2]:

$$\beta = \frac{1}{2}\ln\left(\frac{h}{R}\right) \tag{23.2}$$

for a paraboloidal tip, and

$$\beta = \frac{1}{2}\ln\left(\frac{4h}{R}\right) \tag{23.3}$$

for a hyperboloidal tip.

The field factor is clearly dependent on the tip shape as well as the tip-to-screen distance. Taking h in the range $10^{-5} - 1$ m, and R in the range $10^{-9} - 10^{-6}$ m, the β value lies in the range 2–8. However, no simple relationship exists that takes into account parameters such as the shank angle of the tip or the presence of the sample holder. To take into account these parameters as well as the exact geometry of the atom probe chamber, some investigators modeled the field distribution around the specimen with finite element methods, but these simulations required the exact knowledge of geometric parameters [3–6].

Considering the simple analytical expression in Eqs. (23.1–23.3), a field of several volts per nanometer is induced by the application of a few kilovolts (less than 20 kV in practice) to a sample with R in the range 10–100 nm. To reduce significantly the required voltage, a counter electrode with a circular aperture is placed from several micrometers to about 1 mm in front of the tip (reducing the electrostatic distance h).

In the general case, the shank angle of the end apex is not null. During the controlled field evaporation of the specimen, the radius of curvature gradually increased. By considering a constant half shank angle a, geometric considerations lead to a simple expression for the variation of the curvature radius all along the analysis. As the depth, z, is progressively analyzed, the radius of curvature increases, in a first approximation:

$$\frac{dR}{dz} = K_a = \frac{\sin(a)}{1 - \sin(a)} \tag{23.4}$$

This increase is compensated for by a controlled variation of the voltage applied to the sample. Note that the field factor β also varies slightly during the analysis. Nevertheless, the variation is small so that this parameter is considered as a constant for the sake of simplicity.

23.2.2
Field Evaporation and Pulsed Field Evaporation

At temperature used in APT, that is, T_0 in the range 20–150 K, the evaporation process is mostly thermally activated and is well described, considering the rate of evaporation for a given surface atom submitted to a field F, by the following Arrhenius equation:

$$K_e = v_0 \exp\left(-\frac{Q_n(F)}{k_B T_0}\right) \qquad (23.5)$$

where v_0 is the frequency of vibration of the surface atoms, of the order of 10^{12}–10^{13} s^{-1} [1, 7–14], k_B the Boltzmann constant, and $Q_n(F)$ is the energy barrier that must overpass the atom to become an ion and be ejected from the tip surface. This last barrier is strongly dependant on the field F present at the surface, and the height of the barrier continuously decreases with the applied field F. At a field called "0 K evaporation field," the barrier completely disappears. This law describes well the observed phenomenon in which a critical field is required for efficient field evaporation. It is worth noting here that ion tunneling has been observed at a very low temperature, especially for low-mass atoms [15]. However, this effect is thought to be negligible in most cases and will be ignored in the treatment of the field evaporation process presented here.

In the vicinity of the threshold for field evaporation at a given evaporation rate, the activation energy $Q_n(F)$ can be described, in a rather rough approximation, by a linear relationship with F:

$$Q_n(F) = Q'_0 \left(1 - \frac{F}{F_{evap}}\right) \qquad (23.6)$$

This law can be derived from experimental measurements of the electric field or the voltage necessary to provoke field evaporation at a given rate from the specimen surface as a function of the temperature. Considering this linear relationship, F_{evap} is the threshold electric field for which the barrier disappears at $T = 0$, referred to as the *evaporation field*, and Q'_0 is equivalent to a barrier height when no field is applied. The validity of such an empirical expression is theoretically limited to less than 20% around the field evaporation threshold, but is sufficient in most cases to describe the observed field evaporation behavior.

This empirical expression finds its explanation through several theories that were developed to describe the mechanisms of field evaporation. These theories are based on the assumption that field evaporation is the transition of a surface atom into an *n*-times charged ion under the influence of an external electric field F. The simplest treatments assume the atom as a particle confined inside a potential well that can be approximated in 1D form. The simplest model (image hump or Müller–Schottky model) considers the process as an escape of a charged ion pulled away by the electric field competing with the classical image force induced by its own electrostatic image. The particle must cross a field-dependant barrier to be ionized and finally repelled from the surface. The energy barrier corresponding to this transition can be denoted as $Q_n(F)$.

The ionic curve is shifted by the sum of the sublimation energy (Λ) and ionization energies necessary to transform the atom bonded to the surface into an n-times charged free ion ($\sum_n I_n$). An energy corresponding to its work function ($n\Phi$) is gained by the surface as electrons are transferred into the material. The electric field modifies the ionic potential continuously as the ion is driven away from the surface. As the electric field is increased, the height of the potential barrier that the atom has to overcome to field evaporate is reduced.

When considering the presence of a hump in the potential energy, and by neglecting high-order additional terms in the field (polarization terms), the activation energy is given by the image force theory as

$$Q_n = Q_0 - \left(\frac{n^3 e^3 F}{4\pi \varepsilon_0}\right) \tag{23.7}$$

with

$$Q_0 = \Lambda + \sum_n I_n - n\phi \tag{23.8}$$

Assuming this equation, it is easy to demonstrate that the barrier vanishes for a critical field usually called the *evaporation field* at $T = 0$ K, F_{evap}

$$F_{evap} \approx \frac{4\pi \varepsilon_0}{n^3 e^3} Q_0^2 \tag{23.9}$$

Close to this threshold, the energy barrier can be approximated by Eq. (23.6) with

$$Q_0' \cong \frac{Q_0}{\left(1 + \left(\frac{F}{F_{evap}}\right)^{\frac{1}{2}}\right)} \cong \frac{Q_0}{2} \tag{23.10}$$

The attractive feature of the image hump model is its extreme simplicity, which allows a fully analytical treatment with very few parameters. Relatively good estimates of field evaporation constants can be deduced from tabulated values of physical constants of the different materials. However, the existence of the hump is hypothetical since the position of the hump is in the order of 0.1 nm from the surface, smaller than the nearest neighbor distance for metals. At this distance, the strong repulsive force between ion cores predominates and the model becomes unrealistic. However, the exact calculation of Q_n and F_{evap} is difficult, since explicit knowledge of the atomic and ionic energy-distance curves is required. Note that the exact process by which the atom becomes an ion has been the subject of a long debate, but it now seems clear that the charge is progressively drained from the departing atom as it crosses the energy barrier to become an ion.

The extreme sensitivity of the erosion process to the local amplitude of the surface electric field makes field evaporation the sharpest tool for the atom-by-atom analysis of a surface. Owing to the small field penetration in conductive materials, any roughness at the atomic scale induces field enhancement that provokes local field evaporation of atoms. As a result, under the influence of field evaporation, the tip surface of the specimen is eroded to produce a smooth and regular surface at a nanometer scale. The extremity of the tip develops naturally an end form

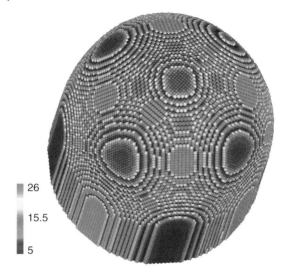

Figure 23.2 Tip shape end form under field evaporation. The color is relative to the surface field (arbitrary unit). A unique evaporation field was used for the modeling of the field evaporation process. (Courtesy of M. Gruber, GPM UMR CNRS 6634, University of Rouen.)

defined by a unique radius of curvature R (at least over a field of view of 40°). Only second-order regional curvatures induced by lattice binding energy variations, work function variations, or field variations are generally observed (Figure 23.2). At the atomic scale, the atoms that are likely to be field evaporated are the atoms situated at the border of lattice terraces of the studied materials (kink sites or ledge sites). These protrusions give rise to field enhancement of several percentages compared to the field present at the center of terraces.

In atom probe experiments, the evaporation is controlled by the combined effect of the field and the temperature. Experimental evaporation fields are generally defined for a given evaporation rate at the temperature of analysis. An increase in the temperature leads generally to the observed decrease in this critical threshold of evaporation. This behavior is theoretically relatively well explained by the models described above.

Let us consider analyzing a surface containing N atoms with a high probability to be field evaporated. These are atoms subjected to high field enhancement and are simply kink site or ledge site atoms. Hence, it is reasonable to assume that the amplitude of the electric field above these atoms is the same. In Eqs. (23.5) and (23.6), the evaporation rate of the surface is given by

$$\Phi_e = N v_0 \exp\left(-\frac{Q_0'(1 - F/F_0)}{k_B T_0}\right) \quad (23.11)$$

Thus, the field F_e^{T,Φ_e} giving rise to the evaporation of an atom-flux Φ_e (in atom per second) is given by

$$F_e^{T,\Phi_e} = F_e \left(1 - \frac{k_B T_0}{Q'_0} \ln\left(\frac{v_0 n}{\Phi_e}\right)\right) \quad (23.12)$$

There is theoretically a linear reduction of the evaporation field with the temperature. This prediction is in excellent agreement with experimental observation. This behavior was observed in most metals via FIM. Note that the slope of the reduction is dependant on the evaporation rate and specific to the material. The higher the evaporation rate, the smaller the dependence. This is crucial in the determination of the proper experimental conditions for composition measurement in APT, as is discussed in Section 23.4.

23.2.3
Image Formation

Once the atoms are field evaporated from the tip surface, the generated ions are accelerated away from the specimen surface by the diverging electric field to the PSD. Note that the use of a time-resolved PSD makes it possible to locate ion impact positions from which atoms originate at the tip surface. Several types of detectors were implemented on the atom probe (multianode detector, the first French generation of APT, and time-resolved charge-coupled device (CCD) detector). Delay-line-based detectors are now most often used. The spatial resolution of PSDs is generally much better than 0.1 nm [16–18].

The classical theory assumes that the ions are projected normally to the surface in the first steps of flight. In the distance equivalent to several radii of curvature and up to the PSD, the ions are influenced by the electrostatic field lines bent by the presence of the shank of the specimen.

As the field factor, the ion trajectories are dependant on the shape of the sample evaporated. In a first approximation, the radial projection is replaced by a point projection relationship P with a center that is different from the tip center O (with $OP = mR$; Figure 23.1) [19, 20].

m is a compression factor that also depends on the electrostatic environment with a value close to 0.6 ($0 < m < 1$).

As a result, in a first approach, the impact angles $(\theta', \varphi'$ are deduced by the relations:

$$\begin{aligned}\varphi &= \varphi' \\ \theta &= (m+1)\theta'\end{aligned} \quad (23.13)$$

Taking into account the spatial resolution of the PSD, the projected angle is evaluated with accuracy better than $0.1°$.

The $(m+1)$ factor is generally named as the image compression factor (ICF). Following these equations, the positions of ion impacts on the detector are used to deduce the atom positions onto the tip with the geometrical relations:

$$x = R\sin(\theta)\sin(\varphi)$$
$$y = R\sin(\theta)\cos(\varphi) \qquad (23.14)$$
$$z = R(1 - \cos(\theta))$$

With these equations, the magnification M is given by

$$M = \frac{L}{(m+1)R} \qquad (23.15)$$

where L signifies the distance between the tip and the detector.

Since the tip is eroded atom by atom, the depth inside the sample is simply deduced from the evaporation order. The z coordinate is incremented for each atom by the ratio between its atomic volume Ω and the surface of analysis S_a. S_a is deduced from the detector surface S_d and the magnification M [21].

This fundamental assumption is directly linked to the nature of the evaporation process. At low temperatures, the evaporation process is mainly influenced by the surface field at the atom scale, and is deterministic. The more protruding atom is always the atom chosen as the next atom to be field evaporated. The tip gradually evolves to keep a smooth radius of curvature that is defined completely by a geometric description. Assuming that only atoms from the very surface of the specimen are likely to be field evaporated, the order in which the atoms are collected by the PSD is a direct indication of the depth of the surface from which the atoms are originating. The reconstruction procedure designed by Bas et al. [21] assumes that the analyzed surface was progressively going down the axis of the specimen of this increment, which is directly proportional to the atomic volume of the detected ion distributed over the whole analyzed surface.

The next step is thus to estimate the analyzed surface. This basically assumes that the apex of the specimen has a perfectly hemispherical shape. The main problem in ensuring the correct reconstruction for both the lateral and the depth scale is thus to obtain all along the experiment a sufficiently good estimate of the radius of curvature $R(n)$, with n the index of the evaporated atom in the list.

There are two generally applied methods. The first considers the field F_e required to field evaporate the tip to be constant. At very low temperatures, this field is generally taken as the field evaporation constant defined previously. The radius of curvature can then be directly deduced from Eq. (23.1), and the voltage $V(n)$ applied to the tip all along the analysis, that is, $R(n) = V(n)/F_e\beta$.

The second method assumes a tip that can be defined as a truncated cone of half shank angle a with an end apex of radius R_0. This shape can be measured by transmission electron microscopy (TEM) or scanning electron microscopy (SEM) before analysis. In this case, the volume evaporated during the analysis was calculated by Walck et al. [22] for 1D atom probe purpose.

The volume v_{probe} defined in Figure 23.1 is given by

$$v_{probe} = S_{a,0} \times h \left[1 - K_a \left(\frac{1 - \cos\gamma}{\sin\gamma}\right)^2\right] \times \left[1 + \frac{h}{R_0}K_\alpha \left(\frac{h}{R_0}K_\alpha\right)^2\right] \qquad (23.16)$$

K_α is defined by $K_\alpha = \frac{\sin\alpha}{1-\sin\alpha}$ and γ the aperture angle on the tip with $\gamma = (m+1)\,\gamma$ if γ defines the physical angle of detection.

The probed depth is h, and $S_{A,0} = \pi R_0^2 \sin^2 \gamma$ the initial analyzed area onto the tip. For a not-too-large shank angle ($<25°$) and a small depth variation δh, which corresponds to an amount of atoms Δn, the volume is well approximated by

$$\delta v = \pi R^2 \sin^2 \gamma \times \delta h \left[1 - K_\alpha \left(\tan\frac{\gamma}{2}\right)^2\right] = \Delta n \times \frac{\Omega}{Q} \qquad (23.17)$$

Note Q is here the detection efficiency, depending on the detection system.

Most of the reconstruction parameters are generally easy to be evaluated or measured. The detector efficiency and the detector field of view are linked to instrumental features. Atomic volumes are deduced from material atomic densities. Evaporation fields F_{evap} are obtained from tabulated values or deduced from measurements of charge states [23, 24].

The reconstruction accuracy is thus dominated by the precision of the two parameters m and b. The ICF, in the general case, is in the range $1.3 < m+1 < 2$. The field factor b is in the range $2 < b < 8$. In pure metals, the ICF can be closely determined using FIM or field desorption images, and the field factor is deduced assuming an initial radius R_0. In addition, the presence of visible atomic planes is used to optimize the parameter with good accuracy. Nevertheless, the visibility of crystallographic features such as depleted poles or zone lines or atomic planes in the reconstruction is not systematic. In these cases, there is no straightforward calibration of the parameters.

It is known that this model, based on a quasi-stereographic point projection, does not necessarily provide the most accurate description of the actual image projection [25], especially for wide field-of-view instruments. When the procedure was developed in the mid-1990s, the field of view of the instrument was rather limited and small angle approximations were made and validated, while the two projections give equivalent results. Since it was the most advanced procedure at the time, it has been implemented in commercial software, and yet yields accurate reconstructions. Nevertheless, the use of correct parameters for the reconstruction enables to achieve extremely high performances in terms of spatial resolution. The depth resolution, that is, the resolution perpendicular to the specimen surface is only limited by the field evaporation mechanism, as discussed previously. Best experimental measurements achieve less than 0.1 nm in depth. The lateral resolution is limited by small trajectory deviations in the first step of field evaporation. Experimental evidences of lateral resolution better than 0.3 nm were demonstrated.

The best spatial resolution is attained in pure metals, at low temperatures. In alloys, the spatial resolution can be strongly degraded when analyzing a sample with elements very different in terms of their evaporation fields. In single-phase disordered alloy, the presence of species with strong differences in the evaporation field ($>20\%$) reduces experimentally the lateral resolution up to several nanometers because of the roughness of the tip surface under the process of field evaporation. Indeed, the species with high (low) evaporation fields are subjected to surface retention (preferential evaporation) compared to their local environments. The observed trajectory aberrations induced by the subtle deviations in the first step of flight depend on the small range neighborhood of the atom leaving the surface and on the local atomic structure and arrangement. This makes the correction of these aberrations impossible, as much of the information is not a priori known [26]. In the case of multiphase materials, the presence of phases enriched with an element with strong evaporation field differences may cause unexpected inhomogeneous field evaporation of the surface. When the phase of interest arises from the tip surface, a local curvature radius, different from the average radius of the tip apex tends to be formed at its surface in order to accommodate the differences in evaporation fields [27]. Following this shape, the local magnification is strongly affected by this global tip shape change. This mechanism is responsible for the presence in reconstructed images of artificial variations of atomic density. In practice, these density artifacts can be corrected using the density correction algorithm in post treatment [28, 29].

The maximum depth that can be explored by APT is a few hundred nanometers. It is experimentally limited by the maximum voltage (V) that can be applied to the tip. A small shank angle makes it possible in principle to analyze a larger depth. However, the main limiting factor is due to tip fracture: the high electrostatic pressure that is exerted leads to fracture and this generally occurs before attaining the maximum voltage (20 kV) that can be applied to the sample.

23.2.4
Pulsed Field Evaporation

Time-controlled field evaporation or pulsed field evaporation is enabled via the application of either voltage pulses (VP) or laser pulses (LP) superimposed on a DC electric field. This enables the elemental identification of the ion by means of time-of-flight mass spectrometry. Assuming that the energy is instantaneously acquired by the departing ion, the mass-to-charge-state ratio can be written as

$$M = \frac{m}{n} = 2eU\frac{t^2}{L^2} \tag{23.18}$$

where U is the voltage applied to the tip, m is the mass of the ion, n the charge state, e the elemental charge, t the time of flight, and L the flight distance from the tip to the PSD [30].

In standard atom probe, the applied voltage is in the range 2–20 kV for a flight length in the range 10–60 cm. With monoatomic species, it gives standard time of flights in the range 10^2–10^4 ns. In order to achieve mass spectrometry with good performances in terms of mass resolving power, atoms must be field evaporated within durations much shorter than the nanosecond.

The process of field evaporation is extremely sensible to the applied field. For instance, at 80 K, using expression (7.11), the evaporation rate varies by several orders of magnitude by changing the applied voltage of only a small percentage. In the same way, the evaporation rate can be controlled by changing the base temperature of a few tens of Kelvin. As a result, the mass spectrometry performance of the instrument is closely related to the control of evaporation conditions at the surface of the tip.

23.2.4.1 Field Pulsing

In HV pulsing mode, the ions are emitted because of the transient increase in the electric field during the VP pulse. Depending on the design of the instrument, the repetition rate can vary from 1 to 100 kHz. Field evaporation occurs in a relatively short period of time during the pulse itself.

However, as the ions are flying away from the surface, they are accelerated by a time-varying electric field. Therefore, depending on the instant in the duration of an HV pulse at which different ions are field evaporated, they are not necessarily accelerated by the same electric field, and hence do not acquire the same total kinetic energy [31]. This spread in energy results in a spread in time of flight, limiting the mass resolving power of the technique. This limitation was overcome by the development of energy compensation devices based on electrostatic mirrors in APT instruments. These devices cause ions with more energy to travel longer flight paths to reach the detector and thus increase their time of flight. The only limitation of this kind of device is the reduction in detection efficiency induced by the presence of a grid in the electrostatic mirrors and spatial resolution losses induced by the scattering of ions close to the grid mesh. Note that obviously, VP pulsing is limited to good electrically conductive materials. In practice, materials with resistivity above about 1 Ω cm are generally difficult to analyze.

The pulsed field is derived from a pulsed voltage. Assuming that the evaporation occurs preferentially at the topmost voltage within each pulse, and assuming a square profile voltage pulse giving rise to a pulsed field ϕ_{VP}, the evaporation rate is well approximated by

$$\phi_{VP}(\text{atom/pulse}) \approx vN\tau_F e^{\frac{Q_n(F_{DC}+F_{VP})}{k_B T_0}} \tag{23.19}$$

τ_F is evaporation duration induced by the pulsed field, and is in the nanosecond range. Note that the square profile pulse approximation is a reasonable description of the pulsed field evaporation. Indeed, the exponential variation of the evaporation rate with the field makes the rising edge and the falling edge of the evaporation pulses extremely abrupt. It is always possible to model the evaporation rate with this simple expression considering a mean evaporation duration τ_F where the evaporation field is constant and equal to $F_{DC} + F_{VP}$.

The amplitude of VP is fundamental to obtain good and reliable composition measurements. Indeed, as summarized, if the DC field is maintained at a relatively high level, species with lower evaporation fields are likely to be field evaporated between evaporation pulses, and will thus be preferentially lost, artificially increasing the concentration in the high evaporation field species. This underpins erroneous compositions in atom probe experiments of alloys and complex materials, as specific species will be preferentially lost during the course of the analysis. In most alloys, the best compromise is a VP amplitude of about 20% of V_{DC}.

23.2.4.2 Thermal Pulsing

The past few years have been witness to a major breakthrough. Formerly limited to metals or good conductors, the implementation of ultrafast pulsed laser to the instrument opened APT to semiconductors or oxides and, as a consequence, to the important domain of microelectronics and nanosciences (e.g., tunnel junctions). In the first generation of voltage pulsed 3D atom probes, materials with low electrical conductivity could not be analyzed properly. For highly resistive materials, it was impossible to get any data. In the last generation of APT, VP have been replaced by ultrafast LP (picosecond or subpicosecond pulse). These pulses give rise to a very rapid thermal pulse that promotes the field evaporation of surface atoms. This made it possible to analyze bad conductors such as semiconductors or oxides that are key materials in microelectronics. As an example, successful analyses of highly resistive materials, such as intrinsic silicon or even silicon oxide, were performed. One additional advantage of laser pulses is that sample fractures are observed to be much less frequent, increasing, therefore, the depth that can be analyzed.

In laser pulsing mode, ions are emitted during the thermal pulse. They are therefore accelerated by the DC field only, and thus acquire the same kinetic energy. However, ions can still be emitted at various instants during the thermal pulse, which in turn induces a small spread in the times of flight. Very high mass resolutions can be reached by simply increasing the flight path, as the relative differences in the time of flight will become negligible [32–34]. In this case, energy deficits are restricted to intrinsic energy deficits related to the field evaporation process. These deficits are limited in practice to a few electronvolts. Mass resolutions up to 4000 were reported by Tsong and coworkers (M/ΔM ~4000 with a flight length of about 7 m; M/ΔM and stands for mass resolving power expressed as mass (M) over width of mass peak (ΔM) at mass M. DB stands for beam diameter).

In pulsed laser APT, a laser beam is applied onto the apex of the specimen to trigger time-controlled field evaporation. TEM00 Gaussian mode lasers are generally used, so that the beam profile can always be described with a Gaussian shape of width DB ~ 2 s with a maximum of energy centered with the tip apex. Note that a standing DC field is still applied on the specimen in order to lower the potential barrier. In the case of a purely thermally activated process, the duration of the pulsed field evaporation T is not controlled by the laser pulse duration itself, but by the temperature response of the tip apex, which can be regarded as a thermal pulse. Converse to the voltage pulsed mode, the duration of this pulse is strongly dependant on the materials illuminated and on the parameters of the laser. The notion of thermal pulsing is based on the concept that under pulsed laser illumination, the specimen temperature is rapidly increased by light absorption, and subsequently quenched as the heat flows along the tip main axis. Hence, laser pulsing can be regarded as subjecting the specimen to an equivalent of a temperature pulse.

As with voltage pulse, the rising edge and the falling edge of the evaporation pulse are generally extremely abrupt (even if the shape is asymmetric in the common case). Assuming again that the temperature $T(t)$ exhibits a square profile with a temporal width of τ_T, and that the tip temperature rise is proportional to the pulse energy U, the evaporation rate can be written as

$$\phi_{TP}\,(\text{atom/pulse}) \approx vN\tau_{T}e^{\dfrac{Q_n(F_{DC})}{k_B(T_0 + \alpha_T U)}} \tag{23.20}$$

T_0 is the base temperature and T the proportionality constant between the temperature rise and the absorbed laser energy U.

During the excitation by the light pulse, energy is first absorbed by the electrons. After a characteristic electron-phonon rising time of the order of a few picoseconds, the hot electrons progressively become coupled to the lattice. After this time, the lattice and the electrons are in thermal equilibrium, at a peak temperature T_{max}. In the model describing field evaporation, the electron temperature has no known direct consequence; only the lattice temperature is of interest and is considered in the calculation. As a result, the temperature rising time is much shorter than the time measurement accuracy of PSD and is clearly not the limiting factor.

After electron-lattice thermalization, the cooling process is governed by the classical laws of heat flows in bulk materials (Fourier equation of heat). In a material, the cooling constant is proportional to $(\lambda T)^2/D$, with λT the heat-affected zone size within the specimen and D the thermal diffusivity. This length is related to the absorption distribution inside the needle specimen. Because of the complex interaction of a laser illumination with a subwavelength specimen, the heated zone size is strongly dependant on the materials properties (mainly the optical index related to the energy band diagram of the considered materials) and the geometric features of the tip. The most prominent parameters regarding the laser illumination are the direction of polarization of the incident light with respect to the specimen main axis (z), the size of the beam (DB), and the wavelength of the electromagnetic wave. Polarizations of the laser light, parallel or perpendicular to the specimen axis, are referred to as *axial* or *transverse polarizations*, respectively.

It was found both experimentally and theoretically that the best mass resolutions (i.e., the shortest temperature pulses) were found in axial polarization with short wavelength illumination in almost all materials [35]. As depicted in Figure 23.3, this result was proven to be related to the confinement of absorption in a small zone situated at the end apex of the sample. This temperature focusing is first induced by the local resonance of the end apex when the radius of the hemispherical end cap approaches $0.15\lambda_{\text{laser}}$ [36, 37]. Second, in axial polarization, the incident wave is strongly diffracted by the presence of the end apex. All along the specimen, the interaction of this diffracted wave with the incident wave creates a standing interference wave that modulated the absorption of light. As a result, the second absorption confinement is observed on a local zone of size about half a wavelength situated close to the end apex, whatever the tip radius [38].

Nevertheless, even though the temperature of the specimen apex can be assumed to be homogeneous across the emitting surface in the case of pure metals, since their thermal diffusivity is high, the presence of defects, alloying elements, or interfaces within the specimen will decrease the mean free path of the electrons

Figure 23.3 3D map of temperature 10 ps after the illumination of a Fe tip of end radius about 50 nm by a planar wave with axial polarization and subpicosecond duration ($\lambda = 500$ nm). Note the presence of a confined zone of absorption at the end apex of the sample related to the interaction between the tip and the laser wave.

and thus the ballistic length of penetration. This can give rise to an asymmetric three-dimensional distribution of the temperature in the specimen apex. The temperature of the side of the tip directly illuminated by the laser becomes higher than that of the so-called shadow side, inducing asymmetric field evaporation across the specimen surface. This effect is therefore enhanced for short wavelength illumination on large tip. The lateral dimensions of the tip are sufficient to prevent diffraction effects that normally contribute to homogenize the heat absorption all around the tip. This is in good agreement with some studies of specimen deformation in laser pulsing mode for pure metals, where distortions are not observed in the vicinity of the illuminated side of the specimen, as in poor heat conductors [39].

It is important to note that the differences existing between the evaporation mechanisms of the laser pulse mode and the voltage pulse mode cannot be reduced simply to a difference in temperature. The field at the specimen surface during evaporation and, therefore, the activation energies are very different. Potentially, this can have significant consequences on the conditions of analyses that are required to ensure high-quality data (e.g., gas-assisted evaporation, molecular ions, irregular tip shapes, etc.). The evaporation duration is also different and could vary from one sample to another. Indeed, it is directly related to the duration of the thermal pulse and thus to the illumination conditions and the specific specimen geometry.

23.3
Field Ion Microscopy

23.3.1
Introduction

As previously discussed in Section 23.2, ion trajectories can be deduced from different projection laws. As a consequence, the FIM and the atom probe are basically projection microscopes, the magnification of which is directly proportional to the specimen to detector over specimen radius of curvature ratio. Under standard operating conditions, this ratio is several millions. With such a magnification, true atomic resolution is easily achievable. Indeed, FIM was the first ever method that allowed individual atoms to be observed. The first observation was made in 1956 by Bahadur and Müller at Penn State University on tungsten (W) atoms [40]. As shown in Figure 23.4, on a field ion micrograph, individual bright spots are clearly resolved. Each bright spot corresponds to a unique atom, indicating that true atomic resolution is achieved. Under simple assumptions, which have been partly discussed previously, invaluable information regarding the atomic scale structure of investigated materials can be derived from FIM.

23.3.2
Image Interpretation

In metallic materials, atoms are periodically arranged in three dimensions. Symmetries arising from the interception of this rigid lattice by the hemispherical end shape of the specimen generate a series of concentric rings called *poles*, each of them being the trace of the emergence of a given crystallographic direction out of the specimen. Between, or in some cases inside, these concentric rings, parts of atomic planes are clearly observed. The micrograph presented in Figure 23.4 is characteristic of a crystalline specimen. Depending on the respective orientation of the crystal and the specimen axis, the observed symmetries can be rotated. In the case of very simple (but common) crystal structures (simple, face or body center cubic, and hexagonal), different symmetries will be observed. The analysis of these

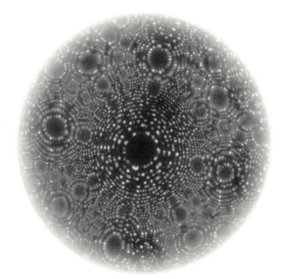

Figure 23.4 Field ion micrograph of a Cu tip. Specimen temperature is 20 K, the imaging gas is 5×10^{-5} Torr Ne, and the tip voltage 5.0 kV, leading to a lateral extent of the micrograph close to 50 nm.

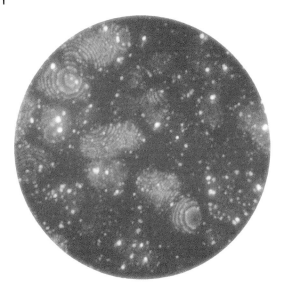

Figure 23.5 Field ion micrograph of a nanocrystallized AlSm tip. Specimen temperature is 20 K, the imaging gas is 5×10^{-5} Torr Ne + He, and the tip voltage 5.5 kV, leading to a lateral extent of the micrograph close to 50 nm.

symmetries can be interpreted in the simple framework of stereographic projection, and the structure of the analyzed crystal structure derived. If the structure can be obtained from field ion micrographs, lattice parameters are usually not derived with this method. Indeed, to achieve a precision similar to what can be obtained by X-ray (or even electron) diffraction would require a perfect knowledge of the projection laws, which is not the case. This issue of projection laws is still a matter of debate, in particular regarding the accuracy of APT reconstructions.

In the case of nanocrystallites, each individual particle will have its own crystal orientation, which may be completely disconnected from the surrounding ones. This is displayed in the micrograph of a partially devitrified AlSm amorphous alloy shown in Figure 23.5. Al nanocrystals are embedded in an amorphous matrix [41]. As the symmetries observed on field ion micrographs arise from the regular arrangement of atoms in the crystal lattice, these symmetries cannot be observed in the amorphous phase, as shown in Figure 23.5. The difficulty in deriving the orientation of such nanocrystals is a consequence of their small size. Symmetries are not straightforward to determine, preventing any indexing when the crystallites are too small.

23.3.3
Field Ion Microscopy in Materials Science

23.3.3.1 Point Defects

As atoms can be individually imaged, one could assume that vacancies would be easy to identify, and therefore numbered, as they would appear as a lack of atom in the field ion image. Unfortunately, there are many artifacts that can result in the absence of a bright spot corresponding to an atomic position in micrographs. First of all, as field evaporation is a statistical process, even if unlikely, evaporation may take place from locations where it is not expected. In worst cases, atoms can be evaporated from inside fully resolved atomic planes. In addition, in alloys, solute atoms may show with a dark contrast, or preferentially evaporate (which may in some cases be assisted by field corrosion), leading to the biased conclusion that a given atomic site is not occupied. Therefore, any attempt to determine the vacancy number density requires carefully controlled experimental conditions, often based on the examination of evaporation sequences. Such an approach has been successfully used by Berger et al. [42], who determined the vacancy concentration in quenched refractory metals. This approach was further developed to study the distribution of vacancies produced by the impact of a high-energy Kr^{2+} ion. This distribution was reconstructed in three dimensions, where the atom-by-atom evaporation sequence is computed and treated with the aid of an early image analysis system [43]. Going further in the study of aggregate of point defect, tentative studies of voids have been conducted, but, because severe distortions in the electric field distribution at the specimen surface affected the observed contrast, only qualitative descriptions of such voids could be obtained by FIM [44, 45]. In contrast with vacancies, autointerstitials are another type of defects in pure metals that FIM should be able to resolve. Indeed, these defects should be characterized by the presence of additional brightly imaging spots in micrographs. As in the case of vacancies, other sources of bright spots, such as preferential retention of solute atoms, field adsorption, and corrosion, exist, and make unambiguous identification of autointerstitials very tedious. An elegant, but indirect, proof of the presence of autointerstitials was given by Seidman and Lie [46], by studying the contract modification at the specimen surface, with the autointerstitial atom lying a few atomic planes under the surface.

23.3.3.2 Linear Defects

The second major type of lattice defects studied by FIM are the linear defects, namely, dislocations. Numerous studies of dislocations have been conducted in the 1950s and 1960s, and it is beyond the scope of this contribution to provide a complete description of these studies. The interested reader should refer to the monograph by Bowkett and Smith [47]. The contrast observed by FIM originates from the atomic displacements generated by the presence of a dislocation. The only possibly observed displacements have to have a component normal to the specimen axis. For example, for a dislocation having a Burgers vector component normal to the specimen surface, a characteristic spiral will replace the set of concentric rings. If the Burgers vector is parallel to the specimen surface, then no modification of the local symmetry will be generated.

Because only components normal to the specimen surface generate atom stacking modifications in the plane of observation (i.e., specimen surface), the complete determination of a Burgers vector is not always possible. It is only if the dislocation is observed to emerge from two different poles that this can be possible. One of the major problems related to dislocation study by FIM is the potential influence of the mechanical stress resulting from the externally applied electric field. TEM observations of FIM specimen before and after FIM imaging have unambiguously shown dislocation glide [48]. It should be noticed that no direct observation of dislocation glide have been reported so far. In addition, it is not clear how the emergence of a dislocation core will be modified when emerging at the free surface of the specimen.

23.3.3.3 Planar Defects

As FIM allows the crystal orientation to be determined, any perturbation of the symmetric arrangement can be detected. In the situation where an internal interface is present in the observed region, any misorientation across this interface will generate a symmetry breaking, as shown in Figure 23.6. Only in the unlikely event of a misorientation corresponding exactly to a crystal symmetry will this breaking clearly appear on the field ion micrograph. Simple geometrical considerations are generally enough to deduce the specific orientations on each part of the interface. Depending on the sophistication of the projection model used, the precision of the orientation can be reduced down to less than 2° [47]. A specific case of internal boundaries is twin boundaries. Because all atom planes are continuous across twin boundaries, very little contrast is visible in FIM, explaining the limited application of FIM to twin boundaries.

23.3.3.4 Phase Contrast

When different phases are present in an alloy, they usually have different compositions. The binding energies between atoms in the different phases are most

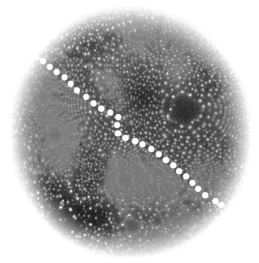

Figure 23.6 Field ion micrograph of a nanocrystallized AlSm tip. Specimen temperature is 20 K, the imaging gas is 5×10^{-5} Torr Ne, and the tip voltage 5.0 kV, leading to a lateral extent of the micrograph close to 50 nm.

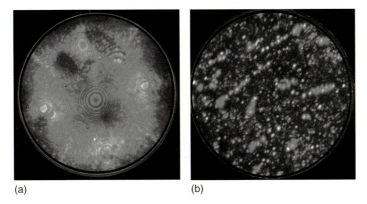

Figure 23.7 Field ion micrographs of (a) CuTi specimen, showing darkly imaging Cu$_4$Ti ordered particles and (b) tempered martensitic steel specimen, showing brightly imaging rod-shaped M$_2$C carbides. Specimen temperature is 80 K, the imaging gas is 5×10^{-5} Torr Ne, and the tip voltages 9.0 kV, leading to a lateral extent of the micrograph close to 100 nm.

likely to be different. The evaporation field (i.e., the field requested to evaporate a given phase) will also be different and generate an imaging contrast at the specimen surface. For example, highly cohesive phases will be more difficult to field evaporate, and will have a higher evaporation field. To generate this local higher field, the specimen has to develop locally a smaller radius of curvature. With a higher local field, the ionization rate of imaging gas atoms will be enhanced, resulting in a brighter contrast on the field ion micrographs. On the contrary, low evaporation field phases will show with a dark contrast. This is the basic mechanism responsible for phase (in its metallurgical sense) contrast in the FIM. Examples of brightly and darkly imaging phases are shown in Figure 23.7. Field ion micrographs may help in estimating characteristics of the second phase, such as particle size, number density, and volume fraction. Even if successfully applied to a number of metallurgical applications, these estimates need to be considered with care.

Two main aspects must be taken into account. First, a field ion micrograph is only a two-dimensional section of the investigated material. The collected data must therefore be corrected in order to account for this geometrical constraint. In addition, as the local radius of curvature of the second phase is different from the average one of the specimen, the projection law is locally affected, and so is the magnification. Brightly imaging phases will be overmagnified and the size of brightly imaging particles, as measured on the field ion micrographs, will be overestimated. The size of particles exhibiting a dark contrast will conversely be underestimated.

23.3.3.5 Three-Dimensional Field Ion Microscopy

When the voltage applied to a specimen at equilibrium is raised, the specimen will need to develop a larger radius of curvature to return to equilibrium. This increase in the radius of curvature can only be achieved by evaporating the specimen, a phenomenon known as *field evaporation*. When carefully controlled, individual atoms can be field evaporated from the specimen surface, atomic layer by atomic layer. Examination of evaporation sequences offers the possibility of investigating not only the surface but also the depth of the specimen, at the ultimate atomic resolution. The analysis then turns from 2D to 3D. As briefly mentioned in the case of point defect, the investigation of evaporation sequences overcomes some of the limitations of the FIM, as true 3D information becomes available. Following the pioneer work of Seidman [43], who processed semimanually individual field ion micrographs, progress in computer power and data analysis makes automatic processing of evaporation sequences possible. With modern computers, thousands of high-resolution field ion micrographs can be processed, with the aim of rendering a true 3D atomic 3D reconstruction. This method, alternatively called three-dimensional field ion microscopy(3DFIM) or *computerized field ion microscopy* (3DFIM [49] or cFIIT [50]), has recently been established as a characterization method providing 3D reconstruction at the atomic scale. It has successfully been used to determine characteristic features of decomposed alloys (precipitate size, number density, and volume fraction). Because this method is still in its early years, it is expected that it can also be used to study lattice defects in materials. Even if based on individual field ion micrographs, it appeared that more information is contained in 3DFIM reconstruction. Indeed, as classical field ion micrographs are recorded at equilibrium, evaporation sequences are dynamic. Surprisingly, the contrasts are slightly different, and may bring complementary information. As shown in Figure 23.8, the core shell structure of $Al_3(ZrSc)$ particles cannot be observed on standard field ion micrographs, whereas it is clearly evidenced in 3DFIM reconstruction. Among the other advantages of

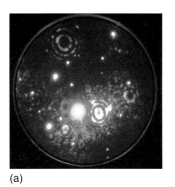
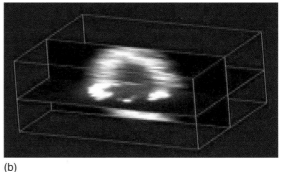

(a) (b)

Figure 23.8 (a) Field ion micrograph of an aged AlZrSc specimen, showing brightly imaging $Al_3(ZrSc)$ intermetallic particles. (Specimen temperature is 20 K, the imaging gas is 5×10^{-5} Torr Ne + He, and the tip voltages 5.0 kV, leading to a lateral extent of the micrograph close to 50 nm.) (b) Three-dimensional field ion micrograph of the same specimen, showing the core shell structure of the intermetallic particles. (The diameter of the selected particle is approximately 5 nm). (Courtesy W. Lefebvre, University of Rouen.)

3DFIM, number density of particles can be directly measured. Indeed, as this method allows atomic resolution, calibration of the reconstructed volume is exact. Determining the number of particles in the volume leads directly to the number density [51]. The precision of the method only depends on the relative characteristic length scales (volume and interparticle distances), but not on any assumption made to deduce 3D information from projected volumes (as in standard FIM and TEM). Another advantage of 3DFIM is that particle size can be measured along the analysis direction, where the bias introduced by local magnification is not active. Particle size along the specimen axis can, for example, be measured by counting the number of atomic planes evaporated during the evaporation of a given particle. This protocol can be applied to many particles, and precipitate size distribution histograms established [52, 53].

23.4
Atom Probe Tomography

23.4.1
State of the Art

One of the major limitations of FIM is that it does not allow chemical identification of the observed atoms. This limitation was overcome in 1968 by Müller and coworkers [54], who designed the first atom probe. The underlying concept is to identify by mass spectrometry the atoms that are field evaporated from the specimen surface. With the aim of combining the spatial resolution of the FIM and single atom identification, the first efficient instrument was the combination of a FIM and a time-of-flight spectrometer. The major challenge was the measurement of the time of flight between the specimen and the detector. High-voltage pulses were superimposed to the specimen DC voltage, allowing field evaporation, and thus departure signal of the surface atoms. When hitting a single particle detection capability detector, the incident ion generates a current pulse, which is used as the arrival signal. The difference between these two times is the time of flight. The mass to charge ratio of the detected ion is derived from the equivalence between the initial potential energy of the atom at rest at the specimen surface, and its kinetic energy when hitting the detector. A simple, but fairly good, assumption has to be made regarding the velocity of ions during their path. Owing to their very small mass, the ions are considered to reach instantly their final velocity, that is, they cruise at a constant speed. Assuming that the detector is grounded, the energy equivalence allows mass to charge ratio to be derived, as shown in Eq. (23.18).

The mass resolution of such instruments is limited by two main factors. The first one is the precision of the time-of-flight measurement. Both departure and detection signals will contribute to the precision of the measurement, and a typical resolution of 1 ns gives a mass resolution full width at tenth of maximum (FWTM) of 150 for a 1 m flight path. The second source of degradation of the mass resolution is the energy deficit of evaporated ions. Indeed, as discussed in Section 23.2, field evaporation is a probabilistic process, and some ions may evaporate before the maximum potential (i.e., during the evaporation pulse raise) is achieved. These ions will, therefore, have lower than assumed potential energy and velocity. The time of flight will be larger than the one corresponding to an ion with the maximum potential energy. Events with larger time of flights, that is, higher apparent masses, will be identified, degrading the mass resolution. With the original instruments, the final mass resolution was about 50, making a mass determination accuracy of about 1 amu.

The trajectories of imaging gas and field evaporated ions are quasi-identical [55]. As a consequence, if the detector is far enough from the specimen, its projection onto the specimen, which will define the instrument resolution, may be reduced down to a few tenths of a nanometer. The resolution of the atom probe can thus theoretically reach the one of the FIM. But because of physical aspects of the field evaporation process itself, the lateral resolution of the atom probe is generally not fully atomic, and so not as good as the one of the FIM. It is only in very specific cases that full atomic resolution, and, for example, lattice reconstruction, can be achieved [56].

Once the concept of atom probe was validated, improved generations of instruments were conceived. Three major improvements can be identified, which currently make APT a well-recognized analytical microscope. The first is the improvement of mass resolution. The major improvement is the development of energy compensators. These energy compensators are electrostatic lenses, located between the specimen and the detector. The general concept is to provide lower energy ions a shorter flight path. As a result, ions will be focused in terms of flight times, but their final energy will not be affected. The term *energy compensator* is somehow misleading; time-of-flight compensation would be more accurate. The two most popular compensators are the Poschenreider [57] toroidal sector, and the reflectron lenses [58]. If the achievable mass resolution with a Poschenreider-type lens is better that with a reflectron (1000–500 respectively), it suffers a major problem of trajectory overlaps, making it unsuitable for three-dimensional atom probes. When time-of-flight compensators are used, the influence of the timing signal resolution becomes increasingly important. For most recent instruments, this resolution has been improved to 100 ps.

The second major improvement with respect to the first generation of instruments is the use of multihit capability PSDs. The advantage of such detectors is that they allow lateral positioning of the incident ions. With such detector capabilities, three-dimensional reconstructions of the analyzed volume are possible. X and Y coordinates (at the specimen surface) are deduced form the impact position and the Z coordinate is derived, at a first approximation, from the temporal sequence of ion arrival [59]. The major drawback, as compared to the first generation of detectors, is that they are based on microchannel plates (MCPs). Because of MCPs, the efficiency of the current instruments is reduced to 40–50%. A second advantage of PSDs is that larger areas of the specimen surface can be analyzed with the same lateral resolution. Indeed, converse to early instruments in which the lateral resolution was the projection of the detector at the specimen surface, for recent atom probes it is the intrinsic resolution of the detector. With a detector intrinsic resolution of 100 μm, the spatial resolution of the instrument is in the order of a tenth of a nanometer [60]. The increase of the analyzed surface is obtained by moving the detectors toward the specimen. But this results in a degradation of the mass resolution, due to the reduction of the flight times, and the increase in the relative importance of the timing errors. The current instruments use the reflectron as a focusing electrostatic lens to focus the emitted ion beam, and thus enlarge the acceptance angle, for a given flight path [61]. Thanks to these developments, the latest generation of instruments allows sections of about 100×100 nm^2 to be analyzed, with a mass resolution of about 500 when a reflectron is used as both a compensator and a focusing lens. Volumes of $100 \times 100 \times 500$ nm^3 can then be mapped out, with virtually all detected ions individually chemically identified.

The last major improvement is the development of laser-assisted atom probes. The use of electric pulses intrinsically limits the application of APT to conductive materials. The use of laser increased the field of application to semiconductor and insulating materials. As previously discussed, laser pulses promote thermal field evaporation of surface atoms at the specimen DC voltage. As the electric field remains constant during operation, no energy deficit can arise, and the mass resolution is limited by the precision of the time-of-flight measurements. In addition to the resolution of the timing electronics, an additional source of error is the exact time-of-field evaporation. This is controlled by the temperature rise, which is not instant. The dispersion in evaporation times is the factor limiting mass resolution in laser-assisted instruments. The optimization of laser atom probe operating conditions is a very active field of research, the better understanding of laser matter interaction being the key to further improvements [62]. The virtual absence of any energy deficit is an alternative to reflectron-type compensators for wide-angle instruments. With a flight path of about 10 cm, laser-assisted instruments reach a mass resolution of about 300 with reconstructed volumes of about $100 \times 100 \times 500$ nm^3.

23.4.2
APT in Materials Science

23.4.2.1 Specimen Preparation
One of the key issues for APT is the preparation of the specimen as a needle, with an end radius of curvature in the range 50–100 nm. Historically, tips were made using salt bath or standard electrochemical polishing [63, 64]. The most popular technique is the two-stage method, where a match is first etched in its middle, and then placed in the electrolytic bath until separation into two parts. This method is well adapted to the study of conductive massive samples, when the analysis site is not of prior importance. When site-specific preparation is required, the method has to be adapted. In particular, if grain boundary (GB) analysis is aimed at, such a boundary must be positioned at a distance of less than 100 nm from the tip apex. The standard electropolishing method has to be combined with regular

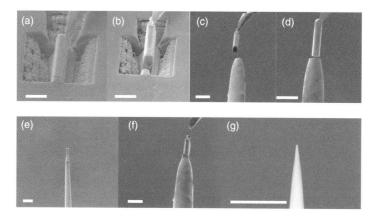

Figure 23.9 Scanning electron microscope micrographs showing the FIB-assisted lift-out preparation of atom probe tomography specimens. (a,b) Lift-out of a preformed pillar. (c–f) Attachment of the post to a presharpened tip. (g) Specimen after FIB sharpening. The final radius of curvature is about 50 nm. (White bar in SEM micrographs is 2 µm). (From P. Jessner, PhD thesis, GPM UMR CNRS 6634, University of Rouen, 2010 [51].)

observation of the specimen by TEM. If a GB is observed a few microns away from the apex, small amounts of material can be removed by pulse electropolishing, until the boundary is placed less than 100 nm from the apex, at a depth that can be reached during analysis [65]. Recently, the introduction of focused ion beam (FIB) milling [66], combined with lift-out techniques, has drastically expanded the specimen preparation technique. It has made possible the preparation of specimen at controlled depths, even at a few nanometers from the surface. Site-specific specimen preparation, as illustrated in Figure 23.9, with an accuracy of a few tens of nanometers, is now routinely conducted [67]. It is also possible to prepare atom probe specimens from powder grains, even as small as 1 µm [68].

23.4.2.2 Phase Composition Measurements

As an analytical microscope, APT provides quantitative analysis of the entire volume analyzed. If it is not worth using this instrument for composition measurements where other standard analytical techniques can be used, some intrinsic characteristics of the instrument can be used in the field of chemical analysis. In particular, its ability to detect all the atomic species with the same efficiency makes it an interesting instrument for direct concentration when light elements are concerned. One of the interesting aspects is that no calibration is needed before a local composition is to be estimated.

Of particular interest is the direct measurement of carbon and nitrogen levels in steels, with a precision in the order of 10 atppm. The measurement of nitrogen levels along the diffusion front during nitriding is one example [69]. Even if APT is able to detect hydrogen, accurate concentrations are more difficult to obtain, as there is always a hydrogen signal present in mass spectra, due to the presence of residual hydrogen in the analysis chamber. The use of deuterium is a viable alternative, and recent studies have shown that deuterium segregation at particle–matrix interfaces can be directly observed in steels by APT [70]. It is expected for the near future that APT will provide invaluable information in the field of hydrogen/deuterium distribution in solid materials.

Even if APT can bring information on specimen composition, its main field of application is when a combination of quantitative composition and spatial resolution is requested. Indeed, it is possible to measure the composition in a volume as small as $1\,\text{nm}^3$. In condensed materials, $1\,\text{nm}^3$ contains about 100 atoms. With a detection efficiency of 50%, 50 atoms can be detected from such a volume. The detection process being independent of the chemical nature of the ions, detection is a random process, and the precision of the measurement can be simply estimated from statistical laws. It can be shown that the precision of the measurement is $\sqrt{\frac{c(100-c)}{n}}(1-Q)$, where c is the concentration (in percentage), n the number of detected ions, and Q the detection efficiency [71, 72]. It is therefore possible to obtain a precision of a small atomic percentage. When analyzed phases are much larger than $1\,\text{nm}^3$, the precision of the local concentration may reach the 10 ppm level.

23.4.2.3 Segregation

Taking advantage of its ultimate spatial resolution, the atom probe has been used to study composition evolution at the nanometer scale, in many cases close to internal defects (dislocations, interfaces, and particles).

Segregation to Linear Defects Linear defects are known to be preferential sites for interstitial atom segregation. Such segregations are known as *Cottrell atmospheres* [73], and play an important role in materials, as they alter the matrix composition. The first reported atom probe analysis of a Cottrell atmosphere concerns carbon segregation along dislocation cores in a steel specimen [74]. Later, another illustration of interstitial atom (in this case boron) segregation to dislocation in intermetallic FeAl alloy was provided by Blavette *et al.* [75]. In this later example, simultaneous observation of the dislocation core and the associated segregation was highlighted, thanks to the depth resolution of the instrument. Since then, many observations of carbon segregation to dislocations in steels, similar to the one showed in Figure 23.10, have been published. Another example of a Cottrell atmosphere is shown in Figure 23.11, where the segregation of As into a dislocation loop in silicon is clearly observed [76]. After heat treatment (annealing at 600°C

Figure 23.10 Distribution of carbon atoms showing Cottrell atmospheres in Cr–P steel. (From Pereloma *et al. Microscopy and Microanalysis* (2005), **11**(Suppl 2):876–877, Ref. [78].)

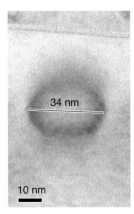 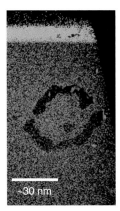

Figure 23.11 Transmission electron micrograph and atom probe tomographic reconstruction showing dislocation loops aligned along (111)Si planes, in As-implanted Si. As-rich regions containing more than 2 at% are delimited by the black surface. Note that the scales are similar for both images, but that it is not the same loop imaged. (Courtesy K. Thompson, T.F. Kelly, Cameca inc.)

for 0.5 h followed by 1000 °C for 30 s), the atomic defects present in silicon have arranged into dislocation loops, and trapped the arsenic-implanted dopant atoms. The impact that these Cottrell atmospheres may have on the concentration of electrically active dopant atoms, and therefore on device performance, remains undetermined but could prove important [77].

Segregation to GB Among the major contribution of APT to materials science is the study of GB segregation. It is well recognized that GBs play a crucial role in the mechanical properties of materials. Depending on the interface nature and the chemical nature of segregating atoms, segregation may either be beneficial or detrimental. In order to reinforce, or to limit, solute segregation, it is therefore essential to characterize the local chemistry at interfaces, and to connect it with the mechanical behavior of the material. The main advantage of APT is that it gives direct access to GB concentration, and therefore to the Gibbsian interfacial excess [79]. Initial investigations of GB chemistry were mostly conducted by Norden and coworkers [80] and by Seidman and coworkers [81] on refractory metal alloys. Because of their industrial interest, GBs in steel have also been investigated. Most of the studies, as reviewed by Thuvander and Andren [82] were devoted to carbon and nitrogen segregation to GBs. The other main topic in steels is the austenite to low temperature product (ferrite and bainite) transformation in model FeMoC [83], FeMnC [84], and industrial steels [85]. The concentration profiles at the transformation interface obtained by APT are used to understand the transformation mode acting during austenite to ferrite transformation. As shown in Figure 23.12, the interstitial and substitutional element distribution across the interface helps discriminate between local equilibrium (with or without

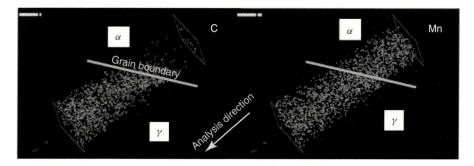

Figure 23.12 3D reconstruction showing the distribution of C and Mn across an austenite-ferrite interface in a Fe-1.8Mn-0.4C after holding 50 s at 700 °C. No obvious manganese segregation accumulation in the vicinity of the α/γ can be seen in the analyzed volume. (From O. Thuillier, PhD thesis, GPM UMR CNRS 6634, University of Rouen $10 \times 10 \times 36$ nm^3, Ref. [90].)

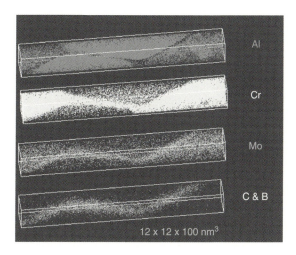

Figure 23.13 Grain boundary segregation in Astroloy® Ni-based superalloy. Segregation of B, C, Mo, and Cr at serrated grain boundaries. (From L. Letellier, PhD thesis, GPM UMR CNRS 6634, University of Rouen 1994, Ref. [91].)

partitioning) and paraequilibrium modes. Nickel-based superalloys are materials in which GB chemistry has been widely investigated. It is well known that small amounts of carbon and boron improve creep rupture properties. One of the major issues concerns the distribution of these solute elements. Buchon et al. [86] and Miller et al. [87] showed that they strongly segregate at the γ/γ' GBs, but that the g/g boundaries are free from segregation. In commercial Astroloy®, Blavette et al. [88] showed that B and C segregate at GBs, as illustrated in Figure 23.13. A general overview of this issue of GB segregation in nickel-based alloys is given by Blavette et al. [89].

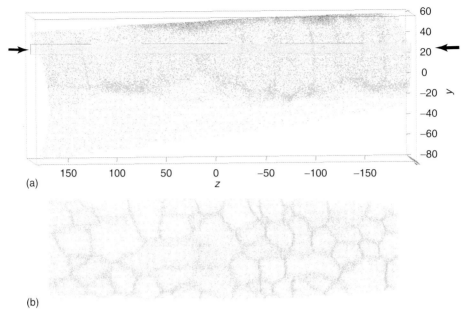

Figure 23.14 (a) Three-dimensional reconstruction of a NiSiPt structure consisting of a 50 nm thick Ni-5 at% Pt film sputter deposited onto Si. (b) Transverse reconstruction (along the slice indicated by the arrow in (a) showing the partitioning of Pt atoms at columnar grain boundaries in the NiSi-rich layer. Scale in nanometer. (Courtesy K. Hoummada, University Paul Cezanne, Marseille.)

Another important system in which GB analysis by APT has brought important information is Ni–Si. This system is currently the best silicide for contact in advanced integrated circuits. Its high temperature stability is of concern, as it may transform to the undesired Ni_2Si. It has been shown that the addition of Pt in the Ni film increases the nucleation temperature of Ni_2Si by approximately 150 °C, and thus stabilizes the NiSi films. In order to understand the role of Pt, the formation behavior of Ni silicide alloyed with Pt was recently analyzed by APT [92–95]. An APT atom map of the distribution of Pt in the Ni_2Si region formed after annealing of a NiSiPt structure is shown in Figure 23.14. Pt segregation to GBs in the Ni_2Si region is clearly seen. Not only the location but also the amount of Pt segregated at GBs can be derived from such an analysis. This is an important element for understanding the nucleation and lateral growth of this phase, and thus the enhanced stability of the NiSi phase. Also, knowledge of this Pt segregation is essential as it has a direct influence on the electrical properties of the NiSi contacts in microelectronics.

23.4.2.4 Early Stages of Phase Transformation

Phase separation, and in particular the early stages, is a field in which APT, thanks to its high spatial resolution, has brought major information over the past decades. Studying the early stages of phase separation, in other words the nucleation stage, has always been challenging for microscopic techniques, as the length scales involved are usually in the subnanometer range and as very limited numbers of atoms are involved. The characterization of early stages of precipitation in engineering alloys has early been recognized as a major element of materials characterization, as it influences the whole precipitation sequence. It is therefore a very important aspect, in particular in the case of age-hardenable alloys, where a careful control of the precipitation microstructure is the key to optimization of mechanical properties.

Aluminum-Based Alloys Because they are known to age even at room temperature, the studies of nucleation stages in aluminum-based alloys have always been challenging. The situation is even more difficult as the first precipitate to form are GP (for Guinier Preston) zones [96, 97], which are fully coherent with the matrix and may have spatial extensions as small as one atomic layer in thickness. The model system for studying the precipitation of GP zones is the Al–Cu system, with Cu contents lower than 5 at%. This system was studied in detail by Hono *et al.* [98], who observed single- and multi-{001}Al layer Cu zones using the FIM. Atom probe analysis of such zones has proved to be very difficult, and accurate atomic scale reconstructions have not been reported yet. On the other hand, these zones appear as clouds in 3D reconstruction, most likely because of a too high difference in the evaporation field between the matrix and the zones [99].

In many other aluminum-based alloys, different types of GP zones have been identified, ranging from rodlike guinier preston bagaryatsky (zones) (GPB) zones in AlCuMg [100, 101], globular (MgSi)-rich clusters in 6xxx AlMgSi [102–104], and (MgZn) rich in AlMgZn 7xxx alloys [105–109]. Because of its destructive nature, APT has not been able to answer the question whether in these aluminum-based alloys, metastable precipitates form at the former GP zone sites, or homogeneously in the matrix, after dissolution of the GP zones. A large review of studies related to particle composition evolution during age hardening in Al-based alloys can be found in Ref. [110].

Steels Another system in which APT has brought original information related to the early stages of precipitation is steels. In these alloys, two different types of precipitates can be distinguished, the ones involving interstitial elements (carbides and nitrides mostly) and the one involving only substitutional elements.

Iron Carbides Iron-based carbides are the one that form at the lower temperatures because of the high diffusivity of carbon. The most common, and most stable, iron carbide is Fe_3C cementite. In the classical view of martensite tempering, epsilon e-$Fe_{2.4}C$ is regarded as the first intermediate carbide to precipitate. Even if extensively investigated, e carbide with the expected stoichiometry has never been observed. Conversely, regions with local carbon contents in the range 10–15 at% are almost systematically observed [111, 112]. In addition, even before the formation of this e carbide, carbon concentration fluctuations have been measured in martensites aged at very low temperatures (i.e., lower than 100 °C). As shown in Figure 23.15,

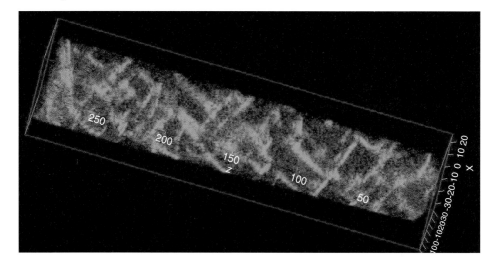

Figure 23.15 Three-dimensional reconstruction showing the distribution of C atoms in the virgin martensite of a Fe-25Ni-0.4C (wt%) aged one month at room temperature. The tweedlike network is produced by spinodal decomposition. Scale in nanometers.

these concentration fluctuations develop as a triaxially oriented network. The evolution of the characteristic wavelength and concentration amplitude is coherent with a spinodal-type decomposition. Interestingly, these concentration fluctuations reach a maximum value in the same 10–15 at% range, similar to what is observed at higher temperatures [113].

In complex steels, substitutional elements partition between matrix and cementite. Babu et al. [114] studied the partitioning behavior of Mn and Si. They showed no partitioning of these elements in the early stages of transformation, neither in the formed phase nor at the transformation interface. It is only after prolonged aging in the temperature range 350–500 °C that partitioning of Mn to cementite and Si to matrix is observed. A similar behavior for Cr and Mo in FeCrMoC steels was observed by Thomson and Miller [115]. Interestingly, identical behavior of Mn was observed in a FeMnC alloy, where cementite was formed during autotempering [116]. The distribution of solute elements in the different phases, and at the interfaces, has been shown to be consistent with a paraequilibrium mechanism for cementite precipitation [117].

Alloy Carbides In contrast with cementite, alloy carbides form directly from the combination of carbon with alloying elements. The most efficient carbide-forming elements are Ti, V, Cr, Nb, and Mo. They are intentionally added in steels for two different reasons. Either they are used in low-carbon alloys to trap carbon, in order to help producing high stength low alloy steels (HSLA) steels, or they are added in higher carbon containing steels to promote the precipitation of secondary hardening carbides. Depending on the solute content, different types of carbides can be formed.

In low-alloy steels, solutes are used to trap carbon as carbides. The most common carbides are VC, NbC, and TiC. The precipitation in the Nb–C system has been extensively studied by APT. As shown in Figure 23.16, the precipitation is shown to be heterogeneous, niobium carbides nucleating both at GBs and dislocation lines. This result is in agreement with TEM observations [118], APT bringing the additional information that homogeneous nucleation can be ruled out.

In Cr- and Mo-rich alloys, M_2C, M_7C_3, and $M_{23}C_7$ carbides form. M_2C carbides are desired in medium carbon martensitic steels (such as hot work tool steels and high-speed steels) in order to improve wear resistance, or in low-alloy ferritic steels to improve creep resistance. They are the first to form in the precipitation sequence,

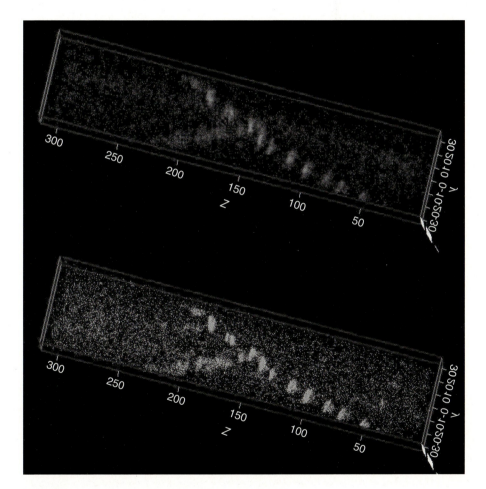

Figure 23.16 Distribution of C (dark gray) and Nb (light gray) atoms in a Fe-500 at ppm Nb-500 atppm C alloy aged 5 min at 650 °C. The alignment of NbC precipitates is due to heterogeneous precipitation along dislocation lines. Scale in nanometers.

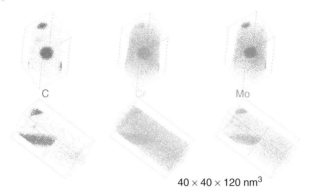

$40 \times 40 \times 120$ nm^3

Figure 23.17 Morphology of M$_2$C carbides in a medium carbon martensitic steel aged 100 h at 500 °C. The precipitate is clearly rod shaped and contains Cr and Mo as metallic elements.

because of their lower M/C ratio. As they usually precipitate with very high number densities (up to 10^{24} part m^{-3}), they greatly affect mechanical properties. As shown in Figure 23.17, they adopt a needle-shaped morphology, with dimensions as small as a few nanometers [119, 120]. They have been found to be responsible for an outstanding combination of strength and toughness observed in Co bearing high Ni AF1410 alloy. This steel has been analyzed by 1D atom probe [121, 122], and more recently APT investigations on a closely related material brought new insight into its microstructure [123, 124]. The M$_2$C carbides were shown to precipitate at temperatures close to 500 °C, and with no apparent connection with prior cementite. The metal elements found in these carbides are Cr and Mo, but the Cr/Mo ratio was shown to be different from one particle to another.

For prolonged aging times, M$_7$C$_3$ (mostly in Mo-rich alloys) and M$_{23}$C$_6$ (mostly in Cr-rich alloys) also form. It is not clear whether these carbides precipitate homogeneously in the matrix, or from heterogeneous nucleation sites, such as dissolving cementite, or preexisting M$_2$C carbides.

Nitrides Modern steels also incorporate controlled amounts of nitrogen. During thermal treatments at temperatures below 600 °C, nitrogen may combine with alloying elements to form either carbonitrides or nitrides, depending on the amount of carbon in the material. The study of niobium nitride precipitation was conducted in parallel with the study of niobium carbides by substituting nitrogen with carbon. Significant differences in the precipitation mechanism resulted in this substitution. The major difference was the modification of the precipitation mode, from purely heterogeneous to mostly homogeneous, as shown in Figure 23.18. Careful examination of the early stages of niobium nitride precipitation indicated that these nitrides are single-layered disc-shaped particles, fully coherent with the ferritic bcc matrix [125]. They can therefore be regarded as GP zones, the first to be identified in Fe-based alloys, even if predicted by Jack [126], back in the 1970s. Their particularity is to include both interstitial and substitutional atoms. To date, only

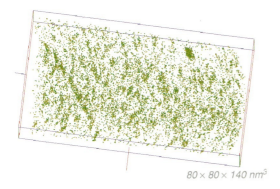

Figure 23.18 Distribution of N (blue) and Nb (green) atoms in a Fe-500 atppm Nb-500 atppm N alloy aged 300 min at 600 °C. The alignment of NbN precipitates is due to heterogeneous precipitation along dislocation lines, but the majority of NbN precipitates have formed homogeneously in the ferritic matrix.

the position of the substitutional atoms (Nb) is known, the position of interstitial atoms (N) is still a matter of debate, and the experimental results are compared with *ab initio* type simulation.

Alloy Nitrides Very limited studies on alloy nitrides have been conducted by APT. The first system investigated was FeMo [127–129], where the disc-shaped molybdenum nitrides were investigated. A more recent study was conducted on FeCr alloys, with the aim of investigating the structure of the nitrided layer as a

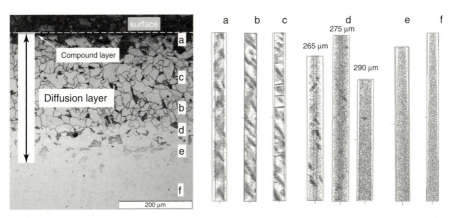

Figure 23.19 APT analysis of nitrided Fe-5 at% Cr at different depths from the surface. For clarity, only Cr (light grey) and N (dark grey) ions are shown. The cross section of all APT volumes is 10×10 nm^2. (From P. Jessner, PhD thesis, GPM UMR CNRS 6634, University of Rouen, 2010, Ref. [51].)

function of the depth. Figure 23.19 shows the distribution of chromium nitrides as a function of the position in the diffusion layer. It clearly shows that the number density and morphology of the precipitates is clearly related to the position in the nitrides layer, and therefore to the local nitrogen content [130]. Taking advantage of the combination of high spatial and mass resolution, it was also possible to study the composition in the matrix. The challenging issue was to access the matrix, with disc-shaped precipitates about 10 nm apart from each other. Because of its spatial resolution in the cubic nanometer range, it was possible to measure the matrix composition for different heat treatments, and prove that in the as-nitrided condition, the nitrogen level was almost twice as high as the equilibrium content at the same temperature. This was the first experimental evidence of excess nitrogen in the matrix, contributing to the so-called excess nitrogen in the material [131]. In addition, it was shown that the nitrides are not stoichiometric, but that they still contain up to 7 at% iron.

Precipitation Precipitation in alloys has always been a field of intensive atom probe activity. It would be unrealistic to provide, if possible, a complete list of all the systems investigated. Detailed illustrations can be found in classical textbooks and atom probe monograms, and in review papers, such as in Ref. [132]. Basically, the contribution of APT, as illustrated in the case of nitrided Fe–Cr alloys, allows composition measurements of both matrix and precipitates, as well as a precise determination of the number densities and volume fraction of the precipitates. These last two parameters can be used to discriminate the different sequences in the precipitation mechanism, namely, nucleation, growth, and coarsening, and compare them with theoretical prediction. Results obtained by APT can also be compared with simulation at the atomic scale, namely, the Monte Carlo simulation and phase field in its last development, atomic phase field. Experimental results can be used as reference states in order to validate calculated structures. On the other hand, simulation, once validated in advanced stages, can be used to track back the early stages of phase separation, which are hardly observable experimentally. This approach has been successfully applied in the case of substitutional alloys [133, 134], and has now to be expanded to interstitial alloys.

Spinodal Decomposition Nucleation and growth is not the only mechanism that can be involved in the early stages of phase separation. Nucleation and growth takes place when the single-phase configuration is metastable with respect to the more stable two-phase configuration. In the particular case when the single-phase configuration is unstable, then the spinodal mechanism, as described by Cahn [135], will be active. As the solution is unstable, composition fluctuations will emerge spontaneously, and develop until the equilibrium two-phase configuration is reached. The decomposition does not require any second-phase nuclei to develop to proceed. In the case of metallic materials, the initial composition fluctuation will develop homogeneously in the material with length scales at the nanometer scale. APT is therefore a very well-suited investigation tool, which provides a direct description of the microstructure, complementary to the small angle diffusion techniques. Spinodal decomposition in metallic systems has been studied in various metallic systems, including Fe- and Al-based alloys. Most of the studies were conducted with 1D atom probe, making the deconvolution between phase transformation and spatial convolution, and therefore the discrimination between spinodal and nucleation and growth mechanisms very difficult, when possible [136, 137]. A conclusion on the concentration fluctuation increase is questionable. In a combined TEM and APT study of the low temperature decomposition of Fe–Be alloys, Miller *et al.* demonstrated that the initial solid solution decomposes into a triaxially aligned B2-ordered Be-rich phase. The preferential orientation of the composition fluctuations is connected to the elastic strain because of the misfit between the Fe-rich bcc matrix and the ordered Be-rich phase. In the case of Fe–Cr alloys aged at intermediate temperatures (i.e., between 300 and 500 °C), two types of bcc Fe- and Cr-rich domains develop from the homogeneous solid solution. Because of the absence of any sensible misfit between the two phases, the decomposition is isotropic, as it would be in liquids. The system was extensively studied, being a model system for both fundamental and applied (as the basis of stainless steels) research. The composition fluctuation and characteristic wavelength, as observed in Figure 23.20, have been measured by APT, and found in good agreement with

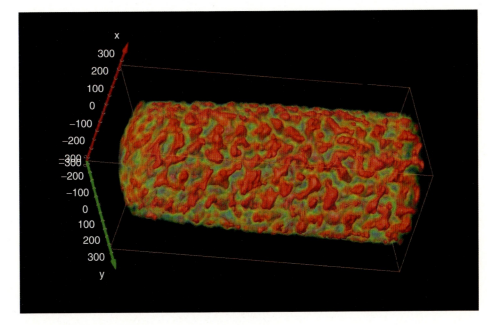

Figure 23.20 Spinodal decomposition in Fe-32%Cr aged 50 h at 450 °C. The red regions correspond to Cr domains, containing more than 50 at% Cr. Their morphology shows the continuity of Cr-rich domains over the reconstructed volume, typically produced by an isotropic spinodal decomposition. $50 \times 50 \times 130$ nm^3.

the different models of spinodal decomposition, and Monte Carlo simulations. In particular, the time exponent for coarsening of the microstructure was measured in the range 0.18–0.2, significantly lower than the 0.33 value predicted from the lifshitz slyozov wagner (theory) (LSW) theory. In addition, the evolution of the hardness of the material was found to vary linearly with concentration fluctuation amplitudes, supporting the model for embrittlement of Fe–Cr-based stainless steels.

23.5
Conclusion

In conclusion, this contribution has presented the basic principles of FIM and APT, from the early developments in the 1950s to the latest ones, aiming to describe the interaction of different types of specimens with a femtosecond laser beam under high electric field conditions. This field is still very active, and it is expected that better understanding of the underlying physics will help in expanding the technique to new fields, in particular, biological sciences.

In the second part, the current fields in which APT has been, historically or more recently, used have been illustrated on the basis of various examples. They cover the field of conductive materials (mainly metals) and the newly investigated area of semiconductors, which can now be routinely analyzed with the help of laser-assisted instruments. Even more recently, insulators have also successfully been analyzed, potentially opening the field of mineralogy.

It can therefore be concluded that APT, despite its spectacular contribution to material science, is still a young technique, in rapid evolution, and that the picture given in this contribution is only an instant snapshot, in need of regular updates in the near future.

References

1. Müller, E.W. (1951) *Z. Phys.*, **131**, 136.
2. Gomer, R. (1961) *Field Emission and Field Ionisation*, Harvard University, Cambridge, pp. 45–46.
3. Gipson, J. (1980) *J. Appl. Phys.*, **51**, 3884.
4. Bajikar, S.S., Kelly, T.F., and Camus, P.P. (1996) *Appl. Surf. Sci.*, **94–95**, 464.
5. Schlesiger, R. and Schmitz, G. (2009) *Ultramicroscopy*, **109**, 497.
6. Nishikawa, O., Ohtani, Y., Maeda, K., Watanabe, M., and Tanaka, K. (2000) *Mater. Charact.*, **44**, 29.
7. Nishikawa, O. and Müller, E.W. (1964) *J. Appl. Phys.*, **35**, 2806.
8. Müller, E.W. (1953) *J. Appl. Phys.*, **24**, 1414.
9. Gomer, R.J. (1959) *Chem. Phys.*, **31**, 341.
10. Gomer, R. and Swanson, L. (1963) *J. Chem. Phys.*, **38**, 1613.
11. Forbes, R.G. (1981) *Surf. Sci.*, **102**, 255.
12. Kellogg, G.L. (1984) *Phys. Rev. B*, **29**, 4304.
13. Wada, M. (1984) *Surf. Sci.*, **145**, 451.
14. Kellogg, G.L.J. (1981) *Appl. Phys.*, **52**, 5320.
15. Tsong, T.T. (1968) *Surf. Sci.*, **10**, 102.
16. (a) Blavette, D., Deconihout, B., Bostel, A., Sarrau, J.M., Bouet, M.,

and Menand, A. (1993) *Rev. Sci. Instrum.*, **64**, 2911; (b) Da Costa, G., Vurpillot, F., Bostel, A., Bouet, M., and Deconihout, B. (2005) *Rev. Sci. Instrum.*, **76**, 013304.

17. Deconihout, B., Renaud, L., Bouet, M., Da Costa, G., Bostel, A., and Blavette, D. (1998) *Ultramicroscopy*, **73**, 253.
18. Da Costa, G., Vurpillot, F., Bostel, A., Bouet, M., and Deconihout, B. (2005) *Rev. Sci. Instrum.*, **76**, 013304.
19. Walls, J.M. and Southworth, H.N. (1979) *J. Phys. D-Appl. Phys.*, **12**, 657.
20. Southworth, H.N. and Walls, J. (1978) *Surf. Sci.*, **75**, 129.
21. Bas, P., Bostel, A., Deconihout, B., and Blavette, D. (1995) *Appl. Surf. Sci.*, **87**, 298.
22. Walck, S.D., Bwuklimanli, T., and Hren, J.J. (1986) *J. Phys. Colloq. C2*, **47**, 451.
23. Haydock, R. and Kingham, D.R. (1980) *Phys. Rev. Lett.*, **44**, 1520.
24. Tsong, T.T. (1978) *Surf. Sci.*, **70**, 211.
25. Cerezo, A., Warren, P.J., and Smith, G.D.W. (1999) *Ultramicroscopy*, **79**, 251.
26. Vurpillot, F., Bostel, A., Cadel, E., and Blavette, D. (2000) *Ultramicroscopy*, **84**, 213.
27. Vurpillot, F., Bostel, A., and Blavette, D. (2001) *Ultramicroscopy*, **89**, 137.
28. De Geuser, F., Lefebvre, W., Danoix, F., Vurpillot, F., and Blavette, D. (2007) *Surf. Interface Anal.*, **39**, 268.
29. Vurpillot, F., Larson, D.J., and Cerezo, A. (2004) *Surf. Interface Anal.*, **36**, 552.
30. Miller, M.K., Cerezo, A., Hetherington, M.G., and Smith, G.D.W. (1996) *Atom Probe Field Ion Microscopy*, Clarendon Press, Oxford.
31. Müller, E.W. and Krishnaswamy, S.V. (1974) *Rev. Sci. Instrum.*, **45**, 1053.
32. Tsong, T.T., McLane, S.B., and Kinkus, T.J. (1982) *Rev. Sci. Instrum.*, **53**, 1442.
33. Tsong, T.T. and Kinkus, T.J. (1974) *Phys. Rev. B*, **29**, 529.
34. Gault, B., Vurpillot, F., Vella, A., Gilbert, M., Menand, A., Blavette, D., and Deconihout, B. (2006) *Rev. Sci. Instrum.*, **77**, 043705.
35. Bunton, J.H., Olson, J.D., Lenz, D.R., and Kelly, T.F. (2007) *Microsc. Microanal.*, **13**, 418.
36. Houard, J., Vella, A., and Vurpillot, F. (2009) *Appl. Phys. Lett.*, **94**, 121905.
37. Vurpillot, F., Houard, J., Vella, A., and Deconihout, B. (2009) *J. Phys. D-Appl. Phys.*, **42**, 125502.
38. Houard, J., Vella, A., Vurpillot, F., and Deconihout, B. (2010) *Phys. Rev. B*, **81**, 125411.
39. Sha, G., Cerezo, A. and Smith, G.D.W. (2008) *Appl. Phys Lett.*, 92.
40. Müller, E.W. and Bahadur, K. (1956) *Phys. Rev.*, **22**, 624.
41. Gloriant, T., Danoix, F., Lefebvre, W., and Greer, A. (2007) *Adv. Eng. Mater.*, **9**, 151.
42. Berger, A.S., Seidman, D.N., and Balluffi, R.W. (1973) *Acta Metall.*, **21**, 123.
43. Wei, C.Y., Current, M.I., and Seidman, D.N. (1981) *Philos. Mag.*, **44**, 459.
44. Brenner, S.S. and Seidman, D.N. (1975) *Radiat. Eff.*, **24**, 73.
45. Godfrey, T.F., Lewis, R.J., Smith, D.A., and Smith, G.D.W. (1976) *J. Less-Common Met.*, **44**, 319.
46. Seidman, D.N. and Lie, K.H. (1972) *Acta Metall.*, **20**, 1045.
47. Bowkett, K.M. and Smith, D.A. (1970) *Field Ion Microscopy*, North Holland, Amsterdam.
48. Loberg, B. and Norden, H. (1970) *Ark. Foer Fys.*, **40**, 413.
49. Vurpillot, F., Gilbert, M., and Deconihout, B. (2007) *Surf. Interface Anal.*, **39**, 273.
50. Semboshi, S., Al-Kassab, T., Gemma, R., and Kirchheim, R. (2009) *Ultramicroscopy*, **109**, 593.
51. Jessner, P. (2010) Nanometer scale precipitation in nitrided alloys: atom probe tomography contribution. PhD Thesis, University of Rouen (France) (in English).
52. Akré, J., Danoix, F., Leitner, H., and Auger, P. (2009) *Ultramicroscopy*, **109**, 518.
53. Cazottes, S. (2008) PhD Thesis, University of Rouen (France) (in French).
54. Müller, E.W., Panitz, J.A., and McLane, S.B. (1968) *Rev. Sci. Instrum.*, **39**, 83.
55. Waugh, A.R., Boyes, E.D., and Southon, M.J. (1976) *Surf. Sci.*, **61**, 109.

56. Vurpillot, F., Da Costa, G., Menand, A., and Blavette, D. (2001) *J. Microsc.*, **203**, 295.
57. Poschenreider, W.P. (1972) *Int. J. Mass Spectrom. Ion Phys.*, **9**, 83.
58. Drachsel, W.V., Alvensleben, K., and Melmed, A.J. (1989) *J. Phys.*, **50–C8**, 541.
59. Bas, P., Bostel, A., Deconihout, B., and Blavette, D. (1995) *Appl. Surf. Sci.*, **87/88**, 298.
60. Deconihout, B. (1993) mise au point de la sonde atomique tomographique. PhD Thesis, University of Rouen (France) (in French).
61. Mamyrin, B.A. (1966) Russian Patent No. 198,034.
62. Proceedings of the 50th, 51st and 52nd International Field Emission Symposia, Guilin (China) 2006, Rouen (France) 2008, and Sydney (Australia) 2010.
63. Müller, E.W. and Tsong, T.T. (1969) *Field Ion Microscopy–Principle and Applications*, Elsevier, New York, NY.
64. Miller, M.K. and Smith, G.D.W. (1989) *Atom Probe Microanalysis: Principle and Applications to Materials Problems*, Materials Research Society, Pittsburg, PA.
65. Henjered, A. and Norden, H. (1983) *J. Phys. E*, **16**, 617.
66. Larson, D.J., Foord, D.T., Petford-Long, A.K., Anthony, T.C., Cerezo, A., and Smith, G.D.W. (1989) *Ultramicroscopy*, **75**, 147.
67. Miller, M.K., Russell, K., Thomson, K., Alvis, R., and Lardon, D.J. (2007) *Microsc. Microanal.*, **13**, 428.
68. Calvo-Dahlborg, M., Chambreland, S., Bao, C.M., Quelennec, X., Cadel, E., Cuvilly, F., and Dahlborg, U. (2009) *Ultramicroscopy*, **109**, 672.
69. Jessner, P., Danoix, R., Hannoyer, B., and Danoix, F. (2009) *Ultramicroscopy*, **109**, 530.
70. Takhahashi, J. Kawakami, K. Kobayashi, Y. Tarui, T. (2010) *Ultramicroscopy, Scripta Mater.*, **63** (3), 261–264.
71. Danoix, F., Grancher, G., Bostel, A., and Blavette, D. (2007) *Ultramicroscopy*, **107**, 734–738.
72. Danoix, F., Grancher, G., Bostel, A., and Blavette, D. (2007) *Ultramicroscopy*, **107**, 739–743.
73. Cottrell, A.H. and Bilby, B.A. (1949) *Proc. Phys. Soc. Lond.*, **A62**, 49.
74. Chang, L., Barnard, S.J., and Smith, G.D.W. (1992) in *Fundamental of Ageing and Tempering in Bainitic and Martensitic Products* (eds G. Krauss and P.E. Repas), Iron and Steel Society, Warrendale, PA, p. 19.
75. Blavette, D., Cadel, E., Fraczkiewicz, A., and Menand, A. (1999) *Science*, **286**, 2317.
76. Thompson, K., Flaitz, P.L., Ronsheim, P., Larson, D.J., and Kelly, T.F. (2007) *Science*, **317**, 1370.
77. Larson, D.J., Prosa, T.J., Geiser, B.P., Lawrence, D., Jones, C.M., and Kelly, T.F. Atom probe tomography for microelectronics, (2011) in *Handbook of Instrumentation and Techniques for Semiconductor Nanostructure Characterization*, (eds R. Haight, F. Ross, and J. Hannon), World Scientific Publishing/Imperial College Press, p. 407.
78. Peremola, E.V. et al. (2005) *Microsc. Microanal.*, **11** (Suppl 2), 876–877.
79. Krakauer, B.W. and Seidman, D.N. (1993) *Phys. Rev. B*, **48**, 6724.
80. Lai, Z.H., Norden, H., and Eaton, H.C. (1986) *J. Phys.*, **47–C7**, 269.
81. Hu, J.G., Kuo, S.M., Seki, A., Krakauer, B.W., and Seidman, D.N. (1989) *Scr. Metall.*, **23**, 2033.
82. Thuvander, M. and Andren, H.O. (2000) *Mater. Charact.*, **44**, 87.
83. Reynolds, W.T., Brenner, S.S., and Aaronson, H.I. Jr. (1988) *Scr. Metall.*, **22**, 1343.
84. Thuillier, O., Danoix, F., Gouné, M., and Blavette, D. (2006) *Scr. Mater.*, **55**, 1071–1074.
85. Stark, I., Smith, G.D.W., and Bhadeshia, H.K.D.H. (1990) *Metall. Trans. A.*, **21A**, 837.
86. Buchon, A., Menand, A., and Blavette, D. (1991) *Surf. Sci.*, **246**, 246.
87. Miller, M.K., Horton, J.A., Cao, W.D., and Kennedy, R.L. (1996) *J. Phys.*, **6-C5**, 241.

88. Blavette, D., Duval, P., Letellier, L., and Guttmann, M. (1996) *Acta Mater.*, **44**, 4995.
89. Blavette, D., Cadel, E., and Deconihout, B. (2000) *Mater. Charact.*, **44**, 133.
90. Thuillier, O. (2007) Transformation austénite-ferrite dans un alliage modéle Fe-C-Mn: modélisation et étude expérimentale à l'échelle nanométrique, PhD Thesis, GPM UMR CNRS 6634, University of Rouen.
91. Letellier, L. (1994 Investigation of grain boundary and interphases in nickel base superalloy Astroloy with electron microscopy and tomographic atom probe) PhD Thesis, GPM UMR CNRS 6634, University of Rouen.
92. Kim, Y.-C., Adusumilli, P., Lauhon, L.J., Seidman, D.N., Jung, S.-Y., Lee, H.-D., Alvis, R.L., Ulfig, R.M., and Olson, J.D. (2007) *Appl. Phys. Lett.*, **91**, 113106/1–113106/-3.
93. Cojocaru-Miredin, O., Mangelinck, D., Hoummada, K., Cadel, E., Blavette, D., Deconihout, B., and Perrin-Pellegrino, C. (2007) *Scr. Mater.*, 1–4.
94. Adusumilli, P., Lauhon, L.J., Seidman, D.N., Murray, C.E., Avayu, O., and Rosenwaks, Y. (2009) *Appl. Phys. Lett.*, **94**, 103113-1.
95. Mangelinck, D., Hoummada, K., Portavoce, A., Perrin, C., Daineche, R., Descoins, M., Larson, D.J., and Clifton, P.H. (2010) *Scr. Mater.*, **62**, 568.
96. Guinier, A. (1938) *Nature*, **142**, 569.
97. Preston, G.D. (1938) *Nature*, **142**, 570.
98. Hono, K., Satoh, T., and Hirano, K. (1986) *Philos. Mag.*, **53A**, 495.
99. Bigot, A., Danoix, F., Auger, P., Blavette, D., and Menand, A. (1996) *Appl. Surf. Sci.*, **94/95**, 261.
100. Ringer, S.P., Hono, K., Polmear, I.J., and Sakurai, T. (1996) *Acta Mater.*, **44**, 1883.
101. Majimel, J., Molenat, G., Danoix, F., Thuillier, O., Blavette, D., Lapasset, G., and Casanove, M.J. (2004) *Philos. Mag.*, **84**, 3263.
102. Edwards, G.A., Stiller, K., Dunlop, G.L., and Cooper, M.J. (1998) *Acta Mater.*, **46**, 3893.
103. Murayama, M. and Hono, K. (1998) *Acta Mater.*, **47**, 1537.
104. Hasting, H.K., Lefebvre, W., Marioara, C., Walmsley, J., Andersen, S., Holmestad, R., and Danoix, F. (2007) *Surf. Interface Anal.*, **39**, 189.
105. Ortner, S.R., Grovenor, C.R.M., and Shollock, B.A. (1988) *Scr. Metall.*, **22**, 843.
106. Brenner, S.S., Kowalik, J., and Hua, M.J. (1991) *Surf. Sci.*, **246**, 210.
107. Warren, P.J., Grovenor, C.R.M., and Cromton, J.S. (1992) *Surf. Sci.*, **266**, 342.
108. Schmuck, C., Auger, P., Danoix, F., and Blavette, D. (1995) *Appl. Surf. Sci.*, **87/88**, 228.
109. Bigot, A., Danoix, F., Auger, P., Blavette, D., and Reeves, A. (1996) *Mater. Sci. Forum*, **217**, 695.
110. Ringer, S.P. and Hono, K. (2000) *Mater. Charact.*, **44**, 101.
111. Miller, M.K., Beaven, P.A., Brenner, S.S., and Smith, G.D.W. (1983) *Metall. Trans. A*, **14A**, 1021.
112. Zhu, C., Cerezo, A., and Smith, G.D.W. (2009) *Ultramicroscopy*, **109**, 545.
113. Taylor, K.A., Chang, L., Olson, G.B., Smith, G.D.W., Cohen, M., and Vander Sande, J.B. (1989) *Metall. Trans.*, **20A**, 2717.
114. Babu, S.S., Hono, K., and Sakurai, T. (1993) *Metall. Trans.*, **25A**, 499.
115. Thomson, R.C. and Miller, M.K. (1998) *Acta Mater.*, **46**, 2203.
116. Thomson, R.C. and Miller, M.K. (1996) *Appl. Surf. Sci.*, **94-95**, 313.
117. Thomson, R.C. (2000) *Mater. Charact.*, **44**, 219.
118. Perrard, F., Donnadieu, P., Deschamps, A., and Barges, P. (2006) *Philos. Mag.*, **86**, 4271.
119. Karagoz, S. and Andren, H.O. (1992) *Z. Metall.*, **83**, 386.
120. Leitner, H., Stiller, K., Andren, H.-O., and Danoix, F. (2004) *Surf. Interface Anal.*, **36**, 540.
121. Liddle, J.A., Smith, G.D.W., and Olson, G.B. (1986) *J. Phys.*, **47-C7**, 211.
122. Olson, G.B., Kinkus, T.J., and Montgomery, J.S. (1991) *Surf. Sci.*, **246**, 238.
123. Danoix, F., Danoix, R., Akre, J., Grellier, A., and Delagnes, D. (2011) *J. Micros.*, **244**, 305–310.

124. Danoix F., Epicier T., Vurpillot F., and Blavette D. (2012) *J Mater. sci.*, **47**, 1567–1571.
125. Epicier T., Danoix F., Vurpillot F., and Blavette D. *Nature*, (submitted).
126. Jack, D.H. (1976) *Acta Metall.*, **24**, 137.
127. Brenner, S.S. and Goodman, S.R. (1971) *Scr. Metall.*, **5**, 865.
128. Driver, J.H. and Papazian, J.M. (1973) *Acta Metall.*, **21**, 1139.
129. Isheim, D., Siem, E.J., and Seidman, D.N. (2001) *Ultramicroscopy*, **89**, 195.
130. Jessner P., Danoix R., Hannoyer B., and Danoix .F. *Ultramicroscopy*, **109**, 530.
131. Jessner, P., Gouné, M., Danoix, R., Hannoyer, B., and Danoix, F. (2010) *Philos. Mag. Lett.*, **90**, 793.
132. Miller, M.K., Cerezo, A., Hetherington, M.G., and Smith, G.D.W. (1996) *Atom Probe Field Ion Microscopy*, Chapter 6, Clarendon Press, Oxford.
133. Pareige, C., Soisson, F., Martin, G., and Blavette, D. (1999) *Acta Mater.*, **47**, 1889–1899.
134. Mao, Z., Sudbrack, C.K., Yoon, K.E., Martin, G., and Seidman, D.N. (2007) *Nat. Mater.*, **6**, 210.
135. Cahn, J.W. (1968) *Trans. AIME*, **242**, 166.
136. Piller, J., Wagner, W., Wollenberger, H., and Mertens, P. (1984) in Acta-Scripta Metallurgica Conference Series, Vol. 2 (eds P. Haasen, V. Gerold, R. Wagner and M.F.Ashby), Pergamon Press, p. 156.
137. Abe, T. (1992) *Acta Metall.*, **40**, 1951.

24
Signal and Noise Maximum Likelihood Estimation in MRI
Jan Sijbers

24.1
Probability Density Functions in MRI

Maximum likelihood (ML) estimation requires the knowledge of the underlying data distribution. This section, therefore, describes various data probability density functions (PDFs) that appear when dealing with magnetic resonance (MR) data.

24.1.1
Gaussian PDF

Raw MR data acquired in *K*-space during an MR acquisition scheme are known to be complex valued. The complex data are composed of noiseless signal components and noise contributions that are assumed to be additive and independent, and characterized by a zero mean Gaussian PDF [1–3]. An MR reconstruction is then obtained by means of an inverse Fourier transform (FT). Owing to the linearity and orthogonality of the FT, the complex data resulting from the transformation are still independent and Gaussian distributed[1] [4–6]. Hence, the PDF of a raw, complex data point $\underline{c} = (\underline{w}_r, \underline{w}_i)$ is given by Sijbers and Van der Linden [7]:

$$p_{\underline{c}}(\omega_r, \omega_i | A, \varphi, \sigma) = \frac{1}{2\pi\sigma^2} e^{-\frac{(\omega_r - A\cos\varphi)^2}{2\sigma^2}} e^{-\frac{(\omega_i - A\sin\varphi)^2}{2\sigma^2}} \qquad (24.1)$$

where σ^2 denotes the noise variance and (ω_r, ω_i) are the real and imaginary variables, respectively, corresponding to the complex observation $(\underline{w}_r, \underline{w}_i)$, with underlying true amplitude and phase value, A and φ, respectively. In Eq. (24.1) and in what follows, stochastic (i.e., random) variables are underlined. In general, a Gaussian PDF is described by

---

1) It is assumed that the MR signals are sampled on a uniform grid in K-space. Furthermore, the variance of the noise is assumed to be equal for each raw data point.

Handbook of Nanoscopy, First Edition. Edited by Gustaaf Van Tendeloo, Dirk Van Dyck, and Stephen J. Pennycook.
© 2012 Wiley-VCH Verlag GmbH & Co. KGaA. Published 2012 by Wiley-VCH Verlag GmbH & Co. KGaA.

$$p_{\underline{x}}(x|\mu,\sigma) = \frac{1}{\sqrt{2\pi\sigma^2}} e^{-\frac{(x-\mu)^2}{2\sigma^2}} \qquad (24.2)$$

with μ and σ denoting the mean and standard deviation of the PDF, respectively.

24.1.1.1 Moments of the Gaussian PDF

Analytical expressions for the moments of a Gaussian PDF are given by Gradshteyn and Ryzhik [8]

$$\mathbb{E}[\underline{x}^\nu] = \sigma^{2(\nu-1)} \left[\frac{d^{\nu-1}}{d\mu^{\nu-1}} \left(\mu e^{\frac{\mu^2}{2\sigma^2}} \right) \right] e^{-\frac{\mu^2}{2\sigma^2}} \qquad (24.3)$$

where $\mathbb{E}[.]$ is the expectation operator and $\nu \in \mathbb{N}_0$. For the central moments, we have

$$\mathbb{E}[(\underline{x}-\mu)^\nu] = \begin{cases} 0 & \text{if } \nu \text{ is odd} \\ \frac{\nu! \sigma^\nu}{(\nu/2)! 2^{\nu/2}} & \text{if } \nu \text{ is even} \end{cases} \qquad (24.4)$$

24.1.2 Rician PDF

During MR data processing, it is common practice to work with magnitude data instead of real and imaginary data because magnitude data have the advantage of being immune to the effects of incidental phase variations due to radio frequency (RF) angle inhomogeneity, system delay, noncentered sampling windows, and so on. In this section, the PDF of the magnitude data is discussed.

To construct a magnitude image from the complex data, the magnitude is computed on a pixel by-pixel-base:

$$m = \sqrt{\omega_r^2 + \omega_i^2} \qquad (24.5)$$

where m is the magnitude variable corresponding to the magnitude observation $\underline{m}$. As root extraction is a nonlinear transformation, the PDF of the magnitude data is no longer expected to be Gaussian [9], but is called a *Rician distribution* after Rice [9] who derived it in the context of the communication theory in 1944:

$$p_{\underline{m}}(m|A,\sigma) = \frac{m}{\sigma^2} e^{-\frac{m^2+A^2}{2\sigma^2}} I_0\left(\frac{Am}{\sigma^2}\right) \epsilon(m) \qquad (24.6)$$

with I_0 denoting the 0th order modified Bessel function of the first kind. The unit step Heaviside function $\epsilon(.)$ is used to indicate that the expression for the PDF of $\underline{m}$ is valid for nonnegative values of m only. Note that the shape of the Rician PDF depends on the signal-to-noise ratio (SNR), which is here defined as the ratio A/σ.

- For low SNR, the modified Bessel function is given by $I_{\nu(z)} \sim (z/2)^\nu \Gamma(\nu+1)$ such that the Rician PDF leads to a *Rayleigh distribution*:

$$p_{\underline{m}}(m|\sigma) = \frac{m}{\sigma^2} e^{-\frac{m^2}{2\sigma^2}} \epsilon(m) \quad \text{for} \quad \text{SNR} \to 0 \qquad (24.7)$$

The Rayleigh PDF characterizes the random intensity distribution of nonsignal background areas such as air.

- At high SNR, the asymptotic approximation of the modified Bessel function is given by $I_{v(z)} \sim e^z/(\sqrt{2\pi z})$. Then, given that $\sqrt{m/A} \to 1$ for SNR $\to \infty$, we find for the Rician PDF

$$p_{\underline{m}}(m|A,\sigma) = \frac{1}{\sqrt{2\pi\sigma^2}} e^{-\frac{(m-A)^2}{2\sigma^2}} \epsilon(m) \quad \text{for} \quad \text{SNR} \to \infty \tag{24.8}$$

Hence, for high SNR, the Rician PDF approaches a Gaussian PDF with mean A and variance σ^2.

Finally, for the PDF of magnitude data in case the Gaussian variables from which it is computed are correlated, we refer to the work of Dharmawansa et al. [10].

24.1.2.1 Moments of the Rician PDF

The moments of the Rician PDF can be analytically expressed as a function of the confluent hypergeometric function of the first kind $_1F_1$ [11]:

$$\mathbb{E}[\underline{m}^v] = (2\sigma^2)^{v/2} \Gamma\left(1 + \frac{v}{2}\right) {}_1F_1\left[-\frac{v}{2}; 1; -\frac{A^2}{2\sigma^2}\right] \tag{24.9}$$

with Γ representing the Gamma function. The even moments of the Rician distribution (i.e., when v is even) are simple polynomials. The expressions for the odd moments are much more complex. However, the confluent hypergeometric function can be expressed in terms of modified Bessel functions, from which an analytic expression of the odd moments can be derived.

The general expression for the *moments of the Rayleigh PDF* to which the Rician PDF tends at low SNR are given by

$$\mathbb{E}[\underline{m}^v] = (2\sigma^2)^{v/2} \Gamma\left(1 + \frac{v}{2}\right) \tag{24.10}$$

24.1.3
Generalized Rician PDF

If magnitude data are computed from more than two Gaussian-distributed variables, the underlying PDF is the *generalized Rician PDF*. Such data are found in phased array magnitude MR images, where use is made of multiple receiver coils [12], and in phase contrast magnetic resonance (PCMR) images [13, 14]. In general, a (random) PCMR pixel variable, denoted by $\underline{m}$, can be written as

$$\underline{m} = \sqrt{\sum_{k=1}^{K} \underline{s}_k^2} \tag{24.11}$$

with K denoting twice the number of orthogonal Cartesian directions in which flow is encoded. The set $\{\underline{s}_k\}$ contains independent Gaussian-distributed variables with means $\{a_k\}$ and variance σ^2. The deterministic signal component of the PCMR pixel variable is given by

$$A = \sqrt{\sum_{k=1}^{K} a_k^2} \qquad (24.12)$$

The PDF of such a PCMR variable is given by

$$p_{\underline{m}}(m|A,\sigma) = \frac{m}{\sigma^2}\left(\frac{m}{A}\right)^{\frac{K}{2}-1} \exp\left(-\frac{m^2+A^2}{2\sigma^2}\right) I_{\frac{K}{2}-1}\left(\frac{mA}{\sigma^2}\right)\epsilon(m) \qquad (24.13)$$

When $A \to 0$, the PDF of the magnitude PCMR variable turns into a *generalized Rayleigh PDF*:

$$p_{\underline{m}}(m|\sigma) = \frac{2m^{K-1}}{(\sigma\sqrt{2})^K \Gamma(K/2)} \exp\left(-\frac{m^2}{2\sigma^2}\right)\epsilon(m) \qquad (24.14)$$

24.1.3.1 Moments of the Generalized Rician PDF

The general expression for the moments of the generalized Rician PDF is written as

$$\mathbb{E}[\underline{m}^\nu] = (2\sigma^2)^{\nu/2}\frac{\Gamma[(K+\nu)/2]}{\Gamma(K/2)}\,{}_1F_1\left(-\frac{\nu}{2},\frac{K}{2};-\frac{A^2}{2\sigma^2}\right) \qquad (24.15)$$

For $SNR = 0$, we obtain the moments of the generalized Rayleigh PDF:

$$\mathbb{E}[\underline{m}^\nu] = (2\sigma^2)^{\nu/2}\frac{\Gamma[(K+\nu)/2]}{\Gamma(K/2)} \qquad (24.16)$$

24.1.4
PDF of Phase Data

Phase data, which are commonly obtained during flow imaging, are constructed from the real and imaginary observations $\{(\underline{w}_{r,n}, \underline{w}_{i,n})\}$ by calculating for each complex data point the arctangent of their ratio for each complex data point:

$$\phi_n = \arctan\left(\frac{\omega_{i,n}}{\omega_{r,n}}\right) \qquad (24.17)$$

The PDF of the phase deviation $\underline{\Delta\phi}$ from the true phase value is given by Gudbjartsson and Patz [15]:

$$p_{\underline{\Delta\phi}}(\Delta\phi) = \frac{1}{2\pi}e^{-A^2/2\sigma^2}\left[1 + \frac{A}{\sigma}\cos\Delta\phi\, e^{A^2\cos^2\Delta\phi/2\sigma^2}\int_{-\infty}^{\frac{A\cos\Delta\phi}{\sigma}} e^{-x^2/2}dx\right] \qquad (24.18)$$

Note that the distribution can be expressed solely in terms of the SNR, defined as A/σ. Although the general expression for the distribution of $\underline{\Delta\phi}$ is complicated, the two limits of the SNR turn out to yield simple distributions.

- In regions where there is only noise, (i.e., where the SNR is zero), Eq. (24.18) reduces to a uniform PDF:

$$p_{\underline{\Delta\phi}}(\Delta\phi) = \begin{cases} \frac{1}{2\pi} & \text{if } -\pi < \Delta\phi < \pi \\ 0 & \text{otherwise} \end{cases} \quad (24.19)$$

Stated in another way, complex data that only consist of noise "point in all directions" with the same probability.

- For high SNR, it is easy to see that the probability of observing large values for the phase deviation will be small. In that case, Eq. (24.18) reduces to

$$p_{\underline{\Delta\phi}}(\Delta\phi) = \frac{1}{\sqrt{2\pi}} \frac{A}{\sigma} \exp\left(-\frac{\Delta\phi^2 A^2}{2\sigma^2}\right) \quad (24.20)$$

Thus, the phase noise $\underline{\Delta\phi}$ is governed by a Gaussian distribution when SNR $\to \infty$.

The standard deviations of the phase noise can, in general, be calculated from Eq. (24.18). However, for the SNR limits, given in Eqs. (24.19) and (24.20), the standard deviation is given by

$$\sigma_{\underline{\Delta\phi}} = \begin{cases} \pi/\sqrt{3} & \text{if SNR} = 0 \\ \sigma/A & \text{if SNR} \gg 1 \end{cases} \quad (24.21)$$

Finally, for the PDF of the phase random variable in case the Gaussian variables from which it is computed are correlated, we refer to the work of Dharmawansa et al. [10].

24.2 Signal Amplitude Estimation

24.2.1 Introduction

In this section, the problem of signal amplitude estimation from MR data is addressed. In particular, we focus on the estimation of the magnitude magnetization values from data acquired during a magnetic resonance imaging (MRI) procedure.

Raw MR data, directly obtained from an MR scanner, represent the FT of a magnetization distribution of a volume at a certain point in time. Such data are generally complex valued and corrupted by zero-mean Gaussian-distributed noise [16]. An inverse FT, which yields the complex magnetization distribution of the object under study, does not change the type of the data PDF because of the linearity and orthogonality of the FT. Hence, these data are as well corrupted by zero-mean Gaussian-distributed noise. The complex data, however, are generally not retained, but transformed into magnitude and phase data. This is because the amplitudes and phase values of the magnetizations are of greater interest than the real and imaginary components of the complex data values. Transformation to a magnitude or a phase MR image, however, is nonlinear. As a consequence of this transformation, the PDF of the data changes to a Rician PDF. Using ML estimation, the Rice distribution can be exploited when estimating signal parameters from magnitude MR data.

Furthermore, consider a data processing application (e.g., noise filtering) that requires estimation of the underlying signal amplitude from a number of noise-corrupted, complex-valued data points. Therefore, for each complex data point, belonging to the data set under concern, the underlying signal amplitude is assumed to be the same. This amplitude can be estimated either by first transforming the complex data points into a set of magnitude data points and afterwards estimating the signal amplitude from the thus obtained data set, or, by directly estimating the signal amplitude from the original complex-valued data points. Hence, questions may rise such as "Should we use the complex data set or the magnitude data set when estimating the unknown signal amplitude?" and "Does it matter whether the true phase values of the complex data points, from which the signal amplitude is estimated, are the same?". These questions have already been addressed in [17] but are summarized in this section as well.

24.2.2
Signal Amplitude Estimation from Complex Data

We start by considering complex, Gaussian-distributed data. The CRLB for unbiased estimation of the underlying amplitude signal as well as the ML estimator of this signal will be derived. This will be done for data with identical underlying phase values, as well as for data with different underlying phase values.

24.2.2.1 Region of Constant Amplitude and Phase

Consider a set of N-independent, Gaussian-distributed, complex data points $\underline{c} = \{(\underline{w}_{r,n}, \underline{w}_{i,n})\}$ with underlying true amplitude and phase values, A and φ, respectively. The CRLB for unbiased estimation of (A, φ) is then given by van den Bos [18]

$$\text{CRLB} = \begin{pmatrix} \frac{\sigma^2}{N} & 0 \\ 0 & \frac{\sigma^2}{NA^2} \end{pmatrix} \tag{24.22}$$

The likelihood function L is obtained by substituting the available observations $\{(w_{r,n}, w_{i,n})\}$ for $\{(\omega_{r,n}, \omega_{i,n})\}$ into the joint PDF:

$$L(A, \varphi | \{(w_{r,n}, w_{i,n})\}) = \left(\frac{1}{2\pi\sigma^2}\right)^N \prod_{n=1}^{N} e^{-\frac{(w_{r,n}-A\cos\varphi)^2}{2\sigma^2}} e^{-\frac{(w_{i,n}-A\sin\varphi)^2}{2\sigma^2}} \tag{24.23}$$

Then, the ML estimates of (A, φ) are found by maximizing this function with respect to A and φ:

$$\widehat{A}_{ML} = \frac{1}{N}\sqrt{\left(\sum_{n=1}^{N} w_{r,n}\right)^2 + \left(\sum_{n=1}^{N} w_{i,n}\right)^2} \tag{24.24}$$

$$\widehat{\varphi}_{ML} = \arctan\left(\frac{\sum_{n=1}^{N} w_{i,n}}{\sum_{n=1}^{N} w_{r,n}}\right) \tag{24.25}$$

Notice that the estimator $\widehat{A}_{ML}$ is obtained by taking the square root of the quadratic sum of two Gaussian-distributed variables. Hence, $\widehat{A}_{ML}$ is Rician distributed [9]. Its mean squared error (MSE) is hence given by

$$\text{MSE} = 2A\left(A - \mathbb{E}\left[\widehat{A}_{ML}\right]\right) + 2\frac{\sigma^2}{N} \tag{24.26}$$

with

$$\mathbb{E}\left[\widehat{A}_{ML}\right] = \sigma\sqrt{\frac{\pi}{2N}} e^{-\frac{NA^2}{4\sigma^2}} \left[\left(1 + \frac{NA^2}{2\sigma^2}\right) I_0\left(\frac{NA^2}{4\sigma^2}\right) + \frac{NA^2}{2\sigma^2} I_1\left(\frac{NA^2}{4\sigma^2}\right)\right] \tag{24.27}$$

24.2.2.2 Region of Constant Amplitude and Different Phases

Now assume that the complex data $\underline{c} = \{(w_{r,n}, w_{i,n})\}$ have an underlying signal amplitude A and arbitrary phase values $\varphi_1, \ldots, \varphi_N$. Then, the joint PDF of the complex data, $p_{\underline{c}}$, is given by

$$p_{\underline{c}}(\{(w_{r,n}, w_{i,n})\}|A, \{\varphi_n\}) = \left(\frac{1}{2\pi\sigma^2}\right)^N \prod_{n=1}^{N} e^{-\frac{(w_{r,n} - A\cos\varphi_n)^2}{2\sigma^2}} e^{-\frac{(w_{i,n} - A\sin\varphi_n)^2}{2\sigma^2}} \tag{24.28}$$

Then, the CRLB for unbiased estimation of $(A, \varphi_1, \ldots, \varphi_N)$ is given by

$$\text{CRLB} = \begin{pmatrix} \frac{\sigma^2}{N} & 0 & \cdots & 0 \\ 0 & \frac{\sigma^2}{A^2} & \cdots & 0 \\ \vdots & \vdots & \ddots & \vdots \\ 0 & 0 & \cdots & \frac{\sigma^2}{A^2} \end{pmatrix} \tag{24.29}$$

From the likelihood function for N statistically independent, Gaussian-distributed complex observations $\underline{c} = \{(w_{r,n}, w_{i,n})\}$ with underlying noiseless signal amplitude A and arbitrary phase values $\varphi_1, \ldots, \varphi_N$ the ML estimators of A and φ_n can be derived as

$$\widehat{A}_{ML} = \frac{1}{N} \sum_{n=1}^{N} \sqrt{w_{r,n}^2 + w_{i,n}^2}, \tag{24.30}$$

$$\widehat{\varphi}_{n,ML} = \arctan\left(\frac{w_{i,n}}{w_{r,n}}\right) \tag{24.31}$$

The ML estimator of the signal amplitude, given by Eq. (24.30), is distributed as the average of N-independent, Rician-distributed variables. Therefore, its mean value is simply given by the average of the mean values of the individual Rician-distributed variables, whereas its variance is given by the sum of their variances, divided by N^2. Hence, we have for the MSE:

$$\text{MSE}\left(\widehat{\underline{A}}_{\text{ML}}\right) = \left(A - \mathbb{E}\left[\widehat{\underline{A}}_{\text{ML}}\right]\right)^2 + \left(A^2 + 2\sigma^2 - \mathbb{E}\left[\widehat{\underline{A}}_{\text{ML}}\right]^2\right)/N, \quad (24.32)$$

where $\mathbb{E}[\widehat{\underline{A}}_{\text{ML}}]$ is now given by

$$\mathbb{E}\left[\widehat{\underline{A}}_{\text{ML}}\right] = \sigma\sqrt{\frac{\pi}{2}} e^{-\frac{A^2}{4\sigma^2}} \left[\left(1 + \frac{A^2}{2\sigma^2}\right) I_0\left(\frac{A^2}{4\sigma^2}\right) + \frac{A^2}{2\sigma^2} I_1\left(\frac{A^2}{4\sigma^2}\right)\right] \quad (24.33)$$

Note that Eq. (24.33) does not depend on N. Furthermore, Eq. (24.33) is identical to Eq. (24.27) in case $N = 1$.

24.2.3
Signal Amplitude Estimation from Magnitude Data

Raw, complex-valued, and Gaussian-distributed MR data are commonly transformed into magnitude MR data, as physiological and anatomical information are more closely related to the length of the magnetization vectors. However, as noted earlier, computing the magnitude results in a change in the underlying data PDF, which has to be accounted for when extracting quantitative information. In this section, we describe the estimation of the underlying, noiseless amplitude signal A from a region where A is assumed to be constant. The region now consists of N-independent, Rician-distributed data points $\underline{m} = (\underline{m}_1, \ldots, \underline{m}_N)$.

Unlike estimation of the signal amplitude from *complex* data, estimation of the signal amplitude from *magnitude* data requires either prior knowledge of the noise variance or simultaneous estimation of signal amplitude and noise variance.

- The noise variance may be estimated separately if a background region is available, that is, a region in which the underlying signal is zero (Section 24.3.3.1). Moreover, if large background areas are available, which is often the case, much more data points are available for the estimation of the noise variance than for the estimation of the signal amplitude. Then, the noise variance can be estimated with much higher precision. Hence, it might be a valid assumption to regard the noise variance as known (i.e., to regard the estimated noise variance as the true noise variance).
- If the noise variance cannot be estimated separately (with sufficient precision), it acts as a nuisance parameter that needs to be estimated simultaneously with the signal amplitude.

Both cases are discussed in this section.

24.2.3.1 Region of Constant Amplitude and Known Noise Variance
The CRLB for unbiased estimation of (A, σ) is given by Karlsen et al. [19]

$$\text{CRLB} = \frac{\sigma^2}{N}\left(Z - \frac{A^2}{\sigma^2}\right)^{-1} \quad (24.34)$$

with

$$Z = \mathbb{E}\left[\frac{\underline{m}^2}{\sigma^2} \frac{I_1^2\left(\frac{A\underline{m}}{\sigma^2}\right)}{I_0^2\left(\frac{A\underline{m}}{\sigma^2}\right)}\right] \qquad (24.35)$$

and $\underline{m}$ a Rician-distributed random variable with true parameters (A, σ). The expectation value in Eq. (24.35) can be evaluated numerically.

For the estimation of the signal A, commonly, $\mathbb{E}\left[\underline{m}^2\right]$ is exploited. Therefore, $\mathbb{E}\left[\underline{m}^2\right]$ is estimated from a simple spatial average of the squared pixel values of a constant assumed region [20–23]:

$$\widehat{\mathbb{E}[\underline{m}^2]} = \langle \underline{m}^2 \rangle = \frac{1}{N}\sum_{n=1}^{N} \underline{m}_n^2 \qquad (24.36)$$

Note that this estimator is unbiased since $\mathbb{E}[\langle \underline{m}^2 \rangle] = \mathbb{E}[\underline{m}^2] = A^2 + 2\sigma^2$. If the noise variance σ^2 is assumed to be known, an unbiased estimator of A^2 is given by

$$\widehat{\underline{A}_c^2} = \langle \underline{m}^2 \rangle - 2\sigma^2 \qquad (24.37)$$

Taking the square root of Eq. (24.37) gives the conventional estimator of A [20–23]:

$$\widehat{\underline{A}_c} = \sqrt{\langle \underline{m}^2 \rangle - 2\sigma^2} \qquad (24.38)$$

Obviously, A is a priori known to be real valued and nonnegative. However, the conventional estimator $\widehat{\underline{A}_c}$, given in Eq. (24.38), may reveal estimates that violate this a priori knowledge and are therefore physically meaningless. This is the case when $\widehat{\underline{A}_c^2}$ becomes negative. Therefore, $\widehat{\underline{A}_c}$ cannot be considered a useful estimator of A if the probability that $\widehat{\underline{A}_c^2}$ is negative differs from zero significantly. In practice, $\widehat{\underline{A}_c}$ will only be a useful estimator if the SNR is high. However, even if the condition of high SNR is met, the use of $\widehat{\underline{A}_c}$ as an estimator of A should still not be recommended, since the results obtained are biased because of the square root operation in Eq. (24.38).

In what follows, we will consider the ML estimator of A from a set N Rician-distributed magnitude data points $\underline{m} = (\underline{m}_1, \ldots, \underline{m}_N)$.

The joint PDF $p_{\underline{m}}$ is given by

$$p_{\underline{m}} = \prod_{n=1}^{N} \frac{m_n}{\sigma^2} e^{-\frac{m_n^2 + A^2}{2\sigma^2}} I_0\left(\frac{Am_n}{\sigma^2}\right) \epsilon\,(m_n) \qquad (24.39)$$

where $\{m_n\}$ are the magnitude variables corresponding to the magnitude observations $\{\underline{m}_n\}$. The ML estimate of A is constructed by substituting the available observations $\{\underline{m}_n\}$ in the expression for the joint PDF (Eq. (24.39)) and maximizing the resulting function $L(A)$, or equivalently $\ln L(A)$, with respect to A. Hence, it follows that

$$\ln L = \sum_{n=1}^{N} \ln\left(\frac{m_n}{\sigma^2}\right) - \sum_{n=1}^{N} \frac{m_n^2 + A^2}{2\sigma^2} + \sum_{n=1}^{N} \ln I_0\left(\frac{Am_n}{\sigma^2}\right) \qquad (24.40)$$

The ML estimate is then found from the global maximum of $\ln L$:

$$\widehat{A}_{ML} = \arg\left\{\max_{A} (\ln L)\right\} \quad (24.41)$$

Notice that Eq. (24.41) cannot be solved analytically. Finding the maximum of the (log-)likelihood is therefore a numerical optimization problem.

Discussion It is not possible to find the maximum of the $\ln L$ function directly because the parameter A enters that function in a nonlinear way. Therefore, finding the maximum of the $\ln L$ function will, in general, be an iterative numerical process. It can be shown that $A = 0$ is a stationary point of $\ln L$, independent of the particular data set and that $A = 0$ is a minimum of $\ln L$ whenever

$$\frac{1}{N}\sum_{n=1}^{N} m_n^2 > 2\sigma^2 \quad (24.42)$$

If this condition is met, the $\ln L$ function will have two further stationary points, being maxima. Notice that condition (24.42) is always met for noise-free data. However, in practice, the data will be corrupted by noise, and for particular realizations of the noise, condition (24.42) may not be met. Then $A = 0$ will be a maximum [24].

24.2.3.2 Region of Constant Amplitude and Unknown Noise Variance

If the noise variance σ^2 is unknown, the signal amplitude and the noise variance have to be estimated simultaneously (i.e., the noise variance is a nuisance parameter). The CRLB for unbiased estimation of (A, σ^2) is given by

$$\text{CRLB} = \frac{1}{\det I}\begin{pmatrix} I(2,2) & -I(2,1) \\ -I(1,2) & I(1,1) \end{pmatrix} \quad (24.43)$$

where $I(i,j)$ denotes the (i,j)th element of the matrix I

$$I(1,1) = \frac{N}{\sigma^2}\left(Z - \frac{A^2}{\sigma^2}\right) \quad (24.44)$$

$$I(1,2) = I(2,1) = \frac{NA}{\sigma^4}\left(1 + \frac{A^2}{\sigma^2} - Z\right) \quad (24.45)$$

$$I(2,2) = \frac{N}{\sigma^4}\left(1 + \frac{A^2}{\sigma^2}(Z-1) - \frac{A^4}{\sigma^4}\right) \quad (24.46)$$

and Z is given by Eq. (24.35).

If a background region is not available for noise variance estimation, the signal A and variance σ^2 have to be estimated simultaneously from the N available data points by maximizing the log-likelihood function with respect to A and σ^2:

$$\{\widehat{A}_{ML}, \widehat{\sigma}^2_{ML}\} = \arg\left\{\max_{A,\sigma^2}(\ln L)\right\} \quad (24.47)$$

where $\ln L$ is given by Eq. (24.40).

24.2.4
Discussion

We first discuss the expressions for the CRLB for unbiased estimation of the signal amplitude from complex and magnitude data separately.

Complex Data. Analytical expressions for the CRLBs for unbiased estimation of A from complex data with identical and different phase values were derived in Section 24.2.2. They are given by Eqs. (24.22) and (24.29), respectively. Note that both CRLB do not depend on the phase values.

Magnitude Data. In contrast to the CRLBs for complex data, no analytical expressions for the CRLB for unbiased estimation of the signal amplitude from *magnitude* MR data with known and unknown noise variance can be derived. However, the lower bounds given by Eqs. (24.34) and (24.43) can be evaluated numerically. Therefore, the expectation values can be evaluated from Monte Carlo simulations.

Note also that all lower bounds are inversely proportional to the number of data points. It follows that

- for low SNR, the CRLB for unbiased estimation of A
  - from complex data is significantly smaller than for estimation from magnitude data.
  - from magnitude data with known noise variance is significantly larger than for estimation from magnitude data with unknown noise variance (i.e., in which the noise variance is a nuisance parameter).

  Recall that knowledge of the noise variance is not required when estimating the signal amplitude from complex data.
- for increasing SNR, the CRLBs for unbiased estimation from magnitude data tend to the CRLB for unbiased estimation from complex data, which equals σ^2/N.

In terms of the MSE, experiments have shown that [17]

- $\widehat{\underline{A}}_{ML}$ for complex data with identical phase values performs best, independent of the SNR.
- $\widehat{\underline{A}}_{ML}$ for magnitude data with known noise variance is significantly better compared to $\widehat{\underline{A}}_{ML}$ for magnitude data with unknown noise variance and $\widehat{\underline{A}}_{ML}$ for complex data with different phase values.

Also, for increasing SNR, the performance difference in terms of the MSE between the MLEs of A based on complex and magnitude data tend to zero. As in practice, the assumption of identical phases for complex data is generally invalid, it may be concluded that the signal amplitude is preferentially estimated from magnitude MR data for which the noise variance is known. The latter requisite is not too restrictive as often, in practice, the noise variance can be estimated with a much higher precision than the signal amplitude.

24.3
Noise Variance Estimation

24.3.1
Introduction

Many image processing methods require the knowledge of the image noise variance. For example, a denoising step is often required before any relevant information can be extracted from an image and, in turn, this denoising step relies on the noise variance [25]. Image processing methods generally assume the data to be Gaussian distributed [26–29]. As magnitude MR data are Rician distributed, these images require noise estimation methods that exploit this knowledge [30].

In this section, several noise estimation methods are discussed, which can be classified as follows:

> **Single Acquisition Methods.** In MRI, the image noise variance is commonly estimated from a large uniform signal region or nonsignal regions within a single magnitude MR image [20, 31]. In this section, we consider ML estimation of the noise variance from complex (Section 24.3.2) as well as magnitude MR data (Section 24.3.3) from a region in which the underlying signal amplitude is nonzero but constant as well as from a background region, that is, a region in which the underlying signal amplitude is zero [32].
> **Double Acquisition Methods.** Furthermore, methods were developed based on two acquisitions of the same image: the so-called double acquisition methods [21, 33, 34]. At the end of Section 24.3.3, we describe in short a robust method based on a double acquisition scheme [15, 35].

24.3.2
Noise Variance Estimation from Complex Data

Suppose the noise variance σ^2 needs to be estimated from N complex-valued observations $\underline{c} = \{(\underline{w}_{r,n}, \underline{w}_{i,n})\}$. We will consider the case of identical underlying phase values, as well as the case of different underlying phase values.

24.3.2.1 Region of Constant Amplitude and Phase
Let us first consider a region with a constant, nonzero underlying signal amplitude and identical underlying phase values.

The CRLB for unbiased estimation of (A, φ, σ^2) is given by

$$\text{CRLB} = \begin{pmatrix} \frac{\sigma^2}{N} & 0 & 0 \\ 0 & \frac{\sigma^2}{NA^2} & 0 \\ 0 & 0 & \frac{\sigma^4}{N} \end{pmatrix} \qquad (24.48)$$

For identical true phase values, the ML estimator of σ^2 can be shown to be given by

$$\widehat{\underline{\sigma}^2}_{ML} = \frac{1}{2N} \sum_{n=1}^{N} \left[\left(\widehat{\underline{A}}_{ML} \cos \widehat{\underline{\varphi}}_{ML} - \underline{w}_{r,n} \right)^2 + \left(\widehat{\underline{A}}_{ML} \sin \widehat{\underline{\varphi}}_{ML} - \underline{w}_{i,n} \right)^2 \right] \quad (24.49)$$

with $\widehat{\underline{A}}_{ML}$ and $\widehat{\underline{\varphi}}_{ML}$ given by Eqs. (24.24) and (24.25), respectively. The MSE of $\widehat{\underline{\sigma}^2}_{ML}$ can be shown to be given by

$$\text{MSE}\left(\widehat{\underline{\sigma}^2}_{ML}\right) \simeq \frac{\sigma^4}{N} \quad (24.50)$$

24.3.2.2 Region of Constant Amplitude and Different Phases

Next, consider a region with a constant nonzero underlying signal amplitude and different underlying phase values.

The CRLB for unbiased estimation of $(A, \varphi_1, \ldots, \varphi_N, \sigma^2)$ is then given by

$$\text{CRLB} = \begin{pmatrix} \frac{\sigma^2}{N} & 0 & \cdots & 0 & 0 \\ 0 & \frac{\sigma^2}{A^2} & \cdots & 0 & 0 \\ \vdots & \vdots & \ddots & \vdots & \vdots \\ 0 & 0 & \cdots & \frac{\sigma^2}{A^2} & 0 \\ 0 & 0 & \cdots & 0 & \frac{\sigma^4}{N} \end{pmatrix} \quad (24.51)$$

The CRLB for unbiased estimation of the noise variance from *complex* data with identical and different phase values is given by Eqs. (24.48) and (24.51), respectively. Note that in both cases, and independent of the signal amplitude of the data points, the CRLB is equal to σ^4/N.

The ML estimator of σ^2 in case of different phase values is

$$\widehat{\underline{\sigma}^2}_{ML} = \frac{1}{2N} \sum_{n=1}^{N} \left[\left(\widehat{\underline{A}}_{ML} \cos \widehat{\underline{\varphi}}_{n,ML} - \underline{w}_{r,n} \right)^2 + \left(\widehat{\underline{A}}_{ML} \sin \widehat{\underline{\varphi}}_{n,ML} - \underline{w}_{i,n} \right)^2 \right] \quad (24.52)$$

with $\widehat{\underline{A}}_{ML}$ and $\widehat{\underline{\varphi}}_{n,ML}$ given by Eqs. (24.30) and (24.31), respectively.

It can be shown that for large N, the MSE of $\widehat{\underline{\sigma}^2}_{ML}$ is given by

$$\text{MSE}\left(\widehat{\underline{\sigma}^2}_{ML}\right) \simeq \frac{\sigma^4}{4} \left(1 + \frac{4}{N} - \frac{1}{N^2} \right) \quad (24.53)$$

24.3.2.3 Background Region

Next, consider the case in which the noise variance is estimated from a background region (i.e., a region where A is known to be zero). It can easily be shown that the CRLB for unbiased estimation of σ^2 is given by

$$\text{CRLB} = \frac{\sigma^4}{N} \quad (24.54)$$

independent of the underlying phase values.

The ML estimator is given by

$$\widehat{\underline{\sigma}^2}_{ML} = \frac{1}{2N} \sum_{n=1}^{N} \left(\underline{w}_{r,n}^2 + \underline{w}_{i,n}^2 \right) \quad (24.55)$$

independent of the underlying phase values. Furthermore, it can easily be shown that the ML estimator (24.55) is unbiased and that its variance equals the CRLB. Therefore, its MSE is simply given by

$$\text{MSE}\left(\widehat{\sigma^2}_{\text{ML}}\right) = \frac{\sigma^4}{N} \tag{24.56}$$

24.3.3
Noise Variance Estimation from Magnitude Data

We will now describe ML estimation of the noise variance (and standard deviation) from magnitude MR data. First, we will consider ML estimation of the noise variance from a so-called background region, that is, a region in which the underlying signal is zero. The CRLB for unbiased estimation of both noise variance and standard deviation will be computed and the ML estimators will be derived. Next, we will consider ML estimation of the noise parameters from a so-called constant region, that is, a region in which the (nonzero) signal amplitude is assumed to be constant. In this case, the noise parameters have to be estimated simultaneously with the signal amplitude.

It will be assumed that the available data is governed by a generalized Rician distribution. The methods described below can also be applied to conventional Rician-distributed magnitude MR data; that would be the case when $K = 2$.

24.3.3.1 Background Region

Suppose that a set of N statistically independent magnitude data points $\underline{m} = \{\underline{m}_n\}$ is available from a region where the true signal value A is zero for each data point (background region). Hence, these data points are governed by a Rayleigh distribution and their joint PDF, $p_{\underline{m}}$, is given by (cf. Eq. (24.14))

$$p_{\underline{m}}(\{m_n\}) = \prod_{n=1}^{N} \frac{2m_n^{K-1}}{(2\sigma^2)^{K/2}\Gamma(K/2)} \exp\left(-\frac{m_n^2}{2\sigma^2}\right) \tag{24.57}$$

where $\{m_n\}$ are the magnitude variables corresponding to the magnitude observations $\{\underline{m}_n\}$.

CRLB (Variance) The CRLB for unbiased estimation of σ^2 is given by Sijbers *et al.* [36]

$$\text{CRLB}_{\sigma^2} = \frac{2\sigma^4}{NK} \tag{24.58}$$

The CRLB for unbiased estimation of σ can be derived from Eq. (24.58)

$$\text{CRLB}_\sigma = \frac{\sigma^2}{2NK} \tag{24.59}$$

If the noise variance is estimated from N magnitude data of a *background* region, the CRLB is equal to σ^4/N (cf. Eq. (24.58)). Then, the CRLB is the same as for estimation from N complex data of a background region. This might be surprising as estimation from N complex data actually exploits $2N$ real-valued (N real and N imaginary) observations, while estimation from N magnitude data only exploits N real-valued observations. However, this is compensated for by the fact that the Rayleigh PDF has a smaller standard deviation.

ML Estimation from Continuous Magnitude MR Data The likelihood function is obtained by substituting the available background data points $\{m_n\}$ for the variables $\{m_n\}$ in Eq. (24.57). Then the log-likelihood function, only as a function of σ^2, is given by

$$\ln L \sim -N \ln \sigma^2 - \frac{1}{K\sigma^2} \sum_{n=1}^{N} m_n^2 \qquad (24.60)$$

Maximizing with respect to σ^2 yields the ML estimator of σ^2

$$\widehat{\sigma}^2_{ML} = \frac{1}{KN} \sum_{n=1}^{N} m_n^2 \qquad (24.61)$$

It can be shown that Eq. (24.61) is an unbiased estimator, that is, its mean is equal to σ^2. Furthermore, the variance of the ML estimator (24.61) is equal to $2\sigma^4/NK$, which equals the CRLB given by Eq. (24.58) for all values of N.

One might as well be interested in the value of the standard deviation σ, for example, to estimate the SNR: A/σ. Simply taking the square root of the ML estimator of σ^2 Eq. (24.61) yields an estimator of σ

$$\widehat{\sigma}_{ML} = \sqrt{\frac{1}{KN} \sum_{n=1}^{N} m_n^2} \qquad (24.62)$$

This estimator is identical to the ML estimator of σ, as the square root operation has a single valued inverse (cf. invariance property of ML estimators [37]). Its variance is approximately equal to

$$\text{Var}\left(\widehat{\sigma}_{ML}\right) \simeq \frac{\sigma^2}{2NK} \qquad (24.63)$$

which equals the CRLB (cf. Eq. (24.59)). The estimator (24.62) is, however, biased because of the square root operation. Its expectation value is approximately equal to

$$\mathbb{E}\left[\widehat{\sigma}_{ML}\right] \simeq \sigma \left(1 - \frac{1}{4NK}\right) \qquad (24.64)$$

Notice that this means that it is possible to apply a bias correction. This, however, would increase the variance of the estimator.

ML Estimation from Discrete Magnitude MR Data Often, magnitude MR values are stored in integer types. Hence, the background mode from the histogram can be used to estimate the noise variance [38].

Let $\{l_i\}$ with $i = 0, \ldots, L$ denote the set of boundaries of histogram bins. Furthermore, let n_i represent the number of observations (counts) within the bin $[l_{i-1}, l_i]$, which are multinomially distributed. Then, the joint PDF of the histogram data is given by Mood et al. [37]

$$p(\{n_i\}|\sigma,\{l_i\}) = \frac{N_K!}{\prod_{i=1}^{K} n_i!} \prod_{i=1}^{K} p_i^{n_i}(\sigma) \tag{24.65}$$

with $N_K = \sum_{i=1}^{K} n_i$ the total number of observations within the partial histogram and p_i the probability that an observation assumes a value in the range $[l_{i-1}, l_i]$. The ML estimate of σ is then given by Sijbers et al. [38]

$$\hat{\sigma}_{\text{ML,L}} = \arg\min_{\sigma} \left[N_L \ln \left(e^{-\frac{l_0^2}{2\sigma^2}} - e^{-\frac{l_L^2}{2\sigma^2}} \right) - \sum_{i=1}^{L} n_i \ln \left(e^{-\frac{l_{i-1}^2}{2\sigma^2}} - e^{-\frac{l_i^2}{2\sigma^2}} \right) \right]. \tag{24.66}$$

Equation (24.66) is the ML estimate of the noise standard deviation from L bins. The selection of the number of bins L is important, since only bins that contain Rayleigh-distributed data should be used. In [38], a method was presented to select the optimal number of bins L. Alternatively, a preprocessing step can be incorporated to segment the background from the foreground before the computation of the histogram [39].

24.3.3.2 Nonbackground Region

In this section, we address the estimation of the noise level in magnitude MR images in the absence of background data. Most of the methods proposed earlier exploit the Rayleigh-distributed background region in the MR images to estimate the noise level. These methods, however, cannot be used for images in which no background information is available.

A first alternative is exploiting a signal region that is assumed to be constant. Suppose that a set of magnitude data points $\{m_n\}$ is available from a region where the true signal value A is the same for each data point. Then, the CRLB for unbiased estimation of σ^2 from magnitude data of a constant region is given by Eq. (24.43), in which the elements are defined by Eqs. (24.44)–(24.46).

If the noise variance is estimated from N magnitude data of a nonzero constant region, the CRLB is given by Eq. (24.43). It can numerically be shown that for magnitude data this CRLB tends to $2\sigma^4/N$ when the SNR increases, which is a factor 2 larger compared to estimation from complex data. This is not surprising since the Rician PDF tends to a Gaussian PDF for high SNR with the same variance as the PDF of the real or imaginary data. Hence, for high SNR, the difference in CRLB between magnitude and complex data can simply be explained by the number of observations available for the estimation of the noise variance.

Furthermore, the value of σ^2 can be estimated using the ML method as follows (cf. Eq. (24.47)):

$$\left(\hat{A}_{\text{ML}}, \hat{\sigma}^2_{\text{ML}}\right) = \arg\left\{\max_{A,\sigma^2}\left(-N\ln\sigma^2 - \sum_{n=1}^{N}\frac{m_n^2 + A^2}{2\sigma^2} + \sum_{n=1}^{N}\ln I_0\left(\frac{Am_n}{\sigma^2}\right)\right)\right\} \tag{24.67}$$

Note that it requires the optimization of a 2D function, which cannot be solved analytically.

Recently, an automatic noise estimation method that does not rely on a substantial proportion of voxels being from the background was proposed [40]. Therefore, the magnitude of the observed signal is modeled as a mixture of Rice distributions with a common noise parameter. The expectation-maximization (EM) algorithm is then used to estimate all parameters, including the common noise parameter. The algorithm needs initializing values and the number of components in the mixture model also needs to be estimated *en route* to noise estimation.

Finally, methods have been proposed based on the local estimation of the noise variance using ML estimation and the second method is based on the local estimation of the skewness of the magnitude data distribution [41, 42].

24.3.3.3 Double Acquisition Method

Estimation of the noise variance from a single magnitude image requires homogeneous regions in the image. However, large homogeneous regions are often hard to find, and, therefore, only a small amount of data points is available for estimation. Also, background data points sometimes suffer from systematic intensity variations. To cover these disadvantages, methods were developed based on two acquisitions of the same image: the so-called double acquisition methods. Therefore, the noise variance is, for example, computed by subtracting two acquisitions of the same object and calculating the standard deviation of the resulting image pixels [21, 33, 34]. Alternatively, the image noise variance can be computed from two magnitude MR images, as follows.

When two conventional MR images ($K = 2$) are acquired under identical imaging conditions, one can solve σ^2 from two equations and two unknowns using the averaged (where averaging is done in K-space, where the data is Gaussian distributed) and single images, since

$$\mathbb{E}\left[\langle m_s^2 \rangle\right] = \frac{1}{N} \sum_{n=1}^{N} A_n^2 + 2\sigma^2 \tag{24.68}$$

$$\mathbb{E}\left[\langle m_a^2 \rangle\right] = \frac{1}{N} \sum_{n=1}^{N} A_n^2 + 2\left(\frac{\sigma}{\sqrt{2}}\right)^2 \tag{24.69}$$

where $\langle \rangle$ denotes the spatial average of the whole image. The subscripts s and a refer to the single and averaged images, respectively. From Eqs. (24.69) and (24.68), an unbiased estimator of the noise variance is derived:

$$\widehat{\sigma^2} = \langle m_s^2 \rangle - \langle m_a^2 \rangle \tag{24.70}$$

This approach has the following advantages:

- It does not require any user interaction, as no background pixels need to be selected.
- It is insensitive to systematic errors such as ghosting, ringing, and DC artifacts, as long as these appear in both images. It is clear that if this type of error appears

in only one of the two images, none of the double acquisition methods yields the correct result.
- The precision of the noise variance estimator (24.70) is drastically increased, compared to the precision of the estimator given in Eq. (24.61), as all the data points (not only those from the background region) are involved in the estimation.
- It is valid for any image SNR.

An obvious disadvantage is the double acquisition itself. However, in MR acquisition schemes, it is common practice to acquire two or more images for averaging. Hence, those images may as well be used for the proposed noise estimation procedure, without additional acquisition time. In addition, the images require proper geometrical registration, that is, no movement of the object during acquisition is allowed.

24.3.4
Discussion

For *complex* data from a region with constant amplitude with identical and different phase values, the ML estimator $\widehat{\underline{\sigma}^2}_{ML}$ of σ^2 was derived, and the following are clear:

- Both noise variance estimators are biased. Also, it can easily be shown that the bias of $\widehat{\underline{\sigma}^2}_{ML}$ is independent of the true signal amplitude.
- For identical phases, the bias of $\widehat{\underline{\sigma}^2}_{ML}$ decreases inversely proportionally with the number of observations (N). In contrast, the bias of $\widehat{\underline{\sigma}^2}_{ML}$ for different phases does not; for large N, it converges to $\sigma^2/2$.

Only for complex data with identical phases, the variance of $\widehat{\underline{\sigma}^2}_{ML}$ asymptotically attains the CRLB. This is because for estimation from complex data with different phases the number of unknown parameters that need to be estimated simultaneously with σ^2, is proportional to N.

The MSE of $\widehat{\underline{\sigma}^2}_{ML}$ for *complex* data from a region with constant amplitude and identical and different phase values is given by Eqs. (24.50) and (24.53), respectively. The MSE of $\widehat{\underline{\sigma}^2}_{ML}$ for *magnitude* data from a region with constant amplitude can be found numerically from Eq. (24.67). Finally, the MSE of $\widehat{\underline{\sigma}^2}_{ML}$ from background MR data is given by σ^4/N, for magnitude as well as for complex data, and independent of the phases. From this, it is clear that the noise variance should be estimated from background data points, whenever possible. If a background region is not available, a similar reasoning for the estimation of σ^2 as for the estimation of A holds. That is, estimation of σ^2 from complex data with identical phases is then preferred to estimation from magnitude data, which, in turn, is preferred to estimation from complex data with different phases.

24.4
Conclusions

In this chapter, the problem of signal and noise ML estimation from MR data was addressed. It was noted that original data, coming from the scanner, are complex and Gaussian distributed. However, because of multiple digital data processing steps, the PDF of the resulting data may change. In this chapter, most of the PDFs one is confronted with when processing MRI data were discussed, along with their moments and asymptotic behavior. Furthermore, it was shown how these PDFs can be exploited in an ML estimation procedure to optimally estimate signal and noise parameters.

Next, the question was addressed whether complex or magnitude data should be used to estimate signal or noise parameters from low SNR data using the ML method. In summary, the following conclusions can be drawn:

1) The image noise variance should preferentially be estimated from background data (i.e., from a region of interest in which the true magnitude values are zero). Therefore, it does not matter whether the noise variance is estimated from magnitude or from complex-valued data.
2) On the other hand, whether or not the signal amplitude should be estimated from magnitude or complex-valued data depends on the underlying phase values:
   a. If the true phase values are known to be constant, the signal amplitude should be estimated from complex-valued data.
   b. If the true phase values are unknown, or if the true phase model deviates from a constant model, it is generally better, with respect to the MSE, to estimate the signal amplitude from magnitude data.

In this chapter, the framework of ML estimation for signal and noise estimation was addressed, which might be useful as a basis to create more advanced parameter estimation methods (as was, e.g., recently shown by He and Greenshields, who proposed a nonlocal ML estimation method for noise reduction [43]).

References

1. Breiman, L. (1968) *Probability*, Addison-Wesley, Reading, MA.
2. Wang, Y. and Lei, T. (1994). Statistical analysis of MR imaging and its applications in image modeling. Proceedings of the IEEE International Conference on Image Processing and Neural Networks, Newport Beach, CA, USA, vol. I, pp. 866–870.
3. Wang, Y., Lei, T., Sewchand, W., and Mun, S.K. (1996) MR imaging statistics and its application in image modeling. Proceedings of the SPIE conference on Medical Imaging, Newport Beach (CA).
4. Ross, S. (1976) A first course in probability, Collier Macmillan Publishers, New York.
5. van den Bos, A. (1989) Estimation of Fourier coefficients. *IEEE Trans. Instrum. Meas.*, 38 (4), 1005–1007.
6. Cunningham, I.A. and Shaw, R. (1999) Signal-to-noise optimization of medical imaging systems. *J. Opt. Soc. Am. A*, 16 (3), 621–631.
7. Sijbers, J. and Van der Linden, A. (2003) Magnetic Resonance Imaging. *Encyclopedia of Optical Engineering*, Marcel Dekker, New York, pp. 1237–1258. ISBN: 0-8247-4258-3.
8. Gradshteyn, I.S. and Ryzhik, I.M. (1965) Table of Integrals, Series and Products, 4th edn, Academic Press, New York, London.

9. Rice, S.O. (1944) Mathematical analysis of random noise. *Bell Syst. Tech.*, **23**, 282–332.
10. Dharmawansa, P., Rajatheva, N., and Tellambura, C. (2009) Envelope and phase distribution of two correlated Gaussian variables. *IEEE Trans. Commun.*, **57** (4), 915–921.
11. Abramowitz, M. and Stegun, I.A. (1970) Handbook of mathematical functions, Dover Publications, New York.
12. Constantinides, C.D., Atalar, E., and McVeigh, E.R. (1997) Signal-to-noise measurements in magnitude images from NMR phased arrays. *Magn. Reson. Med.*, **38**, 852–857.
13. Andersen, A.H. and Kirsch, J.E. (1996) Analysis of noise in phase contrast MR imaging. *Med. Phys.*, **23** (6), 857–869.
14. Pelc, N.J., Bernstein, M.A., Shimakawa, A., and Glover, G.H. (1991) Encoding strategies for three-direction phase-contrast MR imaging of flow. *J. Magn. Reson. Imaging*, **1** (4), 405–413.
15. Gudbjartsson, H. and Patz, S. (1995) The Rician distribution of noisy MRI data. *Magn. Reson. Med.*, **34**, 910–914.
16. Callaghan, P.T. (1995) Principles of Nuclear Magnetic Resonance Microscopy, 2nd edn, Clarendon Press, Oxford.
17. Sijbers, J. and den Dekker, A.J. (2004) Maximum likelihood estimation of signal amplitude and noise variance from MR data. *Magn. Reson. Med.*, **51** (3), 586–594.
18. van den Bos, A. (1982) Parameter estimation, in Handbook of Measurement Science, vol. 1, (P.H. Sydenham), Wiley, Chichester, England, pp. 331-377. Chapter 8.
19. Karlsen, O.T., Verhagen, R., and Bovée, W.M. (1999) Parameter estimation from Rician-distributed data sets using a maximum likelihood estimator: application to T_1 and perfusion measurements. *Magn. Reson. Med.*, **41** (3), 614–623.
20. Kaufman, L., Kramer, D.M., Crooks, L.E., and Ortendahl, D.A. (1989) Measuring signal-to-noise ratios in MR imaging. *Radiology*, **173**, 265–267.
21. Murphy, B.W., Carson, P.L., Ellis, J.H., Zhang, Y.T., Hyde, R.J., and Chenevert, T.L. (1993) Signal-to-noise measures for magnetic resonance imagers. *Magn. Reson. Imaging*, **11**, 425–428.
22. McGibney, G. and Smith, M.R. (1993) An unbiased signal-to-noise ratio measure for magnetic resonance images. *Med. Phys.*, **20** (4), 1077–1078.
23. Miller, A.J. and Joseph, P.M. (1993) The use of power images to perform quantitative analysis on low SNR MR images. *Magn. Reson. Imaging*, **11**, 1051–1056.
24. Sijbers, J., den Dekker, A.J., Scheunders, P., and Van Dyck, D. (1998) Maximum likelihood estimation of Rician distribution parameters. *IEEE Trans. Med. Imag.*, **17** (3), 357–361.
25. Lysaker, M., Lundervold, A., and Tai, X.-C. (2003) Noise removal using fourth-order partial differential equation with applications to medical magnetic resonance images in space and time. *IEEE Imag. Proc.*, **12** (12), 1579–1590.
26. Olsen, S.I. (1993) Estimation of noise in images: an evaluation. *Graph. Model. Image Process.*, **55** (4), 319–323.
27. Close, R.A. and Whiting, J.S. (1996) Maximum likelihood technique for blind noise estimation. In Proceedings of SPIE Medical Imaging, Newport Beach CA.
28. Lee, J.S. (1981) Refined filtering of image noise using local statistics. *Comput. Vis. Graph.*, **15**, 380–389.
29. Meer, P., Jolion, J., and Rosenfeld, A. (1990) A fast parallel algorithm for blind noise estimation of noise variance. *IEEE Trans. Pattern Anal. Mach. Intell.*, **12** (2), 216–223.
30. Kisner, S.J., Talavage, T.M., and Ulmer, J.L. (2002) Testing a model for MR imager noise. In Proceedings of the Second Joint EMBS-BMES Conference, Houston, TX, USA.
31. Henkelman, R.M. (1985) Measurement of signal intensities in the presence of noise in MR images. *Med. Phys.*, **12** (2), 232–233.
32. De Wilde, J.P., Hunt, J.A., and Straughan, K. (1997) Information in magnetic resonance images: evaluation of signal, noise and contrast. *Med. Biol. Eng. Comput.*, **35**, 259–265.
33. Sano, R.M. (1988) MRI: Acceptance Testing and Quality Control–The Role

of the Clinical Medical Physicist, Medical Physics Publishing Corporation, Madison, WI.

34. Firbank, M.J., Coulthard, A., Harrison, R.M., and Williams, E.D. (1999) A comparison of two methods for measuring the signal to noise ratio on MR images. *Phys. Med. Biol.*, **44**, 261–264.

35. Sijbers, J., den Dekker, A.J., Verhoye, M., Van Audekerke, J., and Van Dyck, D. (1998) Estimation of noise from magnitude MR images. *Magn. Reson. Imaging*, **16** (1), 87–90.

36. Sijbers, J., den Dekker, A.J., Raman, E., and Van Dyck, D. (1999) Parameter estimation from magnitude MR images. *Int. J. Imag. Syst. Technol.*, **10** (2), 109–114.

37. Mood, A.M., Graybill, F.A., and Boes, D.C. (1974) Introduction to the Theory of Statistics, 3rd edn, McGraw–Hill, Tokyo.

38. Sijbers, J., Poot, D., den Dekker, A.J., and Pintjens, W. (2007) Automatic estimation of the noise variance from the histogram of a magnetic resonance image. *Phys. Med. Biol.*, **52**, 1335–1348.

39. Rajan, J., Poot, D., Juntu, J., and Sijbers, J. (2010) Segmentation based noise variance estimation. In International Conference on Image Analysis and Recognition, vol. 6111, pp. 62–70.

40. Maitra, R. and Faden, D. (2009) Noise estimation in magnitude MR datasets. *IEEE Trans. Med. Imaging*, **28** (10), 1615–1622.

41. Koay, C.G. and Basser, P.J. (2006) Analytically exact correction scheme for signal extraction from noisy magnitude MR signals. *J. Magn. Reson. Imaging*, **179**, 317–322.

42. Rajan, J., Poot, D., Juntu, J., and Sijbers, J. (2010) Noise measurement from magnitude MRI using local estimates of variance and skewness. *Phys. Med. Biol.*, **55**, 441–449.

43. He, L. and Greenshields, I.R. (2009) A nonlocal maximum likelihood estimation method for Rician noise reduction in MR images. *IEEE Trans. Med. Imaging*, **28** (2), 165–172.

25
3-D Surface Reconstruction from Stereo Scanning Electron Microscopy Images

Shafik Huq, Andreas Koschan, and Mongi Abidi

25.1
Introduction

Three-dimensional (3-D) stereovision allows surface metrology in 3-D for an enhanced inspection. 3-D reconstruction of microscale surface roughness is a useful approach to quality control in many applications. Study of surface roughness in 3-D may provide useful information in such areas as the resistance of a surface to corrosion over time, or friction during repeated use, or how strongly one surface becomes bonded to another. In the case of integrated circuit (IC) fabrication, engineers want to evaluate the performance of fabrication processes; one very effective way to accomplish this is to look at the surfaces in 3-D after various processing steps.

For the 3-D reconstruction of surfaces of an object, multiple images of the object are required, each taken from a different viewpoint. In this chapter, we describe techniques for the 3-D reconstruction of surfaces from two images forming a stereo image pair. A scanning electron microscope (SEM) can be operated in such a way that one can capture stereo images of a subject. In our experiments, we use stereo images captured with a large-chamber scanning electron microscope (LC-SEM) located at the Oak Ridge National Lab (ORNL) (*www.ornl.gov*), Oak Ridge, Tennessee, USA (Figure 25.1). The data we have used could, however, also have been captured with a standard SEM.

Two main problems have to be solved for 3-D modeling using stereo: sensor calibration and stereo matching. While most techniques described in the literature are designed for a special experimental configuration, we introduce in this chapter a more generalized affine calibration technique. Nevertheless, more specialized techniques could be derived from the affine technique for a constrained experimental configuration. The second research area that we focus on in this chapter addresses the automatic search for corresponding pixels in the stereo images. The process is called *stereo matching*. Four different algorithms will be presented in detail for the matching of stereo images: (i) *iterated conditional mode* (*ICM*), (ii) *belief propagation* (*BP*), (iii) *graph cuts*, and (iv) *local support window* (*LSW*). All of these algorithms apply energy minimization to solve an optimization problem.

Handbook of Nanoscopy, First Edition. Edited by Gustaaf Van Tendeloo, Dirk Van Dyck, and Stephen J. Pennycook.
© 2012 Wiley-VCH Verlag GmbH & Co. KGaA. Published 2012 by Wiley-VCH Verlag GmbH & Co. KGaA.

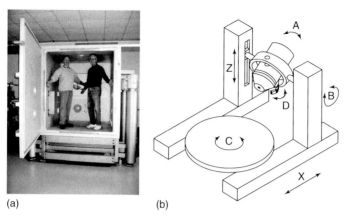

Figure 25.1 (a) LC-SEM chamber and (b) LC-SEM diagram showing the platform and electron gun with rotational and translational axes. (Images are provided by Oak Ridge National Laboratory).

An optimization problem can be represented in the following manner: for a given function f from a set S to the real numbers, we search for an element x_0 in S such that $f(x_0) \leq f(x)$ for all x in S ("minimization"). Such a formulation is called an *optimization problem* and the problem of finding corresponding points in stereo images can be modeled in this general framework. Problems in the field of computer vision formulated using this technique refer to the technique as energy minimization, speaking of the value of the function f as representing the energy of the system being modeled. The function f is called an *objective function, cost function*, or *energy function*. A feasible solution that minimizes the objective function is called an *optimal solution*. While in the literature, the energy parameters used in the minimization process are often chosen manually, we emphasize in this chapter techniques that estimate their stereo matching energy parameters automatically.

In the remainder of this chapter, Section 25.1 presents an introduction to the geometric calibration of an SEM/LC-SEM and the stereo image acquisition procedure using SEM/LC-SEM. Section 25.2.1 presents a local optimization–based algorithm that matches points that are located on an evenly distributed grid. Recently, the Middlebury College in Vermont performed an evaluation of results obtained when applying approximately 100 different stereo matching algorithms to four standard stereo data sets [1]. Section 25.2.2 presents an algorithm based on BP, which is currently ranked as the top performer in this evaluation. The algorithm works by propagating matching information of a point to its neighboring points. Section 25.2.3 describes an LSW-based algorithm. This is based on local optimization where each point works with a neighborhood of a particular shape while trying to maintain smoothness in the solution. Section 25.2.4 presents another popular algorithm called *graph cuts*. Graph cuts determine energy parameters from heuristics. This section also shows the overlay of visual and energy-dispersive X-ray spectroscopy (EDS) textures, captured by a LC-SEM, on the 3-D surface models. Section 25.3 draws conclusions.

25.1.1
Geometric Calibration of SEM/LC-SEM

LC-SEM is a new and emerging class of SEM (Figure 25.1a) with sample chambers sufficiently large such that meter-scale large objects can be scanned at the micro- or nanoscale with ease; the microscope also offers positioning with six degrees of freedom (Figure 25.1b). Each scan delivers 2-D images of surfaces. An additional capability of this microscope is EDS with simultaneous analysis of the elements from boron (5) to uranium (92), allowing for elemental characterization of surface materials.

An SEM/LC-SEM can be operated in such a way that multiple stereo images of an object are captured with known angles between them. In comparison to typical stereo camera systems that often use multiple sensors, an SEM/LC-SEM has only one imaging sensor available. In SEM, the object is placed on a stage and then posed to the scanner at different tilt angles by rotating the stage to obtain multiple scans, with each scan delivering one of the stereo images. In LC-SEM, both the stage and the scanner can be repositioned to perform the scans. For a reconstruction of the object from stereo images, the scanner can be modeled with an affine projection if the objects are scanned at high magnification (300× or more). Owing to the affine assumption, the focal length (intrinsic parameter) becomes linearly proportional to the magnification. Assuming the magnification to be 1.0 allows the reconstruction to happen up to a scale factor. The aspect ratio is assumed to be 1.0; hence, magnification remains the same along both the X- and Y-axes. The sensor skew angle is 90°. The extrinsic parameters of the system represent the orientation and location of the scanner. In the experimental configuration, the orientation is known and the parameters related to the location are unknown. In our modeling, there are four parameters related to the location of the scanner. These parameters are determined from geometric calibration.

To present the theory of geometric calibration, we describe the SEM/LC-SEM imaging system with the following three coordinate systems: one attached to the scanner, another one for the stage, and a third one for the object. We assume that the stage is located directly below the scanner, that is, the principal axis of the scanner, which is also the Z-axis of the scanner-coordinate system, and goes through the origin of the stage. The Z-axis of the stage is parallel to the principal axis of the scanner. XY planes of the scanner and the stage are also parallel. We further assume that the difference between the origins of the stage and the scanner coordinate systems is T_z.

The coordinate systems for the scanner, the stage, and the object are introduced in Figure 25.2. The object coordinate system has an origin located at an arbitrary feature point of the object, also referred to as the *anchor point*, which can be easily detected in all stereo images (those with two or more images). XY planes of all three coordinate systems and their X- and Y-axes are parallel. The Z-axes are vertical and

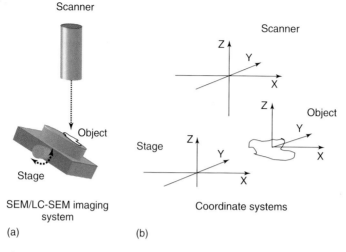

Figure 25.2 (a) SEM/LC-SEM stereo imaging configuration and (b) scanner, stage, and object coordinate systems.

parallel with the principal axis of the scanner. With respect to the object coordinate system, the origin of the scanner coordinate system is (X_c, Y_c, Z_c), and the origin of the stage coordinate system is (X_s, Y_s, Z_s). An object could be reconstructed with respect to the stage coordinate system. However, by utilizing the object coordinate system we gain the advantage of a self-calibrating system.

Assume that at tilt angle 0, the object point (X, Y, Z) is affine projected to (x, y). Before the projection is applied, the object point is transformed into the scanner coordinate system by a translation to (X_c, Y_c, Z_c). An affine transformation with magnification M is described by

$$\begin{bmatrix} x \\ y \\ 0 \end{bmatrix} = \begin{bmatrix} M & 0 & 0 \\ 0 & M & 0 \\ 0 & 0 & 0 \end{bmatrix} \begin{bmatrix} X - X_c \\ Y - Y_c \\ Z - Z_c \end{bmatrix} \quad (25.1)$$

Assume that at a tilt angle β, the object point (X, Y, Z) is affine projected to $(x\prime, y\prime)$. Before the projection is applied, the object point is rotated by the angle β around the Y-axis. Usually, such rotation moves the object out of the field of view of the microscope. A translation $(T_x, 0, 0)$ is applied to bring the object back into the field of view. The rotation changes the relative coordinates of the origins of the stage and the scanner coordinate systems, which are defined with respect to the object coordinate system. The new scanner location after rotation of the object is

$$\begin{bmatrix} X_c - X_s - T_x \\ Y_c - Y_s \\ Z_c - Z_s \end{bmatrix} + R_Y(\beta) \begin{bmatrix} X_s \\ Y_s \\ Z_s \end{bmatrix}$$

which is obtained by first transforming all the coordinates into the stage coordinate system, then applying rotation β and translation $(T_x, 0, 0)$, and finally transforming all the coordinates back into the object coordinate system. Thus, we can establish the following projection relation for the second view.

$$\begin{bmatrix} x' \\ y' \\ 0 \end{bmatrix} = \begin{bmatrix} M & 0 & 0 \\ 0 & M & 0 \\ 0 & 0 & 0 \end{bmatrix}$$

$$\left(R_Y(\beta) \begin{bmatrix} X \\ Y \\ Z \end{bmatrix} - \begin{bmatrix} X_c - X_s - T_x \\ Y_c - Y_s \\ Z_c - Z_s \end{bmatrix} - R_Y(\beta) \begin{bmatrix} X_s \\ Y_s \\ Z_s \end{bmatrix} \right) \quad (25.2)$$

From Eqs. (25.1) and (25.2) we obtain

$$\begin{bmatrix} 1 & 0 & 0 \\ 0 & 1 & 0 \\ \cos\beta - 1 & 0 & -\sin\beta \end{bmatrix} \begin{bmatrix} X \\ Y \\ Z \end{bmatrix}$$

$$= \begin{bmatrix} X_c + x \\ Y_c + y \\ -\sin\beta \cdot Z_s + (\cos\beta - 1) \cdot X_s - T_x + x' - x \end{bmatrix} \quad (25.3)$$

where, $M = 1.0$ is assumed for a reconstruction up to a scale factor. There are four unknowns X_c, Y_c, Z_s, and X_s in Eq. (25.3), which can be estimated from the correspondence of one point in three views or one and a half points in two views. Let image coordinates of the anchor point in the first and second views be (x_a, y_a) and (x'_a, y'_a), respectively. When substituting (X, Y, Z) with $(0, 0, 0)$, (x, y) with (x_a, y_a), and (x', y') with (x'_a, y'_a) in Eq. (25.3), we obtain

$$X_c = -x_a,$$
$$Y_c = -y_a, \text{ and} \quad (25.4)$$
$$\sin\beta \cdot Z_s - (\cos\beta - 1) \cdot X_s + T_x - x'_a + x_a = 0$$

The last equation is obtained by simplification when inserting $X_c = -x_a$. From Eq. (25.4) we see that X_c and Y_c can be calculated from (x_a, y_a) and its match (x'_a, y'_a). To solve for Z_s and X_s, we need one more equation which can be obtained from a third view taken at another tilt angle. For tilt angles β_1 and β_2 we obtain the following two equations:

$$\sin\beta_1 \cdot Z_s - (\cos\beta_1 - 1) \cdot X_s + T_{x\beta_1} - x'_a + x_a = 0$$
$$\sin\beta_2 \cdot Z_s - (\cos\beta_2 - 1) \cdot X_s + T_{x\beta_2} - x''_a + x_a = 0 \quad (25.5)$$

Z_s and X_s are solved from the pair of equations in Eq. (25.5). x_a, x'_a, and x''_a are the x-image coordinates of the anchor point in the first, second, and third views. Finally, we obtain the following reconstruction equations for (X, Y, Z) of a matched pair of points $\{(x, y), (x', y')\}$ in the first and second views (i.e., left and right stereo images) with relative tilt angle β,

$$X = X_c + x$$
$$Y = Y_c + y, \text{ and} \quad (25.6)$$
$$Z = \frac{-\sin\beta \cdot Z_s + (\cos\beta - 1) \cdot X_s - T_x + x' - x - (X_c + x) \cdot (\cos\beta - 1)}{-\sin\beta}$$

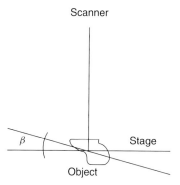

Figure 25.3 Constrained SEM/LC-SEM stereo imaging configuration.

For a constrained configuration (Figure 25.3), the set of equations in Eq. (25.6) can be simplified in the following way. Assume that the stage coordinate system is aligned with the object coordinate system $(X_s, Y_s, Z_s) = (0, 0, 0)$ and the camera coordinate system differs only in Z. In this case, $T_x = 0$, $X_c = 0$, and $Y_c = 0$.

By including these values into Eq. (25.6), (X, Y, Z) is approximated by

$$X = x,$$
$$Y = y, \text{ and} \qquad (25.7)$$
$$Z = -x \cdot \tan\frac{\beta}{2} + \frac{x - x'}{\sin \beta}$$

The equations in Eq. (25.7) have also been used by Hemmleb *et al.* [2]. While generating the stereo images, the working distance of the SEM/LC-SEM system must remain unchanged. When tilted, the object moves out of focus; the stage is then translated along its optical axis (which is parallel to the Z-axis in the experimental configuration) to bring the object back into focus. This movement changes the Z-location of the origin of the scanner coordinate system, (X_c, Y_c, Z_c). However, this change does not affect the reconstruction, since we do not use Z_c in Eq. (25.6) or (25.7).

25.1.2
LC-SEM Stereo Image Acquisition

The LC-SEM stereo image acquisition is performed in a controlled geometry. The principal points in all images are chosen to appear approximately at the centers of the images. After the images are captured, image rectification is applied, which is a transformation process used to project two or more images onto a common image plane. The objective of image rectification in stereovision is to transform the image planes onto a common image plane where corresponding points are located on the same scan line in both images. As mentioned in the calibration section, given that the images are captured at high magnification (300×), the distortion effect on the surface geometry along the axis of rotation is negligible owing to the perspective projection. Therefore, we can assume the image formation is an affine process. Owing to the affine assumption and the fact that the object is tilted only around the X-axis of the LC-SEM, the images become rectified after a 90° rotation, either clockwise or counterclockwise. Note that stereo images from SEM or LC-SEM can be reconstructed only up to duality, which means that the real 3-D surface could be inverted. Figure 25.4 shows an original LC-SEM stereo image pair and their corresponding rectified images.

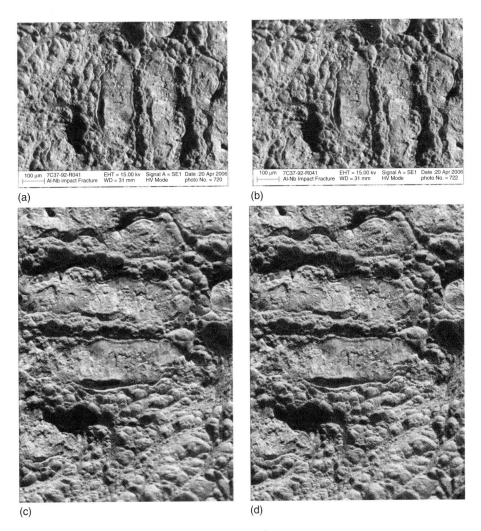

Figure 25.4 (a) and (b) are stereo image pair captured at 400× magnification with 6° of tilt angle apart; the image size is 1024 × 690 with each pixel 1 μm × 1 μm in physical dimension; and (c,d) are stereo image pair cut from images in (a,b) and rectified with a counterclockwise 90° rotation.

25.2
Matching Stereo Images

Stereo image matching is widely considered to be a difficult computer vision problem. In recent years, several state-of-the-art algorithms based on energy minimization have been proposed. These algorithms, including energy-minimizing grid [3], graph cuts [4–7], and BP [8, 9] match stereo images with energy minimization that derives from Bayesian inference and Markov assumption on Gibbs distribution. In this section, we describe four energy minimization–based matching algorithms that apply automatically estimated (either heuristically or statistically) energy parameters to match the two stereo images. For convenience of reference, we call them left and right stereo images.

25.2.1
Matching with Energy-Minimizing Grid

When applying an energy-minimizing grid algorithm, which is an annealing-based two-step energy minimization technique, one of the stereo images is first overlaid with sparse grids of regular shape with equally spaced lines and columns. An application of the sparse-grid concept was used in [3] for 3-D face modeling, where the authors obtained an initial matching from laser scans of the face and then refined the model by applying energy minimization on the grid. Figure 25.5 shows the grid overlaid on the left stereo image. Image points at the grid nodes are the points we expect to match in the right image. Since the images are rectified, points

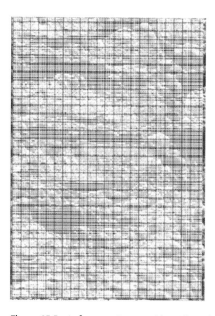

Figure 25.5 Left stereo image with grid overlaid; image size is 1024 × 690.

in the right image corresponding to the grid nodes in the left image have the same y-coordinate values. The algorithm does not address the presence of occlusions in the scene. When occlusions are present, they are implicitly treated as noise. A description of the grid configuration follows.

For a node p, an intersection between a grid line and a column in the grid in the left image, we define two neighborhoods $\mathcal{N}_1(p)$ and $\mathcal{N}_2(p)$. $\mathcal{N}_1(p)$ includes image points surrounding p. $\mathcal{N}_2(p)$ includes neighboring grid nodes. A typical cost function that obeys Bayesian inference is

$$E(p) = -V_d(p, \mathcal{N}_1(p)) - \alpha_p V_s(p, \mathcal{N}_2(p)) \tag{25.8}$$

where the first term is the data likelihood and the second one is the prior with the prior model parameter α_p. Note that in a Markov random field (MRF) framework, Eq. (25.8) assumes the prior models to be independent but not identical. It is defined locally for one node. The data term $V_d(p, \mathcal{N}_1(p))$ is within the interval [0.0, 1.0] and is obtained from shifting and scaling of the normalized cross-correlation coefficient. Shifting is done by adding 1.0 to the original score, which is in [−1.0, 1.0] and scaling is achieved by dividing the summation by 2.0. The size of the cross-correlation window is chosen to be greater than or equal to the space between two columns in the grid to decrease the number of inconsistent matches. $V_s(p, \mathcal{N}_2(p))$ is the prior term, also known as the *smoothness term*. The ability of Eq. (25.8) to reach a global or near-global solution is dependent on the interaction mechanism of p with its neighbors. To avoid local minima, we introduce two techniques. The first one is a mean field annealing–based interaction and the second one is a two-step optimization based on two neighborhoods. For each step we have one equation. In the first equation, the neighborhood $\mathcal{N}_2(p)$ includes all the neighbors around p. We call this neighborhood $\mathcal{N}_{2a}(p)$. In the second equation, $\mathcal{N}_2(p)$ includes only one neighbor that is also a grid node. We call this neighborhood $\mathcal{N}_{2b}(p)$. The two cost functions are defined as follows:

$$E_a(p) = -V_d(p, \mathcal{N}_1(p)) - \alpha_p V_{sa}(p, \mathcal{N}_{2a}(p)) \tag{25.9}$$
$$E_b(p) = -V_d(p, \mathcal{N}_1(p)) - \alpha_p V_{sb}(p, \mathcal{N}_{2b}(p)) \tag{25.10}$$

The functions in Eqs. (25.9) and (25.10) are still defined for one node p and the Markov assumption still holds. The only difference between them is the number of neighbors with which p interacts in the smoothness term. The smoothness term is associated with three annealing parameters which will be described in detail in the following sections. These parameters help in the progressive propagation of matching for global convergence. In analogy with the temperature for potential functions in the Gibbs distribution, α_p is called the *inverse temperature parameter*.

25.2.1.1 Cost Function with Mean Field Annealing

If $d(p)$ is the disparity label of p, we define $\mu_{\mathcal{N}_2(p)}$ with mean field approximation as

$$\mu_{\mathcal{N}_2(p)} = \frac{1}{|\mathcal{N}_2(p)| + 1} \sum_{q \in p \cup \mathcal{N}_2(p)} d(q) \tag{25.11}$$

with $\mathcal{N}_2(p) = \mathcal{N}_{2a}(p)$ and $\mathcal{N}_2(p) = \mathcal{N}_{2b}(p)$ for Eqs. (25.9) and (25.10), respectively. $\mathcal{N}_{2b}(p)$ includes all the first-order neighbors of p. For $\mathcal{N}_{2a}(p)$, we select one of the neighbors in $E_a(p_i)$. Let p_i, $0 \leq i \leq 7$, be the eight neighbors of p. We select the neighbor p_i that has the highest value for $E_a(p_i)$ estimated in the last iteration. In the energy minimization scheme, the smoothness term is defined by

$$V_s(p, \mathcal{N}_2(p)) = 1.0 - |\mu_{\mathcal{N}_2(p)} - d(p)| \tag{25.12}$$

The variable $|\mu_{\mathcal{N}_2(p)} - d(p)|$ in Eq. (25.12) is remotely related to an exponential distribution, $\eta \exp\left(-\eta\left(1.0 - |\mu_{\mathcal{N}_2(p)} - d(p)|\right)\right)$, where η is the decay rate. We assign a value of 1.0 to η based on the assumption that the disparity smoothness constraint, $|\mu_{\mathcal{N}_2(p)} - d(p)| < 1.0$, holds. This constraint is stronger than the disparity gradient constraint $|\Delta d/\Delta x| < 1.0$, used in many stereo matching algorithms as, for instance, by Chen and Medioni [10] for 3-D face modeling. Equation (25.12) points out that p interacts with itself and its neighbors to maintain smoothness locally in the neighborhood. This interaction occurs with the mean field of disparity labels, which is in contrast with pairwise interactions of disparity labels described in the current popular MRF-based algorithms [4, 8]. Naturally, the mean field approximation of Eq. (25.11) plays the role of pulling unmatched grid nodes toward correct matching locations.

25.2.1.2 Annealing with Inverse Temperature

Wrong matches or mismatches are usually observed in image regions with little or no texture. The smoothing energy can distribute the grid nodes evenly in such regions and thus help to avoid mismatches that tend to make the grid's shape overly irregular. The smoothness parameter α_p regularizes these regions by applying high smoothness energy. In this regard, the derivation of α_p is motivated by two objectives: (i) to initially disregard the smoothness energy in order to facilitate rearrangement of the grid nodes and (ii) to obtain a smoothness energy that helps to avoid mismatches in low-texture regions. Heuristically, these two objectives are achieved by defining the following expression for α_p:

$$\alpha_p = (1 - u_p) \times g \tag{25.13}$$

with,

$$g = \frac{1}{|\Omega|} \sum_{p \in \Omega} \left(1 - \frac{|I_l(p) - I_r(p')|}{255}\right) \tag{25.14}$$

and

$$u_p = \frac{1}{1 + \sum_{q \in \mathcal{N}_u} \prod_{j,k} f\left(I_l(p_{j,k}) - I_l(q_{j,k})\right)} \tag{25.15}$$

g is the global matching score and u_p describes the uniqueness of p. In Eq. (25.15), $I_l(p)$ is the intensity of pixel p. $I_r(p')$ is the intensity of p', the point in the right image corresponding to p. Ω is the set of all grid nodes in the left image. Initially, the images are not matched and the value of g remains small, therefore reducing the effect of α_p. At this stage, node points with strong features match. After they match, the value of g becomes large and enforces matching in low-textured regions. At the same time, u_p describes how unique a node is in its neighborhood. Since we use a window-based data term for a node, the uniqueness is also defined on a window. The definition is based on the likelihood of all the pixels of the window centered at p with respect to all the pixels of the windows centered at each point in $p \cup \mathcal{N}_u$. $\mathcal{N}_u$ includes neighboring points located on the horizontal line at the left and at the right of p. $f(\cdot)$ in Eq. (25.15) is defined using σ_N^2, the variance of the Gaussian noise in the left image, with $I_l(p_{j,k})$ and $I_l(q_{j,k})$ being the intensities of image points (j, k) in a window centered at grid node p and neighboring image point q in the left image.

$$f(x) = \exp(\frac{-x^2}{2\sigma_N^2}) \tag{25.16}$$

Note that u_p is estimated solely from the left image in the neighborhood $\mathcal{N}_u$. We typically use $\sigma_N = 5$ and $\mathcal{N}_u$ is chosen to include 10 neighbors; 5 on each side of p.

25.2.1.3 Optimization Algorithm

Equations (25.9) and (25.10) are applied alternatively through an iterative algorithm. Matches for grid nodes are searched in a predefined order in a similar way to the inference algorithm ICM, first proposed by Besag in 1986 [11]. In accordance with the order rule in ICM, a grid node with smaller row and column numbers comes first in the stochastic search process. The two-step optimization process runs in the following way. First, we apply Eq. (25.10) once for the best neighbor-based energy estimation and matching. Then, Eq. (25.9) is applied repeatedly for all neighbor-based energy estimations and matchings until convergence is obtained. When convergence is achieved, Eq. (25.9) has reached an optimal solution, which could be local. At this point, $\mathcal{E}_{i,j} = \sum_{p \in \Omega} E_a(p)$ denotes the aggregated energy over Ω.

Here j indicates the number of iterations in the inner loop where Eq. (25.9) is applied and i indicates the number of iterations in the outer loop. In a next processing step, we apply Eq. (25.10) again so that the new labeling forces the matching out of a local minima. We apply Eq. (25.9) repeatedly again and obtain $\mathcal{E}_{i+1,j} = \sum_{p \in \Omega} E_a(p)$. Both modules run alternatively until convergence in global energy E_i is achieved. Figure 25.6 depicts this alternating optimization algorithm through a flowchart. In the diagram, S is the set of grid nodes.

The inner loop A is designed to achieve local convergence and the outer loop B to achieve global convergence. Local convergence is achieved when two consecutive energies $\mathcal{E}_{i,j-1}$ and $\mathcal{E}_{i,j}$ are close, that is, $|\mathcal{E}_{i,j-1} - \mathcal{E}_{i,j}| < \Delta$, where Δ is an arbitrarily selected small number. To avoid infinite looping, this condition can be tied with a bounded number of iterations that are empirically chosen.

25.2.1.4 Matching Results

Example results obtained when applying grid matching to an SEM stereo image pair are shown in Figure 25.7. The matching was first performed with a grid

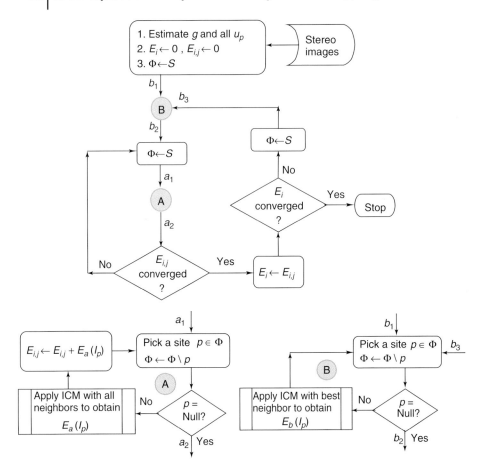

Figure 25.6 Flowchart of the two-step alternating optimization algorithm.

resolution of 150 × 100 on images of resolution 1024 × 690 and then the grid resolution was doubled. The cross-correlation window size was 7 × 7, that is, $|\mathcal{N}_1(p)| = 48$ and $\mathcal{N}_2(p)$ was a first order eight-neighborhood system. The result shows that all the grid nodes match including those in the dark regions. We apply stereo matching in two directions, "left image to right image" and "right image to left image," to further study the performance of the algorithm. The algorithm delivered similar results for both matchings (Figure 25.7a,b and c,d).

25.2.2
Matching with Belief Propagation

BP was formulated independently in 1986 by Pearl [12] and Lauritzen and Spiegelhalter [13]. BP works on finite cycle-free graphs. However, because all of its operations are local, it may also be applied to graphs with loops (hence the alternate

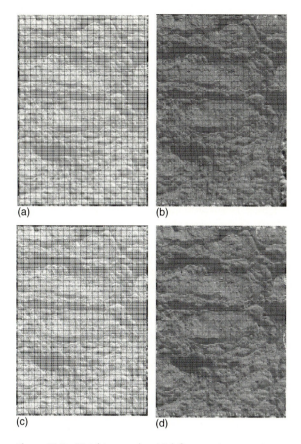

Figure 25.7 Matching results: (a) left stereo image overlaid with grid, (b) matched grid in the right image, (c) right stereo image overlaid with grid, and (d) matched grid in the left stereo image.

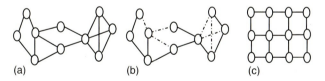

Figure 25.8 (a) Graph with a loop, (b) graph without loop, and (c) loopy graph created by a 3 × 4 stereo image; each matching site is a node in the graph.

name *loopy belief propagation*), such as in stereo images, where BP becomes iterative and approximate (Figure 25.8). In BP, each site (node in the graphs or pixels in stereo images) estimates a maximum a posteriori (MAP) assignment of a discrete probability distribution and passes the MAP information, called *message*, to its neighbors.

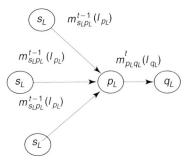

Figure 25.9 Message passing mechanism in belief propagation (BP).

Two algorithmic versions of BP exist: (i) sum-product BP and (ii) max-product BP, where both algorithms produce the same results [9]. We discuss here the max-product BP algorithm that works by passing messages around the graph defined by the four-neighbor systems of order one (Figure 25.9).

Each message is represented by a vector of the dimension given by the number of possible labels. Let $m^t_{q_L p_L}$ be the message that node q_L sends to a neighboring node p_L at time t. s_L are all the neighbors of p_L except q_L, the neighbor to whom the message is sent. All entries in $m^0_{q_L p_L}$ are initialized with zeros. New messages are computed at each iteration in the following way when p_L sends a message to q_L:

$$m^t_{q_L p_L}(l_{p_L}) = \min_{l_{p_L}} \left(D_{q_L}(l_{q_L}) + \lambda V(l_{q_L}, l_{p_L}) + \sum_{s_L \in \mathcal{N}(q_L) \setminus p_L} m^{t-1}_{s_L p_L}(l_{q_L}) \right) \quad (25.17)$$

where $\mathcal{N}(q_L) \setminus p_L$ denotes the neighbors of p_R excluding p_L. After T iterations, a belief vector is computed for each node by

$$b_{p_L}(l_{p_L}) = D_{p_L}(l_{p_L}) + \sum_{q_L \in \mathcal{N}(p_L)} m^T_{q_L p_L}(l_{p_L}) \quad (25.18)$$

$D_{p_L}(l_{p_L})$ is the absolute difference between the intensities of p_L and p_R. $V(l_{q_L}, l_{p_L})$ is the cutoff function as defined in Eq. (25.19) and illustrated in Figure 25.10. The function can also be constructed by truncating linear smoothness cost such as the following:

$$V(l_{p_L}, l_{q_L}) = \min(T_{q_L}, |l_{p_L} - l_{q_L}|) \quad (25.19)$$

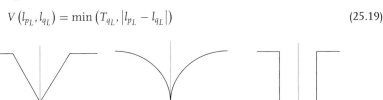

Figure 25.10 Discontinuity-preserving smoothness functions. (a) Truncated linear function, (b) exponential function, and (c) Potts model [14].

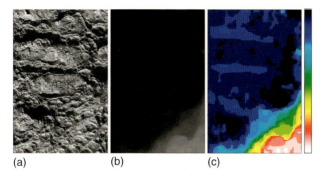

(a) (b) (c)

Figure 25.11 Disparity map generated by belief propagation. (a) One of the stereo images, (b) gray-coded disparity map, and (c) color-coded disparity map.

where T_{q_L} is the threshold parameter handling discontinuity. To allow discontinuities, a cutoff function was introduced by Sun et al. [8], who defined a function that changes smoothly from being almost linear near the origin to a constant value as the cost increases. For the estimation of the smoothness parameter λ, an expectation maximization (EM) algorithm adopted in [15] can be used. The labels $l^*_{p_L}$ that minimize $b_{p_L}(l_{p_L})$ individually at each node are selected as the disparity map. Figure 25.11 shows results of BP stereo matching applied to the LC-SEM stereo image pair.

Several variants of the BP algorithm have been suggested by others. In [16–18], the authors tried to further develop or establish the theoretical aspects of BP. In [19, 20], the performances of BP were improved by using image cues. In [19], the idea was to segment the left image based on color and assume that each segmented part represents a plane. The color segmentation improved matching around discontinuous regions. With the same objective, the data cost was obtained from color-weighted cross correlation. Another application of BP for modeling 3-D faces was presented in [21], where the data cost is obtained from an area-based estimate.

Standard BP propagates the messages around a graph with loops. Therefore, the MAPs are redundant in the estimated cost, that is, the disparity map is not obtained from an optimum on approximated cost. An experiment conducted by Szeliski et al. [22] showed that, at best, BP gives an energy that is 3.4% higher and at worst 30% higher. The top-ranked stereo matching algorithms listed in the performance table of the Middlebury College stereo site are BP and their variants [1].

25.2.3
Matching with Local Support Window

When matching stereo images with LSWs, the images are matched symmetrically to improve performance and also to detect occlusions. The algorithm picks a site p_L from the left image S_L and finds its minimum cost in a winner-take-all (WTA) manner. The cost function and its parameters are defined by

$$E_{p_L}(l_{p_L}) = \sum_{q_L \in \mathcal{N}(p_L)} \lambda_D D_{p_L}(l_{p_L}) + \lambda_{q_L} V(l_{p_L}, l_{q_L})$$

$$D_{p_L}(l_{p_L}) = \min(T_D, |I(p_L) - I(p_R)|), \text{ and}$$

$$V(l_{p_L}, l_{q_L}) = \min(T_{q_L}, |l_{p_L} - l_{q_L}|)$$

The data parameter λ_D, the smoothness parameter λ_{q_L}, and the cutoff threshold T_{q_L} are estimated statistically [1]. The minimum cost of a point can be calculated using WTA by applying

$$l_{p_L} = \underset{l_{p_L} \in \{d_{min}, \ldots, d_{max}\}}{\arg\min} \sum_{q_L \in \mathcal{N}(p_L)} \lambda_D D_{p_L}(l_{p_L}) + \lambda_{q_L} V(l_{p_L}, l_{q_L})$$

A flow diagram of the matching algorithm is shown in Figure 25.12. First, the data parameter is estimated and the smoothness parameter is initialized with a very small number. Then, an alternating optimization algorithm starts. The algorithm has two phases. In the first phase, the smoothness parameter is updated. In the second phase, the candidate sites p_L are picked one by one from the left image and matched to the right. When all the sites have been matched, the algorithm returns to the first phase. The iterations continue in a similar manner until convergence in the aggregated cost is achieved. Convergence is achieved when the difference of two aggregated costs from two consecutive iterations is less than a small number predefined by the user. Alternatively, or at the same time, the iterations can be tied to an empirically determined sufficiently large number to avoid infinite looping.

The flow diagram in Figure 25.12 describes one-way matching only. We developed a pair of symmetric cost functions and matched the images symmetrically, whereby we reduced the matching error by about 1% averaged over all the tested image pairs. In symmetric matching, the smoothness energy is computed from both the stereo images. First, we perform left-to-right matching. In one-way matching, we

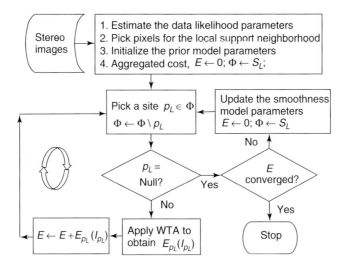

Figure 25.12 Flow diagram of the proposed local support window algorithm.

assumed that disparities of only the neighbors q_L are given. In symmetric matching, additionally we consider that disparities of q_R are also given. Here, q_R in the right image are the correspondences of q_L. Thus, the inference algorithm uses disparity information from both images. In the same way, we perform the right-to-left matching. The symmetric matching equations are described in equation set (25.20) with the first equation for the left-to-right match and the second for the right-to-left one.

$$E_{p_L}(l_{p_L}) = \sum_{q_L \in \mathcal{N}(p_L)} \lambda_D D_{p_L}(l_{p_L}) + \lambda_{q_L} \left(\frac{1}{2} V_{LR}(l_{p_L}, l_{q_L}) + \frac{1}{2} V_{RL}(l_{p_R}, l_{q_R}) \right), \text{where } p_L \in S_L$$

$$E_{p_R}(l_{p_R}) = \sum_{q_R \in \mathcal{N}(p_R)} \lambda_D D_{p_R}(l_{p_R}) + \lambda_{q_R} \left(\frac{1}{2} V_{RL}(l_{p_R}, l_{q_R}) + \frac{1}{2} V_{LR}(l_{p_L}, l_{q_L}) \right), \text{where } p_L \in S_R$$

(25.20)

S_R is the right image. Since both sets of corresponding neighbors are equally likely, each of the smoothness terms are divided by two. Owing to the uniqueness constraint, only one data term is used in both equations. In the symmetric matching algorithm, first the left-to-right matching is performed and then the right-to-left matching is performed. Following this, the algorithm continues with the left-to-right match and the iteration continues in a similar manner. Figure 25.13 shows the disparity map of the stereo image pair obtained when applying the LSW. As described above, all the matching parameters are chosen manually.

25.2.4
Matching with Graph Cuts

Graph cuts as a combinatorial optimization technique was first proposed and applied to stereo matching by Boykov et al. in 2001 [5]. The central idea was to represent an energy function with graphs and then optimize the function through minimum cuts. The theory of minimum cuts of a graph was developed by Karl Menger in 1927 [23]. This theory states that if s and t are distinct, nonadjacent vertices in a connected graph G, then the maximum number of s–t paths in G

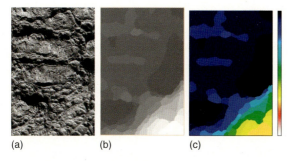

(a)   (b)   (c)

Figure 25.13 Disparity map generated by local support window matching. (a) One of the stereo images, (b) gray-coded disparity map, and (c) color-coded disparity map.

equals the minimum number of vertices needed to separate s and t. The theory derives the max-flow min-cut theory: the value of the max flow is equal to the value of the min cut, which was proved by Elias et al. in 1956 [24], and also independently by Ford and Fulkerson in the same year [25].

Graph cuts operate on graph-representable functions of binary variables. In stereo matching, each site can have more than two labels; hence they are not binary. To treat the sites as binary variables, two choices are given to all the sites in a matching iteration: either the site moves to the label w_i or it remains with whatever label it has. To be graph representable, a function has to be regular. A function is regular if it meets the following constraint:

$$E(0,0) + E(1,1) \leq E(0,1) + E(1,0) \tag{25.21}$$

where $E : \{0,1\}^2 \rightarrow \mathcal{R}$ is a function of the second-order cliques of binary variables. For $E_{p_L}(l_{p_L})$ to be a regular function, only $V(l_{p_L}, l_{q_L})$ needs to be a metric since $D_{p_L}(l_{p_L})$ is a unary function, hence, trivially regular. $V(l_{p_L}, l_{q_L})$ is regular when it is a metric; since then, the triangular relation for the constraint in Eq. (25.21) is true. Graph cuts use $D_{p_L}(l_{p_L})$ as a quadratic of the absolute difference between intensities of p_L and its corresponding site p_R at l_{p_L} and $V(l_{p_L}, l_{q_L})$ as a truncated Euclidean distance, where the Euclidean part makes $V(l_{p_L}, l_{q_L})$ metric and a truncation makes it discontinuity preserving. Thus,

$$V(l_{p_L}, l_{q_L}) = \min\left(K, |l_{p_L} - l_{q_L}|\right) \tag{25.22}$$

where K is a free parameter working as a bound on $V(l_{p_L}, l_{q_L})$. Another choice for $V(l_{p_L}, l_{q_L})$ in graph cuts is Potts' interaction penalty [14], which is also both metric and discontinuity preserving (Eq. (25.23)).

$$V(l_{p_L}, l_{q_L}) = \delta\left(l_{p_L} \neq l_{q_L}\right), \text{where } \delta\,(true) = 1 \text{ and } \delta\,(false) = 0 \tag{25.23}$$

Both Eqs. (25.22) and (25.23) have disadvantages. While an unknown parameter K has to be estimated in Eq. (25.22), Eq. (25.23), on the other hand, sacrifices matching quality by replacing slanted surfaces with stairlike structures.

Algorithms based on graph cuts [26, 27] are typically good at preserving discontinuities (occlusions or discontinuities of surface smoothness). However, they are sensitive to noise, since the algorithms work with data and smoothness terms that are not chosen statistically but rather heuristically. The smoothness term described by Eq. (25.23) imposes piecewise linearity on surfaces, leaving out stair effects in reconstructions. Figure 25.14 shows a disparity map of the stereo image pair obtained by graph cuts [26] where the smoothness parameter λ and the cutoff threshold K are determined from heuristics.

The equation set (Eq. (25.23)) reconstructs the matched point pairs. These equations were proposed in [2] and also derive from the generalized equations in Eq. (25.6).

$$X = x, Y = y, \text{and } Z = -x \cdot \tan\frac{\beta}{2} + \frac{x - x'}{\sin \beta} \tag{25.24}$$

(X, Y, Z) are the 3-D coordinates of the reconstructed model and β is the tilt angle around the Y-axis and between viewing angles of the images. In the experiment,

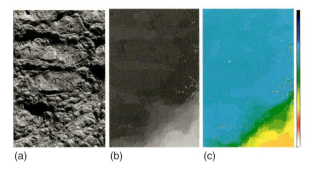

(a) (b) (c)

Figure 25.14 Disparity map generated by graph ss. (a) One of the stereo images, (b) gray-coded disparity map, and (c) color-coded disparity map.

$\beta = 6°$. In Eq. (25.24), (x, y) are the image coordinates of p_L, and x' is the x-coordinate of p_R. Figure 25.15 shows a visualization of a 3-D reconstruction. The 3-D model consists of approximately 534 600 points. Points not lying in the grid are approximated with bilinear interpolation. A comparison with virtual 3-D shows that overall the reconstructed 3-D models are reasonably accurate. Virtual 3-D is obtained by combining the red channel of the left image and the green and blue channels of the right image into a single color image.

An LC-SEM also provides spectral information of surfaces at different energy levels. The spectral information at two energy levels, one level for an Al (Aluminum) composite and the other for a Nb (Niobium) composite, is extracted using EDS. The spectral information that is extracted in the form of intensity values is overlaid on the 3-D surface. EDS images are acquired using the identical stereo image

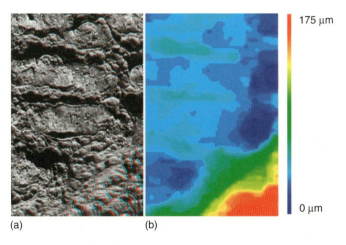

(a) (b)

Figure 25.15 (a) Virtual 3-D combining red channel of the left image and green and blue channels of the right image and (b) color-coded 3-D model.

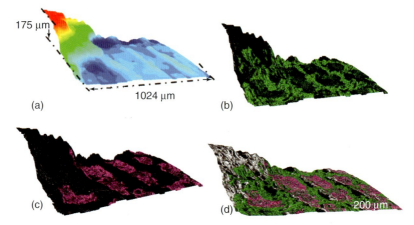

Figure 25.16 (a) Color-coded 3-D model; (b) Al composite texture overlay on the surface; (c) Nb composite texture overlay; and (d) LC-SEM image texture of (b,c) overlaid together on the surface in (a).

capturing geometry. Therefore, the stereo images and EDS images are perfectly registered. Thus, in addition to performing 3-D metrology on the surface, one is able to visually inspect the distribution characteristics of constituent materials and correlate them with surface structure such as creases or dents caused by chemical, physical, or other effects.

Figure 25.16 shows color-coded 3-D views of a surface reconstructed via stereo matching and EDS textures corresponding to the surface constituents. Al and Nb composites are later overlaid on the 3-D surface obtained from stereo matching and reconstruction (Figure 25.16a). Renderings shown in Figure 25.16b–d are from virtual reality modeling language (VRML) files written with shape of the 3-D surface as triangular meshes and texture as color of the corresponding vertices. Visually, it is evident that the textures on the surface are complementary. While the Nb composite is dominant in valley regions, the Al composite is dominant around the hill regions.

25.3
Conclusions

Four energy-minimizing algorithms were described for the 3-D reconstruction of material surfaces from stereo images captured with a SEM/LC-SEM. The surface material distribution is also observed by overlaying EDS images of the LC-SEM for detection of materials of interest on 3-D surfaces. Note that the signal-to-noise ratio (SNR) is relatively high in LC-SEM (or SEM) images [28]. Therefore, algorithms applying statistical estimation for finding stereo matching parameters are well suited for matching LC-SEM (or SEM) stereo images.

3-D reconstructions obtained from graph cuts methods are known to have stair effects on the surface. The graph cuts result presented in this chapter uses heuristic parameters. BP uses stereo matching parameters estimated from statistics. The visual inspection of the disparity map shows more surface details for the 3-D reconstruction with BP as compared to the other three algorithms. A local approach, such as ICM [11], suffers from a lack of global convergence. The version of ICM described in this chapter increases the chance of global convergence by using two alternately iterated cost functions, instead of one. The LSW technique produced over-smoothing effects similar to graph cuts. All these observations are visual. The degree of accuracy of different algorithms can be numerically determined from reconstructions of known shapes.

Acknowledgments

This work was supported by the DOE University Program in Robotics under grant DOE-DE-FG02-86NE37968 and BWXT-Y12 Subcontract #4300056316. The SEM images are provided by the Oak Ridge National Lab.

References

1. http://vision.middlebury.edu/stereo/eval/ (accessed 8 January 2012).
2. Hemmleb, M., Albertz, J., Schubert, M., Gleichmann, A., and Köhler, J. (1996) Digital microphotogrammetry with the scanning electron microscope. *Int. Arch. Photogram. Rem. Sens.*, **XXXI** (Part B5), 225–230.
3. Huq, S., Abidi, B., and Abidi, M. (2007) Stereo-based 3-D face modeling using annealing in local energy minimization, Proceedings of the IEEE 14th International Conference on Image Analysis and Processing (ICIAP), Modena, Italy.
4. Kolmogorov, V. and Zabih, R. (2001) Computing visual correspondence with occlusions using graph cuts. Proceedings of the 8th IEEE International Conference on Computer Vision, Vol. 2, pp. 508–515.
5. Boykov, Y., Veksler, O., and Zabih, R. (2001) Fast approximate energy minimization via graph cuts. *IEEE Trans. Pattern Anal. Mach. Intell.*, **20** (12), 1222–1239.
6. Hong, L. and Chen, G. (2004) Segment-based stereo matching using graph cuts. Proceedings of the IEEE Computer Society Conference on Computer Vision and Pattern Recognition (CVPR), Vol. 1, pp. I-74–I-81.
7. Jang, J.Y., Lee, K.M., and Lee, S.U. (2006) Stereo matching using iterated graph cuts and mean shift filtering. Proceedings of the 7th Asian Conference on Computer Vision, Hyderabad, India, pp. 31–40.
8. Sun, J., Zheng, N., and Shum, H. (2003) Stereo matching using belief propagation. *IEEE Trans. Pattern Anal. Mach. Intell.*, **25** (7), 787–800.
9. Felzenswalb, P. and Huttenlocher, D. (2004) Efficient belief propagation for early vision. Proceedings of the IEEE Computer Society Conference on Computer Vision and Pattern Recognition (CVPR), Vol. 1, pp. 261–268.
10. Chen, Q. and Medioni, G. (2001) Building 3–D human face models from two photographs. *J. VLSI Signal Process.*, **27** (1–2), 127–140.
11. Besag, J. (1986) On the statistical analysis of dirty pictures (with discussion). *J. R. Stat. Soc.*, **B 48**, 259–302.
12. Pearl, J. (1988) *Probabilistic Reasoning in Intelligent Systems: Networks of Plausible*

Inference, Morgan Kaufmann Publishers, Inc.
13. Lauritzen, S.L. and Spiegelhalter, D.J. (1988) Local computations with probabilities on graphical structures and their application to expert systems. *J. R. Stat. Soc.*, **B 50**, 157–224.
14. Potts, R.B. (1952) Some generalized order-disorder transformations. *Proc. Camb. Phil. Soc.*, **48**, 106–109.
15. Zhang, L. and Seitz, S.M. (2007) Estimating optimal parameters for MRF stereo from a single image pair. *IEEE Trans. Pattern Anal. Mach. Intell.*, **29** (2), 331–342.
16. Wainwright, M.J., Jaakkola, T.S., and Willsky, A.S. (2005) MAP estimation via agreement on (Hyper) trees: message-passing and linear-programming approaches. *IEEE Trans. Inform. Theory*, **51** (11), 3697–3717.
17. Meltzer, T., Yanover, C., and Weiss, Y. (2005) Globally optimal solutions for energy minimization in stereo vision using reweighted belief propagation. Proceedings of the International Conference on Computer Vision (ICCV), pp. 428–435.
18. Kolmogorov, V. (2006) Convergent tree-reweighted message passing for energy minimization. *IEEE Trans. Pattern Anal. Mach. Intell.*, **28** (10), 1568–1583.
19. Klaus, A., Sormann, M., and Karner, K. (2006) Segment-based stereo matching using belief propagation and a self-adapting dissimilarity measure. Proceedings of the 18th International Conference on Pattern Recognition, Vol. 3, pp. 15–18.
20. Yang, Q., Wang, L., Yang, R., Stewénius, H., and Nistér, D. (2006) Stereo matching with color-weighted correlation, hierarchical belief propagation and occlusion handling. Proceedings of the IEEE Computer Society Conference on Computer Vision and Pattern Recognition (CVPR), Vol. 2, pp. 2347–2354.
21. Onofrio, D., Sarti, A., and Tubaro, S. (2004) Area matching based on belief propagation with applications to face modeling. Proceedings of the International Conference on Image Processing, pp. 1943–1946.
22. Szeliski, R., Zabih, R., Scharstein, D., Veksler, O., Kolmogorov, V., Agarwala, A., Tappen, M., and Rother, C. (2006) A comparative study of energy minimization methods for Markov random fields. Proceedings of the 9th European Conference on Computer Vision (ECCV), Graz, Austria, Vol. 2, pp. 19–26.
23. Menger, K. (1927) Zur allgemeinen Kurventhoerie. *Fund. Math.*, **10**, 96–115.
24. Elias, P., Feinstein, A., and Shanon, C.E. (1956) Note on maximum flow through a network. *IRE Trans. Inform. Theory*, **IT-2**, 117–119.
25. Ford, L.R. and Fulkerson, D.R. (1956) Maximum flow through a network. *Can. J. Math.*, **8**, 399–404.
26. Hong, L. and Chen, G. (2004) Segment-based stereo matching using graph cuts. Proceedings of the IEEE Computer Society Conference on Computer Vision and Pattern Recognition (CVPR), Vol. 1, pp. I-74–I-81.
27. Jang, J.Y., Lee, K.M., and Lee, S.U. (2006) Stereo matching using iterated graph cuts and mean shift filtering. Proceedings of the 7th Asian Conference on Computer Vision (ACCV), Hyderabad, India, pp. 31–40.
28. Oho, E., Baba, N., Katoh, M., Nagatani, T., Osumi, M., Amako, K., and Kanaya, K. (2005) Application of the Laplacian filter to high-resolution enhancement of SEM images. *J. Electron Microsc. Tech.*, **1** (4), 331–340.

Further Reading

Huq, S., Koschan, A., Abidi, B., and Abidi, M. (2008) MRF stereo with statistical estimation of parameters. Proceedings of the 4th International Symposium on 3-D Data Processing, Visualization, and Transmission (3-DPVT), Atlanta.

Part II
Applications

26
Nanoparticles

Miguel López-Haro, Juan José Delgado, Juan Carlos Hernández-Garrido, Juan de Dios López-Castro, César Mira, Susana Trasobares, Ana Belén Hungría, José Antonio Pérez-Omil, and José Juan Calvino

26.1
Introduction

The past decade has witnessed a steeply increasing interest in the development of applications of nanoparticles in a wide range of fields [1]. From heterogeneous catalysis, where metal and oxide nanoparticles dispersed on different type of supports have been classically employed as active materials to promote the rates of a large number of chemical reactions, most of them of interest in industrial chemistry, the use of nanoparticles have more recently spread over many other areas ranging from medicine [2, 3], construction [4], textiles [5], food processing [6], photonics and electronics [7], or preservation of cultural heritage [8], among others. Thus, it is nowadays quite common to find the word *nano* as prefix in the name of many disciplines, as nanocatalysis, nano(bio)medicine, nanopharmacy, or nanoelectronics.

The benefits of decreasing the size of particles to the nano range stem at least from two separate contributions: (i) the huge increase in surface to volume ratio and (ii) the modification of the electronic structure of the solid. Most of the applications mentioned above take advantage of the perturbation of the physical and chemical properties derived from shrinking the size to the nanoscale. A large surface to volume ratio is a crucial parameter in fields such as catalysis, adhesion or friction, and wear, where not only the total extent of exposed surface is of utmost importance but also its exact chemical and physical properties. On the other hand, quantization effects because of the limited size of these objects do strongly influence their electronic, optical, and magnetic properties, which are key in many other applications.

A simple consideration of the number of reviews related to nanoparticle research published in 2010, a total of 190, clearly demonstrates the huge interest of the scientific community in these objects.

More recently, a great concern has also been raised about the potential threat of nanoparticles to the environment [4, 9, 10] and also about their toxic effects on living organisms [11]. Thus, the word *nanotoxicity* has also been very recently coined.

Handbook of Nanoscopy, First Edition. Edited by Gustaaf Van Tendeloo, Dirk Van Dyck, and Stephen J. Pennycook.
© 2012 Wiley-VCH Verlag GmbH & Co. KGaA. Published 2012 by Wiley-VCH Verlag GmbH & Co. KGaA.

Tuning the macroscopic performance of nanoparticle-based systems to real needs requires tailoring very finely their structural and compositional parameters, among them their size, shape, crystalline structure, or surface chemistry. Such a tuning process requires establishing appropriate correlations between these parameters, synthesis methods, and macroscopic performance. Therefore, their precise and as accurate as possible determination is a requisite to rationalize the behavior of nanoparticle-based systems, to achieve the goal of effectively targeting particular applications as well as properly understanding their potential risks.

Electron microscopy-based techniques are currently the most suitable to cope with the detailed characterization of nanoparticles. This is so not only because they allow to deal with them on an individual basis but also because the current spatial resolution achievable with most imaging and spectroscopic (scanning) transmission electron microscopy ((S)TEM) techniques allow, at the same time, to depict structural and compositional features at the smallest length scales.

Regarding the first point, in contrast to macroscopic characterization techniques, which can only provide average information about the whole ensemble of particles present in the material, (S)TEM techniques provide a direct visualization of individual particles, whose characteristics (size, shape, structure, etc.) can be determined. By studying and characterizing a sufficiently high number of particles, an overall picture, representative and statistically meaningful, of the whole system can be obtained. For some properties, statistical distributions can even be obtained. Moreover, given that a lot of properties do not show a linear relationship with size, average values may not be valid to understand properly the macroscopic behavior of a set of particles.

Possibly, the property that could help to illustrate these ideas is *particle size*. Of course, size is one of the most basic features of a nanoparticle, but we have to admit that it is in fact, for most applications, a crucial parameter. Most synthesis methods lead to systems in which particles of varying size are present, that is, in general the relevant feature concerning size is not a particle size but a particle size distribution.

The most widely used technique to achieve information about particle size from crystalline nanoparticle-based materials is X-ray diffraction (XRD). Thus, by appropriate analysis methods, XRD data allow determining, for samples containing nanoparticles in the nanometer-size range, an average "particle" diameter. This average diameter corresponds, in fact, to the "volume-averaged" diameter of the crystalline domains present in the sample. In a completely homogeneous sample, made up of particles constituted by only one domain, this diameter would correspond to the diameter of the particles, but in a sample in which there is a dispersion of the particle diameter, the value of the XRD-determined diameter will be more influenced by the larger particles. In some cases, even if a significant number of small particles are present, the average XRD diameter may significantly deviate from the diameter of these smaller particles. Moreover, in samples made up of particles constituted by aggregates of smaller domains, particle size would clearly deviate from the domain size.

All these comments only intend to highlight that just by having access to the whole range of particle size, the necessary information about the real structure of the material, required to rationalize its macroscopic behavior, can be obtained. And this can only be done on the basis of the characterization of individual particles. It is also important to emphasize that in this type of analysis, a large enough number of particles have to be considered to guarantee that the final conclusions are representative of the whole sample.

Regarding the second of the aspects commented on above, the high spatial resolution characteristic of (S)TEM-based techniques, let us simply recall that currently, with conventional equipment, the structure and chemistry of materials can be probed at resolutions in the 0.15–0.17 nm. Moreover, the recent development of aberration correctors for electronoptical lenses has allowed surpassing the angstrom limit, and modern advanced electron microscopes provide structural and analytical information at resolutions better than 0.1 nm. The characterization of the local average structure and defects, both at the bulk and boundaries (surfaces and interfaces) of nanoparticles, on which their chemical and physical properties depend, can only be performed with (S)TEM, as we will illustrate with a number of examples.

This chapter does not intend to review the applications of electron microscopy or even (S)TEM to the characterization of nanoparticles. A simple bibliographic analysis reveals not only an increasing interest for these techniques (Figure 26.1) but also, for example, a total of 2083 papers in which the keyword high-resolution electron microscopy (HREM) is used as a characterization tool of nanoparticles (second keyword in the search) in the last 20 years, 1990–2010 period. Likewise, a total number of 358 reviews about applications of electron microscopy in

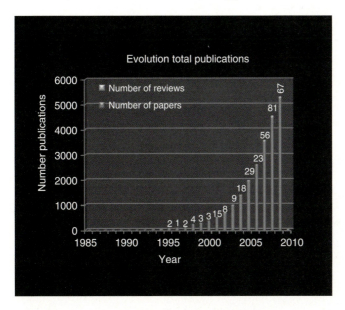

Figure 26.1 Evolution in the past decades of the number of publications and reviews (figures on top of each bar) related to nanoparticle characterization by electron microscopy techniques.

nanoparticle research are reported by the ISI Web of Science search engine in the period 1985–2010.

Instead, we will focus in this chapter on how (S)TEM techniques can nowadays be used to determine specific aspects of nanoparticles as well as on the detailed description of the procedures followed to extract the information from the experimental recordings. With this approach, the reader will gain a comprehensive overview on how the whole set of current (S)TEM techniques can be used for an in-depth characterization of nanoparticles with applications in a wide variety of fields.

The chapter has been organized on the basis of the different techniques. Section 26.2 is devoted to the applications of imaging techniques, including both atomic resolution and other complementary imaging modes, as diffraction contrast in weak-beam dark field (WBDF). Imaging analytical modes are included in Section 26.4, which focuses on nanoanalytical techniques. Section 26.3 gathers information about the use of electron tomography in nanoparticle characterization. Finally, Section 26.5 is devoted to reactivity aspects. In this section, the applications of *"in situ"* TEM to nanoparticle investigation is considered.

Let us finally mention that not only the study of isolated particles is illustrated but also the cases of supported, embedded, or encapsulated (core-shell or surface functionalized) particles.

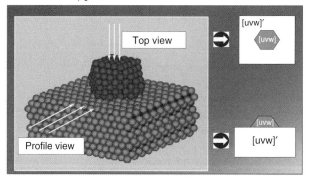

Figure 26.2 Schemes depicting the two principal imaging modes in the case of supported or embedded nanoparticle systems: profile view and plane or top view.

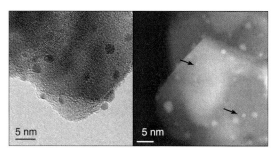

Figure 26.3 HREM (a) and HAADF-STEM (b) images of a Au (2.5% wt)/$Ce_{0.5}Zr_{0.5}O_2$ catalyst.

26.2
Imaging Nanoparticles

Currently, a number of electron microscopy techniques may be used to image different properties of nanoparticles at spatial resolutions ranging from the nanometer down to the subangstrom level. Some of them, that is, HREM, may picture the atomic structure of the particles projected along some crystallographic directions. Others may provide information about the spatial distribution of elements in the nanoparticles, that is, energy-filtered transmission electron microscopy (EFTEM) images. Finally, others are able to combine structural and chemical information, as in the case of high-angle annular dark field (HAADF) images recorded in the scanning transmission electron microscopy (STEM) mode or atomic-resolution chemical imaging. In the case of HAADF-STEM, the contrasts in the images arise from variations in the value of the average atomic number in the atomic columns; this being the reason why these images are also called *Z-contrast images* [12]. On the other hand, atomically resolved chemical images are formed from the information contained in the electron energy loss spectroscopy (EELS) spectra recorded on the imaged area working in the so-called spectrum-imaging (SI) mode [13]. The application of both EFTEM and atomically resolved chemical images are presented in Section 26.4. Therefore, here we focus on the use of HREM and HAADF-STEM, two techniques that nowadays provide in most modern microscopes atomically resolved images, with resolutions roughly in the range 0.08–0.20 nm.

The electrons contributing to these two types of images and their image formation mechanisms are totally different. In the case of HREM images, the electrons involved in the imaging process are those dynamically scattered by the sample at low angles (usually <20 mrad). The image is formed by a nonlinear, partially

(a)

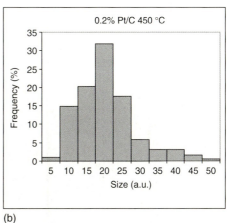

(b)

Figure 26.4 (a) HAADF-STEM image recorded on a Pt(0.2% wt)/C catalyst and (b) Pt particle size distribution.

coherent, interference process between all the diffracted beams allowed in the aperture used to synthesize the image [14]. In contrast, the electrons involved in HAADF images are those scattered at very large angles (usually >50 mrad). No interference between diffracted electrons takes place in this case [13]. These comments are only made to point out that we cannot expect to get exactly the same type of information from them. As we will see, to a large extent these two techniques can be considered complementary and their combined use open access to a very detailed structural and, directly or indirectly, chemical information about both the bulk and the surface of nanoparticles.

In the following sections, the application of HREM and HAADF-STEM as tools to reveal different features of nanoparticle materials is considered. Although treated in different sections, most of these parameters cannot, in fact, be separated because usually the validation of the results requires performing image simulation, and the input models required for this task have to include all the necessary details: size, shape, bulk, and surface structure or interface structure.

26.2.1
Nanoparticle Size

As mentioned in Section 26.1, the most basic but at the same time one of the most interesting parameters of a nanoparticle, which can be determined from the analysis of both HREM and HAADF-STEM images, is their projected size. In most cases, image formation processes provide the necessary contrast to discriminate the nanoparticles from the material onto which these nanoparticles are supported. The latter in many cases is the integral, and usually relevant, part of the material which is under analysis. This is the case, for example, in metal- and oxide-supported catalysts or in materials consisting of particles embedded in a matrix. In other cases, the supporting material is just a medium to allow the observation inside the microscope. In this case, very thin and light materials (usually amorphous carbon, silicon or, more recently, silicon nitride), either in the form of a continuous thin film or as a holey film, are used. Amorphous films are preferred, so as to avoid the contribution of diffraction effects to the images.

Supported and embedded nanoparticles constitute possibly the most complicated cases. So, we refer mainly to this type of materials in the following to provide a view of the analysis capabilities of HREM and HAADF-STEM in the least favorable situation. For this kind of materials, two different type of imaging conditions do occur, namely profile view and plane view imaging (Figure 26.2). In the first case, the electron beam impinges on the sample along a direction contained in the particle/support interface, whereas in plane or top view imaging, the electron beam direction is inclined with respect to the metal/support interface. The imaging process becomes more complex in the top view because of the overlap of both components of the system. As we will see, such overlap gives rise to particular effects, which can be fruitfully exploited to gain very rich structural information about the system.

Figure 26.3 illustrates HREM (a) and HAADF-STEM (b) imaging of supported nanoparticle materials, in particular a catalyst made up of gold nanoparticles dispersed over the surface of a $Ce_{0.5}Zr_{0.5}O_2$ oxide. Observe how, in both cases, the Au nanoparticles can be discriminated from the underlying support, even in the case where they are imaged on a heavy, high atomic number material.

Both images were recorded at similar magnifications, but note how the visibility is slightly improved in the HAADF-STEM mode, especially in the case of the smallest particles, as are those marked with arrows on the HAADF-STEM image, which are just a few angstroms wide.

Given that the intensity in HAADF-STEM images is related to roughly Z^2 values [12, 15], this technique is especially well suited to detect nanoparticles made up of heavy elements distributed on or within low atomic number materials. Thus, it has been extensively applied to characterize metal catalysts supported on high surface area C, MgO, Al_2O_3, or SiO_2. In these cases, the large difference in atomic number between the metallic component and the supporting material allows quite a clear discrimination of the two components as well as the detection of very small entities, down to the angstrom level [16, 17].

Figure 26.4 illustrates the case of a Pt (0.2% wt)/C. The very intense areas correspond to Pt nanoparticles ($Z = 78$). Note how a large number of particles can be detected in spite of the very low loading of the metallic component, just 0.2% in this case. Likewise, Figure 26.5 illustrates the identification of highly dispersed species on alumina, in particular of isolated La atoms in a La-doped alumina support [18] (Figure 26.5a) and Pt trimers on alumina [19] (Figure 26.5).

It is clear, therefore, that HAADF imaging provides a wider particle size visibility window than HREM. Visibility limits, in profile and plane view imaging conditions

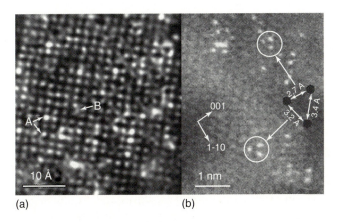

Figure 26.5 HAADF images of (A) La-doped γ-Al_2O_3; (b) a Pt/Al_2O_3 catalyst. Bright dots (A and B) in (a) locate the position of atomically dispersed La atoms in the doped oxide. The dots encircled in (b) demonstrate the presence of Pt trimers over the surface of the alumina support. (Adapted from Refs. [18, 19].)

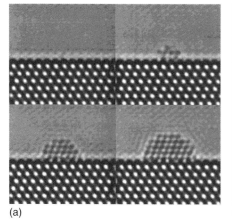 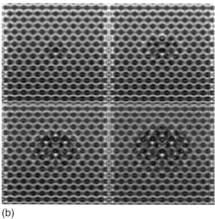

(a)   (b)

Figure 26.6 Particle size visibility study in a Rh/CeO$_2$ catalyst. Structural models of rhodium particles of increasing size supported over a CeO$_2$ crystallite 10 nm thick. Particle size: 0.27 nm (1 atom), 0.8 nm (28 atoms), 1.4 nm (133 atoms), 1.9 nm. (a) Profile view HREM images calculated along the CeO$_2$-[110] zone axis. (b) Top view calculated images. The Rhodius [20] code was used to build the metal/support supercells.

for any particular system, may in any case be established from image simulation studies [20, 21], as shown in Figure 26.6 for HREM images in profile and plane view.

Building appropriate structural models of particles of increasing diameter, both HREM and HAADF images may be calculated. From the comparative analysis of models and calculated images, semiquantitative conclusions can be drawn about the minimum detectable size in each type of image. Note how in this case, where Rh particles supported over CeO$_2$ were modeled; only those particles with diameters larger than about 1 nm produce contrast enough to allow their identification in the image.

The great advantage of both HREM and HAADF, against other macroscopic techniques that provide information about average particle size, is that by analyzing a sufficiently high number of images particle size distributions, such as that in Figure 26.4, can be obtained. These distributions are of utmost importance to rationalize the macroscopic performance. Actually, in many cases the change in physical or chemical behavior does not linearly relate to particle size. When this happens, an average particle size would never represent the behavior of the whole ensemble of particles. Instead, the macroscopic response would be that of the added contribution of every particle size present in the distribution, their frequency acting in that case as the weighing factor.

Datye et al. [22] have compared the use of HREM and HAADF-STEM images to characterize supported particles and concluded that HAADF-STEM is better suited to determine particle size distributions. A similar conclusion is reported by Treacy et al. [23] in the characterization of very small Pt clusters (<1 nm) in zeolites. In these systems, HREM images do not allow detection of particles with less than 20 atoms when the zeolite crystallite is thicker than 10 nm. In contrast, HAADF images obtained using a 0.2 nm probe are shown to be capable of detecting single Pt atoms against a 20 nm thick zeolite support. Similarly, using an aberration-corrected (AC) microscope, Kiely et al. have reported [24] the detection of isolated Au species dispersed over Fe_2O_3, which is in fact an oxide heavier ($Z_{Fe} = 26$) than magnesia, alumina, silica, or carbon.

Regarding HREM, aberration correction provides two fundamental benefits in relation to the determination of particle size:

a. A better definition of the particles on the background support. This is due to the strong influence on the image contrasts of even very small defocus variations when images are recorded in this type of instruments and the fact that nanoparticles are at least a few nanometers in height along the electron beam incidence direction. Hutchison et al. have demonstrated this feature in the case of nanoparticles supported on amorphous carbon [25], and showed that a substantial increase in the ratio of the object contrast to the substrate contrast can be achieved in an AC microscope.
b. Simultaneous elimination of delocalization effects [26, 27] for a range of spatial frequencies at a defocus close to zero [28, 29]. This minimizes the artifacts at interfaces, as in the case of surfaces, and, therefore, improved imaging.

Concerning the determination of particle size, image analysis strategies have been developed to improve the segmentation of nanoparticles from the underlying background both in HREM and HAADF images [30]. Thus, routines based on a so-called local adaptive threshold have been proposed to improve particle size analysis [31, 32]. This image processing method applies a procedure, possibly close to the one used by our eyes when performing particle detection, which involves dividing the whole image into a bidimensional array of square subimages. For each subimage, a threshold value is defined using some predefined criteria, for example, the median or the average value between the minimum and maximum intensity in the subimage. This local threshold, which varies with spatial coordinates, is then applied to discriminate between background and foreground (particle) pixels.

By using this digital image processing method, which also has the advantage that it can be automated, not only a much higher number of particles can be detected in a single image but also a more accurate determination of particle size can be obtained. In fact, once the pixels corresponding to the particle objects have been identified, different definitions of *size* can be applied (maximum distance, Feret's diameter, circular diameter, etc.) [33].

26.2.2
Nanoparticle Bulk Structure

It is evident that both physical and chemical properties of nanoparticles are largely influenced by their exact chemical nature and, in the case of crystalline materials, by the polymorph in which the corresponding composition grows. In the synthesis of nanomaterials, a particular chemical composition and a crystalline structure are usually targeted to allow for a specific macroscopic response. However, answering whether the goal has been achieved, that is, whether the actual composition and structure of the prepared material is as expected, or not, is most of the times one of the basic characterization concerns.

Given that in most cases a particular composition is related to a specific crystalline structure or a series of crystalline structures when polymorphism is possible, determining the latter gives access indirectly to information on chemical composition. XRD is one of the routine techniques to check for bulk structure but its use in the case of only a few nanometer-sized crystallites is limited by peak broadening effects. As we show in this section, HREM and HAADF-STEM images can provide rich information about the internal structure of nanocrystals even in complicated situations in which large similarities between different polymorphs complicate the assignment of a structure. Limitations and complementarities of both techniques are discussed and the key role of image simulation is demonstrated with examples.

To get information about bulk structure from atomically resolved images, either HREM or HAADF-STEM, the contrasts have to be analyzed in two steps. The first step consists in the characterization of the spatial frequencies contained in the images, in terms of lattice plane spacing and angles between lattice planes that have been transferred into the image. Such information can be drawn from the analysis of the diffractograms, which can be digitally calculated from the entire image or parts of the images by applying (fast) Fourier transform (FFT) operations. Comparing this information with that corresponding to the different crystalline phases that are expected to be present in the synthesized material, an assignment can be done to zone axis orientations of one, or several, of the expected phases. In the second step, this assignment must be confirmed on the basis of image simulation, that is, a calculated image reproducing the details of the contrasts observed in the experimental one has to be obtained. In a number of cases where assignment to zone axis orientations of different phases is possible, only this second step allows a definite assignment of the images to a particular phase.

Figure 26.7 illustrates the first step of the analysis process using as an example an HREM image recorded on a Rh (3% wt)/CeO_2 system. Note that the image in the center shows the atomic column contrasts characteristic of this type of images. Different regions can be recognized; the first corresponds to a large crystallite, several tens of nanometers wide, showing a square dot contrast pattern. Calculating the FFT of the region marked with a, the digital diffraction pattern (DDP) shown at the right is obtained. Let us first indicate that this DDP corresponds to the log-scaled representation of the power spectrum of the FFT. Note how in this DDP two spots at 0.27 nm, forming an angle of 90°, are observed. These lattice spacing values and angles are those expected for ceria, CeO_2, along the [001] zone axis. Therefore, we can confirm that this part of the image corresponds to the support component.

On the surface of the ceria crystallite, some smaller, nanometer-sized regions with a different contrast pattern are also observed (those marked with b and c). DDPs corresponding to these areas, shown on the left panel, contain two sets of

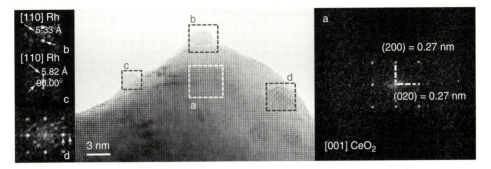

Figure 26.7 (Center) HREM image recorded on a Rh (3% wt)/CeO$_2$ catalyst; (right) DDP corresponding to the region marked with a; and (left) DDPs from areas marked with b, c, and d.

0.22 nm reflections at 70° (marked with white arrows), which can be assigned to the [110] orientation of metallic Rh. These two nanoparticles are imaged in this case in profile view.

The area marked with d deserves a comment. Note that the contrast pattern in this area is clearly more complex that in those already commented on. Such complexity arises from the overlap of a Rh nanoparticle and the CeO$_2$ support crystallite, that is, we are observing the nanoparticle in top view conditions. In fact, in the DDP of this area, both the {200} type reflections of the support at 0.27 nm (yellow arrow) and the {111} reflections of Rh at 0.22 nm (white arrow) are observed aligned along the same direction. In addition, other reflections (those marked with gray arrows), which do not correspond either to the support or to the supported Rh particles, are also observed. These extra reflections, due to double diffraction effects, are called *Moiré reflections*. As we will see in greater detail in one of the following sections, these Moiré reflections are linear combinations of those characteristic of the two crystalline phases that overlap along the electron beam direction. Care must be exercised not to misinterpret them as being due to the presence in the imaged area of a phase different from those that are overlapping. In spite of the intrinsic difficulties involved in their analysis, these reflections provide, as we will see in Section 26.2.4, quite interesting information about the structure of the system.

In summary, spatial frequency analysis from DDPs offer quite a convenient way to characterize the crystalline structure and changes in chemical composition in nanoparticles due to synthesis or postsynthesis treatments. To illustrate this last idea, Figure 26.8 presents a series of HREM images of metal/ceria catalysts submitted to different thermochemical treatments. In the case of Figure 26.8a, a rhodium/ceria catalyst was reduced in H$_2$ at a low temperature (350 °C) after depositing a rhodium precursor onto ceria; Figure 26.8c corresponds to the same catalyst but treated in O$_2$ at 250 °C; finally, Figure 26.8e was recorded on a platinum/ceria catalyst reduced at a very high temperature, 900 °C, after deposition of the platinum precursor.

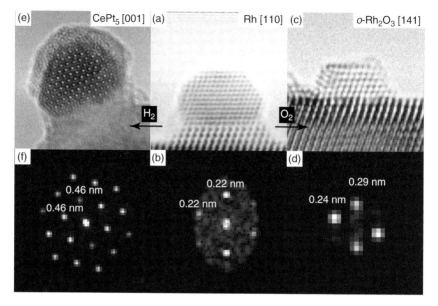

Figure 26.8 HREM images recorded on: (a) a rhodium/ceria catalyst reduced at 200 °C; (c) a rhodium/ceria catalyst oxidized at 250 °C; and (e) a platinum/ceria catalyst reduced at 950 °C. DDPs of each sample are shown just below (b,d,f).

The DDP corresponding to Figure 26.8a shows the same type of reflections in Figure 26.7b,c, which can be assigned to {111} fcc metallic rhodium in [110] orientation. This suggests that the treatment in hydrogen transforms the rhodium oxidized precursor into metallic rhodium nanoparticles.

The reflections observed in the DDP of the sample treated in oxygen cannot be assigned to metallic rhodium. No Rh zone axis contains reflections at 0.24 and 0.29 nm. Instead, they can be interpreted as being due to a Rh_2O_3 oxide with orthorhombic structure along the [141] zone axis.

In the case of the platinum catalyst, increasing the treatment in hydrogen at 900 °C also gives rise to a phase change. Thus, the 0.46 nm × 0.44 nm dot pattern cannot be assigned to any of the zone axis orientations of metallic platinum. The high-temperature treatment promoted in this case the transformation of Pt into a Ce–Pt intermetallic compound [34]. From the five intermetallics described in this system ($CePt$, $CePt_2$, $CePt_5$, Ce_3Pt_2, Ce_7Pt_3), the pattern observed in the DDP can only be interpreted in terms of the [010] zone axis of the $CePt_5$ phase.

Figure 26.9 illustrates the second step of the image interpretation process described above, confirmation by image simulation of the assignment carried out on the basis of DDP analysis. In this case, confirmation of Ce–Pt alloying into a $CePt_5$ stoichiometry is proven. For this purpose, a model of a $CePt_5$ nanoparticle supported on a CeO_2 crystallite was built and submitted to image calculation.

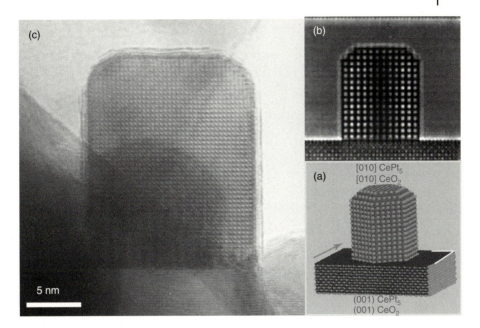

Figure 26.9 (a) Structural model of a CePt$_5$ nanoparticle supported on a CeO$_2$ crystallite; (b) simulated image; and (c) experimental HREM image of the Pt/CeO$_2$ catalyst reduced at 900 °C.

The calculated image reproduces both the 0.46 nm × 0.44 nm dot pattern and the projected shape of the experimental image.

The last two examples illustrate how atomically resolved images allow not only detecting the bulk structure of particles only a few nanometers large but also discriminating between different possible phases. In the case of the rhodium catalyst treated in oxygen, not only the occurrence of oxidation of the nanoparticles could be confirmed but also the formation, specifically, of the orthorhombic phase. This polymorph could be distinguished among four different Rh–O phases [35]. Likewise, in the case of the platinum catalyst treated in hydrogen, a specific intermetallic stoichiometry, CePt$_5$, could be confirmed among five different possibilities.

Detailed analysis of the imaging process allows establishing reliable criteria to distinguish even quite similar structures. This is the case, for example, in the discrimination between the cubic and tetragonal polymorphs of ceria-zirconia mixed oxides [36] (Figure 26.10). As shown in Figure 26.11, the unit cell parameters of these two phases are quite close to each other. This shows that the differences in lattice spacing values and angles between the lattice planes between these two phases are within the usual experimental errors. Thus, by simply measuring lattice spacing values or angles, a reliable assignment of an image to the cubic or tetragonal polymorph is not possible.

Note also from Figure 26.11 that the major difference between these two phases is the slight displacement, <0.02 nm, of the oxygen atoms along the c direction

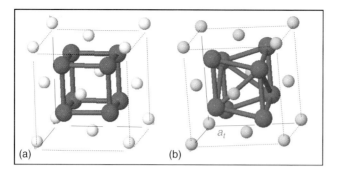

Figure 26.10 The light and medium gray spheres correspond to the cation, Ce/Zr, positions. The dark spheres correspond to oxygen. The major difference between the two structures corresponds to the displacement of oxygens.

of the original cubic cell. Four. oxygens shift downwards, whereas the other four shift by the same amount but upwards. This displacement shows that in the tetragonal phase the oxygens are no longer on the body diagonals of the cell and destroys the ternary axes characteristic of the cubic phase. This reduces the space group symmetry from $Fm\text{-}3m$ down to $P4_2/nmc$. This symmetry lowering involves the activation in the tetragonal phase of some reflections, which were systematic absences in the $Fm\text{-}3m$. This is the case of the (102) tetragonal reflection. In the cubic unit cell, this reflection is the (112), which in the Bravais cell is a systematic absence (Figure 26.11a). The $(102)_t$ reflection gains intensity even under kinematic diffraction conditions (Figure 26.11b). Other systematic absences in the cubic space group, as in the case of the $(001)_c = (001)_t$ or $(1-10)_c = (100)_t$, do not activate under kinematic conditions, but for thicker crystals, under dynamic diffraction conditions, they do also become active (Figure 26.11c). The presence of any of these reflections in the DDPs of the experimental images constitutes the key feature to detect the tetragonal phase. This is the case of the images shown in Figure 26.11d,e.

Given that the intensities of these reflections, which become active in the tetragonal phase, strongly depend on crystal thickness and also that in the case of nanoparticle thickness changes quite steeply, because of the particle shape, it becomes important to study the influence of particle size on their activation. Such a study can be carried out using image simulation. Thus, Figure 26.12 shows a set of calculated HREM images of ceria-zirconia nanoparticles with tetragonal structure of increasing size. According to the DDPs from the simulated images, the detection of the tetragonal phase seems feasible, on the basis of both the $\{102\}_t$ and the $\{100\}_t$ reflections, in nanoparticles with a diameter down to 4 nm. For particles equal to or smaller than 4 nm, the intensity of these reflections in the DDP would not be high enough to allow a clear discrimination from the cubic phase in an experimental image containing noise. Therefore, 4 nm is roughly the size limit to distinguish the c from the t phases in the case of nanoparticles.

This last example does not only provide a clear illustration about the particularities of electron microscopy studies of nanoparticles but also about the key role of

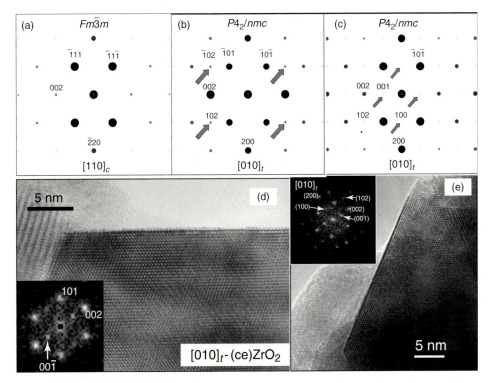

Figure 26.11 (a–c) Calculated electron diffraction patterns corresponding to (a) [110] cubic zone axis; (b) [010] tetragonal zone axis under kinematic diffraction conditions; and (c) [010] tetragonal zone axis under dynamic conditions. (d,e) Experimental HREM images recorded on a CeO_2-ZrO_2 mixed oxide.

simulation in the interpretation of electron microscopy data recorded on this particular type of materials. It allows us, in fact, to define accurate criteria to avoid misinterpretation of experimental results and to fix the limits of each technique. Thus, according to the analysis just exposed, assigning a DDP of a ceria-zirconia crystallite smaller than 4 nm in which no extra reflections are observed to the cubic phase would be simply incorrect.

Concerning the limits of techniques, it would be convenient to talk about the possibilities of HAADF-STEM to carry out the same crystallographic phase, tetragonal versus cubic, discrimination. For this purpose, Figure 26.13a shows a simulated HAADF image expected for a tetragonal crystallite. Note that the DDP (Figure 26.13b) shows only the reflections expected for the cubic structure. No extra reflection is present. This can be understood if we take into account that the intensity in the HAADF is related to the atomic number. Thus, the HAADF image shows maxima at the positions of the cationic (Ce and Zr) atomic columns. The positions of the oxygen atoms are not reproduced in the image. Since the major

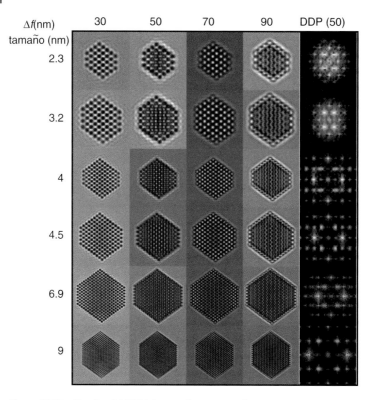

Figure 26.12 Simulated HREM images for tetragonal ceria-zirconia nanoparticles of increasing size at different defocus values. The DDPs at the right were calculated from the images at $\Delta f = 50$ nm. Both the $(112)_t$ and the $(001)_t$ spots are clearly visible down to the particle of 4 nm of diameter.

difference between tetragonal and cubic phases is related to displacements in the oxygen sublattice, the image of the tetragonal phase cannot be discriminated from that of the cubic. As depicted in Figure 26.13d, even HAADF images recorded in an AC microscope do only show contrasts attributable to the position of the cationic columns. In this case, a model of a wedge-shaped crystal of varying thickness, as is usual in the case of nanocrystals, has been modeled (Figure 26.13c). According to the image intensity profiles obtained on the simulated image, maxima only peak at the cation column positions, at distances 0.532 nm from each other along the [101] direction (Figure 26.13e).

It becomes clear that in this particular case, HREM images offer advantages over HAADF-STEM imaging. Similarly, as we will see just now, HAADF-STEM is better suited in those cases where structural changes are linked to modifications in the cation, high atomic number, sublattices. To illustrate this case, we will refer to the characterization of disorder–order transitions in mixed oxides. This is the case, for example, of ceria-zirconia oxides when treated in hydrogen at high temperature.

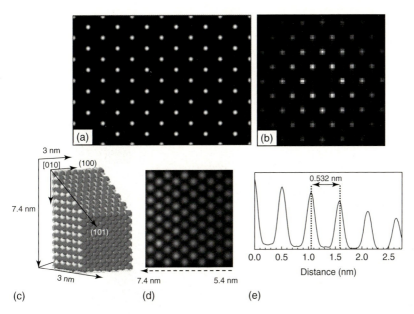

Figure 26.13 (a) Simulated HAADF-STEM image of the tetragonal ceria-zirconia phase down the [100] zone axis; (b) corresponding DDP; (c) model of a wedge-shaped tetragonal ceria-zirconia crystallite used as input to calculate the HAADF image shown in (d); and (e) image intensity profile of the calculated image along the [101] direction.

This kind of thermochemical treatments give rise to the transformation of oxides with cubic, fluorite-type, (Fm-$3m$), or tetragonal ($P4_2/nmc$) solid solution structures into oxides with a pyrochlore ($Fd3m$)-related structure. This pyrochlore structure can be considered a $2 \times 2 \times 2$ superstructure of the fluorite that arises when Ce and Zr, which are randomly distributed in the cationic sublattice of the cubic and tetragonal solid solutions, become ordered into an arrangement in which they alternate along the three major <100> directions.

As shown in Figure 26.14, this structural transition can be detected in the DDPs of both HREM and HAADF images recorded along the [110] axis of the original cubic structure. In this case, extra $1/2\{111\}_F$ and $1/2\{002\}_F$ (F stands for fluorite) appear, which can be considered a fingerprint of the structural transformation.

We can then wonder what the advantage of HAADF-STEM in this case is. To answer this question and to clarify the importance of this technique, we have to analyze the simulated HREM images shown in Figure 26.15. They correspond to two different models of a ceria-zirconia oxide with $2 \times 2 \times 2$ superstructure features. In Figure 26.15a, the superstructure was formed by displacing slightly the oxygen atoms along the body diagonals of the original fluorite unit cell, that is, the <111> directions. The Ce and Zr atoms were kept fully disordered in the model used to calculate this image. In other words, this is an anion superstructure.

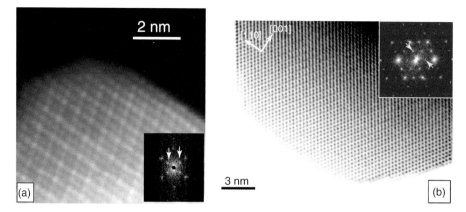

Figure 26.14 (a) HAADF-STEM and (b) HREM images of a ceria-zirconia oxide after treatment in hydrogen at 900 °C. Note in both cases the presence of $1/2\{111\}_F$ superstructure spots in the corresponding DDPs (marked with arrows).

In the model used to calculate Figure 26.15b, the oxygens were kept at the original, undistorted, fluorite positions but Ce and Zr were ordered into the pyrochlore-type arrangement. Therefore, a cation superstructure is considered in this second case.

Note that the DDPs of both images (Figure 26.15c,d) show the same $\frac{1}{2}\{111\}_F$ superstructure reflections. This clearly indicates that although HREM is sensitive to the formation of a superstructure, it is not able to discriminate the real origin of the superstructure effects. The superstructure features cannot be unambiguously assigned to a transition either in the oxygen or the cation sublattices. It is not therefore a specific technique.

On the contrary, given that the contrasts in HAADF images are dominated by the cationic columns, the appearance of a superstructure can be reliably linked to the ordering of the cations. Thus, Figure 26.15c,g shows the simulated HAADF-STEM images corresponding to the same models previously described for the cation and anion $2 \times 2 \times 2$ superstructures. Note that in this case, in contrast to that described for HREM images, the superstructure spots are only visible in the DDP of the model corresponding to the cation ordered mixed oxide (Figure 26.15e). In the DDP of the image corresponding to the model in which the superstructure is created by oxygen displacements, only the reflections due to the basic fluorite structure are observed (Figure 26.15h).

Atomic imaging techniques do also allow the detection of highly localized structural defects occurring in the bulk of the nanoparticles. To illustrate this point let us consider the image recorded on a Ru/CeO$_2$-ZrO$_2$ catalyst shown in Figure 26.16. Ruthenium crystallizes in a hcp structure. The analysis of the stacking sequence of the close-packed planes (the {001} planes in the hexagonal unit cell) in the ruthenium particle observed in this image clearly shows a local structural deviation, which extends over a nanometer-sized region. Note how, as depicted in the enlargement in Figure 26.16b, in the lower part of the image the stacking

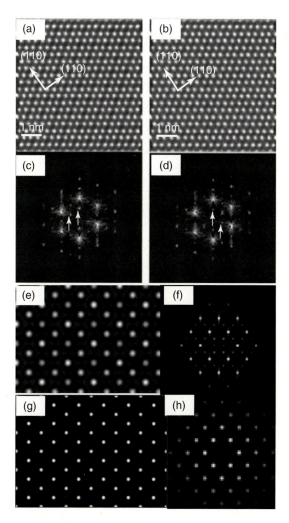

Figure 26.15 Simulated HREM images corresponding to a 2 × 2 × 2 superstructure generated by (a) oxygen displacements along the <111> directions and (b) cation ordering. DDPs of these images are shown in (c) and (d) respectively. Simulated HAADF-STEM images for the cation ordered and oxygen displacement models are shown in (e) and (g) together with their DDPs, (f) and (h).

sequence corresponds to that expected for the hcp phase, ABABA... However, in the upper part of the particle, the stacking sequence changes to an ABCABC... pattern, which is expected for a cubic close-packed, fcc-type phase which is not the expected for Ru. Structural modifications taking place at such local levels, as the intergrowth of two types of structure within a single particle such as the one evidenced in this example, can only be evidenced by atomic imaging techniques. They could never be unveiled by macroscopic structural analysis, which provided

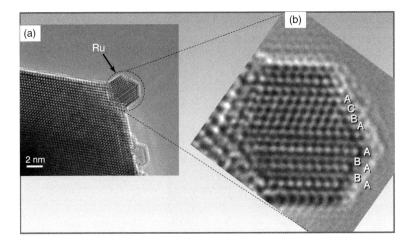

Figure 26.16 (a) HREM image recorded on a Ru/CeO$_2$-ZrO$_2$ catalyst and (b) enlargement of the nanoparticle encircled in (a). The stacking sequence of close-packed planes ({111} in the cubic close packing or {001} in the hexagonal close packing) are marked with the A, B, and C.

average information about the whole set of particles. Given the great impact that such structural defects could play on the electronic properties of the nanoparticles and, hence, on their physical and chemical properties, this example clearly shows the high characterization potential of atomic imaging techniques, both HREM and HAADF.

Summarizing all the results presented in this section, we have shown different examples that demonstrate how HREM and HAADF-STEM are complementary techniques to characterize in detail the bulk structure of nanoparticles. Analysis in the spatial frequency space, by means of the calculation and geometric characterization of DDPs, constitute an appropriate tool to assign crystallographic phases even from nanosized regions. Likewise, image simulation is completely necessary to establish the limits of each technique as well as to get rid of possible misinterpretations.

Finally, we will consider the case in which the structure of the system does not refer to the internal structure of the particles themselves, but, instead, to that corresponding to the assembly of different nanoparticles. A lot of synthetic efforts have been addressed to prepare materials in which more than one type of nanoparticles are assembled into different type of arrangements, which could be termed either as *molecular particles* or *molecular crystals* [1]. The idea behind this strategy is not only to combine at the nanoscale the specific properties of the individual components, for example, optical + magnetic properties but also to benefit from the interactions that could arise once the individual nanoobjects are arranged together into a new superstructure [1, 37]. (S)TEM imaging techniques provide the necessary clues to verify that the synthetic goal has been achieved, that is, the assembly has taken place in the expected arrangement. Figure 26.17

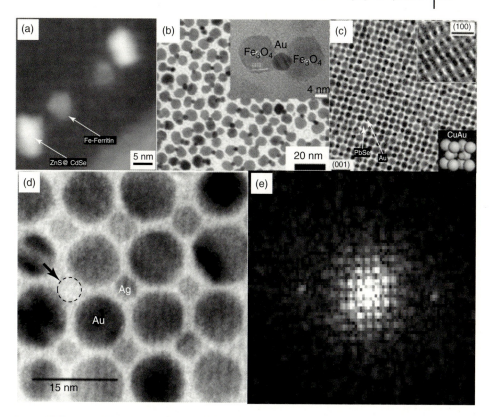

Figure 26.17 (a) HAADF image of a ZnS@CdSe–Fe ferritin assemblies [38]; (b) low- and high-magnification TEM images of Fe_2O_3-Au-Fe_2O_3 trimers [37]; (c) TEM image of a PbSe-Au supercrystal with CuAu structure [39]; (d) Ag–Au supercrystal [40]; and (e) DDP from (d). (Adapted from Refs. [37–40].)

illustrates some characterization examples of molecular and supercrystal structures. In the case of Figure 26.17a, the goal was to assemble into pairs quantum dots consisting of ZnS@CdSe core-shell nanoparticles and Fe-Ferritin nanoparticles. HAADF images recorded on this system clearly reveal the formation of these pairs. The large difference between the atomic number of the elements involved in the quantum dot ($Z_{Zn} = 30$, $Z_S = 16$, $Z_{Cd} = 48$, $Z_{Se} = 34$) and the iron ferritin nanoparticle ($Z_{Fe} = 26$, $Z_O = 8$) allows to directly identify the two components in the image. Thus, as marked on Figure 26.17a, the quantum dot would correspond with the high-intensity particle, whereas the ferritin would correspond with the low-intensity one.

Fringe analysis of HREM images was employed to identify the metallic and oxide components in the case of the iron oxide–gold molecular assemblies shown in Figure 26.17b as well as in the PbSe-Au supercrystal in Figure 26.17c. Note how DDPs from low-magnification images of these superlattices, in which the internal structure of each nanoparticle is not resolved, can be used to characterize the crystallographic features of the superlattice itself, that is, its symmetry and superlattice parameters. Thus, by measuring the diffraction spots observed in Figure 26.17e, the distances characteristic of the squarelike arrangement of Ag–Au nanoparticles in Figure 26.17d could be determined.

Of course, in all the examples illustrated above, the structural analysis of the material could be further improved by the detailed characterization of the structure of the individual nanoparticle components, using the methods already described for isolated or supported nanoparticles.

Likewise, the defects in the superlattices would also be characterized. Note, for example, the identification of a *vacancy* (marked with an arrow) in the Ag particles sublattice in Figure 26.17c.

In any case, these last examples highlight the possibilities of electron microscopy imaging techniques in the characterization of more complex, nanoparticle-based materials.

26.2.3
Nanoparticle Shape and Surface Structure

Determining the shape of nanoparticles, which, on the other hand, is intimately related to the determination of some of its surface features, becomes key to understanding both their physical and chemical properties. Both crystallography and chemical composition are crucial aspects.

It is clear that a determination of shape would certainly require a 3D description and, therefore, a 3D characterization technique. In Section 26.3, we consider the possibilities of electron tomography, but we have to advance here that for particles smaller than about 5 nm, given the actual resolution limits of electron tomography, the best approaches to describe the shape of very small nanoparticles are based on 2D images.

One of these approaches is that based on WBDF imaging. In this imaging mode, the crystal to be imaged is tilted a few degrees out of a predefined zone axis so as to put it into kinematic scattering conditions. Then, one (hkl) beam is selected with the objective lens aperture to form the image. Figure 26.18 illustrates WBDF imaging of Pd nanoparticles using the (002) beam. As shown in Figure 26.18a, the intensity of the diffracted beam selected to form the image changes periodically because of its coupling with the rest of the beams in the diffraction plane.

Thus, minima appear at multiples of the so-called extinction distance (ε_{002}) (roughly 2.2 nm in this case) followed by maxima at multiples of $\frac{1}{2}\varepsilon_{002}$. Thus, when the image is formed using this beam, it finally shows a series of black and white fringes centered at the position of the minima and maxima in the intensity versus thickness plot. As shown in Figure 26.18b, these fringes are similar to contour lines in a thickness map. Therefore, from the outline of these contours a particle shape can be deduced (Figure 26.18c). Note that in this case, the particle diameter is about 10 nm.

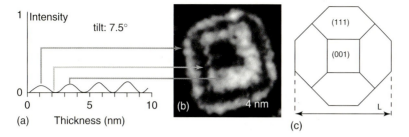

Figure 26.18 (a) Evolution with thickness of the intensity of the (002) beam in Pd for a crystal tilted 7.5° off the [001] zone axis; (b) (002)-WBDF image of a few nanometers Pd particle; and (c) sketch of the Pd particle deduced from (b). (Adapted from S. Giorgio et al. [41].)

This technique has been fruitfully applied to the morphological characterization of a variety of metal particles [42–44] such as Au, Pd, Ag, Rh, among others. Both isolated and supported nanoparticles can be characterized using this one-beam one-tilt technique. Note, however, that the resolution of the images is rather limited.

An alternative way to derive the representative morphology of the nanoparticles present in a sample is by considering a series of atomically resolved images (HREM or HAADF) in profile view.

Once the bulk structure of the nanoparticle has been determined using the procedures described in Section 26.2.2, the crystallographic planes defining the surfaces of the particles can be assigned, as shown in Figure 26.19a,d. Once this assignment has been made on a series of images recorded on different particles, a specific shape can be guessed, as depicted in Figure 26.19b,e, fixing the growth of the different planes. Then, these models can be submitted to image simulation (Figure 26.19c,f). A good match between the contrasts observed in the experimental and simulated images can be considered the figure of merit to validate the proposed morphology. Data presented previously in Figure 26.19 also illustrated this procedure.

In general, although this approach is applicable to any particle size, it becomes most important for the smallest sizes for which tomographic reconstructions are out of reach. Moreover, the resolution in the definition of the surface is that characteristic of atomic imaging, currently in some microscopes below the angstrom. Thus, any step at the surface which is imaged in profile will be detected in the corresponding image, thereby allowing incorporation of this information into the structural model.

Following the discussion by Flüeli and coworkers [45, 46] about the effects of thickness variation across nanoparticles on the displacements of the atomic column contrasts in HREM images with respect to their exact position in the structure, Yacaman et al. proposed [47] that the number of metal atoms along the incident beam direction could be determined, thus opening the possibility of estimating the roughness of the particle surface, that is, variations in the occupancy

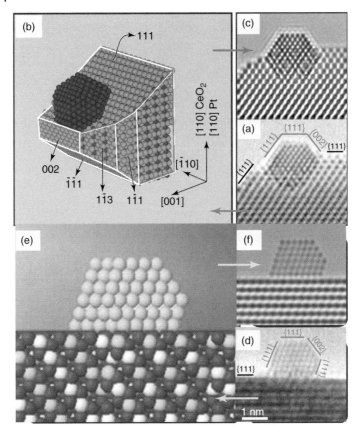

Figure 26.19 (a) Experimental HREM image recorded in a Pt/CeO$_2$ catalyst; (b) structural model built using the information contained in (a); (c) simulated HREM image corresponding to the model in (b); (d) experimental HREM image recorded in a Au/CeO$_2$–ZrO$_2$ catalyst; (e) structural model built using the information contained in (d); and (f) simulated HREM image corresponding to the model in (e).

of surface sites with respect to that corresponding to perfectly flat surfaces. The procedure proposed by these authors consisted in detailed comparisons of the intensities and exact distances between adjacent columns, as measured on digital images of experimental and simulated HREM images of nanoparticles. Following this procedure, a description of the shape of the particle not only in directions perpendicular to the incident beam direction but also along the electron beam direction could be gained. In any case, the application of this procedure requires very detailed calculations considering not only the influence of the particle shape but also of residual crystal or beam tilts, which could also contribute to a shift in the position of the maxima in the experimental images with respect to that of the projected columns. Considering that full correction of very small crystal tilts in the case of nanoparticles is a difficult task, care must be exercised when applying the proposals of Yacaman *et al.*

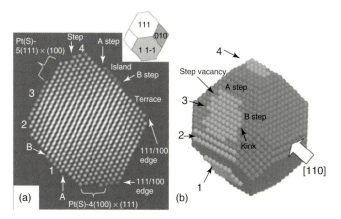

Figure 26.20 (a) Phase image of a Pt nanoparticle restored by EWR from a series of AC-HREM images at different defocus values. Different types of surface defects are marked and (b) detailed structural model built from the analysis of the reconstructed phase image. (Adapted from A. Kirkland et al. [51].)

With the development of AC microscopes and through focus exit wave restoration (EWR) algorithms [48], the detailed analysis of the contrasts in AC-HREM images, actually on the reconstructed phase images, allows a very detailed and reliable detection of surface defects lying both along and perpendicular to the incident beam direction [49, 50]. An example of this analysis is illustrated in Figure 26.20. A detailed comparison between the phases corresponding to the different atomic columns in the restored image and those corresponding to trial structural models is necessary to fix the details of the surface defects.

HAADF-STEM images recorded in AC microscopes are particularly well suited for this kind of analysis. Two properties of these images are of large interest. First, the direct relationship between the intensity of the signal and the values of Z in the atomic columns and thickness. Thus, with appropriate calibration procedures, the intensities can be quantified to yield the approximate number of atoms in a particle and, moreover, the number of atoms in individual atomic columns [52–54]. Second, they do not suffer from delocalization effects, so by appropriate procedures, the position of atomic columns in the projected image can be very precisely defined.

The first property has been exploited, in the case of monometallic particles, to monitor the change in particle thickness with position inside the nanoparticle. Thus, Terasaki et al. [55] used this approach to determine the number of atomic layers stacked along the [001] direction of the Au nanoparticles present in a Au/Fe_2O_3 catalyst (Figure 26.21a). The intensity of the HAADF signal observed in image intensity profiles was compared to that corresponding to simulated images in which the number of gold atoms stacked along the [001] was varied between 1 and 14 (Figure 26.21b). The linearity observed in the relationship between image

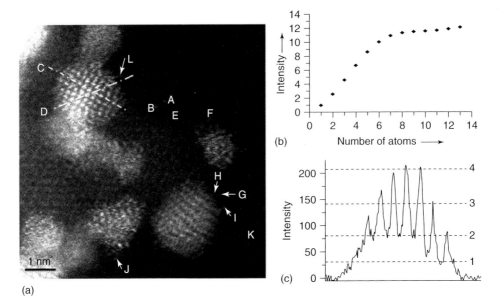

Figure 26.21 (a) AC-HAADF-STEM image recorded on a Au/Fe$_2$O$_3$ catalyst; (b) correlation between simulated atomic column intensity along the Au [001] direction and the number of gold atoms; and (c) intensity profile across the line marked with D in the experimental image. (Adapted from O. Terasaki et al. [55].)

intensity and number of stacked atoms in the range between 1 and 6 atoms, allowed the authors to interpret directly the intensity in the experimental images in terms of number of atoms in the atomic columns (Figure 26.21c). Particles in which up to four atomic layers were stacked could be detected. Similar studies have been performed by Pennycook et al. on Au/TiO$_2$ catalysts [56, 57].

A similar approach has been fruitfully applied to more complicated systems, as in the case of bimetallic nanoparticles. Thus, from the analysis of experimental intensity profiles recorded on Pt–Pd bimetallic nanoparticles and their comparison with those obtained from simulated images, Wang et al. [58] determined with large accuracy the formation of Pt@Pd core-shell nanoparticles with shell thickness varying from just one (Figure 26.22a,c) up to four atomic Pt layers (Figure 26.21b,d). Even the structure of nanoparticles composed of three different layers has been determined in AuPd bimetallic nanoparticles [59]. As shown in Figure 26.21e,f, the quantitative analysis of HAADF-STEM images in this case allowed the authors to identify Pd@Au@Pd nanoparticles. Finally, Figure 26.22g,h illustrates quite elegant results reported by Sánchez et al. [53] in the analysis of Pd–Pt bimetallics. Both Pd@Pt core-shell nanoparticles, as the one shown in Figure 26.22g, and Pt@Pd could be detected and clearly discriminated by comparison with calculated intensities obtained from model nanoparticles, as that shown in Figure 26.22h.

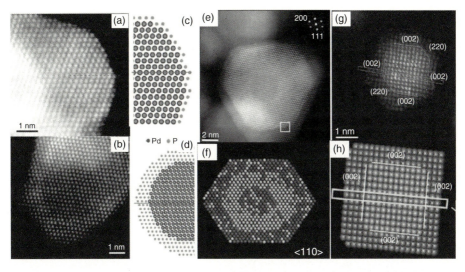

Figure 26.22 HAADF-STEM images of bimetallic particles: (a,b) PtPd; (e) AuPd, and (g) PdPt. Core-shell type structural models of the different systems which explain the contrasts observed in the experimental images are shown in (c) and (d); (f) and (h) respectively. (Adapted from Refs. [53, 58, 59].)

To finish the comments on bimetallic nanoparticles, let us mention that using specific image analysis routines, Shannon *et al.* have detected the presence of isolated dopant Re atoms randomly distributed into the columns of Co nanoparticles in a Fischer–Tropsch catalyst [60].

The last examples commented on above illustrate that by the appropriate analysis of HAADF-STEM images not only shape but also bulk structure and surface nature can be simultaneously determined.

As depicted in Figure 26.23, HREM and STEM-HAADF images do also allow the detection of less conventional morphologies. Thus, Figure 26.23a corresponds to a Au nanoparticle with the shape of a hollow sphere, that is, an egg-shell structure [61]. A number of features are worth being commented on about this HREM image. First, note how the width of the shell can be easily estimated from the measurement of the darker circular region in the image, a value close to 3 nm could be thus estimated. More important, the DDP shown as an inset at the bottom right, which was obtained from the whole area inside the large circle, shows only one set of {002} reflections of metallic gold. This clearly means that the whole nanoparticle, in spite of its peculiar shape, is a single crystal. In this case, this single crystal hollow sphere is imaged down the [001] zone axis. Finally, the HREM image indicates that it is a closed structure. Note in this respect how, although in the image the contrasts are hardly visible, the DDP obtained from the area encircled within the small diameter circle at the center, shown as an inset in the upper right corner, shows exactly the same {002}-type reflections.

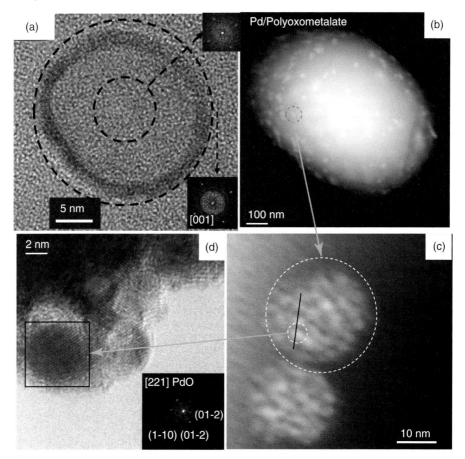

Figure 26.23 (a) HREM image recorded on a Au nanoparticle with hollow sphere structure. DDPs of the areas encircled are shown as insets; (b) HAADF-STEM image recorded in a Pd catalyst supported on a polyoxometallate support. An enlargement of an area as the encircled is shown in (c); and (d) HREM image of one of the small particles as that marked with the small diameter circle in (c). The DDP that is shown as inset corresponds to this particle. (Adapted from Refs. [61, 62].)

Therefore, the Au particle is covering the whole area within the larger diameter circle and not only the area within the darker circular ring, which is easily identified in the image. The darker aspect of the annulus is simply due to the larger thickness, in projection, of this area. This would, in fact, be the thickness variation expected for a hollow sphere structure.

On the other hand, the HAADF-STEM image shown in Figure 26.23b was recorded on a Pd catalyst dispersed on a polyoxometalate support, POM: [bmim]4HPO4OV$_2$Mo10, bmim$^+$ = butylmethyl-imidazolium. At first glance, the brighter areas in the low-magnification image can be interpreted as Pd-containing nanoparticles, given the higher atomic number of Pd (Z_{Pd} = 46), in comparison to those of molybdenum (($Z_{Mo} = 42$), vanadium ($Z_V = 23$), or phosphorus ($Z_P = 15$)). These particles, with diameter of a few tens of nanometers, seem to be single particles, but images recorded at higher magnification (Figure 26.23c) reveal that they are in fact berrylike particles constituted by aggregation of smaller, about 2 nm particles. One of these smaller particles is marked with a small diameter circle in Figure 26.23c. Moreover, the analysis of one of these small particles by HREM indicates that they are, in fact, not metallic Pd particles but, instead, PdO nanoparticles. The DDPs of the HREM images clearly revealed the presence of reflections characteristic of this oxide stoichiometry.

HREM can also sometimes be used to detect the presence of monolayer structures on the surface of nanoparticles. This is particularly possible when the structure of the atomic layer covering, partially or totally, the surface of the nanoparticle has a structure that differs significantly from that of the nanoparticle itself. To illustrate this point with an example, let us consider the images presented in Figure 26.24. The experimental HREM image was recorded on a Pt/CeO$_2$ catalyst, which was treated in hydrogen at high temperature, 700 °C. DDP analysis of the structure of the particle, which is observed clearly, indicates that it corresponds to metallic, fcc, Pt. Both the support and metal particles are imaged down the [110] zone axis (Figure 26.24a). An inspection of the image reveals the presence of a series of

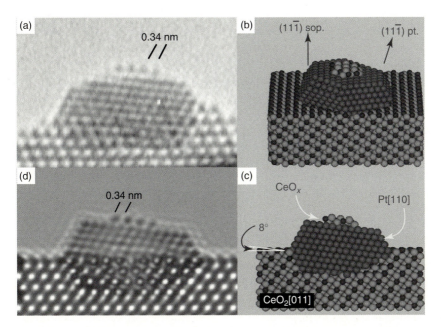

Figure 26.24 (a) Experimental HREM image recorded on a Pt/CeO$_2$ catalyst; (b) structural model of a Pt nanoparticle partially buried into the ceria support and with the topmost {111} surface covered with a one atomic layer high CeO$_x$ patch. Perspective view; (c) cross-section projection of the model along the [110] direction of the metal; and (d) simulated image corresponding to the model in (c). (Adapted from R.T. Baker et al. [63].)

dark dots on the topmost {111} surface of the nanoparticle separated 0.34 nm from each other, as marked on the image. These dots should correspond to projected atomic columns. Nevertheless, there is no way to get, in any projection, a 0.34 nm distance between adjacent atomic columns in the structure of Pt. Therefore, we must admit that these contrasts cannot be associated with Pt atoms. Considering that the chemical treatment to which the catalyst was submitted could promote the migration of support moieties onto the metal particles, a model considering this effect was built (Figure 26.24b). As shown in Figure 26.24c, a Pt particle partially buried, its lower part, into the ceria support and covered by a one-monolayer height circular patch of support was considered. The simulated HREM image shows the contrast features observed in the experimental one. In fact, the dots at 0.34 nm on the top surface are very accurately reproduced, which confirms that the coverage by the support can be detected in the HREM recordings.

Detailed structure modeling and image simulation studies have revealed that the observation of this type of subnanometer thick overlayers depends strongly on the defocus value at which the image is recorded [64].Thus, Figure 26.25 shows a model of a metallic rhodium particle fully encapsulated within a thin CeO_2 layer epitaxially grown on top of the nanoparticle surfaces. The cross section of the model (Figure 26.25a) shows how thin the overlayer is, whereas the top view of the model shown in Figure 26.25b indicates that the particle is fully covered by ceria. Note how in the image calculated for a 54 nm defocus (Figure 26.25c), the contrasts due to the support overlayer are appreciated both along the particle edges and in the area of the particle, in the form of double-diffraction Moiré-type contrasts. However, in the image calculated for a defocus of 104 nm (Figure 26.25d), the support contrasts are only observed along the perimeter of the particle. The particle area shows the [110] zone axis contrasts of the metal particle. This would erroneously lead us to think that the particle is only partially covered.

Care must then be exercised to make conclusions about the extent of the coverage in these cases. Two examples of this apparent partial coverage are included in Figure 26.26. In AC-HREM images, which are even more sensitive to small changes in defocus value, detailed simulation studies, as that illustrated in Figure 26.25, would be completely necessary to make precise assessments on this aspect.

Let us dedicate some final comments to the possibilities of analysis of the topmost surface structure using HAADF-STEM imaging. We will concentrate on two aspects: (i) the displacement of surface atoms in this layer associated with relaxation phenomena and (ii) the use of this technique to characterize compositional aspects.

For the first question, we will refer again to metal nanoparticle study cases. Thus, on the basis of phase images obtained by restoring the exit wave function from a focal series of AC-HREM images (a total of 20 images) of a 6 nm Pt nanoparticle, Dunin-Borkowski *et al.* have estimated with great accuracy the displacement of atomic columns around different surface defects (Figure 26.27). The displacements were measured with respect to the position of bulk columns in the particle. This analysis showed that the surface atomic sites on (111) and (100) facets

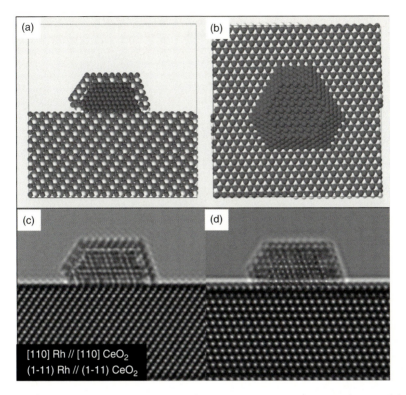

Figure 26.25 Image simulation study of nanoparticle coverage effects; (a) cross section of an structural model where a Rh metal particle is supported and decorated by ceria in a parallel epitaxial relationship; (b) top view of the model showing that the particle is fully encapsulated by the support. HREM image simulations of the model for a defocus value of 54 nm (c) and 104 nm (d) using the following parameters: HT= 200kV; Cs = 0.7 mm; θ_c = 1.2 mrad; and Δ = 10 nm. The support thickness and particle size considered in the model were 9 and 3.2 nm, respectively.

generally shifted outward with largely varying magnitudes. The largest atomic displacement was observed on the (111)/(100) edge. Also important, significant lateral displacements were detected at the surfaces of the nanoparticle.

To illustrate point (ii) above we will come back to the HAADF-STEM characterization of ceria-zirconia mixed oxides. In particular, to the study of the cation-ordered, pyrochlore-type phase of this type of oxides. Figure 26.28a shows an HAADF-STEM image recorded on a $Ce_{0.68}Zr_{0.32}O_2$ mixed oxide with pyrochlore structure. The crystal was slightly tilted out of the [110] zone around the (111) axis so as to produce a fringelike image. The cation ordered structure is revealed by the 1/2{111} reflections, 0.61 nm, observed in the DDP shown as an inset. Each fringe corresponds to one of the (111) planes in this structure.

The image intensity profile along the direction perpendicular to the (111) planes shows peaks of alternating low and high intensity (Figure 26.28b). As shown in the model included in Figure 26.28c, such an alternation in the intensity of the

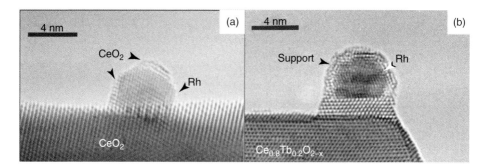

Figure 26.26 HREM images of encapsulated nanoparticles in a Rh/CeO$_2$ catalyst [63] (a) and a Rh/CeTbO$_x$ catalyst (b) At the defocus values at which the images were recorded, part of the covering layers are not properly imaged. (Adapted from Refs. [63, 64].)

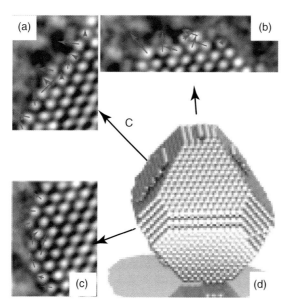

Figure 26.27 (a) 3D model of a Pt nanoparticle showing surface defect features such as steps or kinks and (b–d) AC-HREM images corresponding to the surface locations indicated by arrows. The red arrows in these images indicate the direction and magnitude of the atomic column displacements around the surface defect sites. (Adapted from R. Dunin-Borkowski et al. [65].)

fringes is due to the presence in the structure of two types of planes; a Ce-rich variant and a Zr-rich one. These two types of chemically distinct (111) planes stack in an alternate manner along the <111> direction. The HAADF-STEM simulated image and its corresponding intensity profile shown in Figure 26.28d,e support this interpretation.

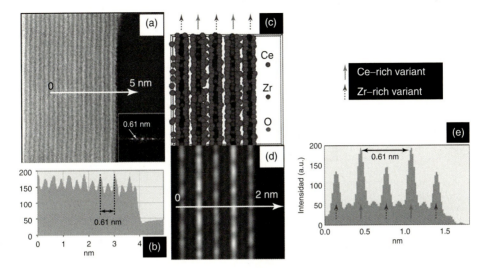

Figure 26.28 (a) HAADF-STEM image of a Ce$_{0.68}$Zr$_{0.32}$O$_2$ mixed oxide with pyrochlore structure; (b) intensity profile along the line perpendicular to the {111} planes marked on (a); (c) structural model of the oxide showing the stacking of the two types of {111} variants; (d) simulated HAADF-STEM image of the model shown in (c); and (e) intensity profile on the line marked on the simulated image. Note that the pattern of alternating low- and high-intensity peaks observed in the experimental image is reproduced. (Adapted from J.C. Hernández et al. [66].)

The most important feature of this image refers to the nature of the first plane exposed at the surface of the nanocrystallite. The image intensity profile shown in Figure 26.28b indicates that it corresponds to a low-intensity line and, therefore, to the Zr-rich variant. It becomes, therefore, evident from this study that using appropriate interpretation procedures, HAADF-STEM images can be exploited to retrieve chemical information about the surface of nanocrystals.

A quite important message derives from the last two examples: atomic imaging techniques provide unique capabilities to characterize the structural and compositional features of nanoparticle surfaces.

26.2.4
Nanoparticle Lattice Distortions and Interface Structure

We devote this last section of the application of 2D atomic imaging techniques to the analysis of two aspects of nanoparticles, which are, in fact, connected through the procedure that can be used to characterize them: (i) the deviation of the lattice parameters with respect to those of the bulk phases and some features of the growth of nanoparticles onto support or matrix materials; in particular, the orientation relationships between nanoparticles and matrix or support component. They both are aspects that can influence the behavior of the nanoparticle through modifications either of their morphology or of their electronic properties and that, as we will now see, can be very finely investigated by both HREM and HAADF.

Atomically resolved images provide information on lattice spacings. This information can be processed either in real space, by direct measurements of distances either on the images or on intensity profiles, or in reciprocal space, by detecting the position of maxima in DDPs of the images and transforming the position of maxima in d values. In the last case, the information is averaged over the whole area from which the DDP is calculated.

Hofmeister *et al.* have routinely used the reciprocal space approach to study the influence of particle size on lattice distortion phenomena in different nanoparticle systems. They used this approach in the investigation of both free [67–69] and embedded [70, 71] nanoparticles. Figure 26.29 illustrates some of these studies.

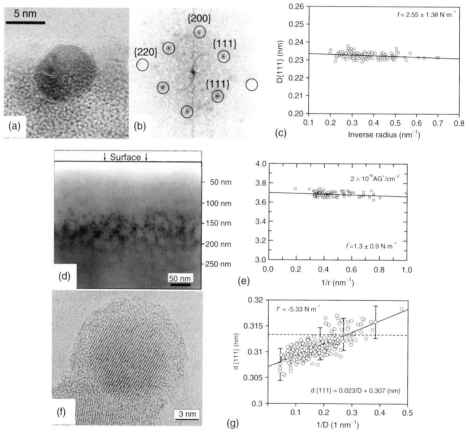

Figure 26.29 (a) HREM image of a silver nanoparticle supported on carbon; (b) DDP from the HREM image; (c) evolution of the {111} lattice spacing with the reciprocal particle size; (d) TEM image of the cross section of a sample consisting of Ag nanoparticles embedded in a silicate glass; (e) evolution of the {111} lattice spacing with the reciprocal particle size in sample (d); (f) HREM image of a Si nanoparticle produced by laser pyrolysis of silane; and (g) evolution of the {111} lattice spacing with the reciprocal particle size in sample (f). (Adapted from H. Hofmeister *et al.* [67–71].)

In the case of silver nanoparticles, either isolated (Figure 26.29a) or embedded (Figure 26.29d), they find an increase in the average lattice parameter with particle size (Figure 26.29c,e). In contrast, Si nanoparticles (Figure 26.29f) did show a contraction effect with increasing particle size (Figure 26.29g).

The work of Giorgio et al. [72] on the structure of 4 nm Ni nanoparticles supported on MgO microcubes illustrates an analysis in real space. These authors determined, from direct measurements on intensity profiles of HREM images, the distance between neighboring (002) planes lying parallel to the metal/support interface. They found lattice distances different from bulk-Ni metal values for the first two planes closer to the interface. The values remained close to the bulk value for planes further from the support.

In a much more recent work, based on AC-HREM images and using an approach similar to that used by Dunin-Borkowski et al. [65], Yacaman et al. have analyzed with a precision of about 0.002 nm the evolution of the {111} lattice spacings in a complex nanoparticle system, icosahedral FePt bimetallic nanoparticles smaller than 10 nm [73].

In addition to an enrichment of Pt at the surface of the nanoparticles (Figure 26.30a), they found that the {111} lattice spacing (Figure 26.30b) decreases exponentially from the surface to the core of this type of multiply twinned particles (Figure 26.30c).

Similarly, combining AC-HREM images and strain mapping measurements, Hÿtch et al. [74] have characterized in detail the internal structure of another type of metallic multiple twin, decahedral Au nanoparticles. Their measurements confirm the presence of a disclination consistent with the commonly accepted strain model. However, they also observed shear gradients, which are not usually considered in these models.

In the case of supported or embedded particles, the occurrence of double diffraction effects, that is, diffraction in the first component (nanoparticle) followed by further diffraction in the second component of the system (matrix or support), opens up new possibilities to determine the lattice parameters of the nanoparticles.

As shown in Figure 26.31a, the double diffraction process gives rise to additional reflections in the diffraction plane, which are linear combinations of those corresponding to the nanoparticle and the matrix or support. In general, given any reflections from the nanoparticle and the support, different linear combinations could be expected to occur; some of them will be of very high frequency (those corresponding to additions of the basic nanoparticle and support material), whereas others will be of low frequency (differences). In general, all these new reflections are called *Moiré reflections* because they produce in the images, as we will see now, low-frequency fringes resembling Moiré-type patterns.

As Figure 26.31b shows, these Moiré reflections can be detected at the diffraction stage, for example, in selected area diffraction patterns. The one shown in this figure was recorded on a Rh/CeO$_2$ catalyst. Different linear combinations (μ_1 to μ_4) could be assigned on this pattern. In HREM images, the Moiré reflections give rise to peculiar low-frequency details, which overlap with the contrasts due to the nanoparticle. Figure 26.31c illustrates one bidimensional Moiré pattern occurring

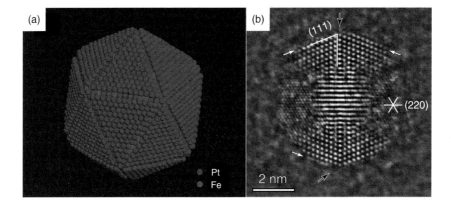
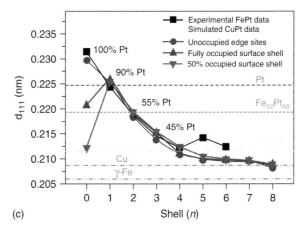

Figure 26.30 (a) Structural model of a FePt bimetallic icosahedral nanoparticle; (b) AC-HREM restored phase image; and (c) plot of the evolution of {111} lattice spacing from the outer surface of the nanoparticle ($n = 0$) toward the particle core ($n = 8$). (Adapted from M.J. Yacaman et al. [73].)

in the same Rh/CeO$_2$ system. In this case, the rhodium nanoparticle is observed partially in profile view conditions. The portion of the particle imaged in the top view, lower part, gives rise to the Moiré pattern.

As shown in Figure 26.32, double diffraction effects can be clearly identified on the DDPs of the HREM images. Two examples are illustrated in this case, one corresponding to Pb nanoinclusions within a metallic aluminum matrix and the second related to metallic platinum nanoparticles supported on a ceria support. Although in both cases Moiré contrasts are easily identified in the images, their precise characterization requires analysis in reciprocal space.

In the two cases, bidimensional Moiré patterns are observed. Correspondingly, μ-type diffraction spots are observed along two directions of the DDPs, marked

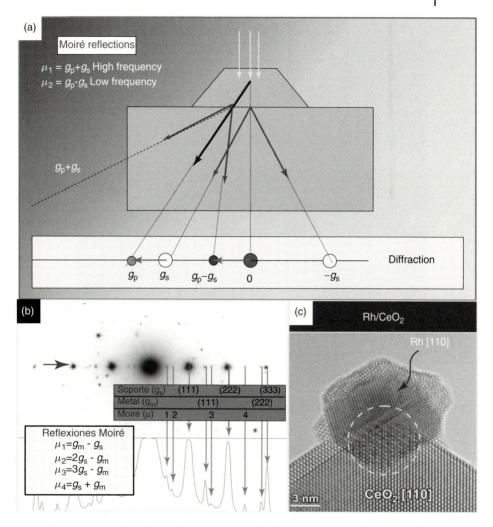

Figure 26.31 (a) Sketch explaining the origin of the Moiré reflections by double diffraction in a supported nanoparticle system (g_p and g_s correspond to the diffracted beam of the particle and the support respectively); (b) selected area electron diffraction pattern recorded on a Rh/CeO$_2$ catalyst. Diffraction spots due to the metal (g_m), the support (g_s) and different Moiré-type reflections (μ_n) are assigned on the intensity profile of the pattern recorded along the direction marked by the blue arrow; and (c) HREM image recorded on a Rh/CeO$_2$ catalyst showing Moiré-type contrasts (area encircled in yellow).

in both cases with yellow and white arrows in Figure 26.32b,d. The reflections corresponding to the isolated components, support (matrix) and particle (inclusion) are marked with red and blue arrows.

In the Pb-in-Al DDP, only one type of Moiré reflection is observed, $\mu_1 = g_{Al}(111) - g_{Pb}(111)$ (yellow arrows). In the Pt catalysts, two different Moirés can

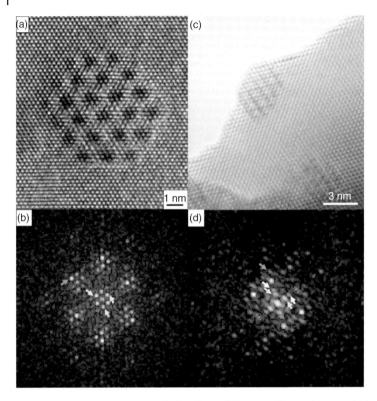

Figure 26.32 (a) HREM image of Pb inclusions in Al; (b) DDP from (a); (c) HREM image of a Pt/CeO$_2$ catalyst; and (d) DDP from (c). The reflections marked in both cases with arrows correspond to double diffraction effects. Those marked with red arrows to the Al matrix or CeO$_2$ support. Blue arrows correspond to Pb nanoinclusions or Pt nanoparticles. (Adapted from U. Dahmen et al. [75].)

be detected in the DDP, $\mu_1 = g_{Pt}(111) - g_{CeO_2}(111)$ (yellow arrows) and $\mu_2 = 2\, g_{CeO_2}(111) - g_{Pt}(111) = g_{CeO2}(222) - g_{Pt}(111)$ (white arrows). Note also how in both cases the Moiré reflections are aligned with those of the individual components. This indicates the existence of particular orientation relationships between the structures of the two components.

As any other reflection in the electron diffraction pattern or in the DDP, each μ spot is characterized by its modulus and its orientation in reciprocal space. By proper quantification of these parameters, using digital image processing techniques, quite a lot of interesting information can be obtained regarding the two aspects: (i) as just mentioned, the orientation relationship between the lattices of the two components and (ii) the modifications in the lattice parameter of the nanoparticle, if we assume that the support or matrix does not change its lattice parameters, which seems a reasonable assumption.

In relation to the first point, it is quite frequent that when nanosized particles grow either on the surfaces or inside voids of another material, their growth is influenced by the intrinsic structure of the latter. The template effect, which exerts the matrix or the support on the nanoparticles, may finally determine both their exact orientation with respect to the lattice onto which or inside which they are growing and the degree of structural modification needed to accomplish such structural interaction. These in turn will determine the exact features of the nanoparticle/support or nanoparticle/matrix interface.

It is not our goal here to present detailed analysis of the interface structure in specific materials, but instead to illustrate how double diffraction effects can be easily exploited to obtain information about these interfaces.

Concerning orientation relationships between different components, profile view images can provide, as shown in Figure 26.33, evidence of their occurrence and the necessary crystallographic details. As shown in all the examples discussed up to this point, image processing, DDP analysis, and image simulation tasks are required in order to reach reliable conclusions, no matter that the images are recorded in the HREM or in HAADF-STEM mode.

In these profile view images, the crystallographic nature of the planes contacting at the interface can be determined from the analysis of the DDPs. This is not the case in top view images in which Moiré effects are recorded. Nevertheless, these latter images present some other advantages, as illustrated in Figure 26.34.

This figure contains an image simulation study of the influence on both profile and top view images of small rotations of the metal nanoparticle out of a perfect orientation relationship. Thus, Figure 26.34a,c are simulations corresponding to the situation in which a Rh nanoparticle is grown under a perfect parallel orientation relationship on a {111} surface of ceria. Note in this case how the exact alignment of metal and support is clearly observed in the Moiré-type contrasts of the plan view image (Figure 26.34c). In the DDP of this image (Figure 26.34e), the reflections from the metal, the support, and the two Moires are perfectly aligned.

If the particle is rotated only by 2° out of this perfect orientation relationship, around the axis perpendicular to the metal/support interface, the profile view image does not change significantly (Figure 26.34b). In fact, it looks even clearer than that corresponding to the perfect orientation. We could not tell from the comparison of the two images that any rotation has occurred. In contrast, the Moiré fringes change significantly (Figure 26.34d). Even a naked eye inspection of the simulated image suggests that the metal and support are not aligned in this case. The misalignment is better appreciated on the DDP (Figure 26.34f). The rotation between the reflections of the metal and the support is too small, 2°, but note how the angle between the line that connects the two Moiré reflections, μ_1 and μ_2, and the line from the center of the diffraction pattern to the support reflection is pretty large (dashed white lines), larger than 20°. In other words, this angle, which is just 0° for the perfect alignment situation, changes up to more than 20° after a 2° rotation of the particle on the surface of the support. This proves that the Moiré reflections amplify by a factor larger than 10, the effect of the rotation.

Figure 26.34g shows an experimental top view image recorded on a Rh/CeO$_2$ system in which the Moiré contrasts suggest a rotated situation. Note the clear

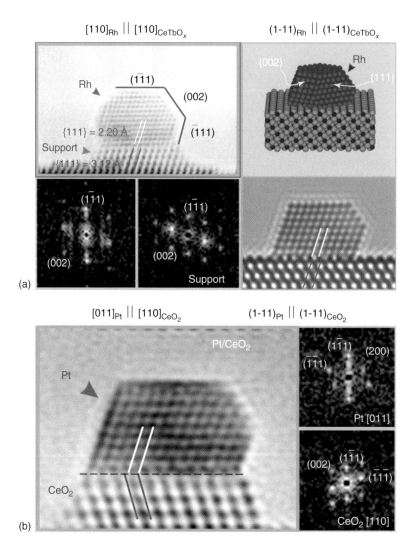

Figure 26.33 HREM studies of metal/support orientation relationships in supported catalysts on the basis of profile view images. In the case illustrated in (a), the Rh particles were grown under a parallel-type orientation relationship. In the case of the Pt nanoparticle shown in (b), there is a twin-type orientation relationship. In the first case, the {111} planes of the support keep aligned across the interface into the nanoparticle, whereas in the second they are mirrored into the nanoparticle {111} planes. The interface (dashed lines) acts as the mirror. In both cases, the planes in contact at the interface are both {111} for the two components. The exact crystallographic relationships between the two components of the interface are shown above the images. (Adapted from S. Bernal et al. [76].)

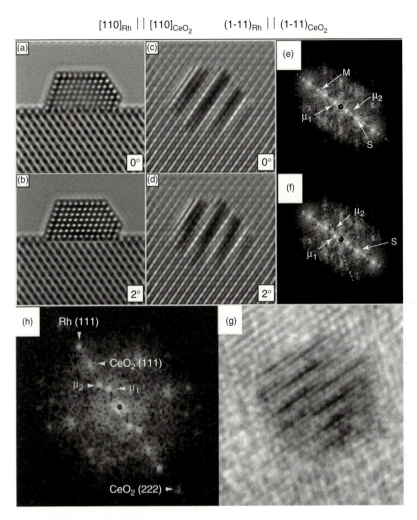

Figure 26.34 (a–h) Image simulation and experimental study of the influence of small misalignments between supported particle and support on profile and top view images. (Adapted from R.T. Baker et al. [77].)

misalignment between the metal, support, and Moiré reflections in the DDP (Figure 26.34h).

Therefore, we can conclude that top view images are much more sensitive than profile views to the effect of very small crystal misalignments between the two components of an interface. Its quantification by the analysis of the DDPs can provide very accurate information on the amplitude of the misalignment [77]. As shown in Figure 26.35, the angle between the two lines mentioned above, that connecting the μ reflections and that connecting the support reflection with the support (marked with Φ in the image) is the key parameter in this quantification.

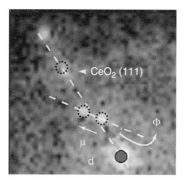

Figure 26.35 Enlargement of the DDP of Figure 26.34h. The dashed circles locate the position of the Moiré and the support reflections. The solid circle corresponds to the center of the DDP. (Adapted from R.T. Baker et al. [77].)

Double diffraction effects also play a magnifying effect on the modifications of the lattice parameter of the nanoparticle [78]. Thus, small changes in this parameter make a strong difference on the exact position of, μ_1 and μ_2, that is, on the modulus of these reciprocal vectors. An accurate procedure to make quantifications of the particle lattice parameter is by measuring the magnitude of the vector that connects these two Moiré reflections. We can call this vector the difference Moiré vector, μ_d (marked in Figure 26.35). By using appropriate formulae [77], both the value of the nanoparticle lattice parameter and its orientation with respect to the support can be accurately estimated.

This type of analysis can be performed even with pretty small particles, close to the visibility limit of the system (e.g., 1 nm). Care must in any case be exercised because the exact orientation of the metal/support system with respect to the electron beam plays a major influence on the Moiré contrasts [64].

Data presented also showed that both profile and top view images are required for a complete characterization of nanosized interfaces such as the ones we have illustrated in the different examples shown in this section. Figure 26.36 emphasizes in any case the importance of top view images. Note in fact how in this image, recorded on a Pt/CeO$_2$ system, most of the nanoparticles are observed under plane view conditions. If we intend that any conclusion obtained from the analysis of these systems has statistical significance, it is clear that the interpretation of this type of images is imperative.

26.3
Electron Tomography of Nanoparticles

Electron microscopy techniques commonly used to image nanoparticles (HREM and HAADF-STEM) provide extremely useful information, but they are limited to the projection of the structures. However, the design and characterization of

Figure 26.36 HREM image of a Pt/CeO$_2$ catalyst.

nanoparticles have to be accomplished considering their three-dimensional shape in order to fully understand and optimize their particular properties compared to bulk materials. In many cases, besides the size and the shape of the nanoparticles, their three-dimensional distribution within or over their support becomes of major importance.

Electron tomography allows determining of the three-dimensional structure of a nanoparticle by reconstructing the object from a series of images acquired in the microscope at different tilt angles. Tomographic reconstruction is based on the assumption that the images acquired are "true projections of the structure." This implies that the intensity in the images show, at least, a monotonic relationship with some function of the thickness or density of the structure [79].

Traditionally used in life sciences, the first attempt in materials science was that of Koster et al. [80], who used TEM electron tomography to get unequivocal information about the location of silver particles (10–40 nm in diameter) in a metal/zeolite crystal (Ag/NaY). It is worth noting that metallic nanoparticles inside a heterogeneous catalyst (Pd–Ru bimetallic nanoparticles supported on mesoporous silica) were as well the material used to perform the first tomographic reconstruction using the HAADF-STEM imaging mode [81]. The intensity in HAADF-STEM images is monotonically dependent on the sample thickness and proportional to the square of the atomic number, Z, of the material [82, 83]. HAADF-STEM tomography has been successfully applied over the last decade to the characterization of small particles dispersed over high surface area supports of low atomic number elements, [81, 84–89]. As an example of this kind of characterization, Figure 26.37 shows an axial projection of a tomogram of a trimetallic catalyst (Ru$_5$PtSn) supported on Davison 38 Å silica, together with successive slices through the reconstructed tomogram [85]. This reconstruction allows clear differentiation between the silica support, which has low intensity in the reconstruction, because of the low Z of Si and O, and the active nanoparticles, which correspond to the high-intensity areas

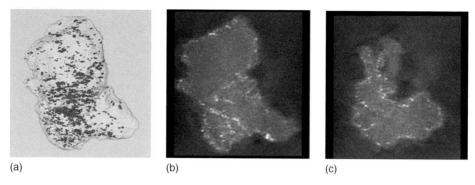

Figure 26.37 Axial projection of a specimen tomogram (a), with successive slices (b–c) through the tomogram of Ru$_5$PtSn supported on silica.

in the reconstruction because of the high Z of the metallic particles (Ru$_5$PtSn). Note that the reconstruction resolution (1 nm^3) allows a direct visualization of the position of the nanoparticles inside the mesoporous support, close to its surface.

A more challenging situation is the description by means of HAADF-STEM electron tomography of catalysts based on heavy element (transition metal or lanthanide) supports [90, 91]. In these systems, image conditions need to be optimized in order to get the highest possible contrast difference between the nanoparticles and the support, given their smaller Z difference. Figure 26.38 reports results obtained on a Au catalyst supported on a heavy Ce$_{0.50}$Tb$_{0.12}$Zr$_{0.38}$O$_{2-x}$ support. Thus, Figure 26.38a shows a HAADF-STEM image of the tilt series, where it can be seen that the contrast between the gold particles and the support is high enough to discriminate the presence of small (about 1 nm) particles even in relatively thick regions of the support. Figure 26.38b displays a surface-rendered representation of the segmented reconstructed volumes of the gold nanoparticles

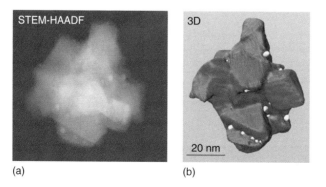

Figure 26.38 (a) HAADF-STEM image of the tilt series of sample Au/Ce$_{0.50}$Tb$_{0.12}$Zr$_{0.38}$O$_{2-x}$. (b) Surface-rendered visualization of the reconstructed tomogram after segmentation of gold nanoparticles (white) and the support (gray).

(colored in white and corresponding to the high-intensity areas of the HAADF images) and to the mixed oxide support (colored in blue).

A very interesting piece of information that can be exclusively obtained from reconstruction such as the one shown in this figure is the location of the gold nanoparticles. In this particular case, it becomes remarkable that the gold nanoparticle locate either at the nanocrystals boundaries or at sites where a stepped surface is present. It is clear from the reconstruction displayed in Figure 26.38b that few or only a small fraction of the metal nanoparticles are located on the facets themselves.

Besides information about distribution and location of metallic nanoparticles on the support, HAADF-STEM tomography can provide particularly valuable information about the surface crystallography of nanocystallites [66, 91]. As we have already mentioned, this knowledge is important to fully understand their chemical properties, for example, the catalytic behavior of metal/support systems. Studies performed in Ref. [66] over a Ce–Zr mixed oxide sample illustrate this approach. In that case, 3D reconstructions revealed that the support consisted mainly of octahedral crystallites of several tenths of nanometers in size, as depicted in Figure 26.39a. A crystallographic analysis of the particles showed that the surface of the nanoparticles was dominated by {111} facets. The angles between the facets measured in the orthoslice through the reconstructed volume displayed in Figure 26.39b are close to 110°, as expected for the angle between {111}planes.

Besides determining the morphology and exposed facets of oxide particles used in the field of heterogeneous catalysis [92], HAADF-STEM tomography has unveiled the 3D shape of nanoparticles with applications in several fields. Some interesting examples are the analysis of the shape of the magnetite (Fe_3O_4) crystals from a single magnetotactic bacterial cell [93]; the study of the position, size, and shape of Sn-rich quantum dots embedded in a silicon matrix [94]; and the characterization of CdSe quantum dots grown on carbon nanotubes (CNTs) [95]. These CdSe change from rodlike (growing along the 001 direction) to dihexagonal pyramids, with the (001) planes bonded to the (002) graphitic planes of the nanotubes [95].

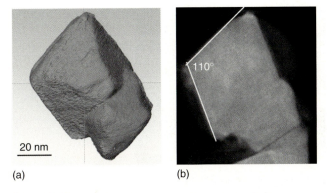

(a) (b)

Figure 26.39 (a) Surface visualization of a reconstructed tomogram of a treated $Ce_{0.62}Zr_{0.38}O_2$ nanocrystal. (b) Central slice through the reconstructed tomogram.

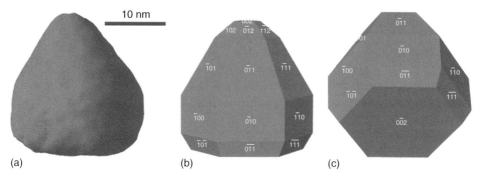

Figure 26.40 (a) Surface-rendered reconstructed nanoparticle. (b,c) Model particle at different orientations.

The epitaxial ordering of the CdSe nanocrystals with the hexagonal periodicity of the graphite surface described could be responsible for the stabilization of the (001) planes on the surface of the nanotube, leading to particles with a wider basal plane than those of a rodlike nanoparticle. One of the CdSe reconstructed nanoparticles is shown in Figure 26.40a, together with a model of the nanoparticle.

The precise 3D characterization of nanoparticles achievable by HAADF-STEM tomography can be useful to rationalize their physicochemical properties. The 3D shapes derived from the tomographic reconstructions can be used as models much more realistic than those inferred from 2D microscopy images, which are usually simplified. For example, the 3D study of noble metal nanoparticles is crucial for understanding their optical response, which it is determined by the collective excitation of the localized surface plasmons. Figure 26.41 shows results from the combination of electron tomography with electrodynamic simulations, which allows generation of an accurate optical model of a highly irregular gold nanoparticle [96].

3D reconstruction of nanoparticles can take advantage of a new reconstruction algorithm, DART (discrete algebraic reconstruction technique), based on discrete

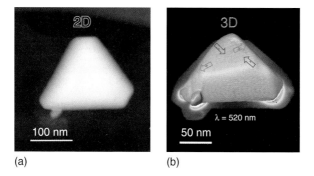

Figure 26.41 Single images acquired at different tilt angles (a) allow the reconstruction of the 3D shape needed to calculate the optical response (b).

tomography [97, 98]. Discrete tomography focuses on the reconstruction of samples that consist of only a few different materials. Ideally, each material should correspond to a unique gray level in the reconstruction. Using this *a priori* knowledge as an additional reconstruction parameter, discrete tomography provides a method for reducing artifacts, including missing wedge artifacts, without the need to extend the tilt range, to increase the number of tilted images, or to improve image acquisition. The quality of the reconstruction can be enhanced even with a reduced number of projection images [98]. DART reconstructions are typically less noisy and the boundaries between the different compositions are clearly visible. The technique is particularly interesting if an additional analysis (e.g., quantification) needs to be performed after the reconstruction, as the discrete reconstruction is already segmented.

26.4
Nanoanalytical Characterization of Nanoparticles

The performance of nanoparticles is strongly influenced by their chemical and structural properties as well as their interaction with the surroundings (nanoparticle coupling, nanoparticle-support/matrix interaction). The ability to combine different signals, as high-resolution imaging techniques (HREM and HAADF), with spectroscopic techniques, such as X-ray energy dispersive spectroscopy (XEDS) and EELS, has demonstrated to be a powerful tool to provide information on individual nanoparticles at the nanometric scale as well as information related to the interaction of the nanoparticle with the medium (support or surrounding nanoparticles). In this section, after introducing the spectroscopic techniques, we illustrate several examples that demonstrate the information that can be obtained when combining HAADF, XEDS, and EELS signals.

EELS involves measurement of the energy distribution of electrons that have interacted with a specimen and lost energy because of the inelastic scattering. In the EELS spectrometer, the electrons are distributed as a function of their energy; those electrons with similar energy will be focused on the same point. In addition, as a result of the interaction of the high-energy electron beam with the sample, an atom of the nanoparticle can be ionized, causing the ejection of an inner shell electron. A subsequent relaxation process of the excited atom can occur, generating a vacancy that is further refilled with an electron of a higher energy level, which loses a portion of its energy by the emission of an X-ray photon. The number and energy of the emitted photons are measured by XEDS. The observed energy values (either in EELS or XEDS) are characteristic of the analyzed elements, providing chemical information of the characterized sample (Figure 26.42).

When analyzing individual nanoparticles, the XEDS or EELS spectrum can be acquired on a specific area of the sample, the analyzed volume defined by the beam size of the probe; this is the so called *point analysis mode*. Alternatively, the whole individual nanoparticle can be analyzed using the *SI mode* [99]. This acquisition mode consists in acquiring a series of spectra, while the probe is scanned over the

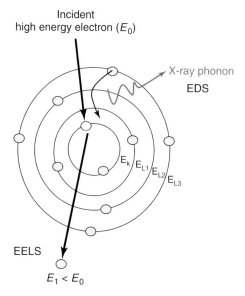

Figure 26.42 Scheme to illustrate the XEDS and EELS signals obtained when a fast electron (E_0 incident energy) interacts with an inner-shell electron of an atom. After the interaction, the fast electron energy is E_1 ($E_1 < E_0$).

sample following a digitally controlled 1D or 2D pattern. This technique provides an accurate correlation between spectroscopic information and a specific (sub) nanometer area of the sample defined by the probe size (Figure 26.43).

In the case of XEDS, the energy of the X-ray emitted peaks can be used as a fingerprint to identify the different elements present in the sample. Moreover, if the intensity under the peaks is quantified using standard methods [100, 101], the composition of the analyzed area can be extracted. In this context, combining HAADF imaging with XEDS, the core-shell morphology of Au–Ag nanoparticles has been determined and the evolution of the Au–Pd nanoparticle morphology as a function of the temperature treatment investigated. In particular, it has been observed that the initial Au–Pd alloy nanoparticles supported on Al_2O_3 evolves toward the formation of a core-shell morphology with a Pd-rich shell and a Au-rich core [102] Figure 26.44. Recently, atomic-resolved STEM-EDX chemical maps of In, Ga, and As for the 1.47-Å dumbbell structure of InGaAs have been achieved using AC electron microscopy [103].

The information contained in an EELS spectrum (Figure 26.45) is manifold and can be extracted after more or less refined spectrum analysis. In particular, three different zones, which will provide different information, can be found; the zero-loss peak, the low-loss, and core-loss zones. The zero-loss peak corresponds to electrons that have not lost energy when passing through the sample and those electrons whose energy loss is too small to be detected. There is no analytical information to be extracted from this part of the spectrum. The low-loss area (0–50 eV)

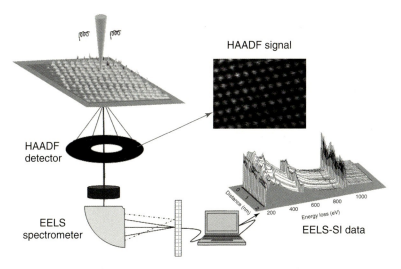

Figure 26.43 Spectrum imaging acquisition mode. The electron beam is scanned across the sample and the HAADF and EELS signals are acquired in every pixel of the analyzed area. EELS-SI data corresponds to a series of EELS spectra extracted from the whole dataset.

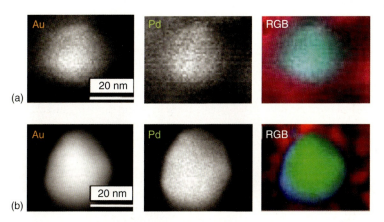

Figure 26.44 Au and Pd XEDS maps of a AuPd/Al$_2$O$_3$ sample before (a) and after calcination at 400 °C (b) (RGB image Au in green and Pd in blue). (Image adapted from Ref. [103].)

corresponds to transitions involving single excitations of outer-shell electrons and collective excitations of the valence electrons (plasmons). A refined analysis of the low-loss area can provide information about the sample thickness, the electron density of the material. As we will illustrate, optical information about individual nanoparticles can also be extracted from this energy range of the spectrum. In certain cases, the low-loss region of the spectrum can be used as a fingerprint to

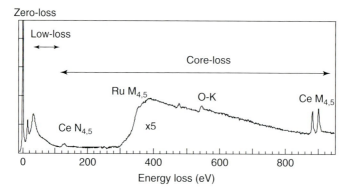

Figure 26.45 EELS spectrum from individual Ru nanoparticle supported on cerium oxide. Three areas of the spectrum can be distinguished; zero-loss peak, low-loss, and core-loss area.

identify the elements present on the sample. Thus, Kim et al. [104] have used the low-loss EELS spectra from pure amorphous ice, pure poly(dimethyl siloxane), and a pure copolymer made of methylacrylate (MA), methyl methacrylate (MMA), and vinyl acetate (Vac) to characterize the spatial distribution, at a molecular level, in nanocolloids. In particular, using the references mentioned above and multiple least square (MLS) fitting of the experimental data allowed them to gain information about the spatial distribution of water, PDMS polymer, and copolymer on byphasic polymer nanocolloids (see chemical maps illustrated in Figure 26.46).

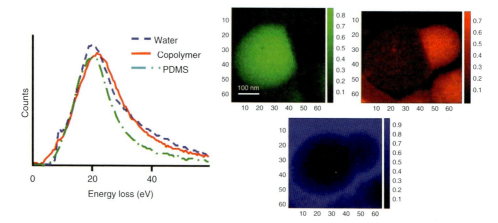

Figure 26.46 Low-loss EELS spectra from pure amorphous ice (water reference), pure poly(dimethyl siloxane) (PDMS reference), and pure copolymer made of methylacrylate (MA), methyl methacrylate (MMA) (copolymer reference), and chemical maps obtained using the indicated references (water in blue, copolymer in red, and PDMS in green). (Image adapted from Kim et al. [104].)

The core-loss area (>100 eV) corresponds to transitions of the inner-shell electrons toward the unoccupied states, which are above the Fermi level. Such transitions are reflected in the EELS spectrum by the formation of core edges, which rise at the corresponding binding energy of the core electron. Therefore, identification of the elements present in the analyzed volume is possible, together with the quantification of the edge by measuring an area under the core edge. Finally, considering the fine structure of a core edge, one can also get information about the electronic structure of the analyzed sample, such as the unoccupied density of states. Although in the following we mainly focus on the analysis of the core-loss energy range, we also discuss first an example of how optical information about individual nanoparticles using the low-loss area of the spectrum can be extracted.

26.4.1
Optical Information on Individual Nanoparticles

During the past years the improvements in the EELS spectrometer design, which have allowed a higher energy resolution and shorter acquisition times, jointly with the development of more powerful data analysis methods, have pushed forward the characterization of the optical properties of individual nanoparticles on the basis of EELS. Since the first measurements of surface plasmons on individual nanostructures [105, 106], a huge effort has been carried out to develop a technique and methodology that would allow mapping the optical properties of individual nanoparticles at the nanometer scale and, therefore, opening the possibility of carrying out studies to establish relationships between the energy and intensity of the surface plasmons and the size and shape of the metal nanostructures. Such information could be crucial to tailor the metal nanoparticle properties for applications on medical treatment of tumors [107], single-molecule detection [108], and subwavelength manipulation of light [109]. In this context, combining the subnanometer electron probe available on STEMs, SI EELS, and mathematical methods developed to extract the zero-loss peak tail of the low-loss EELS spectra, several authors have mapped the surface plasmon resonance on individual nanostructures [110–112]. In the case of gold nanorods [110] of 50 nm length, two distinct plasmon resonances were observed at 1.70 and 2.40 eV. The map distribution and intensity indicated that when the electron beam passes at the end of the rod, the electric field is primarily directed along the long axis of the rod, which is equivalent to polarization of light along this direction (Figure 26.47). In the case of silver nanoparticles of triangular shape [111, 112], the spectra vary considerably from one position to the next, the energy of the three lower modes being size dependent. Figure 26.47 illustrates the EELS amplitude distributions of the 1.75, 2.70, and 3.20 eV modes on an individual nanoprism. Analysis performed on triangles of different sizes has concluded that the optical properties of the analyzed nanostructures are mainly driven by their aspect ratio. This methodology has been further applied to explore the electromagnetic coupling

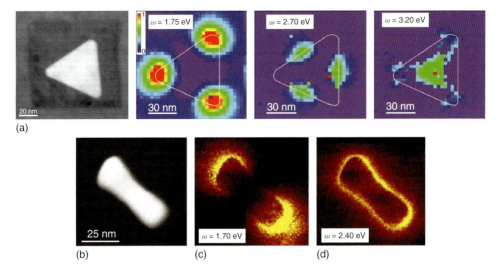

Figure 26.47 (a) HAADF–STEM image of the Ag particle and EELS amplitude distributions of the 1.75, 2.70, and 3.20 modes on the nanoprism displayed in the HAADF image. (b) HAADF image of Au nanorod and (c,d) are the intensity maps of the 1.70 and 2.40 eV plasmons resonance on the Au nanorod displayed in (b). (Image adapted from Refs. [110, 112].)

between nanoparticles [113] and to investigate the zeptomol detection limit by ultrasensitive surface-enhanced Raman scattering [114].

26.4.2
Chemical Identification and Quantification of Individual Nanoparticles

Focusing on the core-loss range of the spectrum, the onset of the inner shell ionization edges correspond to the binding energy of the core electrons (about 285 eV for C and 399 eV for N). Thus, an identification of the different elements present in the individual nanostructure can be obtained from an EELS spectrum. Moreover, using the capabilities of the SI mode, previously described, a detailed map of the element distribution is nowadays available at the nanometric and atomic scale [115, 116]. In this section, we illustrate the capabilities of the technique to identify the elements on the sample. In particular, we show how the technique can be used to identify the presence of bimetallic and core-shell structures.

The properties of nanoparticles are strongly influenced by their nanostructure, size, and chemical composition. Thus, when synthesizing bimetallic nanoparticles, differences are found as we go from metal alloy systems to core-shell structures, and the first question to be answered is whether the two metals are combined in every particle or whether, instead, a core-shell structure is formed. Thus Co–Ni nanoparticles synthesized within the apoferritin cavity were characterized using the EELS SI mode [117]. In this mode, a series of EELS spectra were acquired, while a 0.5 nm beam probe with a current of 0.1–0.3 nA was scanned across

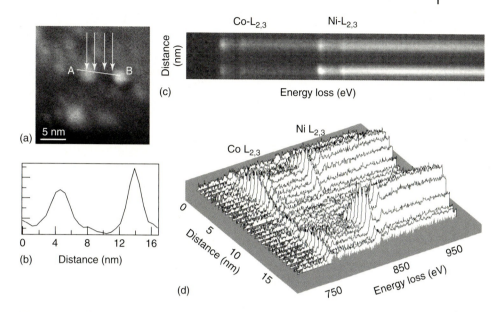

Figure 26.48 EELS Spectrum imaging analysis on Co–Ni nanoparticles; (a) STEM HAADF image of the two nanoparticles illustrating the A-B line where the analysis was performed. (b) HAADF signal acquired simultaneously with the EELS signal. (d) 3D illustration of a set of 30 spectrum in the energy range of 700–950 eV taking along the line A-B, the data was extracted from the full spectrum imaging data set (c).

two individual nanoparticles (Figure 26.48). Every 0.6 nm, the HAADF and EELS signals were simultaneously recorded. In order to evaluate the Ni, Co, and O distribution, the EELS spectra were acquired in the 500–950 eV range (only Co and Ni distribution are illustrated in Figure 26.48). The HAADF signal, a signal that can be related to the projected sample mass, confirms that the analysis has been performed across the two nanoparticles. Moreover, since both signals have been acquired simultaneously, we can correlate the HAADF and EELS signals at every position of the analysis. The grayscale image of the Co–Ni signals (Figure 26.48c) illustrates the variation of the Co-$L_{2,3}$ and Ni-$L_{2,3}$ (780 and 853 eV respectively) signals as we go across the two particles. From Figure 26.48d, it is clear that Co and Ni metals are associated in both nanoparticles, confirming the formation of bimetallic nanoparticles.

In the case of core-shell structures, as reported in Section 26.2.3 (Figure 26.22), HAADF imaging using an AC microscope has allowed the identification of Pt@Pd core-shell nanostructures. In this context, combining HAADF imaging techniques with EELS in the *point analysis mode*, the core-shell morphology of Au–Fe nanoparticles has been reported by Cho et al. [118] (Figure 26.49). Fe $L_{2,3}$ edge is only present when analyzing the core area of the particle, whereas the Fe $L_{2,3}$ signal vanished on the nanoparticle shell. Using the *SI* acquisition mode, Shooneveld et al. [119] have been able to visualize and quantify the subnanometer morphology of hybrid organic–inorganic nanostructures made of a quantum dot-containing

932 | 26 Nanoparticles

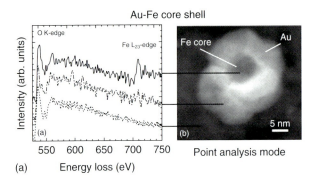

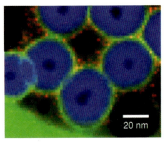

Figure 26.49 Core-shell nanostructures identified using three different acquisition modes; (a) HAADF image indicating the three areas where EELS spectrum has been acquired using the point analysis mode and EELS spectra corresponding to the core and shell area of the Au–Fe core-shell nanoparticle. (b) Spectrum imaging mode to analyze hybrid inorganic–organic nanostructures. (c) Colored (Ag in blue and Cu in yellow) EFTEM image of CuAg core-shell nanoparticles. (Image adapted from Refs. [118, 119, 122].)

silica particle coated by a lipid layer (Figure 26.49). As it will be shown further, the SI mode can also provide quantitative information about the elements present in the nanostructure. In addition, EELS information can also be obtained through the *EFTEM acquisition mode*. In this mode, a parallel beam illuminates the sample and images are registered with the electrons that fit within a predefined energy window of the EEL spectrum. In the EFTEM imaging mode, chemical maps of the elements present in the sample can be obtained by using the three-window elemental mapping [120, 121] procedure. In these maps, the intensity of each pixel is proportional to the integrated intensity under the ionization edge in the EELS spectrum recorded at that point. Using this acquisition mode and the Cu $L_{2,3}$ and

Ag M$_{4,5}$ (920 and 1100 eV respectively) ionization edges, Langlois et al. [122] have imaged CuAg core-shell nanoparticles (Figure 26.49).

In the case of nanoparticles with applications in catalysis, their properties are strongly influenced not only by the size but also by their surface. Moreover, the chemistry and structure of the nanoparticle can be modified during the catalytic process. In case of supported nanoparticles, strong interaction between the support and the metal nanoparticle can also influence the catalytic properties of the system. As indicated in Section 26.2, HREM and HAADF imaging can provide information about the nanoparticle surface. However, to obtain direct chemical information of the nanoparticle surface, spectroscopic techniques with high spatial resolution are required. In this context, EELS, either in the point analysis or in SI modes, has demonstrated to be a powerful technique. Thus, EELS using the point analysis mode has been crucial to confirm the growth of Au nanoparticles inside cucurbituril 7 (CB7) molecules. HREM images have indicated the presence of fcc gold nanoparticles, but only by analyzing the nanoparticle surface can the presence or absence of C in the system be detected. In order to analyze the surface of individual nanoparticles, Au nanoparticles synthesized using cucurbituril as template for the encapsulation of the metal atoms were deposited onto CNTs. Such deposition methods allowed isolating individual nanoparticles on the nanotube surface (Figure 26.50a), whose surface composition could be properly analyzed by local EELS analysis. In this context, combining HAADF and EELS spectroscopy the C-K edge was detected on an EELS spectrum acquired at the nanoparticle surface (Figure 26.50a). This result, together with the nanoparticle size distribution obtained by HAADF, confirmed that gold nanoparticles did grow inside the cucurbituril host molecule [123].

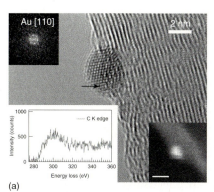
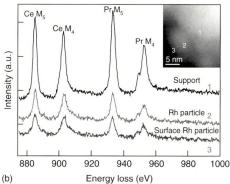

(a) (b)

Figure 26.50 (a) HREM image of an fcc gold nanoparticle where the EELS analysis was performed at the particle surface that projects into the vacuum. Inset HAADF image of a 1 nm nanoparticle and EELS spectrum obtained at the surface where the C-K edge signal is present. (b) HAADF image of a Rh nanoparticle supported on Ce$_x$Pr$_{1x}$O$_2$ and EELS spectra recorded on the points indicated in the image; at the support (1), nanoparticle–support interface (2), and nanoparticle surface (3).

As indicated in Section 26.2.1, supported and embedded nanoparticles are one of the most complicated cases to characterize. Moreover, during the catalyst preparation, structural and chemical changes can occur at the nanoparticle and support surface. In-depth characterization of a series of catalysts made of Rh nanoparticles supported on $Ce_xPr_{1-x}O_2$ mixed oxides have shown significant chemical changes in the system during the preparation of a rhodium-supported catalyst by wet impregnation techniques. Focusing on the Rh metal nanoparticles, HREM images [124] showed the presence of a surface layer on the Rh nanoparticles (Figure 26.50b). EELS spectra acquired on the nanoparticle surfaces (point 3), nanoparticle-support interface (point 2), and substrate (point 1) indicated the presence of Ce $M_{4,5}$ and Pr $M_{4,5}$ signals, in the 870–970 eV energy range. Both signals are characteristic of the catalytic $Ce_xPr_{1-x}O_2$ support. Taking into account that the analyzed points were away from the support, in fact, at a distance a few times larger than the diameter of the electron probe, the contribution to spectra 3 of some signal coming from the underlying support crystallite could be ruled out. Hence, these results could be considered a clear evidence of the decoration of the metal nanocrystallites by patches of the oxide support. This example illustrates how combining HAADF and EELS spectroscopy information of relevance to understand catalyst deactivation by surface decoration, that is, information on the most external surface key to catalyst performance, can be obtained from EELS experiments.

Apart from identifying the elements present in the sample, by analyzing the signal under the characteristic edges one can obtain quantitative chemical information about the sample. In this chapter, we cannot cover EELS theory; however, some basic ideas concerning EELS spectrum quantification will be introduced before illustrating the capabilities of the technique to quantify the elements distribution on individual nanoparticles.

At the core-loss area of the spectrum, the ionization edges are superimposed on a monotonically decreasing background, which is due to the excitation of core levels of lower binding energy. In order to quantify the edge, the noncharacteristic signal (background) has to be removed first through standard methods [120]. Usually, the background signal is fitted using a power law function, which depends on the energy as follows: AE^{-r}. Such a fitting has to be done for every edge to proceed with quantification, as indicated in Figure 26.51. Once the background is removed, the quantification of the edge is performed by integrating the core-loss intensity over a defined energy window (ΔE). The characteristic signal (S), for a given collection angle (β), is defined by Eq. (26.1), where I_0 is the incident beam intensity, N is the number of atoms per unit area (atoms/nm^2), $S(\beta, \Delta E)$ is the integrated intensity of a core-loss event measured over a ΔE window for electrons scattered under a β acceptance semiangle, $\sigma(\beta, \Delta E)$ is the partial inelastic scattering cross section dependent on β and ΔE [120, 125].

$$S(\beta, \Delta E) = I_0(\beta, \Delta E) N \sigma(\beta, \Delta E) \tag{26.1}$$

The partial cross section is a measurement of the ionization probability, whereby the fast electron loses energy within an integration energy window (fitting window ΔE) above the threshold energy and with scattered angles less than β (collection

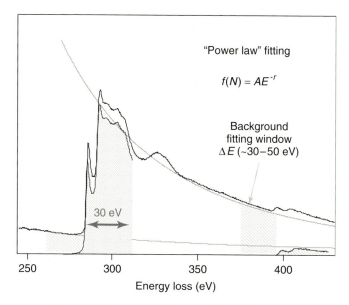

Figure 26.51 Spectrum from a CN_x nanotube sample recorded at $E_0 = 100$ keV. The spectrum has been recorded with a collection angle of 24 mrad and an acquisition time of 3s.

angle). When more than one element is present in the sample, for example, C and N (Figure 26.51), one can be interested in measuring relative concentrations of different elements, and thus the N/C ratio is defined by Eq.(26.2). Taking into account the experimental conditions of the spectrum acquisition (100 kV, collection angle of 24 mrad, and convergence angle of 15 mrad) and following the procedure described in Figure 26.51, the estimated N/C atomic ratio is 0.11 ± 0.02. In addition, mathematical methods based on model fitting and principal component analysis have been further developed [126–128].

$$\frac{N_N}{N_C} = \frac{S_N(\Delta E)}{S_C(\Delta E)} \frac{\sigma_C(\Delta E')}{\sigma_N(\Delta E)} \tag{26.2}$$

The quantification methodology described for an individual spectrum can be also applied to a series of data obtained through the SI mode using processing routines adapted to this type of experiment. Some routines are simple extensions of the techniques previously developed for processing individual spectra and can be straightforwardly used to provide elemental maps. Using the described methodology, the Ce $M_{4,5}$ and Pr $M_{4,5}$ have been quantified to characterize the Ce/Pr ratio in individual $Ce_xPr_{1-x}O_2$ nanoparticles. The overlap of the Pr-$M_{4,5}$ edge, especially that of the Pr-M5 one with the extended region of the Ce-M_4, and the lack of precisely determined Ce and Pr $M_{4,5}$ cross sections, makes quantification a delicate task. In the case illustrated in Figure 26.52, the Ce/Pr molar ratio was measured from the EELS data by determining the integrated intensity in the Ce-M_5 and Pr-M_4 ionization edges using an energy window of 14 eV. Before integration,

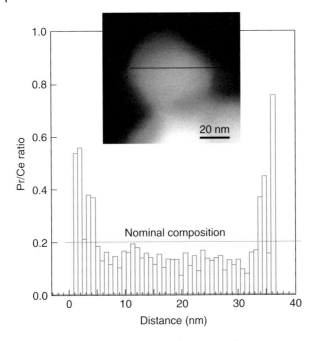

Figure 26.52 Pr/Ce ratio obtained after treating the spectra acquired across the line indicated on the HAADF image.

a power law fitting model (in a 10 eV energy window) was used to remove the background before Ce-M_5 and Pr-M_5 edges. A Hartree-Slater model was used for estimating the M-edge cross sections. In particular, 64 EELS spectra were acquired using the *SI mode*; an EELS spectrum and the corresponding HAADF signal were acquired every 0.78 nm while the beam was scanned across a $Ce_xPr_{1-x}O_2$ nanoparticle. From the collection of spectra, the Pr/Ce ratio was estimated at each point. The results indicate that the nanoparticle surface has higher Pr concentrations than the bulk. The Pr atomic percent (%Pr) at the first atomic layers, down to about 2 nm, is in the range of 28–37%, above the nominal expected value [126]. A similar quantification procedure has been used by Shooneveld *et al.* [119] to quantify the relative quantities of Cd, Gd, C, and Si on hydrophilic quantum dots containing silica nanoparticles before and after being coated with octadecane and lipid layers. More precisely, the number of gadolinium atoms in the lipid patches at the surface of the nanoparticles was determined [119].

26.4.3
Monitoring the Chemical Bonding and Oxidation State; Analysis of the EELS Fine Structure

EELS allows to also obtain information about the local electronic structure by appropriate analysis of the fine structure of the spectra. The characteristic edges arise from the excitation of inner shell electrons. The shapes of the basic edges are determined by atomic physics and, therefore, they are independent of the environment or bonding of the atom. Superimposed on the basic shape, oscillations do also occur, often called *fine structures*, which are strongly dependent on bonding, coordination, or nearest neighbor distances. Two areas are defined in the fine structure, strong oscillations up to 20–30 eV above the threshold called energy loss near edge structures (ELNESs) and weaker oscillations beyond about 30 eV, called the extended energy loss fine structures (EXELFSs). In this section, we focus our attention on the ELNES part of the spectrum. The near edge structure provides information about the bonding configuration in the system. However, the interpretation of the ELNES is not trivial. In the simplest case, the useful signal is directly observed in the EELS spectrum and it is identified by comparing the recorded ELNES with fingerprints (reference previously acquired from standards, Figure 26.53) available in libraries [129, 130].

To illustrate this approach, we will refer to the investigation of oxidation states of cerium and iron oxides. When trying to investigate the distribution of oxidation states in individual nanoparticles made of these oxides, given the high sensitivity of both materials to suffer reduction and the potential reducing power of high-energy electron beams on metal oxide materials, electron beam damage effects are a major concern. Thus, the analysis has to be performed under experimental conditions,

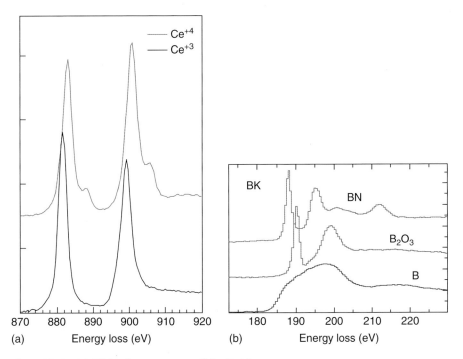

Figure 26.53 (a) EELS reference spectra of the Ce M [4, 5] signal of Ce^{+3} (black line) and Ce^{+4} (red line). (b) EELS reference spectra of B-K signal on boron, boron nitride, and boron oxide samples.

which minimize the influence of the beam. Thus, in addition to high-energy resolution, very short acquisition times are required to limit as much as possible the total electron dose. Taking into account these experimental requirements, the evolution of the oxidation state of elements as Ce can be followed by monitoring the Ce $M_{4,5}$ signal across individual nanoparticles. In particular, change in the oxidation state of cerium from +3 to +4 introduces measurable changes in the $M_{4,5}$ signal, among these being the variation of the M_4/M_5 white line intensity ratio, the shift in energy position of these lines, and the disappearance of the small shoulder at the right side of the two white lines. With these oxidation state fingerprints in mind, the evolution of cerium oxidation state across individual $Ce_xPr_{1-x}O_2$ nanoparticles after its impregnation with Rh metal catalyst was carried out in Ref. [124]. In general, both Ce^{3+} and Ce^{4+} species were present on the catalyst and the relative amount of these two species varied from one position to another in the individual particle. Using a similar procedure, the Fe removal mechanism on ferritine systems has been investigated [131]. In this case, the Fe oxidation state distribution was followed by acquiring the Fe $L_{2,3}$ signal across individual nanoparticles, as that displayed in Figure 26.54. In particular, using the SI mode, a 0.5 nm probe was scanned across the nanoparticle and the HAADF signal and the corresponding EELS spectrum were acquired every 0.4 nm. In total,

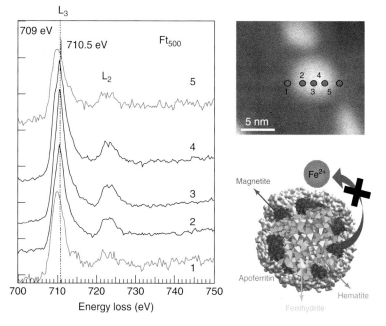

Figure 26.54 HAADF image of a ferritine particle indicating the position from where the EEL spectra in the energy loss region of Fe $L_{2,3}$ edges were acquired. Positions 1 and 5 correspond to the surface of the particle (the thinner regions), and positions 2–4, to inner bulk sites.

64 spectra with an energy dispersion of 0.3 eV and an acquisition time of 1 s per spectrum were recorded. A series of five spectra extracted from different positions on the nanoparticle (one and five at the particle surface, and two to four at the nanoparticle bulk) show that the position of the L_3 peak shifts slightly from the surface to the bulk. Thus, the peak is situated at 709.9 eV in spectra acquired at the surface (points 1 and 5) and at 710.5 eV in those recorded inside the particle (points 2–4), indicating that there is a higher fraction of iron(II) species at the surface of the ferritin cores.

In some cases, the experimental ELNES signal is not trivially obtained and sophisticated adapted methods are needed to extract the characteristic signal. The Nonnegative linear square (NNLS) fitting method and multivariate statistical analysis (MSA) are some of the available methods. NNLS is an MLS fitting method, which consists in simulating the experimental ELNES as a linear combination of weighted elemental reference components. It is extremely important to select properly the reference signals: identical instrumental recording conditions and close thickness values are requisites. To illustrate this approach, using the boron references displayed in Figure 26.53, Arenal *et al.* [132] have combined HAADF imaging, EELS in the SI mode, and mathematical methods to determine bonding maps on individual nanoparticles. Thus, using the EELS spectrum acquired from pure boron, boron oxide, and h-BN as references, a model consisting of a linear combination of the ELNES references was built to picture the spatial distribution of the different ELNES structures on the acquired data. As a result, bonding maps as that displayed in Figure 26.55 were obtained. The bonding maps clearly identified a particle consisting of a pure boron core encapsulated within a BN shell but which is oxidized into boron sesquioxide, B_2O_3, at the surface.

26.4.4
Chemical Imaging at the Atomic Scale

The last part of this section dedicated to the analytical characterization of nanoparticles is focused on chemical imaging of nanostructures at the atomic scale. Even

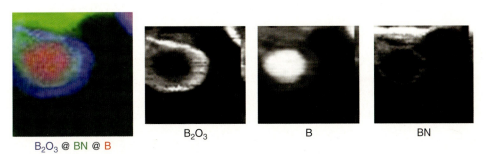

B_2O_3 @ BN @ B B_2O_3 B   BN

Figure 26.55 B_2O_3, BN, B, and color code (boron oxide in blue, boron nitride in green, and boron in red) bonding chemical maps extracted for a nanoparticle presenting a core-shell morphology.

though visualizing individual atoms through HREM and HAADF is now possible with AC electron microscopes, spectroscopic techniques are still required to obtain the chemical information of the sample. In this section, we illustrate an example of how combining HAADF imaging and EELS using an AC microscope, makes chemical imaging of individual nanocrystals at atomic scale resolution possible.

The combination of subangström probes formed by high brightness sources coupled with the improvement in EELS spectrometers has simultaneously increased the signal-to-noise ratio and very significantly decreased the acquisition times [133]. This has allowed imaging based on STEM EELS to be a powerful tool to obtain two-dimensional (2D) atomic-resolution chemical information. Thus, systems such as $Bi_{0.5}Sr_{0.5}MnO_3$ [134] and $La_{1.2}Sr_{1.8}Mn_2O_7$ [135] crystals or $(La,Sr)MnO_3/SrTiO_3$ multilayers [136] have been characterized with this approach. Recently, combining the capabilities of an AC STEM (Nion UltraSTEM operated at 100 kV) with sophisticated calculations based on the Bloch wave method, the first atomically resolved chemical maps on nanocrystals with applications in catalysis have been obtained, in particular on ceria-zirconia mixed oxide nanocrystals. Using the SI mode, as illustrated in Figure 26.43, a series of EELS spectra were acquired while scanning a probe of nominally 1 Å full-width at half-maximum along a defined area of the crystal (2.44 nm × 2.16 nm). The EELS spectra showed the characteristic signals for $Ce-N_{2,3}$, $Zr-M_{3,2}$, O-K, and $Ce-M_{4,5}$ edges at 207, 330, 530, and 881 eV, respectively. Following the methodology described for an individual spectrum (Figure 26.51), atomically resolved $Ce-M_{4,5}$ and $Zr-M_{2,3}$ maps have been obtained on individual $Ce_{0.5}Zr_{0.5}O_2$ nanocrystals (Figure 26.56). The maximum signals for Ce and Zr are found at the vertices and center of the marked rhombus respectively [137].

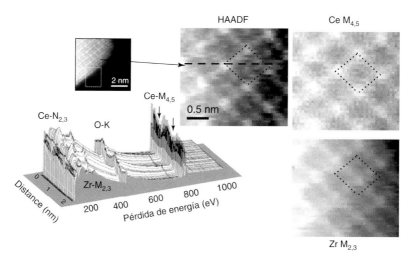

Figure 26.56 HAADF image of a $Ce_2Zr_2O_8$ nanocrystal indicating the area analyzed by spectrum imaging EELS, 3D plot of the EELS spectra extracted from the whole data set. HAADF image and atomic-resolution chemical maps of $Zr-M_{2,3}$ and $Ce-M_{4,5}$.

26.5
In situ TEM Characterization of Nanoparticles

Over the past two decades, the use of *in situ* electron microscopy techniques has dramatically increased in materials science [138, 139]. The development of new electron microscopes and sample holders makes possible the direct observation of microstructural evolution under well-controlled gas environments [140–142], temperature [143], mechanical stress [144], and magnetic/electrical fields [145]. Although conventional characterization before and after exposure to the operation conditions have greatly contributed to the progress of materials science, the dynamic observation of the sample is a powerful tool that allows us to study the material evolution and eventually the process mechanism. In addition, physical and chemical properties of the nanoparticles depend on their size, shape, microstructure, atomic surface structure, and defect organization. Therefore, the advantage of gaining reliable structural and compositional information at the atomic scale under working conditions is evident when we try to find out structure–property relationships. There are several good examples in the literature regarding the *in situ* observation of melting processes [146], deformation-induced growth during nanoindentation [147], catalyst [148, 149] and electrochemical chemical deposition cells [150], fuel cell anodes [151] under working conditions, and biased and unbiased semiconductor p–n junctions [152].

The aim of this section is to give the reader a brief overview of the advantages of using *in situ* TEM techniques. We specifically focus on nanoparticle characterization under well-controlled temperatures and chemical environments using environmental transmission electron microscopy (ETEM). This approach is extremely useful in different fields in which gas–solid interactions take place, such as synthesis, catalysis, and gas sensors. We describe below several relevant examples related to dynamic studies of supported nanoparticle mobility and sintering, their shape evolution, and interaction with the substrate.

26.5.1
Environmental Cells for *In situ* Investigations of Gas–Solid Interaction

One of the major experimental disadvantages of ETEM is the high vacuum requirements of conventional TEM to prevent the electron source degradation, as well as to minimize the effect of inelastic scattering interaction with the gas molecules. This inconvenience can be avoided by introducing an environmental cell (E-cell), which would restrict the gas path length around the sample. The early attempts of building E-cell systems were based in the use of two thin windows, which kept the sample and the gases sandwiched between them. Several modified specimen holders were designed considering this approach and they were successfully applied to the study of wet samples and gas–solid interaction [153, 154].

It should be pointed out that the operation conditions of these windowed E-cell holders strongly depend on the nature of the window. Obviously, these windows should be made of electron transparent and nondiffracting material, to avoid artifacts in the diffraction pattern of the specimen. Amorphous silica and carbon are the most common materials used in this case. Jeol Ltd has manufactured one E-cell sample holder with carbon windows that can operate up to 350 °C and a maximum H_2 or O_2 pressure of 10 mbar [155]. However, new E-cell designs based on innovative nanoreactor designs using microelectromechanical systems (MEMS) have been recently developed [156, 157]. This promising approach is based on 1.2 μm thick SiN_x windows hosted in a silicon substrate, which allow to operate up to 500 °C and 1.2 bar of H_2. Figure 26.57a illustrates the cross-sectional view of the nanoreactor developed by Zandbergen et al. [156], where the gas inlet, gas outlet, and the window membrane can be easily identified. The MEMS methodology allows miniaturizing the reactor, as well as integrating the resistance heater and temperature sensor, Figure 26.57b. This specimen holder has been successfully used in the dynamic characterization of Cu/ZnO catalyst under reducing atmospheres, being possible to observe lattice spacing of 0.18 nm.

A reliable alternative to the windowed cells consist in introducing a complex differential pumping aperture system in the microscope. This approach was initially

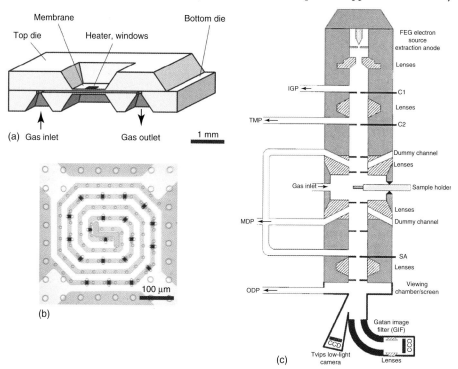

Figure 26.57 MEMS-based E-cell (a,b) and differential pumping ETEM (c) diagrams. (a) Cross section of the nanoreactor. (b) Optical view of the nanoreactor membrane where we can observe the platinum heater (bright spiral) and the electron-transparent windows. (c) Schematic of the in situ microscope showing the two additional apertures located above and below the sample.

developed in the 1950s and it is well documented by Ito, Hashimoto, and Baker [158–160]. This work was continued by Sharma and Crozier [161–167] and Boyes and Gai [168–172], who managed to record high-resolution images under different gas environments (H_2, O_2, and hydrocarbons) using a successfully modified Philips microscope. On the basis of their work, Philips commercialized one modified CM 300 ETEM and recently FEI (former Philips) introduced in the market the Titan^{TM} ETEM, which combines E-cell with the image Cs-corrector and monochromator technology. The actual design of this approach incorporates two additional apertures, above and below the sample, to reduce the gas diffusion into the rest of the microscope. An additional molecular drag pump is connected by two channels to each set of apertures in order to ensure the high vacuum requirements in the rest of the column. A schematic of a standard *in situ* microscope setup is included in Figure 26.57c. It should be pointed out that during the past two decades most works, and in fact the most remarkable ones, reported in the literature have used this approach. However, the promising new generation of windowed holders is a really appealing alternative considering the high cost of the ETEM modified microscopes.

Nevertheless, although environmental transmission microscopes allow a live monitoring of the dynamics of gas–solid interactions, the currently available *in situ* stages can only operate at pressure values in the millibar range. Thus, a challenging gap to be addressed is to study the gas–solid interactions under more realistic operating conditions. On the other hand, although new windowed holders allow the use of high pressure, the operation temperature cannot exceed 500 °C. Therefore, the use of an *ex situ* environmental reaction chamber specific to an anaerobic-transfer TEM holder is an alternative approach that allows carrying out the pretreatment of the sample under more realistic conditions and subsequently transferring it to the TEM, under conditions that prevent any ulterior sample modification. Although the dynamic aspects of the gas–solid interactions are lost in this alternative approach, the characterization of those structural and compositional features that are induced by realistic thermochemical treatments, which could be lost by interaction with air components but which are not reversed by evacuation or cooling, can be investigated.

The quasi *in situ* methodology was initially used by Parkinson and White [173] and Colloso-Davila *et al.* [174]. More recently, Inceesungvorn *et al.* used a quasi-*in situ* TEM approach based on a high-resolution vacuum transfer holder (VTST4006, GATAN ltd.) to study a Ag/γ-Al_2O_3 catalyst. In particular, a structural investigation of these catalysts under redox reaction conditions has been performed by means of *in situ* diffuse reflectance UV spectroscopy (DRUVS) and quasi *in situ* TEM techniques. A good correlation between the UV–vis results and TEM studies were obtained. These two techniques were used to investigate the nature of the silver active sites present in the Ag/γ-Al_2O_3 catalyst during the selective reduction of oxygen by hydrogen in the presence of hydrocarbons. Their results suggest that redispersion of silver clusters/particles takes place after contact with the oxidizing environment, which justified the use of the quasi *in situ* approach after thermal treatment under reducing conditions [175].

26.5.2
Synthesis of Nanostructured Materials

The *in situ* ETEM technique has greatly contributed to understanding the synthesis of different nanostructured materials, such as highly dispersed metal nanoparticles supported on oxides and CNTs. It is commonly accepted that the shape, size, and structure of supported metal nanoparticles determine their chemical and catalytic properties [177, 178]. Therefore, the attention that the scientific community has paid to the nucleation/formation mechanisms of particles on high surface area supports from different precursors is justified. Thus, Crozier *et al.* have extensively characterized the formation of Ni nanoparticles during the reduction of a $Ni(NO_3)_2 6H_2O$ precursor supported on a commercial titania substrate via an incipient wetness technique [176, 179]. Figure 26.58 shows the particle nucleation and growth during reduction under CO flow at 350 °C [176]. In the case of low-metal loadings, the Ni nanoparticles nucleate from the nickel precursor that is uniformly covering the support (Figure 26.58a). The dynamic characterization indicates that the particle remains at the same position after nucleation, although its size increases by the migration of the Ni atoms on the titania support surface leading to a three-dimensional growth mode (Figure 26.58b). Parallel chemical analysis performed by EELS showed that about 20% of the nickel nanoparticles were covered on the top by titania patches. In this case, time-resolved HREM images proved an extremely useful tool to study the growth mechanism, which follows a layer-by-layer mode. The importance of these studies in catalysis should be pointed out, since they can be related to the strong metal support interaction (SMSI) phenomena, which have been proposed as the main reason for dramatic modification of chemical properties observed for catalysts based on highly dispersed noble metals on reducible oxides, such as titania. Among these chemical properties, we can cite the

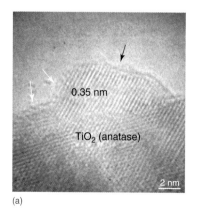
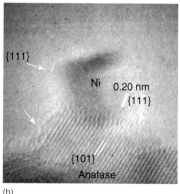

(a)   (b)

Figure 26.58 Time-resolved HREM images showing the nucleation and growth mechanism of Ni particles on anatase. (a) Before the reduction and (b) after 20 min using 0.2 Torr of CO in the ETEM at 350 °C. (Adapted from Ref. [176].)

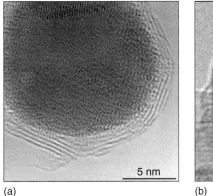 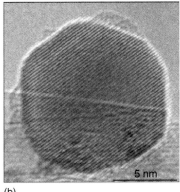

(a)   (b)

Figure 26.59 (a) Representative *ex situ* TEM images of Ba-promoted Ru catalyst supported on BN after a reducing pretreatment. (b) HREM image obtained in the ETEM at 552 °C and 5.2 mbar in a gas mixture of H_2/N_2 of 3 : 1. (Adapted from Ref. [185].)

reducibility and the chemisorption of probe molecules such as hydrogen and carbon monoxide. This metal decoration affects also the catalytic properties, and could eventually lead to a total deactivation [180–184]. The opportunity to evaluate the support coverage and the evolution under real reaction or pretreatment conditions opens new possibilities to understand catalyst deactivation mechanisms such as the SMSI phenomena.

Hansen *et al.* have applied the ETEM technique to the characterization of Ba-promoted Ru catalyst supported on BN [185] for ammonia synthesis. Their results clearly show strong divergences between traditional TEM and *in situ* ETEM characterization. The samples pretreated under reducing conditions in a conventional laboratory-scale reactor and transferred into the TEM equipment under air showed the total coverage of the metal nanoparticles by the support (Figure 26.59a). Owing to this, it was not possible to identify the location of the promoter in the catalyst by HREM images, although spectroscopic analysis by EDS corroborated its presence. ETEM allowed studying the evolution of the sample during the activation pretreatment under hydrogen. Under 331 °C and 3 mbar of a 3 : 1 H_2/N_2 mixture, the metal surface was not covered by the BN and the authors were able to observe the high mobility of barium under operation conditions (Figure 26.59b). The combination of HREM with EELS analysis determined the presence of single Ba atoms close to ruthenium crystal edges and corners, which were considered by the author as those responsible for the catalytic activity enhancement. The crucial contribution of similar studies for a deep understanding of the role of promoters in catalysis is obvious.

Dynamic *in situ* ETEM observations have been also proven as a powerful tool to reveal the growth mechanism of silicon nanowires [140, 186, 187], CNTs, and carbon nanofibers (CNFs) [149, 188–190]. The in-depth analysis of time-resolved HREM

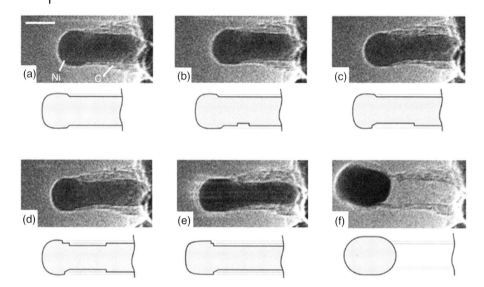

Figure 26.60 (a–f) Representative time-resolved TEM images of CNF growth using a Ni-based catalyst. The images were recorded at 536 °C using a total pressure of 2.1 mbar and a reaction mixture of $CH_4:H_2 = 1:1$. The scale bar is 5 nm. (Adapted from Ref. [191].)

images of nickel-based catalysts for CNT and CNF synthesis by catalytic chemical vapor decomposition (CCVD) allowed to perform kinetic measurements and to determine the growth mechanism. The Ni nanoparticles, which act as a catalyst and a template of the final multiwalled CNTs, change their shape dramatically during the growth of the CNTs. The reshaping of the Ni nanoparticles as well as the graphene growth and alignment during the reaction are summarized in Figure 26.60. These measurements revealed the importance of the spontaneously induced Ni step edges, which seemed to be the origin of layer growth. This idea implies the diffusion of carbon atoms from the particle surface toward the nucleation center located at the Ni steps. The images also showed the dramatic elongation of the Ni nanoparticle during the process. The authors used theoretical calculations of atom transportation and C–Ni binding energies for a deep understanding of the *in situ* observations [191, 192]. The intensive research of Helveg *et al.* demonstrated that the combination of reliable high-resolution characterization under operation conditions and theoretical calculation of physical-chemistry properties is nowadays one of the most appealing research challenges.

In situ ETEM observations have also greatly contributed to understanding CNT/CNF growth rates and the effect of the synthesis conditions on the CNT–CNF morphology [140, 188, 189, 193]. In this respect, Sharma *et al.* have observed that at low temperatures (480 °C) waves, loops, and spiral-like CNTs are formed because of a zigzag growth mechanism. However, the synthesis of straight single-walled CNTs can be achieved at higher temperatures (600 °C) and low reactant partial pressures. These authors also estimated the growth rate using the projected length of the CNTs in time-resolved HREM image series. According to their results, the CNT growth rate is lineal and it depends on the tube diameter and the number of carbon walls.

26.5.3
Nanoparticle Mobility

The traditional *ex situ* TEM characterization shows static nanoparticles, although structural and morphological changes may be induced by temperature and chemical environment. As it was mentioned previously, the particle size and shape strongly influences nanoparticle properties. Therefore, an exhaustive characterization at nanometric scale during operation conditions is needed if we want to reach reliable correlations between the properties and nanostructure. In this respect, we should highlight that *in situ* TEM has showed dynamic reshaping of nanoparticles under reaction conditions [194]. Figure 26.61 shows, as an example, the dramatic shape changes observed in Cu particles supported on ZnO under different chemical environments at 220 °C [194]. The high-resolution images obtained under well-controlled conditions are useful to reveal the changes in particle faceting and

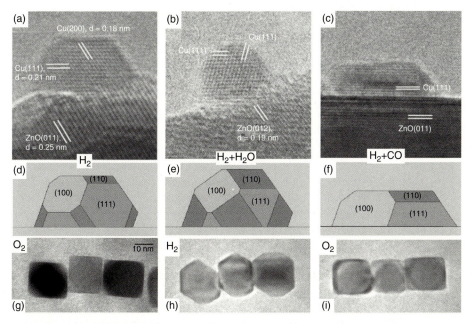

Figure 26.61 ETEM images of a Cu/ZnO catalyst at 220 °C and under the following environment: (a) 1.5 mbar of H_2, (b) 1.5 mbar of $H_2:H_2O = 3:1$, and (c) 5 mbar of $H_2:CO = 19:1$. The corresponding Wulff models of each image are included below them. It is also included in the shape evolution of Pt nanoparticles under oxygen (g), hidrógeno (H_2) (h), and reoxidation (i). (Adapted from Ref. [191, 196].)

the interaction with the support and to establish accurate particle morphology models. Observe how in this case, well-faceted particles were formed under hydrogen and how the particle faceting changed to a more spherical shape after introducing water vapor in the gas stream. The metal morphology modeling provided important information about the active site exposed on the metal particles. As it can be deduced from the Wulff constructions included in Figure 26.61, the fraction of {111} facets decreased after introducing water into the gas mixture, whereas the extent of interfacial surface between the metal and the support was not affected. The introduction of a stronger reducing gas, such as CO, into the E-cell resulted in more flattened particles and an increase in the metal–support contact surface.

Similarly, Giorgio et al. have intensively studied the shape stability of different metal particles, such as Au and Pt, under different chemical environments [155, 195, 196] using a carbon windowed E-cell holder. Figure 26.61g–i illustrates the shape evolution of three square-shaped Pt particles during an oxidation–reduction cycle. The original particle shape was limited by six {001} facets truncated by {111} facets at the corners. After exposure to hydrogen, the {111}/{001} ratio increases significantly. However, this transformation was shown to be reversible and the particles recovered their original shape after reexposure to oxygen. This results were explained considering the stronger adsorption of oxygen on the more open {100} faces and highlighted the importance of determining the particle faceting under operation conditions to establish correlation between performance and nanostructure.

26.5.4
Nanoparticles under Working Conditions

One of the most appealing characteristics of *in situ* ETEM is the possibility to perform dynamic observations of the specimen under operation conditions. This approach gives us the unique opportunity to evaluate the effect of thermochemical environment on structure, and subsequently to relate those with macroscopic performance. Moreover, these observations would eventually show how the nanoparticles work. A good example of this can be found in the work performed by Simonsen et al. on soot oxidation by ceria catalyst nanoparticles [197] using a CM300 FEG ETEM. These authors monitored the combustion reaction of the carbon black (CB) located in direct contact with ceria nanoparticles. Figure 26.62a,b shows consecutive images within a time interval of about 12 min. Note how the CB area close to the ceria nanoparticles gradually disappears because of catalytic combustion. A deeper quantitative analysis of the TEM images made it possible to estimate kinetic parameters such as the apparent activation energy of the reaction. In fact, the authors measured the evolution with time of the projected CB-CeO_2 distance to calculate the average projected rate of the CB relative to the CeO_2 nanoparticles. The activation energy was obtained by the Arrhenius representation of the projected rate at several temperatures (300–600 °C).

Gai et al. have also worked intensively to evaluate the new opportunities of ETEM studies to study specimen stability under working conditions [170, 198–201].

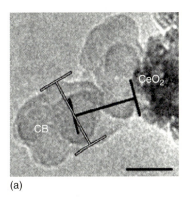 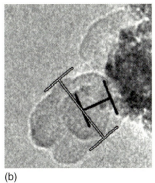

Figure 26.62 (a,b) *In situ* HREM images of soot combustion using ceria as catalyst at 500 °C and 2 mbar of O_2. (Adapted from Ref. [197].)

One of their most relevant contributions to this field has been the development of a wet ETEM holder for *in situ* studies of liquid–solid and gas–liquid–solid catalytic reactions [142, 171]. Initial experiments allowed them to follow the *in situ* polymerization reaction over Co-Ru/titania catalyst system in hexamethylene diamine and adipic acid solutions at different temperatures, from room temperature up to 188 °C.

References

1. Goesmann, H. and Feldmann, C. (2010) Nanoparticulate functional materials. *Angew. Chem. Int. Ed.*, **49** (8), 1362–1395.
2. Arvizo, R., Bhattacharya, R., and Mukherjee, P. (2010) Gold nanoparticles: opportunities and challenges in nanomedicine. *Expert Opin. Drug Deliv.*, **7** (6), 753–763.
3. Giljohann, D.A., Seferos, D.S., Daniel, W.L., Massich, M.D., Patel, P.C., and Mirkin, C.A. (2010) Gold nanoparticles for biology and medicine. *Angew. Chem. Int. Ed.*, **49** (19), 3280–3294.
4. Lee, J., Mahendra, S., and Alvarez, P.J.J. (2010) Nanomaterials in the construction industry: a review of their applications and environmental health and safety considerations. *ACS Nano*, **4** (7), 3580–3590.
5. Dastjerdi, R., Montazer, M., Dastjerdi, R., and Montazer, M. (2010) A review on the application of inorganic nano-structured materials in the modification of textiles: focus on anti-microbial properties. *Colloids Surf. B Biointerfaces*, **79** (1), 5–18.
6. Vinayaka, A.C. and Thakur, M.S. (2010) Focus on quantum dots as potential fluorescent probes for monitoring food toxicants and foodborne pathogens. *Anal. Bioanal. Chem.*, **397** (4), 1445–1455.
7. Homberger, M. and Simon, U. (2010) On the application potential of gold nanoparticles in nanoelectronics and biomedicine. *Philos. Trans. R. Soc. A*, **368** (1915), 1405–1453.
8. Giorgi, R., Baglioni, M., Berti, D., and Baglioni, P. (2010) New methodologies for the conservation of cultural heritage: micellar solutions, microemulsions, and hydroxide nanoparticles. *Acc. Chem. Res.*, **43** (6), 695–704.

9. Scown, T.M., van Aerle, R., and Tyler, C.R. (2010) Do engineered nanoparticles pose a significant threat to the aquatic environment? *Crit. Rev. Toxicol.*, **40** (7), 653–670.
10. Paschoalino, M.P., Marcone, G.P.S., and Jardim, W.F. (2010) Nanomaterials and the Environment. *Quim. Nova*, **33** (2), 421–430.
11. Hillegass, J.M., Shukla, A., Lathrop, S.A., MacPherson, M.B., Fukagawa, N.K., and Mossman, B.T. (2010) Assessing nanotoxicity in cells in vitro. *Wiley Interdiscipl. Rev. Nanomed. Nanobiotechnol.*, **2** (3), 219–231.
12. Nellist, P.D. and Pennycook, S.J. (2000) The principles and interpretation of annular dark field Z-contrast imaging. *Adv. Imaging Electron Phys.*, **113**, 147–203.
13. Muller, D.A., Fitting Kourkoutis, L., Murfitt, M., Song, J.H., Hwang, H.Y., Silcox, J., Dellby, N., and Krivanek, O.L. (2008) Atomic-scale chemical imaging of composition and bonding by aberration-corrected microscopy. *Science*, **319**, 1073–1076.
14. J.C.H., Spence (ed.) (2003) *High Resolution Electron Microscopy*, 3rd edn, Oxford University Press, Inc., New York, ISBN: 978-0-19-955275-7.
15. Wang, Z.L. and Cowley, J.M. (1990) Theory of high-angle annular dark field stem images of Ge/Si interfaces. *Ultramicroscopy*, **32**, 275.
16. Uzun, A. Ortalan, V. Browning, N.D. Gates, B.C. (2010) A site-isolated mononuclear iridium complex catalyst supported on MgO: characterization by spectroscopy and aberration-corrected scanning transmission electron microscopy, *J. Catal.*, **269**, 318–328.
17. Blom, D.A., Allard, L.F., Narula, C.K., and Moses-DeBusk, M.J. (2008) Aberration-corrected STEM imaging of Ag on γ-Al_2O_3. *Microsc. Microanal.*, **14**, 98–103.
18. Varela, M., Lupini, A.R., van Benthem, K., Borisevich, A.Y., Chisholm, M.F., Shibata, N., Abe, E., and Pennycook, S.J. (2005) Materials characterization in the aberration-corrected scanning transmission electron microscope. *Annu. Rev. Mater. Res.*, **35**, 539–569.
19. Sohlberg, K., Rashkeev, S., Borisevich, A.Y., Pennycook, S.J., and Pantelides, S.T. (2004) Origin of the anomalous Pt-Pt distances in the Pt/Alumina catalytic system. *Chem. Phys. Chem.*, **5**, 1893–1897.
20. Bernal, S., Botana, F.J., Calvino, J.J., López Cartes, C., Pérez Omil, J.A., and Rodríguez-Izquierdo, J.M. (1998) The interpretation of HREM images of supported metal catalysts using image simulation: Profile view images. *Ultramicroscopy*, **72**, 135–164.
21. Avalos-Borja, M. and Yacamán, M.J. (1982) On the visibility of small metallic particles on crystalline substrates. *Ultramicroscopy*, **10**, 211–216.
22. Yao, M.H., Smith, D.J., and Datye, A.K. (1993) Comparative study of supported catalyst particles by electron microscopy methods. *Ultramicroscopy*, **52**, 282–288.
23. Rice, S.B., Koo, J.Y., Disko, M.M., and Treacy, M.M.J. (1990) On the imaging of Pt atoms in zeolite frameworks. *Ultramicroscopy*, **34**, 108–118.
24. Herzing, A.A., Kiely, C.J., Carley, A.F., Landon, P., and Hutchings, G.J. (2008) Identification of active gold nanoclusters on iron oxide supports for CO oxidation. *Science*, **321**, 1331.
25. Allsop, N.A., Dobson, P.J., and Hutchison, J.L. (2006) Optimized HREM imaging of objects supported by amorphous substrates using spherical aberration adjustment. *J. Microsc.*, **223**, 1–8.
26. ThustA., A., Coene, W.M.J., Op de Beek, M., and Van Dyck, D. (1996) Focal-series reconstruction in HRTEM: simulation studies on non-periodic Objects. *Ultramicroscopy*, **64**, 211–223.
27. Zandbergen, H.W., Tang, D., and Van Dyck, D. (1996) Non-linear interference in relation to strong delocalisation. *Ultramicroscopy*, **64**, 185–198.
28. Urban, K., Kabius, B., Haider, M., and Ross, H. (1999) A way to higher resolution: spherical-aberration correction in a 200 kV transmission electron microscope. *J. Electron Microsc.*, **48**, 821–826.

29. Gontard, L.C., Dunin-Borkowski, R.E., Htch, M.J., and Ozkaya, D. (2006) Delocalisation in images of Pt nanoparticles. *J. Phys. Conf. Ser.*, **26**, 292–295.
30. Gontard, L.C., Ozkaya, D., and Dunin-Borkowski, R.E. (2011) A simple algorithm for measuring particle size distributions on an uneven background from TEM images. *Ultramicroscopy*, **111**, 101–106.
31. Gontard, L.C., Dunin-Borkowski, R.E., Ozkaya, D., Hyde, T., Midgley, P.A., and Ash, P. (2006) Crystal size and shape analysis of Pt nanoparticles in two and three dimensions. *J. Phys. Conf. Ser.*, **26**, 367–370.
32. Bele, P., Jäger, F., and Stimming, U. (2007) *Microsc. Anal.*, **21**, S5.
33. Matyi, R.J., Schwartz, L.H., and Butt, J.B. (1987) Particle size, particle size distribution, and related measurements of supported metal catalysts. *Catal. Rev. Sci. Eng.*, **29** (1), 41–99.
34. Blanco, G., Calvino, J.J., Cauqui, M.A., Corchado, P., and López-Cartes, C. (1999) Nanostructural evolution under reducing conditions of a Pt/CeTbOx catalyst: a new alternative system as a TWC component. *Chem. Mater.*, **11**, 3610–3619.
35. Blanco, G., Calvino, J.J., Cauqui, M.A., Corchado, P., López-Cartes, C., and Pérez-Omil, J.A. (1999) HREM analysis of the nanostructure/oxygen buffering capacity relationship in a Rh/CeTbOx catalyst. *Inst. Phys. Conf. Ser.*, **161** (Section 10), 541–544.
36. López-Haro, M., Pérez-Omil, J.A., Hernández-Garrido, J.C., Trasobares, S., Hungría, A.B., Cíes, J.M., Midgley, P.A., Bayle-Guillemaud, P., Martínez-Arias, A., Bernal, S., Delgado, J.J., and Calvino, J.J. (2010) Advanced electron microscopy investigation of Ceria–Zirconia-based catalysts. *ChemCatChem*, doi: 10.1002/cctc.201000306
37. Zeng, H. and Sun, S. (2008) Syntheses, properties, and potential applications of multicomponent magnetic nanoparticles. *Adv. Funct. Mater.*, **18**, 391–400.
38. Fernández, B., Gálvez, N., Cuesta, R., Hungría, A.B., Calvino, J.J., and Domínguez-Vera, J.M. (2008) Quantum dots decorated with magnetic bionanoparticles. *Adv. Funct. Mat.*, **18**, 3931–3935.
39. Shevchenko, E.V., Talapin, D.V., Kotov, N.A., O'Brien, S., and Murray, C.B. (2006) Structural diversity in binary nanoparticle superlattices. *Nature*, **439**, 55.
40. Kiely, C.J., Fink, J., Zheng, J.G., Brust, M., Bethell, D., and Schiffrin, D.J. (2000) Ordered colloidal nanoalloys. *Adv. Funct. Mat.*, **12** (9), 640–643.
41. Giorgio, S., Graoui, H., Chapon, C., and Henry, C.R. (1998) Epitaxial growth of clusters on defined oxide surfaces-HRTEM studies. *Cryst. Res. Technol.*, **33**, 1061–1074.
42. Rupprechter, G., Hayek, K., and Hofmeister, H. (1998) Electron microscopy of thin-film model catalysts: activation of alumina-supported rhodium nanoparticles. *J. Catal.*, **173**, 409–422.
43. Hofmeister, H. (1984) Habit and internal structure of multiply twinned gold particles on silver bromide films. *Thin Sohd Films*, **116**, 151–162.
44. Pérez, O.L., Romeu, D., and Yacamán, M.J. (1982) Distribution of surface sites on small metallic particles. *Appl. Surf. Sci.*, **13**, 402–413.
45. Fluei, M. (1989) Observation des structures anormales de petites particulesd'or et d'argent par microscopieélectronique à haute résolution et diffraction d'électrons pat un jet d'agrégatsd'argent. PhD Thesis. No. 796, Ecole Polytechnique Federale de Lausanne, Lausanne EPL.
46. Spycher, R.P. Stadelmann, P. Buffat P.H., and Fluei, M. (1988) High-resolution electron microscopy image simulation on a Cray 1S/2300 computer. *J. Electron Microsc. Tech.*, **10**, 369.
47. Yacamán, M.J., Herrera, R., Tehuacanero, S., Gómez, A., and Beltrán del Río, L. (1990) Surface roughness studies in small particles using HREM. *Ultramicroscopy*, **33**, 133–141.

48. Kirkland, A.I. and Meyer, R.R. (2004) "Indirect" high-resolution transmission electron microscopy: aberration measurement and wavefunction reconstruction *Microsc. Microanal.*, **10**, 401–413.
49. Gontard, L.C., Chang, L.-Y., Dunin-Borkowski, R.E., Kirkland, A.I., Herethington, C.J.D., and Ozkaya, D. (2006) The application of spherical aberration correction and focal series restoration to high-resolution images of platinum nanocatalyst particles. *J. Phys.: Conf. Ser.*, **26**, 25–28.
50. Gontard, L.C., Chang, L.Y., Hetherington, C.J.D., Kirkland, A.I., Ozkaya, D., and Dunin-Borkowski, R.E. (2007) Aberration-corrected imaging of active sites on industrial catalyst nanoparticles. *Angew. Chem. Int. Ed.*, **46**, 3683–3685.
51. Gontard, L.C., Chang, L.Y., Hetherington, C.J.D., Kirkland, A.I., Ozkaya, D., and Dunin-Borkowski, R.E. (2007) Aberration-corrected imaging of active sites on industrial catalyst nanoparticles. *Angew. Chem. Int. Ed.*, **46**, 3683–3685.
52. LeBeau, J.M. and Stemmer, S. (2008) Experimental quantification of annular dark-field images in scanning transmission electron microscopy. *Ultramicroscopy*, **108**, 1653–1658.
53. Sanchez, S.I., Small, M.W., Zuo, J.M., and Nuzzo, R.G. (2009) Structural characterization of Pt-Pd and Pd-Pt core-shell nanoclusters at atomic resolution. *J. Am. Chem. Soc.*, **131**, 8683–8689.
54. Singhal, A., Yang, J.C., and Gibson, J.M. (1997) STEM-based mass spectroscopy of supported Re clusters. *Ultramicroscopy*, **67**, 191–206.
55. Liu, Y., Jia, C.-J., Yamasaki, J., Terasaki, O., and Schüth, F. (2010) Highly active iron oxide supported gold catalysts for CO oxidation: how small must the gold nanoparticles be? *Angew. Chem. Int. Ed.*, **49**, 5771–5775.
56. Rashkeev, S.N., Lupini, A.R., Overbury, S.H., Pennycook, S.J., and Pantelides, S.T. (2007) Role of the nanoscale in catalytic CO oxidation by supported Au and Pt nanostructures. *Phys. Rev. B*, **76**, 035438-1–035438-7.
57. Lupini, A.R., Franceschetti, A.G., Pantelides, S.T., Dai, S., and Chen, B. (2004) Aberration corrected STEM analysis of gold nanoparticle catalytic activity. *Microsc. Microanal.*, **10**, 462–463.
58. Wang, J.X., Inada, H., Wu, L., Zhu, Y., Choi, Y.M., Liu, P., Zhou, W.-P., and Adzic, R.R. (2009) Oxygen reduction on well-defined core-shell nanocatalysts: particle size, facet, and Pt shell thickness effects. *J. Am. Chem. Soc.*, **131**, 17298–17302.
59. Ferrer, D., Blom, D.A., Allard, L.F., Mejía, S., Pérez-Tijerina, E., and Yacamán, M.J. (2008) Atomic structure of three-layer Au/Pd nanoparticles revealed by aberration-corrected scanning transmission electron microscopy. *J. Mater. Chem.*, **18**, 2442–2446.
60. Shannon, M.D., Lokc, C.M., and Casci, J.L. (2007) Imaging promoter atoms in Fischer–Tropsch cobalt catalysts by aberration-corrected scanning transmission electron microscopy. *J. Catal.*, **249**, 41–51.
61. Gröger, H., Gyger, F., Leidinger, P., Zurmühl, C., and Feldmann, C. (2009) Microemulsion approach to nanocontainers and its variability in composition and filling. *Adv. Mater.*, **21**, 1586–1590.
62. Corma, A., Iborra, S., Xamena, F.X., Montón, R., Calvino, J.J., and Prestipino, C. (2010) Nanoparticles of Pd on hybrid polyoxometalate-ionic liquid material: synthesis, characterization, and catalytic activity for heck reaction. *J. Phys. Chem. C*, **114**, 8828–8836.
63. Baker, R.T., Bernal, S., Calvino, J.J., Pérez-Omil, J.A., and López-Cartes, C. (2004) in *Nanotechnology in Catalysis*, vol. **2**, Chapter 19 (eds B. Zhou, S. Hermans, and G.A. Somorja), Springer-Verlag, Berlin, pp. 403–429.
64. Bernal, S., Botana, F.J., Calvino, J.J., Cifredo, G.A., Pérez-Omil, J.A., and Pintado, J.M. (1995) HREM study of the behaviour of a Rh/CeO$_2$ catalyst under high temperature reducing and

oxidizing conditions. *Catal. Today*, **23**, 219–250.

65. Chang, L.Y., Barnard, A.S., Gontard, L.C., and Dunin-Borkowski, R.E. (2010) Resolving the structure of active sites on platinum catalytic nanoparticles. *Nano Lett.*, **10**, 3073–3076.

66. Hernández, J.C., Hungría, A.B., Pérez-Omil, J.A., Trasobares, S., Bernal, S., Midgley, P.A., Alavi, A., and Calvino, J.J. (2007) Structural surface investigations of Cerium-Zirconium mixed oxide nanocrystals with enhanced reducibility. *J. Phys. Chem. C*, **111** (6), 9001–9004.

67. Hofmeister, H., Miclea, P.-T., Steen, M., Mörke, W., and Drevs, H. (2007) Structural characteristics of oxide nanosphere supported metal nanoparticles. *Top. Catal.*, **46** (1–2), 11–21.

68. Cai, W., Hofmeister, H., and Dubiel, M. (2001) Importance of lattice contraction in surface plasmon resonance shift for free and embedded silver particles. *Eur. Phys. J. D*, **13**, 245–253.

69. Hofmeister, H., Huisken, F., and Kohn, B. (1999) Lattice contraction in nanosized silicon particles produced by laser pyrolysis of silane. *Eur. Phys. J. D*, **9**, 137–140.

70. Dubiel, M., Hofmeister, H., Schurig, E., Wendler, E., and Wesch, W. (2000) On the stress state of silver nanoparticles in ion-implanted silicate glasses. *Nucl. Instrum. Methods Phys. Res. B*, **166–167**, 871–876.

71. Dubiel, M., Hofmeister, H., and Schurig, E. (1997) Compressive stresses in Ag nanoparticle doped glasses by ion implantation. *Phys. Stat. Sol. B*, **203**, R5–R6.

72. Sao-Joao, S., Giorgio, S., Mottet, C., Goniakowski, J., and Henry, C.R. (2006) Interface structure of Ni nanoparticles on MgO (100): a combined HRTEM and molecular dynamic study. *Surf. Sci.*, **600**, L86–L90.

73. Wang, R., Dmitrieva, O., Farle, M., Dumpich, G., Acet, M., Mejia-Rosales, S., Perez-Tijerina, E., Yacaman, M.J., and Kisielowski, C. (2009) FePt Icosahedra with magnetic cores and catalytic shells. *J. Phys. Chem. C*, **113**, 4395–4400.

74. Johnson, C.L., Snoeck, E., Ezcurdia, M., Rodríguez-González, B., Pastoriza-Santos, I., Liz-Marzán, L.M., and Hÿtch, M.J. (2007) Effects of elastic anisotropy on strain distributions in decahedral gold nanoparticles. *Nat. Mater.*, **7**, 120–124.

75. Johson, E., Andersen, H.H., and Dahmen, U. (2004) Nanoscale lead and noble gas inclusions in Aluminum: structures and properties. *Micros. Res. Tech.*, **64**, 356–372.

76. Bernal, S., Baker, R.T., Burrows, A., Calvino, J.J., Kiely, C.J., López-Cartes, C., Pérez-Omil, J.A., and Rodríguez-Izquierdo, J.M. (2000) Structure of highly dispersed metals and oxides: exploring the capabilities of high-resolution electron microscopy. *Surf. Interface Anal.*, **29**, 411–421.

77. Baker, R.T., Calvino, J.J., López-Cartes, C., Pérez-Omil, J.A., and Rodríguez-Izquierdo, J.M. (1999) Detection and quantification of small misalignments on nanometer-sized particles on oxide support systems by the analysis plan view HREM images. *Inst. Phys. Conf. Ser.*, **161** (Section 10), 537–540.

78. Woltersdorf, J., Nepijko, A.S., and Pippel, E. (1981) Dependence of lattice parameter of small particles on the size of the nuclei. *Surf. Sci.*, **106**, 64–69.

79. Hawkes, P.W. (1992) in *Electron Tomography: Three-Dimensional Imaging With the Transmission Electron Microscope* (ed. J. Frank), Plenum Press, New York, London, pp. 17–38.

80. Koster, A.J., Ziese, U., Verklej, A.J., Hansen, A.H., and de Jong, K.P. (2000) Three-dimensional transmission electron microscopy: a novel imaging and characterization technique with nanometer scale resolution for materials science. *J. Phys. Chem. B*, **104**, 9368.

81. Midgley, P.A., Weyland, M., Thomas, J.M., and Johnson, B.F.G. (2001) Z-Contrast tomography: a technique in three-dimensional nanostructural analysis based on Rutherford scattering. *Chem. Commun.*, 907–908.

82. Howie, A. (1979) Image-contrast and localized signal selection techniques. *J. Microsc.*, **117**, 11.

83. Pennycook, S.J. (1989) Z-Contrast STEM for materials science. *Ultramicroscopy*, **30**, 58.

84. Midgley, P.A. and Weyland, M. (2003) 3D electron microscopy in the physical sciences: the development of Z-contrast and EFTEM tomography. *Ultramicroscopy*, **96**, 413–431.

85. Hungria, A.B., Raja, R., Adams, R.D., Captain, B., Thomas, J.M., Midgley, P.A., Golovko, V., and Johnson, B.F.G. (2006) Single-step conversion of dimethyl terephthalate into cyclohexanedimethanol with Ru_5PtSn, a trimetallic nanoparticle catalyst. *Angew. Chem. Int. Ed.*, **45**, 4782–4785.

86. Weyland, M., Midgley, P.A., and Thomas, J.M. (2001) Electron tomography of nanoparticle catalysts on porous supports: a new technique based on Rutherford scattering. *J. Phys. Chem. B*, **105** (33), 7882.

87. Wikander, K., Hungria, A.B., Midgley, P.A., Palmqvist, A.E.C., Holmberg, K., and Thomas, J.M. (2007) Incorporation of platinum nanoparticles in ordered mesoporous carbon. *J. Colloid Interface Sci.*, **305**, 204–208.

88. Arslan, I., Walmsley, J.C., Rytter, E., Bergene, E., and Midgley, P.A. (2008) Toward three-dimensional nanoengineering of heterogeneous catalysts. *J. Am. Chem. Soc.*, **130**, 5716–5719.

89. Kaneko, K., Furuya, K., Hungria, A.B., Hernandez-Garrido, J.C., Midgley, P.A., Onodera, T., Kasai, H., Yaguchi, Y., Oikawa, H., Nomura, Y., Harada, H., Ishihara, T., and Baba, N. (2009) Nanostructural characterization and catalytic analysis of hybridized platinumphthalocyanine nanocomposites. *J. Electron Microsc.*, **58**, 289–294.

90. Hernández-Garrido, J.C., Yoshida, K., Gai, P.L., Boyes, E.D., Christensen, C.H., and Midgley, P.A. (2011) The location of gold nanoparticles on titania: a study by high resolution aberration-corrected electron microscopy and 3D electron tomography. *Catal. Today*, **160** (1), 165–169.

91. González, J.C., Hernández, J.C., López-Haro, M., Del Río, E., Delgado, J.J., Hungría, A.B., Trasobares, S., Bernal, S., Midgley, P.A., and Calvino, J.J. (2009) 3D characterization of gold nanoparticles supported on heavy metal oxide catalysts by HAADF-STEM electron tomography. *Angew. Chem. Int. Ed.*, **48**, 5313–5315.

92. Kaneko, K., Inoke, K., Freitag, B., Hungria, A.B., Midgley, P.A., Hansen, T.W., Zhang, J., Ohara, S., and Adschiri, T. (2007) Structural and morphological characterization of cerium oxide nanocrystals prepared by hydrothermal synthesis. *Nanoletters*, **7** (2), 421–425.

93. Weyland, M., Yates, T.J.V., Dunin-Borkowski, R.E., Laffont, L., and Midgley, P.A. (2006) Nanoscale analysis of three-dimensional structures by electron tomography. *Scr. Mater.*, **55**, 29–33.

94. Arslan, I., Yates, T.J.V., Browning, N.D., and Midgley, P.A. (2005) Embedded nanostructures revealed in three dimensions. *Science*, **309**, 2195.

95. Hungria, A.B., Juarez, B.H., Klinke, C., Weller, H., and Midgley, P.A. (2008) 3-D characterization of CdSe nanoparticles attached to carbon nanotubes. *Nano Res.*, **1**, 89–97.

96. Perassi, E.M., Hernández-Garrido, J.C., Moreno, M.S., Encina, E.R., Coronado, E.A., and Midgley, P.A. (2010) Using highly accurate 3D nanometrology to model the optical properties of highly irregular nanoparticles: a powerful tool for rational design of plasmonic devices. *Nano Lett.*, **10**, 2097–2104.

97. Jinschek, J.R., Batenburg, K.J., Calderon, H.A., Kilaas, R., Radmilovic, V., and Kisielowski, C. (2008) 3-D reconstruction of the atomic positions in a simulated gold nanocrystal based on discrete tomography: prospects of atomic resolution electron tomography. *Ultramicroscopy*, **108**, 589–604.

98. Batenburg, K.J., Bals, S., Sijbers, J., Kübel, C., Midgley, P.A., Hernandez, J.C., Kaiser, U., Encina, E.R., Coronado, E.A., and VanTendeloo, G. (2009) 3D imaging of nanomaterials by

discrete tomography. *Ultramicroscopy*, **109**, 730–740.

99. Jeanguillaume, C. and Colliex, C. (1989) Spectrum-image – the next step in EELS digital acquisition and processing. *Ultramicroscopy*, **28** (1–4), 252–257.

100. Lorimer, G.W. (1983) in *Quantitative Electron Microscopy* (ed. J.N. Chapman and A.J. Craven), Scottish Universities Summer School in Physics, p. 305.

101. Watanabe, M. and Williams, D.B. (2006) The quantitative analysis of thin specimens: a review of progress from the Cliff-Lorimer to the new zeta-factor methods. *J. Microsc. (Oxford)*, **221** (2), 89–109.

102. Herzing, A.A., Watanabe, M., Edwards, J.K., Conte, M., Tong, Z.R., Hutchings, G.H., and Kiely, C.J. (2008) Energy dispersive X-ray spectroscopy of bimetallic nanoparticles in an aberration corrected scanning transmission electron microscope. *Faraday Discuss.*, **138**, 337–351.

103. Chu, M.-W., Liou, S.C., Chang, C.-P., Choa, F.-S., and Chen, C.H. (2010) Emergent chemical mapping at atomic-column resolution by energy-dispersive x-ray spectroscopy in an aberration-corrected electron microscope. *Phys. Rev. Lett.*, **104**, 196101–196104.

104. Kim, G., Sousa, A., Meyers, D., and Libera, M. (2008) Nanoscale composition of biphasic polymer nanocolloids in aqueous suspension. *Microsc. Microanal.*, **14**, 459–468.

105. Batson, P.E. (1980) Damping of bulk plasmons in small aluminum spheres. *Solid State Commun.*, **34** (6), 477–480.

106. Batson, P.E. (1982) Surface-plasmon coupling in clusters of small spheres. *Phys. Rev. Lett.*, **49** (13), 936–940.

107. Hirsch, L.R., Stafford, R.J., Bankson, J.A., Sershen, S.R., Rivera, B., Price, R.E., Hazle, J.D., Halas, N.J., and West, J.L. (2003) Nanoshell-mediated near-infrared thermal therapy of tumors under magnetic resonance guidance. *Proc. Natl. Acad. Sci. U.S.A.*, **100** (23), 13549–13554.

108. Nie, S., and Emory, S.R. (1997) Probing single molecules and single nanoparticles by surface-enhanced Raman scattering. *Science*, **275** (5303), 1102–1106.

109. Ebbesen, T.W., Lezec, H.J., Ghaemi, H.F., Thio, T., and Wolff, P.A. (1998) Extraordinary optical transmission through sub-wavelength hole arrays. *Nature*, **391**, 667–669.

110. Bosman, M., Keast, V.J., Watanabe, M., Maaroof, A.I., and Cortie, M.B. (2007) Mapping surface plasmons at the nanometre scale with an electron beam. *Nanotechnology*, **18**, 165505–165505.

111. Nelayah, J., Kociak, M., Stephan, O., Geuquet, N., Henrard, L., de Abajo, F.J.G., Pastoriza-Santos, I., Liz-Marzan, L.M., and Colliex, C. (2010) Two-dimensional quasistatic stationary short range surface plasmons in flat nanoprisms. *Nano Lett.*, **10** (3), 902–907.

112. Nelayah, J., Kociak, M., Stephan, O., de Abajo, F.J.G., Tence, M., Henrard, L., Taverna, D., Pastoriza-Santos, I., Liz-Marzan, L.M., and Colliex, C. (2007) Mapping surface plasmons on a single metallic nanoparticle. *Nat. Phys.*, **3** (5), 348–353.

113. Chu, M.W., Myroshnychenko, V., Chen, C.H., Deng, J.P., Mou, C.Y., and de Abajo, F.J.G. (2009) Probing bright and dark surface-plasmon modes in individual and coupled noble metal nanoparticles using an electron beam. *Nano Lett.*, **9** (1), 399–404.

114. Rodriguez-Lorenzo, L., Alvarez-Puebla, R.A., Pastoriza-Santos, I., Mazzucco, S., Stephan, O., Kociak, M., Liz-Marzan, L.M., and Garcia de Abajo, F.J. (2009) Zeptomol detection through controlled ultrasensitive surface-enhanced Raman scattering. *J. Am. Che. Soc.*, **131** (131), 4616–4618.

115. Colliex, C., Kociak, M., Stephan, O., Suenaga, K., and Trasobares S. (2001) in *Nanostructured Carbon for Advanced Applications*, NATO Science Series II, Vol. 24, Editorial (si libro), (eds G. Benedek et al.), Kluwer Academic Publishers, Netherlands, pp. 201–232.

116. Maigne, A. and Twesten, R.D. (2009) Review of recent advances in spectrum imaging and its extension of reciprocal space. *J. Electron Microsc.*, **58** (3), 99–109.
117. Gálvez, N., Valero, E., Ceolin, M., Trasobares, S., López-Haro, M., Calvino, J.J., and Domínguez-Vera, J.M. (2010) A bioinspired approach to the synthesis of bimetallic CoNi nanoparticles. *Inorg. Chem.*, **49**, 1705–1711.
118. Cho, S.J., Idrobo, J.C., Olamit, J., Liu, K., Browning, N.D., and Kauzlarich, S.M. (2005) Growth mechanisms and oxidation resistance of gold-coated iron nanoparticles. *Chem. Mater.*, **17**, 3181–3186.
119. van Schooneveld, M.M., Gloter, A., Stephan, O., Zagonel, L.F., Koole, R., Meijerink, A., Mulder, W.J.M., and de Groot, F.M.F. (2010) Imaging and quantifying the morphology of an organic–inorganic nanoparticle at the sub-nanometre level. *Nat. Nanotechnol.*, **5**, 538–544.
120. Egerton, R.F. (1996) *Electron Energy-Loss Spectroscopy in the Electron Microscope*, Plenum Press, New York, pp. 134–135.
121. Berger, A., and Kohl, H. (1992) Elemental mapping using an imaging energy filter-image-formation and resolution limits. *Microsc., Microanal. Microstruct.*, **3** (2–3), 159–174.
122. Langlois, C.T., Oikawa, T., Bayle-Guillemaud, P., and Ricolleau, C. (2008) Energy-filtered electron microscopy for imaging core–shell nanostructures. *J. Nanopart. Res.*, **10**, 997–1007.
123. Corma, A., Garc, H., Montes-Navajas, P., Primo, A., Calvino, J.J., and Trasobares, S. (2007) Gold nanoparticles in organic capsules: a supramolecular assembly of gold nanoparticles and cucurbituril. *Chem. Eur. J.*, **13**, 6359–6364.
124. Rodriguez-Luque, M.P., Hernandez, J.C., Yeste, M.P., Bernal, S., Cauqui, M.A., Pintado, J.M., Perez-Omil, J.A., Stephan, O., Calvino, J.J., and Trasobares, S. (2008) Preparation of Rhodium/$Ce_xPr_{1-x}O_2$ catalysts: a nanostructural and nanoanalytical investigation of surface modifications by transmission and scanning-transmission electron microscopy. *J. Phys. Chem. C*, **112**, 5900–5910.
125. Leapman, R.D. (1992) in *Transmission Electron Energy Loss Spectrometry in Material Science* (eds M.M. Disko, C.C. Ahn, and B. Fultz), The Mineral, Metals & Materials Society.
126. Bonnet, N., Brun, N., and Colliex, C. (1999) Extracting information from sequences of spatially resolved EELS spectra using multivariate statistical analysis. *Ultramicroscopy*, **77**, 97.
127. Keenan, M.R. and Kotula, P.G. (2004) Accounting for Poisson noise in the multivariate analysis of ToF-SIMS spectrum images. *Surf. Interface Anal.*, **36**, 203–212.
128. Verbeeck, J., and Van Aert, S. (2004) Model based quantification of EELS spectra. *Ultramicroscopy*, **101** (2–4), 207–224.
129. Rez, P. and Muller, D.A. (2008) The theory and interpretation of electron energy loss near-edge fine structures. *Annu. Rev Mater. Res*, **38**, 535–558.
130. Ahn, C.C. (ed.) *Transmission Electron Energy Loss Spectrometry in Materials Science and The EELS Atlas*, Wiley-VCH Verlag GmbH & Co. KGaA. ISBN: 9783527405657.
131. Galvez, N., Fernandez, B., Sanchez, P., Cuesta, R., Ceolin, M., Clemente-Leon, M., Trasobares, S., Lopez-Haro, M., Calvino, J.J., Stephan, O., and Dominguez-Vera, J.M. (2008) Comparative structural and chemical studies of ferritin cores with gradual removal of their iron contents. *J. Am. Chem. Soc.*, **130**, 8062–8068.
132. Arenal, R., de la Peña, F., Stephan, O., Walls, M., Tence, M., Loiseau, A., and Colliex, C. (2008) Extending the analysis of EELS spectrum-imaging data, from elemental to bond mapping in complex nanostructures. *Ultramicroscopy*, **109**, 32–38.
133. Allen, L.J. (2008) Electron microscopy: new directions for chemical maps. *Nat. Nanotechnol.*, **3**, 255–256.

134. Bosman, M., Keast, V.J., Garcia-Munoz, J.L., D'Alfonso, A.J., Findlay, S.D., and Allen, L.J. (2007) Two-dimensional mapping of chemical information at atomic resolution. *Phys. Rev. Lett.*, **99**, 086102.
135. Kimoto, K., Asaka, T., Nagai, T., Saito, T., Matsui, Y., and Ishizuka, K. (2007) Element-selective imaging of atomic columns in a crystal using STEM and EELS. *Nature*, **450**, 702–704.
136. Muller, D.A., Kourkoutis, L.F., Murfitt, M., Song, J.H., Hwang, H.Y., Silcox, J., Dellby, N., and Krivanek, O.L. (2008) Atomic-scale chemical imaging of composition and bonding by aberration-corrected microscopy. *Science*, **319**, 1073–1076.
137. Trasobares, S., López-Haro, M., Kociak, M., March, K., de La Peña, F., Perez-Omil, J.A., Calvino, J.J., Lugg, N.R., D'Alfonso, A.J., Allen, L.J., and Colliex, C. (2011) Chemical imaging at atomic resolution as a technique to refine the local structure of nanocrystals. *Angew. Chem. Int. Ed.*, **50** (4), 868–872.
138. Ross, F.M. (2007) in *Science of Microscopy* (eds P.W. Hawkes and J.C.H. Spence), Springer, New York, pp. 445–534.
139. Banhart, F. (ed.) (2008) *In-Situ Electron Microscopy at High Resolution*, World Scientific Publishing Co. Pte. Ltd.
140. Sharma, R. (2009) Kinetic measurements from in situ TEM observations. *Microsc. Res. Tech.*, **72** (3), 144–152.
141. Sharma, R. (2005) An environmental transmission electron microscope for in situ synthesis and characterization of nanomaterials. *J. Mater. Res.*, **20** (7), 1695–1707.
142. Gai, P.L. (2002) Developments in in situ environmental cell high-resolution electron microscopy and applications to catalysis. *Top. Catal.*, **21** (4), 161–173.
143. Walsh, M.J., Yoshida, K., Gai, P.L., and Boyes, E.D. (2010) In-situ heating studies of gold nanoparticles in an aberration corrected transmission electron microscope. *J. Phys.: Conf. Ser.*, **241**, 012058.
144. Legros, M., Dehm, G., Arzt, E., and Balk, T.J. (2008) Observation of giant diffusivity along dislocation cores. *Science*, **319**, 1646–1649.
145. Cumings, J., Olsson, E., Petford-Long, A.K., and Zhu, Y. (2008) Electric and magnetic phenomena studied by in situ transmission electron microscopy. *MRS Bull.*, **33**, 101–106.
146. Arai, S., Tsukimoto, S., and Saka, H. (1998) In situ transmission electron microscope observation of melting of aluminum particles. *Microsc. Microanal.*, **4** (3), 264–268.
147. Ohmura, T., Minor, A.M., Stach, E.A., and Morris, J.W. Jr. (2004) Dislocation-grain boundary interactions in martensitic steel observed through in situ nanoindentation in a transmission electron microscope. *J. Mater. Res.*, **19**, 3626–3632.
148. Gai, P.L., and Calvino, J.J. (2005) Electron microscopy in the catalysis of alkane oxidation, environmental control, and alternative energy sources. *Annu. Rev. Mater. Res.*, **35**, 465–504.
149. Molenbroek, A.M., Helveg, S., Topsoe, H., and Clausen, B.S. (2009) Nano-Particles in heterogeneous catalysis. *Top. Catal.*, **52** (10), 1303–1311.
150. Williamson, M.J., Tromp, R.M., Vereecken, P.M., Hull, R., and Ross, F.M. (2003) Dynamic microscopy of nanoscale cluster growth at the solid-liquid interface. *Nat. Mater.*, **2**, 532–536.
151. Jeangros, Q., Faes, A., Wagner, J.B., Hansen, T.W., Aschauer, U., Van herle, J., Hessler-Wyser, A., and Dunin-Borkowski, R.E. In situ redox cycle of a nickel-YSZ fuel cell anode in an environmental transmission electron microscope, *Acta Mater.*, **58**(14), 4578–4589.
152. Twitchett, A.C., Dunin-Borkowski, R.E., Broom, R.F., and Midgley, P.A. (2004) Quantitative electron holography of biased semiconductor devices. *J. Phys.: Condens. Matter.*, **16**, S181–S192.
153. Butler, P. and Hale, K. (1981) *Practical Methods in Electron Microscopy* (ed. A.M.G. Glauert), Elsevier, Amsterdam, pp. 239–309.
154. D.L., A. (1975) in *Principles and Techniques in Electron Microscopy, Biological*

Applications (ed. M.A. Hayat), Van Nostrand Reinhold, New York, p. 52.
155. Giorgio, S., Sao, J.S., Nitsche, S., Chaudanson, D., Sitja, G., and Henry, C.R. (2006) Environmental electron microscopy (ETEM) for catalysts with a closed E-cell with carbon windows. *Ultramicroscopy*, **106** (6), 503–507.
156. Creemer, J.F., Helveg, S., Hoveling, G.H., Ullmann, S., Molenbroek, A.M., Sarro, P.M., and Zandbergen, H.W. (2008) Atomic-scale electron microscopy at ambient pressure. *Ultramicroscopy*, **108** (9), 993–998.
157. Allard, L.F., Bigelow, W.C., Jose-Yacaman, M., Nackashi, D.P., Damiano, J., and Mick, S.E. (2009) A new MEMS-based system for ultra-high-resolution imaging at elevated temperatures. *Microsc. Res. Tech.*, **72**, 208–215.
158. Ito, T. and Hiziya, K. (1958) Specimen reaction device for the electron microscope and its applications. *J. Electron Microsc.*, **6**, 4.
159. Hashimoto, H., Naiki, T., Etoch, T., and Fujiwara, K. (1968) High temperature gas reaction specimen chamber for an electron microscope. *Jpn. J. Appl. Phys.*, **7**, 6.
160. Baker, R.T.K. and Harris, P.S. (1972) Controlled atmosphere electron microscopy. *J. Phys. E*, **5**, 793–797.
161. Atzmon, Z., Sharma, R., Mayer, J.W., and Hong, S.Q. (1994) An in situ transmission electron microscopy study during NH_3 ambient annealing of Cu-Cr thin films. *Mater. Res. Soc. Symp. Proc.*, **317**, 245–250.
162. Atzmon, Z., Sharma, R., Russell, S.W., and Mayer, J.W. (1994) Kinetics of copper grain growth during nitridation of Cu-Cr and Cu-Ti thin films by in situ TEM. *Mater. Res. Soc. Symp. Proc.*, **337**, 625–630.
163. Drucker, J., Sharma, R., Weiss, K., and Kouvetakis, J. (1994) In situ, real-time observation of Al chemical-vapor deposition on SiO_2 in an environmental transmission electron microscope. *J. Appl. Phys.*, **76** (12), 8198–8200.
164. Sharma, R., Atzmon, Z., Mayer, J., and Hong, S.Q. (1994) In situ studies on nitridation of Cu/Ti thin films using environmental cell in transmission electron microscopy. *Mater. Res. Soc. Symp. Proc.*, **317**, 251–256.
165. Crozier, P.A., Sharma, R., and Datye, A.K. (1998) Oxidation and reduction of small palladium particles on silica. *Microsc. Microanal.*, **4** (3), 278–285.
166. Sharma, R. and Weiss, K. (1998) Development of a TEM to study in situ structural and chemical changes at an atomic level during gas-solid interactions at elevated temperatures. *Microsc. Res. Tech.*, **42** (4), 270–280.
167. Crozier, P.A., Wang, R., and Sharma, R. (2008) In situ environmental TEM studies of dynamic changes in cerium-based oxides nanoparticles during redox processes. *Ultramicroscopy*, **108** (11), 1432–1440.
168. Gai, P.L. and Boyes, E.D. (2009) Advances in atomic resolution in situ environmental transmission electron microscopy and 1 Å, aberration corrected in situ electron microscopy. *Microsc. Res. Tech.*, **72** (3), 153–164.
169. Gai, P.L., Boyes, E.D., Helveg, S., Hansen, P.L., Giorgio, S., and Henry, C.R. (2007) Atomic-resolution environmental transmission electron microscopy for probing gas-solid reactions in heterogeneous catalysis. *MRS Bull.*, **32** (12), 1044–1050.
170. Gai, P.L. and Boyes, E.D. (2005) Pioneering development of atomic resolution in situ environmental transmission electron microscopy for probing gas-solid reactions and in situ nanosynthesis. *Mater. Res. Soc. Symp. Proc.*, **876E**, R9.1/P6.1.
171. Gai, P.L. (2002) Development of wet environment TEM (wet-ETEM) for in situ studies of liquid-catalyst reactions on the nanoscale. *Microsc. Microanal.*, **8** (1), 21–28.
172. Gai, P.L. and Goringe, M.J. (1981) Application of in-situ electron microscopy in catalysis. *Proc. Annu. Meet., Electron Microsc. Soc. Am.*, **39**, 68–71.
173. Parkinson, G.R. and White, D. (1986) The application of a controlled atmosphere reaction cell for studying electroactive polymers by TEM. Proceedings of the XI International

Congress on Electron Microscopy, Kyoto, p. 331.
174. Collazo-Davila, C., Landree, E., Grozea, D., Jayaram, G., Plass, R., Stair, P.C., and Marks, L.D. (1995) Design and initial performance of an ultrahigh vacuum sample preparation evaluation analysis and reaction (SPEAR) system. *Microsc. Microanal.*, **1** (06), 267–279.
175. Inceesungvorn, B. López-Castro, J. Calvino, J.J. Bernal, S. Meunier, F.C. Hardacre, C. Griffin, K., and Delgado, J.J. (2011) Nano-structural investigation of Ag/Al_2O_3 catalyst for selective removal of O_2 with excess H_2 in the presence of C_2H_4, *Appl. Catal. A: Gen.*, **391** (1–2), 187–193.
176. Li, P., Liu, J., Nag, N., and Crozier, P.A. (2006) Dynamic nucleation and growth of Ni nanoparticles on high-surface area titania. *Surf. Sci*, **600** (3), 693–702.
177. López-Haro, M., Delgado, J.J., Cies, J.M., del Rio, E., Bernal, S., Burch, R., Cauqui, M.A., Trasobares, S., Pérez-Omil, J.A., Bayle-Guillemaud, P., and Calvino, J.J. (2010) Bridging the gap between CO adsorption studies on gold model surfaces and supported nanoparticles. *Angew. Chem. Int. Ed.*, **49** (11), 1981–1985.
178. Boccuzzi, F., Chiorino, A., Manzoli, M., Lu, P., Akita, T., Ichikawa, S., and Haruta, M. (2001) Au/TiO_2 nanosized samples: a catalytic, TEM, and FTIR study of the effect of calcination temperature on the CO oxidation. *J. Catal.*, **202** (2), 256–267.
179. Li, P., Liu, J., Nag, N., and Crozier, P.A. (2005) Atomic-scale study of in situ metal nanoparticle synthesis in a Ni/TiO_2 system. *J. Phys. Chem. B*, **109** (29), 13883–13890.
180. Bernal, S., Calvino, J.J., Cauqui, M.A., Gatica, J.M., López Cartes, C., Pérez Omil, J.A., and Pintado, J.M. (2003) Some contributions of electron microscopy to the characterisation of the strong metal-support interaction effect. *Catal. Today*, **77** (4), 385–406.
181. Dulub, O., Hebenstreit, W., and Diebold, U. (2000) Imaging cluster surfaces with atomic resolution: the strong metal-support interaction state of Pt supported on TiO_2(110). *Phys. Rev. Lett.*, **84** (16), 3646.
182. Haller, G.L. and Resasco, D.E. (1989) Metal-support interaction: group VIII metals and reducible oxides. *Adv. Catal.*, **36**, 173–235.
183. Tauster, S.J. and Fung, S.C. (1978) Strong metal-support interactions: occurrence among the binary oxides of Groups IIA-VB. *J. Catal.*, **55**, 29–35.
184. Tauster, S.J., Fung, S.C., and Garten, R.L. (1978) Strong metal-support interactions. Group 8 noble metals supported on titanium dioxide. *J. Am. Chem. Soc.*, **100**, 170–175.
185. Hansen, T.W., Wagner, J.B., Hansen, P.L., Dahl, S., Topsoe, H., and Jacobsen, C.J. (2001) Atomic-resolution in situ transmission electron microscopy of a promoter of a heterogeneous catalyst. *Science*, **294** (5546), 1508–1510.
186. Ross, F.M., Tersoff, J., and Reuter, M.C. (2005) Sawtooth faceting in silicon nanowires. *Phys. Rev. Lett.*, **95** (14), 146104.
187. Dick, K.A., Kodambaka, S., Reuter, M.C., Deppert, K., Samuelson, L., Seifert, W., Wallenberg, L.R., and Ross, F.M. (2007) The morphology of axial and branched nanowire heterostructures. *Nano Lett.*, **7** (6), 1817–1822.
188. Sharma, R., Rez, P., and Treacy, M.M.J. (2006) Direct observations of the growth of carbon nanotubes using in situ transmission electron microscopy. *e-J. Surf. Sci. Nanotechnol.*, **4**, 460–463.
189. Sharma, R., Rez, P., Brown, M., Du, G., and Treacy, M.M.J. (2007) Dynamic observations of the effect of pressure and temperature conditions on the selective synthesis of carbon nanotubes. *Nanotechnology*, **18** (12), 125602/125601–125602/125608.
190. Hofmann, S., Sharma, R., Ducati, C., Du, G., Mattevi, C., Cepek, C., Cantoro, M., Pisana, S., Parvez, A., Cervantes-Sodi, F., Ferrari, A.C., Dunin-Borkowski, R., Lizzit, S., Petaccia, L., Goldoni, A., and Robertson, J. (2007) In situ observations of catalyst dynamics during

surface-bound carbon nanotube nucleation. *Nano Lett.*, **7** (3), 602–608.
191. Helveg, S., Lopez-Cartes, C., Sehested, J., Hansen, P.L., Clausen, B.S., Rostrup-Nielsen, J.R., Abild-Pedersen, F., and Norskov, J.K. (2004) Atomic-scale imaging of carbon nanofibre growth. *Nature*, **427** (6973), 426–429.
192. Abild-Pedersen, F., Nørskov J.K., Rostrup-Nielsen, J.R., Sehested, J., and Helveg, S. (2006) Mechanisms for catalytic carbon nanofiber growth studied by ab initio density functional theory calculations. *Phys. Rev. B*, **73** (11), 115419.
193. Sharma, R., Rez, P., Treacy, M.M.J., and Stuart, S.J. (2005) In situ observation of the growth mechanisms of carbon nanotubes under diverse reaction conditions. *J. Electron Microsc.*, **54** (3), 231–237.
194. Helveg, S. and Hansen, P.L. (2006) Atomic-scale studies of metallic nanocluster catalysts by in situ high-resolution transmission electron microscopy. *Catal. Today*, **111** (1–2), 68–73.
195. Cabie, M., Giorgio, S., Henry, C.R., Axet, M.R., Philippot, K., and Chaudret, B. (2010) Direct observation of the reversible changes of the morphology of Pt nanoparticles under gas environment *J. Phys. Chem. C*, **114** (5), 2160–2163.
196. Giorgio, S., Cabié, M., and Henry, C.R. (2008) Dynamic observations of Au catalysts by environmental electron microscopy. *Gold Bull.*, **41** (2), 7.
197. Simonsen, S.B., Dahl, S., Johnson, E., and Helveg, S. (2008) Ceria-catalyzed soot oxidation studied by environmental transmission electron microscopy. *J. Catal.*, **255** (1), 1–5.
198. Gai, P.L. (2007) in *Nanocharacterisation*. (eds A. Kirkland, and J.L. Hutchison, Royal Society of Chemistry, pp. 268–290.
199. Gai, P.L. (1999) Environmental high resolution electron microscopy of gas-catalyst reactions. *Top. Catal.*, **8** (1,2), 97–113.
200. Gai, P.L. (2001) Developments of electron microscopy methods in the study of catalysts. *Curr. Opin. Solid State Mater. Sci.*, **5** (5), 371–380.
201. Boyes, E.D., Gai, P.L., and Hanna, L.G. (1996) Controlled environment [ECELL] TEM for dynamic in-situ reaction studies with HREM lattice imaging. *Mater. Res. Soc. Symp. Proc.*, **404**, 53–60.

27
Nanowires and Nanotubes

Yong Ding and Zhong Lin Wang

27.1
Introduction

In the past two decades, intensive research interest has been focused on nanowires, nanotubes, and nanobelts, which are the foundation of today's nanoscience and nanotechnology [1–20]. Transmission electron microscopy (TEM) and associated techniques have been widely used to understand and explore the structures and properties of these one-dimensional (1D) nanomaterials. The objective of this chapter is to give a brief introduction about the applications of TEM techniques for characterizing nanowires and nanotubes. The contents are organized as follows. First, we demonstrate how to identify the geometric configuration of nanowires and nanotubes based on TEM images and electron diffraction patterns; such geometric information includes the growth direction and shape, surface reconstruction, the defects inside nanowires, and the chiral indices of carbon nanotubes. Then, the applications of modern *in situ* TEM techniques for studying the properties of nanowires and nanotubes, including mechanical, electrical, electromechanical, and electrochemical properties, are reviewed.

27.2
Structures of Nanowires and Nanotubes

In the literature, there are a few names being used to describe 1D nanostructures, such as nanorods, nanowires, nanofibers, nanobelts, and nanoribbons, and so on. Nanowires usually mean a linear structure that has a specific growth direction, but its side surfaces and cross-section shape may not be well defined or uniform (Figure 27.1a) [21]. A nanorod is a nanowire with a short length (Figure 27.1b). A nanotube can be a 1D nanostructure with a hollow interior channel (Figure 27.1c). Nanobelts/nanoribbons are 1D nanostructures with well-defined side facets (Figure 27.1d), and they have more restrictive shape and uniformity compared to the nanowires. Less restriction in the shape of the cross section and its uniformity can include a lot of nanostructures in the family of nanowires.

Handbook of Nanoscopy, First Edition. Edited by Gustaaf Van Tendeloo, Dirk Van Dyck, and Stephen J. Pennycook.
© 2012 Wiley-VCH Verlag GmbH & Co. KGaA. Published 2012 by Wiley-VCH Verlag GmbH & Co. KGaA.

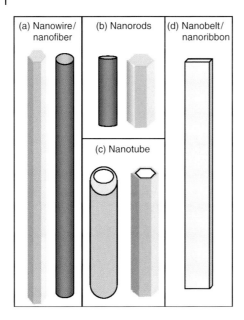

Figure 27.1 Typical morphologies of one-dimensional nanostructures: (a) nanowire, (b) nanorod, (c) nanotube, and (d) nanobelt [21]. (With permission from ACS.)

27.2.1
Determination of the Growth Directions of Nanowires

In order to determine the growth direction of 1D nanowires based on TEM images and diffractions, it is important to ensure that the incident electron beam is perpendicular to the growth direction. Just by depositing 1D nanostructures onto a carbon-film-coated TEM grid, there is no guarantee that the nanowires will remain flat on the substrate; bending could happen as illustrated in Figure 27.2c. A reliable way is to measure dozens of nanowires to see whether the determined growth directions are unique and self-consistent. If not, one must carefully check the data to judge whether there is more than one growth direction.

Just as we know that the operation of TEM relies on magnetic lenses, the high-speed electrons traveling in magnetic field follow a helical trajectory. Between the diffraction and image modes, the electron-traveling paths are normally different. As a result, there is a rotation angle between the diffraction pattern and images showing on the screen (same situation at the film position). For newly built equipment manufactured in the past 10 years at least, such a rotation effect has been automatically corrected by compensating optics, but for the older instrument, this rotation does exist. Figure 27.2a is a bright-field TEM image of a ZnO nanowire recorded at 400 kV by using a JEOL 4000EX, which has such a rotation effect. After tilting the specimen perpendicular to the electron beam, the selected-area electron diffraction (SAED) pattern from the nanowire is recorded and shown in Figure 27.2b

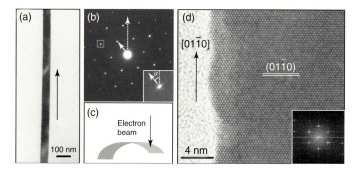

Figure 27.2 (a) TEM image of a ZnO nanobelt and (b) its corresponding SAED pattern. The streaking of the spot in the inset of (b) is due to the sharp edge of the nanowire, which can be used to correct the magnetic rotation angle between the image and the diffraction pattern. (c) Schematic diagram showing the folded shape of a nanobelt and the incident electron beam, showing the possibility of incorrect determination of growth direction. (d) HRTEM image from the nanobelt shown in (a), and its diffractogram pattern [21]. (With permission from ACS.)

without any correction in the rotation angle. The incident electron beam direction in the diffraction pattern is parallel to [0001], and it looks like the growth direction of the nanowire does not match any low-index diffraction spots in the pattern. This is due to the magnetic rotation between the image and the diffraction pattern.

If the magnetic rotation angle between the image and the SAED pattern is known, then, after rotating the angle back, the growth direction can be identified based on the SAED pattern and the real-space image. However, if the rotation angle is not known, we still have a lot of methods to determine its value. The easiest one for nanowires is based on the shape factor in the electron diffraction pattern introduced by the fine size of the nanowires in their width direction. Such size-related effect would add a streaking around each diffraction spot in the diffraction pattern along the direction perpendicular to the side surface of the nanowire (see the magnified diffraction spot shown in Figure 27.2b). The growth direction of the nanowire is thus determined as [01$\bar{1}$0], while the rotation angle α between the real-space image and the diffraction pattern is identified in Figure 27.2b.

If we can record a high-resolution transmission electron microscopy (HRTEM) image of the nanowire, then its diffractogram (the Fourier transform of the HRTEM image) has the same function as the diffraction pattern but without the rotation angle. Figure 27.2d is an HRTEM image of the nanowire shown in Figure 27.2a. The diffractogram shown directly gives the growth direction of the nanowire as [01$\bar{1}$0], which is consistent with the result retrieved through the shape factor.

When the diffraction pattern is along a high-indexed axis, the HRTEM image normally shows line fringes instead of 2D lattices due to the limited resolution of the microscope. In such a situation, the diffractogram cannot serve as the diffraction pattern because of the information loss. In order to identify the growth direction, we can use the shadow imaging technique. We know that we get a convergent beam electron diffraction (CBED) pattern in diffraction mode when the incident electron beam converges to set the beam crossover on sample. However, in the shadow imaging process, the electron beam is slightly expanded to move the beam crossover below the sample plane, as illustrated in Figure 27.3a. In the diffraction mode, we get the shadow image instead of the CBED pattern. Actually, it contains a bright-field image (center beam) and a set of dark-field images (diffraction beams) at the same time. In the shadow image, we have the diffraction pattern and real-space images simultaneously. Therefore, there was no rotation angle between them. What remains is to index the diffraction spot g, which is parallel to the growth direction of the nanowire. Figure 27.3b is a TEM image and the corresponding SAED pattern from a ZnO nanowire. With a converged incident electron beam, a shadow image was recorded, as displayed in Figure 27.3c. The nanowire growth direction points to the $[2\bar{1}\bar{1}0]$ direction.

Another method to determine the growth direction as illustrated in Figure 27.4 is slightly complicated compared with the above ones. In this method, we recorded the image and "diffraction pattern" in two steps but both in image mode. The first step is to record a bright-field TEM image of a well-focused nanowire using a parallel beam (Figure 27.4a). In the following step, we still keep in image mode, but converging the beam onto the specimen (normally we need to switch to large

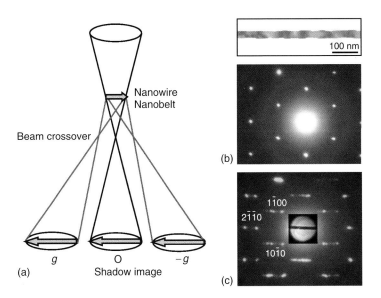

Figure 27.3 Shadow image technique for determining the growth direction of a nanowire/nanobelt. (a) Optical diagram for shadow imaging. (b) TEM image and the SAED pattern from a nanobelt. (c) The corresponding shadow image recorded in diffraction mode [21]. (With permission from ACS.)

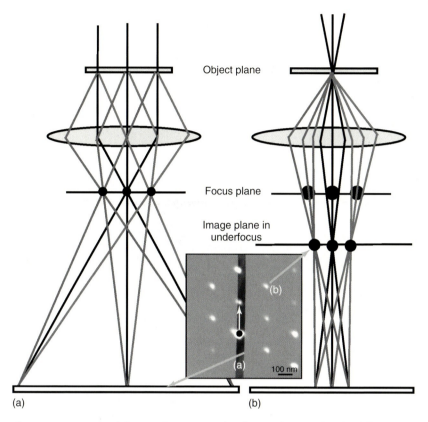

Figure 27.4 (a) Optical diagram for receiving the imaging of a nanowire under parallel beam illumination. (b) Optical diagram for recording the diffraction information from the nanowire in the *imaging mode* by converging the electron beam and underfocusing the objective lens, so that the diffraction pattern appears in the image mode. The image shown is a double-exposed image and "diffraction pattern" recorded at the optical configurations of (a) and (b) [21]. (With permission from ACS.)

spot size, small beam radius), with underfocusing the objective lens heavily, we can bring the diffraction spots to show in the image mode (Figure 27.4b). Different from the case of shadow imaging, which is operated in diffraction mode, here we get the "diffraction pattern" and image both in the image mode. Therefore, there was no rotation between the image and diffraction pattern as well; the index of the diffraction spot parallel to the nanowire will give the growth direction.

Sometimes, the wurtzite-structured ZnO, [0001] and [000$\bar{1}$] directions are not equivalent to each other because of the missing inversion symmetry in the structure. Solely basing them on the electron diffraction pattern cannot separate them. Such problems can be resolved by the CBED technique. An example is shown in Figure 27.5. The ZnO nanowires were grown on Si substrate by the wet chemical method. The shape with a small tip and a big root revealed by the SEM image in Figure 27.5a can guide us to identify the direction of the growth front.

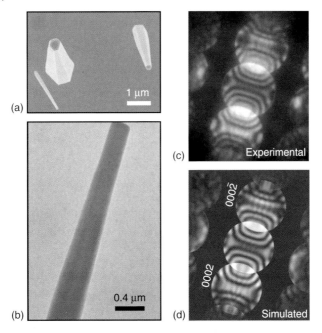

Figure 27.5 (a,b) SEM and TEM images of ZnO nanowires. (c,d) The experimental and simulated CBED pattern corresponding to the ZnO nanowires in (b).

We first need to tilt the sample to get the precise zone-axis condition. Then, with the convergence of the incident electron beam to set the beam crossover on the sample, we get the CBED pattern in diffraction mode, as shown in Figure 27.5c. After the experimental pattern is acquired, we need to do the simulation based on the structural model. There are several software that can perform such simulations, for example, Java version electron microscopy image simulation package (JEMS) and MacTempas. Readers can also use the web electron microscopy application software online at: *http://emaps.mrl.uiuc.edu/*. By verifying the sample thickness, the best matching to the CBED pattern, shown in Figure 27.5c, happens at the thickness close to 180 nm, which is shown in Figure 27.5d. On the basis of the simulation results, [0001] and [000$\bar{1}$] directions can be successfully separated. The drawback of the CBED technique is that the sample thickness cannot be too thin. Otherwise, the dynamic effect is so weak that the diffraction discs have less contrast inside. Later, we will see that HRTEM images can resolve this problem in a thin sample, but the quality of the image must be guaranteed.

It is worthwhile to emphasize that some anomalous or odd electron diffraction patterns may exist in nanowires [22–24]. Most cases are due to the existence of nanotwins and/or stacking faults in them. Figure 27.6 gives an example observed in Si nanowires with incident electron beams along [111] directions. Figure 27.6a is an experimental result [24]; besides the fundamental diffraction spots included in [111] zone axis, there are some extra weak spots, such as the gray arrowheads marked

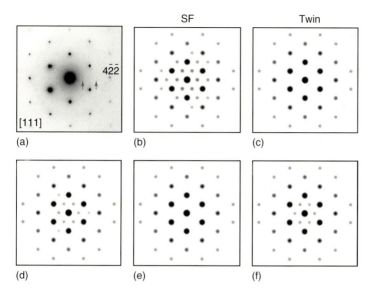

Figure 27.6 (a) Experimental [111] axis SAED pattern from a Si nanowire showing abnormal diffraction spots. (b,c) Simulated diffraction patterns with a stacking fault, twin plane inside, respectively. (d,f) Simulated diffraction patterns corresponding to a C plane missing, equal A, B, C planes, and an extra A plane in the crystal, respectively [24]. (With permission from International Union of Crystallography.)

$1/3[4\bar{2}\bar{2}]$, $2/3[4\bar{2}\bar{2}]$ in Figure 27.6a. In order to better understand the formation mechanism of these anomalous diffractions, we can do some simulations using well-developed software. Here, we used the MacTempas software, and considered three different configurations. The first is due to the existence of a stacking fault, with one of the (111) close-packed planes missed. The simulated diffraction pattern is shown in Figure 27.6b with sample thickness ∼10 nm. The extra diffraction spots are well reproduced. The second configuration is to consider the twin structure. There was no extra diffraction spot in the simulated pattern, as displayed in Figure 27.6c, while the sample thickness is the same as in the case including a stacking fault. The third configuration is to consider the unequal number of the A, B, and C hexagonal planes perpendicular to the [111] direction. As we know, the face-centered cubic (FCC) structure can be considered as the close-packed A, B, and C planes successively stacked along <111> directions. Each unit cell has one of the A, B, and C planes. If viewed along the beam direction, the number of A, B, and C planes are equal, for example, as in Figure 27.6e, there are a total of 11 A, 11 B, 11 C planes, and no anomalous diffraction spots appear in the simulated pattern. While removing a C plane or adding an A plane, those anomalous diffraction spots appear as shown in Figure 27.6d–f, respectively [25]. Considering the round-shaped cross section of Si nanowires, the third configuration is easily satisfied. On the basis of

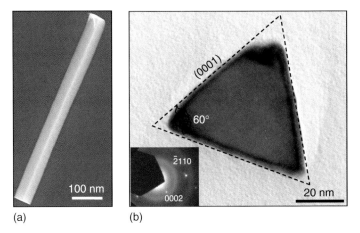

Figure 27.7 (a) An SEM image of a GaN nanowire deposited on a Si substrate. (b) Cross-section TEM image of a GaN nanowire to show the triangular cross section, the SAED pattern inserted [26]. (With permission from ACS.)

the simulation results, the anomalous diffraction spots can come from the stacking fault or the unequal number of the A, B, and C planes.

27.2.2
The Three-Dimensional Structure of Nanowires

In order to reveal the 3D structure of nanowires, traditionally, we combined the plane-view and cross-section electron microscopy images of the nanowires to reconstruct the 3D structure. Figure 27.7a,b gives the scanning electron microscopy (SEM) image and cross-section TEM image of a GaN nanowire [26]. The cross-section sample was prepared by embedding the GaN nanowires in epoxy, followed by the slicing using ultramicrotone. On the basis of the images in Figure 27.7, the side surfaces of the GaN nanowires can be identified as (0001), ($\bar{2}$112), and ($2\bar{1}\bar{1}$2).

If nanowires have complicated 3D structures, the newly developed electron tomography technique is more suitable to resolve their 3D structures compared to the traditional method. Electron tomography is a method of reconstructing the 3D morphologies from a series of 2D TEM images or projections obtained by high-angle annular dark-field (HAADF) scanning transmission electron microscopy (STEM). An example from Zn_3P_2 nanowires is given in Figure 27.8 [27]. The SEM and TEM images in Figure 27.8a–d reveal that the structure is entirely zigzagged over the whole nanowire. The zigzag period is 140 nm on average and the zigzag angle is uniformly 140°. Figure 27.8e displays the tomographic 3D reconstruction images of the nanowire, (i–iv) images for the sequential 30° turns. The sliced views along the nanowire (v–viii) reveal the triangular cross section of the segments. The videos of the 3D reconstruction can be found at the *Nano Letters* web site [27].

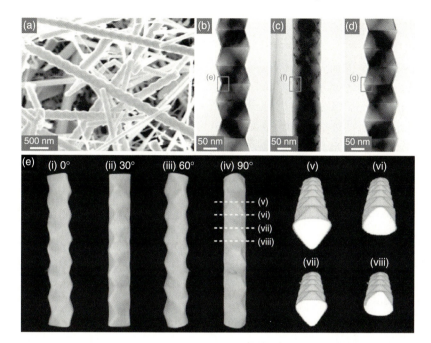

Figure 27.8 (a) SEM micrograph of high-density Zn_3P_2 NWs homogeneously grown on the substrate. (b) TEM image reveals the zigzagged morphology of the Zn_3P_2 NW having a zigzagged period of 140 nm. Consecutive 30° tilt around the wire axis changes the morphology to (c) straight and (d) zigzagged. (e) Images of the Zn_3P_2 NW obtained through tomographic 3D reconstruction; (i–iv) correspond to a series of 0°, 30°, 60°, and 90° tilt around the axial direction, respectively; (v–viii) images show the sliced views along the NWs (as marked in (iv)), having a triangular cross section [27]. (With permission from ACS.)

27.2.3
Surface Reconstruction of Nanowires

Some unique properties of nanowires compared to bulk counterparts are due to their large surface-to-bulk atomic ratio; in other words, most of the atoms are located at or close to the surface in nanowires. Because the dangling bonds of the surface atoms make the system unstable, the relaxation or reconstruction of surface atoms is energetically preferred, and as a result, the surface structure may be much different from the bulk. The HRTEM profile imaging is a good technique to characterize the surface reconstruction. Figure 27.9 gives such a profile image of the $(01\bar{1}0)$ surface acquired from a ZnO nanowire [28].

Before discussing the surface reconstruction, let us discuss another piece of useful information contained in Figure 27.9. The positive and negative (0001) planes of the wurtzite-structured ZnO are terminated with pure Zn and oxygen ions, respectively; they are the so-called polar surfaces. If the sample thickness is larger than 50 nm, we can use the CBED technique to separate the positive

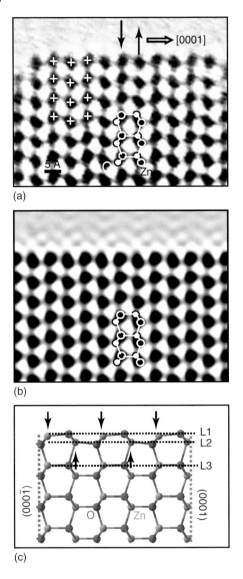

Figure 27.9 (a,b) Experimental and simulated HRTEM profile images of a reconstructed ZnO (0̄110) surface. The atomic model used for the simulation is displayed in (c) [28]. (With Permission from Elsevier.)

and negative [0001] directions as discussed in the previous section. In the thin samples, such as tiny ZnO nanowires, however, the CBED technique is not effective. However, along the special projection direction [2̄1̄10], solely based on the high-quality HRTEM images as the one displayed in Figure 27.9a, we still can identify the positive [0001] direction. The atomic model of ZnO projected in the same direction is shown in the image as well. Keeping the different scattering powers of heavy atom Zn and light atom O in mind, the contrast of the dark rods in Figure 27.9a is mostly contributed by the heavy Zn ions, while the O ions only introduce a slim tail. As each of these dark rods comes from a Zn-O dimer, the two Zn atoms in the two horizontal nearby dimers are not kept in the same line parallel to the $(01\bar{1}0)$ surface. Then, we can proximately identify the Zn and oxygen locations in each dark rod or dimer, as indicated by the atomic model. Using this method, the polar directions [0001] in Figure 27.9a have been marked.

Surface reconstruction is presented in Figure 27.9a. Some centers of the dark rods are marked by "+." Those centers in the first surface layer are much closer to the horizontal line compared to the inside layers. Considering that the dark contrast of these dots is mainly contributed by the Zn ions, we assume several configurations of the surface reconstruction. The first possibility is that of the first layer Zn ions moving inward. The second possibility is that of the Zn ions in the second layer moving outward. The third case is that of the first layer Zn moving inward plus the second layer Zn moving outward. In order to verify each assumption, HRTEM simulation has been carried out by using MacTempas software. The best-matched simulation result is shown in Figure 27.9b. It suggests that the real case belongs to the third assumption. The atomic model is illustrated in Figure 27.9c. On the basis of the simulation, the displacement of the Zn ions in the first and second surface layers can be estimated as −13% (inward) and 17% (outward) of $d_{(01\bar{1}0)}$.

27.2.4
Chiral Indices of Carbon Nanotubes

Carbon nanotubes have a lot of unique properties. But most of the properties are structure dependent. For example, some carbon nanotubes behave like a semiconductor, while some of them act as metals. From the structural point of view, cutting the monolayered graphite (also named as *graphene*) into a rectangular shape, as depicted in Figure 27.10a, and then rolling it up with the A as the perimeter, the seamless cylinder formed is just a single-walled carbon nanotube. The structure of a single-walled carbon nanotube can be uniquely characterized by its chiral indices (u, v). On the basis of the chiral indices, its perimeter can be calculated as $\vec{A} = u\vec{a}_1 + v\vec{a}_2$, and the helical angle as $\alpha = \tan^{-1}(\frac{\sqrt{3}v}{2u+v})$.

As we know, the distance between the two nearby carbon atoms inside the graphene plane is close to 0.14 nm, which is beyond the resolution of most traditional TEMs. Further, the carbon nanotubes cannot be exposed for a long time under high-energy electron beam in order to avoid beam damage. Therefore, only when one has a subangstrom, low-voltage aberration-corrected TEM does the recording of the atomic resolution HRTEM images of the carbon nanotube become possible [30]. In the past two decades, the determination of the chiral indices of carbon nanotubes has mostly relied on their electron diffraction patterns.

The strongest intensity peaks in the electron diffraction pattern from a single-walled carbon nanotube belong to the primary graphene diffractions. These

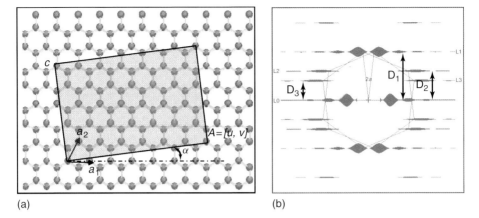

Figure 27.10 (a) Schematic structure of graphene with basis vectors $\vec{a}_1$ and $\vec{a}_2$. (b) Simulated diffraction pattern from a (7, 1) single-walled carbon nanotube [29]. (With permission from IOP.)

diffractions form three pairs of lines, labeled as l_1 (linked to the graphene (010) diffraction), l_2 (linked to the graphene ($\bar{1}$00) diffraction), and l_3 (linked to the graphene (110) diffraction) with layer line spacing with respect to the equatorial layer line denoted by D_1, D_2, and D_3, and they are referred to as the *principal layer lines* and are schematically illustrated in Figure 27.10b, which is a simulated diffraction pattern from a carbon nanotube with chiral indices as (7,1). The diffraction pattern, as shown in Figure 27.10b, is determined by the structure factor of the carbon nanotube, while the structure factor of the carbon nanotube can be simply considered as the structure factor of graphene multiplied with a modified Bessel function. The diffraction intensities of the three principal layer lines are related to the Bessel functions in different orders. Separately, the l_1, l_2, and l_3 layers correspond to the v, u, and $u+v$ ordered Bessel functions. The order of the Bessel function, and then the chiral indices (u, v) of the carbon nanotube, can be determined by examining the positions of the peaks in each principal layer. An example of how to determine the chiral indices based on the peak positions in principal layers is shown in Figure 27.11 [29]. Figure 27.11a gives the electron diffraction pattern recorded from a single-walled carbon nanotube; the inset gives the HRTEM image of the carbon nanotube. The line-scan intensity profiles of the l_1 and l_2 principal layers are shown in Figure 27.11b,c, respectively. On the basis of the X_2/X_1 ratios measured from Figure 27.11b,c and the database shown in Table 27.1, the orders of Bessel functions, and thus the chiral indices of the nanotube, can be determined to be $v = 2$ and $u = 17$.

Another easy way to determine the chiral indices is based on the lengths D_1 and D_2, which are the distances between the principal layers l_1, l_2, and the equatorial layer. The chiral indices u and v have the relationship as $\frac{v}{u} = \frac{2D_2 - D_1}{2D_1 - D_2}$. With the precisely measured D_1 and D_2, the ratio between v and u can be determined.

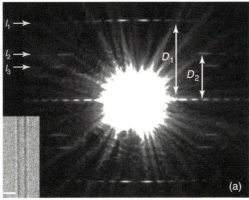
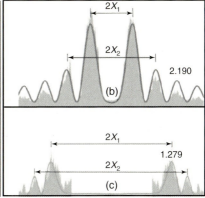

Figure 27.11 (a) Electron diffraction pattern of a carbon nanotube (17, 2). Inset is a high-resolution electron microscope image of the nanotube. The three principal layer lines, l_1, l_2, and l_3, are indicated in the figure. (b) Intensity profile of principal layer line l_1. The ratio of the positions of the second peak (X_2) and the first peak (X_1) is 2.190. (c) Intensity profile of principal layer line l_2. The ratio of the positions of the second peak (X_2) and the first peak (X_1) is 1.279 [29]. (With permission from IOP.)

Table 27.1 Ratio of the second and the first peak positions of Bessel functions [29].

N	X_2/X_1	n	X_2/X_1	n	X_2/X_1
1	2.892	11	1.373	21	1.239
2	2.197	12	1.350	22	1.232
3	1.907	13	1.332	23	1.226
4	1.751	14	1.315	24	1.218
5	1.639	15	1.301	25	1.211
6	1.565	16	1.287	26	1.206
7	1.507	17	1.275	27	1.201
8	1.465	18	1.266	28	1.196
9	1.428	19	1.256	29	1.192
10	1.398	20	1.247	30	1.188

Combined with the diameter measured in the real-space TEM images, the chiral indices can be resolved as well.

Of course, we can perform the simulation. By comparing the experimental diffraction with the simulated ones, we can retrieve the chiral indices. Figure 27.10b shows the simulated electron pattern of the (7,1) carbon nanotube. Such simulation can be carried out online at *http://www.physics.unc.edu/project/lcqin/www/nds/hdsh/hdsh.html*.

For a multiwalled carbon nanotube, it is necessary to determine the chiral indices for each individual shell. The detailed method can be found in Ref. [29].

27.3
Defects in Nanowires

27.3.1
Point Defects

Imaging of point defects is a challenge for TEM. TEM imaging relies on diffraction and interference effect, and it is much less sensitive to point defects distributed randomly or with a short-range ordering. Oxygen-deficient oxide nanostructures are of great interest because their conductivity depends on oxygen contents. Although TEM cannot be directly applied to determine oxygen deficiency in ZnO nanowires, it may help in other cases. We have applied TEM to directly image defects created in ZnO nanobelts by Mn implantation [21, 31]. A high density of ministacking faults was created as shown in Figure 27.12a. Taking a Fourier transform of the HRTEM image recorded from the nanowire followed by an inverse Fourier transform only using ±0002 diffraction spots, we get the distribution of the (0002) fringes (Figure 27.12b), which clearly displays the ministacking fault.

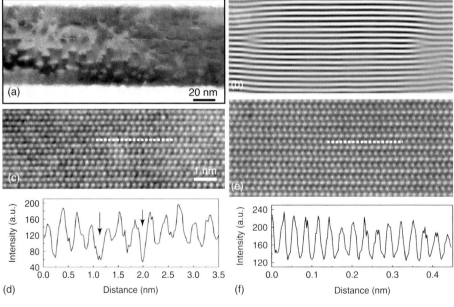

Figure 27.12 Imaging point defects in nanobelts. (a) Low-magnification TEM image of a ZnO nanobelt implanted with Mn, showing numerous ministacking faults produced by ion implantation. (b) Fourier filtered image showing a dislocation loop created by the ministacking fault. (c) HRTEM image of the as-implanted sample. (d) Intensity profile parallel to the (0001) plane, showing fluctuation in image contrast due to ion-implantation-induced point defects. (e) HRTEM image of the implanted ZnO nanobelt after being annealed in oxygen to compensate for the lost oxygen and the displaced cations. (f) Intensity profile parallel to the (0001) plane, showing the disappearance of the fluctuation contrast in the image [21]. (With permission from AC.)

Figure 27.12c is an HRTEM image recorded from an as-implanted ZnO nanowire. A line-scan profile along the dashed line gives the intensity variation across the atomic columns in a (0001) plane. An ideal profile should be periodic with equal amplitude. The large-scale variation in the image contrast from column to column (Figure 27.12d) reflects the existence of point defects as well as lattice distortion due to ion implantation. Possibly, the point defects can be oxygen deficiency and/or Zn interstitials. After annealing the sample in air, the image contrast becomes very uniform for less point defects (Figure 27.12e,f).

27.3.2
Dislocations

Dislocations are rarely observed in nanowires, especially for oxide nanostructures. It may be due to the size effect, which makes the dislocations unstable inside them. In the past, almost all the oxide nanostructures we have reported were free of dislocations. But recently, we have surprisingly observed dislocations in ultrathin ZnO nanowires of 6 nm in width [32]. Figure 27.13a,b shows, respectively, bright-field and dark-field images of these fine ZnO nanowires. The variation in the contrast (especially in the dark-field image) reveals the nonuniform strain field

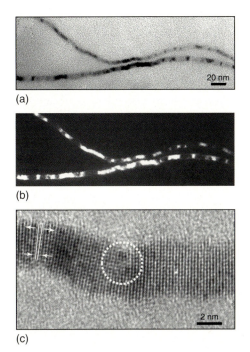

Figure 27.13 Dislocations in ultranarrow nanobelts. (a,b) Bright-field and dark-field TEM images of ultrafine ZnO nanobelts. (c) HRTEM image indicating the existence of a dislocation [21]. (With permission from ACS.)

distribution along the nanowires, which may be created by the large deformation. The HRTEM image in Figure 27.13c reveals the existence of an edge dislocation with Burgers vector as $1/2[0001] + 1/3[01\bar{1}0]$ at the twisted area indicated by a circle. At the left-hand side of the nanowire, no dislocation is formed, but the interplanar distance is expanded to ~5.4 Å at the exterior arc and compressed to ~4.8 Å at the inner arc to accommodate the local strain. Such a gigantic change of ~6% in interplanar distance without fracture for oxide seems only possible at the nanoscale. The nonuniform deformation along these fine nanowires corresponds to a nonuniform strain field. When the accumulated strain in some local areas is beyond the structural endurance, dislocations are introduced. This metallike structural characteristic for oxide is unusual.

27.3.3
Planar Defects: Twins

The most common defects in 1D nanostructures are planar defects, such as twins [21], stacking faults [33], and inversion domain walls [34]. These defects are not only essential for the growth of the nanostructures but also strongly affect their optical, electrical, and possibly chemical properties. Therefore, we describe some planar defects mostly observed in ZnO 1D nanostructures using HRTEM.

From the literature, it can be seen that twin structure is most common in FCC-structured metallic nanoparticles and silicon-based nanowires [35]. For the wurtzite structure, from the theoretical point of view, the predicted possible twin boundaries are $(01\bar{1}1)$, $(01\bar{1}2)$, and $(01\bar{1}3)$ [36]. Theoretical calculations revealed that the $(01\bar{1}3)$ twin has a relatively low energy. However, such a twin boundary has not been observed in bulk crystals so far. When the size of materials shrinks into nanometer scale, not only the $(01\bar{1}3)$ twin but also the $(01\bar{1}1)$ and $(01\bar{1}2)$ twins have been observed in ZnO 1D nanostructures [33]. The population of $(01\bar{1}1)$ and $(01\bar{1}2)$ twins is much lower than that of the $(01\bar{1}3)$ twin, in agreement with theoretical energy calculation. Besides the above three dominant twin structures, the $(\bar{2}112)$ twin was also observed in 1D ZnO nanostructures. Here, we only give the example of the $(01\bar{1}3)$ twin structure in ZnO nanowires.

Figure 27.14a shows a typical ZnO $(01\bar{1}3)$ twin structure. Its SAED pattern is displayed in Figure 27.14b, which is composed of two sets of diffraction spots that have symmetrical geometrical distribution but possibly variable intensity. The two set patterns are labeled using subscripts L and R specifying the left and right crystals, respectively. The common spot corresponds to the twin boundary plane $(01\bar{1}3)$ as indexed in the SAED pattern, and the incident beam direction is $[\bar{2}110]$. The existence of high-density stacking faults in the dark-field image (Figure 27.14c) indicates the large local strain. The optimum orientation to image twins is parallel to the twin plane, so that the diffraction pattern shows mirror symmetry between the two sets of diffraction spots. Figure 27.14d, an HRTEM image recorded from the twin boundary, clearly displays the mirror symmetry between the two crystals. HRTEM image simulation was carried out to clarify the Zn arrangement at the

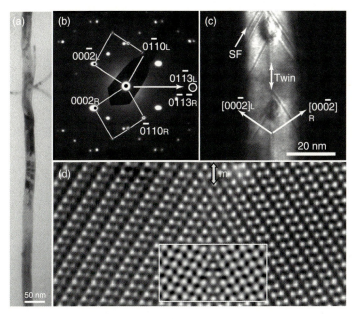

Figure 27.14 (a,b) TEM image and corresponding SAED pattern of a ZnO nanobelt that has a (01$\bar{1}$3) twin parallel to the growth direction. (c) A dark-field TEM image showing the existence of stacking faults in the nanobelt. (d) HRTEM image of the (01$\bar{1}$3) twin with the incident electron beam along [2$\bar{1}\bar{1}$0] direction, the simulation image is shown [33]. (With permission from Elsevier.)

boundary. The best-matched simulated image is shown in Figure 27.14d. In such focus conditions identified by the simulation, the dark contrast corresponds to atom positions, mostly Zn atoms, for the contrast of O is very weak. The (01$\bar{1}$3) twin structure is not only identified in ZnO nanowires but also in ZnS and CdSe nanowires [37], which belong to the wurtzite structure as well.

27.3.4
Planar Defects: Stacking Faults

Stacking faults are the most frequently observed planar defects in 1D ZnO nanostructures. They are formed by the changes in stacking sequence of close-packed planes. Depending on the stacking faults lying in or perpendicular to their closed-packed plane, they are named basal-plane and prismatic-plane stacking faults, respectively.

In the wurtzite structure, there are three possible types of basal-plane stacking faults, I_1, I_2, and E in bulk materials, which are produced by extracting one or two periodic layers, and inserting a layer, respectively [38, 39]. Combining HRTEM images with simulation or diffraction contrast extinguishing conditions, we can uniquely identify the different types of stacking faults. For imaging the contrast of stacking faults, the diffraction beam g has to avoid the extinguishing condition as

$g \cdot R = 0$, or n where R is the displacement crossing the defect and n is an integer. As we know, the R of stacking faults in wurtzite structures is $1/3[01\bar{1}0]+1/2[0001]$ for type I_1 case and $2/3[01\bar{1}0]$ fro type I_2 case.

A stacking fault in a ZnO nanobelt is displayed in Figure 27.15a. Its HRTEM image is recorded in Figure 27.15b. An arrowhead highlights the c direction in the image. In spite of the perfect stacking sequence of the wurtzite structure as ABABAB along the [0001] direction, a new consequence identified as ... ABABACAC ... in the defect area suggests the existence of an I_1 stacking fault. The displacement across the I_1 basal-plane stacking fault can be identified as $R = 1/3[01\bar{1}0] + 1/2[000\bar{1}] = 1/6[02\bar{2}3]$.

Figure 27.16a is a low-magnification TEM image of another ZnO $[01\bar{1}0]$ growth nanobelt. There are two different basal-plane stacking faults in Figure 27.16a; the top one is the I_1 stacking fault as discussed in Figure 27.15. An HRTEM image of the bottom defect is shown in Figure 27.16b. In the defect area, the stacking sequence can be identified as ... ABABCACAC ... It can be classified as an intrinsic type I_2 basal-plane stacking fault since it is equivalent to extracting two periodic layers, and a $2/3[01\bar{1}0]$ translation displacement has been measured across the defect.

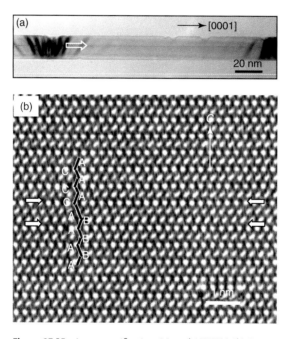

Figure 27.15 Low-magnification (a) and HRTEM (b) image in ZnO nanobelt showing the type I_1 intrinsic basal-plane stacking fault. The incident electron beam in the HRTEM image is along the $[2\bar{1}\bar{1}0]$ direction [33]. (With permission from Elsevier.)

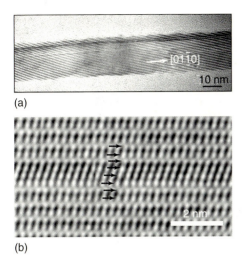

Figure 27.16 Low-magnification (a) and HRTEM (b) images of a [01$\bar{1}$0] growth ZnO nanobelt with incident electron beam along the [2$\bar{1}\bar{1}$0] direction. The defect in (b) belongs to the type I_2 stacking fault [33]. (With permission from Elsevier.)

From the crystal structural point of view, the hexagonal wurtzite structure takes space group as $P6_3mc$. It can be described as a number of alternating planes composed of tetrahedrally coordinated cations and anions, stacking alternatively along the c-axis. The oppositely charged ions produce positively charged (0001) and negatively charged (000$\bar{1}$) polar surfaces, resulting in a normal dipole moment and sometimes even spontaneous polarization along the c-axis as well as a divergence in surface energy. As a general case, 1D wurtzite nanostructures usually grow along the c-axis and their side surfaces are {01$\bar{1}$0} and/or {2$\bar{1}\bar{1}$0} because of their lower energies than that of (0001), resulting in an almost neutralized dipole moment. I_1 type of basal-plane stacking faults can be easily formed in those [0001] growth nanowires. The 1D nanostructures growing normal to the [0001] direction, as in the cases shown in Figures 27.15 and 27.16, are normally accompanied by at least a stacking fault from beginning to end. The surface kink at the intersection between the stacking fault and the growth front can be a preferred position to adsorb the vapor molecules to accelerate the growth along such a direction. As a result, the existence of basal-plane stacking faults can span the energy barrier set by the polar surfaces, and then lead the 1D nanostructures to grow along the [01$\bar{1}$0] direction.

Besides basal-plane stacking faults, we also observed prismatic-plane stacking faults in wider ZnO nanobelts (~1 μm) [33]. Prismatic-plane stacking faults have two configurations classified by the displacements across them. Of those displacements, one is 1/2<01$\bar{1}$1>, and the other is 1/6<02$\bar{2}$3> [40, 41]. In order to determine the structural nature of the defect in the nanobelts, a series of

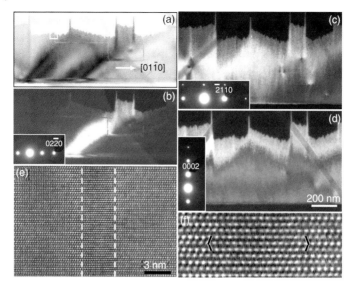

Figure 27.17 Bright-field (a) and dark-field (b) images of a [01$\bar{1}$0] growth ZnO nanobelt recorded under two-beam conditions with g = (02$\bar{2}$0). Prismatic- and basal-plane stacking faults coexist in the belt. In the dark-field images recorded using g = ($\bar{2}$110) (c) and 0002 (d), the contrast of the defect vanishes. (e,f) HRTEM images showing the prismatic-plane stacking fault [33]. (With permission from Elsevier.)

dark-field images have been recorded as shown in Figure 27.17. The defect contrast can be seen clearly in Figure 27.17a,b, which are the corresponding bright-field and dark-field images under two-beam condition with g = (02$\bar{2}$0). The diffraction condition is shown in Figure 27.17b. After identifying the [01$\bar{1}$0] direction by the diffraction pattern, we know that part of the defect lies in the basal plane and part in the prismatic plane. The extended contrast of defects with tilting the belt suggests that they are planar defects and rules out the possibility of dislocation loops. Only one set of SAED patterns was recorded from the defects area; thus, they cannot be twin structures. They must be stacking faults, part in the basal plane and part in the prismatic plane. By further tilting the nanobelt, we got two extinguishing conditions of the defects as shown in Figure 27.17c,d. They correspond to diffraction g as ($\bar{2}$110) and (0002), respectively. Both displacements 1/2<01$\bar{1}\bar{1}$> and 1/6<02$\bar{2}$3> can be perpendicular to ($\bar{2}$110), or when multiplied with (0002) will be an integer. However, if the displacement is 1/2<01$\bar{1}\bar{1}$>, the stacking faults may lose contrast in the dark-field image formed by the (02$\bar{2}$0) diffraction. As a conclusion, the stacking faults should take the displacement of 1/6<02$\bar{2}$3>, which are the same as that of the I_1 basal-plane stacking fault. Figure 27.17e is an HRTEM image recorded from the prismatic plane part of the defect shown in Figure 27.17a. Figure 27.17f gives a magnified defect area in Figure 27.17e. The displacement along the c-axis can be easily identified. The projected width of the I_1 basal-plane stacking fault along the [2$\bar{1}\bar{1}$0] direction is just an atomic plane thickness, as shown in Figure 27.15. However, the projection of the prismatic-plane stacking fault in the same direction is nearly 3 nm in Figure 27.17e. A possible explanation is that the defect is not edgy-on in Figure 27.17. The prismatic stacking fault most likely lies in the $(\bar{2}110)$ plane instead of the $(01\bar{1}0)$ plane. There is a $1/3[01\bar{1}0] + 1/2[000\bar{1}] = 1/6{<}202\bar{3}{>}$ translation across the defect, which is self-consistent with the result obtained from the diffraction contrast analysis presented above.

The formation of the prismatic-plane stacking faults seems be related to the secondary growth led by the self-catalysis effect [42]. We can find basal-plane stacking faults at the bottom of the nanobelt in Figure 27.17 and they cross the whole nanobelt from beginning to end, and are presumed to guide the dominant growth of the nanobelt along $[01\bar{1}0]$. The exposed top and low surfaces of the nanobelt then are Zn-terminated (0001) and oxygen terminated $(000\bar{1})$ planes, respectively. The self-catalysis growth at the Zn-terminated surface will lead the secondary growth at the (0001) surface, which is the upper surface of the nanobelt in Figure 27.17. The domains growing from different nucleii at the (0001) surface will merge together at the end; however, the stacking faults nucleated in each domain do not easily cross the boundary to elongate into the nearby domain. Further, the defect properties require that they cannot be terminated inside the body; therefore, in order to minimize the system energy, the defects lying in the domain boundaries, in this case those prismatic planes, will be favorable. It corresponds to the fold of the stacking faults from basal plane to prismatic plane, and then we can observe those prismatic-plane stacking faults.

27.3.5
Planar Defects: Inversion Domain Walls

Planar defects, such as basal-plane stacking faults, can guide ZnO nanowires/nanobelts growing along the $[01\bar{1}0]$ direction to expose the polar surface as side surfaces. Inversion domain walls have the same function, which are introduced by doping. The doped impurity we chose was indium [34]. As a result, the $[01\bar{1}0]$ growth nanobelts were formed. Because the nanobelts have large exposed polar surfaces, they can even self-coil to form nanosprings or nanorings [43].

Figure 27.18a shows a ZnO nanobelt with inversion domain walls parallel to the (0001) polar surfaces. The X-ray energy-dispersive spectrum (EDS) required from the nanobelt shows the presence of minor indium besides majority of Zn and O (Cu and Si signals come from copper grid and the substrate) (Figure 27.18b). Quantitative analysis indicates the atomic ratio of In:Zn $\approx 1 : 15$, suggesting that indium ions are successfully doped in the ZnO lattice. The HRTEM image in Figure 27.18c presents a nanobelt with two sharp-contrast plane defects. After constructing several possible structural models, we performed the detailed image simulations in reference to the experimental data. If we considered the two dark layers (labeled I and II in the image) as two In-O octahedral layers, we found the best-matched structural model is the one displayed in Figure 27.18e. If we define the (0001) surface as Zn-terminated and the $(000\bar{1})$ surface as oxygen terminated, so that the polarization is along the c-axis, the two slabs of ZnO on both sides of the In-O octahedral layer must have opposite polarization, which means that the

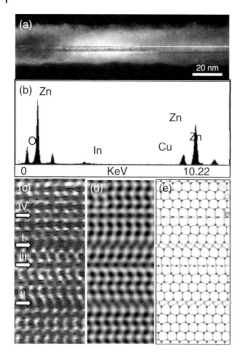

Figure 27.18 (a) Dark-field TEM image of a ZnO nanobelt with inversion domain boundaries. (b) EDS spectrum showing the presence of In ions in the structure. Experimental (c) and simulated (d) HRTEM images of the inversion domain boundaries with incident electron beam along the [$2\bar{1}\bar{1}0$] direction. (e) The structural model used in the simulation. The simulation is for a sample thickness of 2.924 nm, defocus −28.67 nm, for JEOL 4000EX at 400 kV [33]. (With permission from Elsevier.)

In-O layer effectively induces a "head-to-head" polarization domain, the so-called inversion domain boundary. To configure the two sharply contrasted plane defects (I and II), there must exist another type of defect (labeled as III) between the I and II layers, and it should correspond to a "tail-to-tail" inversion domain wall. The III layer does exit in the image and the adjacent bright spots forms a rectangle pattern. On the other hand, based on the structural model of In_2O_3, the two slabs of ZnO on either side of the In-O octahedral layer can take not only head-to-head polarity but also tail-to-tail polarity as presented in Figure 27.18e. In the first case, the fourfold symmetry axis of the In-O octahedra lies in the ZnO c-plane to form a head-to-head inversion domain wall. In the second case, the fourfold symmetry axis of the In-O octahedron is parallel to the c-axis to form a tail-to-tail inversion domain wall. Such a tail-to-tail layer also exists above the I layer, labeled to be the IV layer. The transverse translation of the IV layer may have resulted from the relocation of the doped indium ions. It is noteworthy that the translation in the [0001] direction across the tail-to-tail inversion domain wall is very small, and there is no translation along the [01$\bar{1}$0] direction. A simulated image based on the model in Figure 27.18e is shown in Figure 27.18d. An excellent match between the simulated and the experimental images supports our model.

27.4
In situ Observation of the Growth Process of Nanowires and Nanotubes

Chemical vapor deposition is one of the most important approaches in the growth of 1D nanowires and nanotubes. In the growth process, there are several sequential processes: (i) gas precursor transport, (ii) precursor adsorption and dissociation at the catalyst surface, (iii) material diffusion across the catalyst particles, and (iv) precipitation of the nanowire materials. Depending on whether the catalyst is in liquid state or not, the growth mechanism is classified as vapor–liquid–solid (VLS) and vapor–solid–solid (VSS) mechanisms [44, 45]. In situ TEM under environmentally controlled conditions provides the best way for directly imaging the growth of nanowires.

27.4.1
Si Nanowires Catalyzed by Au and Pd Nanoparticles

The Si nanowires grown from Au seeds follow the VLS process. The as-evaporated Au nanoparticles deposited on SiO_x membranes show well-developed facets in Figure 27.19a [46]. Increasing the system temperature to 590 °C and then

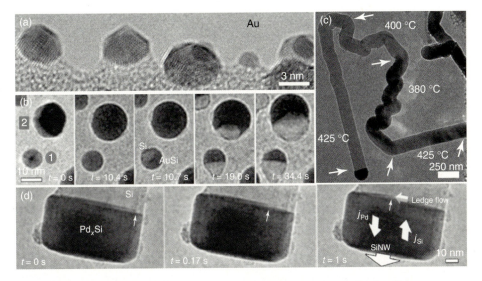

Figure 27.19 (a) As-evaporated Au on SiO_x. (b) Plane-view environmental TEM image sequence if Au-SiO_x system exposed to Si_2H_6 gas at 590 °C. (c) Frame grabbed from the end of an in situ TEM movie of a Si nanowire grown using 2.5 m torr disilane acquired along the [110] azimuth. (d) Environmental TEM image sequence of growing Si nanowire with Pd as catalyst [46, 47]. (With permission from ACS and NPG.)

introducing Si_2H_6 gas, the Au particles show contrast fluctuations and swell and liquefy as Si is incorporated into the particles (Figure 27.19b). A video showing the growth process can be downloaded from the journal web site of Nature Materials. The Si nucleation occurs burstlike, by growing into the Au–Si droplets. The smaller catalyst particles nucleate Si first, for in the droplets, the Si concentration reaches the supersaturation threshold faster. Si(111) is the slowest growing face with the lowest free energy and thus dominates Si nanowire growth. The temperature influence on the growth can be seen in Figure 27.19c [47]. The Si nanowire grows straight at 425 °C with the first segment along [1$\bar{1}$1] and the last along [$\bar{1}$12]. The segment grown at 400 °C changes its growth axis from one to another crystallographic direction at localized kinks. The segment growth at 380 °C changes its growth direction smoothly and continuously along random directions, following a wormy trajectory. While changing the temperature back to 425 °C, the Si nanowire becomes straight and along <111> again.

The growth of Si nanowires using Pd nanoparticles as catalyst follows the VSS process, for the catalyst alloy was in solid state throughout the growth (Figure 27.19d). The growth temperature was kept at 560 °C, while the Pd_xSi catalyst alloy shows well-developed facets instead of the liquid droplets. Further, lattice fringes can be observed. With time elapses, Pd leaves the Pd_xSi/Si interface, diffuses through or around the Pd_sSi particle, and reacts at its free surface with the excess Si, generated by decomposition of Si_2H_6 there. Stress buildup at the Si nanowire–catalyst interface is prevented by Si diffusion in the other direction. The driving force is the Si concentration gradient and gas decomposition on the catalyst surface as the rate-limiting step.

27.4.2
In situ Observation of the Growth of Carbon Nanotubes

The crucial questions on the growth of carbon nanotubes are as follows [48]. (i) Are catalyst nanoparticles liquid, fluctuating crystalline, or crystalline? (ii) Are catalyst nanoparticles carbide particles? (iii) Do carbon atoms migrate on the surface and/or through the bulk of catalyst nanoparticles? The *in situ* TEM observation has the answers to all of the above questions.

Figure 27.20 presents the nucleation and growth process of a single-walled carbon nanotube in time sequence. Before the nucleation, the catalyst nanoparticle (Fe) exhibits different facets in every frame. Furthermore, the various carbon cages frequently expelled from the catalyst particle disappear in a few seconds. The shapes of both unstable carbon cages and the catalyst particle change rapidly. After the incubation period, the stable dome, or the nucleus of a single-walled carbon nanotube appears. The nucleus grows gradually into a well-defined carbon nanotube of 1.5 nm in diameter and 3.6 nm in length, as seen in Figure 27.20. In the nucleation process, graphene layers are first observed on a facet of the catalyst particle. The graphene layers gradually extend and bend along the facets of the catalyst particle and additional graphene layers nucleate beneath the extend ones. The lattice images revealed by the HRTEM indicate that the catalyst particles

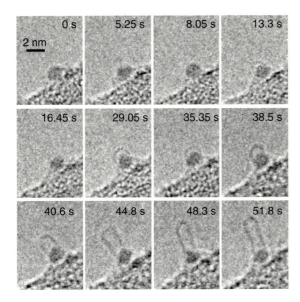

Figure 27.20 Nucleation and growth process of a single-walled carbon nanotube from a catalyst nanoparticle on a substrate [48]. (With permission from ACS.)

are actually below the Fe-carbide structure. Volume diffusion of carbon inside of carbide catalyst particles is very likely based on the above prescriptions of the growth process.

27.5
In situ Mechanical Properties of Nanotubes and Nanowires

The main challenge in measuring the mechanical properties of carbon nanotube is the ultrasmall size, which prohibits the application of conventional techniques. We need to see the nanotube first, and then measure its properties. Apparently, we must rely on electron microscopy. There are two techniques that have been developed for this application. The first technique relies on the thermal vibration of the carbon nanotube and the second depends on the electrically induced resonance of the nanotubes; both of them were performed in TEM.

27.5.1
Young's Modulus Measured by Quantifying Thermal Vibration Amplitude

Figure 27.21a shows a TEM image recorded from carbon nanotubes grown by arc-discharge. There are many nanotubes with a wide range of distributions in length and diameter. The as-synthesized sample is not pure, so the measured result, based on conventional technique, is an average value contributed by all of

 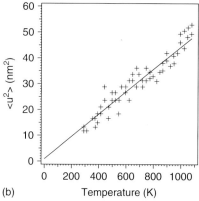

Figure 27.21 (a) A low-magnification TEM image of carbon nanotubes grown by arc discharging. The thermal vibration can be directly seen at the tips of the nanotubes with a large aspect ratio, as indicated by an arrowhead. A high-resolution TEM image is seen, which shows the inner and outer diameters of a nanotube. (b) Plot of the mean square vibration amplitude of a carbon nanotube measured in TEM as a function of temperature. The length of the nanotube is 5.1 μm and diameter is 16.6 nm. The slope of the solid line is 0.044 nm^2 K^{-1}, which gives an effective Young's modulus of 3.7 ± 0.2 TPa [49]. (With permission from NPG.)

the phases and the nanotubes with a wide range of structures. To characterize the mechanical properties of individual nanotubes and correlate them with the observed microstructure, one must do it using *in situ* electron microscopy. A striking feature noticed in Figure 27.21a is that a few nanotubes, indicated by arrowheads, show blurring contrast especially at their ends. A systematic measurement of the vibration amplitude as a function of temperature shows that the blurring contrast is due to the thermal-induced vibration at the tip, as shown in Figure 27.21b, and a quantitative analysis of the vibration amplitude gives the Young's modulus [49].

This technique has been extended to measure the Young's modulus of individual single-wall carbon nanotubes (SWNTs). By measuring the vibration amplitude from electron micrographs and assuming that the vibration modes are driven stochastically, an average Young's modulus of $E = 1.25$ TPa was found for SWNTs, about 25% higher than the currently accepted value of the in-plane modulus of graphite [50]. However, this technique is only appropriate for nanotubes with small diameters and long lengths, as the ones with larger diameters show no visible thermal vibration. The inaccuracy of this type of measurement arises from the inaccuracy in determining the vibration amplitude experimentally, and in some cases, an error of more than 100% can be yielded.

27.5.2
Bending Modulus by Electric-Field-Induced Mechanical Resonance in TEM

In this *in situ* experiment, we need a special specimen holder, which can apply a voltage between a nanotube and its counter electrode. In the area where a specimen is loaded in conventional TEM, an electromechanical system that allows not only the lateral movement of the tip is built but also applies a voltage across the nanotube with the counter electrode [51]. This setup is similar to the integration of the scanning probe microscopy (SPM) technique with TEM. The static and dynamic properties of the nanotubes can be obtained by applying a controllable static and alternating electric field.

The carbon nanotubes in Figure 27.22a is produced by an arc-discharge technique, and the as-prepared nanotubes are agglomerated into a fiberlike rod. The carbon nanotubes have diameters of 5–50 nm and lengths of 1–20 μm and most of them are nearly defect-free. The fiber is glued using silver paste onto a gold wire, through which the electric contact was made. The counter electrode is an Au/Pt ball of diameter ∼0.25 mm. The information provided by TEM directly reveals both the surface and the intrinsic structure of the nanotube. This is a unique advantage over the SPM technique. The distance from the nanotube to the counter electrode is controllable.

After applying an electric potential between the carbon fiber and the counter electrode, the induced charge on the carbon nanotubes mostly locate at their tips and the electrostatic force results in the deflection of the nanotubes. Alternatively, if an applied voltage is an alternating voltage, the charge on the tip of the nanotube is also oscillating, same as the force. If the applied frequency matches the natural resonance frequency of the nanotube, mechanical resonance will be triggered. By tuning the applied frequency, the first and the second harmonic resonances can be achieved as shown in Figure 27.22c,d. The geometric parameters of the carbon

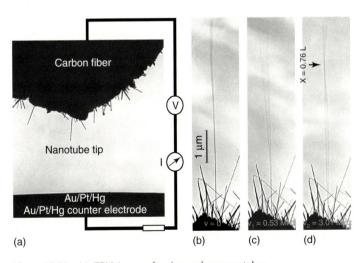

Figure 27.22 (a) TEM image showing carbon nanotubes at the end of the electrode and the other counter electrode. A selected carbon nanotube at (b) stationary, (c) the first harmonic resonance ($v_1 = 1.21$ MHz), and (d) the second harmonic resonance ($v_2 = 5.06$ MHz) [52]. (With permission from Elsevier.)

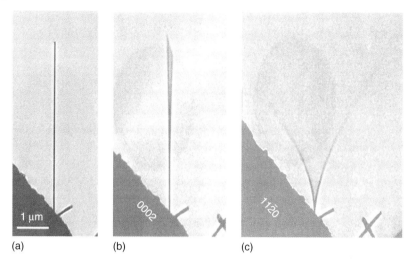

Figure 27.23 A selected ZnO nanobelt at (a) stationary, (b) the first harmonic resonance in x direction, $v_{x1} = 622$ kHz, and (c) the first harmonic resonance in y direction, $v_{y1} = 691$ kHz [54]. (With permission from AIP.)

nanotube can be directly measured from TEM images at a high accuracy. The determination of length has to consider the 2D projection effect of the nanotube. It is essential to tilt the tube and to catch its maximum length in TEM, which is likely to be the true length. Combined with the theoretical modeling, the bending modulus of each carbon nanotube can be retrieved [52, 53].

Figure 27.23 gives another example of how to measure the bending modulus of a ZnO nanobelt based on the alternative electric field induced resonance [54]. Two fundamental frequencies in the two orthogonal transverse vibration directions have been found. On the basis of the experimentally measured data, the bending modulus of the ZnO nanobelts is calculated as ∼52 GPa.

27.6
In situ Electric Transport Property of Carbon Nanotubes

In classical physics, conductance is a quantity defined to specify the capability that an object can transport current. Under an applied external electric field, the movement of the electrons in a wire is accelerated, but the moving speed of the electron is limited by the mean-free-path-length of the electron in the wire, which is typically a few nanometers for metals. Inelastic collision causes the electron to lose energy, and this is the primary source of heat. The main source of inelastic collision is phonon–electron interaction. The electron moves at a mean speed on average, which is called the *Fermi velocity*. The resistance of the wire is related to its cross section A, length L, and resistivity ρ by $R = \rho L/A$.

27.6 In situ Electric Transport Property of Carbon Nanotubes

The inverse of the conductance G is the resistance, which is a quantity for characterizing the heat generation in the object under externally applied voltage. Taking a metal wire as an example, its conductance is proportional to the cross section of the wire and inversely proportional to its length. The concept holds if the wire length is much longer than the mean-free-path-length of the electron in the wire, which is typically a few nanometers for most metals. Therefore, heat production is a very common phenomenon in classical electric transport.

In quantum mechanics, the conductance of a system is related to its scattering property because electrons are waves. One of the main assumptions in quantum conductance is that the electron mean-free-path-length is so long that the electrons experience no inelastic collision during transport through the wire, which is possible only if the temperature is extremely low or the system is so special to have a long mean-free-path-length. The conductance defined in quantum mechanics is to characterize the wave scattering property rather than heat generation. Quantum conductance occurs if the electron mean-free-path-length due to inelastic scattering is larger than the length of the wire. Therefore, quantum conductance, if observed, does not have heat dissipation.

The conductance measurement of individual carbon nanotubes has also been carried out using the same *in situ* technique introduced in Figure 27.22 [55]. The specimen holder used for the mechanical property measurements was used for the conductance measurement except that a mercury droplet, which served as the soft contact for the measurement, replaces the solid counter electrode. Figure 27.24 shows the contact of a carbon nanotube with the mercury electrode, and the conductance of G_0 was observed. It is also interesting to note that the contact area between the nanotube and the mercury surface is curved mostly due to the difference in surface work function between the carbon nanotube and the mercury.

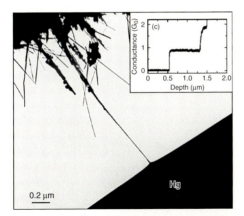

Figure 27.24 *In situ* TEM image showing the conductance measurement through a single carbon nanotube. The inset is the contact area of the nanotube with the mercury surface. The inset gives the measured conductance [55]. (With permission from AAAS.)

In addition, electrostatic attraction could distort the mercury surface as well. These effects effectively reduce the contact resistance between the nanotube and the Hg electrode.

In the original experiment, nanotube fibers were conditioned by many (up to several thousand) dipping cycles until the quantized (flat) plateaus developed. The initial cycles do not show the effect. After many cycles, a pattern of reproducible steps appears (each about 1 G_0 high), with cycle-to-cycle step height variations of the order of ± 0.05 G_0. The plateaus are remarkably flat (typically $|dG/dL| < 0.04$ G_0 μm^{-1}), which implies conductivity greater than for copper. Usually positive, but sometimes slightly negative, plateau slopes are observed, as seen in Figure 27.24. This may be due to a slight reduction in the electronic transmission with increasing strain. Flat plateaus (ideally $\rho = 0$) and plateau conductance G in the range 0.8 $G_0 < G < 1$ G_0, are consistent with conductance quantization and ballistic conduction, allowing for a slight residual contact resistances at the fiber/nanotube contact.

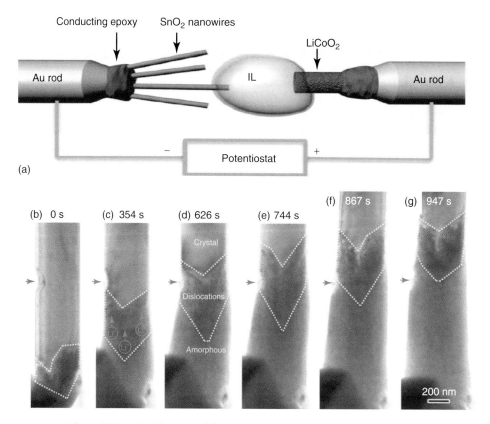

Figure 27.25 (a) Schematic of the experimental setup. (b–g) TEM images revealed a high density of dislocations emerging from the reaction front [56]. (With permission from AAAS.)

27.7
In situ TEM Investigation of Electrochemical Properties of Nanowires

With a special design of the TEM sample holder, even some electrochemical reaction can be visualized real-time in TEM [56]. Recently, by combining a SnO_2 nanowire anode, an ionic liquid-based electrolyte (ILE) and a $LiCoO_2$ cathode inside an HRTEM, Huang and coworkers successfully recorded the whole process of the lithiation of the SnO_2 nanowire during electrochemical charging. The experimental design can be found in Figure 27.25a. After contacting the nanowire with the ILE, a -3.5 V potential was applied to the nanowire with respect to the $LiCoO_2$ counter electrode. This initiated an electrochemical reaction at the point of contact between the nanowire and the ILE where the SnO_2 reduction was observed. On charging, a reaction front propagated progressively along the nanowire, causing the nanowire to swell, elongate, and spiral. The reaction front contains a high density of mobile dislocations, which are continuously nucleated and absorbed at the moving front (Figure 27.25b–g). This dislocation cloud indicates large in-plane misfit stresses and is a structural precursor to electrochemically driven solid-state amorphization. The electron energy loss spectroscopy (EELS) indicated that, after reaction, the nanowire contained metallic Sn, Li, and Li_2O. The electron diffraction patterns further confirmed the existence of Sn and Li_xSn nanoparticles, while Li_2O took the amorphous state.

In the discharge process, the TEM showed that the Li_xSn alloy nanoparticles were converted back to pure Sn and the diameter of the nanowire decreased. However, the Li_2O did not participate in the discharging process. Such an observation provides an important insight into the design of advanced batteries.

27.8
Summary

In this chapter, we have presented some examples to demonstrate the applications of TEM techniques in the research related to nanowires and nanotubes. The examples demonstrated are routine techniques for characterizing their structures and properties. It is worth emphasizing that, with the high spatial and image resolution provided by TEM, the *in situ* technique provides a direct method for imaging and measuring the properties of the same nanostructure. The TEM technique can be used not only to identify the geometric information of nanowires and nanotube but also to measure their mechanical, electrical, electromechanical, and electrochemical properties. We anticipate that many more unique techniques related to TEM will be explored in the near future.

References

1. Lieber, C.M. and Wang, Z.L. (2007) *MRS Bull.*, **32**, 99.
2. Wang, Z.L. (2009) *Mater. Sci. Eng., R.*, **64**, 33.

3. Chou, Y.C., Lu, K.C., and Tu, K.N. (2010) *Mat. Sci. Eng., R.*, **70**, 112.
4. Wang, Z.L. (2010) *Nano Today*, **5**, 540.
5. Barth, S., Hernandez-Ramirez, F., Holmes, J.D., and Romano-Rodriguez, A. (2010) *Prog. Mater. Sci.*, **55**, 563.
6. Ando, Y. (2010) *J. Nanosci. Nanotechnol.*, **10**, 3726.
7. Fan, Z.Y., Ho, J.C., Takahashi, T., Yerushalmi, R., Takei, K., Ford, A.C., Chueh, Y.L., and Javey, A. (2009) *Adv. Mater.*, **21**, 3730.
8. Lee, M., Baik, K.Y., Noah, M., Kwon, Y.K., Lee, J.O., and Hong, S. (2009) *Lab Chip*, **9**, 2267.
9. Schmidt, V., Wittemann, J.V., Senz, S., and Gosele, U. (2009) *Adv. Mater.*, **21**, 2681.
10. Lu, W., Xie, P., and Lieber, C.M. (2008) *IEEE Trans. Electron Dev.*, **55**, 2859.
11. Wang, N., Cai, Y., and Zhang, R.Q. (2008) *Mat. Sci. Eng. R.*, **60**, 1.
12. Briseno, A.L., Mannsfeld, S.C.B., Jenekhe, S.A., Bao, Z., and Xia, Y. (2008) *Mater. Today*, **11**, 38.
13. Lu, W. and Lieber, C.M. (2007) *Nat. Mater.*, **6**, 841.
14. Allen, B.L., Kichambare, P.D., and Star, A. (2007) *Adv. Mater.*, **19**, 1439.
15. Huang, X.J. and Choi, Y.K. (2007) *Sens. Actuators B Chem.*, **122**, 659.
16. Li, Y., Qian, F., Xiang, J., and Lieber, C.M. (2006) *Mater. Today*, **9**, 18.
17. Lu, W. and Lieber, C.M. (2006) *J. Phys. D Appl. Phys.*, **39**, R387.
18. Agarwal, R. and Lieber, C.M. (2006) *Appl. Phys. A Mater.*, **85**, 209.
19. Wanekaya, A.K., Chen, W., Myung, N.V., and Mulchandani, A. (2006) *Electroanal*, **18**, 533.
20. Law, M., Goldberger, J., and Yang, P.D. (2004) *Annu. Rev. Mater. Res.*, **34**, 83.
21. Ding, Y. and Wang, Z.L. (2004) *J. Phys. Chem. B*, **108**, 12280.
22. Gibson, J.M., Lanzerotti, M.Y., and Elser, V. (1989) *Appl. Phys. Lett.*, **55**, 1394.
23. Wu, Y., Cui, Y., Huynh, L., Barrelet, C.J., Bell, D.C., and Lieber, C.M. (2004) *Nano Lett.*, **4**, 433.
24. Cayron, C., Den Hertog, M., Latu-Romain, L., Mouchet, C., Secouard, C., Rouviere, J.L., Rouviere, E., and Simonato, J.P. (2009) *J. Appl. Crystallogr.*, **42**, 242.
25. Bell, D.C., Wu, Y., Barrelet, C.J., Gradecak, S., Xiang, J., Timko, B.P., and Lieber, C.M. (2004) *Microsc. Res. Tech.*, **64**, 373.
26. Huang, C.T., Song, J.H., Lee, W.F., Ding, Y., Gao, Z.Y., Hao, Y., Chen, L.J., and Wang, Z.L. (2010) *J. Am. Chem. Soc.*, **132**, 4766.
27. Kim, H.S., Myung, Y., Cho, Y.J., Jang, D.M., Jung, C.S., Park, J., and Ahn, J.P. (2010) *Nano Lett.*, **10**, 1682.
28. Ding, Y. and Wang, Z.L. (2007) *Surf. Sci.*, **601**, 425.
29. Qin, L.C. (2006) *Rep. Prog. Phys.*, **69**, 2761.
30. Bell, D.C., Russo, C.J., and Benner, G. (2010) *Microsc. Microanal.*, **16**, 386.
31. Ronning, C., Gao, P.X., Ding, Y., Wang, Z.L., and Schwen, D. (2004) *Appl. Phys. Lett.*, **84**, 783.
32. Wang, X.D., Ding, Y., Summers, C.J., and Wang, Z.L. (2004) *J. Phys. Chem. B*, **108**, 8773.
33. Ding, Y. and Wang, Z.L. (2009) *Micron*, **40**, 335.
34. Ding, Y., Kong, X.Y., and Wang, Z.L. (2004) *Phys. Rev. B*, **70**, 235408.
35. Wiley, B.J., Xiong, Y.J., Li, Z.Y., Yin, Y.D., and Xia, Y.A. (2006) *Nano Lett.*, **6**, 765.
36. Bere, A. and Serra, A. (2003) *Phys. Rev. B*, **68**, 033305.
37. Moore, D., Ding, Y., and Wang, Z.L. (2006) *Angew. Chem. Int. Ed.*, **45**, 5150.
38. Potin, V., Ruterana, P., and Nouet, G. (2000) *J. Phys. Condens. Matter*, **12**, 10301.
39. Stampfl, C. and Van de Walle, C.G. (1998) *Phys. Rev. B*, **57**, 15052.
40. Vermaut, P., Nouet, G., and Ruterana, P. (1999) *Appl. Phys. Lett.*, **74**, 694.
41. Northrup, J.E. (1998) *Appl. Phys. Lett.*, **72**, 2316.
42. Wang, Z.L., Kong, X.Y., and Zuo, J.M. (2003) *Phys. Rev. Lett.*, **91**, 185502.
43. Kong, X.Y., Ding, Y., Yang, R., and Wang, Z.L. (2004) *Science*, **303**, 1348.
44. Wagner, R.S. and Ellis, W.C. (1964) *Appl. Phys. Lett.*, **4**, 89.
45. Kodambaka, S., Tersoff, J., Reuter, M.C., and Ross, F.M. (2007) *Science*, **316**, 729.

46. Hofmann, S., Sharma, R., Wirth, C.T., Cervantes-Sodi, F., Ducati, C., Kasama, T., Dunin-Borkowski, R.E., Drucker, J., Bennett, P., and Robertson, J. (2008) *Nat. Mater.*, **7**, 372.
47. Madras, P., Dailey, E., and Drucker, J. (2009) *Nano Lett.*, **9**, 3826.
48. Yoshida, H., Takeda, S., Uchiyama, T., Kohno, H., and Homma, Y. (2008) *Nano Lett.*, **8**, 2082.
49. Treacy, M.M.J., Ebbesen, T.W., and Gibson, J.M. (1996) *Nature*, **381**, 678.
50. Krishnan, A., Dujardin, E., Ebbesen, T.W., Yianilos, P.N., and Treacy, M.M.J. (1998) *Phys. Rev. B*, **58**, 14013.
51. Poncharal, P., Wang, Z.L., Ugarte, D., and de Heer, W.A. (1999) *Science*, **283**, 1513.
52. Wang, Z.L., Poncharal, P., and de Heer, W.A. (2000) *J. Phys. Chem. Solids*, **61**, 1025.
53. Wang, Z.L., Poncharal, P., and de Heer, W.A. (2000) *Pure Appl. Chem.*, **72**, 209.
54. Bai, X.D., Gao, P.X., Wang, Z.L., and Wang, E.G. (2003) *Appl. Phys. Lett.*, **82**, 4806.
55. Frank, S., Poncharal, P., Wang, Z.L., and de Heer, W.A. (1998) *Science*, **280**, 1744.
56. Huang, J.Y., Zhong, L., Wang, C.M., Sullivan, J.P., Xu, W., Zhang, L.Q., Mao, S.X., Hudak, N.S., Liu, X.H., Subramanian, A., Fan, H.Y., Qi, L.A., Kushima, A., and Li, J. (2010) *Science*, **330**, 1515.

28
Carbon Nanoforms
Carla Bittencourt and Gustaaf Van Tendeloo

> ... every element says something to someone (something different to each) .... One must perhaps make an exception for carbon, because it says everything to everyone ...
>
> Primo Levi

Our personage ... the carbon atom is amazing; it holds the combined experience of our evolution, and our history. Castro Neto [1] added importance to this element, describing a new carbon age with carbon taking the place of silicon as active element in new electronic devices – thus, it may also hold our future. The carbon atom is present in about 16 million compounds, and the number of possible allotropes is almost infinite. It bonds to both electronegative and electropositive elements. The electronic configuration of $1s^2\ 2s^2\ 2p^2$ allows carbon atoms to form three different types of bonding, that is, single, double, and triple bonds. This bonding versatility of the carbon occurs because it can hybridize its 2s and 2p atomic orbitals in three different manners: sp^3 (for single bonding, tetrahedral), sp^2 (for double bonding, trigonal planar), and sp (for triple bonding, linear) [2]; it can form sigma (σ)-bonds and/or pi (π)-bonds. Carbon allotropes or carbon-containing compounds can have pure sp^1, sp^2, sp^3 hybridizations or a mixture of them. The properties of the carbon-based materials are influenced by the sp^2/sp^3 ratio, the morphology, the structure, the presence of defects, and impurities. Diamond and graphite are considered the main allotropic forms of carbon. In diamond, each sp^3-hybridized carbon atom covalently bonds to four others, extending three-dimensionally as a network of tetrahedral. Diamond has two forms: cubic and hexagonal, which can be interconverted under specific conditions [3]. Graphite comprises infinite layers of sp^2-hybridized carbon. Within a layer, each carbon atom bonds to three others, forming a two-dimensional network of hexagons. The unhybridized $2p_z$ orbital that accommodates the fourth electron forms a delocalized orbital of π symmetry that stabilizes the in-plane bonds. The noncovalent π-interaction keeps the layers stacked together, 3.354 Å

Handbook of Nanoscopy, First Edition. Edited by Gustaaf Van Tendeloo, Dirk Van Dyck, and Stephen J. Pennycook.
© 2012 Wiley-VCH Verlag GmbH & Co. KGaA. Published 2012 by Wiley-VCH Verlag GmbH & Co. KGaA.

apart [4]. The most common form of graphite is hexagonal, while rhombohedral graphite is much rarer. Amorphous carbons consist of nearly pure carbon with varying $sp^2:sp^3$ ratios and have been considered as a carbon allotropic "form" (although lacking any crystallinity and often containing significant amounts of hydrogen) [5]. These three forms of carbon compose the traditional carbon family tree. Nevertheless, improvements in the spatial resolution of traditional imaging techniques and development of new experimental techniques combined with theoretical analysis tools have led to the discovery of nanoscale carbon forms, such as fullerenes [6–10], carbon nanotubes (CNTs) [11–16], and graphene [11], [17–19], thus enlarging the traditional carbon tree. Recently, Suarez-Martinez proposed an extended carbon tree, adding a large number of carbon nanoforms [20]. Like their 3D counterparts, the carbon nanoforms can have pure sp^1, sp^2, sp^3 hybridizations or a mixture of them; however, because of the atomic rearrangement for forming the nanoform, further "rehybridization" of electronic states may occur. Saito et al. noted that hybridization of the graphitic-associated $\sigma, \pi, \pi^*, \sigma^*$ electronic states should occur in CNTs because of the curvature [18]. The importance of this effect was reported by Blase et al.; these authors showed that in small-radius CNTs, hybridization of the π^* and σ^* states of the graphene network occurs, drastically changing the electronic band structure from that obtained by simply "folding" the graphene band structure [21]. In (6,0) and (9,0) CNTs, hybridization effects change the energy and character of the lowest lying conduction band states, with important consequences to the metallicity and transport properties of the nanotube [21]. In fullerenes, the geometric and electronic structures of carbon atoms and bonds primarily differ from those of graphite [22]. For example, the surface of the C_{60} contains 20 hexagons and 12 pentagons and two different types of bonds: "short bonds" or 6,6 junctions, shared by two adjacent hexagons (circa 1.38 Å long) and "long bonds," or 5,6 junctions, fusing a pentagon and a hexagon (circa 1.45 Å long) [23]. The geometric demand of the spherical cage is such that all the double bonds in C_{60} deviate from planarity [24]. This pyramidalization of the sp^2-hybridized carbon atoms confers an excess of strain to C_{60}, which is responsible for the enhanced reactivity of the fullerene [23]. Different from the 3D counterparts, in nanoscale graphene flakes (nanosized hexagonal networks) and nanodiamonds (nanosized tetrahedral networks), the presence of edges and surfaces strongly impacts on the electronic structure of the carbon nanoform [25]. Nanoscale graphene flakes have unconventional nonbonding π-states (edge states) localized around its edges that depend on the edge geometry, these states having localized spins, impart a molecular magnetic character to the nanoscale graphene flake [25–27]. Similar to the edge states in nanoscale graphene, the chemical composition of the surface in a nanodiamond particle influences the electronic structure [28]. If at the surface of the nanodiamond all the carbon atoms are unsaturated, the surface is subjected to structural reconstruction at the expenses of σ-dangling bonds, being covered with shells consisting of a π-bond network. Hydrogenation of the surface can passivate the dangling bonds, thus stabilizing the surface structure. The nanodiamond particles, with their surface hydrogen-terminated, are expected to work as electron reservoirs owing to the presence of a surface electric bilayer

[28–30]. The incompletely hydrogenated surface has localized spins [28, 31, 32]. The above examples illustrate the importance of the knowledge on the morphology of the carbon nanoform as well as the possibility to perform analysis at selected regions of an isolated carbon nanoform – determine and differenciate the features of the electronic states resulting from edge or the "bulk" atoms. To perform these analyses, it is important to "see" the carbon nanoform and to be able to select the desired region of analysis. Therefore, traditional spectroscopic techniques were combined with microscopy while the need to determine the number and location of atoms pushed the development of new microscopes that are capable of imaging with reduced impact on the carbon nanoform structure.

This chapter aims at providing a reference that will help researchers to gain an overview of different techniques based on imaging presently used to study carbon nanoforms. First, we present carbon nanoforms that were identified using electron microscopy, with the exception for graphene that was "put in evidence" via optical microscopy [11] (while writing this chapter, we came across a report published in 1962 showing a beautiful and comprehensive electron micrograph of a graphene oxide flake). The selection of the nanoforms described in this chapter was based on the curiosity of the authors and on the possibility to show the development of different imaging techniques. It is outside the scope of this chapter to provide a description of synthesis techniques and applications of each carbon nanoform. However, appropriate references are indicated for the reader who desires to learn more about a particular nanoform. For a good description of the wider world of carbon allotropes (including bulk forms such as graphite, diamond, and amorphous carbons), see E. Falcao *et al.* [5], I. Suarez-Martinez *et al.* [2], and P. Harris [33].

In the second part, we briefly show less conventional image techniques such as photon-induced near-field electron microscopy (PINEM) [34–36], conductance imaging atomic force microscopy (CI-AFM) [37–39], or scanning transmission X-ray microscopy combined with near edge X-ray absorption fine structure (STXM-NEXAFS) [40–43] and their use to "image" CNTs. Others imaging techniques such as optical microscopy [44–47], not conventionally used in the analysis of nanoscale objects, fluorescence quenching microscopy (FQM) [48–50], and scanning photocurrent microscopy (SPCM) [51, 52] are considered when applied to study graphene. The great majority of the "nonconventional" imaging techniques combine spectroscopy with microscopy, allowing spectroscopic studies of an isolated carbon nanoform, removing the uncertainties of probing different carbon nanoforms at the same time. In certain cases, a selected region of the carbon nanoform can be studied, such as the tips or the walls of a CNT.

One of the most prominent imaging technique for identifying carbon nanoforms is electron microscopy. Information about morphology, structure and defects, crystal phases, and elemental composition can be obtained. In order to illustrate the importance of electron microscopy imaging, we present different carbon nanoforms that were studied via electron imaging together with brief remarks on the context of the related research. This section is divided in electron microscopy at high-resolution using high-energy acceleration voltages and electron microscopy

28.1
Imaging Carbon Nanoforms Using Conventional Electron Microscopy

Traditional electron microscopes, although they lacked resolution, had a key role in the observation of different carbon nanoforms. A great talent and unique imaging strategies were requirements for the earliest researchers working in electron microscopy. In this context, Figure 28.1 reproduces an electron microscopic image recorded in 1960 by K. Heideklang (Eduard Zintl Institute, Technical University at Darmstadt, Germany) [53]. It shows a "particle of graphite oxide soot" obtained from reduction of graphite oxide with hydrazine or hydroxyl-amine by H.-P. Boehm and colleagues [53]. By modeling the observed intensity on the microscope's photographic plate, calibrated to the known thickness of the film used as support, they extracted the flake thickness in different points. The average thickness of the thinnest sites was 0.46 nm; however, regions with two, three, and four times this value were also observed. Presented at the "Fifth Conference on Carbon" at Penn State University in 1961, no one took interest at that time [53].

28.1.1
Carbon Nanotubes

Three historical high-resolution electron micrographs reported by Sumio Iijima in 1991 are reproduced in Figure 28.2 – these were the first reported high-resolution images of a carbon nanoform [54]. In this image, the three structures show evenly spaced lattice fringes with an equal number of fringes on either side of the central core, they were associated with different CNTs. Structurally, CNTs can be

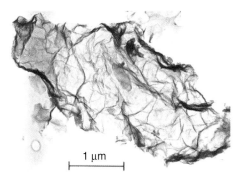

Figure 28.1 Electron microscopic image of "a particle of graphite oxide soot" recorded in 1960 by K. Heideklang using a Siemens Übermikroskop 100 e. Folds can be observed [53].

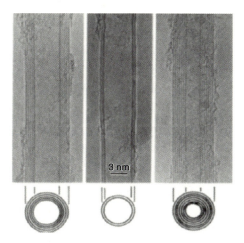

Figure 28.2 High-resolution electron micrographs: multi-wall carbon nanotubes with respectively, five, two, and seven walls by Sumio Iijima in 1991 [54].

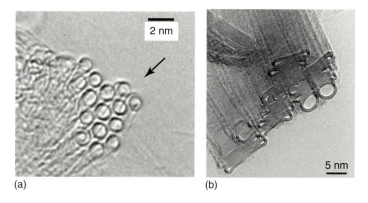

Figure 28.3 Bundles of (a) single-walled nanotubes [56] and (b) collapsed "dog-bone" double-walled nanotubes [57].

visualized as seamless tubes made of graphene. The topological transformation required to obtain a single-walled carbon nanotube (SWCNT) from a graphene sheet of defined size is to roll it and bond the two edges together. Nanotubes are classified based on the number of walls: SWCNTs, double-walled carbon nanotubes (DWCNTs), triple-walled carbon nanotubes (TWCNTs), and multiwalled carbon nanotubes (MWCNTs). All consist of concentric cylinders with spacing between nanotube walls approximately the interlayer distance in turbostratic graphite, that is, 0.34 nm. The number of walls can be determined – if the nanotube is isolated – by the number of lines (fringes) in a transmission electron microscopic image.

SWCNTs commonly form bundles because of van der Waals interactions between neighboring tubes (Figure 28.3a). Although normally considered weak forces, van der Waals interactions between two neighboring tubes can be high, ~1.2 eV/nm along a nanotube interface [55, 56]. Motta *et al.* reported that large-diameter SWCNTs and DWCNTs collapse to give "dog-bone" cross sections (Figure 28.3b) [57, 58]. Theoretical modelling shows that single-wall tubes under atmospheric pressure auto-collapse when the diameter is greater than a critical value in the range of 4.2–6.9 nm.

From the electron diffraction pattern, it is possible to derive the exact lattice structure of an individual SWCNT [59]. However, diffraction experiments on individual SWCNTs require sufficiently long, straight sections of individual nanotubes that are stable throughout the exposure. SWCNTs tend to curve and combine into bundles – a curved nanotube is not a periodic one-dimensional structure. In order to avoid the bundling of SWCNTs, Meyer and colleagues used free-standing structures as support to synthesize their samples resulting in well-separated, long, and straight individual SWCNTs (Figure 28.4) [59]. These authors presented a detailed electron diffraction study of individual SWCNTs synthesized via chemical vapor deposition (CVD) [59]. Their diffraction experiments were carried out at 60 kV, that is, below the threshold for knock-on damage in CNTs, using a thermal emission transmission electron microscope [60]. Figure 28.5 shows a diffraction pattern obtained on a (16,9) nanotube. The smallest SWCNT identified by these authors was a (7,7) nanotube, which has a diameter of 0.94 nm. Their statistical analysis on the distribution of the indices (n,m) showed that in a batch of SWCNTs synthesized under certain parameters [61], the rolling angle is not randomly distributed but grouped toward the armchair orientation.

The presence of diffraction resulting from tubes with different chiralities observed on the diffraction pattern recorded on MWCNTs makes their interpretation very difficult. A detailed work in this subject was reported by Amelinckx and coworkers in 1993–1994, who found evidence for zigzag, armchair, and chiral tubes within the same MWCNT [15, 33, 62, 63].

Figure 28.4 Dark-field mode TEM image showing long straight individual SWCNTs. The dark-field mode was used for imaging since SWCNTs are not seen in bright-field mode at lower magnifications. Scale bar = 1 mm [59].

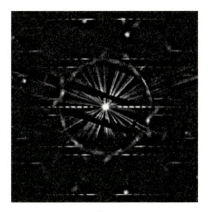

Figure 28.5 Diffraction patterns of a (16,9) nanotube. A background subtraction was performed [59].

SWCNTs have been reported to reach lengths of up to 18.5 cm, although more typically, lengths range from microns to millimeters [64]. Owing to their entangled nature, in general, the length of few-walls CNTs is rather challenging to determine, unless they are grown perpendicular to substrates (so-called nanotube forests) (Figure 28.6) [65]. The length of MWCNTs grown on substrates via chemical vapor deposition, on the contrary, is easily measured in a standard cross-sectional scanning electron microscopic image [66]. With a growth process similar to that of carbon nanotube forests, synthesis of vertically aligned few-layered graphite can be performed using microwave-enhanced plasma CVD (Figure 28.7) [67, 68]. The "graphite nanowalls" oriented almost perpendicularly to the substrate surface, are a few nanometers thick (typically less than 10 nm), and typically a micron long [67].

The use of carbon nanoforms in applications such as nanoelectronics, metal composites, and catalysis requires knowledge on the characteristics of the interface

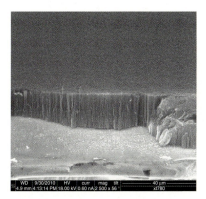

Figure 28.6 Vertically aligned MWCNTs. The length can be evaluated by acquiring a lateral image. (Courtesy: Ke and van Tendeloo, Antwerp).

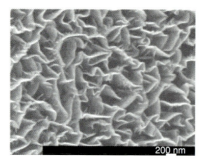

Figure 28.7 Vertically oriented few-layers graphite [67].

between nanoforms and metals. Investigating metals with different wetting characteristics at the CNT surface, Cha and colleagues reported that islanding rather than wetting of a metal at the surface of CNTs can deform the tube at the contact region [69]. These authors showed that under a gold island or a Au-Pd island the nanotube can undergo significant deformation (Figure 28.8) [69]. Pd has a greater tendency to wet the CNT surface; however, for the given Au–Pd alloy composition, islands formed at the CNT surface (Figure 28.8b). A cross section through a tomographic reconstruction of a CNT covered with Au-Pd reveals a relatively flat contact surface and the nanotube deformation (Figure 28.8c).

A electron tomographic analysis performed on a CNT covered with Ti indicated that Ti covers the tube uniformly, indicating good wetting (Figure 28.9a) [69]. A cross-sectional view revealed that Ti wraps around the nanotube and preserves its round shape (Figure 28.9b). The wetting by Ti was observed regardless of the nanotube diameter [69].

The chemistry at the interface of Ti/CNT was studied by Felten and colleagues [70]. These authors found that, different from the Pd/CNT that do not form strong chemical bonds, in the Ti/CNT interface, chemical bonds are formed. Gloter and Bittencourt (unpublished) investigated the interface formation between a Ti layer and a MWCNT using scanning transmission electron microscopy high angle

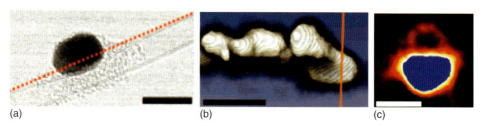

Figure 28.8 (a) Au island on a triple-wall carbon nanotube. The nanotube is kinked at the contact region with a Au island. The red dotted line shows the deviation from a straight line. (b) Carbon nanotubes with Au-Pd islands; gray and black isointensity surface and volume render denote the Au-Pd islands and the nanotube, respectively. (c) A cross-sectional view of the orange slice in (b); the radial deformation of the nanotube (in black) is evident. Scale bars, 10 nm in (a,b) and 4 nm in (c) [69].

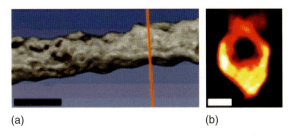

Figure 28.9 (a) Volume render of a carbon nanotube (black) with isointensity surface of Ti wetting layer (gray). (b) Cross-sectional view of the orange slice in (a). Scale bars, 10 nm in (a) and 4 nm in (b) [69].

annular dark field (STEM-HAADF). The Ti layer was thermally evaporated onto the CNT. These researchers observed the formation of a continuous layer at the CNT surface. The unidirectional evaporation led to the deposition of large amounts of Ti on the left side of the tube as shown in the line profile (Figure 28.10) [70–72].

Depending on the synthesis parameters or applied postgrowth treatments, the structure and morphology of the CNTs can be changed. Trasobares and colleagues reported the engineering of MWCNTs by exposure to an Ar/CH_4 microwave plasma [73]. These authors reported the formation of graphitic platelets (like wings) covalently attached to the side wall of the nanotubes (Figure 28.11a) [73]. Different synthesis techniques were reported to lead to a peculiar CNT called *bamboo tubes* [74–76]. These tubes are tubular structures with a compartmented hollow core.

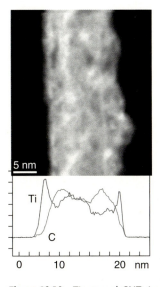

Figure 28.10 Ti covered CNT, imaged (STEM-HAADF) and analyzed (EELS carbon and titanium chemical profile). (Courtesy of Alexander Gloter, Orsay).

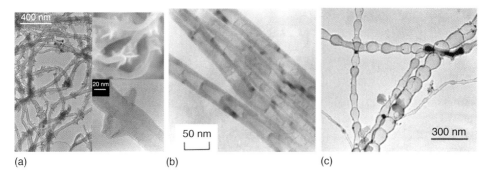

Figure 28.11 (a) MWCNTs showing nanowings at both sides. (Courtesy of Trasobares, Cadiz.) (b) Bamboolike nanotubes [75]. (c) Carbon necklaces presenting the repetition of an elementary unit looking like a nanobell, sometimes empty or filled with a metal particle [76].

Depending on the structure of the outer walls, there are two possible bamboo structures: bamboo nanotubes in which the external walls are almost parallel to the tube axis and herringbone-bamboo tubes in which the layers are at an angle to the tube axis (Figure 28.11b). Figure 28.11c shows "necklacelike" carbon nanostructures obtained by thermal plasma technology [76].

Nanobuds are synthesized via vapor decomposition or hot wire generator [77, 78]. These carbon nanoforms consist of fullerene units covalently attached to SWCNTs. The fullerenes can have a spherical structure and can be covalently bonded to CNTs via sp^3-hybridized carbon atoms. Otherwise, all carbon atoms are sp^2-hybridized, and the nanobud is as a fully hybrid structure, where a fullerene is a continuous part of a CNT as shown in Figure 28.12. Careful investigation of this material by transmission electron microscopy (TEM) revealed the presence of C_{42} and C_{60} fullerene units covalently bonded to the outer surface of the SWCNTs [79].

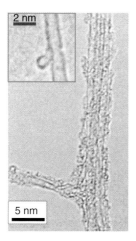

Figure 28.12 Nanobud. Spherical structures on the surface of SWCNTs [78].

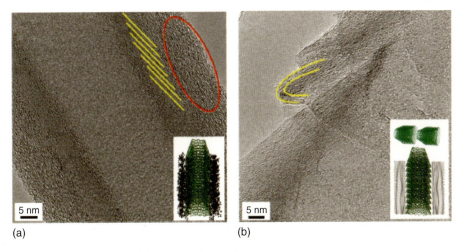

Figure 28.13 Pristine tubes before (a) and after (b) exfoliation. The yellow lines indicate the catalytically grown truncated cones stacked along the tube length, whereas the red circle and lines indicate the deposited amorphous carbon [81].

Cup-stacked CNTs are characterized by truncated conical graphene layers along the tube length, with open edges at the outer surface. In vapor-phase catalytically grown pristine cup-stacked tubes, the edges of the truncated cones are covered with amorphous carbon (Figure 28.13a). This unavoidable amorphous carbon layer degrades the electrochemical function of the edges of the truncated cones [80, 81]. Hence, for optimal performance of this nanoform, the amorphous layer covering the cones must be removed. Jang and colleagues reported a strategy based on the chemical exfoliation method that exposes the active edges of the cones by simply detaching the amorphous carbon deposited on the surface of the cup-stacked nanotube (Figure 28.13b) [81].

Helically coiled CNTs can be observed in samples produced catalytically (Figure 28.14). Amelinckx and colleagues discussed the growth of these tubes in terms of an elliptical locus of active sites around the periphery of the catalytic particle. They showed that a catalytic activity that varies around an ellipse will produce helical growth [82, 83].

28.1.2
Crop Circles

Circular ropes of SWCNTs are a by-product of laser-based synthesis of SWCNTs (Figure 28.15) [84]. This carbon nanoform was called *"crop circles,"* even if many of the individual tubes in these ropes are perfect tori. The ropes consist of between 10 and 100 individual nanotubes aligned over the entire length, with diameters typically ranging between 300 and 500 nm [84]. Transmission electron

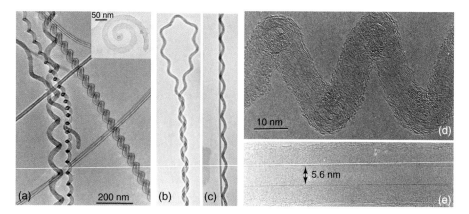

Figure 28.14 Examples of helical carbon nanotubes (a–c). A high-resolution micrograph of a helical tube (d) and the inner tube (e).

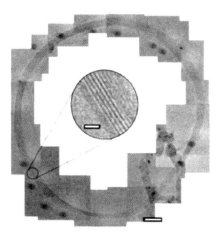

Figure 28.15 Transmission electron micrograph of a "crop circle," showing fringes with spacing typical of SWCNTs in normal ropes. Scale bars, 15 nm; inset, 5 nm [84].

micrographs of "crop circles" clearly show fringes corresponding to individual tubes (Figure 28.15). Liu and colleagues used a Kekuléan image of a growing nanotube eating its own tail to explain the formation of the "crop circles": first the nanotube would bend around to touch near its edges [83, 84]. Afterward, they would align to maximize van der Waals interaction but would slide over another nanotube to alleviate bending strain. If there was a metal particle (catalyst) at the edges of a tube, the two edges of the same tube would become welded together by the metal "plug." The metal particle could then be shunted either to the exterior or to the interior. With no edges, there would be no further growth [84].

28.1.3
Carbon Beads

Carbon beads form during the synthesis of MWCNTs via arc discharge. De Heer and colleagues studied the formation of CNTs in a carbon arc, analyzing cylindrical fragments of carbonaceous materials formed on the cathode during the synthesis [85]. These authors observed that the morphology of the MWCNTs formed at the inner and at the outer surface of the fragments had a similar structure; however, beads were often observed on the nanotubes formed at the outer surface (Figure 28.16). Analysis of TEM images recorded on the beads indicated their wetting ability on the CNT surface and lower density when compared to MWCNTs (Figure 28.17). Using a dynamical theory developed for studying the instability of a liquid cylinder under the action of a capillary force [86], De Heer et al. [85] suggested that carbon-arc-produced MWCNTs form by homogeneous nucleation in liquid carbon, inside elongated carbon droplets coated with a thin layer of carbon glass.

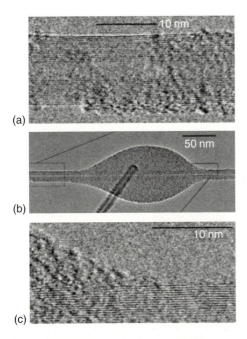

Figure 28.16 TEM of a bead on a MWCNT. The nanotube is visible inside the bead, a clear extended meniscus is seen at the nanotube-bead interface, indicating adhesive wetting. Note its cylindrical symmetry, the absence of facets, and a very well formed meniscus. There is no evidence for graphitic layering in the bead, nor does the bead distort the nanotube. (a second nanotube is in the background). Image of the meniscus on the left side (a) and on the right side (c) of the bead. (b) Elongated bead [85].

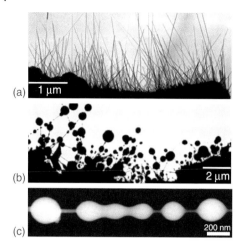

Figure 28.17 (a) Nanotubes are straight and relatively free of amorphous carbon particles [85]. (b) TEM image of the surface of a column, showing nanotube beads on most of the tubes [85]. (c) ADF-SEM image of beads on the CNT surface. (Courtesy of D. Ugarte and P. Silva, Campinas.) The undulations on the second bead are caused by the Rayleigh instability, which ordinarily would cause the structure to separate into three beads. The authors associated the arrested beading to the cooling of the carbon glass.

28.1.4
Carbon Nanoscrolls

Nanoscrolls can be formed by rolling more than one graphene sheet (multiscrolls) or by wrapping around conventional nanotubes [87–91]. Figure 28.18a,b, show nanoscrolls that were formed upon sonication of exfoliated graphite sheets that curled onto themselves [88]. Current theories suggest that multiwall nanotubes may form through a scrolling mechanism [16, 92]. Lavin *et al.* reported that in arc-discharge MWCNTs defects equivalent to edge dislocations in a bulk crystal

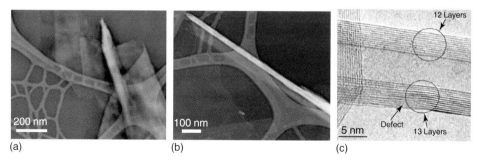

Figure 28.18 (a) A thin plate of graphitic sheets in the process of scrolling, (b) an isolated carbon nanoscroll with open ends, and (c) extra layer plane in the vicinity of a slip-plane defect [87]. (a) and (b) [88].

might initiate the transition from scrolls to nested tubes [87]. These authors hypothesized that graphene monolayers would initially form scrolls and subsequently transform into multiwall nanotubes through a series of defects. Scrolls and nested tubes thus would coexist within a single MWCNT (Figure 28.18c).

28.1.5
Herringbone Carbon Nanofibers

Herringbone-type carbon nanofibers (CNFs) are a special kind of CNF with angles between the graphene layers direction and the main axis of fiber in the range of 0–90° [93]. Figure 28.19 shows some typical high-resolution transmission electron microscopy (HRTEM) images of herringbone-type CNFs synthesized via catalytic decomposition of methane [94]. In these fibers, some adjacent graphene layers can be closed at the edge of the outer surface, forming a loop structure (Figure 28.19a), or they can remain open (Figure 28.19b). The interlayer spacing can be constant (Figure 28.19c) or vary along the graphene layer direction (Figure 28.19d).

28.1.6
Carbon Nanocones

Carbon nanocones (CNCs) are also referred to as *nanohorns*, and their aggregation (tips outward) is referred to as a *dahlialike aggregate* [20, 95, 96]. Iijima *et al.* reported

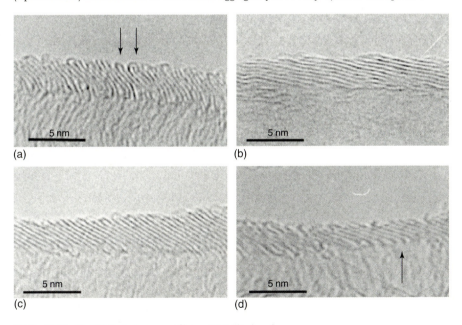

Figure 28.19 Herringbone-type nanofibers: (a) with closed edge sites, (b) with open edge sites, (c) with a constant interlayer spacing, and (d) with a varying interlayer spacing [94].

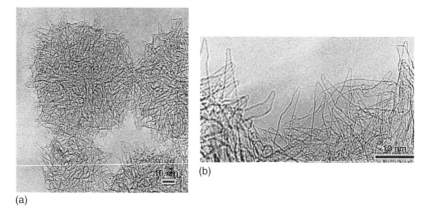

Figure 28.20 (a) Dahlia aggregates of nanocones and (b) a magnified image of an edge of a dahlia aggregate showing single graphene sheet carbon nanocones with closed caps whose diameter is similar to that of the fullerene molecules [96].

that aggregates of graphitic carbon nanoforms obtained by CO_2 laser ablation of carbon were tubules with cone caps (Figure 28.20) [95, 96]. The average cone angle was 20°, indicating that the cone cap contained five carbon pentagon rings together with many carbon hexagons [97].

Using a carbon-arc plasma generator (torch configuration) under a continuous flow of hydrocarbon (heavy oil), Krishnan and collaborators [98] synthesized turbostratic microstructures (where the graphitic sheets are not ordered relative to each other in the plane) that have total disclination (TD) that were multiples of 60° (Figure 28.21). These authors showed that only five types of cones can be made from a continuous sheet of graphite; the angles of these cones, θ, are given by $\sin(\theta/2) = 1 - (P/6)$, where P is the number of pentagons necessary to produce a particular disclination ($TD = P \times 60° = 300°, 240°, 180°, 120°, 60°$). An interesting feature is the presence of facets at the rim of the open end of the cones (Figure 28.21a,b).

28.1.7
Carbon Onionlike Particles

The interaction of high-energy electrons with carbonaceous materials can induce the formation of symmetrical and spherical onionlike particles [99, 100]. Ugarte showed that two major structural aspects characterize the onionlike particles formed *in situ* in a electron microscope: the concentrically arranged graphitic layers are nearly perfect spheres and the size of the innermost shell is close to that of the C_{60} (Figure 28.22) [99]. The carbon onionlike particles represent the transition from fullerene to macroscopic graphite. These particles present both the closed graphitic surface of fullerenes and the stacked layers interacting by π-stacking interactions, as in graphite.

28.1 Imaging Carbon Nanoforms Using Conventional Electron Microscopy

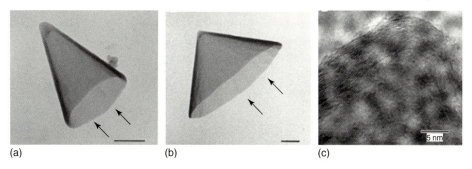

Figure 28.21 Examples of two types of nanocones (scale bars in (a,b), 200 nm). (a) $TD = 240°$, $\theta = 38.9°$. (b) $TD = 120°$, $\theta = 84.6°$. The arrows indicate the faceting. (c) Image showing the graphitic sheets at a cone tip [98].

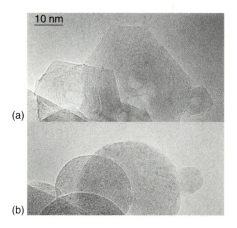

Figure 28.22 Electron micrographs of graphitic particles: (a) polygonal onion displaying a well-defined faceted structure and a large inner hollow. These onions are typical by-products of arc-electric carbon nanotube growth. (b) The very same particles after being subjected to intense electron irradiation; dark lines represent graphitic layers [99].

28.1.8
Carbon Spiroid

Ozawa and coworkers studied the conversion of commercial furnace black particles into carbon onions by electron beam irradiation (Figure 28.23) [101]. They suggested that the onion configuration is reached through a reversible transformation, with a spiral intermediate playing the role of shell adjuster [101]. This "intermediary" carbon nanoform was called *spiroid*. Archimedean-type spiroids with equal spacing between shells often appear preceding the formation of the onions independent of the synthesis method used to produce the carbon sample – a particle having a spiral

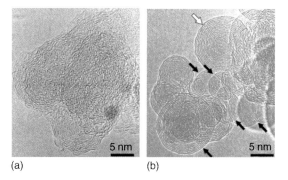

Figure 28.23 Conversion of commercial furnace black particles into carbon onions by irradiation of focused electron beam. (a) Before irradiation and (b) after irradiation. The arrow indicates a carbon spiroid [101].

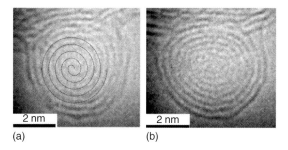

Figure 28.24 Archimedean spiral observed during electron irradiation of commercial furnace black particles. (a) A spiral with more than seven layers and (b) after complete transformation into an onion [101].

core in marked with a white arrow (Figure 28.23b). On continued irradiation by the electron beam, the spiroid particles were found to change into onions (Figure 28.24).

28.1.9
Nanodiamonds

Carbon onions can be converted into nanodiamonds [102, 103]. Banhart and colleagues reported that during electron irradiation of the carbon precursor the sputtering-induced loss of atoms and closure of the shells around vacancies lead to an extreme surface tension, which is responsible for the spherical shape of the onion. The exchange of carbon atoms between the shells and the cumulative action of all shells results in a decreasing interlayer spacing toward the core regions. Under continued intense irradiation, the structure evolves to a boundary state when the highly compressed onion cores eventually transform into cubic diamond crystals (Figure 28.25) [79, 102, 103].

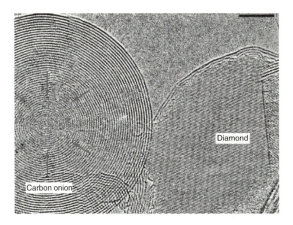

Figure 28.25 Carbon onions (left) self-compress under electron irradiation until, under favorable conditions, a diamond crystal nucleates in their core. The diamond continues to grow until the onion has wholly transformed to diamond. Scale bars, 5 nm [79].

28.1.10
Carbon Nanospheres

Well-shaped multishell carbon spheres with encapsulated copper nanocrystals can be synthesized via pyrolysis using copper(II)-phthalocyanine (CuPc) as precursor [104]. Using this method Schaper and colleagues synthesized Cu particles of 50 nm encapsulated in carbon nanospheres with an average size of 200 nm and a shell thickness in the range 60–80 nm. Figure 28.26 shows the copper-filled carbon spheres (a) and the energy-filtered transmission electron microscopy (EFTEM) mapping mode using the K-edge intensities of C (b) and Cu (c); the corresponding

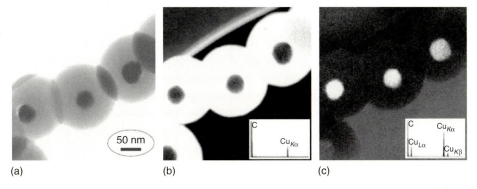

Figure 28.26 Copper-filled carbon nanospheres in a TEM micrograph (a), and in EFTEM images showing the C map (b), and the Cu map (c) with the corresponding EDX spectra [104].

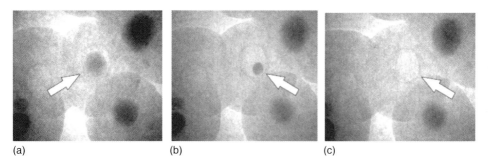

(a) (b) (c)

Figure 28.27 *In situ* melting and displacement at 850 °C under high-voltage electron irradiation of a copper nanocrystal (arrows) encapsulated in a graphitic carbon cage. Micrographs recorded before heating (a) after 7 s (b) and 21 s (c) [105].

energy-dispersive X-ray (EDX) spectra are inserted [104]. The encapsulated Cu material usually fills in the hollow carbon sphere entirely. After *in situ* high-temperature annealing, an increase in the degree of order was observed. The shell of the Cu-CNSs was found to be formed by a concentric carbon structure [104]. Under high-voltage electron irradiation and heating, the melting point of the enclosed nanosized copper reduced more than 200 K. During the melting, the copper migrated through the carbon shells, leaving intact the carbon cages with a central hole (Figure 28.27) [104].

28.1.11
Fullerides

Fullerenes can be fused when exposed to high temperatures and pressure forming ultrahard fullerene crystals (fullerites). In these crystals, the structural arrangement of the fullerene will depend on the type of the fullerene and temperature [105]. Van Tendeloo and colleagues studied the structure and the phase transitions in C_{70} fullerites at different temperatures [105, 106]. These authors observed that at room temperature the structure of C_{70} fullerites is a mixture of hexagonal and face-centered cubic arrangement of C_{70} fullerenes (Figure 28.28). They showed that around room temperature, the c/a ratio of the hexagonal phase depends on the sample temperature. Decreasing the temperature to ∼276 K, an ordering transformation occurs toward a monoclinic distorted HCP phase. At higher temperatures when the instantaneous rotation axis of the C_{70} fullerenes becomes almost isotropically distributed, the crystals undergo a shear transformation into a slightly rhombohedral crystal. At still higher temperatures, the crystal gradually becomes FCC.

Taking in account that the sample was cooled to liquid nitrogen temperature to be sliced for the analysis, these authors suggested that a shear transformation in the sense HCP → FCC might take place below the observation temperature, indicating that below the transition temperature the stable phase is HCP and above this temperature it is FCC [105].

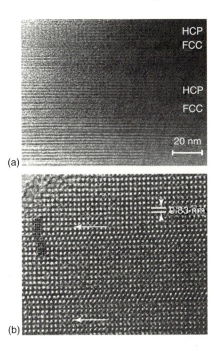

Figure 28.28 (a) C_{70} fulleride. Image along $[110]_{FCC}$ showing the interweaving of FCC and HCP bands. (b) HCP stacked region; stacking faults are indicated by arrows [106].

28.1.12
Carbon Nanoform Spatial Distribution in a Cell

Cheng and colleagues [107] combined cell viability assays with nanometer scale imaging to elucidate the physiological and structural effects of cellular exposure to carbon nanoforms. These researchers used EFTEM, HAADF-STEM, and tomography and confocal microscopy to generate 3D images enabling the determination of a carbon nanoform spatial distribution in a cell. They found that MWCNTs can penetrate the nuclear membranes, whereas C_{60} aggregates along the nuclear membrane (Figure 28.29). Viability tests showed increase in cell death via necrosis after exposure to MWCNTs, whereas the high-resolution images associate the spatial distribution of the carbon nanoforms within the cell with an increased incidence of necrosis. This combined approach might enable to probe the mechanism of particle uptaking and subsequent chemical changes within the cell, these are essential for identifying the toxicological profiles of carbon nanoform particles [107].

28.1.13
In situ Evaluation of Tensile Loading of Carbon Nanotubes

Muoth and colleagues developed a TEM-compatible platform that allows access for investigations on CNTs integrated in microelectromechanical system (MEMS)

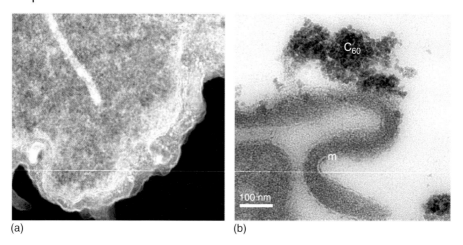

Figure 28.29 (a) HAADF-STEM tomographic projection of a MWCNT in the nucleus of a human monocyte-derived macrophage cell (HMM) and (b) zero-loss EFTEM. C_{60} clusters along plasma membrane of a HMM [107].

[108–110]. Using this platform, they determined the chirality and the displacement of an individual SWCNT under an applied actuation voltage in a transistor (Figure 28.30).

28.1.14
Observation of Field Emission Sites in MWCNTs

Using a Lorenz microscopy in conventional transmission electron microscopy mode, Fujieda and colleagues studied emission sites near the tip end of a single MWCNT while measuring the field emission currents. [111] They observed a marked fluctuation in the emission current above 20–30 μA, which was associated to structural changes at the tip of the MWCNT. Figure 28.31 shows that near the tip of the MWCNT the layers peeled off during field emission when under high emission current – the bright spots observed in the underfocused condition correspond to emission sites. When a new bright spot was observed near the main bright spot (Figure 28.31c) the emission current increased suddenly from 20 to 30 μA, suddenly decreasing when this spot disappeared (Figure 28.31d). Comparing the in-focus and underfocused images (Figure 28.31), these authors suggested that the peeled-off layers may function as the second emission site.

28.2
Analysis of Carbon Nanoforms Using Aberration-Corrected Electron Microscopes

The high operating energy necessary for atomic resolution imaging in conventional electron microscopes produces considerable knock-on damage in low Z materials

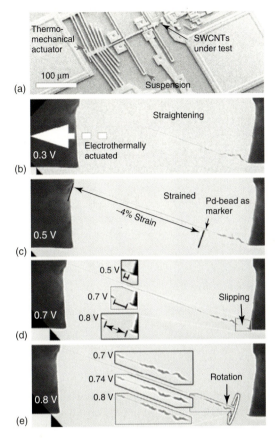

Figure 28.30 (a) Electrically heated thermomechanical MEMS actuator for TEM-compatible integration of SWCNTs. (b–e) TEM images recorded during *in situ* tensile loading of nanotubes with Pd markers that allow detecting slipping by extracting distances between markers and anchors. The nanotube started slipping at a strain of ∼5%. At 0.9 V actuator bias, resulting in a displacement of 0.38 μm in plane, which corresponds to an elongation in tube direction of 19% (initial slack), the nanotube slipped off the anchor. Thereafter, vibration of the now single-clamped nanotube was observed. The chirality of a tube rupturing after an increase in strain of ∼3% (not shown) was determined from electron diffraction to be (24,0). The corresponding diameter of 1.88 nm matches with the bright-field image 1.8 ± 0.2 nm [108].

such as carbon [112, 113], limiting the usable electron doses and hence the counting statistics of experimental data [114]. Before aberration correction, operation below the carbon knock-on threshold [115], that is, in low operating energy, implied a large loss of spatial resolution and, thus, prohibited atomic resolution imaging of closely packed lattices [114]. This situation changed with the development of microscopes equipped with aberration-corrected lenses. Aberration correction has opened new frontiers in electron microscopy by overcoming the limitations of conventional electron microscopy, providing subangstrom-sized probes and extending the information limit. The impact of the use of aberration correction

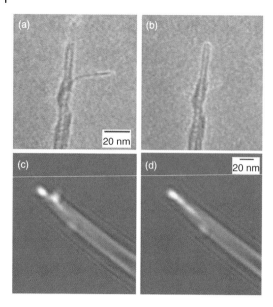

Figure 28.31 Structural changes of the tip of a MWCNT during field emission under a high emission current. (a) – (b) in-focus images and (c) – (d) underfocused images. The emission currents and applied voltages are: (a) I = 30 μA, V = −450V, (b) I=18 μA, V= −540.

in imaging carbon nanostructures can be easily seen from Figure 28.32; note that the images recorded using aberration correction have high-quality structural information. Figure 28.32a,b reproduces images of a DWCNT reported by the Tsukuba group, respectively, without and with aberration correction. The chiral indices of the small inner tube (4,3) (∼0.47 nm in diameter) were determined from the image recorded using aberration correction [54, 116, 117]. Figure 28.32c shows a STEM image where several carbon nanoforms can be observed.

28.2.1
Nanodiamond

Large-scale production of nanodiamonds can be achieved via detonation synthesis using high-energy explosives. Turner and colleagues [118, 119] showed that the product of this synthesis is a highly aggregated cluster of primary particles with sizes depending on the synthesis parameters (Figure 28.33). Using scanning transmission electron microscopy electron energy-loss spectroscopy (STM-EELS), these authors showed that during the detonation synthesis nitrogen is incorporated preferentially at the central part of the nanodiamond or at twin boundaries [118, 119]. The advances in microscopy on imaging a sp^3 carbon nanoform (detonation nanodiamond) can be seen in Figure 28.33, in which characteristics features of the nanoform are much easily identified from the image recorded using aberration correction.

28.2 Analysis of Carbon Nanoforms Using Aberration-Corrected Electron Microscopes

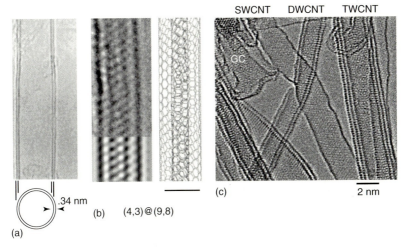

Figure 28.32 (a,b) Images of a DWCNT reported by the Tsukuba group (a) 1991 [54] and (b) 2008 Image (up), simulation (down) and schematic model (right) for a (4,3)@(9,8) [117]. (c) STEM image showing several carbon nanoforms: an SWCNT (18,18 armchair), DWCNT, TWCNT, and a graphitic cone (GC). (Courtesy, J.A. Rodriguez-Manzo and F. Banhart, Strassbourg).

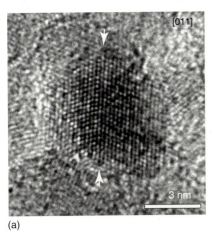
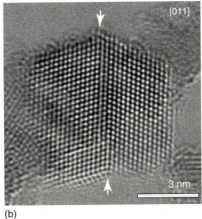

Figure 28.33 Detonation nanodiamond. (a) The main fraction of DND particles has a particle size close to 6 nm. Twinning is frequently observed in these small particles, with all visible twinning occurring over a {111} twin boundary (twin planes are indicated by white arrows in (a,b). Similar diamond cluster, but recorded after aberration correction. The diamond surfaces show minimal presence of graphitic material. (Courtesy of Turner and van Tendeloo, Antwerp).

28.2.2
Carbon Peapods

Peapods, can be described as fullerenes within CNTs (Figure 28.34). Their naming convention follows the fullerene community that use x@y to specify that x lies inside y [2]. Fullerenes are inserted in SWCNTs because they induce changes in the Fermi energy level and in the band gap of the host SWCNTs [120]. Similar to pristine SWCNTs, peapods are insoluble, which prevents their easy manipulation, processing, and integration. A common strategy to overcome this drawback is the grafting of functional groups on the outer wall of the nanotubes. However, doubts existed in the stability of the peapods during the functionalization due to their exposure to harsh chemical environments. Karousis *et al.* reported a strategy to covalently graft aryl moieties onto the external wall of C_{60}@SWCNT peapods, and based on TEM observation, they showed that the morphological characteristics of the C_{60}@SWCNTs remained unaffected [121]. Figure 28.34 shows the peapods that were functionalized by these authors [121]. The soft material at the SWCNT surface was attributed to organic moieties that are covalently attached onto the sidewalls of the peapods.

28.2.3
Determination of the Chiral Angle of SWCNTs

Sato and colleagues reported on a statistical analysis of the chiral indices (n,m) of metallic and semiconducting SWCNTs selectively separated via the density gradient ultracentrifugation process [122]. Their statistical analysis revealed that armchair (n,n) and chiral (n,n − 3) SWNTs with large chiral angles (>20°) are dominantly metallic nanotubes in the separated samples, whereas such a noticeable preference of particular indices was not observed for semiconducting nanotubes.

Figure 28.34 Encapsulated fullerene spheres inside SWCNTs and aryl addends as soft matter decorating the outer surface of SWCNTs. Inset: simulated image and an atomic model of aryl-functionalized C_{60}@SWCNT peapods [121].

28.2 Analysis of Carbon Nanoforms Using Aberration-Corrected Electron Microscopes

Using an aberration-corrected electron microscope operated at 80 kV, these authors visualized, without pronounced electron irradiation damages, the graphene sidewall of the SWCNTs as a moiré pattern. Figure 28.35 shows TEM images recorded on two separated SWCNTs of a batch that showed mainly optical absorption peaks characteristic of metallic nanotubes. The chiral angle (α) was determined from the characteristic bright spots forming hexagons in the fast Fourier transform (FFT) pattern, and the diameter (d) of the nanotubes directly measured on the TEM image. Simulated TEM images, as well as their FFT patterns, show fine agreement with the experimentally observed images (Figure 28.35). Furthermore, these authors showed that SWCNTs with identical structures but lying at different rotational angles about their axes can be distinguished Figure 28.36 [122].

28.2.4
Deformation of sp^2-Carbon Nanoforms during TEM Observation

Deformation of the sp^2-carbon nanoforms has been extensively reported [99, 100, 102, 103, 123, 124]. Suenaga and colleagues explained the deformation that occurs in SWCNTs during TEM observation analyzing the displacement of topological defects [116]. By means of high-resolution TEM with atomic sensitivity, these

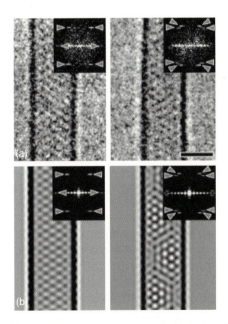

Figure 28.35 (a) TEM images of SWCNTs S1 (left) and S2 (right) and their FFT images (inset). (b) Simulated TEM images considering S1 (10,10) and S2 (11,8) and their FFT images (inset). Characteristic bright spots in the FFT images are indicated by triangles. Scale bar = 1 nm [122].

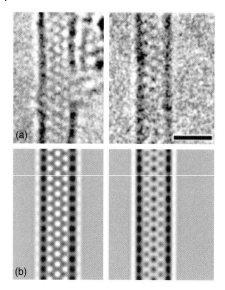

Figure 28.36 (a) TEM images of (6,6) SWCNTs and (b) simulated TEM images based on different orientations with respect to the incident beam. Scale bar = 1 nm [122].

authors imaged at 120 kV the displacement of pentagon–heptagon pair defects near a kink region in a SWCNT (Figure 28.37). Observing that these topological defects tend to gather around the kink region, they suggested that dislocation motions or active topological defects are responsible for plastic deformation of SWCNTs. However, they stated that in drastic transformations with extreme structural changes the motion of topological defects is not the principal mechanism and that the loss of carbon atoms by means of vaporization (due to local heating caused by the electron beam) or the knock-on [125] effect must be taken in account [116].

Aiming at understanding the effect of the electron beam irradiation on fullerenes, Liu and colleagues analyzed a hybrid carbon nanoform composed of a C_{60} fullerene molecule attached to the surface of a SWCNT via a pyrrolidine (Figure 28.38) [10, 126, 127].

Analyzing intact C_{60} molecules with different orientations at the CNT surface, these authors determined the structure of the C_{60} molecules. They observed that the pyrolidine functional groups were stable during observations at 120 kV, while for longer irradiation times (~30 s) or high electron doses the C_{60} molecule deforms [10]. Simulations suggested that the deformation appears after the transformation from C_{60} to C_{58}. The structural instability, continuous deformation, and displacement of the molecules during the observation prevent the determination of the deformation process of the C_{60} molecule (Figure 28.39). To overcome this drawback, a recording media with higher sensibility or lower accelerating voltage was suggested.

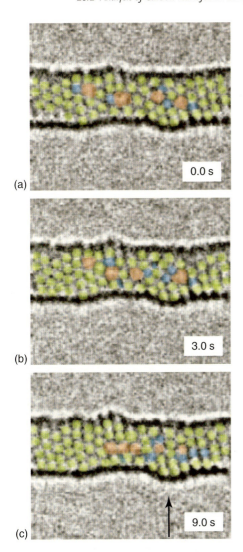

Figure 28.37 Sequential images of SWCNT active topological defects. (a–c) Heptagons or higher membered rings (red), hexagons (green), and pentagons or smaller rings (blue) of carbon atoms tend to gather around the kink structure (arrow) during observations [116].

28.2.5
Visualization of Chemical Reactions

A step further in electron microscopy is the visualization of chemical reactions. Koshino and colleagues showed that high-resolution electron microscopy can visualize chemical reactions by monitoring time-dependent changes in the atomic

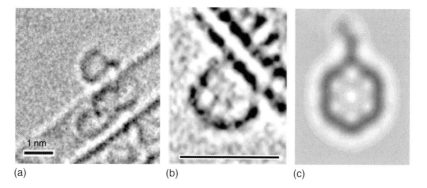

Figure 28.38 Functionalized fullerene molecules with pyrrolidine (C_{60}-C_3NH_7) attached to a SWCNT surface. (a) Image taken under 120 kV without Cs-corrector. (b) Image taken with corrector. The intramolecular structures are clearly visible for each fullerene. (c) Image simulation of the C_{60} fullerene [10].

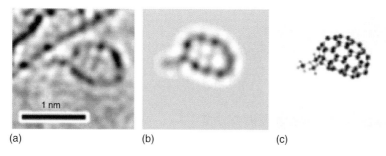

Figure 28.39 (a) HRTEM images of the degraded (or deformed) fullerene structures. (b) The simulated images are based on the assumption that C_{58} molecules probably form after losing two carbon atoms (or a carbon dimer). (c) Corresponding atomic model. The pyrrolidine type functional groups are attached arbitrarily in the image simulations [10].

positions of reacting molecules [128]. These authors report a study on bimolecular reactions of fullerene and metallofullerene molecules inside CNTs. Two distinct stages of electron-irradiated C_{60} transformations in a nanotube were observed: reversible bond formation – a bond is formed between two C_{60} molecules (dark contrast single line in Figure 28.40a), followed by an irreversible fusion – fusion of the molecules with multiple-bond formation (indicated by the red arrow heads, Figure 28.40b). By observing the displacement of the fullerenes during the dimerization process, these researchers determined the specific orientations in which two molecules interact and how bond reorganization occurs after their initial interaction [128]. Their studies on the concentration, specimen temperature, effect

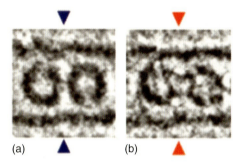

Figure 28.40 C_{60} dimerization: (a) reversible bond formation – single line between two C_{60} (blue arrow) and (b) irreversible bond formation multiple lines between C_{60} (red arrow) [128].

of catalyst (by endohedral metallofullerene), and accelerating voltage indicate that chemical reactions can be imaged under a variety of conditions.

28.2.6
Imaging at 30 and 60 kV

Although imaging at 80–120 kV reduces the probability of damage by irradiation, from the examples above it is obvious that deformation and motion of topological defects can take place. Recently, Sasaki and colleagues reported on the observation of carbon nanoforms at electron acceleration energies as low as 30 kV [129]. Figure 28.41 shows HAADF images of single Er atoms in a (Er@C_{82})@SWNT sample recorded at 30 and 60 kV [130]. The intensity in HAADF images is approximately proportional to the square of the atomic number Z, thus the Er atoms ($Z = 68$) have a brighter contrast than the lighter carbon atoms ($Z = 6$). In the image recorded at 30 kV (Figure 28.41a), single Er atoms in C_{82} fullerene cages are clearly visible (bright dots), but the fullerene structure is rather indistinct. At 60 eV,

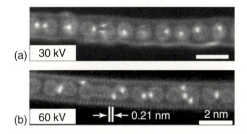

Figure 28.41 HAADF images of (Er@C_{82})@SWNTs taken at (a) 30 kV and (b) 60 kV. Bright dots in the images correspond to single Er atoms in C_{82} fullerene molecules; 0.21-nm-separated lattice fringes of an SWCNT are clearly visible in (b) [130].

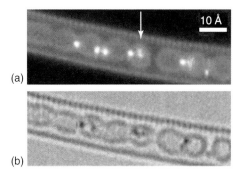

Figure 28.42 (a) Medium angle annular dark-field images (MAADF) (inner half-angle ~60 mr, outer half-angle ~200 mr). (b) BF image (collection half-angle ~5 mr) [114].

details of the structure of the carbon nanoforms are more easily observed – note the lattice fringes separated by 0.21 nm in the SWCNT wall.

Krivanek et al. [114] reported that the movement of single Er atoms in peapods can be observed in medium angle annular dark field (MAADF) images recorded at 60 kV (Figure 28.42). These authors associated the apparent irregular shape of the Er atoms in MAADF images to their displacement during the imaging. In Figure 28.42a, the white arrow indicates one Er atom that oscillates between two positions during the measurements. The image intensities of the atom at its two different positions are similar; this similarity was associated to an equal probability of the atom of spending about the same time in each position during the imaging.

28.2.7
Graphene

A single layer of graphite is called *graphene*. In graphene, the topological arrangement of three equally spaced planar carbon neighbors gives 120° bond angles, resulting in a planar array of tessellated hexagons. Depending on the lateral size of the graphene layer, the electronic, thermal, and mechanical properties are influenced by the atomic arrangement at the edge, the presence of impurity atoms as well as by the type and the number of defects [11, 20, 131]. Transmission electron microscopy is a unique technique to investigate the graphene topology, the nature of the defects, and the characteristics of the edges. Figure 28.43 shows a typical unfiltered image of a graphene recorded at 80 kV using an aberration-corrected, monochromated transmission electron microscope. The carbon atoms are in white, for better visualization, red points are used to label three equally spaced neighbors and one hexagon of the planar array. The additional structures delimited by high-intensity lines are adsorbates; holes can also be observed [132].

The electronic properties of a graphene flake depend on the number of layers that compose the flake [133]. Meyer and colleagues reported that the number of layers in a graphene flake is better determined by electron-diffraction analysis

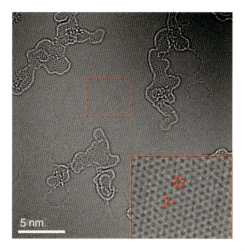

Figure 28.43 HRTEM image of a graphene sheet. The carbon atoms are in white. Red points are used to label three equally spaced neighbors and one hexagon of the planar array. Roles and adsorbates (structures delimitated by high-intensity lines) can also be observed. (Courtesy of R. Erni, Zurich).

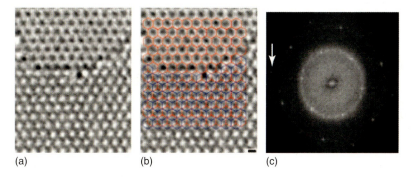

(a) (b) (c)

Figure 28.44 (a,b) Direct image of a graphene bilayer (atoms in white). Step from a monolayer to a bilayer, showing the unique appearance of the monolayer. In the bilayer region, white dots appear where two carbon atoms align the projection. Overlay of the graphene lattice (red) and the second layer (blue) (b). (c) Numerical diffractogram, calculated from an image of the bilayer region. The outermost peaks, one of them indicated by the arrow, correspond to a resolution of 1.06 Å. The scale bar = 2 Å [136].

because of ambiguities that can arise in direct TEM images [134–136]. These authors show that although the atomic patterns arising from areas with different layers are qualitatively different, they can be easily misinterpreted. Figure 28.44 shows the step from a single to a bilayer region, the patterns of the bilayer and the single layer area are different; however, at a different defocus, the pattern of the single-layer region appears like a bilayer (and vice versa). Figure 28.44c shows the Fourier transform of a bilayer region. The outermost set of peaks corresponds to an information transfer of 1.06 Å. The two innermost sets of hexagons correspond to 2.13 and 1.23 Å. It must be noted that although resolving the 2.13 Å reflection provides lattice images of graphene already in moderate resolution microscopes, the individual carbon atoms in graphene are resolved only if the second reflection at 1.23 Å is transferred [136].

28.2.8
Defects in Graphene

Real-time atomic-scale observation of the formation and dynamics of defects in graphene was also reported by Meyer and colleagues [136]. During sequential imaging at 80 kV of an unperturbed graphene lattice, they observed the formation of an isolated Stone-Wales (SW) defect (during one exposure of 1 s) that relaxed to the unperturbed lattice in the next exposure (4 s later) (Figure 28.45a–d). Conversely, in Figure 28.46, an unperturbed lattice was formed after a few seconds of observation. Most likely, electron interaction with a mobile adsorbate at the graphene surface induced its diffusion and replaced the missing carbon in the vacancy [136].

28.2.9
Graphene Edges

Koskinen and colleagues reported on the existence of reconstructed graphene edges other than pure zigzag or armchair edges [132, 137]. Theoretical modeling had predicted the existence of a reconstructed zigzag resulting from a lowering in the edge energy (reczag edge) (Figure 28.47) [137]. Besides lowering the energy, the reconstruction changes the edges' chemical properties. The dangling bonds, responsible for the reactivity of the zigzag edge, are removed by the reconstruction [138].

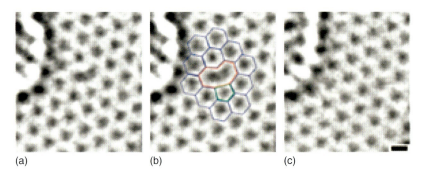

Figure 28.45 Metastable defects found in image sequences. Stone-Wales (SW) defect: (a) unperturbed lattice before appearance of the defect, (b) a SW defect (c) image in (b) with atomic configuration superimposed, and (d) relaxation to unperturbed lattice (after circa 4 s). The scale bar = 2 Å [136].

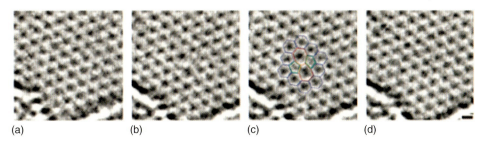

(a)　　　　　　　(b)　　　　　　　(c)　　　　　　　(d)

Figure 28.46 Metastable defects found in image sequences. Reconstructed vacancy: (a) original image and (b) with atomic configuration; a pentagon is indicated in green and vacancy in red. (c) Unperturbed lattice, 4 s later. This defect returned to the unperturbed lattice after 8 s. The scale bar = 2 Å [136].

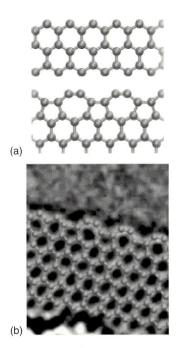

Figure 28.47 (a) Zigzag edge (schema up) and reczag (schema below). (b) Reczag with highlighted structure assignment [139].

During the TEM observation, electrons may kick out doubly coordinated atoms at the edges and the zigzag edges are repaired rapidly by diffusing single carbon atoms, while for armchair the repair requires two atoms; hence, its reparation is slower [132]. The reczag edge has an edge profile similar to armchair and is therefore less probable during TEM imaging. Furthermore, during TEM observation, the

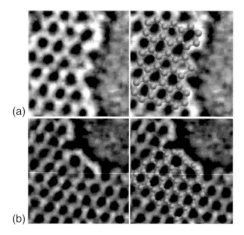

Figure 28.48 (a) Armchair edge (left) with structure assignment (right), revealing an armchair edge with pentagons. (b) Armchair edge reconstruction (left) with structure assignment (right), revealing an armchair edge with hexagons and two heptagons [139].

edge can be eroded so that graphene with pristine hexagons is revealed, favoring a zigzag instead of a reczag edge [137].

Huang and colleagues predicted a partial metastability for the armchair edge; the edge stress could be relieved by forming heptagons at the edge and this decreases the energy, attaining an energy minimum with ~30% heptagon concentration [137, 140]. In fact, Figure 28.48 shows an armchair edge with pentagons and an armchair edge with two adjacent heptagons [139].

28.2.10
Electronic and Bonding Structure of Edge Atoms

Site-specific single-atom spectroscopy was used by Kazu Suenaga and Masanori Koshino to investigate the electronic and bonding structure of the edge atoms localized at a graphene boundary with atomic resolution [114, 141]. Analyzing the carbon K-edge recorded on atoms at different positions these authors determined the spectroscopic fingerprint of single-, double-, and triple-coordinated carbon atoms (Figure 28.49). The absence of oxygen at the investigated edges was associated to *in situ* etching with continuous removal of the carbon edge atoms because of the electron beam irradiation used for the TEM observation.

28.2.11
Elemental Analysis of the Graphene Edges

Krivanek and colleagues [114] reported that annular dark field (ADF) imaging can be used to support the elemental analysis by EELS of impurities at graphene

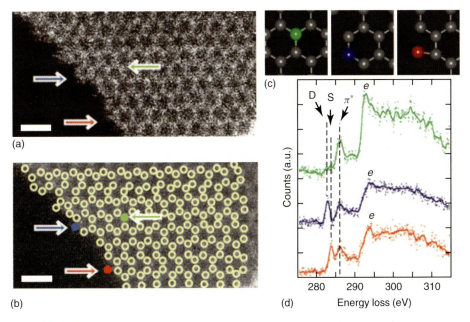

Figure 28.49 Graphene edge spectroscopy. (a) ADF image of single graphene layer at the edge region (without any image processing). Atomic positions are marked by circles in a smoothed image (b). Scale bars, 0.5 nm. (d) energy loss near edge structure (ELNES) of carbon K (1s) spectra taken at the color-coded atoms indicated in (b). Green, blue, and red spectra correspond, respectively, to carbon atoms with different states of atomic coordination: an sp² carbon atom, a double-coordinated atom, and a single-coordinated atom. These different states of atomic coordination are marked by colored arrows in (a,b) and illustrated in (c). The green spectrum exhibits the features of typical sp² coordinated carbon atoms, such as the sharp π^* peak around 286 eV and the exciton peak of σ^* at 292 eV. The spectrum in blue, recorded from an edge atom with two-coordination, has an extra peak labeled D, with both the π^* and the exciton (*e*) peaks having reduced intensities. The spectrum in red shows similar features, also with weaker π^* peak and broadened σ^* peak. Its extra peak (labeled S) occurs at a different energy position. The frequent change of edge morphology and the damages due to electron irradiation make the assignment of the atomic position not fully reproducible. Therefore, this energy state was related to damage occurring during the observations [141].

edges. They showed that the graphene edges are characterized by a variety of nonstationary atomic arrangements and impurity adatoms that depend on the atomic number Z can give much stronger contrast in the image than the carbon atoms (Figure 28.50a). In an intensity line profile recorded through an adatom that proved to be stationary during an MAADF imaging experiment, the intensity of the adatoms was 3.6 times larger than that of the C intensity (inset in Figure 28.50b). Using the dependence of the atomic intensity I on the atomic number Z reported in Ref. [142], these authors evaluated the $Z_{impurity}$ equal to 13.1, thus, allowing to identify the atom as an aluminum atom. They stated that their calculation results from an extrapolation so errors are expected. The tendency of the adatoms at the

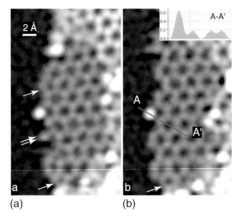

Figure 28.50 (a) MAADF image of the edge of monolayer graphene; the single arrows point to fivefold rings at the graphene edge, the double arrow to a single atom of carbon dangling off the graphene edge and (b) same area imaged about 2 min later. Inset: A-A′ line profile between the points A and A′ having an impurity atom near the point A. The impurity atom is localized at the edge. Both images were processed by a defogging filter and displayed slightly nonlinearly [114].

graphene surface to diffuse away under the electron beam irradiation prevented these authors, to record EEL spectra on the studied atom and on other impurity atoms with similar intensity.

28.2.12
Graphene Grain Structure

The large difference in size between grain boundaries and atoms prevents an easy analysis of the graphene grain structure. At a grain boundary, different grains stitch together predominantly through pentagon–heptagon pairs (Figure 28.51). Huang and colleagues reported that diffraction-filtered imaging can be used to rapidly map the location, orientation, and shape of several hundred grains and boundaries [143]. Figure 28.52 describes the imaging strategy proposed by these authors. In dark-field TEM individual grains are imaged using an objective aperture filter in the back focal plane to collect electrons diffracted through a small range of angles. The resulting real-space image (Figure 28.52c) will contain only the grains corresponding to the selected in-plane lattice orientations. Repeating this process using several different aperture filters, then coloring and overlaying these dark-field images (Figure 28.52d,e), Huang *et al.* created complete maps of the graphene grain structure, color-coded by lattice orientation showing grains connected by tilt boundaries as shown in Figure 28.52e. By correlating grain imaging with scanning probe and transport measurements, they showed that the grain boundaries severely weaken the mechanical strength of graphene but do not drastically alter their electrical properties [143].

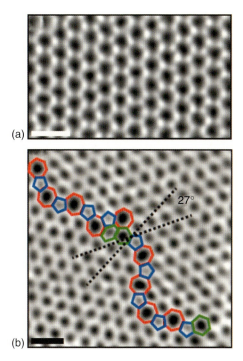

Figure 28.51 (a) ADF-STEM image showing the defect-free hexagonal lattice inside a graphene grain. (b) Two grains intersect with a 27° relative rotation: pentagons (blue), heptagons (red), and distorted hexagons (green). Low-pass-filtered to remove noise; scale bars, 5 Å [143].

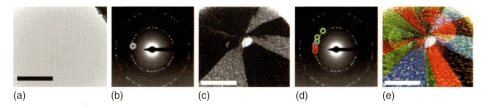

Figure 28.52 Large-scale grain imaging using DFTEM. (a) Samples appear uniform in bright-field TEM images. (b) Diffraction pattern taken from a region in (a) reveals that this area is polycrystalline. Placing an aperture in the diffraction plane filters the scattered electrons forming (c), a corresponding dark-field image showing the real-space shape of these grains. (d) Using several different aperture locations and color-coding them produces (e), a false-color, dark-field image overlay depicting the shapes and lattice orientations of several grains [143].

28.2.13
Engineering of Carbon Nanoforms: *In situ* TEM

28.2.13.1 Graphene

The *in situ* growth of graphene by a metal-catalyzed solid-state transformation of amorphous carbon was reported by Rodriguez-Manzo and colleagues [144]. Monitoring the formation and migration of metal islands (Ni, Co, or Fe) during the annealing of thin metal layers on amorphous carbon films, they observed the transformation of the underlying amorphous carbon film to graphene. Figure 28.53 illustrates the graphene grown from Ni after heating-induced metal migration. These authors showed that the nucleation and growth of graphene on the metal surfaces happen after the amorphous carbon film has been dissolved. The transformation of the energetically less favorable amorphous carbon to the more favorable phase of graphene occurs by diffusion of carbon atoms through the catalytically active metal [144–147].

Extending their work, the same authors showed that site-selective growth of graphene ribbons (GRs) can be implemented by metal lamellas deposited at selected sites. Figure 28.54 shows the retraction of a Ni metal lamella grown by *in situ* electron irradiation over an amorphous carbon layer. After retraction, the amorphous carbon layer has vanished and the vacated traces are occupied by graphene, whereas the surrounding layer is still amorphous.

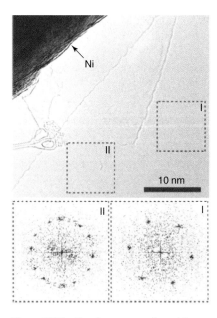

Figure 28.53 Graphene grown from Ni at 730 °C. Single- (I) and double-layer (II) graphene regions at the edge of a retracting Ni crystal. Diffractograms I and II (obtained by Fourier transformation from the dashed squares I and II) show the presence of a single layer (I) and a double layer (II), where the two planes are rotated by 25° relative to each other [144].

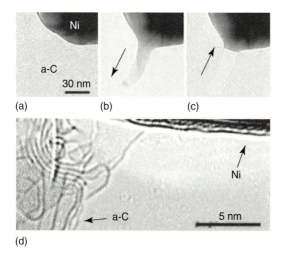

Figure 28.54 Growth and retraction of metal lamellae on amorphous carbon by electron irradiation and subsequent growth of graphene. (a) Ni crystal before generation of the lamella. (b) Lamella as created by electron irradiation. (c) Retraction of the lamella at 560 °C. (d) Double-layer graphene after retraction of a Ni lamella at 720 °C. Graphitic layers originating from carbon segregation on top of the Ni crystal are visible on the left [144].

28.2.13.2 Fullerene

The shape of the edges of the graphene sheet continuously changes while exposed to an 80-keV electron beam during TEM observations. In addition, as often observed, small flakes adsorbed at the graphene surface are fragmented and transformed in singular carbon nanoforms. Combining quantum chemical modeling with TEM visualization, Chuvilin et al. determined four critical steps in the mechanism of the transformation of a small graphene flake adsorbed at the graphene surface to a fullerene: first a loss of carbon atoms at the edge of graphene leads to the formation of pentagons; this triggers the curving of graphene into a bowl-shaped structure, and it subsequently zips up its open edges to finally form a closed fullerene structure (Figure 28.55) [148]. The authors showed that transformation occurs for flakes within a size range, because in large flakes (∼several hundred of atoms) the van der Waals interactions between the graphene flake and the underlying graphene refrain the curving step, while in very small flakes (less than 60 atoms) the excessive strain on C-C bonds imposed by the high curvature of small fullerene cage and the violation of the isolated pentagon rule suppresses the transformation.

28.2.13.3 Standing Carbon Chain

Carbon atomic chains were engineered from graphene by Jin and colleagues [149]. These authors showed that stable and rigid carbon atomic chains can be experimentally engineered by removing carbon row by row from graphene using electron beam irradiation inside a transmission electron microscope [149]. The dynamics of the formation of freestanding carbon atomic chains through electron

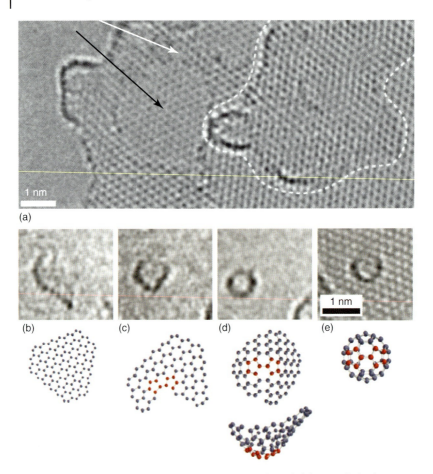

Figure 28.55 Experimental TEM images showing stages of fullerene formation directly from graphene. (a) The black arrow indicates a double layer of graphene, which serves as the substrate. The white arrow indicates a strip of graphene (monolayer) adsorbed on this substrate. The dashed white line outlines a more extended island of graphene mono- or bilayer, which has its edges slightly curved. (b–d) Consecutive steps showing the gradual transformation of a small graphene flake into fullerene (e). The graphene lattice is filtered out of images (b–d) for clarity. (e) The final product of graphene wrapping: a fullerene molecule on the surface of graphene. Quantum modeling of stages of fullerene formation [148].

beam irradiation is resumed in Figure 28.56. The stability of the carbon chains was associated to the sp hybridization of the carbon atoms in the chains – compared to sp^2, in sp hybridized atoms, there are more valence electrons distributed around the bonds corroborating to the bond strength [149]. While the edge carbon atoms of the graphene nanororibbon (GNR) (mainly sp^2 hybridization) were reported to be easier to be sputtered through local bond breakages, the relatively higher stability of the carbon chain may help it to sustain under the electron beam

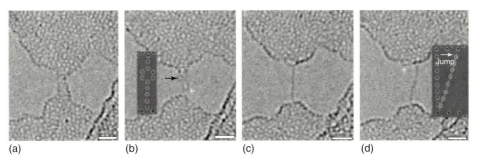

(a) (b) (c) (d)

Figure 28.56 Formation of freestanding carbon atomic chains. (a) A graphene ribbon formed between two holes was thinned row by row under continuous irradiation. (b) A carbon chain consisting of double strands was formed, and a knot remained on the left chain (marked by the black arrow). The right chain broke from its bottom end (marked by a white arrow) and migrated along the left chain to connect with the edge of the upper graphene. (c) The carbon chain was found to be linear and flexible. (d) The carbon chain made a jump along the graphene edge with a changing of edge bonding. Scale bar = 1 nm [149].

irradiation. The formation energies, migration, and breakage behavior of these chains were investigated based on DFT calculations [149].

28.3
Ultrafast Electron Microscopy

The main difference between the well-established spectroscopic pump-probe techniques and the ultrafast electron microscopy (UEM) is that electron, rather than photons are used to probe the excited sample. The use of electrons as probes allows the analysis of structural changes induced by the pump laser pulse at the atomic-scale. In a conventional electron microscope, the electrons are produced by heating the source of by field emission, resulting in randomly distributed burst of a continuous electron beam. In UEM, images are obtained stroboscopically with single-electron coherent packets that are timed with femtosecond precision, each electron has a unique coherence volume. Different variants of the UEM have been reported, here we present their use to visualize motion and study polarization properties of CNTs, further details can be found in Ref. [150].

28.3.1
Four-Dimensional Ultrafast Electron Microscopy (4D-UEM)

The four-dimensional ultrafast electron microscopy (4D-UEM) provides the ability to image structures with the spatial resolution of TEM, but as snapshots captures with ultrafast electron packets derived from a train of femtosecond pulses. This technique enables the study of nonequilibrium structures and transient processes. Using the 4D-UEM, Kwon and Zewail evaluated the structural changes induced in

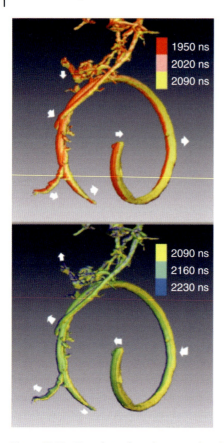

Figure 28.57 Time-dependent changes induced in a braceletlike carbon bundle by heating. Colors indicate temporal evolution. The wiggling motion of the whole bracelet is highlighted by arrows [151].

a bracelet-like bundle of carbon nanotubes by a heating (pump laser) pulse [151]. Upon absorption of the heating (pump laser) pulse lattice phonons are formed in a few picoseconds [152]. The irradiated region heats up rapidly (tens of picoseconds) inducing motion of the sample [151]. Kwon and Zewail showed that there are two prominent time scales in this motion. At early-times (less than 75 ns) a resonance "breathing-type" motion was identified. At longer times, the bracelet resonates on a slower time scale; the reversal of the motion was observed. On the microsecond time scale (from 1950 ns to 2090 ns), the bracelet wiggles as evidenced from changes in the volume density, which shows displacements in the same direction (Figure 28.57a). In the following 140 ns, the direction of the motion is reversed, revealing that the resonance motion is a wiggling of the whole bracelet around the anchored position (Figure 28.57b).

28.3.2
Photon-Induced Near-Field Electron Microscopy (PINEM)

Barwick and colleagues showed that the evanescent electromagnetic fields can be imaged with electron pulses when such fields are resolved in both space (nanometer and below) and time (femtosecond), establishing the PINEM [35, 36]. This variant of the UEM allows the probing of interfacial evanescent and ephemeral fields in a nanostructure. In PINEN the gain of electron energy is observed when electron-photon interaction takes place at the nanostructure, in this case they will show unique spatial, temporal and polarization properties of the nanomaterial. Barwic and collegueas showed that the precise spatiotemporal overlap of femtosecond single-electron packets with intense optical pulses at an individual CNT results in the direct absorption of integer multiples of photon quanta by the relativistic electrons, in this case, accelerated to 200 keV. By energy filtering only those electrons resulting from this absorption, it is possible to image directly in space the near-field electric field distribution, obtaining the temporal behavior of this field on the femtosecond timescale (Figure 28.58), and mapping its spatial polarization dependence (Figure 28.59).

28.4
Scanning Tunneling Microscopy (STM)

Scanning tunneling microscopy is often used when atomic scale resolution is needed. By combining scanning tunneling microscopy (STM) with scanning tunneling spectroscopy (STS) it is possible to probe both the structure and the electronic properties of carbon nanoforms. The chiral indices of the external wall of MWCNTs can be easily determined (Figure 28.60). In the specific case of DWCNTs, Giusca and colleagues [153] reported that by combining STS and STM the chiral index for the inner tube can be extracted from the additional van Hove singularities in the tunneling spectra.

The imaging of a solvent molecule (N-methyl-pyrrolidone, NMP) at the surface of CNTs was reported by Bergin and colleagues Figure 28.61 [154]. These authors reported that the solvent molecules observed at the CNT surface migrates and become irreversibly bound to the Si(100) surface, leaving behind defect-free tubes. On the basis of this observation, they suggested that the role of the sonication in the dispersion of the CNTs in solvents is to promote the insertion of solvent molecules into CNT bundles. Solvent intercalation weakens the intertube interaction, thus facilitating the formation of solvated tubes.

Cai and colleagues used an STM to follow the bottom-up fabrication of GNRs [155]. These authors propose a method for the production of atomically defined GRs, which uses surface-assisted coupling of molecular precursors into linear polyphenylenes and subsequent cyclohydrogenation [155–157]. They show that the topology, width, and edge periphery of the GNRs can be designed by

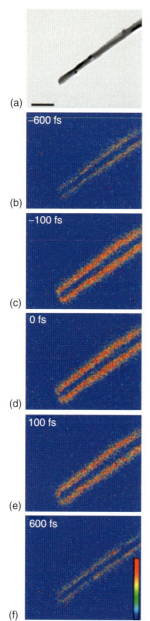

Figure 28.58 (a) Bright-field image; scale bar = 500 nm. (b–f) Five energy-filtered UEM images. False color: blue regions – no counts were recorded. The time of arrival of the electron packet at the nanotube relative to the clocking laser pulse is shown in the upper left corner of each image. The electric field of the clocking pulse was linearly polarized perpendicular to the CNT long axis [35].

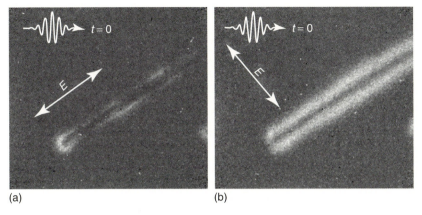

Figure 28.59 Polarization dependence of the imaged interfacial fields. Images taken when the E-field polarization of the femtosecond laser pulse is parallel to (a) and perpendicular to (b) the CNT long axis. Both polarization frames were taken at $t = 0$ (time of arrival of the electron packet at the nanostructure relative to the clocking laser pulse) [35].

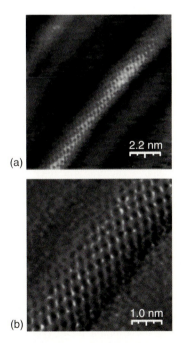

Figure 28.60 (a) STM image showing atomic resolution of a DWCNT. (b) Derivative image of a zoomed region of the DWCNT [153].

 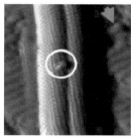 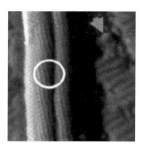

Figure 28.61 STM current images of SWCNTs deposited on Si(100) surfaces. Note the evidence of molecular migration on SWCNTs (circles). The grey arrow is a reference point on the surface [154].

engineering monomers [155]. Figures 28.62 and 28.63 show GNRs with different morphologies obtained from 10,10′-dibromo-9,9′-bianthryl (**1**), 6,11-dibromo-1,2,3,4-tetraphenyltriphenylene (**2**), and/or 1,3,5-tris(4″-iodo-2′-biphenyl)benzene (**3**) monomers. Using the monomer **1**, straight GNRs are produced (Figure 28.62). First, intermolecular colligation through radical addition are thermally activated – the dehalogenated intermediates diffuses along the Au(111) surface and form single covalent C-C bonds between each monomer; deviation from planarity explains the large apparent height (Figure 28.62b) [155]. The GNRs are obtained when the sample is annealed at 400 °C (Figure 28.62c). By using/combining different monomers, ribbons with other morphologies than the simple straight morphology is obtained (Figure 28.63).

The present bottom-up route to atomically precise GNRs involves modest temperatures (<450 °C) compatible with the current complementary metal–oxide semiconductor (CMOS) technology, and all fabrication steps are performed *in situ* so that the intrinsic properties of the GNRs can be probed on clean and well-defined substrates.

In addition to characterization of the electronic properties of pristine CNTs and synthesis of GRs, STM, and STS can be used in structural and electronic characterization of CNT hybrids. The structure of a DNA–CNT complex was reported by Yaratski and colleagues (Figure 28.64) [158]. These authors reported that strands of DNA wrapped around (6,5) nanotubes at ∼63° angle with a coiling period of 3.3 nm.

28.5
Scanning Photocurrent Microscopy (SPCM)

Park and colleagues used SPCM to investigate the spatial mapping of photocurrent (PC) generation and collection in graphene in a multielectrode geometry [51]. In SPCM, a focused beam induces localized generation of the PC, and its collection at different electrodes and gate potentials allows probing how the transport and collection of the PC is influenced by an electric potential profile. The authors

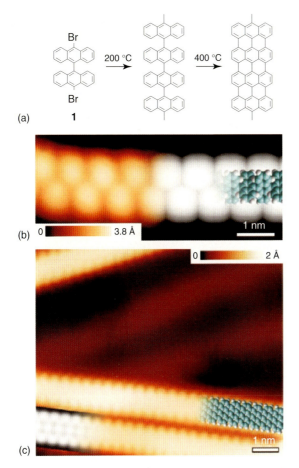

Figure 28.62 (a) Reaction scheme from precursor 1 to straight $N = 7$ GNRs. (b) STM image taken after surface-assisted C-C coupling at 200 °C but before the final cyclodehydrogenation step, showing a polyanthrylene chain, and DFT-based simulation of the STM image (right) with partially overlaid model of the polymer (blue, carbon; white, hydrogen). (c) High-resolution STM image after cyclodehydrogenation at 400 °C, with partly overlaid molecular model (blue) of the ribbon. At the bottom left is a DFT-based STM simulation of the $N = 7$ ribbon shown as a grayscale image [155].

observe that a strong electric field near metal-graphene contacts leads to an efficient PC generation, resulting in >30% efficiency for electron-hole separation. The polarity and magnitude of the contact PC are used to study the band alignment and graphene electrical potential near contacts. The graphene device used by these authors is shown in Figure 28.66. The graphene, which is visible as a dark shade, was contacted by eight Cr/Au electrodes. The SPCM image measured with a green laser ($\lambda = 532$ nm) is shown in Figure 28.66b. The electrical and optical images are shown overlaid. The gray light-reflection image shows the positions of all eight electrodes, while the color scale displays the PC $I_{C \to A}$ measured

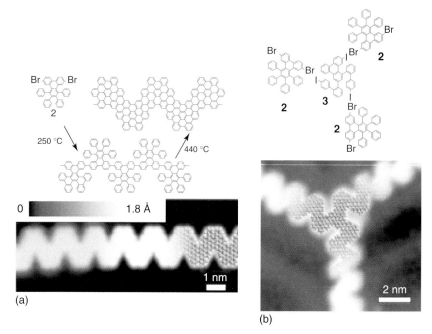

Figure 28.63 (a) Chevron-type GNR from tetraphenyl-triphenylene monomers (**2**). Up: Reaction scheme from monomer **2** to chevron-type GNRs. Bottom: STM image of chevron-type GNR fabricated on a Au(111) surface and a DFT-based simulation of the STM image (grayscale) with partly overlaid molecular model of the ribbon (blue, carbon; white, hydrogen). (b) Threefold GNR junction obtained from a 1,3,5-tris(4''-iodo-2'-biphenyl)benzene monomer **3** at the nodal point and monomer **2** for the ribbon arms. Up: Schematic model of the junction fabrication process with components **3** and **2** monomers. Bottom: STM image on Au(111) model (blue, carbon; white, hydrogen) of the colligated and dehydrogenated molecules forming the threefold junction overlaid on the image [155].

while only two electrodes (**C** and **A**) are connected to an external measurement circuit at zero bias. A strong PC signal with alternating polarity is observed along the edges of several electrodes, most notably **A–C**. The presence of the strong PC spots here is due to electron band bending and local electric fields located near electrode-graphene contacts. On laser illumination, electron-hole pairs are separated and then transported to opposite directions because of local electric fields, resulting in a net electrical current. The polarity of the PC is dependent on the direction of the band bending.

Figure 28.65 shows the total PC generation map of the graphene device when all eight electrodes are connected. The electrodes are faintly colored to show the positions of PC generation. As expected, the PC generation is strong along the edges of each electrode; however, the PC generation is stronger at the corners of electrodes, compared to the straight edges. These authors associated this behavior to a 50% increase in graphene area directly exposed to the light near the corners and to the stronger electric field near the sharp corners of the metal electrodes.

28.5 Scanning Photocurrent Microscopy (SPCM)

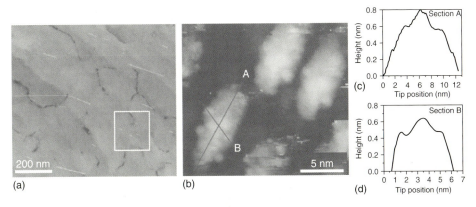

Figure 28.64 (a) AFM image of the same CNT solution deposited onto a Au(111) substrate (b) 21 × 21 nm STM topographic image of CNT-DNA hybrids on a Si(110) substrate acquired. Height profiles along the lines A (c) and B (d) [158].

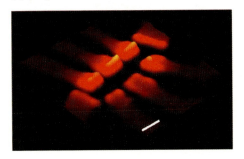

Figure 28.65 Map of PC generation efficiency when all eight electrodes are connected for PC measurement. Scale bar = 2 μm [51].

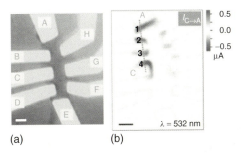

Figure 28.66 Photocurrent imaging of a graphene device. (a) Optical image of the graphene device (the channel defined by electrodes is 12 μm long and 2.5 μm wide); scale bar = 2 μm. (b) Combined light reflection and photocurrent image (ICfA; scale bar = 4 μm) measured with a 532 nm laser (230 μW) and a 60× objective with NA 0.7 [51].

28.6
X-Ray Electrostatic Force Microscopy (X-EFM)

The imaging of electron trapping centers on distorted CNTs using X-ray excited scanning probe microscopy in the X-ray electrostatic force microscopy configuration (X-EFM) was reported by Ishii and colleagues [159]. The probe detects the Coulomb force that results from the relaxation of an electron bound to a defect site into the core-hole state created by X-ray photon absorption. Figure 28.67 shows the AFM and X-EFM images of a CNT branch. The X-EFM image was obtained at $h\nu = 290$ eV. The X-EFM image reveals the distribution of electron trapping centers, while the AFM image shows the topography of the branch. On overlapping the AFM and X-EFM images, it is found that the electron trapping centers are at the branches on the torsional CNT. This finding indicates that an external force that deforms the CNTs degrades their electric properties.

28.7
Atomic Force Microscopy

An important feature of the atomic force microscopy (AFM) is it association to other techniques such as scanning tunnelling microscopy or Raman spectroscopy. AFM was developed to obtain surface topography at the nano/atomic scale. However, gradually, the application of AFM has been greatly expanded to realize functional characterization; this is made possible through modifications of the probe tip that change or augment its interaction mechanism with the target surface. Modified probes open the possibility to explore a range of surface physical and chemical properties at the nanoscale.

28.7.1
Conductance Atomic Force Microscopy

In conductance atomic force microscopy (C-AFM), the current passing between a conducting cantilever and the sample generates a conductance image. Figure 28.68

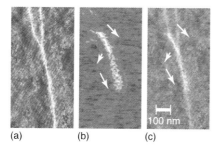

(a)   (b)   (c)

Figure 28.67 (a) AFM and (b) X-EFM images of CNT branch. These images are overlapped in the frame (c) [159].

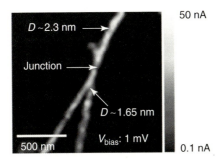

Figure 28.68 C-AFM image on a pristine sparse carbon nanotube network. Current map of a nanotube bundle (D ∼ 2.3 nm), which is intersected by an individual tube (D ∼ 1.65 nm). The chromatic scale shows a decrease in the conductance at the junction [38].

shows a conductance mapping image recorded from a sparse network of SWCNTs studied by Nirmalraj and colleagues [38]. They found that the transport in the sparse CNT network is dominated by the resistance at the network junctions [38].

28.7.2
Chemical Force Microscopy

Chemical force microscopy (CFM) overcomes the chemical nonspecificity of AFM by covering the tips with molecular groups having specific interactions with the surface species to be studied. Poggi and colleagues, used CFM to evaluate the adhesion forces between alkanethiols modified AFM cantilevers and SWCNTs [160]. These authors showed that adhesive interactions between the sidewall of SWCNTs and self-assembled alkanethiol monolayers (SAMs) depend on the identity of the terminal group of the SAM (NH_2 > OH > CH_3) [160]. Topographical and adhesive force measurements were acquired simultaneously on SWCNT paper [161] using chemically modified AFM probe tips with a series of ω-substituted alkanethiols or para-substituted arylthiols. They showed that there is a direct correlation between the topographical and adhesion force images (Figure 28.69).

28.7.3
Magnetic Force Microscopy

Taking into account that CNTs interact with a magnetic field, Lillehei and colleagues used a magnetic force microscope (MFM) for imaging the dispersion of CNTs in polymers composites (Figure 28.70) [162, 163]. The nanotubes were often found as agglomerates throughout the film. Individual nanotube bundles and regions of localized high nanotube concentration are clearly visible only in the MFM

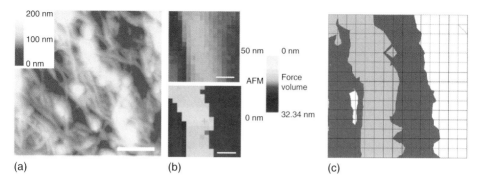

Figure 28.69 (a) AFM topographic image of the SWCNT paper acquired in contact mode using a clean, gold-coated tip showing individual and bundles of SWCNTs. Scale bar = 250 nm; scan domain: 1.0 × 1.0 µm. (b) Topographical (up) and force volume (bottom) images acquired using a hydroxyl-terminated, alkanethiol-functionalized gold-coated tip. Scale bar = 12 nm. Scan domain: 50 nm × 50 nm. (c) Adhesion force map of the data presented in (b).

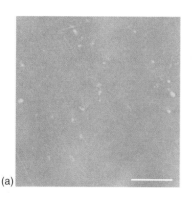

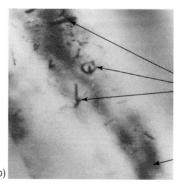

Figure 28.70 MFM image on a carbon nanotube polymer composite film: (a) topography and (b) phase showing localized SWCNT bundles via mapping the magnetic field gradient. Scale bar = 2.0 µm [162].

phase image (Figure 28.70). The ability to track the nanotubes in the polymer was found to be a function of the magnetic field strength used to induce the magnetic field in the nanotubes. Qualitatively, the contrast between the nanotubes and the background was lost after 15 nm of lift height. Using nonmagnetic tips, the images were not reproduced, indicating that the response is indeed due to mapping the magnetic field gradient above the sample and not due to contact interactions.

28.8
Scanning Near-Field Optical Microscope

Schreiber and colleagues recorded the first Raman spectrum of an isolated SWCNT using near-field optical Raman spectroscopy in 2003 [165]. In order to overcome the drawback of low spatial resolution ~500 nm associated with Raman spectrometers combined with common optical microscopes, these authors combined a Raman spectrometer with a scanning near-field optical microscope (SNOM). In a simple comparison to another near-field microscope, the AFM, the SNOM used by these authors had a metallized fiber-optical taper as a probe and a "shear-force mechanism" to lock the distance tip-sample. The sample was excited by a far-field focused laser beam and the Raman signal detection of the optical near field was collected by the metallized fiber through an aperture of near 50 nm. This aperture was used to select the sub-wavelength signals only and reject the far-field signals. To enhance the Raman signal a silver SERS surface was used; the elastic optical intensity is recorded during the scan. They observed that in the near-field set-up, the Raman spectra can be recorded only when the SWNT coincided with "hot spots", i.e., with sites of local field enhancement (Figure 28.71). Still in 2003, another approach, perhaps more pertinent to obtain an image rather than to obtain a localized spectrum, was proposed by Hartschuh and colleagues.[161] In their approach, a metallic tip without aperture was used allowing excitation and collection in the far-field mode. Using the tip to enhance locally the electric local fields, the excitation as well as the inelastic Raman signals (TERS for tip enhancement Raman spectroscopy) these authors recorded AFM images correlated with their respective hyperspectral Raman maps. Nowadays this approach is more used due to the straightforward combination of an AFM microscope with a Raman confocal device.

28.9
Tip-Enhanced Raman and Confocal Microscopy

Hartschuh and colleagues analyzed different types of SWCNTs via TERS [164, 166]. Chemical vapor deposition and arc-discharge-synthesized CNTs show different response to the laser excitation at 633 nm (Figure 28.72). TERS images of

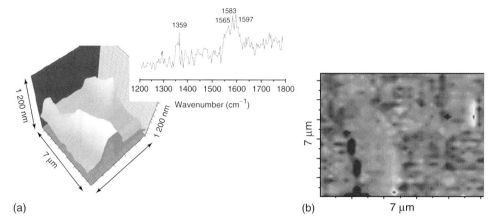

Figure 28.71 (a) Sample of SWCNTs deposited on a silver SERS surface, characterized by SNOM, and (b) the corresponding optical image. Schreiber and colleagues [165] proposed that the spectrum recorded when the intensity was maximum was generated from either one or two individual nanotubes. The observed shift in the G mode was associated with the use of the SERS surface. These surfaces induce an up-shift of all the Raman bands of well-dispersed SWCNTs. They suggested that these bands may not have a different origin as those observed on SWCNTs in the far-field setup [165].

arc-discharge-grown SWCNTs showed dependence of the intensity on the laser polarization: tubes oriented perpendicular to the laser polarization did not produce any signal (see arrows in Figure 28.72a,b). This polarization contrast was associated to polarization properties of the G and G'Raman bands [170, 171]. On the contrary, confocal Raman imaging of SWCNTs grown by CVD was not possible. However, as shown in Figure 28.72c for CVD grown SWCNTs, the contrast and resolution of Raman images are greatly increased when a sharp silver tip is near the laser focus. Besides the SWCNTs, the topographic image shows a large number of small circular features with a height of ∼2 nm, which are associated with condensating water (Figure 28.72c,d). It is important to notice that vertically *and* horizontally oriented SWCNTs are observed in the Raman image with similar signal intensity even though the incident laser light is oriented vertically (Figure 28.72c).

Localized defects in CNTs were imaged by Georgi and Hartshcuh [168]. They used tip-enhanced Raman spectroscopy (TERS) to study defect induced D-band Raman scattering in metallic SWCNTs. Figure 28.73 shows simultaneously acquired topography and near-field Raman G-band and D-band images. A localized defect site was photoinduced in the lower part of the nanotube (B). The D-band intensity I_D strongly increased around position B, whereas the G-band remained rather uniform along the SWCNT. The intensity ratio I_D/I_G, was used for the quantification of defect densities in sp^2 carbon structures [168, 172].

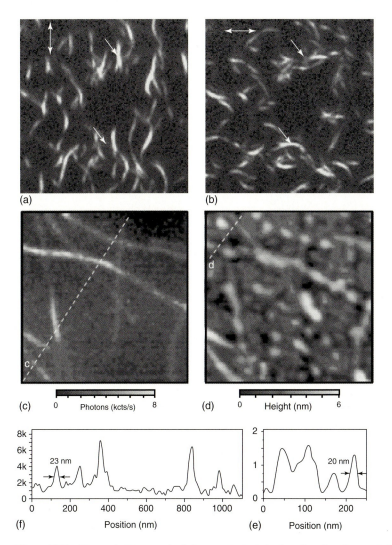

Figure 28.72 Raman images acquired detecting the intensity of the G'band on laser excitation at 633 nm. (a,b) Arc-discharge SWCNT bundles: confocal Raman images (raster scanning without metal tip, scan area 15 × 15 μm) for vertical (a) and horizontal polarization (b) of the laser excitation at 633 nm. The arrows indicate tubes that are oriented completely perpendicular to one polarization direction demonstrating the polarization contrast. (c,d) Simultaneous near-field Raman image (c) and topographic image (d) of CVD SWNTs (scan area 1 × 1 μm). Cross section taken along the indicated dashed line in the Raman image (e) and topographic image (f). Vertical units are photon counts per second for (e) and nanometer for (f) [166].

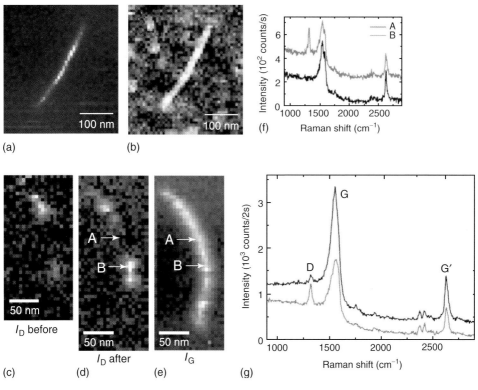

Figure 28.73 (a) TERS Raman G-band image of a metallic SWCNT detected by an avalanche photodiode (APD) after band pass filtering at 1580 cm^{-1}. (b) Simultaneously acquired topography. (c–e) Spectroscopic images of a single SWCNT before and after generation of an additional defect. (f) Spectra from positions A and B, representing defect-free and highly defective positions. (g) Spectrum at position B before (upper) and after (lower curve) the near-field-induced generation of a localized defective region [168].

28.10
Tip-Enhanced Photoluminescence Microscopy

The energy transfer in pairs of semiconducting nanotubes was studied by Qian and colleagues [173]. They investigated the distance-dependent energy transfer between two nanotubes of different (n,m) forming a bundle. Chirality (n,m)-specific PL imaging of SWCNTs allows identification and location of different nanotubes. Figure 28.74 shows the simultaneously detected, energy resolved image data for (9,1) and (6,5) nanotubes forming a bundle. The energy transfer efficiency for different nanotube–nanotube distances was evaluated from the intensity of the (9,1) nanotube acting as energy donor – an efficient transfer was observed only inside a few nanometers [173].

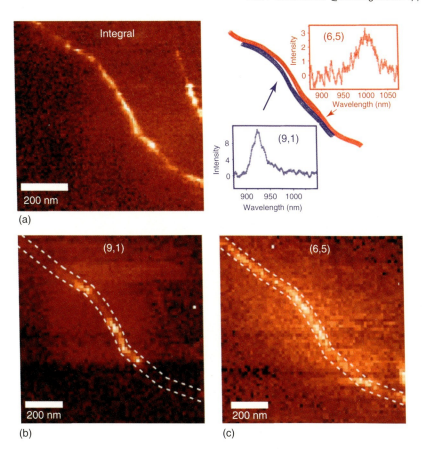

Figure 28.74 (a,b) Near-field PL image (a) of DNA-wrapped single-walled carbon nanotubes created by detecting all emission between 860 and 1050 nm on laser excitation at 632.8 nm. Near-field PL images of the same sample area obtained by measuring spectra at each pixel reveal that the structure is a nanotube bundle composed of two nanotubes with different chiralities: blue spectrum (9,1) and red spectrum (6,5) nanotubes (c) [173].

28.11
Fluorescence Quenching Microscopy

Fluorescence labels can be used to enhance or enable the visibility of objects of interest in fluorescence microscopy [174]. However, graphitic systems are naturally efficient fluorescence quenchers for dye molecules [175, 176]. Taking advantage of the fluorescence quenching effect, Kim and colleagues reported on the use of the FQM to analyze GRs [48, 49]. Instead of being labeled to be bright, the GRs are made dark utilizing their strong quenching effect of fluorescent dyes. Since the fluorescence quenching effect by GRs is strong enough to generate dark contrast

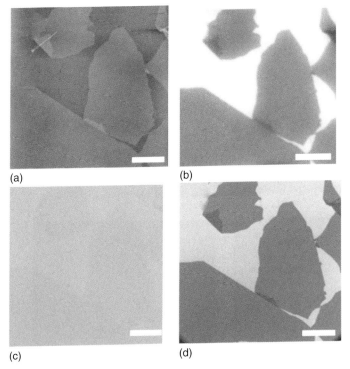

Figure 28.75 (a) AFM image of GO sheets deposited on SiO$_2$/Si substrate showing single layers and folded double layers. (b) FQM image of the same area after spin-coating a fluorescein/PVP thin film. (c) The reflectance optical microscopic image. FQM image can offer much improved contrast and layer resolution comparable to AFM. (d) SEM images. All scale bars = 10 μm [49].

against the bright background, FQM produces high-contrast images comparable to those taken by a scanning electron microscope (SEM) or AFM (Figure 28.75); the contrast in FQM allows layer counting [49]. Figure 28.75 shows the same sample areas acquired by (a) AFM, (b) FQM, (c) reflectance optical microscopy [46], and (d) SEM. It can be clearly seen that FQM offers a drastic improvement over reflectance optical microscopy.

28.12
Fluorescence Microscopy

Fluorescence microscopy can be used to identify the positions of target objects in both solid and liquid form, which has been especially useful for biological samples [49, 177]. Using selected fluorescent labels that prevent fluorescence quenching, fluorescence microscopy can be used to visualize SWCNTs in aqueous

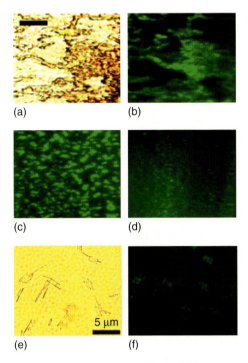

Figure 28.76 (a) Transmitted bright-field image of c-SWCNT-QD; SWCNT bundles are the black features. (b) Fluorescent image of same region in c-SWCNT-QD. (c) Fluorescent image of d-SWCNT-QD. (d) Fluorescent image of QD suspended in SDS alone. Fluorescent images show the decreasing SWCNT cluster size from (b) to (c). Almost no features can be seen in (d) because QDs are monodispersively distributed. Scale bar ~10 μm. Spin-coated d-SWCNT-QD on glass slides: (e) fluorescent and (f) transmitted bright-field images of individual SWCNTs and their ropes. [176].

solutions. Visualization of the CNT dispersion is an important step toward their optimal integration in actual applications. Chaudhay and colleagues reported the use of fluorescent core-shell CdSe-ZnS nanocrystals (QD) as label for the visualization of CNTs in aqueous solution [177]. Mercaptoacetic-acid-stabilized QDs were utilized for fluorescent visualization of SWCNTs dispersed in 1% solution of SDS. These authors studied two kinds of dispersion of SWCNTs: c-SWCNT-QDs (concentrated uncentrifuged solution) and d-SWCNT-QDs (supernatant containing mostly individual SWCNTs/SWCNT ropes). Figure 28.76a–d shows transmitted bright-field and fluorescence images of SWCNTs in solution. The size of fluorescent clusters decreases from Figure 28.76b,c. The cluster size of the bundled SWCNTs (i.e., dispersion) can be evaluated from the bright areas in Figure 28.76b,c. In another approach, the d-SWNT-QD solution was spin-coated on glass slides, resulting in individual SWCNTs/SWCNT ropes labeled with fluorescent QDs across their length spread across the slide (Figure 28.76e,f) [177].

28.13
Single-Shot Extreme Ultraviolet Laser Imaging

Recent advances in the development of high peak brightness extreme ultraviolet (EUV) and soft X-ray (SRX) lasers have opened new opportunities for compact full-field EUV/SRX microscopes capable of capturing images with short exposures down to a single laser shot [178–182]. Brizuela and colleagues using a zone-plate EUV microscope captured with a single laser pulse of ∼1 ns duration the image of a 50 nm CNT (Figure 28.77) [183, 184]. The combination of near-wavelength spatial resolution and high temporal resolution opens up numerous opportunities in imaging, such as the ability to directly investigate dynamics of nanoscale structures.

28.14
Nanoscale Soft X-Ray Imaging

The foundations for X-ray imaging at the nanoscale were laid by the work Schmahl and colleagues, who pursued the development of full-field transmission X-ray microscopy (TXM) [187, 188], and Kirz and colleagues, who investigate scanning transmission X-ray microscopy (STXM) [189, 190]. Both groups employed zone plate optics and synchrotron radiation [191–193]. Spatial resolution were around 100–200 nm in the early works, moving towards 50 nm in the late 1980 and finally towards 10 nm today [194]. Combined with near edge absorption X ray fine spectroscopy (NEXAFS), both set-ups, i.e. TXM and STXM, have been used to study carbon nanostructures [40–43].

28.14.1
Near-Edge Absorption X-Ray Spectromicroscopy

Felten and colleagues reported on the use of scanning transmission X-ray microscopy (STXM) to study the electronic and structural properties of CNTs [40–43].

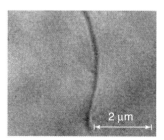

Figure 28.77 Transmission images of a carbon nanotube on a Si membrane obtained with a single laser shot [183].

28.14 Nanoscale Soft X-Ray Imaging

STXM combines both NEXAFS and microscopy with a spatial resolution better than 30 nm. NEXAFS spectroscopy is a well-adapted probe of CNTs, as it is able to investigate both electronic and structural properties of carbon-based systems and can provide information about the composition of organic materials.

An STXM analysis of a MWCNT powder synthesized by arc discharge is shown in Figure 28.78. The TEM image (Figure 28.78a) shows that the powder mainly contains bundles of straight nanotubes and impurities. The quantitative chemical maps of the three different forms of carbon present in the sample can be seen in Figure 28.78a–d. These images are obtained by fitting the C 1s stack to spectra of the three components. The gray scale of each component map gives the thicknesses in nanometers, assuming a density of the materials equal to $1.4\,g/cm^3$. The color composite map is constructed to visualize their respective spatial distributions. Here the red, green, and blue colors correspond to the spatial distributions of MWCNT, onionlike carbon particles [40–43].

The spatial resolution of the technique enables the study of individual CNTs by avoiding contributions from other components of the raw powder, while the high-energy resolution (~ 0.1 eV) allows differentiation among very similar nanostructures.

Felten and colleagues evaluated the defect density in an individual CNT that was irradiated in a focused ion beam system [40–43]. By measuring the polarization dependence (linear dichroism) of the C 1s $\to \pi^*$ transition at specific locations along individual CNT with an STXM, they determined a quantitative relationship between ion dose, nanotube diameter, and defect density. Figure 28.79 shows a TEM image of a CNT after site-selective irradiation; STXM stacks (a sequence of images over a range of photon energies) and HRTEM images were recorded for the fully irradiated (site 3), partially irradiated (site 2), and pristine (site 1) segments as indicated by the black boxes and circles, respectively. For each region, C 1s stacks were recorded with both parallel and perpendicular orientations of the E-vector relative to the tubes. The corresponding NEXAFS spectra and HRTEM images of fully irradiated (site 3), partially irradiated (site 2), and pristine segments (site 1)

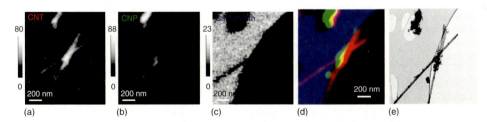

Figure 28.78 CNT, CNP, and carbon film quantitative chemical maps derived from the C 1s stack. (a–c) The gray scales for each component are given in nanometers. (d) A colored composite map is also included with a single thickness scale (0–88 nm) for all three components/colors. The red, green, and blue colors correspond respectively to nanotube, carbon nanoparticles, and carbon grid rich region. (e) Corresponding TEM image [44].

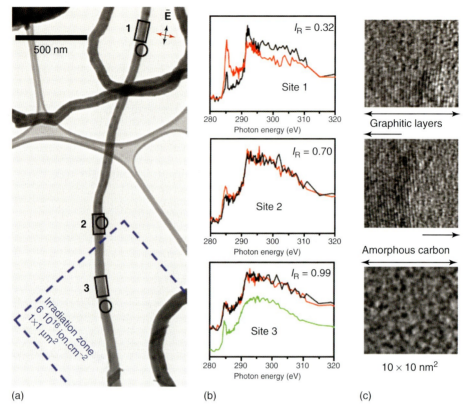

Figure 28.79 (a) Transmission electron microscopic image of a MWCNT after site-selective irradiation of a 1 × 1 μm² zone (indicated in blue). STXM stacks and HRTEM images were recorded for the fully irradiated (site 3), partially irradiated (site 2), and pristine (site 1) segments. (b) C 1s NEXAFS spectra extracted from the three different sites. The red and black curves correspond, respectively, to the nanotube oriented perpendicular and parallel to the E-vector. The ratio I_R between parallel and perpendicular π^* intensities is calculated for each region. The green curve at the bottom was recorded from the amorphous carbon of the supporting TEM grid for comparison purpose. (c) High-resolution TEM images of the three regions indicated by the black circles. The images are 10 × 10 nm² [40].

are shown on Figure 28.79b,c. The red spectra correspond to the nanotube oriented perpendicular to the E-vector and the black spectra to the nanotube oriented parallel to the E-vector. The dichroic ratio ($I_R = I_\parallel/I_\perp$, $I_\parallel(I_\perp)$ π^* intensity when the long axis of the nanotube is parallel (perpendicular) to the E-vector of the incident polarized light) calculated for each site is 0.99, 0.70, and 0.32 respectively. The I_R value for the irradiated section is very close to 1, which indicates that almost all sp² character is lost: this segment was transformed into an amorphous carbon nanowire by ion bombardment, as confirmed by the HRTEM analysis [40–43].

28.15 Scanning Photoelectron Microscopy

In scanning photoelectron microscopy (SPEM), the high spatial resolution is achieved by demagnifying the photon beam to submicrometer dimensions [185]. Such microscopes use the power of the surface-sensitive synchrotron-based photoelectron spectroscopy to obtain chemical characterization of the systems under investigation with submicron spatial resolution and probing depth up to a few nanometers [185, 186].

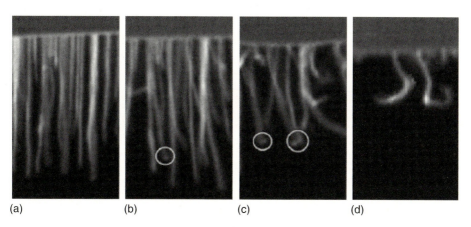

Figure 28.80 Sequence of C 1s images (6.4 × 12.8 µm) of an aligned CNT array (a) before and after exposure to increasing oxygen doses at 200 °C: (b) 2.7×10^{17}, (c) 5.4×10^{17}, and (d) 8.1×10^{17} atoms cm^{-2}. The most intense signal (bright) originates from the MWCNTs. The silicon substrate on top appears dark gray (signal coming from the secondary photoelectron emission), whereas the vacuum around appears dark (no signal) [186].

Using SPEM, Barinov and colleagues studied the formation of oxygen functional groups and the morphological changes of aligned MWCNTs on exposure to atomic oxygen at different temperatures [185]. The C 1s images shown in Figure 28.80 illustrate the impact of the interaction of atomic oxygen with the CNT at 200 °C. The exposure to oxygen atoms leads to wide tips and shorter CNTs. Further exposure causes thinning and bending, until only CNT fractures remain. The type and abundance of oxygen functional groups (ether, epoxy, and carbonyl) are strongly influenced by the density and type of the defects at the CNT surface, thus suggesting an approach for tailoring CNTs via controlled introduction of defects, which could favor the formation of a preferred functional group [185, 186].

> .... Therefore carbon is the key element of living substance: but its promotion, its entry into the living world, is not easy and must follow an obligatory, intricate path ....
>
> Primo Levi

Acknowledgments

We thank our colleagues who have provided material for this review: H.-P Boehm, Rolf Erni, Florian Banhart, Susana Trasobares, Daniel Ugarte, Kazu Suenaga, Alexander Gloter, Mathieu Muoth, Christophe Hierold, John Boland, Renata Papaleo, Chris Ewels, Nikos Tagmatarchis, Pascal Ruffieux, Bernard Humbert, Stuart Turner, Xiaoxing Ke, and Irene Suarez-Martinez. We especially thank Maureen Lagos for the animated discussions on electron microscopy. We are grateful to André De Munck for his help in editing and adapting the images. The support of MP901 and the European Commission under contract 026019 (ESTEEM) is gratefully acknowledged.

References

1. Castro Neto, A.H. (2010) The carbon new age. *Mater. Today* **13** (3), 12–17.
2. Suarez-Martinez, I., Golbert, N., and Ewels, C.P. (2011) *Handbook of Carbon Nanomaterials: Synthesis and Applications*, Panstanford Press, Singapore.
3. He, H.K., Sekine, T., and Kobayashi, T. (2002) Direct transformation of cubic diamond to hexagonal diamond. *Appl. Phys. Lett.*, **81** (4), 610–612.
4. Chung, D.D.L. (2002) Review graphite. *J. Mater. Sci.*, **37** (8), 1475–1489.
5. Falcao, E.H.L. and Wudl, F. (2007) Carbon allotropes: beyond graphite and diamond. *J. Chem. Technol. Biotechnol.*, **82** (6), 524–531.
6. Kroto, H.W., Heath, J.R., Obrien, S.C., Curl, R.F., and Smalley, R.E. (1985) C-60 – Buckminsterfullerene. *Nature*, **318** (6042), 162–163.
7. Kroto, H.W., Allaf, A.W., and Balm, S.P. (1991) C60 – Buckminsterfullerene. *Chem. Rev.*, **91** (6), 1213–1235.
8. Dresselhaus, M.S., Dresselhaus, G., and Eklund, P.C. (1993) Fullerenes. *J. Mater. Res.*, **8** (8), 2054–2097.
9. Wudl, F. (2002) Fullerene materials. *J. Mater. Chem.*, **12** (7), 1959–1963.
10. Liu, Z., Suenaga, K., and Iijima, S. (2007) Imaging the structure of an individual C-60 fullerene molecule and its deformation process using HRTEM with atomic sensitivity. *J. Am. Chem. Soc.*, **129** (21), 6666–6667.
11. Novoselov, K.S., Geim, A.K., Morozov, S.V., Jiang, D., Zhang, Y., Dubonos, S.V., Grigorieva, I.V., and Firsov, A.A. (2004) Electric field effect in atomically thin carbon films. *Science*, **306** (5696), 666–669.
12. Xia, Y.N., Yang, P.D., Sun, Y.G., Wu, Y.Y., Mayers, B., Gates, B., Yin, Y.D., Kim, F., and Yan, Y.Q. (2003) One-dimensional nanostructures: synthesis, characterization, and applications. *Adv. Mater.*, **15** (5), 353–389.
13. Iijima, S. and Ichihashi, T. (1993) Single-shell carbon nanotubes of 1-nm diameter. *Nature*, **363** (6430), 603–605.
14. Colomer, J.F., Stephan, C., Lefrant, S., Van Tendeloo, G., Willems, I., Konya, Z., Fonseca, A., Laurent, C., and Nagy, J.B. (2000) Large-scale synthesis of single-wall carbon nanotubes by catalytic chemical vapor deposition (CCVD) method. *Chem. Phys. Lett.*, **317** (1–2), 83–89.
15. Zhang, X.F., Zhang, X.B., Vantendeloo, G., Amelinckx, S., Debeeck, M.O., and Vanlanduyt, J. (1993) Carbon nano-tubes – their formation process and observation by electron-microscopy. *J. Cryst. Growth*, **130** (3–4), 368–382.
16. Amelinckx, S., Bernaerts, D., Zhang, X.B., Vantendeloo, G., and Vanlanduyt, J. (1995) A structure model and growth-mechanism for multishell carbon nanotubes. *Science*, **267** (5202), 1334–1338.
17. Geim, A.K. and Novoselov, K.S. (2007) The rise of graphene. *Nat. Mater.*, **6** (3), 183–191.

18. Saito, R., Fujita, M., Dresselhaus, G., and Dresselhaus, M.S. (1992) Electronic-structure of chiral graphene tubules. *Appl. Phys. Lett.*, **60** (18), 2204–2206.
19. Castro Neto, A.H., Guinea, F., Peres, N.M.R., Novoselov, K.S., and Geim, A.K. (2009) The electronic properties of graphene. *Rev. Mod. Phys.*, **81** (1), 109–162.
20. Suarez-Marti, I. and Grobert, N., and Ewels, C.P. (2012) Nomenclature of sp^2 carbon nanoforms. *Carbon*, **50** (3), 741–747.
21. Blase, X., Benedict, L.X., Shirley, E.L., and Louie, S.G. (1994) Hybridization effects and metallicity in small radius carbon nanotubes. *Phys. Rev. Lett.*, **72** (12), 1878–1881.
22. Haddon, R.C. (1993) Chemistry of the fullerenes – the manifestation of strain in a class of continuous aromatic-molecules. *Science*, **261** (5128), 1545–1550.
23. Prato, M. (1997) [60]Fullerene chemistry for materials science applications. *J. Mater. Chem.*, **7** (7), 1097–1109.
24. Haddon, R.C. (1992) Electronic-structure, conductivity, and superconductivity of alkali-metal doped C-60. *Acc. Chem. Res.*, **25** (3), 127–133.
25. Enoki, T., Tsuge, S., Hao, S., Takai, K., Nakajima, K., Hara, M., and Kang, F.Y. (2010) Magnetism of nanographene network influenced by the interaction with acid species. *J. Phys. Chem. Solids*, **71** (4), 534–538.
26. Joly, V.L.J., Takahara, K., Takai, K., Sugihara, K., Enoki, T., Koshino, M., and Tanaka, H. (2010) Effect of electron localization on the edge-state spins in a disordered network of nanographene sheets. *Phys. Rev. B (Condens. Matter Mater. Phys.)*, **81**, (6), 115408.
27. Joly, V.L.J., Kiguchi, M., Hao, S.J., Takai, K., Enoki, T., Sumii, R., Amemiya, K., Muramatsu, H., Hayashi, T., Kim, Y.A., Endo, M., Campos-Delgado, J., Lopez-Urias, F., Botello-Mendez, A., Terrones, H., Terrones, M., and Dresselhaus, M.S. (2010) Observation of magnetic edge state in graphene nanoribbons. *Phys. Rev. B*, **81** (24), 245428.
28. Enoki, T., Takai, K., Osipov, V., Baidakova, M., and Vul, A. (2009) Nanographene and nanodiamond; new members in the nanocarbon family. *Chem. Asian J.*, **4** (6), 796–804.
29. Raty, J.Y. and Galli, G. (2003) Ultradispersity of diamond at the nanoscale. *Nat. Mater.*, **2** (12), 792–795.
30. Raty, J.Y., Galli, G., Bostedt, C., van Buuren, T.W., Terminello, L.J. (2003) Quantum confinement and fullerenelike surface reconstructions in nanodiamonds. *Phys. Rev. Lett.*, **90** (3), 037401.
31. Schoenfeld, R.S. and Harneit, W. Real time magnetic field sensing and imaging using a single spin in diamond. *Phys. Rev. Lett.*, **106** (3), 030802.
32. Enoki, T. and Takai, K. (2009) The edge state of nanographene and the magnetism of the edge-state spins. *Solid State Commun.*, **149** (27–28), 1144–1150.
33. Harris, P. (2009) *Carbon Nanotube Science*, Cambridge Unviversity Press, Cambridge.
34. Carbone, F., Barwick, B., Kwon, O.H., Park, H.S., Baskin, J.S., and Zewail, A.H. (2009) EELS femtosecond resolved in 4D ultrafast electron microscopy. *Chem. Phys. Lett.*, **468** (4–6), 107–111.
35. Barwick, B., Flannigan, D.J., and Zewail, A.H. (2009) Photon-induced near-field electron microscopy. *Nature*, **462** (7275), 902–906.
36. Park, H.S., Baskin, J.S., Barwick, B., Kwon, O.H., and Zewail, A.H. (2009) 4D ultrafast electron microscopy: imaging of atomic motions, acoustic resonances, and moire fringe dynamics. *Ultramicroscopy*, **110** (1), 7–19.
37. Garg, A. and Sinnott, S.B. (1998) Effect of chemical functionalization on the mechanical properties of carbon nanotubes. *Chem. Phys. Lett.*, **295** (4), 273–278.
38. Nirmalraj, P.N., Lyons, P.E., De, S., Coleman, J.N., and Boland, J.J. (2009) Electrical connectivity in single-walled carbon nanotube networks. *Nano Lett.*, **9** (11), 3890–3895.

39. Nirmalraj, P.N. and Boland, J.J. (2010) Selective tuning and optimization of the contacts to metallic and semiconducting single-walled carbon nanotubes. *ACS Nano*, **4** (7), 3801–3806.
40. Felten, A., Gillon, X., Gulas, M., Pireaux, J.J., Ke, X.X., Van Tendeloo, G., Bittencourt, C., Najafi, E., and Hitchcock, A.P. (2010) Measuring point defect density in individual carbon nanotubes using polarization-dependent X-ray microscopy. *ACS Nano*, **4** (8), 4431–4436.
41. Najafi, E., Cruz, D.H., Obst, M., Hitchcock, A.P., Douhard, B., Pireaux, J.J., and Felten, A. (2008) Polarization dependence of the C 1s X-ray absorption spectra of individual multi-walled carbon nanotubes. *Small*, **4** (12), 2279–2285.
42. Felten, A., Bittencourt, C., Pireaux, J.J., Reichelt, M., Mayer, J., Hernandez-Cruz, D., and Hitchcock, A.P. (2007) Individual multiwall carbon nanotubes spectroscopy by scanning transmission X-ray microscopy. *Nano Lett.*, **7** (8), 2435–2440.
43. Felten, A., Hody, H., Bittencourt, C., Pireaux, J.J., Cruz, D.H., and Hitchcock, A.P. (2006) Scanning transmission X-ray microscopy of isolated multiwall carbon nanotubes. *Appl. Phys. Lett.*, **89** (9), 093123.
44. Nair, R.R., Blake, P., Grigorenko, A.N., Novoselov, K.S., Booth, T.J., Stauber, T., Peres, N.M.R., and Geim, A.K. (2008) Fine structure constant defines visual transparency of graphene. *Science*, **320** (5881), 1308–1308.
45. Roddaro, S., Pingue, P., Piazza, V., Pellegrini, V., and Beltram, F. (2007) The optical visibility of graphene: interference colors of ultrathin graphite on SiO2. *Nano Lett.*, **7** (9), 2707–2710.
46. Jung, I., Pelton, M., Piner, R., Dikin, D.A., Stankovich, S., Watcharotone, S., Hausner, M., and Ruoff, R.S. (2007) Simple approach for high-contrast optical imaging and characterization of graphene-based sheets. *Nano Lett.*, **7** (12), 3569–3575.
47. Blake, P., Hill, E.W., Neto, A.H.C., Novoselov, K.S., Jiang, D., Yang, R., Booth, T.J., and Geim, A.K. (2007) Making graphene visible. *Appl. Phys. Lett.*, **91** (6), 063124.
48. Kim, J., Cote, L.J., Kim, F., and Huang, J.X. (2010) Visualizing graphene based sheets by fluorescence quenching microscopy. *J. Am. Chem. Soc.*, **132** (1), 260–267.
49. Kim, J., Kim, F., and Huang, J.X. (2010) Seeing graphene-based sheets. *Mater. Today*, **13** (3), 28–38.
50. Treossi, E., Melucci, M., Liscio, A., Gazzano, M., Samori, P., and Palermo, V. (2009) High-contrast visualization of graphene oxide on dye-sensitized glass, quartz, and silicon by fluorescence quenching. *J. Am. Chem. Soc.*, **131** (43), 15576–15577.
51. Park, J., Ahn, Y.H., and Ruiz-Vargas, C. (2009) Imaging of photocurrent generation and collection in single-layer graphene. *Nano Lett.*, **9** (5), 1742–1746.
52. Lee, E.J.H., Balasubramanian, K., Weitz, R.T., Burghard, M., and Kern, K. (2008) Contact and edge effects in graphene devices. *Nat. Nanotechnol.*, **3** (8), 486–490.
53. Boehm, H.P. (2010) Graphene-how a laboratory curiosity suddenly became extremely interesting. *Angew. Chem. Int. Ed.*, **49** (49), 9332–9335.
54. Iijima, S. (1991) Helical microtubules of graphitic carbon. *Nature*, **354** (6348), 56–58.
55. Wang, Y.H., Maspoch, D., Zou, S.L., Schatz, G.C., Smalley, R.E., and Mirkin, C.A. (2006) Controlling the shape, orientation, and linkage of carbon nanotube features with nano affinity templates. *Proc. Natl. Acad. Sci. U.S.A.*, **103** (7), 2026–2031.
56. Terrones, M., Terrones, H., Banhart, F., Charlier, J.C., and Ajayan, P.M. (2000) Coalescence of single-walled carbon nanotubes. *Science*, **288** (5469), 1226–1229.
57. Motta, M., Moisala, A., Kinloch, I.A., and Windle, A.H. (2007) High performance fibres from 'Dog bone' carbon nanotubes. *Adv. Mater.*, **19** (21), 3721–3726.
58. Kim, U.J., Gutierrez, H.R., Kim, J.P, and Eklund, P.C. (2005) Effect of the

tube diameter distribution on the high-temperature structural modification of bundled single-walled carbon nanotubes. *J. Phys. Chem. B*, **109** (49), 23358–23365.
59. Meyer, J.C., Paillet, M., Duesberg, G.S., and Roth, S. (2006) Electron diffraction analysis of individual single-walled carbon nanotubes. *Ultramicroscopy*, **106** (3), 176–190.
60. Smith, B.W. and Luzzi, D.E. (1999) Knock-on damage in single wall carbon nanotubes by electron irradiation. *AIP Conf. Proc.*, (486), 360–363.
61. Paillet, M., Jourdain, V., Poncharal, P., Sauvajol, J.L., Zahab, A., Meyer, J.C., Roth, S., Cordente, N., Amiens, C., and Chaudret, B. (2004) Versatile synthesis of individual single-walled carbon nanotubes from nickel nanoparticles for the study of their physical properties. *J. Phys. Chem. B*, **108**, 17112–17118.
62. Zhang, X.B., Zhang, X.F., Bernaerts, D., Vantendeloo, G.T., Amelinckx, S., Vanlanduyt, J., Ivanov, V., Nagy, J.B., Lambin, P., and Lucas, A.A. (1994) The texture of catalytically grown coil-shaped carbon nanotubules. *Europhys. Lett.*, **27** (2), 141–146.
63. Zhang, X.B., Zhang, X.F., Amelinckx, S., Vantendeloo, G., and Vanlanduyt, J. (1994) The reciprocal space of carbon tubes – a detailed interpretation of the electron-diffraction effects. *Ultramicroscopy*, **54** (2–4), 237–249.
64. Wang, X.S., Li, Q.Q., Xie, J., Jin, Z., Wang, J.Y., Li, Y., Jiang, K.L., and Fan, S.S. (2009) Fabrication of ultralong and electrically uniform single-walled carbon nanotubes on clean substrates. *Nano Lett.*, **9** (9), 3137–3141.
65. Lau, K.K.S., Bico, J., Teo, K.B.K., Chhowalla, M., Amaratunga, G.A.J., Milne, W.I., McKinley, G.H., and Gleason, K.K. (2003) Superhydrophobic carbon nanotube forests. *Nano Lett.*, **3** (12), 1701–1705.
66. Otieno, G., Koos, A.A., Dillon, F., Wallwork, A., Grobert, N., and Todd, R.I. (2010) Processing and properties of aligned multi-walled carbon nanotube/aluminoborosilicate glass composites made by sol-gel processing. *Carbon*, **48** (8), 2212–2217.
67. Wu, Y.H., Qiao, P.W., Chong, T.C., and Shen, Z.X. (2002) Carbon nanowalls grown by microwave plasma enhanced chemical vapor deposition. *Adv. Mater.*, **14** (1), 64–67.
68. Louchev, O.A., Sato, Y., and Kanda, H. (2002) Growth mechanism of carbon nanotube forests by chemical vapor deposition. *Appl. Phys. Lett.*, **80** (15), 2752–2754.
69. Cha, J.J., Weyland, M., Briere, J.F., Daykov, I.P., Arias, T.A., and Muller, D.A. (2007) Three-dimensional imaging of carbon nanotubes deformed by metal islands. *Nano Lett.*, **7** (12), 3770–3773.
70. Felten, A., Suarez-Martinez, I., Ke, X.X., Van Tendeloo, G., Ghijsen, J., Pireaux, J.J., Drube, W., Bittencourt, C., and Ewels, C.P. (2009) The role of oxygen at the interface between titanium and carbon nanotubes. *Chemphyschem*, **10** (11), 1799–1804.
71. Suarez-Martinez, I., Felten, A., Pireaux, J.J., Bittencourt, C., and Ewels, C.P. (2009) Transition metal deposition on graphene and carbon nanotubes. *J. Nanosci. Nanotechnol.*, **9** (10), 6171–6175.
72. Colliex, C., Brun, N., Gloter, A., Imhoff, D., Kociak, M., March, K., Mory, C., Stephan, O., Tence, M., and Walls, M. (2009) Multi-dimensional and multi-signal approaches in scanning transmission electron microscopes. *Philos. Trans. R. Soc. A: Math. Phys. Eng. Sci.*, **367** (1903), 3845–3858.
73. Trasobares, S., Ewels, C.P., Birrell, J., Stephan, D., Wei, B.Q.Q., Carlisle, J.A., Miller, D., Keblinski, P., and Ajayan, P.M. (2004) Carbon nanotubes with graphitic wings. *Adv. Mater.*, **16** (7), 610–613.
74. Koos, A.A., Dowling, M., Jurkschat, K., Crossley, A., and Grobert, N. (2009) Effect of the experimental parameters on the structure of nitrogen-doped carbon nanotubes produced by aerosol chemical vapour deposition. *Carbon*, **47** (1), 30–37.
75. Chen, J.L., Li, Y.D., Ma, Y.M., Qin, Y.N., and Chang, L. (2001) Formation of bamboo-shaped carbon filaments

and dependence of their morphology on catalyst composition and reaction conditions. *Carbon*, **39** (10), 1467–1475.

76. Okuno, H., Grivei, E., Fabry, F., Gruenberger, T.M., Gonzalez-Aguilar, J., Palnichenko, A., Fulcheri, L., Probst, N., and Charlier, J.C. (2004) Synthesis of carbon nanotubes and nano-necklaces by thermal plasma process. *Carbon*, **42** (12–13), 2543–2549.
77. Nasibulin, A.G., Anisimov, A.S., Pikhitsa, P.V., Jiang, H., Brown, D.P., Choi, M., and Kauppinen, E.I. (2007) Investigations of NanoBud formation. *Chem. Phys. Lett.*, **446**, 109–114.
78. Nasibulin, A.G., Pikhitsa, P.V., Jiang, H., Brown, D.P., Krasheninnikov, A.V., Anisimov, A.S., Queipo, P., Moisala, A., Gonzalez, D., Lientschnig, G., Hassanien, A., Shandakov, S.D., Lolli, G., Resasco, D.E., Choi, M., Tomanek, D., and Kauppinen, E.I. (2007) A novel hybrid carbon material. *Nat. Nanotechnol.*, **2** (3), 156–161.
79. Krasheninnikov, A.V. and Banhart, F. (2007) Engineering of nanostructured carbon materials with electron or ion beams. *Nat. Mater.*, **6**, 723–733.
80. Endo, M., Kim, Y.A., Hayashi, T., Fukai, Y., Oshida, K., Terrones, M., Yanagisawa, T., Higaki, S., and Dresselhaus, M.S. (2002) Structural characterization of cup-stacked-type nanofibers with an entirely hollow core. *Appl. Phys. Lett.*, **80** (7), 1267–1269.
81. Jang, I.Y., Ogata, H., Park, K.C., Lee, S.H., Park, J.S., Jung, Y.C., Kim, Y.J., Kim, Y.A., and Endo, M. (2010) Exposed edge planes of cup-stacked carbon nanotubes for an electrochemical capacitor. *J. Phys. Chem. Lett.*, **1** (14), 2099–2103.
82. Bernaerts, D., Zhang, X.B., Zhang, X.F., Amelinckx, S., Vantendeloo, G., Vanlanduyt, J., Ivanov, V., and Nagy, J.B. (1995) Electron-microscopy study of coiled carbon tubules. *Philos. Mag. A: Phys. Condens. Matter Struct. Defects Mech. Prop.*, **71** (3), 605–630.
83. Amelinckx, S., Zhang, X.B., Bernaerts, D., Zhang, X.F., Ivanov, V., and Nagy, J.B. (1994) A formation mechanism for catalytically grown helix-shaped graphite nanotubes. *Science*, **265** (5172), 635–639.
84. Liu, J., Dai, H.J., Hafner, J.H., Colbert, D.T., Smalley, R.E., Tans, S.J., and Dekker, C. (1997) Fullerene 'crop circles'. *Nature*, **385** (6619), 780–781.
85. de Heer, W.A., Poncharal, P., Berger, C., Gezo, J., Song, Z.M., Bettini, J., and Ugarte, D. (2005) Liquid carbon, carbon-glass beads, and the crystallization of carbon nanotubes. *Science*, **307** (5711), 907–910.
86. Rayleigh, L. (1982) On the instability of a cylinder of viscous liquid under capillary force. *Philos. Mag.*, **34** (207), 145–154.
87. Lavin, J.G., Subramoney, S., Ruoff, R.S., Berber, S., and Tomanek, D. (2002) Scrolls and nested tubes in multiwall carbon nanotubes. *Carbon*, **40** (7), 1123–1130.
88. Viculis, L.M., Mack, J.J., and Kaner, R.B. (2003) A chemical route to carbon nanoscrolls. *Science*, **299** (5611), 1361–1361.
89. Suarez-Martinez, I., Savini, G., Zobelli, A., and Heggie, M. (2007) Dislocations in carbon nanotube walls. *J. Nanosci. Nanotechnol.*, **7** (10), 3417–3420.
90. Yudanov, N.F., Okotrub, A.V., Shubin, Y.V., Yudanova, L.I., Bulusheva, L.G., Chuvilin, A.L., and Bonard, J.M. (2002) Fluorination of arc-produced carbon material containing multiwall nanotubes. *Chem. Mater.*, **14** (4), 1472–1476.
91. Berber, S. and Tomanek, D. (2004) Stability differences and conversion mechanism between nanotubes and scrolls. *Phys. Rev. B*, **69** (23), 233404.
92. Zhou, O., Fleming, R.M., Murphy, D.W., Chen, C.H., Haddon, R.C., Ramirez, A.P., and Glarum, S.H. (1994) Defects in carbon nanostructures. *Science*, **263** (5154), 1744–1747.
93. Rodriguez, N.M., Chambers, A., and Baker, R.T.K. (1995) Catalytic engineering of carbon nanostructures. *Langmuir*, **11** (10), 3862–3866.
94. Ren, W.C. and Cheng, H.M. (2003) Herringbone-type carbon nanofibers with a small diameter and large hollow core synthesized by the catalytic

decomposition of methane. *Carbon*, **41** (8), 1657–1660.
95. Naess, S.N., Elgsaeter, A., Helgesen, G., and Knudsen, K.D. (2009) Carbon nanocones: wall structure and morphology. *Sci. Technol. Adv. Mater.*, **10** (6), 065002.
96. Iijima, S., Yudasaka, M., Yamada, R., Bandow, S., Suenaga, K., Kokai, F., and Takahashi, K. (1999) Nano-aggregates of single-walled graphitic carbon nano-horns. *Chem. Phys. Lett.*, **309** (3–4), 165–170.
97. Iijima, S., Ichihashi, T., and Ando, Y. (1992) Pentagons, heptagons and negative curvature in graphite microtubule growth. *Nature*, **356** (6372), 776–778.
98. Krishnan, A., Dujardin, E., Treacy, M.M.J., Hugdahl, J., Lynum, S., and Ebbesen, T.W. (1997) Graphitic cones and the nucleation of curved carbon surfaces. *Nature*, **388** (6641), 451–454.
99. Ugarte, D. (1995) Onion-like graphitic particles. *Carbon*, **33** (7), 989–993.
100. Ugarte, D. (1992) Curling and closure of graphitic networks under electron-beam irradiation. *Nature*, **359** (6397), 707–709.
101. Ozawa, M., Goto, H., Kusunoki, M., and Osawa, E. (2002) Continuously growing spiral carbon nanoparticles as the intermediates in the formation of fullerenes and nanoonions. *J. Phys. Chem. B*, **106** (29), 7135–7138.
102. Banhart, F. and Ajayan, P.M. (1997) Self-compression and diamond formation in carbon onions. *Adv. Mater.*, **9** (3), 261–263.
103. Banhart, F. and Ajayan, P.M. (1996) Carbon onions as nanoscopic pressure cells for diamond formation. *Nature*, **382** (6590), 433–435.
104. Schaper, A.K., Hou, H., Greiner, A., Schneider, R., and Phillipp, F. (2004) Copper nanoparticles encapsulated in multi-shell carbon cages. *Appl. Phys. A: Mater. Sci. Process.*, **78** (1), 73–77.
105. Muto, S., Vantendeloo, G., and Amelinckx, S. (1993) High-resolution electron-microscopy of structural defects in crystalline C-60 and C-70. *Philos. Mag. B: Phys. Condens. Matter Stat. Mech. Electron. Opt. Magn. Prop.*, **67** (4), 443–463.
106. Verheijen, M.A., Meekes, H., Meijer, G., Bennema, P., Deboer, J.L., Vansmaalen, S., Vantendeloo, G., Amelinckx, S., Muto, S., and Vanlanduyt, J. (1992) The structure of different phases of pure C-70 crystals. *Chem. Phys.*, **166** (1–2), 287–297.
107. Cheng, C., Porter, A.E., Muller, K., Koziol, K., Skepper, J.N., Midgley, P., and Welland, M. (2009) Imaging carbon nanoparticles and related cytotoxicity. *J. Phys.: Conf. Ser.*, **151**, 012030.
108. Muoth, M., Lee, S.W., and Hierold, C. (2011) Platform for strainable, TEMP-compatible, MEMS-embedded carbon nanotube transistors. IEEE 24th International Conference on Micro Electro Mechanical Systems (MEMS 2011), pp. 83–86.
109. Muoth, M., Helbling, T., Durrer, L., Lee, S.W., Roman, C., and Hierold, C. (2010) Hysteresis-free operation of suspended carbon nanotube transistors. *Nat. Nanotechnol.*, **5** (8), 589–592.
110. Muoth, M., Gramm, F., Asaka, K., Durrer, L., Helbling, T., Roman, C., Lee, S.W., and Hierold, C. (2009) Tilted-view transmission electron microscopy-access for chirality assignment to carbon nanotubes integrated in MEMS. *Proc. Eurosens. XXIII Conf.*, **1** (1), 601–604.
111. Fujieda, T., Hidaka, K., Hayashibara, M., Kamino, T., Ose, Y., Abe, H., Shimizu, T., and Tokumoto, H. (2005) Direct observation of field emission sites in a single multiwalled carbon nanotube by Lorenz microscopy. *Jpn. J. Appl. Phys. Part 1-Regular Pap. Brief Commun. Rev. Pap.*, **44** (4A), 1661–1664.
112. Smith, B.W. and Luzzi, D.E. (2001) Electron irradiation effects in single wall carbon nanotubes. *J. Appl. Phys.*, **90** (7), 3509–3515.
113. Crespi, V.H., Chopra, N.G., Cohen, M.L., Zettl, A., and Louie, S.G. (1996) Anisotropic electron-beam damage and the collapse of carbon nanotubes. *Phys. Rev. B*, **54** (8), 5927–5931.
114. Krivanek, O.L., Dellby, N., Murfitt, M.F., Chisholm, M.F., Pennycook, T.J., Suenaga, K., and Nicolosi, V.

(2010) Gentle STEM: ADF imaging and EELS at low primary energies. *Ultramicroscopy*, **110** (8), 935–945.

115. Zobelli, A., Gloter, A., Ewels, C.P., Seifert, G., and Colliex, C. (2007) Electron knock-on cross section of carbon and boron nitride nanotubes. *Phys. Rev. B*, **75**, 245402, [9 pages].

116. Suenaga, K., Wakabayashi, H., Koshino, M., Sato, Y., Urita, K., and Iijima, S. (2007) Imaging active topological defects in carbon nanotubes. *Nat. Nanotechnol.*, **2** (6), 358–360.

117. Guan, L.H., Suenaga, K., and Iijima, S. (2008) Smallest carbon nanotube assigned with atomic resolution accuracy. *Nano Lett.*, **8** (2), 459–462.

118. Vlasov, I.I., Shenderova, O., Turner, S., Lebedev, O.I., Basov, A.A., Sildos, I., Rahn, M., Shiryaev, A.A., and Van Tendeloo, G. (2010) Nitrogen and luminescent nitrogen-vacancy defects in detonation nanodiamond. *Small*, **6** (5), 687–694.

119. Turner, S., Lebedev, O.I., Shenderova, O., Vlasov, I.I., Verbeeck, J., and Van Tendeloo, G. (2009) Determination of size, morphology, and nitrogen impurity location in treated detonation nanodiamond by transmission electron microscopy. *Adv. Funct. Mater.*, **19** (13), 2116–2124.

120. Ge, L., Jefferson, J.H., Montanari, B., Harrison, N.M., Pettifor, D.G., and Briggs, G.A.D. (2009) Effects of doping on electronic structure and correlations in carbon peapods. *ACS Nano*, **3** (5), 1069–1076.

121. Karousis, N., Economopoulos, S.P., Iizumi, Y., Okazaki, T., Liu, Z., Suenaga, K., and Tagmatarchis, N. (2010) Microwave assisted covalent functionalization of C-60@SWCNT peapods. *Chem. Commun.*, **46** (48), 9110–9112.

122. Sato, Y., Yanagi, K., Miyata, Y., Suenaga, K., Kataura, H., and Iijima, S. (2008) Chiral-angle distribution for separated single-walled carbon nanotubes. *Nano Lett.*, **8** (10), 3151–3154.

123. Krasheninnikov, A.V. and Banhart, F. (2007) Engineering of nanostructured carbon materials with electron or ion beams. *Nat. Mater.*, **6** (10), 723–733.

124. Yoon, M., Seungwu, H., Gunn, K., Sang Bong, L., Berber, S., Osawa, E., Ihm, J., Terrones, M., Banhart, F., Charlier, J.C., Grobert, N., Terrones, H., Ajayan, P.M., and Tomanek, D. (2004) Zipper mechanism of nanotube fusion: theory and experiment. *Phys. Rev. Lett.*, **92** (7), 075504 / 1–075504/4.

125. Ding, F., Jiao, K., Lin, Y., and Yakobson, B.I. (2007) How evaporating carbon nanotubes retain their perfection? *Nano Lett.*, **7** (3), 681–684.

126. Prato, M., Maggini, M., Giacometti, C., Scorrano, G., Sandona, G., and Farnia, G. (1996) Synthesis and electrochemical properties of substituted fulleropyrrolidines. *Tetrahedron*, **52** (14), 5221–5234.

127. Liu, Z., Koshino, M., Suenaga, K., Mrzel, A., Kataura, H., and Iijima, S. (2006) Transmission electron microscopy imaging of individual functional groups of fullerene derivatives. *Phys. Rev. Lett.*, **96** (8), 088304.

128. Koshino, M., Niimi, Y., Nakamura, E., Kataura, H., Okazaki, T., Suenaga, K., and Iijima, S. (2010) Analysis of the reactivity and selectivity of fullerene dimerization reactions at the atomic level. *Nat. Chem.*, **2** (2), 117–124.

129. Sasaki, T., Sawada, H., Hosokawa, F., Kohno, Y., Tomita, T., Kaneyama, T., Kondo, Y., Kimoto, K., Sato, Y., and Suenaga, K. (2010) Performance of low-voltage STEM/TEM with delta corrector and cold field emission gun. *J. Electron Microsc.*, **59**, S7–S13.

130. Suenaga, K., Sato, Y., Liu, Z., Kataura, H., Okazaki, T., Kimoto, K., Sawada, H., Sasaki, T., Omoto, K., Tomita, T., Kaneyama, T., and Kondo, Y. (2009) Visualizing and identifying single atoms using electron energy-loss spectroscopy with low accelerating voltage. *Nat. Chem.*, **1** (5), 415–418.

131. Novoselov, K.S., Jiang, D., Schedin, F., Booth, T.J., Khotkevich, V.V., Morozov, S.V., and Geim, A.K. (2005) Two-dimensional atomic crystals. *Proc. Natl. Acad. Sci. U.S.A.*, **102** (30), 10451–10453.

132. Girit, C.O., Meyer, J.C., Erni, R., Rossell, M.D., Kisielowski, C., Yang, L.,

Park, C.H., Crommie, M.F., Cohen, M.L., Louie, S.G., and Zettl, A. (2009) Graphene at the edge: stability and dynamics. *Science*, **323** (5922), 1705–1708.

133. Hibino, H., Kageshima, H., Kotsugi, M., Maeda, F., Guo, F.Z., and Watanabe, Y. (2009) Dependence of electronic properties of epitaxial few-layer graphene on the number of layers investigated by photoelectron emission microscopy. *Phys. Rev. B (Condens. Matter Mater. Phys.)*, **79**, 125437.

134. Meyer, J.C., Geim, A.K., Katsnelson, M.I., Novoselov, K.S., Booth, T.J., and Roth, S. (2007) The structure of suspended graphene sheets. *Nature*, **446** (7131), 60–63.

135. Meyer, J.C., Geim, A.K., Katsnelson, M.I., Novoselov, K.S., Obergfell, D., Roth, S., Girit, C., and Zettl, A. (2007) On the roughness of single- and bi-layer graphene membranes. *Solid State Commun.*, **143** (1–2), 101–109.

136. Meyer, J.C., Kisielowski, C., Erni, R., Rossell, M.D., Crommie, M.F., and Zettl, A. (2008) Direct imaging of lattice atoms and topological defects in graphene membranes. *Nano Lett.*, **8** (11), 3582–3586.

137. Koskinen, P., Malola, S., and Hakkinen, H. (2009) Evidence for graphene edges beyond zigzag and armchair. *Phys. Rev. B*, **80** (7), 073401.

138. Kawai, T., Miyamoto, Y., Sugino, O., and Koga, Y. (2000) Graphitic ribbons without hydrogen-termination: Electronic structures and stabilities. *Phys. Rev. B*, **62** (24), R16349–R16352.

139. Koskinen, P., Malola, S., and Hakkinen, H. (2008) Self-passivating edge reconstructions of graphene. *Phys. Rev. Lett.*, **101** (11).

140. Huang, B., Liu, M., Su, N.H., Wu, J., Duan, W.H., Gu, B.L., and Liu, F. (2009) Quantum manifestations of graphene edge stress and edge instability: a first-principles study. *Phys. Rev. Lett.*, **102** (16).

141. Suenaga, K. and Koshino, M. (2010) Atom-by-atom spectroscopy at graphene edge. *Nature*, **468** (7327), 1088–1090.

142. Krivanek, O.L., Chisholm, M.F., Nicolosi, V., Pennycook, T.J., Corbin, G.J., Dellby, N., Murfitt, M.F., Own, C.S., Szilagyi, Z.S., Oxley, M.P., Pantelides, S.T., and Pennycook, S.J. (2010) Atom-by-atom structural and chemical analysis by annular dark-field electron microscopy. *Nature*, **464** (7288), 571–574.

143. Huang, P.Y., Ruiz-Vargas, C.S., van der Zande, A.M., Whitney, W.S., Levendorf, M.P., Kevek, J.W., Garg, S., Alden, J.S., Hustedt, C.J., Ye, Z., Jiwoong, P., McEuen, P.L., and Muller, D.A. (2011) Grains and grain boundaries in single-layer graphene atomic patchwork quilts. *Nature*, **469** (7330), 389–393.

144. Rodriguez-Manzo, J.A., Pham-Huu, C., and Banhart, F. (2011) Graphene growth by a metal-catalyzed solid-state transformation of amorphous carbon. *ACS Nano*, **5** (2), 1529–1534.

145. Murata, Y., Petrova, V., Kappes, B.B., Ebnonnasir, A., Petrov, I., Xie, Y.H., Ciobanu, C.V., and Kodambaka, S. (2010) Moire superstructures of graphene on faceted nickel islands. *ACS Nano* **4** (11), 6509–6514.

146. Eizenberg, M. and Blakely, J.M. (1979) Carbon monolayer phase condensation on Ni(111). *Surf. Sci.*, **82** (1), 228–236.

147. Eizenberg, M. and Blakely, J.M. (1979) Carbon interaction with nickel surfaces-monolayer formation and structural stability. *J. Chem. Phys.*, **71** (8), 3467–3477.

148. Chuvilin, A., Kaiser, U., Bichoutskaia, E., Besley, N.A., and Khlobystov, A.N. (2010) Direct transformation of graphene to fullerene. *Nat. Chem.* **2** (6), 450–453.

149. Jin, C., Lan, H., Peng, L., Suenaga, K., and Iijima, S. (2009) Deriving carbon atomic chains from graphene. *Phys. Rev. Lett.*, **102** (20), 205501.

150. Zewail, A. H. Four-dimensional electron microscopy. *Science*, **328** (5975), 187–193.

151. Kwon, O.H. and Zewail, A.H. (2010) 4D electron tomography. *Science*, **328**, (5986), 1668–1673.

152. Hartland, G.V. (2006) Coherent excitation of vibrational modes in metallic

nanoparticles. *Annu. Rev. Phys. Chem.*, **57**, 403–430.
153. Giusca, C.E., Tison, Y., Stolojan, V., Borowiak-Palen, E., and Silva, S.R.P. (2007) Inner-tube chirality determination for double-walled carbon nanotubes by scanning tunneling microscopy. *Nano Lett.*, **7** (5), 1232–1239.
154. Bergin, S.D., Nicolosi, V., Streich, P.V., Giordani, S., Sun, Z.Y., Windle, A.H., Ryan, P., Niraj, N.P.P., Wang, Z.T.T., Carpenter, L., Blau, W.J., Boland, J.J., Hamilton, J.P., and Coleman, J.N. (2008) Towards solutions of single-walled carbon nanotubes in common solvents. *Adv. Mater.*, **20** (10), 1876–1187.
155. Cai, J.M., Ruffieux, P., Jaafar, R., Bieri, M., Braun, T., Blankenburg, S., Muoth, M., Seitsonen, A.P., Saleh, M., Feng, X.L., Mullen, K., and Fasel, R. (2010) Atomically precise bottom-up fabrication of graphene nanoribbons. *Nature*, **466**, (7305), 470–473.
156. Rim, K.T., Siaj, M., Xiao, S.X., Myers, M., Carpentier, V.D., Liu, L., Su, C.C., Steigerwald, M.L., Hybertsen, M.S., McBreen, P.H., Flynn, G.W., and Nuckolls, C. (2007) Forming aromatic hemispheres on transition-metal surfaces. *Angew. Chem. Int. Ed.*, **46**, 7891–7895.
157. Otero, G., Biddau, G., Sanchez-Sanchez, C., Caillard, R., Lopez, M.F., Rogero, C., Palomares, F.J., Cabello, N., Basanta, M.A., Ortega, J., Mendez, J., Echavarren, A.M., Perez, R., Gomez-Lor, B., and Martin-Gago, J.A. (2008) Fullerenes from aromatic precursors by surface-catalysed cyclodehydrogenation. *Nature*, **454** (7206), 865–U19.
158. Yarotski, D.A., Kilina, S.V., Talin, A.A., Tretiak, S., Prezhdo, O.V., Balatsky, A.V., and Taylor, A.J. (2009) Scanning tunneling microscopy of DNA-wrapped carbon nanotubes. *Nano Lett.*, **9** (1), 12–17.
159. Ishii, M., Hamilton, B., and Poolton, N. (2008) Imaging of charge trapping in distorted carbon nanotubes by X-ray excited scanning probe microscopy. *J. Appl. Phys.*, **104** (10).
160. Poggi, M.A., Bottomley, L.A., and Lillehei, P.T. (2004) Measuring the adhesion forces between alkanethiol-modified AFM cantilevers and single walled carbon nanotubes. *Nano Lett.*, **4** (1), 61–64.
161. Zheng, F., Baldwin, D.L., Fifield, L.S., Anheier, N.C., Aardahl, C.L., and Grate, J.W. (2006) Single-walled carbon nanotube paper as a sorbent for organic vapor preconcentration. *Anal. Chem.*, **78** (7), 2442–2446.
162. Lillehei, P.T., Park, C., Rouse, J.H., and Siochi, E.J. (2002) Imaging carbon nanotubes in high performance polymer composites via magnetic force microscopy. *Nano Lett.*, **2** (8), 827–829.
163. Cui, H., Kalinin, S.V., Yang, X., and Lowndes, D.H. (2004) Growth of carbon nanofibers on tipless cantilevers for high resolution topography and magnetic force imaging. *Nano Lett.*, **4** (11), 2157–2161.
164. Hartschuh, A., Qian, H., Georgi, C., Bohmler, M., and Novotny, L. (2009) Tip-enhanced near-field optical microscopy of carbon nanotubes. *Anal. Bioanal. Chem.*, **394** (7), 1787–1795.
165. Schreiber, J., Demoisson, F., Humbert, B., Louarn, G., Chauvet, O., and Lefrant, S. (2003) First observation of the Raman spectrum of isolated single-wall carbon nanotubes by near-field optical Raman spectroscopy. *Mol. Nanostruct.*, **685**, 181–184.
166. Hartschuh, A., Sanchez, E.J., Xie, X.S., and Novotny, L. (2003) High-resolution near-field Raman microscopy of single-walled carbon nanotubes. *Phys. Rev. Lett.*, **90** (9).
167. Cancado, L.G., Hartschuh, A., and Novotny, L. (2009) Tip-enhanced Raman spectroscopy of carbon nanotubes. *J. Raman Spectrosc.*, **40** (10), 1420–1426.
168. Georgi, C. and Hartschuh, A. Tip-enhanced Raman spectroscopic imaging of localized defects in carbon nanotubes. *Appl. Phys. Lett.*, **97** (14).
169. Scolari, M., Mews, A., Fu, N., Myalitsin, A., Assmus, T., Balasubramanian, K., Burghard, M., and Kern, K. (2008) Surface enhanced Raman scattering of carbon nanotubes

decorated by individual fluorescent gold particles. *J. Phys. Chem. C*, **112** (2), 391–396.

170. Jorio, A., Dresselhaus, G., Dresselhaus, M.S., Souza, M., Dantas, M.S.S., Pimenta, M.A., Rao, A.M., Saito, R., Liu, C., and Cheng, H.M. (2000) Polarized Raman study of single-wall semiconducting carbon nanotubes. *Phys. Rev. Lett.*, **85** (12), 2617–2620.

171. Rao, A.M., Jorio, A., Pimenta, M.A., Dantas, M.S.S., Saito, R., Dresselhaus, G., and Dresselhaus, M.S. (2000) Polarized Raman study of aligned multiwalled carbon nanotubes. *Phys. Rev. Lett.*, **84** (8), 1820–1823.

172. Casiraghi, C., Hartschuh, A., Qian, H., Piscanec, S., Georgi, C., Fasoli, A., Novoselov, K.S., Basko, D.M., and Ferrari, A.C. (2009) Raman spectroscopy of graphene edges. *Nano Lett.*, **9** (4), 1433–1441.

173. Qian, H.H., Georgi, C., Anderson, N., Green, A.A., Hersam, M.C., Novotny, L., and Hartschuh, A. (2008) Exciton energy transfer in pairs of single-walled carbon nanotubes. *Nano Lett.*, **8** (5), 1363–1367.

174. Llopis, J., McCaffery, J.M., Miyawaki, A., Farquhar, M.G., and Tsien, R.Y. (1998) Measurement of cytosolic, mitochondrial, and Golgi pH in single living cells with green fluorescent proteins. *Proc. Natl. Acad. Sci. U.S.A.*, **95** (12), 6803–6808.

175. Nakayama-Ratchford, N., Bangsaruntip, S., Sun, X.M., Welsher, K., and Dai, H.J. (2007) Noncovalent functionalization of carbon nanotubes by fluorescein-polyethylene glycol: supramolecular conjugates with pH-dependent absorbance and fluorescence. *J. Am. Chem. Soc.*, **129** (9), 2448–244+.

176. Kagan, M.R. and McCreery, R.L. (1994) Reduction of fluorescence interference in Raman-spectroscopy via analyte adsorption on graphitic carbon. *Anal. Chem.*, **66** (23), 4159–4165.

177. Chaudhary, S., Kim, J.H., Singh, K.V., and Ozkan, M. (2004) Fluorescence microscopy visualization of single-walled carbon nanotubes using semiconductor nanocrystals. *Nano Lett.*, **4** (12), 2415–2419.

178. Chao, W.L., Harteneck, B.D., Liddle, J.A., Anderson, E.H., and Attwood, D.T. (2005) Soft X-ray microscopy at a spatial resolution better than 15nm. *Nature*, **435**, 1210–1213.

179. Bertilson, M.C., Takman, P.A.C., Holmberg, A., Vogt, U., and Hertz, H.M. (2007) Laboratory arrangement for soft x-ray zone plate efficiency measurements. *Rev. Sci. Instrum.*, **78** (2), 026103.

180. Takman, P.A.C., Stollberg, H., Johansson, G.A., Holmberg, A., Lindblom, M., and Hertz, H.M. (2007) High-resolution compact X-ray microscopy. *J. Microsc. (Oxford)*, **226** (2), 175–181.

181. von Hofsten, O., Takman, P.A.C., and Vogt, U. (2007) Simulation of partially coherent image formation in a compact soft X-ray microscope. *Ultramicroscopy*, **107** (8), 604–609.

182. Bertilson, M.C., Takman, P.A.C., Holmberg, A., Vogt, U., and Hertz, H.M. (2007) Zone plate efficiency measurements with a laser-plasma source. Advances in X-Ray/Euv Optics and Components II, Vol. 6705, pp. F7050–F7050.

183. Brizuela, F., Brewer, C., Fernandez, S., Martz, D., Marconi, M., Chao, W., Anderson, E.H., Vinogradov, A.V., Artyukov, I.A., Ponomareko, A.G., Kondratenko, V.V., Attwood, D.T., Bertness, K.A., Sanford, N.A., Rocca, J.J., and Menoni, C.S. (2009) High resolution full-field imaging of nanostructures using compact extreme ultraviolet lasers. *J. Phys.: Conf. Ser.*, **186**, 012026 (3 pp).

184. Brewer, C.A., Brizuela, F., Wachulak, P., Martz, D.H., Chao, W., Anderson, E.H., Attwood, D.T., Vinogradov, A.V., Artyukov, I.A., Ponomareko, A.G., Kondratenko, V.V., Marconi, M.C., Rocca, J.J., and Menonil, C.S. (2008) Single-shot extreme ultraviolet laser imaging of nanostructures with wavelength resolution. *Opt. Lett.*, **33** (5), 518–520.

185. Barinov, A., Dudin, P., Gregoratti, L., Locatelli, A., Onur Mentes, T., Ángel Niño, M., and Kiskinova, M. (2009) Synchrotron-based photoelectron microscopy. *Nucl. Instrum. Methods Phys. Res. A: Accel. Spectrom. Detect. Assoc. Equip.*, **601** (1–2), 195–202.

186. Barinov, A., Gregoratti, L., Dudin, P., La Rosa, S., and Kiskinova, M. (2009) Imaging and spectroscopy of multiwalled carbon nanotubes during oxidation: defects and oxygen bonding. *Adv. Mater.*, **21** (19), 1916–1920.

187. Schmahl, G., and Rudolph, D. (1969) High power zone plates as image forming systems for soft X-rays. *Optik*, **29** (6), 577–585.

188. Niemann, B., Rudolph, D., and Schmahl, G. (1976) X-ray microscopy with synchrotron radiation. *Appl. Opt.*, **15** (8), 1883–1884.

189. Rarback, H., Shu, D., Feng, S.C., Ade, H., Kirz, J., McNulty, I., Kern, D.P., Chang, T.H.P., Vladimirsky, Y., Iskander, N., Attwood, D., McQuaid, K., Rothman, S. (1976) scanning-X-ray microscope with 75-nm resolution. *Rev. Sci. Instrum.*, **59** (1), 52–59.

190. Kirz, J., Ade, H., Anderson, E., Attwood, D., Buckley, C., Hellman, S., Howells, M., Jacobsen, C., Kern, D., Lindaas, S., McNulty, I., Oversluizen, M., Rarback, H., Rivers, M., Rothman, S., Sayre, D., and Shu, D. (1990) X-ray microscopy with the nsls soft-X-ray Undulator. *Phys. Scr.*, **T31**, 12–17.

191. Schneider, G., Wilhein, T., Niemann, B., Guttmann, P., Schliebe, T., Lehr, J., Aschoff, H., Thieme, J., Rudolph, D., and Schmahl, G. (1995) X-ray microscopy with high resolution zone plates – recent developments. *Proc. SPIE*, **2516**, 90–101.

192. Schmahl, G., Rudolph, D., and Niemann, B. (1984) High-resolution X-ray microscopy with zone plate microscopes. *J. Phys.*, **45** (NC-2), 77–81.

193. Sakdinawat, A., and Attwood, D. Nanoscale X-ray imaging. *Nat. Photonics*, **4** (12), 840–848.

194. Rehbein, S., Heim, S., Guttmann, P., Werner, S., and Schneider, G. (2009) Ultrahigh-resolution soft-X-ray microscopy with zone plates in high orders of diffraction. *Phys. Rev. Lett.*, **103** (11), 110801.

29
Metals and Alloys

Dominique Schryvers

In this chapter, a few applications of electron nanoscopic techniques and the extension of microscopy toward the nanoscale in the field of metals and alloys are reviewed. For the definition of nanoscale, the conventional understanding of any feature below 100 nm is considered. The focus is on the enhanced understanding of the behavior of the different materials, with the techniques used being discussed in Volume I of this publication. Moreover, the emphasis is on quantitative nanoscopy, in which accurate and precise numerical data is retrieved from images, diffraction patterns, and spectra. The examples cover topics from deformation in steels over precipitation by diffusion driven transformations to characteristics of displacive systems such as those exhibiting martensitic transformations. All samples are bulky or thin film in nature; for results on nanoparticles, the reader is referred to the chapter 26 on small particles.

29.1
Formation of Nanoscale Deformation Twins by Shockley Partial Dislocation Passage

Twinning induced plasticity (TWIP) steels have recently received considerable attention [1–3]. The acronym "TWIP" was proposed by Frommeyer and coworkers in 1998 [4, 5] when presenting the excellent mechanical properties of Fe-Mn-Si-Al grades exhibiting either twinning by martensitic transformation or mechanical twinning during straining. Other grades related to the older Fe-Mn-C Hadfield steels were also shown to exhibit a TWIP effect, which proved to be very effective for the increase in the work hardening rate [6, 7]. TWIP steels are thus essentially Fe-Mn-based austenitic steels presenting a huge work hardening rate and in which mechanical twins occur during straining. It is argued that these twins are responsible for the large work hardening rate since they play the role of planar obstacles to the dislocation glide [6–9]. Furthermore, in the Fe-Mn-C grade, the created mechanical twins internally present a huge density of sessile dislocations resulting from the mechanism of twin nucleation and growth [10], so that these twins can be considered as strong inclusions. Both dynamic Hall–Petch

Handbook of Nanoscopy, First Edition. Edited by Gustaaf Van Tendeloo, Dirk Van Dyck, and Stephen J. Pennycook.
© 2012 Wiley-VCH Verlag GmbH & Co. KGaA. Published 2012 by Wiley-VCH Verlag GmbH & Co. KGaA.

strengthening and a composite strengthening thus seem to be responsible for the work hardening rate.

The actual nucleation and growth mechanism of such deformation twins, however, is still under discussion. Several perfect dislocations present in the matrix of an Fe – 20 wt% Mn – 1.2 wt% C alloy were observed to split into two Shockley partial dislocations (SPDs) under low stress (2%) bounding narrow intrinsic stacking faults (SFs) as commonly observed in fcc materials with low stacking fault energy (SFE). The micrograph in Figure 29.1a shows an example of this dissociation in a conventional weak-beam dark field mode with a reflecting plane $g = 2\text{–}20$. Two pairs of dissociated SPDs (marked by small white arrows) with a separation distance of around 10 nm can clearly be observed in the primary plane. The Burgers vectors of these partial dislocations were measured as $b = a/6[-211]$ and $b = a/6[-12-1]$. They thus result from the dissociation of a $b = a/2[-110]$ perfect dislocation in the (111) plane [10].

Since the width separating the leading and trailing partials (i.e., the width of the induced SF) depends on the SFE, this distance was measured for the above material and compared to the SFE calculated using the following equation [11]:

$$d = \frac{G b_1 b_2 (2-\nu)}{8\pi \, SFE (1-\nu)} \left(1 - \frac{2\nu \cos 2\alpha}{2-\nu}\right)$$

where d is the distance between the two Shockley partials, G is the glide modulus, b_1 and b_2 are the absolute values of the Burgers vectors of the two Shockley partials, ν is the Poisson ratio, and α is the angle between the Burgers vector of the undissociated dislocation and the dislocation line.

The elastic constants used for the calculations were $C_{11} = 235.50$ GPa, $C_{12} = 138.55$ GPa, and $C_{44} = 117.00$ GPa [12]. Figure 29.1b shows a good fit between the measured distance between SPDs and the calculated distance for a SFE of

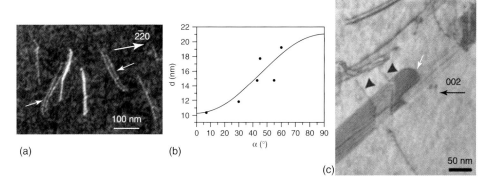

Figure 29.1 (a) Weak-beam dark field (WBDF) micrograph of perfect dislocations dissociated into two Shockley partial dislocations. (b) Comparison between the measured and calculated distances of separation of the Shockley partial dislocations as a function of the angle between the Burgers vector of the undissociated dislocation and the dislocation line. (c) Shockley partial dislocation in the conjugate twinning plane observed with $g = 002$. (Reproduced from [10].)

15 mJ m^{-2}. This value is in good agreement with reported experimentally measured [13–16] and calculated [17] SFE values of austenitic stainless steels.

The micrograph in Figure 29.1c shows bright field (BF) contrast of a large SF in the conjugate twinning plane. A curved dislocation line separating two different faulted regions in the twinning plane is indicated by a white arrow. This dislocation corresponds to a twinning SPD gliding in the (−1−11) plane, dragging a wide SF behind it, eventually creating twin bands. The black arrowheads point at two perfect dislocations connected to sessile dislocations.

Still, under identical deformation conditions, the two families of Fe-Mn-Si-Al and Fe-Mn-C steels exhibit clearly different mechanical behavior, resulting in a considerable difference in the work hardening rate. Figure 29.2a shows true stress–true strain curves for two TWIP steels (i) Fe – 20 wt% Mn – 1.2wt% C (Fe-Mn-C) and (ii) Fe – 28 wt% Mn – 3.5 wt% Si – 2.8 wt% Al (Fe-Mn-Si-Al). The respective strengths at necking are very different, while the yield strengths are similar and the true strains at necking are almost identical. As shown by the inset, this difference results from a large difference in the work hardening rate ($\theta = d\sigma/d\varepsilon$) exhibited by both steels. It is worth noting that the work hardening rate of the Fe-Mn-C grade is rather close to the theoretical limit of $\mu/20$ [18].

The large difference in the work hardening rate of common Fe-Mn-C and Fe-Mn-Si-Al TWIP steels can be explained by considering the structure of the mechanically induced twins. The finer twins in the Fe-Mn-C steel, full of sessile dislocations as seen in Figure 29.2b, are stronger and more effective for the improvement of the work hardening rate than the larger twin bands with less sessile dislocations at the matrix–twin interface in the Fe-Mn-Si-Al steel (Figure 29.2c) by which extended defects can form during further straining [10].

29.2
Minimal Strain at Austenite–Martensite Interface in Ti-Ni-Pd

Systems exhibiting a martensitic transformation, a prototype for first-order displacive transformations, are often used in special applications. Ancient and modern steel makers use the high strength of Fe-C martensite [19], whereas the shape memory and superelastic properties of Ni-Ti are used in many of today's functional materials going from simple actuators to high-grade medical devices [20].

One important feature in the application of shape memory and superelastic alloys is the existence of a hysteresis between transformation temperatures observed on cooling or heating or during stress loading and unloading. Understanding and controlling this hysteresis is of prime importance for any proper application of these materials. Recent studies have coupled the existence of this hysteresis to the energy dissipation during progress of the austenite–martensite interface, the so-called habit plane [21]. As a result of the lattice deformation occurring during transformation, this habit plane is in general not completely free of local strains. The system will try to minimize or compensate these local strains by introducing lamellar twinning or slip in the product phase. The twin width ratio

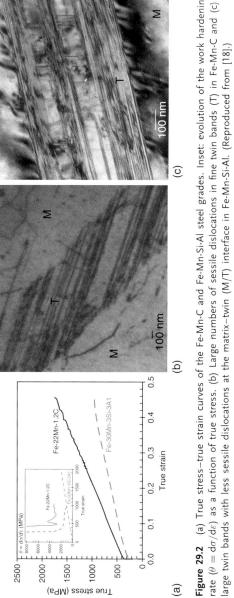

Figure 29.2 (a) True stress–true strain curves of the Fe-Mn-C and Fe-Mn-Si-Al steel grades. Inset: evolution of the work hardening rate ($\theta = d\sigma/d\varepsilon$) as a function of true stress. (b) Large numbers of sessile dislocations in fine twin bands (T) in Fe-Mn-C and (c) large twin bands with less sessile dislocations at the matrix–twin (M/T) interface in Fe-Mn-Si-Al. (Reproduced from [18].)

between twin variants depends on the ratios between the parent and product lattice parameters, while the actual twin width is a function of a competition between surface and volume energies, as a function of grain size. Generally, however, using twinning (or the alternative of slip), a perfect atomic match at the habit plane is never reached, which is believed to be related to energy dissipation mechanisms revealed as hysteresis and the formation of interface trailing dislocation when the transformation proceeds. However, it was recently shown that, next to the condition of no volume change, under certain particular conditions of the lattice parameters of the parent and product phases such a perfect match can exist and no twinning or slip is needed to accommodate any lattice mismatch [21, 22]. This is the case when the middle eigenvector λ_2 of the transformation matrix describing the lattice deformation from austenite to martensite equals 1 (with $\lambda_1 \leq 1$ and $\lambda_3 \geq 1$), which leads to a twin width ratio ω of 0, implying the disappearance of one of the twin variants following the equation

$$\omega = \frac{1}{2} - \frac{1}{2}\sqrt{1 - \frac{4|\lambda_2 - 1|}{|a_2 n_2|}}$$

with $a_2 n_2$ the y-component of the $\hat{a} \otimes n$ vector describing the twin system with n the normal to the twin plane [23, 24]. As an immediate consequence, the hysteresis H decreases to very low values for the condition $\lambda_2 = 1$ following

$$H = \frac{2\omega\theta_c |a|}{L} \sqrt{\frac{kE(1-\omega)(1+\overline{n \cdot m})}{(\overline{m \cdot n^\perp})l_c}}$$

with θ_c the transformation temperature defined as the average of the four characteristic transformation temperatures, that is, (As + Af + Ms + Mf)/4, |a| the length of vector a, L the total length of the strip selected for the calculation of the total energy (i.e., including martensite and austenite), κ the interfacial energy per unit area on the twins, E the elastic shear modulus, n the normal to the twin plane (thus with $n^\perp$ in the twin plane), m the normal to the habit plane, and l_c the length of twin domain for fully developed interfaces ($l_c \ll L$) [22].

The narrow (111) Type I twin in the high-resolution transmission electron microscopic (HRTEM) image in Figure 29.3a is an enlargement of part of the martensite plate in Figure 29.3b with a very low twin width ratio as found in the B19 orthorhombic martensite in a $Ti_{50}Ni_{30}Pd_{20}$ sample with a λ_2 value of 1.0060 [25–27]. Increasing Pd at the expense of Ni yields plates with more periodic lamellae of Type I microtwins, while decreasing Pd toward $Ti_{50}Ni_{39}Pd_{11}$ yields $\lambda_2 = 1.0001$. In the latter case, atomic resolution at the habit plane does not reveal any remaining strains or dislocations, confirming the perfect fit between both structures [25, 27]. This is shown in Figure 29.4, where some remaining austenite platelets are observed surrounded by untwinned martensite. Owing to the high speed of the martensitic transformation, the latter is not a trivial observation and can only be attained when the transformation temperature is close to ambient temperature, so that some but not all material has transformed at room temperature. The location of the habit plane in the HRTEM image is best recognized by a geometric phase analysis (GPA) approach [28] as seen on the right, revealing the changes in

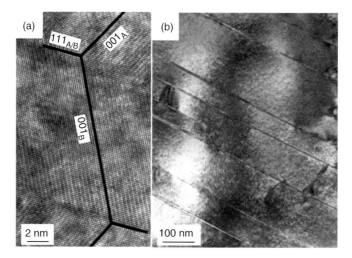

Figure 29.3 (a) HRTEM of a thin twin variant inside larger plates as seen from (b) in B19 martensite in $Ti_{50}Ni_{30}Pd_{20}$. (Reproduced from [25].)

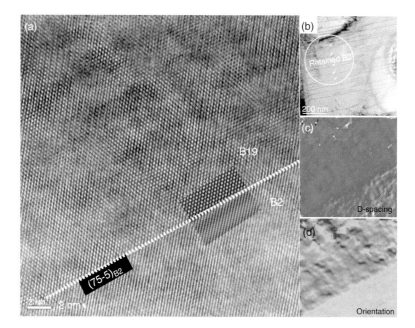

Figure 29.4 (a) HRTEM and (b) overview and of the austenite/martensite habit plane in $Ti_{50}Ni_{39}Pd_{11}$ ($\lambda_2 = 1.0001$) showing perfect fit without the need for dislocations or twins. The location of the habit plane is best recognized from (c) and (d) revealing the changes in d-spacing and lattice plane orientation obtained by GPA. (Reproduced from [27].)

d-spacing and lattice plane orientation. HRTEM multislice simulations, including the proper crystal tilt on both sides of the interface, further confirm the observed atomic lattice images. Also, the observed orientation of the habit plane fits with the predicted $(75\text{-}5)_{B2}$ indices, while groups of untwinned martensite plates form self-accommodating structures following the symmetry relations deducted earlier by Watanabe et al. [29].

Although the above-mentioned tendency of average twin width changing with nominal composition in the Ti-Ni-Pd series is clear and was also confirmed in Ti-Ni-Au samples, the large miscibility of the different elements in these metallic alloy systems implies that obtaining a perfect compositional homogeneity over large regions of the material is virtually impossible, certainly in polycrystalline samples. As a result, the intrinsically strong dependence of the λ_2 parameter on composition implies that small local changes in composition result in local differences in twin ratio. In the present material, it is moreover very hard to avoid precipitation of binary or even ternary phases, which will yield local changes in matrix composition and even compositional gradients near the precipitates. An example of this is shown in Figure 29.5a, where a two-step precipitation sequence occurring during quench in a homogenized $Ti_{50}Ni_{30}Pd_{20}$ alloy has produced an area with one large cubic (Fd3m) but incoherent $Ti_2Ni(Pd)$ precipitate surrounded by coherent tetragonal (C_{11b}) $Ti_2Pd(Ni)$ platelets with different orientation variants. Figure 29.5b shows a HRTEM image of a single $Ti_2Pd(Ni)$ platelet embedded in the B2 matrix and observed along a $[100]_{B2/Ti2Pd}$ zone orientation. GPA was applied to the HRTEM image in order to evaluate the interplanar spacing variations. No interplanar spacing variations above 0.5% (precision of GPA) could be detected for the $(010)_{B2}/(010)_{Ti2Pd}$ planes, which was attributed to a slight enrichment of Ti in the matrix surrounding the central $Ti_2Ni(Pd)$ precipitate leading to an increase in B2 lattice parameters [30]. GPA also allows an easy visualization of the interfaces between the two phases when selecting the $(011)_{B2}/(031)_{Ti2Pd}$ planes. Figure 29.5c,d reveals the changes in d-spacing and lattice rotation, respectively, from which a straight lower interface is seen while the upper interface exhibits a step at which a dislocation is found, as detailed in Figure 29.5e, which is an inverse Fourier transform of the same area using the $(011)_{B2}$ reflections highlighting the shift in lattice planes. The extra half plane terminates in a $(01\text{-}1)_{B2}$ plane, which is the glide plane containing the Burgers vector and the dislocation line, which is in agreement with a glissile dislocation. We further note that the d-spacing for the $(011)_{B2}/(031)_{Ti2Pd}$ planes is larger in the precipitate than in the matrix (Figure 29.5c) and that, as a consequence, the extra half plane at the dislocation is introduced in the B2 matrix on the right side of the step configuration.

29.3
Atomic Structure of Ni_4Ti_3 Precipitates in Ni-Ti

Although the above-mentioned precipitates can disturb the martensitic transformation by affecting the composition of the austenite matrix, the opposite is true for Ni_4Ti_3 precipitates, which are often intentionally grown in Ni-Ti shape

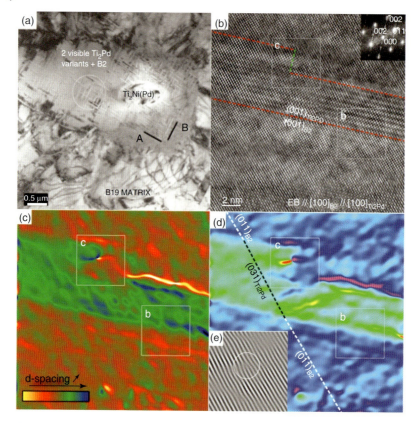

Figure 29.5 (a) BF image showing a precipitation zone containing a central $Ti_2Ni(Pd)$ precipitate surrounded by two sets of perpendicular $Ti_2Pd(Ni)$ platelets embedded in a B2 matrix in homogenized $Ti_{50}Ni_{30}Pd_{20}$. (b) HRTEM image of a $Ti_2Pd(Ni)$ precipitate in $[100]_{Ti2Pd}/[100]_{B2}$ orientation. The interface between the two phases is indicated by a dashed line. (c–d) GPA maps of the interplanar space variations and $(011)_{B2}/(031)_{Ti2Pd}$ lattice rotation, respectively. (e) Fast Fourier transform (FFT) revealing an interface dislocation in the circled area. (Reproduced from [30].)

memory material to stabilize the matrix composition and to provide the needed pinning points for the microscopic martensite variant selection. These precipitates are formed when binary Ni-Ti is annealed at moderate temperatures (around 400–500 °C); they have an ordered crystal structure, are enriched in Ni, and have an estimated Ni : Ti = 4 : 3 composition ratio. The structure of these precipitates was first proposed by Tadaki *et al.* (1986) to be rhombohedral with the space group R-3 [31]. A morphological study by the same authors reveals a lens shape with eight (4 × 2) variants, which was confirmed by Nishida *et al.* [32] and Nishida and Wayman [33]. Each of the four orientation variants have a different $\{111\}_{B2}$ plane as an interface with the cubic B2 matrix and further consist of two ordering variants. The replacement of some B2 Ti atoms by Ni results in a rhombohedral distortion since the Ni atom radius is smaller than that of Ti. The atomic positions in the unit cell proposed by Tadaki are those of the cubic structure projected into the rhombohedral one, which could be expected to be a first-order approximation of the real atomic positions. Since this structure is metastable and exists only in a solid solution with the B2 matrix and since the size of the precipitates is in the range of a few nanometers up to a few microns, it is impossible to perform X-ray or neutron diffraction on a single Ni_4Ti_3 crystal in order to refine its structure via conventional kinematical refinement methods. In transmission electron microscopy (TEM), however, it is possible to use a fine electron probe and obtain diffraction measurements of a single nanoscale precipitate. With the use of the multislice least squares (MSLS) method developed by Zandbergen and Jansen [34] and Jansen and Zandbergen [35], it is possible to refine the crystal structure from these dynamical diffraction patterns in much the same way as is done for kinematic X-ray or neutron refinements.

The experimental part of the MSLS procedure consists of making a series of electron diffraction pattern recordings in different crystal-zone orientations and preferably for different thicknesses of the specimen. A CM30 microscope equipped with a field emission gun and 1 K charge-coupled device (CCD) camera was used to make digital recordings of the diffraction patterns. The probe size of the electron beam has a diameter of around 40 nm, and the microscope was operated at 300 kV. In order to avoid an overlap between the matrix and precipitate and yet have sufficient number of ordering reflections from the precipitate, two suitable zone orientations, shown in Figure 29.6, were chosen: $[100]_{4:3}$ and $[111]_{4:3}$, which are parallel with the $[20\text{-}1]_{B2}$ and $[111]_{B2}$ matrix orientations, respectively [36].

The refinement procedure then consists in simulating these diffraction intensities and improving the fit with the experimental data by optimizing different parameters such as symmetry, atom positions, concentration, sample thickness, sample orientation, and so on, based on a well-chosen starting structure. In the

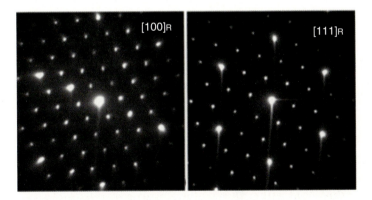

Figure 29.6 Examples of the two Ni_4Ti_3 diffraction patterns used for the MSLS refinement procedure (the streaks are due to the reading of the CCD). Especially the intensities of the super reflections are of interest. (Reproduced from [36].)

Table 29.1 Atomic positions of the Tadaki and MSLS refined unit cells, both with R-3 symmetry.

	x	y	z	Occupancy
Tadaki				
Ni1	0	0	0	1
Ni2	0.5	0.5	0.5	1
Ni3	0.0714 (= 1/14)	0.6429 (= 9/14)	0.7857 (= 11/14)	1
Ti	0.5714 (= 4/7)	0.1429 (= 1/7)	0.2857 (= 2/7)	1
MSLS				
Ni1	0	0	0	1
Ni2	0.5	0.5	0.5	1
Ni3	0.0605 (17)	0.5931 (11)	0.7574 (20)	1
Ti	0.4989 (10)	0.1125 (16)	0.2513 (16)	1

[a] Reproduced from [36].

present case, the latter was the result of a density functional theory (DFT) calculation, while changes in symmetry or concentration did not provide an improvement in the R-factor. The latter is 0.082 for the refined structure, whereas the original Tadaki structure yields a value of 0.125, about 50% worse. In Table 29.1, the resulting atomic positions are compared with the original ones provided by Tadaki et al., from which it is concluded that the absolute positions and distances between neighboring atoms differ significantly.

When the refined structure is projected along the $[111]_{4:3}$ direction, as shown in Figure 29.7, it is seen that the columns consisting of the sequence Ni–Ti–Ni–Ti– ... are distorted; the atomic centers no longer perfectly overlap as is the case for the Tadaki structure. The position of Ti is moved over 0.035 nm, which is 5.2% of the lattice constant, while Ni3 is moved over a distance of 0.027 nm,

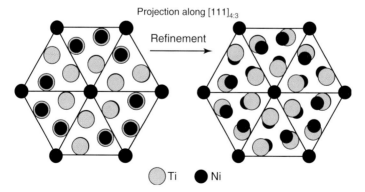

Figure 29.7 Projected structure of Ni_4Ti_3 in the $[111]_{4:3}$ direction before (Tadaki) and after MSLS refinement. (Reproduced from [36].)

equal to 4.0% of the lattice constant. As a result, the Ti atoms are now more widely spaced, since the distance between the two nearest neighbor Ti atoms increases from 0.298 to 0.308 nm. In the case of the Ni3 atoms, this distance decreases from 0.298 to 0.269 nm, while for the Ni2 atoms, the nearest neighbor distance remains 0.254 nm. The reason that the Ni and Ti atom centers are no longer on a straight line is related to the shrinking of the unit cell. In the $[111]_{4:3}$ direction, the rhombohedral structure contracts 2.9% in reference to the matrix, which is normally explained by the substitution of a large Ti atom by a smaller Ni atom. Owing to this substitution, some atom columns in the $[111]_{4:3}$ direction contain only Ni. For these columns, it is evident that the interatomic distance is shorter, but the other columns have the same composition and sequence as in the B2 structure and it would not be energetically favorable to contract these in the same direction. By moving away in different directions from the central $[111]_{4:3}$ axis, the component of the interatomic distance in the $[111]_{4:3}$ direction can become smaller. This explains the contraction of almost 3% in this direction in a more complete way than only giving credit to the smaller size of the Ni atom in one of the columns [36].

A similar refinement procedure was conducted on the so-called R-phase, an intermediate rhombohedral structure appearing during the martensitic transformation in Ni-Ti (in order to stabilize this R-phase, some Fe was added to the alloy). The obtained space group is P-3, where the centrosymmetric nature relates to symmetric shifts of Ti and Ni atoms away from the original {111} planes in the parent B2 structure [37]. Mirror planes parallel to the z-direction and observed in earlier convergent-beam electron diffraction (CBED) patterns, leading to a suggestion of a higher symmetry of P-31m, are explained as an artifact due to the presence of antiphase-like nanodomains.

29.4
Ni-Ti Matrix Deformation and Concentration Gradients in the Vicinity of Ni_4Ti_3 Precipitates

The difference in crystal structure between the Ni_4Ti_3 precipitates and the B2 matrix induces a small lattice mismatch between the cubic matrix and the rhombic precipitate along the interface. As mentioned above, the Ni_4Ti_3 structure is compressed by about 3% in the $<111>_{B2}$ direction perpendicular to the central plane when compared to the B2 matrix. When the precipitates remain small enough to be coherent (diameter below approximately 300 nm), the surrounding matrix is strained to accommodate this lattice mismatch [38]. Moreover, because of the higher Ni content of the precipitates with respect to the B2 matrix, which is close to the stoichiometric 50 : 50 composition, there exists a Ni depletion region surrounding each precipitate [39, 40].

Quantifying a nanoscale strain field around a precipitate can be done by measuring atomic positions from HRTEM images directly captured on CCD or via a scanning procedure from conventional photographic plates. In the present case, such images were obtained with a LaB$_6$ top-entry JEOL 4000EX microscope, avoiding serious image delocalization effects. In a first step, the strain field was characterized for a <110>$_{B2}$ viewing orientation in which the precipitate–matrix interface is viewed edge-on. As a result, no overlapping between both structures exists, and displacement measurements can be obtained with a variety of quantification techniques, mostly including some form of Fourier transform. The results show that the nanoscale precipitates induce a nanoscale deformation region, confirming the expected lattice extension close to the large interface plane and lattice compression at the precipitate edge area. Since in this case the observation is limited to a single orientation, however, only two-dimensional (2D) information could be obtained [38].

Combining two different orientations with independent lattice directions we were able to obtain three-dimensional data on the strain field in the matrix [41]. In the present case, the [10-1]$_{B2}$ and [1-11]$_{B2}$ directions were used, with the remark that, for the latter, the precipitate–matrix interface is inclined over an angle of 19°, leading to a small overlapping region (a schematic of both viewing orientations with respect to the position of the precipitate in the matrix is shown in Figure 29.8a). Consequently, the present results concern a small region about 10 nm inside the matrix (white squares in lower images of Figure 29.8b) in a direction along the central normal to the basal plane of the precipitate, that is, not including the overlapping region. The high-resolution images shown in Figure 29.8b were

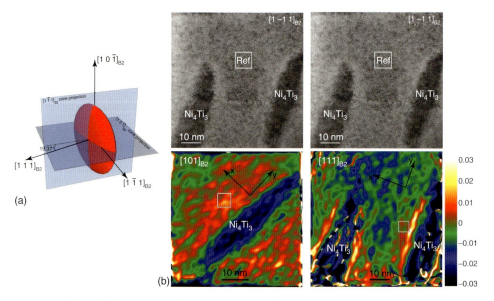

Figure 29.8 (a) Schematic representation of the precipitate with respect to the matrix orientations and HRTEM observation directions [10-1]$_{B2}$ and [1-11]$_{B2}$ (b) HRTEM images and corresponding ε_{xx} strain component with the undeformed matrix used as reference. (Reproduced from [41].)

Table 29.2 Principal strains and directions as measured from the combined HRTEM lattice images of the strained matrix and those calculated for the B2 → R transformation (single variant).

		Principal strains			Principal directions		
E	$E =$	$\begin{pmatrix} -0.0056 \\ 0 \\ 0 \end{pmatrix}$	$\begin{pmatrix} 0 \\ -0.0033 \\ 0 \end{pmatrix}$	$\begin{pmatrix} 0 \\ 0 \\ 0.0110 \end{pmatrix}$	$\begin{pmatrix} -0.2618 \\ -0.4903 \\ 0.8313 \end{pmatrix}$	$\begin{pmatrix} -0.7553 \\ 0.6403 \\ 0.1398 \end{pmatrix}$	$\begin{pmatrix} 0.6008 \\ 0.5913 \\ 0.5380 \end{pmatrix}$
R1	$E =$	$\begin{pmatrix} -0.0059 \\ 0 \\ 0 \end{pmatrix}$	$\begin{pmatrix} 0 \\ -0.0059 \\ 0 \end{pmatrix}$	$\begin{pmatrix} 0 \\ 0 \\ 0.0121 \end{pmatrix}$	$\begin{pmatrix} -0.2991 \\ -0.5084 \\ 0.8075 \end{pmatrix}$	$\begin{pmatrix} -0.7598 \\ 0.6389 \\ 0.1209 \end{pmatrix}$	$\begin{pmatrix} 0.5774 \\ 0.5774 \\ 0.5574 \end{pmatrix}$

Reproduced from [41].

treated with the GPA method [28], yielding a precision of 0.6% on the atomic displacements. In Figure 29.8b also the color maps of both GPA analyses showing plots of ε_{xx} are presented using the undeformed matrix lattice as reference. From these (and similar data sets for ε_{yy} and ε_{zz}) the principal strain components of the chosen deformation region can be calculated, yielding the E values in Table 29.2. Also, the precipitates are seen to retain their fixed lattice parameter, confirming an earlier observation by electron energy low-loss spectroscopy revealing larger elastic moduli for the precipitate than for the matrix [40].

Table 29.2 also shows the transformation strain values for a single variant of the R-phase, a structure that is often seen to appear before the martensitic transformation and in the vicinity of the Ni_4Ti_3 precipitates [42]. The correspondence between both sets of values is apparent, with the same signs and order of magnitude for the strains indicating proper correspondence between lattice compression and tension and only a difference of maximum 2.5° between the respective principal directions. In reality, however, the nucleating R-phase will appear with at least two variants, which is explained by the competition of the above correspondence with the need for energy minimization and the fit at the habit plane with the matrix. Moreover, it is believed that the scale of the distance between adjacent precipitates can further inhibit the martensitic transformation.

The Ni-depletion zone surrounding the precipitates can be measured by different local spectroscopy techniques in a TEM such as energy-dispersive X-ray spectroscopy (EDX) or electron energy loss spectroscopy (EELS), even by using mapping techniques such as EFTEM [39, 40]. Quantification by EELSMODEL [43] confirms the proposed 4 : 3 composition ratio of the precipitates and reveals a dip in the Ni content of the matrix close to the precipitate by about 4%, from which a good matching between the concentrations in the precipitate and the depletion zone can be concluded. The latter is shown in Figure 29.9, in which the contamination cones indicate the measurement positions of the nanoprobe.

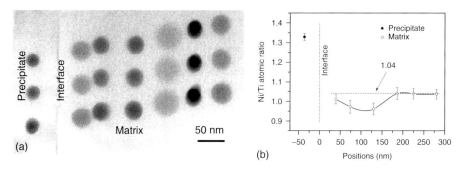

Figure 29.9 (a) Zero-loss image of a large Ni$_4$Ti$_3$ precipitate with contamination cones in the B2 matrix marking the positions for compositional quantification. (b) Measured Ni/Ti ratio indicating a Ni-depletion zone close to the precipitate. (Reproduced from [39].)

29.5
Elastic Constant Measurements of Ni$_4$Ti$_3$ Precipitates

Aside from structural and chemical information that can be obtained by TEM on nanoscale precipitates, the fine structure in a core-loss EELS spectrum, the so-called electron energy near-edge structure (ELNES) can provide detail on the bonding characteristics of the atoms, while plasmon losses can be translated into dielectric properties, nanoparticle size and shape, band-gap determination, and also elastic constants [44]. Figure 29.10 shows the plasmon loss spectra of a Ni$_4$Ti$_3$ precipitate compared with that of the B2 matrix. The shift of the precipitate volume

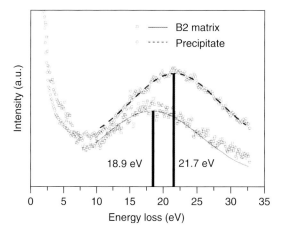

Figure 29.10 Plasmon loss spectra from B2 Ni-Ti ($E_p = 18.9$ eV) and Ni$_4$Ti$_3$ precipitates ($E_p = 21.7$ eV) indicating an increased hardness for the precipitates. (Reproduced from [40].)

plasmon peak E_p toward higher energies with respect to the one from the B2 matrix indicates a higher bulk (B) or Young's (Y) modulus, following the $B = \alpha E_p^\beta$ relation as discussed by Oleshko et al. [45]. From Figure 29.10, the plasmon energies for the B2 phase and the Ni_4Ti_3 precipitate were determined to be 18.9 ± 0.5 and 21.7 ± 0.6 eV, respectively, and after fitting with Gaussian functions. The bulk modulus for the precipitates is evaluated as 141 ± 3 GPa and that of the matrix as 106 ± 3 GPa. This correlates with the fact that the precipitates have an atomic structure denser than the B2 matrix and that they do not undergo a displacive transformation on cooling, while the surrounding matrix will transform martensitically to a more closely packed structure [40]. Again, as with the MSLS refinement discussed above, because of the metastable nature of the precipitates, this information can be obtained only using a nanoprobe energy loss spectroscopy approach in a TEM.

29.6
New APB-Like Defect in Ti-Pd Martensite Determined by HRSTEM

The addition of a third element such as Pd (see also Section 29.2) or Cu at the expense of Ni yields changes in the binary transformation sequence of cubic B2 to monoclinic B19' into a more complex one, including an orthorhombic B19 structure in between. In the extreme case of completely replacing Ni by Pd, a new type of antiphase boundary (APB)-like defect in the B19 martensite was recently described using high-resolution and high-angle annular dark field scanning transmission electron microscopy (HAADF STEM) [46]. These APB-like defects lie primarily on the $(001)_{B19}$ basal plane and occur within a single variant of martensite and can thus not be attributed to the nucleation of different orientation or twin variants. Also, the contrast is not inherited from any conventional APBs in the B2 parent phase, which have a displacement vector of the type $R = (1/2)<111>$. The extinction conditions together with the Z-contrast images, shown in Figure 29.11, result in a displacement vector $\boldsymbol{R}$ of $[uvw] = [1/3\ 0\ -1/2]$ over the interface. As a result, this defect is referred to as a *displacive transformation-induced APB*, which originates in a local heterogeneity of atomic movements due to the existence of two possible sublattices for the nucleation of the shuffling during the B19 martensitic transformation. Recently, similar APB-like defects have also been observed in the ternary Ti-Ni-Pd series [26].

29.7
Strain Effects in Metallic Nanobeams

Nanocrystalline metallic thin films often suffer from a lack of ductility because of a poor strain hardening capacity (see, e.g., Ref. 47). This low ductility impedes the use of these structures in a wide range of applications such as flexible electronics, microelectromechanical systems (MEMS) devices, and thin functional coatings, in which the ability of the materials to deform, stretch, or permanently change shape

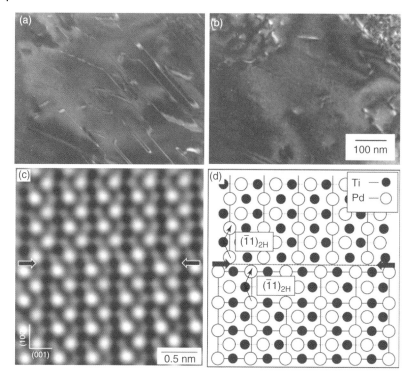

Figure 29.11 DF images of an area showing APB-like defects in Ti-Pd B19 martensite viewed with (a) $g = 100$ and (b) $g = 002$ near to the $[010]_{B19}$ zone. (c) Fourier-filtered HAADF STEM image along $[010]_{B19}$ together with a schematic in (d) revealing the positions of the Pd atom columns as white dots/circles, immediately showing the displacement vector at the defect. (Courtesy of M. Matsuda et al. [46].)

without cracking must be controlled and optimized to improve manufacturability and reliability. However, using proper preparation conditions, very high strain hardening capacities can be found in particular metallic nanocrystalline thin films such as Al and Pd, sometimes strongly depending on specimen size and film thickness [48].

The present example refers to nanocrystalline Pd films that have been deposited by an electron-gun vacuum system on (111) Si with a 1 µm thick SiO_2 intermediate layer and a 5 nm Cr adhesion layer. The thickness of the films ranges from 80 to 310 nm, and they have been reshaped by lithography to form parallel beams of 1–6 µm width. The mechanical tests have been performed using a novel concept of on-chip nanomechanical testing by which internal stresses, present in an actuating beam deposited and shaped during the same procedure (here a 100 nm thick Si_3N_4 layer), deform the attached material (here Pd) by removing the underneath sacrificial layer separating the two materials from the substrate (a detailed description of this procedure can be found elsewhere [49]). Using a high-temperature deposition process, large internal stresses up to about 1 GPa

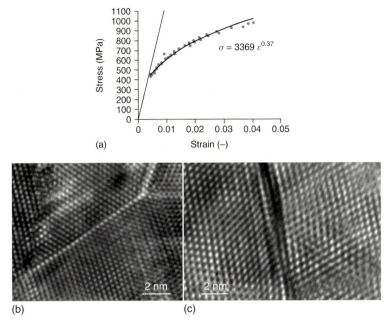

Figure 29.12 (a) Stress–strain curve of sputter-deposited Pd thin film of 310 nm thickness. HRTEM images of twin interfaces (b) before and (c) after application of mechanical stress, clearly showing the difference between coherent twin planes in the former and a strained twin interface and lattice in the latter. (Courtesy of H. Idrissi et al. [52])

can be reached. A single point in the stress–strain curve of the deformed material is provided by the measurement of one given displacement after the release step, coupled with additional experiments used to determine the elastic properties of both the actuator and the Pd thin film. A complete stress-strain curve is then obtained by varying the length ratio between the actuator and the Pd specimen [50, 51]. An example of such a stress–strain curve for a Pd film with a thickness of 310 nm is given in Figure 29.12a, from which a yield stress of about 460 MPa and a large strain hardening coefficient of 0.37 are measured.

TEM of cross-section focused-ion-beam (FIB) samples of the Pd beams reveals a columnar growth of nanoscale grains without clear texture. The lateral diameter of the grains appears independent of the thickness of the films, whereas the height of the columns increases with increasing film thickness. Roughly speaking, the grain height is slightly larger than half the film thickness. Irrespective of film thickness and despite the nanoscale size of the grains, a number of coherent growth twins are observed throughout the sample. An example of such a growth twin observed in unreleased material is shown in Figure 29.12b. Such coherent growth twins offer multiple barriers to dislocation motion as well as sources for dislocation storage and multiplication. The coherency of the twin boundaries is seen to decrease after deformation with the accumulation of sessile dislocations, as shown in Figure 29.12c. The dislocation/twin boundary interactions are thought to be responsible for the remarkable mechanical properties of the Pd thin films with a long elastoplastic transition [52].

29.8
Adiabatic Shear Bands in Ti6Al4V

Conventional diffraction experiments in a TEM have long been limited to a few well-defined crystallographic zones. Recently, however, new automated tools for obtaining large numbers of diffraction patterns have been developed. Computational optimization tools already provide automatic recognition software for the long-standing kinematic diffraction experiments such as X-ray diffraction and more recently also for electron backscattered diffraction (EBSD) in a scanning electron microscope (SEM), including inelastic and dynamic scattering at relatively low voltages. In a TEM, strong dynamic diffraction features lead to more complex diffraction intensities (depending, e.g., on the thickness and orientation of the sample), but these can be minimized by novel spinning methods averaging out diffraction intensities or, as mentioned above, be incorporated in the optimization software, leading to an increase in parameter space as done in MSLS [34, 53]. Alternative applications of these novel diffraction techniques can result in more detailed nanostructural information using so-called fluctuation and conical dark field (CDF) imaging procedures [54, 55]. In Figure 29.13, an example of the use of CDF imaging, in which a series of dark field images is produced by tilting the incident electron beam around a conical surface in such a way that the ring of diffracted intensity, due to a region with nanoscale grains, passes through the objective aperture [55]. This yields a 3D data box from which the orientation of each

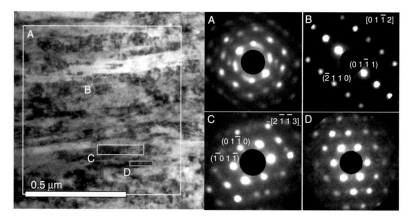

Figure 29.13 CDF image and selected pixel diffraction patterns from an adiabatic shear region in Ti6Al4V. The entire data set is 256 × 256 × 4076 pixels large. (Courtesy of W. Tirry.)

single nanograin can be retrieved. The present example shows an adiabatic shear band (ASB) in hcp α Ti6Al4V of around 1 μm width. The presented diffraction patterns are compiled from the large CDF dataset and indicate local orientation changes in elongated subgrains. The entire region of the image, labeled a, does not show a single zone diffraction pattern, whereas in the smaller elongated regions b and c, two different zone axes can be recognized. Region d appears to have the same zone as region c, but with a different in-plane orientation. Conventional selected area electron diffraction or even microdiffraction cannot obtain the needed lateral resolution for these types of problems. Ultimately diffraction information from a single pixel can be obtained, which, in the present example, corresponds with an area of 36×36 nm^2.

29.9
Electron Tomography

The evolution toward real three-dimensional (3D) TEM is one of the promising paths for future materials characterization. Different tomographic techniques, including dedicated reconstruction procedures, are available or under development. Present applications include conventional dark field (DF) imaging and direct Z-contrast imaging in HAADF STEM. The combination with Cs-corrected machines will ultimately lead to atomic resolution tomography for chemical as well as structural information.

Diffusional processes leading to dual-phase microstructures or precipitation in alloys provide a prime research field for TEM and SEM tomography techniques. Although the examples shown here are at present still on the upper limit of nanoscopy, they already show the capabilities and nanoscale promises of these techniques. The first example is a Ni-based superalloy containing Al and Ti (Ni–8.5at%Al–5.4at%Ti), which shows a typical two-phase microstructure composed of L1$_2$-ordered γ' domains with cuboidal shapes embedded in an A1-disordered γ matrix. On further aging, this system exhibits a secondary phase separation in which precipitation of fine γ occurs in each γ' domain [56]. Depending on the alloy composition, a variety of precipitation behaviors and microstructural evolutions have been observed in the second stage of phase separation. Using tomographic dark field transmission electron microscopic (DFTEM) and energy-filtered transmission electron microscopic (EFTEM) observations, Hata et al. clarified the 3D morphology of the secondary γ precipitates and the compositional distribution in the phase-separated microstructure in this alloy [57]. Figure 29.14 shows conventional dark field images of two consecutive stages after aging at 1213 K for 45 minutes (Figure 29.14a) followed by an aging step at 1023 K for 48 h (Figure 29.14b). The small γ precipitates formed inside the larger γ' ones are clearly seen after the second aging. Keeping the DF diffraction conditions for a selected γ' precipitate, a series of such DF images along different orientations was taken (maximum specimen-tilt angle was $\pm 60°$ and images were taken every $2°$). The tomographic reconstruction followed by volume and surface rendering for a

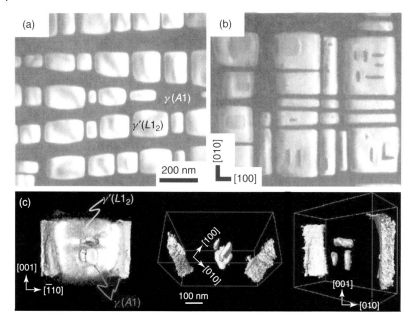

Figure 29.14 Conventional DF images of Ni–8.5at%Al–5.4at%Ti annealed at (a) 1213 K for 45 min followed by (b) 1023 K for 48 h. (c) The rendered volume of three small γ precipitates inside a single γ' precipitate. (Courtesy of S. Hata et al. [57].)

single γ' precipitate is shown in Figure 29.14c, from which the through rectangular plate shape, parallel to one of the {100} planes, of the nanoscale γ precipitates inside the cuboidal γ' ones can be observed. Moreover, the internal orthogonal γ plates are close to each other at their edges, which indicates the influence of elastic interactions between γ precipitates in γ' domains [58, 59].

When larger volumes need to be investigated for 3D information, slice-and-view techniques in a FIB/SEM can be employed. An example of such a problem in an alloy system is that of the distribution of Ni_4Ti_3 precipitates inside the Ni-Ti austenite matrix. When homogenized Ni-Ti material is annealed at the proper conditions of temperature and time, these precipitates will nucleate and grow to form a 3D configuration depending on internal and external parameters such as whether single or polycrystalline, grain size, existence of external compressive or elongation forces, and so on [60–63]. In Figure 29.15a, a 3D view of nanoscale precipitates located near to a grain boundary is shown. As shown in Figure 29.15b, a clear gradient in precipitate size can be recognized from this volume, which is attributed to small concentration differences due to the grain boundary proximity and plays an important role in the multiple-step martensitic transformation in this system [64]. On the basis of the original pixel data, treated to yield binary images, and using appropriate measuring and fitting techniques, large numbers of precipitates can be characterized for shape and dimension. By comparing small

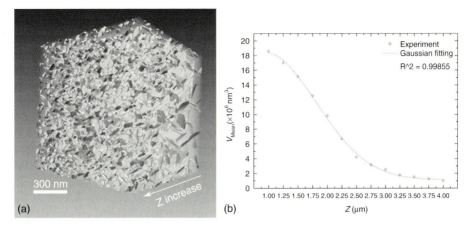

Figure 29.15 (a) Four variants of Ni_4Ti_3 precipitates distributed in a grain of Ni-Ti with a grain boundary close to the front edge. (b) Changing volume of the average precipitate when approaching the grain boundary. (Reproduced from [64].)

and larger precipitates, it is, for example, concluded that the sphericity for the smallest precipitates is larger than for the bigger ones, confirming the expected shape evolution from sphere to plate because of surface-volume energy competition for coherent precipitates.

A similar study on compression-annealed material reveals a stacking of precipitates of a single variant, the one with the normal to the central plane of the precipitate parallel to the compression direction. These precipitates enclose small boxes of austenite, leading to a burstlike martensitic transformation [65].

29.10
The Ultimate Resolution

Using an aberration-corrected TEM and working with monochromated beams, resolutions down to 50 pm can be reached with nanosized beams probing in between the atoms. Even when lowering the acceleration voltage to, for example, 80 kV in order to avoid radiation damage in soft materials, the resolution of the instruments remains below 0.1 nm. In combination with phase retrieval methods, this allows the accurate measurement of atomic positions at, for example, interfaces; location of low Z elements or even vacancies; and so on (see also Chapters 2, 3 and 8). However, not many examples at this level of resolution (and related precision) on metals or alloys have been conducted so far, possibly since such materials are hardly ever used as model systems to test the limits of new techniques and instrumentation. One example is the recent work by Rossell et al., who have reported the imaging of individual atom columns of Li in coherent core-shell precipitates in which the shell has the ordered Al_3Li $L1_2$ structure, whereas the core strongly resembles the surrounding Al matrix [66]. Aside from the obtained structural information on the core-shell structure, the effect of electron energy on the imaging characteristics of Li was investigated by comparing measurements at 80 kV employing a monochromated electron beam with an energy spread ΔE of 0.2 eV and at 300 kV with ΔE of 0.8 eV. These settings enable similar information transfer at both microscope operation conditions and allow a direct comparison between the 80 and 300 kV measurements. In Figure 29.16a,b, the atomic resolution in exit-plane-wave images from a focal series along the [100] $L1_2$ zone is shown for both imaging conditions. Although in both cases the Li columns

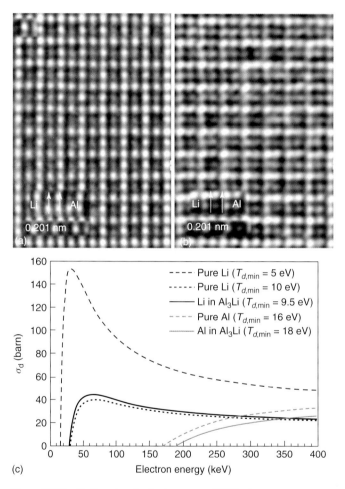

Figure 29.16 Al_3Li retrieved exit-plane wave (EPW) phase images from the shell of core-shell precipitates obtained at (a) 300 kV and (b) 80 kV. (c) Total scattering cross-section σ_d for atom displacement for Al and Li as a function of the primary electron energy and for different phases and critical recoil energy. (Courtesy of M. Rossell et al. [66].)

can be recognized, a considerable improvement with respect to conventional atomic resolution images, the chemical contrast is clearly reduced at 80 kV and local intensity variations strongly affect the Li signal. In other words, the Al/Li intensity ratio is markedly smaller at 80 kV than at 300 kV, which was shown to be in contradiction with predictions based on multislice simulations. The latter was attributed to unexpectedly strong radiation effects at 80 kV, confirmed by a time sequence of HRTEM images revealing structural changes in the Al_3Li shell while the Al core remains unaffected. This behavior can be understood in terms of the total scattering cross-section for atomic displacement σ_d based on relativistic elastic-scattering cross-section between electrons and atoms [66]. The result as a function of accelerating voltage for both elements Al and Li in different phases and for different estimates of the critical recoil energy $T_{d,min}$ to initiate atomic displacements is shown in Figure 29.16c. While σ_d for Al decreases monotonically with decreasing electron energy to disappear around 190 kV, Li displays an anomalous behavior with a maximum for σ_d around 40–60 kV and decreasing with increasing electron energy thus making the Al_3Li phase more susceptible to radiation damage at 80 kV than at 300 kV.

Another example is the characterization by aberration-corrected ADF-STEM of the defect core of radiation-induced dislocation loops in an Al film after Ga^+ irradiation in an FIB instrument [67]. The images were obtained on an FEI Titan 80-300 "cubed" microscope equipped with an aberration corrector for the imaging and probe forming lenses as well as a monochromator, resulting in a resolution of 50 pm in TEM mode and of 80 pm in STEM mode. The high-resolution annular dark field scanning transmission electron microscopic (ADF-STEM) image in Figure 29.17a

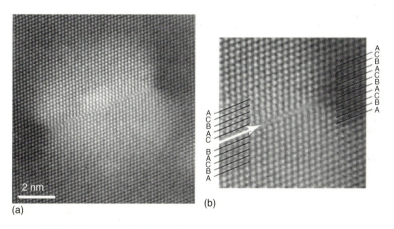

Figure 29.17 High-resolution aberration-corrected ADF-STEM image of a Frank loop induced during Ga^+ FIB thinning of an Al thin film. (b) Enlarged image of one termination point of the loop: the white arrow indicates the extra half plane. (Courtesy of H. Idrissi et al. [67].)

shows the presence of the defect core – a Frank loop – which is easily recognizable through the surrounding white strain-field contrast (the contrast in HAADF STEM appears to be much weaker because of a missing strain field contrast). The enlarged region displayed in Figure 29.17b clearly shows one atomic layer inserted in the Al matrix (white arrow), in agreement with the model of an interstitial Frank loop. The stacking sequence crossing the Frank loop plane is denoted as ABCAB/LOOP/CABCA, with the center loop plane being the extrinsic plane. The extra half plane of the dislocation is parallel to the fault plane in agreement with a Frank dislocation with Burgers vector perpendicular to the fault plane.

Acknowledgments

The author thanks R. Delville, W. Tirry, H. Idrissi, S. Turner, B. Wang, and Z.Q. Yang for support with the TEM observations and S. Cao for the SEM observations. Part of this work was performed in the framework of a European FP6 project "Multi-scale modeling and characterization for phase transformations in advanced materials" (MRTN-CT-2004-505226) and an IAP program of the Belgian State Federal Office for Scientific, Technical, and Cultural Affairs (Belspo), under Contract No. P6/24. Support was also provided by FWO projects G.0465.05 "The functional properties of SMA: a fundamental approach," G.0576.09 "3D characterization of precipitates in Ni–Ti SMA by slice-and-view in a FIB-SEM dual-beam microscope" and G.0180.08 "Optimization of Focused Ion Beam (FIB) sample preparation for transmission electron microscopy of alloys."

References

1. Grässel, O., Krüger, L., Frommeyer, G., and Meyer, L.W. (2000) High strength Fe-Mn-(Al, Si) TRIP/TWIP steels development – properties – application. *Int. J. Plast.*, **16**, 1391–1409.
2. Mi, Z.L., Tang, D., Yan, L., and Guo, J. (2005) High-strength and high-plasticity TWIP steel for modern vehicle. *J. Mater. Sci. Technol.*, **21**, 451–454.
3. Scott, C., Allain, S., Faral, M., and Guelton, N. (2006) The development of a new Fe-Mn-C austenitic steel for automotive applications. *Rev. Metall. Cah. Inf. Techn.*, **103**, 293.
4. Frommeyer, G. and Grassel, O. (1998) High strength TRIP/TWIP and super-plastic steels: development, properties, application. *Rev. Metall. Cah. Inf. Techn.*, **95**, 1299.
5. Grässel, O. and Frommeyer, G. (1998) Effect of martensitic phase transformation and deformation twinning on mechanical properties of Fe-Mn-Si-Al steels. *Mater. Sci. Technol.*, **14**, 1213.
6. Allain, S., Chateau, J.P., and Bouaziz, O. (2002) Constitutive model of the TWIP effect in a polycrystalline high manganese content austenitic steel. *Steel Res.*, **73**, 299.
7. Bouaziz, O. and Guelton, N. (2001) Modelling of TWIP effect on work-hardening. *Mater. Sci. Eng., A*, **319–321**, 246–249.
8. Bouaziz, O., Allain, S., and Scott, C. (2008) Effect of grain and twin boundaries on the hardening mechanisms of twinning-induced plasticity steels. *Scr. Mater.*, **58**, 484–487.
9. Raghavan, K.S., Sastri, A.S., and Marcinkowski, M.J. (1969) Nature of

work-hardening behavior in Hadfields manganese steel. *Trans. Metall. Soc. AIME*, **245**, 1569.
10. Idrissi, H., Renard, K., Ryelandt, L., Schryvers, D., and Jacques, P.J. (2010) On the mechanism of twin formation in FeMnC TWIP steels. *Acta Mater.*, **58**, 2464–2476.
11. Li, J.C., Zheng, W., and Jiang, Q. (1999) Stacking fault energy of iron-base shape memory alloys. *Mater. Lett.*, **38**, 275–277.
12. Bampton, C.C., Jones, I.P., and Loretto, M.H. (1978) Stacking fault energy measurements in some austenitic stainless steels. *Acta Metall.*, **26**, 39–51.
13. Sinclair, C.W., Poole, W.J., and Bréchet, Y. (2006) A model for the grain size dependent work hardening of copper. *Scr. Mater.*, **55**, 739–742.
14. Barlat, F., Glazov, M.V., Brem, J.C., and Lege, D.J. (2002) A simple model for dislocation behavior, strain and strain rate hardening evolution in deforming aluminum alloys. *Int. J. Plast.*, **18**, 919–939.
15. Kireeva, I.V., Luzginova, N.V., Chumlyakov, Y.I., Karaman, I., and Lichter, B.D. (2004) Plastic deformation of nitrogen-containing austenitic stainless steel single crystals with low stacking fault energy. *J. Phys. IV*, **115**, 223.
16. Bracke, L. (2006) *Deformation Behaviour of Austenitic Fe-Mn alloys by Twinning and Martensitic Transformation* University of Ghent.
17. Byun, T.S. (2003) On the stress dependence of partial dislocation separation and deformation microstructure in austenitic stainless steels. *Acta Mater.*, **51**, 3063–3071.
18. Idrissi, H., Renard, K., Schryvers, D., and Jacques, P. (2010) On the relationship between the twin internal structure and the work-hardening rate of TWIP steels. *Scr. Mater.*, **63**, 961–964.
19. Olson, G.B. and Owen, W.S. (eds) (1992) *Martensite: A Tribute to Morris Cohen*, ASM International.
20. Duerig, T., Pelton, A., and Stockel, D. (1999) An overview of nitinol medical applications. *Mater. Sci. Eng., A*, **273**, 149–160.
21. Cui, J., Chu, Y.S., Famodu, O.O., Furuya, Y. et al. (2006) Combinatorial search of thermoelastic shape-memory alloys with extremely small hysteresis width. *Nat. Mater.*, **5**, 286–290.
22. Zhang, Z.Y. (2007) *Special Lattice Parameters and the Design of Low Hysteresis Materials* University of Minnesota, Minnesota.
23. Ball, J.M. and James, R.D. (1987) Fine phase mixtures as minimizers of energy. *Arch. Ration. Mech. Anal.*, **100**, 13–52.
24. Ball, J.M. and James, R.D. (1992) Proposed experimental tests of a theory of fine microstructure and the 2-well problem. *Philos. Trans. R. Soc. London, Ser. A Math. Phys. Eng. Sci.*, **338**, 389–450.
25. Delville, R., Schryvers, D., Zhang, Z., and James, R.D. (2009) Transmission electron microscopy investigation of microstructures in low-hysteresis alloys with special lattice. *Scr. Mater.*, **60**, 293–296.
26. Delville, R. (2010) *From Functional Properties to Micro/Nano-structures: A TEM Study of TiNi(X) Shape Memory Alloys* Antwerp.
27. Delville, R., Kasinathan, S., Zhang, Z., van Humbeeck, J. et al. (2010) Transmission electron microscopy study of phase compatibility in low hysteresis shape memory alloys. *Philos. Mag.*, **90**, 177–195.
28. Hytch, M.J., Snoeck, E., and Kilaas, R. (1998) Quantitative measurement of displacement and strain fields from HREM micrographs. *Ultramicroscopy*, **74**, 131–146.
29. Watanabe, Y., Saburi, T., Nakagawa, Y., and Nenno, S. (1990) Self-accommodation structure in the Ti-Ni-Cu orthorhombic martensite. *Jpn. Inst. Met.*, **54**, 861.
30. Delville, R. and Schryvers, D. (2010) Transmission electron microscopy study of combined precipitation of $Ti_2Ni(Pd)$ and $Ti_2Pd(Ni)$ in a $Ti_{50}Ni_{30}Pd_{20}$ alloy. *Intermetallics*, **18**, 2353–2360.
31. Tadaki, T., Nakata, Y., Shimizu, K.I., and Otsuka, K. (1986) Crystal structure, composition and morphology of a precipitate in an aged Ti-51 at.% Ni shape memory alloy. *Trans. Jpn. Inst. Met.*, **27**, 731–740.

32. Nishida, M., Wayman, C.M., and Honma, T. (1986) Precipitation processes in near-equiatomic TiNi shape memory alloys. *Metall. Trans. A*, **17**, 1505–1515.
33. Nishida, M. and Wayman, C.M. (1987) Electron microscopy studies of precipitation processes in near-equiatomic TiNi shape memory alloys. *Mater. Sci. Eng.*, **93**, 191–203.
34. Zandbergen, H. and Jansen, J. (1998) Accurate structure determinations of very small (4-20 nm) areas, using refinement of dynamic electron diffraction data. *J. Microsc.*, **190**, 222–237.
35. Jansen, J. and Zandbergen, H.W. (2002) Determination of absolute configurations of crystal structures using electron diffraction patterns by means of least-squares refinement. *Ultramicroscopy*, **90**, 291–300.
36. Tirry, W., Schryvers, D., Jorissen, K., and Lamoen, D. (2006) Electron-diffraction structure refinement of Ni4Ti3 precipitates in Ni52Ti48. *Acta Crystallogr., Sect. B*, **62**, 966–971.
37. Schryvers, D. and Potapov, P.L. (2002) Ti-Ni SMAs-R-phase structure refinement using electron diffraction data. *Mater. Trans. (Jpn. Inst. Met.)*, **43**, 774–779.
38. Tirry, W. and Schryvers, D. (2005) Quantitative determination of strain fields around Ni4Ti3 precipitates in NiTi. *Acta Mater.*, **53**, 1041–1049.
39. Yang, Z.Q., Tirry, W., and Schryvers, D. (2005) Analytical TEM investigations on concentration gradients surrounding Ni4Ti3 precipitates in Ni-Ti shape memory material. *Scr. Mater.*, **52**, 1129–1134.
40. Yang, Z.Q., Tirry, W., Lamoen, D., Kulkova, S., and Schryvers, D. (2008) Electron energy-loss spectroscopy and first-principles calculation studies on a Ni-Ti shape memory alloy. *Acta Mater.*, **56**, 395–404.
41. Tirry, W. and Schryvers, D. (2009) Linking a completely three-dimensional nanostrain to a structural transformation eigenstrain. *Nat. Mater.*, **8**, 752–757.
42. Bataillard, L., Bidaux, J.E., and Gotthardt, R. (1998) Interaction between microstructure and multiple-step transformation in binary NiTi alloys using in-situ transmission electron microscopy observations. *Philos. Mag. A*, **78**, 327–344.
43. Verbeeck, J. and Bertoni, G. (2008) Model-based quantification of EELS spectra: treating the effect of correlated noise. *Ultramicroscopy*, **108**, 74–83.
44. Egerton, R.F. (1996) *Electron Energy-loss Spectroscopy in the Electron Microscope*, Plenum Press, New York.
45. Oleshko, V.P., Murayama, M., and Howe, J.M. (2002) Use of plasmon spectroscopy to evaluate the mechanical properties of materials at the nanoscale. *Microsc. Microanal.*, **8**, 350–364.
46. Matsuda, M., Hara, T., and Nishida, M. (2008) Crystallography and morphology of antiphase boundary-like structure induced by martensitic transformation in Ti-Pd shape memory alloy. *Mater. Trans. (Jpn. Inst. Met.)*, **49**, 461–465.
47. Haque, M.A. and Saif, M.T.A. (2003) Strain gradient effect in nanoscale thin films. *Acta Mater.*, **51**, 3053–3061.
48. Coulombier, M., Boe, A., Brugger, C., Raskin, J.P., and Pardoen, T. (2010) Imperfection sensitive ductility of aluminium thin films. *Scr. Mater.*, **62**, 742.
49. Gravier, S., Coulombier, M., Safi, A., André, N. et al. (2009) New on-chip nanomechanical testing laboratory – applications to aluminum and polysilicon thin films. *J. Microelectromech. Syst.*, **18**, 555.
50. Boé, A., Safi, A., Coulombier, M., Fabrègue, D. et al. (2009) MEMS-based microstructures for nanomechanical characterization of thin films. *Smart Mater. Struct.*, **18**, 115018–115018.
51. Boé, A., Safi, A., Coulombier, M., Pardoen, T., and Raskin, J.P. (2009) Internal stress relaxation based method for elastic stiffness characterization of very thin films. *Thin Solid Films*, **518**, 260–264.
52. Idrissi, H., Wang, B., Colla, M.S., Raskin, J.P., Schryvers, D., and Pardoen, T., (2011) Ultrahigh Strain Hardening in Thin Palladium Films with Nanoscale Twins. *Advanced Materials*, **23**, 2119–2122.

53. Jansen, J., Tang, D., Zandbergen, H.W., and Schenk, H. (1998) MSLS, a least-squares procedure for accurate crystal structure refinement from dynamical electron diffraction patterns. *Acta Crystallogr., Sect. A*, **54**, 91–101.
54. Voyles, P.M., Gibson, J.M., and Treacy, M.M. (2000) Fluctuation microscopy: a probe of atomic correlations in disordered materials. *J. Electron. Microsc.*, **49**, 259–266.
55. Wu, X.P., Kalidindi, S.R., Necker, C., and Salem, A.A. (2008) Modeling anisotropic stress-strain response and crystallographic texture evolution in alpha-titanium during large plastic deformation using taylor-type models: influence of initial texture and purity. *Metall. Mater. Trans., A Phys. Metall. Mater. Sci.*, **39A**, 3046–3054.
56. Tian, W.H., Sano, T., and Nemoto, M. (1989) Precipitation of disordered γ phase in an L12 ordered γ'-Ni3(Al,Ti) phase. *J. Jpn. Inst. Met.*, **53**, 1013.
57. Hata, S., Kimura, K., Gao, H., Matsumura, S. et al. (2008) Electron tomography imaging and analysis of γ' and γ domains in Ni-based Superalloys. *Adv. Mater.*, **20**, 1905–1909.
58. Doi, M. (1996) Elasticity effects on the microstructure of alloys containing coherent precipitates. *Prog. Mater. Sci.*, **40**, 79.
59. Doi, M., Moritani, T., Kozakai, T., and Wakano, M. (2006) Transmission electron microscopy observations of the phase separation of $D0_3$ precipitates in an elastically constrained Fe-Si-V alloy. *ISIJ Int.*, **46**, 155–160.
60. Fujishima, K., Nishida, M., Morizono, Y., Yamaguchi, K. et al. (2006) Effect of heat treatment atmosphere on the multistage martensitic transformation in aged Ni-rich Ti-Ni alloys. *Mat. Sci. Eng., A*, **438–440**, 489–494.
61. Khalil Allafi, J., Ren, X., and Eggeler, G. (2002) The mechanism of multistage martensitic transformations in aged Ni-rich NiTi shape memory alloys. *Acta Mater.*, **50**, 793–803.
62. Khalil-Allafi, J., Schmahl, W.W., Wagner, M., Sitepu, H. et al. (2004) The influence of temperature on lattice parameters of coexisting phases in NiTi shape memory alloys - a neutron diffraction study. *Mater. Sci. Eng., A*, **378**, 161–164.
63. Nishida, M., Hara, T., Ohba, T., Yamaguchi, K. et al. (2003) Experimental consideration of multistage martensitic transformation and precipitation behavior in aged Ni-rich Ti-Ni shape memory alloys. *Mater. Trans.*, **44**, 2631–2636.
64. Cao, S., Nishida, M., and Schryvers, D. (2011) Quantitative 3D Analysis of Ni_4Ti_3 Precipitate Morphology and Distribution in Polycrystalline Ni-Ti, *Acta Materialia*, **59**, 1780–1789.
65. Cao, S., Somsen, C., Croitoru, M., Schryvers, D., and Eggeler, G. (2010) Focused ion beam/scanning electron microscopy tomography and conventional transmission electron microscopy assessment of Ni_4Ti_3 morphology in compression-aged Ni-rich Ni–Ti single crystals. *Scr. Mater.*, **62**, 399–402.
66. Rossell, M., Erni, R., Asta, M., Radmilovic, V., and Dahmen, U. (2009) Atomic-resolution imaging of lithium in Al3Li precipitates. *Phys. Rev., B*, **80**, 024110.
67. Idrissi, H., Turner, S., Mitsuhara, M., Wang, B., Hata, S. Coulombier, M., Raskin, J.P., Pardoen, T., Van Tendeloo, G., and Schryvers, D. (2011) Point Defect Clusters and Dislocations in FIB Irradiated Nanocrystalline Aluminum Films: An Electron Tomography and Aberration-Corrected High-Resolution ADF-STEM Study. *Microscopy AND Microanalysis*, **17**, 983–990.

30
In situ Transmission Electron Microscopy on Metals
J.Th.M. De Hosson

30.1
Introduction

Undisputedly, microscopy plays a predominant role in unraveling the mechanisms that underpin the physical properties of metallic systems. Unfortunately, a straightforward correlation between structural information obtained by electron microscopy and physical properties of metallic systems is still hampered by fundamental and practical reasons. First, the defects affecting these properties, such as dislocations and interfaces, are in fact not in thermodynamic equilibrium and their behavior is very much nonlinear [1, 2]. Second, a quantitative evaluation of the structure–property relationship can be rather difficult because of statistics. In particular, situations where there is only a small volume fraction of defects present or a very inhomogeneous distribution statistical sampling may be a problem. A major drawback of experimental and theoretical research in the field of crystalline defects is that most of the microscopy work has been concentrated on static structures. Obviously, the dynamics of moving dislocations are more relevant to the deformation mechanisms in metals. To this end, we have developed nuclear spin relaxation methods in the past as a complementary tool to transmission electron microscopy (TEM) for studying dislocation dynamics in metals [3]. A strong advantage of this technique is that it detects dislocation motion in the bulk of the material, as opposed to *in situ* TEM, where the behavior of dislocations may be affected by image forces because of the proximity of free surfaces. However, information about the local response of dislocations to an applied stress cannot be obtained by the nuclear spin relaxation technique and therefore *in situ* TEM remains a valuable tool in the study of dynamical properties of defects. Direct observation of dislocation behavior during indentation has recently become possible through *in situ* nanoindentation in a TEM. In this contribution, we have chosen to concentrate on the dynamic effects of dislocations and cracks in crystalline and amorphous metals observed with *in situ* TEM nanoindentations and *in situ* TEM straining experiments.

The objective of this contribution is not to address all the various deformation mechanisms in metallic systems, but rather to discuss the various recent advances

Handbook of Nanoscopy, First Edition. Edited by Gustaaf Van Tendeloo, Dirk Van Dyck, and Stephen J. Pennycook.
© 2012 Wiley-VCH Verlag GmbH & Co. KGaA. Published 2012 by Wiley-VCH Verlag GmbH & Co. KGaA.

in *in situ* TEM techniques that can be helpful in attaining a more quantitative understanding of the deformation mechanisms of crystalline and amorphous metals.

As far as fracture is concerned, significant progress has been made in recent years in the understanding of the associated deformation of amorphous metals together with possible control of shear band propagation (SBP) by virtue of (nano-)crystalline additions in order to suppress the tendency for instantaneous catastrophic failure [4–10]. However, it is apparent that there is still much inconsistency, and while many sound hypotheses and proofs abound, clarity is often lacking when comparing published results. The shear band (SB) thickness lies in the range of 10–20 nm for several bulk metallic glasses (BMGs) [11]. TEM should be a suitable tool for this kind of analysis of SB formation in metallic glasses (MGs), since their thicknesses are small and it may be expected that SBs may lead to (nanoscale) structural changes in amorphous materials.

30.2
In situ TEM Experiments

Ex situ nanoindentation followed by postmortem TEM does not allow for direct observation of the microstructure during indentation and thus lacks the possibility of monitoring deformation events and the evolution of dislocation structures as the indentation proceeds. Also, the microstructure observed after indentation is generally different from that of the material under load because of recovery during and after unloading. In several cases of postmortem analyses by TEM, the preparation of the indented surface in the form of a thin foil often leads to mechanical damage to the specimen or relaxation of the stored deformation because of the proximity of free surfaces, thereby further obscuring the indentation-induced deformation. The recently developed technique of *in situ* nanoindentation in a TEM [12–17] does not suffer from these limitations and allows for direct observation of indentation phenomena. Furthermore, as the indenter can be positioned on the specimen accurately by guidance of the TEM, regions of interest such as particular crystal orientations or grain boundaries can be specifically selected for indentation. *In situ* nanoindentation measurements [17] on polycrystalline aluminum films have provided experimental evidence that grain boundary motion is an important deformation mechanism when indenting thin films with a grain size of several hundreds of nanometers. This is a remarkable observation, since stress-induced grain boundary motion is not commonly observed at room temperature in this range of grain sizes.

Grain boundary motion in metals typically occurs at elevated temperatures driven by a free energy gradient across the boundary, which may be presented by the curvature of the boundary or stored deformation energy on either side of the boundary [18]. In the presence of an externally applied shear stress, it was found [19] that migration of both low-angle and high-angle grain boundaries in pure Al occurs at temperatures above 200 °C. This type of stress-induced grain boundary motion (known as *dynamic grain growth*) is considered by many researchers to be the mechanism responsible for the extended elongations obtained in superplastic deformation of fine-grained materials. The occurrence of grain boundary motion in room temperature deformation of nanocrystalline face-centered cubic (fcc) metals was anticipated recently by molecular dynamics simulations [20] and a simple bubble raft model [21]. Experimental observations of such grain boundary motion have subsequently been provided by *in situ* straining experiments of nanocrystalline Ni thin films [22] and *in situ* nanoindentation of nanocrystalline Al thin films [23]. In both the simulations and the experiments, grain boundary motion was observed for grain sizes below 20 nm. The dislocation mobility is greatly restricted at such grain sizes and other deformation mechanisms become more relevant. In contrast, the grain size for which grain boundary motion was found by *in situ* nanoindentation [17] was of the order of 200 nm.

In simple deformation modes such as uniform tension or compression, dislocation-based plasticity is still predominant and grain boundary motion generally does not occur. In the case of nanoindentation, however, the stress field is highly inhomogeneous and consequently involves large stress gradients [24]. These stress gradients are thought to be the primary factor responsible for the observed grain boundary motion at room temperature. Since the properties of high-purity metals such as pure Al are less relevant for the design of advanced materials, we have focused on the indentation behavior of Al–Mg films and the effect of Mg on the deformation mechanisms described above. To this end, *in situ* nanoindentation experiments have been conducted on ultrafine-grained Al and Al–Mg films with varying Mg contents [25–27].

30.2.1
Stage Design

In situ nanoindentation inside a TEM requires a special specimen stage designed to move an indenter toward an electron-transparent specimen on the optic axis of the microscope. The first indentation holder was developed in the late 1990s by Wall and Dahmen [12, 13] for a high-voltage microscope at the National Center for Electron Microscopy (NCEM) in Berkeley, California. In the following years, several other stages were constructed at NCEM with improvements made to the control of the indenter movement and the ability to measure load and displacement. In the work described in this contribution, two of these stages were used: a homemade holder for a JEOL 200CX microscope [14] and a holder for a JEOL 3010 microscope and JEOL2010F with dedicated load and displacement sensors, developed in collaboration with Hysitron (Hysitron Inc., Minneapolis, MN, USA).

The principal design of both holders is roughly the same. The indenter tip is mounted on a piezoceramic tube as illustrated in [14]. This type of actuator allows high-precision movement of the tip in three dimensions, the indentation direction being perpendicular to the electron beam. Coarse positioning is provided by manual screw drives that move the indenter assembly against the vacuum bellows. The indenter itself is a Berkovich-type diamond tip, which is boron-doped in order to be electrically conductive in the TEM. The goniometer of the TEM provides a single tilt axis, so that suitable diffraction conditions can be set up before indentation.

The motion of the indenter into the specimen during indentation is controlled by the piezoceramic tube. In the holder for the JEOL 200CX, the voltage applied to the tube is controlled manually and recorded together with the TEM image. Since the compliance of the load frame is relatively high, the actual displacement of the indenter into the material depends not only on the applied voltage but also to a certain extent on the response of the material. Consequently, this indentation mode is neither load controlled nor displacement controlled. In the holder for the JEOL 3010 and JEOL2010F, a capacitive sensor monitors the load and displacement during indentation. The displacement signal is used as input for a feedback system that controls the voltage on the piezoceramic tube based on a proportional-integral-derivative (PID) algorithm [28]. The indentation is therefore displacement controlled and can be programmed to follow a predefined displacement profile as a function of time.

The need for a separate load and displacement sensor as implemented in the holder is mainly due to the complex response of the piezo tube. If the response was fully known, the load could be calculated at any time during indentation from the displacement (which can be determined directly from the TEM image) and the characteristics of the load frame [29]. Ideally, the correlation between the applied voltage and the displacement of the piezo element is linear. However, hysteresis and saturation effects lead to significant nonlinearities. Moreover, as lateral motion is achieved by bending the tube, the state of deflection strongly affects the response in the indentation direction as well. Calibration measurements of the piezo response in vacuo at 12 points across the lateral range showed an average proportionality constant of $0.12\ \mu m\ V^{-1}$ with a standard deviation as large as $0.04\ \mu m\ V^{-1}$. Although during indentation the deflection of the tube is approximately constant and the response becomes more reproducible, the above-mentioned hysteresis and saturation effects still complicate the measurement of the load. The implementation of a dedicated load sensor, as in the new holder, is therefore essential for obtaining reliable quantitative indentation data. Section 30.2.3 highlights the differences between the *in situ* indentation load displacement and displacement-controlled holders.

30.2.2
Specimen Geometry and Specimen Preparation

The geometry of the specimens used for *in situ* nanoindentation has to comply with two basic requirements: (i) an electron-transparent area of the specimen must be accessible to the indenter in a direction perpendicular to the electron beam and (ii) this area of the specimen must be rigid enough to support indentation without bending or breaking. A geometry that fulfills both these requirements is a wedge that is truncated to a cap width large enough to provide the necessary rigidity while still allowing the electron beam to pass through. For the present investigation, wedge specimens were used as prepared by bulk silicon micromachining. Using this

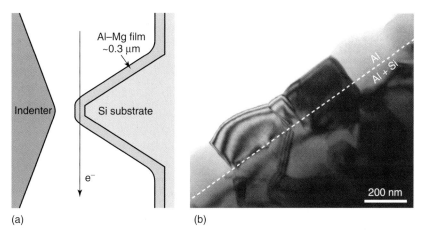

Figure 30.1 (a) Schematic of *in-situ* TEM indentation setup. The deposited Al–Mg film is electron-transparent and accessible to the indenter at the tip of the Si wedge. (b) Typical bright-field image of a deposited film. The dashed line shows the top of the Si ridge.

technique, wedge-shaped protrusions are routinely prepared on Si (001) substrates with a resolution of the order of 1 μm. The side planes of the ridge are aligned with {111} planes of the silicon crystal, so that repeated annealing and oxide removal subsequently lead to sharpening of the wedge driven by a reduction of the surface energy. The silicon ridge specimen geometry provides a means to investigate any material that can be deposited as a thin film onto the silicon substrate. Metals with a low atomic number such as aluminum are particularly suitable for this purpose, since films of these metals can be made to several hundreds of nanometers in thickness and still be transparent at the cap of the wedge to electrons with typical energies of 200–300 kV, as schematically depicted in Figure 30.1a. An example of a resulting TEM image is shown in Figure 30.1b.

The Al and Al–Mg films that are discussed were deposited by thermal evaporation. The substrate was kept at 300 °C to establish a grain size of the order of the layer thickness, which was 200–300 nm for all specimens. After evaporation, the substrate heating was switched off, allowing the specimen to cool down to room temperature in approximately 1 h. One pure Al film was prepared by evaporating a high-purity (5N) aluminum source. Deposition of the Al–Mg alloy films was achieved by evaporating alloys with varying Mg contents. Since Al and Mg have different melting temperatures and vapor pressures, the Mg content of the deposited film is not necessarily equal to that of the evaporated material. Moreover, the actual evaporation rates depend on the quality of the vacuum and the time profile of the crucible temperature. The composition of the deposited alloy films was therefore determined by energy-dispersive spectrometry (EDS) in a scanning electron microscope at 5 kV. The measured Mg concentrations of the four Al–Mg films prepared were 1.1, 1.8, 2.6, and 5.0 wt% [30].

While Al deposited on a clean Si (001) surface may give rise to a characteristic mazed bicrystal structure because of two heteroepitaxial relationships [31], the Si substrates used in the present experiments were invariably covered with a native oxide film. Therefore, the orientations of the Al and Al–Mg grains of the film show no relation to that of the Si surface. An electron backscatter diffraction (EBSD) scan on the evaporated Al film showed a significant <111> texture, which can be explained by the fact that the surface energy of fcc materials has a minimum for this orientation.

On each of the evaporated films, three to four *in situ* experiments were carried out with maximum depths ranging from 50 to 150 nm, using the indentation stage for the JEOL 200CX. The indentation rate, being controlled manually through the piezo voltage, is usually of the order of 5 nm s^{-1}. In addition, several quantitative *in situ* indentation experiments were conducted with the holder for the JEOL 3010/JEOL2010F microscope on the Al and Al–2.6%Mg films. These displacement-controlled indentations are made to a depth of approximately 150 nm with a loading time of 20 s. In order to be able to resolve grain boundary phenomena during each *in situ* indentation, the specimen was tilted to such an orientation that two adjacent grains were both in (different) two-beam conditions.

30.2.3
Load Rate Control versus Displacement Rate Control

The onset of macroscopic plastic deformation during indentation is thought to correspond to the first deviation from elastic response in the load versus displacement curve. For load-controlled indentation of crystalline materials, this deviation commonly has the form of a displacement burst at constant indentation load [32]. The elastic shear stress sustained before this excursion is often much higher than that predicted by conventional yield criteria and can even attain values close to the theoretical shear strength, as was initially observed by Gane and Bowden [33]. The physical origin of the enhanced elastic loading and the subsequent displacement burst has been the subject of extensive discussions in literature [34]. While various mechanisms may be relevant in particular situations, many researchers [35–38] agree that the onset of macroscopic plastic deformation is primarily controlled by dislocation nucleation and/or multiplication, although the presence of an oxide film may significantly affect the value of the yield point.

The initial yield behavior of metals is in some cases characterized by a series of discontinuous yield events rather than a single one [32, 39, 40]. Because of the characteristic steps that result from these yield events during load-controlled indentation, this phenomenon is commonly referred to as "*staircase yielding*." The excursions are separated by loading portions that are predominantly elastic, and the plasticity is thus confined to the yield excursions at this stage of deformation. Staircase yielding may be explained in terms of the balance between the applied stress and the back stress on the indenter exerted by piled-up dislocations that are generated during a yield event at constant load. When the forces sum to zero, the source that generated the dislocations stops operating, and loading continues until another source is activated. This process repeats until fully plastic loading is established. Bahr *et al.* [38] suggested that staircase yielding occurs if the shear stress before the yield point is only slightly higher than the yield stress, so that on yielding, the shear stress drops below the nucleation shear stress and further elastic loading is needed to activate the same or another dislocation source. From these viewpoints, the load at which each excursion occurs depends on the availability of dislocation sources under the indenter and on the shear stress required to nucleate dislocations from them. This accounts for the variation that is commonly observed between indentations in the number of excursions and their size.

While extensive staircase yielding occurs during load-controlled indentation of pure Al thin films, it was found that Al–Mg thin films show essentially continuous loading behavior under otherwise identical conditions [26]. The apparent attenuation of yield excursions was attributed to solute drag on dislocations. It is shown that the effect of solute drag on the resolvability of discrete yield behavior depends strongly on the indentation parameters, in particular on the indentation mode, being either load controlled or displacement controlled.

Conventional *ex situ* nanoindentation measurements were conducted both under load control and displacement control using a TriboIndenter (Hysitron, Inc., Minneapolis, MN) system equipped with a Berkovich indenter with an end radius of curvature of approximately 120 nm. Scanning probe microscopy (SPM) was used to image the surface before each indentation to select a target location on a smooth, flat area of the specimen away from the wedge. The displacement-controlled experiments were performed at a displacement rate of 10 nm s^{-1}; the loading rate in the load-controlled experiments was 10 μN s^{-1}, which, under the present circumstances, corresponds to about 10 nm s^{-1} during the first tens of nanometers of loading.

In situ nanoindentation experiments under displacement control were performed in a TEM using a quantitative indentation stage [39]. The stage has an end radius of approximately 150 nm as measured by direct imaging in the TEM. The *in situ* indentations were carried out on the Al and Al–Mg films at the cap of the wedge, where the surface has a lateral width of the order of 300 nm. The displacement rate during indentation was 7.5 nm s^{-1}.

Given the significant rounding of the indenter in both types of experiments, the initial loading is well described by spherical contact up to a depth of the order of 10 nm. In Tabor's approximation, the elastoplastic strain due to spherical loading is proportional to $\sqrt{\delta/R}$, where δ is the indentation depth and R the indenter radius; the equivalent strain rate is therefore proportional to $1/\sqrt{(4\delta R)}\, d\delta/dt$. Using the above-mentioned values it is easily seen that at a depth of 10 nm the initial strain rates in both types of experiments compare reasonably well to one another, with values of 0.14 and 0.10 s^{-1} for the *ex situ* and *in situ* experiments, respectively.

The load-controlled indentation measurements show displacement bursts during loading on both the Al and the Al–Mg films, as illustrated in Figure 30.2. The curvature of the loading portion before the first excursion is well described by elastic Hertzian contact, as indicated by the dashed curves. With the tip radii used, the depth over which the tip is rounded is larger than the depth over which the

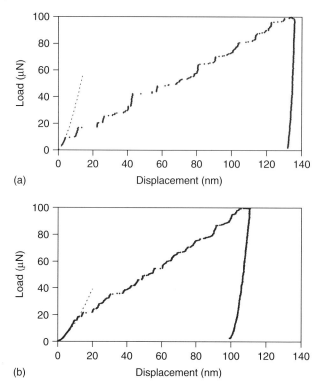

Figure 30.2 *Ex situ* load-controlled indentation response of (a) pure Al and (b) Al–2.6%Mg. The dashed lines represent elastic indentation by a spherical indenter with a radius of 120 nm, with the respective elastic moduli calculated from the slope of the unloading curve.

initial elastic behavior is expected; therefore, the expression for a spherical indenter is used. The subsequent yield behavior is classified as staircase yielding because of the aforementioned dislocation-based mechanisms. Staircase yielding has been reported for indentation of both single crystal and ultrafine-grained polycrystalline Al thin films [15, 27] and, therefore, its occurrence is not expected to depend strongly on the presence of grain boundaries in our case.

The displacement bursts encountered in Al–Mg have a magnitude of up to 7 nm, which is substantially smaller than those observed in pure Al, being up to 15 nm in size. In fact, earlier load-controlled indentation measurements on Al–Mg [26] did not show clearly identifiable discrete yield events whatsoever, because of factors such as lower accuracy of the displacement measurement and more sluggish instrument dynamics with a resonance frequency of only 12 Hz as compared to about 125 Hz in the present instrument. In those experiments, it was observed that the attenuation of displacement bursts occurs for Mg concentrations both below and above the solubility limit in Al, from which it was inferred that the effect is due to solute Mg, which impedes the propagation of dislocation bursts through the crystal. Consequently, at a constant indentation load and for a given amount of stored elastic energy, fewer dislocations can be pushed through the solute atmosphere of Al–Mg than through a pure Al crystal, which accounts for the observed difference in the size of the yield excursions. A comparison of Figure 30.2a,b furthermore reveals that the loading portions in between consecutive yield events in Al–Mg show significant plastic behavior, whereas in Al they are well described by elastic loading, which, at these higher indentation depths, is attested by a slope that is intermediate between spherical [41] and Berkovich [42] elastic contact. The plasticity observed in Al–Mg can be explained in terms of the solute pinning of dislocations that were already nucleated during the preceding yield excursion. As the load increases further, some of the available dislocations are able to overcome the force associated with solute pinning, thereby allowing plastic relaxation to proceed more smoothly. Since dislocation motion is less collective than in pure Al, the measured loading response has a more continuous appearance.

When carried out under displacement control, the *ex situ* indentations show a much more evident effect of the solute drag on the initial yielding behavior, as illustrated in Figure 30.3. The loading curves of both Al and Al–Mg show pronounced load drops, which have the same physical origin as the displacement

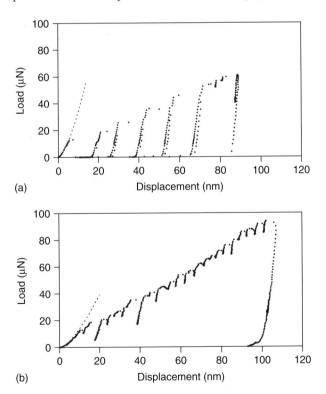

Figure 30.3 *Ex situ* displacement-controlled indentation response of (a) pure Al and (b) Al–2.6%Mg.

excursions in load-controlled indentation, that is, stress relaxation by bursts of dislocation activity. Also, in this case, the loading behavior up to the first load drop closely follows the elastic response under spherical contact. However, the appearance of the load drops is very different: in pure Al, the load drops are large and mostly result in loss of contact, while in Al–Mg they are smaller and more frequent, and contact is maintained during the entire loading segment. The forward surges occurring with each load drop are a result of the finite bandwidth of the feedback system. In the case of pure Al, the observations of complete load relaxation and loss of contact are indicative of the stored elastic energy being fully released in the forward surge before the feedback system is able to reduce the load. In Al–Mg, however, solute pinning strongly reduces the dislocation velocity, which enables the feedback system to respond fast enough to maintain elastic contact. Thus, not all of the stored elastic energy is inputted back into the specimen.

The comparison between the load-controlled and displacement-controlled TEM experiments shows that discrete yield events are far more resolvable under displacement control. This is particularly the case for solid-solution-strengthened alloys, in which the somewhat ragged appearance of the loading portions in between clearly identifiable strain bursts in the load-controlled data is clarified as a series of small but easily distinguishable discrete yield events in the displacement-controlled data. This enhanced sensitivity may be rationalized as follows. When the critical shear stress for a dislocation source under the indenter is reached under load control, a discernible strain burst results only if the source is able to generate many dislocations at constant load, that is, the load–displacement curve must shift from a positive slope to an extended range of zero slope for the slope change to be readily detected. This again is possible only if the newly nucleated dislocations can freely propagate through the lattice, as in pure Al. Under displacement control, however, provided that the feedback bandwidth is sufficiently high, the system may respond to the decrease in contact stiffness when only a few dislocations are nucleated, causing a distinct shift from a positive to a steeply negative slope in the load–displacement curve. Therefore, a detectable load drop can occur without collective propagation of many dislocations and as such may easily be observed even under solute drag conditions. This result cautions against using only load-controlled indentation to determine whether yielding proceeds continuously.

The *in situ* TEM indentations on both Al and Al–Mg show a considerable amount of dislocation activity before the first macroscopic yield point. This is a remarkable observation, as the initial contact would typically be interpreted as purely elastic from the measured loading response. The observations of incipient plasticity are illustrated in Figure 30.4 by the TEM images and load–displacement data recorded during an *in situ* displacement-controlled indentation on Al–Mg. While the indented grain is free of dislocations at the onset of loading (Figure 30.4a), the first dislocations are already nucleated within the first few nanometers of the indentation (Figure 30.4b), that is, well before the apparent initial yield point that would be inferred from the load versus displacement data only. At the inception of the first macroscopic yield event, dislocations are present throughout the entire

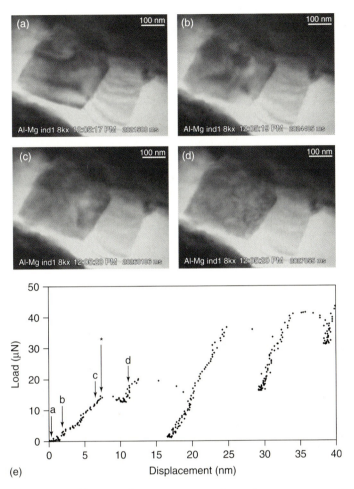

Figure 30.4 TEM bright-field image sequence (a–d) from the initial loading portion (e) of the indentation on Al–2.6%Mg. The first dislocations are nucleated between (a) and (b), that is, before the apparent yield point. The nucleation is evidenced by an abrupt change in image contrast: before nucleation, only thickness fringes can be seen, whereas more complex contrast features become visible at the instant of nucleation. See http://www.dehosson.fmns.rug.nl/ and/or http://materials-science.phys.rug.nl/.

grain (Figure 30.4c). The yield event itself is associated with a rearrangement of these dislocations, which significantly changes the appearance of the dislocation structure (Figure 30.4d). However, the number of newly nucleated dislocations between (c) and (d) is relatively small, as also becomes clear from the limited increase in indentation depth (3 nm corresponds to approximately 10 Burgers vectors). This supports our perception that only a small number of dislocations need to be nucleated in order for a yield event to be detected under displacement control, although the first dislocations nucleated between (a) and (b) do not provide an obvious signature in the load–displacement curve. In the case of *in situ* displacement-controlled indentation of pure Al [16], the onset of dislocation nucleation/propagation coincides with a barely detectable, yet unambiguous, load drop that occurs well before the initial macroscopic yield event, which is further evidence of more collective dislocation motion in Al in comparison to Al–Mg. The *in situ* observations of Al–Mg furthermore provide a self-consistency check for the dynamics of a yield event. With solute drag preventing full load relaxation, the size of a forward surge Δh is essentially determined by the dislocation velocity v and the mechanical bandwidth of the transducer f. Therefore, by ignoring the drag exerted by the feedback system, the dislocation velocity may to a first approximation be estimated as $v \sim \Delta h \times f$, which, using $\Delta h = 7$ nm and $f = 125$ Hz, yields a velocity of the order of $1~\mu m\,s^{-1}$. This is of the same order as observed *in situ* for the initial dislocations in Figure 30.4b, which traversed the 300 nm film thickness in about 130 ms (four video frames at a frame rate of 30 frames per second, see video at *http://www.dehosson.fmns.rug.nl/* or *http://materials-science.phys.rug.nl*).

30.3
Grain Boundary Dislocation Dynamics Metals

Significant grain boundary movement was observed for both low- and high-angle boundaries. This is illustrated in Figure 30.5 by image frames of subsequent stages of the loading part of an indentation near a high-angle boundary. After initial contact (Figure 30.5a) and plastic deformation of grain B (Figure 30.5b), both grain boundaries outlining grain B move substantially (Figure 30.5c,d). By comparing dark-field images taken before and after the indentation, as shown in Figure 30.6, the grain boundary shifts are measured to be 0.04 μm for the left boundary and 0.22 μm for the right boundary.

It should be emphasized that the observed grain boundary motion is not simply a displacement of the boundary together with the indented material as a whole; the boundary actually moves through the crystal lattice and the volume of the indented grain changes accordingly at the expense of the volume of neighboring grains [43]. The trends observed throughout the indentations suggest that grain boundary motion becomes more pronounced with decreasing grain size and decreasing distance from the indenter to the boundary. Moreover, grain boundary motion occurs less frequently as the end radius of the indenter increases because of tip blunting or contamination. Both these observations are consistent with the view that the motion of grain boundaries is promoted by high local stress gradients as put forward in Section 30.1. The direction of grain boundary movement can be both away from and toward the indenter, and small grains may even completely disappear under indentation [23]. Presumably, the grain boundary parameters play an important role in the mobility of an individual boundary, since the coupling of the indenter-induced stress with the grain boundary strain field depends strongly on the particular structure of the boundary.

The quantitative *in situ* indentation technique offers the possibility of directly relating the observed grain boundary motion to features in the load–displacement

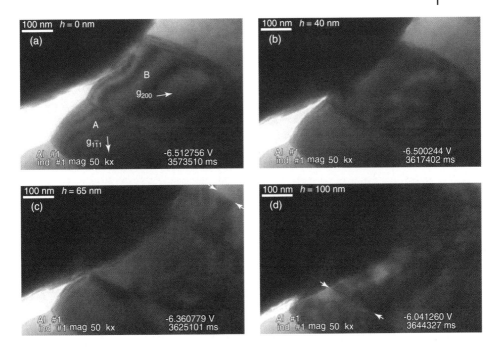

Figure 30.5 Series of bright-field images from an indentation on Al, which is accommodated by movement of the grain boundaries (marked with arrows). The approximate indentation depth h is given in each image.

curve. While this relation has not been thoroughly studied, these results suggest that the grain boundary motion is associated with softening in the loading response. Softening can physically be accounted for by the stress relaxation that occurs on grain boundary motion. However, the quantification of overall mechanical behavior is complicated by the frequent load drops at this stage of indentation, and further *in situ* indentation experiments are needed to investigate this phenomenon more systematically and quantitatively.

The movement of grain boundaries as observed in Al was never found for high-angle boundaries in any of the Al–Mg specimens, even when indented to a depth greater than half of the film thickness. Figure 30.7 shows a sequence of images from an indentation on an Al–1.8%Mg layer. At an indentation depth of approximately 85 nm into grain B (Figure 30.7c), plastic deformation is initiated in grain A by transmission across the grain boundary. However, no substantial grain boundary movement occurs; small grain boundary shifts ($\sim$10 nm) that were measured occasionally can be attributed to displacement of the material under the indenter as a whole, with conservation of grain volume, rather than to actual grain boundary motion (Figure 30.8). Our observations as such indicate a significant pinning effect of Mg on high-angle grain boundaries in these alloys.

In contrast to high-angle grain boundaries, the mobility of low-angle boundaries in Al–Mg was found to be less affected by the presence of Mg. This is illustrated by

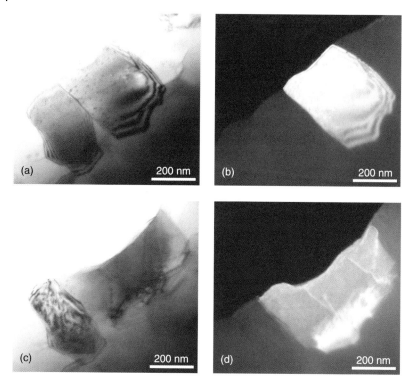

Figure 30.6 Bright- and dark-field images of the indented grain (a,b) before and (c,d) after the indentation. Grain boundary motion leads to a significant volume increase of the indented grain.

the rapid disintegration of a low-angle tilt boundary in Al–5.0%Mg, as shown in Figure 30.9. At a relatively low indentation depth of about 20 nm, the dislocations that were initially confined to the indented grain spread across both grains without being visibly obstructed by the tilt boundary. The boundary effectively disappears at this point with the end result of the two grains becoming one. Figure 30.10a–c shows the orientation of the two grains before indentation. The grains share the same <112> zone axis, but are in different two-beam conditions because of their slight misorientation (∼0.7°). Figure 30.10d shows the grains after the indentation to be both in the same diffracting condition as the grain in Figure 30.10a.

Ideally, in order to compare the observed grain boundary behavior between different measurements, the indenter-induced stress at the boundary should be known. However, because of surface roughness, tip imperfections, and the complicated specimen geometry, it is difficult to accurately measure or calculate the local stress fields. Comparisons between different measurements are therefore mainly based on indentation depth. Our observation of grain boundary pinning in Al–Mg in this context means that no motion of high-angle boundaries was observed in Al–Mg in more than 15 indentations to a depth of the order of

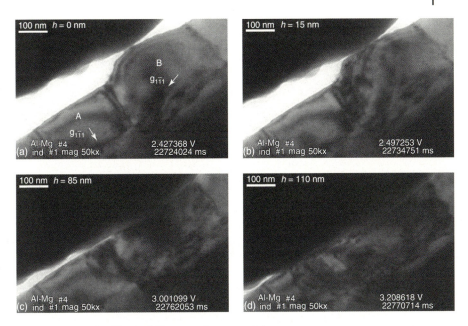

Figure 30.7 Series of bright-field images from an indentation on Al–1.8%Mg. No movement of the high-angle grain boundaries is observed.

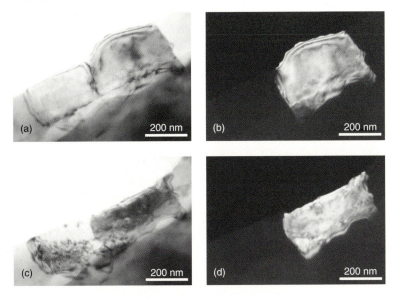

Figure 30.8 Bright- and dark-field images of the indented grain (a,b) before and (c,d) after the indentation. Apart from a slight displacement of the boundaries due to the shape change of the indented grain, no significant grain boundary motion is detected.

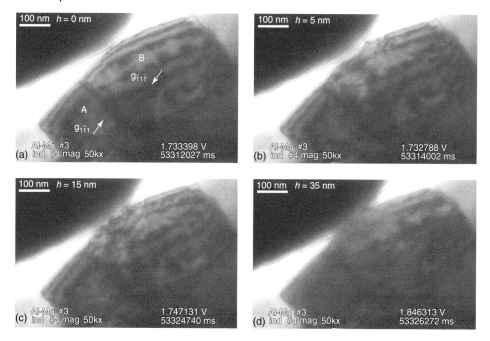

Figure 30.9 Series of bright-field images from an indentation on Al–5.0%Mg, (a) non-contact (b) onset of contact. showing the disintegration of a low-angle <110> tilt boundary between (c) and (d).

100 nm, while in pure Al, grain boundary motion was frequently observed at indentation depths of 50 nm or less. The Al–Mg films presented here include compositions both below and above the solubility limit of Mg in Al. However, no differences in indentation behavior between the solid solution and the precipitated microstructures were observed. Consequently, the observed pinning of high-angle boundaries in Al–Mg is attributed to solute Mg. The pinning is presumably due to a change in grain boundary structure or strain fields caused by solute Mg atoms on the grain boundaries. Relatively few direct experimental observations have been reported of this type of interaction. Sass and coworkers observed that the addition of Au and Sb impurities to bcc Fe changes the dislocation structure of <100> twist boundaries of both low-angle [44] and high-angle [45] misorientation. Rittner and Seidman [46] calculated solute distributions at <110> symmetric tilt boundaries with different boundary structures in an fcc binary alloy using atomistic simulations. However, the influence of solutes on the structure of such boundaries has not been experimentally identified.

Possible changes in atomic boundary structure due to solute atoms may be observed by high-resolution transmission electron microscopy (HRTEM). Atomic-scale observation of grain boundaries using this technique requires that the crystals on both sides share a close-packed direction so that both lattices can be atomically resolved at the same time. The mazed bicrystal structure that forms

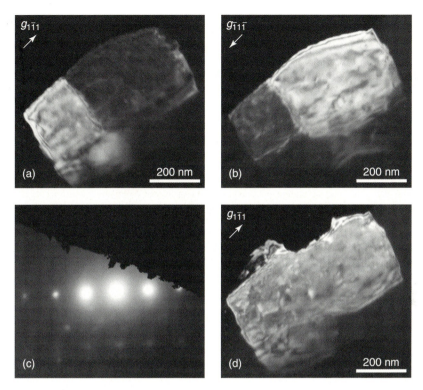

Figure 30.10 (a,b) Dark-field images of the two Al–5.0%Mg grains before indentation. (c) Diffraction pattern showing the <112> orientation of both grains; the cutoff is due to the *in situ* specimen geometry. (d) Dark-field image after indentation.

when an Al film is deposited epitaxially onto a Si (001) surface meets this condition. The epitaxial relationships Al (110)//Si (001), Al [001]//Si [110], and Al (110)//Si (001), Al [001]//Si [1$\bar{1}$0] lead to two possible orientations that are separated exclusively by 90° <110> tilt boundaries [47]. The structure of such boundaries has been successfully studied in HRTEM studies of Al films on Si substrates [48, 49] and Au films on Ge substrates [50–53], which exhibit the same epitaxial relationships. Moreover, the effect of alloying elements in Al has been explored by evaporating alloys such as Al–Cu and Al–Ag [54].

In order to study the effect of Mg on these tilt boundaries, Al and Al–Mg films were deposited onto Si (001) substrates that had been stripped from their native oxide film. It was found that in epitaxial films evaporated from pure Al, the 90° <110> tilt grain boundaries were faceted on $\{100\}_A//\{110\}_B$ and $\{557\}_A//\{557\}_B$ planes, which can be atomically resolved (Figure 30.11a). The addition of Mg, however, drastically changes the microstructure of the deposited film: evaporation of Al–Mg on a Si substrate heated to 300 °C (which is necessary to reduce the lattice

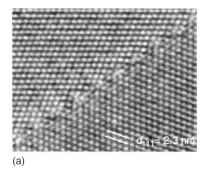
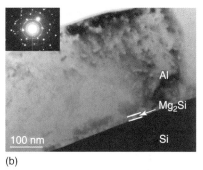

Figure 30.11 (a) High-resolution micrograph of a 90° <110> asymmetrical tilt boundary in an epitaxial Al thin film, showing a periodic structure along the boundary plane. The orientation of the boundary plane is $\{100\}_A//\{110\}_B$. (b) Cross section of a film deposited from an Al–2.2 wt%Mg source onto a Si (001) substrate; the intermetallic compound Mg_2Si, identified by its diffraction ring pattern (inset), forms a 15 nm thick layer on the interface.

mismatch between Al and Si) leads to the formation of the intermetallic compound Mg_2Si, which prohibits any further epitaxial growth (Figure 30.11b). Even in a two-step evaporation consisting of a pure Al deposition to provide a basis for the bicrystal structure and a subsequent Al–Mg deposition to introduce the Mg, the Mg diffuses to the substrate driven by the reaction with the Si substrate. This method could therefore not be used to study the effect of Mg on the atomic structure of the grain boundaries. Another effect that may contribute to the pinning of special boundaries is solute drag on extrinsic grain boundary dislocations (EGBDs) as reported by Song *et al.* [55], who showed that the dissociation rate of EGBDs in Al alloys is reduced by the addition of Mg. This implies that the indenter-induced deformation is accommodated more easily by these boundaries in pure Al than by those in Al–Mg.

The fact that low-angle grain boundaries were found to be mobile regardless of the Mg content can be explained by their different boundary structure. Up to a misorientation of 10–15°, low-angle boundaries can be described as a periodic array of edge and screw dislocations by Frank's rule [56]. In such an arrangement, the strain fields of the dislocations are approximated well by individual isolated dislocations and their interaction with an external stress field can be calculated accordingly. Since there is no significant interaction between the individual grain boundary dislocations, the stress required to move a low-angle boundary is much lower than for a high-angle boundary. Low-angle pure tilt boundaries consisting entirely of parallel edge dislocations are fully glissile and therefore particularly mobile. In general, a combination of glide and climb is required to move a low-angle boundary [57]. As a corollary, the structural difference between low- and high-angle boundaries also affects the extent of solute segregation. Because solutes generally segregate more strongly to high-angle boundaries [58], the observed difference in mobility may partly be a compositional effect.

30.4
In situ TEM Tensile Experiments

MGs are currently the focus of intense studies in view of their great potential in applications [59, 60]. Tremendous efforts have been dedicated to characterize, understand, and possibly control the atomic- and microscale mechanisms underlying the unique properties of MGs, for example, the very high yield strength but near-zero plasticity at room temperature. The basic carriers of plasticity in MGs are widely accepted to be shear transformation zones (STZs), each of which is a local atomic region containing tens to hundreds of atoms and confined within an elastic medium [4, 61]. However, the dynamics and correlation of the local flow units, and their spatiotemporal evolution into macroscopic SBs remain mysterious.

Study on size effects in the mechanical response of MGs is interesting as it may provide hints on these fundamental issues. Also, it may guide us to the practical design of small-sized MG-incorporated materials and devices [62–64]. Consequently, these basic questions gave an impetus to considerable theoretical [65–67] and experimental [68–77] attention. From experiments, it is getting clear that at the micrometer scale SB is still the dominant behavior. [78–80]. Test of submicron- and nanometer-sized samples is highly desirable, but these are extremely challenging because of the experimental difficulties in performing neat experiments, that is, fabricating extremely small, nanosized specimens, which are desired to be free-of geometrical imperfections and surface contaminations, [79–82], and testing them quantitatively. Recent literature hinting at not only homogeneous plastic flow but also remarkable artifacts arising from geometrical tapering and surface modification because of the use of focused ion beam (FIB) have not been clarified. In addition, the driving force and microscopic basic concepts of such transitions in deformation modes remain rather obscure and elusive.

Analysis of inhomogeneous deformation of MGs in compression and their response thereto has been extensively studied [83, 84], since it is believed that this is the easiest manner in which to avoid catastrophic instability [85]. This consequently means that tensile fracture in BMGs has received less attention, mainly due to the high instability associated with plane stress and plane strain investigations. Some interesting tensile investigations have been reported [86], and *in situ* TEM has been shown to be an excellent tool for observing the formation of nanocrystals and their effects (such as crack blunting).

It has been widely believed that shear in MGs can develop high local temperatures [87, 88], leading to a range of fracture surface features in MGs [89], which can be simple, flat, and almost featureless shear planes; characteristic vein patterns formed by the meniscus instability [90] initiated from the surface edges or homogeneously inside a deformed layer; or wide and very elongated veins extending along the shear surface from the edge of the shear step of bulk samples. Again, there remains uncertainty over the differences in fracture surfaces, although Zhang et al. attempt to compare differences in compressive and tensile fracture surfaces in [91]. Analysis of these features, however, suggests that the heat developed during deformation in

MGs means that much more material is involved in the formation of a so-called "liquidlike layer" (LLL) than the accepted width of an SB (10–20 nm).

Direct proofs of those temperature rises have been published, with the analytical methods including high-speed cameras in compressive and impact testing [92, 93]. Recently, Lewandowski and Greer [94] found a very elegant fusible-coating method to confirm experimentally the local temperature increase in the vicinity of an SB formed in amorphous material, using a thin tin coating on their surface, which melts up to some distance from a developed SB. In the model calculation, they consider an SB to be a planar source of heat with density Q (J/m^2) at time $t = 0$. The temperature increase at a distance x to both sides of the SB located at $x = 0$ is given by

$$\Delta T = \left(\frac{Q}{2\rho C \sqrt{\pi \alpha}} \right) \frac{1}{\sqrt{t}} \exp \left(\frac{-x^2}{4\alpha t} \right) \tag{30.1}$$

where ρ is the material density, C the specific heat of the material, and α the material thermal diffusivity. Their measurements approved this relatively simple model as well as an estimation of the amount of released heat Q calculated from the uniaxial yield stress σ_Y and shear displacement Δ [88]:

$$Q = \beta \sigma_Y \delta \tag{30.2}$$

where β is a constant of value <0.35 in a case of 45° shear.

Given that heat is developed during shear in MG, it may be safe to assume that crystallization may occur within that heat-affected area. Published data shows that it may or may not occur[86]; however, to the best of our knowledge, no explanation is offered for these differing findings.

In the following, the problems outlined above are addressed by adopting the model engaged for the tin-coating investigations [94] and applying it to the glass transition temperature of a given material. Owing to the nature of *in situ* TEM testing, the samples must be very thin and therefore the results should not be automatically assumed valid for normal "bulk" samples, although the compositions chosen are from BMG compositions. Several BMGs, namely $Cu_{47}Ti_{33}Zr_{11}Ni_6Sn_2Si_1$ [95], $Zr_{50}Cu_{30}Ni_{10}Al_{10}$ [96], and $Zr_{52.5}Cu_{17.9}Ni_{14.6}Al_{10}Ti_5$ (VIT105) [97] were investigated. The alloys were prepared according to the procedure detailed in [98] and investigated using HRTEM (FEG Jeol 2010) with *in situ* heating and straining, along with EDS.

In the *in situ* setup of the JEOL 2010 (Figure 30.12) one end is "clamped," while the other is extended at a desired extension rate. In these experiments, that strain rate was varied, but most results are shown for 0.1 μm s^{-1} (corresponding to 10^{-5} s) extension rates, as this permits a reasonable propagation time without losing imaging capabilities. Before TEM investigation, all samples were confirmed X-ray amorphous using X-ray diffraction (XRD) technology and information regarding TEM heating parameters was attained using thermo(dynamic) testing.

As an initial cautionary note, the *in situ* TEM straining observation of MG ribbons has been found to be a far from easy experiment, since, in nonnotched

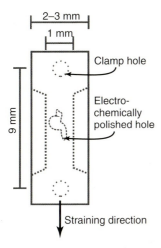

Figure 30.12 Schematic representation of the sample dimensions used in *in situ* TEM straining.

samples, the prediction of where an SB may initiate must be fortunate. However, *in situ* TEM has been found to provide a means by which the propagation of an SB and its associated fracture tip may be followed and analyzed. Figure 30.13 relays a typical result for the $Cu_{47}Ti_{33}Zr_{11}Ni_6Sn_2Si_1$ alloy. In Figure 30.13a, the "state" of the crack tip at 368 μm extension (= 4.08% strain) is shown. The fracture opening is preceded by material thinning induced through the progression of

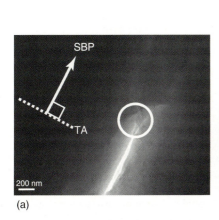
(a)

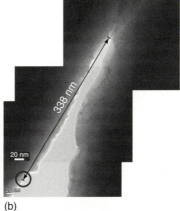
(b)

Figure 30.13 (a) TEM micrograph of a shear band and its crack opening after 4.08% strain in $Cu_{47}Ti_{33}Zr_{11}Ni_6Sn_2Si_1$ alloy. The inset shows that the shear band propagation (SBP) direction relative to the tensile axis (TA) was found to be (∼)90°. The white circle corresponds with the black circle in (b), a TEM micrograph of the crack-opening progress at 4.12% strain.

the SB. The SB itself was initiated at around 3% strain, as may be expected for amorphous materials. After imaging at 368 μm extension, the sample was then re-strained for a further 3–371 μm extension (4.12% strain) accompanied by video recording. The time taken for this displacement to occur was 30 s (extension rate = 0.1 μm s^{-1}). No jerky motion was seen in this case, but instead the "opening" of the crack tip was seen to be "constant." However, in some cases, applied load resulted in no discernible change at the crack tip for up to 5 μm normal extension. This is complementary to macroscale observations in SBP. Imaging the "new" crack tip allows measurement of the crack-tip displacement and crack-tip characteristics (Figure 30.13b). Little change in the "microstructural appearance" of the crack-opening was seen. Characteristics of the area in and around the projected SB include narrow (10–20 nm) featureless bands, at the edge of the fracture surface. The next region is shown to be a "speckled" band 20–50 nm in width. The features in the speckled region are 2–5 nm in size, but could not be resolved as crystals in HRTEM investigations. It may, therefore, be reasonable to consider the smooth (bright) bands to be characteristic of the SB, while in the speckled region, the nanoscale features can be attributed to very local meniscus instability.

An extension of 3 μm in the tensile loading direction (Δl_n) is found to translate to a crack propagation displacement (Δl_p) of 338 nm, heralding an approximate relationship of $\Delta l_n \approx 10 \Delta l_p$ during plastic deformation. This relationship was confirmed over further observations, although instances whereby application of load led to no propagation in the direction of the SB were also seen.

In one special case, crystallization due to SB formation was found. In this case, the propagation rate was much greater, since, during the image collection, 25 full frames were recorded per second. Figure 30.14b reveals the rapid crack propagation

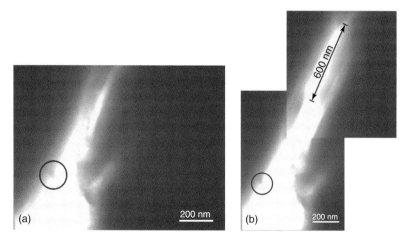

Figure 30.14 Rapid fracture in $Cu_{47}Ti_{33}Zr_{11}Ni_6Sn_2Si_1$ ribbon. The black rings highlight corresponding areas in (a) and (b).

that occurred in one full frame with respect to Figure 30.14a. On sighting this rapid propagation, the progressive loading was stopped immediately and the full extent of the crack movement was imaged. The length over which the crack existed was larger than 600 nm.

At the head of this "crack," a meniscus was found, which signifies liquidlike behavior, and all resolvable material ahead of this tip was shown to be amorphous (HRTEM, Figure 30.15a). The edges of the sample close to the meniscus revealed crystallinity, which must have developed because of the rapid segregation of the shear surfaces and a heat associated with that. Figure 30.15b shows the crack tip developed in the $Cu_{47}Ti_{33}Zr_{11}Ni_6Sn_2Si_1$ ribbon shown in Figure 30.13b. The difference is clear and the area around this crack tip remains amorphous.

The SB detected in $Zr_{52.5}Cu_{17.9}Ni_{14.6}Al_{10}Ti_5$ (HR)TEM (Figure 30.16a) shows a typical crack-opening behavior, as seen already with other amorphous alloys, and

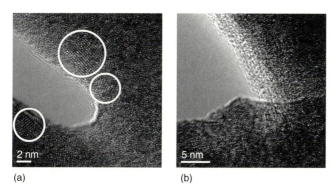

Figure 30.15 HRTEM micrographs revealing (a) a rapid-propagation-induced meniscus at the crack-tip and (b) no meniscus in "controlled" propagation for $Cu_{47}Ti_{33}Zr_{11}Ni_6Sn_2Si_1$ ribbon.

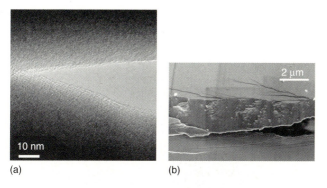

Figure 30.16 (a) HRTEM image showing the crack opening and (b) is an SEM micrograph revealing no evidence of surface feature heating/melting despite full surface separation for $Zr_{52.5}Cu_{17.9}Ni_{14.6}Al_{10}Ti_5$ ribbon.

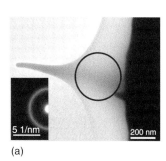 

Figure 30.17 (a) TEM micrograph revealing a triple point vein formed in $Zr_{52.5}Cu_{17.9}Ni_{14.6}Al_{10}Ti_5$ alloy. The inserted selected-area electron diffraction (SAED) ring is taken in the area highlighted by the black circle and shows the vein to be fully amorphous. (b) This was confirmed at the vein tip by HRTEM and fast Fourier transform (FFT).

remained amorphous (HRTEM). The fracture surface did not reveal any liquidlike behavior. The material involved in the shear process does not have an appearance that would suggest it has experienced significant heating, but it has clearly been sheared across the fracture surface and remains intact with the surface. The formation of veins in the full fracture sample was again seen because of meniscus instability and their length is spread over a wider scale, and much longer veins than those seen in the previous case. This suggests that more material is involved in the formation of an LLL during deformation. Figure 30.17a shows one of these triple point veins. Their lengths range from 200 to 600 nm and could be found across the whole length of the fracture surface. Figure 30.17b confirms these vein nipples to be amorphous for a sample heated to 703 K in a TEM chamber.

We used the combination of TEM and scanning electron microscopy (SEM) observations of features on the fracture surfaces exhibiting vein patterns to estimate the lowest limit of the amount of material, which is active during the last moments of failure. Quantitative analysis of a group of TEM observations of veins similar to Figure 30.17a allowed us to estimate the average height of a vein as 0.31 µm and to approximate its profile as an isosceles triangle with a same base in accordance with scanning tunneling microscopic observations [99]. A linear density of veins on the fracture surface may be directly estimated using one of the statistical methods of quantitative stereology [100] applied to SEM micrographs. By simply counting the number of intersections of veins with randomly applied grid lines inside the vein pattern area, calculating the average number of these intersections per unit length, and multiplying by a constant $\pi/2$ gives us the average linear density of veins on a fracture surface of VIT105 alloy $= 1.25\,\mu m\,\mu m^{-2}$. Multiplying this average linear vein density by the average vein profile estimated from TEM observations, we get the volume of $0.061\,\mu m^3$ per each μm^2 of fracture surface. Assuming that the major part of the LLL is, during failure, moving into the veins and that the vein pattern is mirrored at the opposite side of fractured surface, we estimated the lowest limit of thickness of LLL to be 120 nm.

The fact that veins are seen to develop on some fracture surfaces and not on others, and the fact that the appearance of the fracture surfaces can be so very varied, means that the local temperature and/or active stress components must be different in the various cases. There is clear evidence of elevated temperatures, since the (last point of contact) failure becomes ductile in appearance because of necking. The question, of course, is how high this local temperature is. In order for ductile fracture to occur in MGs, viscoplastic deformation must be induced. One manner in which viscoplastic deformation is induced [101] is when the material temperature is raised. The temperature regime in which the material behaves viscoplastically can be determined through thermomechanical analyses. The behavior is associated with the glass transition temperature and this temperature is shown to be 672 K for VIT105 alloy. The temperature at which the crystallization starts, T_x is shown to be 714 K; however, high heating rates can substantially increase this value.

Bengus et al. [102] pioneered and hypothesized as early as 1993 that extreme local heating and melting at the shear crack front developed during tensile tests of amorphous alloys ribbon may occur. Their deductions were supported by the measurement of kinetics of shear crack propagation, by estimation of the amount of elastic energy released and transformed to the local heat in an extremely short time and by the observations of spheroid drops on the fracture surface of $Fe_{83}B_{17}$ amorphous alloy ribbon. Fractographic evidences of the local hot state of the material on the shear failure surfaces were also observed for amorphous ribbons tested in tension at very low temperatures (4.2 K) [103] and for bulk amorphous alloys tested in compression.

According to the contemporary concept of SB formation in amorphous metals, immediately after instantaneous SB development, the surrounding amorphous material is heated up from the environment temperature T_0 because of heat diffusion. The volume where the local temperature is higher than the glass transition temperature T_g may be deformed easily because of substantially decreased viscosity, when both tensile and shear stress components exist.

The size of this volume for alloy VIT105 may be estimated using Eq. (30.1). We assumed $\sigma_y = 1850$ MPa, $\rho = 6600$ kg m^{-3} [104], $C = 420$ J kg^{-1} K^{-1}, $\alpha = 3 \times 10^{-6}$ m^2 s^{-1} and $T_g = 674$ K [105]. Figure 30.18 relays our calculations. Figure 30.18a shows one half of the temperature increase profile in the vicinity of a single SB with a shear displacement, $\delta = 5$ μm (a value near to that observed in Figure 30.19) at three different times 3, 12, and 50 ns, respectively. The horizontal line made at $\Delta T = 376$ K corresponds to the limit when the temperature increase from room temperature reaches the glass transition temperature T_g. From Figure 30.18a, it can be concluded that the volume that may be considered to be at the temperatures over T_g is dependent on time. At a time shortly after SB development (3 ns), the temperature increase is relatively high; however, the "heat-affected volume" is small. After 50 ns, all material is cooled down under T_g. If one would like to know the maximal distance to which the material is overheated over the glass transition temperature, the differentiation of Eq. (30.1) gives the

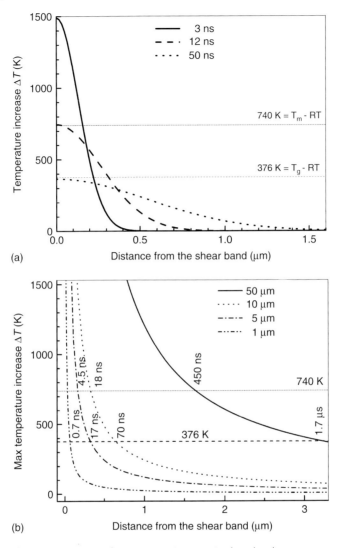

Figure 30.18 (a) Local temperature increase in shear band vicinity calculated for a shear displacement, δ of 5 μm ($Q = 1.34$ kJ m^{-2}) and $t = 10$, 50, and 130 ns for VIT105 alloy and (b) maximum temperature increase calculated for shear displacements, $\delta = 1, 5, 10$, and 50 μm.

value of maximum temperature increase together with the time at which this maximum is reached. This calculation is shown in Figure 30.18b for four different shear displacements −1, 5, 10, and 50 μm. It is seen that 5 μm shear displacement may induce, at time 17 ns, the viscoplastic layer with a thickness slightly exceeding 0.6 μm. A 1 μm shear displacement offers only a 0.13 μm layer for a much shorter time and the larger displacements such as 10 and 50 µm, usually observed on the fracture surfaces of bulk amorphous samples, may form much thicker "LLLs" (1.3 and 6.4 µm) for substantially longer times (70 ns and 1.7 µs).

The horizontal line $\Delta T = 740$ K that corresponds to the temperature increase over the melting temperature of VIT105 glass is also present in Figure 30.18a,b to indicate that a large displacement offers enough heat for local melting. In the examples outlined in this work, where crystallization processes derive, they are driven by temperature increases and not by deformation processes. The maximum temperature is not the most important influencing factor in this case, but the time during which the heated volume stays at an elevated temperature, which must, because of the extreme heating rates, be much higher than conventionally measured T_x (~500 °C). This time scales with the square of shear displacement, δ. However, for large shear displacements (~50 µm), the time at which the material is elevated above the measured T_x is still less than 1 µs, which may be adequate for crystal growth, but insufficient for crystal nucleation, since the activation energy for nucleation is larger than the activation energy for crystal growth in amorphous alloys [106]. This explains why crystals only appear in a very narrow band at the SB edge and also why those crystals are of nanometer size. SBP has been found to be jumplike by *in situ* straining in TEM microscopy. The secondary SBs form a branch when the primary SB stops and one of them will become the new primary SB with the largest shear displacement, δ. Both TEM and SEM observations confirm the presence of a LLL on or near the fracture surface of the amorphous ribbons.

30.5
In situ TEM Compression Experiments

In the case of polycrystalline materials and MGs with surface-dominated geometries (i.e., nanopillars), their mechanical properties can be tailored by altering the extent to which dislocations and SBs, respectively, can nucleate, propagate, and interact. Since metals and alloys are most common in their polycrystalline form, the interaction between defects and grain boundaries is of particular interest. In bulk, grain boundaries act as obstacles to dislocation motion as conveyed through the classical Hall–Petch relation describing the characteristic increase in yield strength of polycrystalline metals with decreasing obstacle distance, down to the grain sizes of ~40 nm [107–110]. An extensive review on intrinsic and extrinsic size effects has been published in [111]. However, plastic deformation of such materials involves a wide range of interaction phenomena between dislocations and grain boundaries, which are still subject to extensive research. Furthermore, when the grains are reduced to ~40 nm, they cannot accommodate multiple lattice dislocations, which engage alternative plastic deformation mechanisms such as grain boundary sliding, partial dislocation emission, and absorption at grain boundaries. At grain sizes below 20 nm, the Hall–Petch relation gives way to the so-called "inverse" Hall–Petch, manifested through softening with decreased grain size because of the activation of grain-boundary-assisted deformation. Although the specific plasticity mechanisms in nanocrystalline metals are controversial, there is a general agreement about the grain size regimes where dislocation-mediated versus grain boundary mechanisms prevail.

In addition to grain-size-dependent plasticity in metals, a different kind of size effect has been reported in the past decade: external geometric size in single crystals, where the attained material strength inversely scales with sample dimensions, that is, thin film thickness or nanopillar diameter due to defect interactions with free surfaces. One such example is single crystalline Au nanopillars subjected to uniaxial-compression-attained strengths ~50 times higher than their bulk form [110, 111]. Moreover, these single-crystalline nanopillars exhibit an unforeseen macroscopic stress–strain response, populated with intermittent discrete strain bursts, indicative of a fundamentally different deformation mechanism compared with bulk. In fcc crystals, one of the possible explanations for increased strength in the nano- (rather than micro-) pillars is "dislocation starvation," a state in which initial mobile dislocations annihilate at the free surface, leaving the crystal dislocation starved, and in order to sustain higher loads new dislocations have to be nucleated, requiring nearly theoretically applied stresses.

These examples of size effects – microstructural in boundary-containing materials and extrinsic in single crystals – utilize a single characteristic length scale parameter, grain size, or specimen size, to explain and model the size effect and to postulate operating deformation mechanisms. The interplay of the length scales in combining both the microstructural and geometrical constraints has the potential of revealing new deformation mechanisms and of significantly extending our knowledge of materials. With the ongoing miniaturization of devices and materials, length scales have come within reach at which the mechanisms by which deformation proceeds change drastically. A thorough understanding of such mechanisms is required to improve the mechanical properties of advanced materials. Ng and Ngan [27, 111] studied the effects of trapping the mobile dislocations by compressing Al micropillars of relatively large diameters – ~1.2 up to 6 µm either coated or center filled with a tungsten-based compound. Consistent with the proposed models, these passivated samples exhibited a notably higher strain-hardening rate and a much smoother stress–strain signature, suggesting the suppression of dislocation avalanches and a lack of nucleation-controlled plasticity. Their TEM analysis reveals much greater dislocation densities and even dislocation cell formation in the post-deformed specimens, confirming the trapping of the dislocations inside the deforming micropillars instead of annihilating at the free surface as is the case in the as-fabricated small-scale specimens [111]. Most recently, the results of *in situ* compression of annealed-to-pristine-state Mo nanopillars inside of TEM were reported by Lowry *et al.* [112]. Owing to the lack of defects in these samples, these authors observed initial elastic loading up to catastrophic failure at nearly theoretical strengths. Ye *et al.* [113] conducted *in situ* uniaxial compression experiments on single-crystalline Mg and Mg–0.2%Ce nanopillars inside of a TEM and revealed the presence of strong size effects associated with two distinct mechanisms: basal plane sliding and extension twinning in both materials. These researchers clearly observed the dependence of twin nucleation stress on pillar size and deemed the availability of preexisting dislocations to be the key factor that enables twin formation. Interestingly, while the nucleation stress for the basal slip was similar between Mg and its alloy, the nucleation stress for extension twinning in Mg–0.2%Ce was 85% lower than that for pure Mg, causing the stresses for both plasticity mechanisms in the former to be nearly equivalent, and therefore competing. Another interesting metallic system studied via microcompression by Withey et al. [114] in an *in situ* TEM is gun metal, or a β-Ti alloy characterized by exceptional elastic elongation. These researchers studied the deformation of 80–250 nm diameter nanopillars and discovered that most samples approached the ideal strength regardless of pillar size and that deformation occurred via a dislocation-free mechanism prompted by instability at the elastic limit. Here, the intrinsic effects are clearly dominant as compared with the external sizes as even in the 100 nm diameter pillars the deformation mode is fine scale. An interesting observation was made that in some tests these materials underwent a martensitic transformation to an orthorhombic structure and in others formed SBs, always occurring post-yield, and therefore not affecting the yield strength.

As far as size effects are concerned, contradicting conclusions have been drawn by different research groups, based on various *in situ* TEM experiments on different metallic glasses. On the basis of *in situ* compression TEM experiments on $Zr_{41}Ti_{14}Cu_{12.5}Ni_{10}Be_{22.5}$ nanopillars, evidence was presented that SB still prevails at specimen length scales as small as 150 nm in diameter. Finite element modeling (FEM) of the stress state within the pillar indicates that the unavoidable geometry constraints accompanying such experiments impart a strong effect on the experimental results, including nonuniform stress distributions and high-level hydrostatic pressures. From these observations it is concluded that the strength of at least this particular MG is size independent. Several other BMGs, namely $Cu_{47}Ti_{33}Zr_{11}Ni_6Sn_2Si_1$ [115], $Zr_{50}Cu_{30}Ni_{10}Al_{10}$ [116], $Zr_{52.5}Cu_{17.9}Ni_{14.6}Al_{10}Ti_5$ (VIT105) [117], and $Zr_{35}Ti_{30}Co_6Be_{29}$ [118] were investigated through *in situ* microscopy. In order to directly observe the evolution of SBs during tensile deformation, *in situ* tensile experiments have also been performed at both room and elevated temperatures. In these tensile experiments, the strain rate corresponded to 10^{-5} s^{-1} extension rates as this allows a reasonable propagation time without losing imaging capabilities.

In situ TEM compression experiments on MGs were performed under load-controlled and displacement-controlled Cu-based $Cu_{47}Ti_{33}Zr_{11}Ni_6Sn_2Si_1$ and Zr-based $Zr_{50}Ti_{16.5}Cu_{15}Ni_{18.5}$ prepared by melt spinning [70, 119, 120]. Micropillars are usually made by FIB from the same kinds of MG ribbons prepared by melt spinning. There are practical challenges in FIB milling of micropillars, especially making a uniform gauge with diameter smaller than 1 µm. Special efforts have to be made to reduce the taper angle of pillars. The pillars have a rather small taper angle between 2.0 and 3.5°, and, importantly, well-defined gauge lengths. The aspect ratios (height/diameter) of the pillars are typically between 3 and 8.

Sometimes, the tapered geometry of the microfabricated pillars causes inhomogeneous distribution of the compressive stress along the axial direction. As a result, yielding and local shear deformation start at the top part of a pillar where the highest compressive stress is concentrated, while the part underneath experiences only elastic deformation. Because of the nonuniform strain and stress distributions along the gauge length of the tapered pillar, engineering stress can be approximated by $4P/[\pi(D + u\tan\beta)^2]$ where P is the load and D is the tip diameter, β is the taper angle, and u is the measured tip displacement. The yield stress can be accurately determined by recording the peak load value before each unloading and measuring the minimum diameter at the elastically deformed region after unloading. The precise measurement of effective diameters permits an accurate determination of the yield stress. Nowadays, picoindenters have several unique features, which are particularly crucial to these kinds of studies. First, it is integrated with a miniature capacitive load–displacement transducer permitting high-resolution load and displacement measurements (resolution of ~ 0.3 μN in load, ~ 1 nm in displacement). In addition, rapid instrument response and data acquisition rates (up to the order of 10^4 s^{-1}, the controller operating in a continuous loop and sampling data at 20 kHz) allow discrete flow events to be well resolved. The compression experiments under displacement rate control exhibit a greater sensitivity to transient load drops, and load rate control has an advantage in evaluating sudden displacement jumps (SB offset). At present, together with the high data acquisition rate, it becomes feasible to evaluate the development of a single SB event. The *in situ* TEM structure monitoring provides information about individual SB evolution both spatially and temporally.

In many crystalline metallic systems, deformation behavior is the same in tension and compression. Exceptions comprise bcc metals and ordered compounds because of dislocation core effect. Also, the mechanical behavior of MGs shows differences and we review some results of compression experiments in relation to size effects. The latter is, in particular, relevant for applications, such as in various complex patterns with smooth surfaces on a nanometer scale were fabricated recently in MGs [121] with highly functional characteristics.

As shown in the previous section, the flow stresses and yield strengths of crystalline metals have been found to be highly size dependent. In particular, fcc single crystals exhibit strengths several times higher than their bulk counter part at micrometer scale and even one or two orders higher at submicrometer or sub-100 nm range, respectively. The strengthening is attributed to dislocation source-limited behavior in small volumes. Such a mechanism does not operate in MGs because of a lack of lattice dislocation in the intrinsically disordered environment. At present, there is no general consensus on how sample size at small scales may influence the strength of MGs. Clearly, a better understanding of yield strength is critical before the microscale components can be successfully incorporated into technological applications. Measurements on even smaller specimens, for example, with characteristic dimensions down to submicrometer or even few tens of nanometer scale will help to elucidate the underpinning mechanisms [122–125]. Accurate measurements of individual, small metallic specimens are still a challenge mainly because of the difficulties in achieving uniform gauge lengths of pillars with diameters smaller than 1 μm. This is in fact one of the origins of the discrepancy in size effects reported in the literature. Also, the positioning of the specimen and alignment of the test system is a critical issue for extremely small specimens.

Only very few quantitative *in situ* TEM compression of micro-/nano-MG pillar experiments that enable measuring the response of pillars of diameters down to tens of nanometer scale have been presented [126]. The load response is correlated with dynamic observation of yielding evolution, which allows real-time measurement of effective diameter supporting the applied load. To illustrate this, Figure 30.19 shows a sequence of frames from a video of the compression of a Cu-based pillar. The pillar has a relatively large tip diameter of 645 nm and was compressed under load rate control. The pillar shows elastic deformation followed by a jerky-type deformation with transient SB events registered with displacement bursts in the displacement–load curve. The SBs 1 and 2 (marked in the figure) initiated sequentially from the top of the pillar, presumably due to the combined effect of the slight tapering and the initial imperfect pillar-punch contact, causing

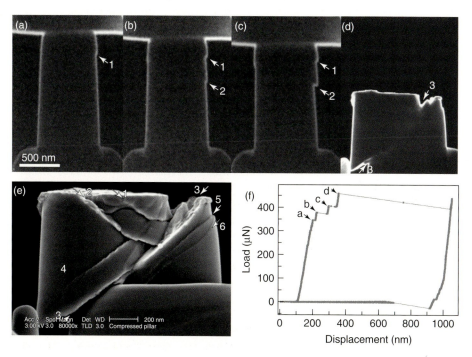

Figure 30.19 Grabbed dark-field TEM video frames recording the evolving deformation of a ø645 nm Cu-based MGs pillar compressed under load-control: (a–d) morphologies immediately after jerky events (a–d) marked in the load–displacement curve in (f); and (e) postmortem SEM micrograph. The initiation of SBs 1 and 2 produced the jerky events (a) and (b) respectively, whereas the subsequent growth/reactivation of the SB 2 causes the event (c). The final large event (d) is associated with a major shear process, involving simultaneous operation of the major SB 3, and minor SBs 4–6 as indicated in image (e).

the sequential appearance of bursts *a* and *b* in the curve. In contrast, the burst *c* is the result of a subsequent jerky-type growth of SB 2. Interestingly, the occurrence of the local SBs 1 and 2 does not seem to weaken the pillar. On the contrary, it enables the pillar to bear a higher load on further loading because of the increase of the load-bearing diameter. With increasing load, the pillar finally deforms with a catastrophic major shear process. From the SEM image, this process is associated with the operation of multiple SBs, that is, the primary SB 3 carrying the majority of the shear displacement (strain) and a few neighboring secondary SBs (4–6). The secondary SBs with minor shear displacements may not traverse the pillar but die out gradually inside the pillar, for example, the SB 4. The duration of the fast running process is very short (<0.04 s) and the sequence of the initiation of SBs 3–6 cannot be resolved from the video frames and the load–displacement curve, indicating simultaneous occurrence of these multiple SBs. The short duration, together with the large shear displacement ($\sim$690 nm in axial direction, and $\sim$1 μm resolved to the shear plane), results in a rather high shear velocity (>25 000 nm s^{-1}). Markedly, after the major shear process the pillar is not completely fractured. On further loading, the load recovers quickly back to the previous value. Similar fracture tolerance and large engineering plastic strain has already been noticed at the micrometer scale. It can be explained by an "extrinsic" size effect, since the critical shear offset at which failure happens (typically taken as 10–20 μm) in BMGs [127, 128] cannot be reached in such small pillars.

On decreasing the diameter, the deformation shows more pronounced intermittent characteristics. This is due either to more frequent initiation of new SBs or to repeated post-initiation growth (reactivation) of preexisting SBs in multiple steps. Figure 30.20 shows frames snapped from a video recording the compression of a 440 nm tip diameter pillar, which was compressed under displacement control and subjected to three loading–unloading cycles. Remarkably, none of these SBs run over a long distance on initiation. Instead, they are arrested after a small distance of propagation ($\sim$20 nm) and subsequently grow (or reactivate) intermittently in many steps, each of them carrying a limited shear offset. Therefore, a higher frequency of transient events in the load–displacement curve is detected. Compared to the predominant SB in the aforementioned thicker pillar, the deformation here is more distributed into many events of SB initiation and subsequent propagation. This tendency increases on decreasing diameter, as will be revealed by a statistical analysis in the following section.

Representative deformation morphologies of Cu-based pillars over the total investigated size range are shown in Figure 30.21, and the representative engineering stress versus displacement curves are shown in Figure 30.22. According to the SEM micrographs, it is interesting to note that as the pillar diameter is reduced from 650 to <100 nm, the apparent deformation mode shows a gradual transition from highly inhomogeneous to relatively more homogeneous. A few well-developed major SBs, with sharp leading fronts at larger diameters in Figure 30.21a–e, are gradually transformed into a larger number of diffuse shear processes with more rounded leading fronts and seemingly torsionlike morphologies (see the 255 and 178 nm pillars shown in (f) and (g)). This phenomenon becomes increasingly

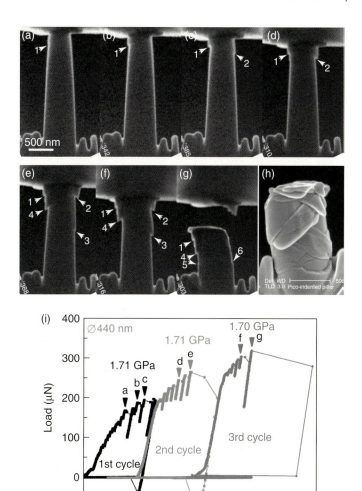

Figure 30.20 Video frames recording the deformation of a ø440 nm Cu-based MG pillar compressed under displacement control and subjected to three loading–unloading cycles. Frames (a–g) correspond respectively to deformation structures immediately after jerky events (a–g) marked on the load–displacement curve in (i). (h) Post-mortem SEM micrograph. The initiation of SBs 1 and 2 cause respectively jerky events (a) and (c). Subsequently, simultaneous initiation of shear bands 3 and 4 causes the larger event (e). The other jumping events following the events (a, c, and e) in the curve are associated, respectively, with intermittent growth (or repeated reactivation) of these preexisting SBs. Image frames (b), (d), and (f) highlight the markedly increased shear displacement of these SBs after the intermittent growth, respectively. The yield stresses measured in each loading cycle is also indicated in (i).

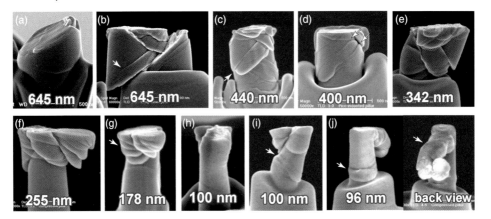

Figure 30.21 SEM micrographs showing a global evolution of deformation morphologies of Cu-based MG pillars with tip-diameters varying from 645 to 96 nm.

significant on further decreasing pillar diameters down to ~100 nm. The inhomogeneity in the deformation of the 96 nm tip diameter pillar is hardly detectable from the morphology (Figure 30.21j), although the stress–response curve still clearly records profuse SB events (Figure 30.22d).

This tendency is also clear from the engineering stress versus displacement curves in Figure 30.22. It can be seen that the frequency of bursts increases rapidly with decreasing pillar diameter. This likely corresponds to the more frequently activated SB events, whereas the shear displacement/strain carried by individual SB events decreases fast, indicating suppression of SBP. At the same time, a close-up examination of individual events (insets in Figure 30.22) reveals that, with decreasing pillars diameter, the abrupt shear processes become increasingly gradual in thinner pillars. The shear displacement in thinner pillars proceeds more smoothly, while the stress decreases more steadily, in agreement with the observation of diffuse SBs. These tendencies can be described by a quantitative analysis of individual shear events versus diameter (Figure 30.23). According to the data, the fast increasing frequency f of bursts with decreasing diameter D follows roughly a power law decay of $f \propto D^{-1.3}$. The amplitude of the burst s can be expressed as $s_{max} \propto D^{2.51}$ and $s_{av} \propto D^{1.2}$, where the subscripts denote the maximum burst and the average of rest bursts, respectively. The velocity of shear displacement jump (SDJ) (Figure 30.23c) shows a similar tendency as the amplitude and can be expressed as $v_{max} \propto D^{2.6}$ and $v_{av} \propto D^{1.3}$, respectively.

From Figure 30.23 at the smallest diameter of ~100 nm, the amplitude of the SDJs is very small (at the order of 10 nm), carrying a shear strain at the order of 1 assuming 10 nm SB thickness according to TEM observations [129]. The propagation of the SBs takes a time lapse of 10 ms, producing a very small shear velocity less than 200 nm s^{-1}. A low propagation speed of SBs in small-sized MG pillars has also been noticed recently in another publication [130]. By compression measurements on MG pillars with effective diameters ranging between 200 and

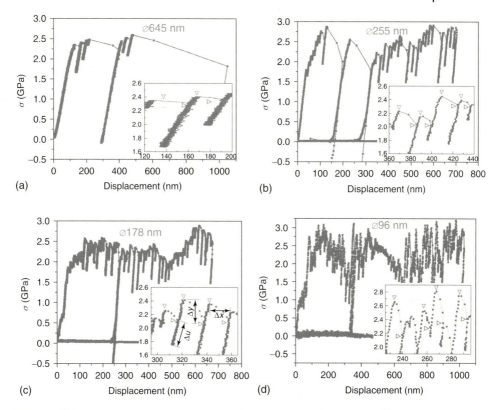

Figure 30.22 Engineering stress σ versus displacement curves of the Cu-based glass pillars under compression of tip diameters: (a) 645 nm, (b) 255 nm, (c) 178 nm, (d) 96 nm. The engineering stress is approximated by: $4P/[\pi(D+2u\tan\beta)^2]$ where P is the load and D tip diameter, β taper angle, and u measured tip displacement. The insets are the enlarged plots showing the individual bursts. Δy denotes the stress drop and Δx the strain carried by an individual burst. Δu is post-burst elastic unloading due to transient withdrawal of indenter to match the predefined displacement rate.

300 nm, the authors concluded that "the shear displacement rate in each jerky motion was only a small fraction of $1\,\mu m\,s^{-1}$" [131]. This lower propagation speed is in sharp contrast to the SBs in unconstrained BMGs, which, according to several investigations, propagate largely to tens of micrometer distance within a few nanoseconds to a couple of microseconds [94, 128, 132, 133] resulting in an extremely high velocity, that is, near the speed of sound. The relatively low speed indicates an incipient and immature nature of the SBs in these small pillars compared to bulk. Also, the acceleration and deceleration transients may contribute, on average, to a lower velocity compared to phenomena in bulk material.

When inspecting literature about the yield strength of MG pillars, one should be mindful that by assuming a tip diameter or an average diameter as the pillar size without taking into account the dynamical evolution of the effective

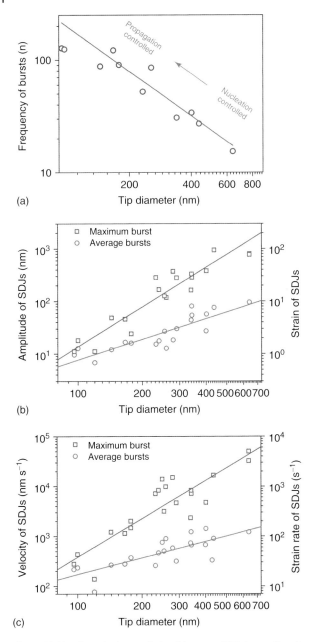

Figure 30.23 Quantitative analysis of individual burst events as a function of pillar tip diameter: (a) *frequency of bursts* defined as the number of bursts per unit engineering strain; (b) amplitude, and (c) velocity of shear displacement jumps (SDJs) transferred to 45° shear planes with the solid trend lines being power law fittings. The right axis reads the strain and strain rate of SDJs assuming 10 nm shear bands thickness. Solid lines are power law fittings.

diameter during compression, an artificial size effect will inevitably be attained, since a constant tapering angle has larger influence on thinner pillars. This kind of error, however, can be circumvented by *in situ* dynamically measuring the effective diameter as mentioned before. Schuster *et al.* [80] evaluated with FEM the effect of tapering (particularly the stress gradient) on the measured apparent yield strength. They concluded that this effect should not have much influence provided the MG follows a Mohr–Coulomb yield criterion. However, if the MG yields according to another criterion, that is, shear plane criterion [134], the reported size-dependent yield strength can be understood as an artifact arising from the taper. Ye *et al.* [135] extracted yield strengths using tapered micropillars in microcompression experiments and derived a formula based on the Mohr–Coulomb law accounting for the effects of a variety of geometrical parameters arising from the tapered micropillar geometries. The experimental results on three classes of BMGs, comprising the Zr-, Mg-, and Fe-based BMGs, show that the yield strengths of the BMG micropillars exhibit a very weak size effect as compared with the literature data obtained from the microcrystals. From an experimental viewpoint, the following remarks have to be made: to accurately determine the yield stress, a test is normally interrupted at a certain stage to measure the effective load-bearing diameter that is the minimum diameter at the purely elastically deformed region measured ahead of the most forefront SB. Together with the real-time recorded load corresponding to this diameter, the procedure leads to an accurate determination of the yield stress (SB initiation stress). One of the examples is displayed in Figure 30.20, with the measured values in each loading–unloading cycle indicated. It can be seen that, by measuring the instantaneous load-bearing diameter, the yield stresses recorded in different test cycles of the same pillar are quite reproducible.

The results so far dealing with microcompression have demonstrated a size dependence in individual SB events, that is, spatially from sharp to diffuse, or temporally from abrupt to gradual, with decreasing size. However, according to the above analysis, we have to clearly distinguish a homogeneous component in the deformation of the pillars subjected to compression. Although the pillar has apparently homogeneous deformation morphology, the deformation is still intermittent plastic flow accommodated by inhomogeneous SB. The apparent homogeneous flow is involved in the characteristic of individual SB which is size dependent. This is still the case even with a diameter down to the sub-100 nm scale, which is the experimentally accessible diameter limit for FIB milling. Nevertheless, we noticed that a deviation from the uniaxial loading condition in the deformation of some extremely thin pillars may provide interesting hints of homogeneous flow. These sub-100 nm diameter pillars invariably have a rounded tip as a side product of FIB milling, while the initial deformation at the tip always flattens the rounded tips out by a bulging (mushrooming) effect, characteristic of homogeneous flow (Figure 30.24). Presumably, this phenomenon is promoted by a combined effect of the extremely small size and the local stress triaxiality arising from the rounded geometry. However, after this initial mushrooming stage, SB takes over. In an attempt to explore the possibility of a fully homogeneous deformation with

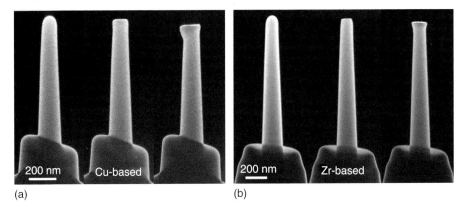

Figure 30.24 Homogeneous deformation at the rounded top part of a Cubased pillar with 100 nm tip diameter (a) and a Zrbased pillar with 116 nm tip diameter (b). The mushrooming bulge near the tip indicative of homogeneous deformation is prompted by the local multiaxial stress state.

decreasing size, microbending tests were explored. It will be demonstrated that microbending tests have particular advantages by minimizing various artifacts and getting the size effect more pronounced in an experimentally accessible regime.

There are a couple of practical difficulties in designing a standard microbending test inside a TEM because of the clamped-free boundary conditions. Simple deflection also suffers from the high stress concentration at the corner near the base. A practical solution that is proposed in this study is to intentionally introduce a small misalignment (3°) between the tip and the pillar. Considering the nonstandard nature of the microbending test, a three-dimensional FEM approach was carried out to assist the stress–strain analysis by assuming an isotropic and ideally elastic–plastic material behavior. The distribution of stress and strain is not fully symmetric, but becomes more symmetric with increasing tip displacement and finally close to symmetric bending. Importantly, the maximum deformation occurs around the region at half height rather than near the tip-pillar contact (as is the case for monotonic compression) or at the base of the pillar (case in simple deflection). Thus, the complexity due to local stress concentrations is avoided and various related artifacts can be minimized. Notice that no details of SB or cracking can be observed from the FEM as isotropic plastic deformation is assumed and no microstructural features are involved. Microbending tests carried out on pillars with relatively large diameters ($D > \sim 500$ nm) confirm that localized shear deformation (SBs) is still the dominant deformation mode, in agreement with the compression test. The deformation evolution of a 505 nm tip diameter pillar is shown in Figure 30.25. The shear displacements associated with SBs at the surface of the compressive side emerge almost simultaneously. Subsequently, the first SB at the tensile side of the pillar nucleates, as shown in Figure 30.25a. This sequence is consistent with the initial asymmetrical stress distribution due to a compression component overlapping with the bending. When deformation

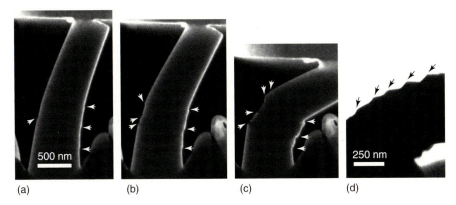

Figure 30.25 *In situ* TEM observation of SBs evolution (arrows annotated) during bending test of a 505 nm tip-diameter pillar (a–c). Another bent pillar of 480 nm tip diameter showing typical stairlike SBs at the surface is shown in (d).

proceeds, more SBs are observed to nucleate at the tensile side and the shear offsets at both sides increase continuously. Figure 30.25d shows typical SBs with stairlike morphology due to shear offsets at the surface. Fracture is not observed in these pillars.

Additional remarks that should be made are the possibilities of "extrinsic" factors taking part in the size effects. One general concern about FIB-milled pillars for mechanical tests is Ga^+ implantation and general ion beam damage, and the other one is the already mentioned tapering angle. Ga^+ implantation has been recognized to have only a minor effect on the deformation behavior of submicrometer samples, especially of amorphous materials that are radiation tolerant [69]. An FIB-introduced disordered surface layer of 3–4 nm thickness [136] and containing less than 1 at% Ga concentration is unlikely to have remarkable effects on the behavior of intrinsically disordered MGs. The tapering angle of 2–3.5° is rather small compared to the 7–15° in other studies and also capable of producing MG pillars with diameters down to submicrometer scale. In addition, the *in situ* TEM observation correlates the structure evolution and load response effectively, which further minimizes possible artifacts. Although the slight tapering still causes the deformation starting from the top and proceeding toward the base, an SB generally traverses through the pillar without constraint at either the leading or the rear front. Nevertheless, the unconstrained SBs do not run away largely but develops intermittently, with the energy released and reaccumulated repeatedly in the limited volume, indicating intrinsic effects. Moreover, with the designed bending test, the limitations of the compression test were circumvented and consequently a robust size effect appeared.

Another influencing factor is the effective strain rate. A high strain rate may lead to apparent homogeneous deformation, however, which is obvious only when the strain rate scales over 3–4 orders of magnitude. In this study, all the compressions are controlled at the same nominal strain rate of $\sim 10^{-2}\,\text{s}^{-1}$, which is unlikely to cause noticeable effects of strain rate on the deformation mode. An exception is the 240 nm pillar compressed at a relatively lower rate (Figure 30.26), which, however, shows frequent arrest of the SB instead of fast and large propagation preferred by a low strain rate, further pointing to an intrinsic size effect.

The study on size effects in the mechanical response of MGs is interesting as it may provide hints on these fundamental issues. Also, it may guide us to the practical design of small-sized MG-incorporated materials and devices [62–64, 137]. Consequently, these basic questions gave an impetus to considerable theoretical [65–67, 129] and experimental [78–82, 138–140] attention. From experiments, it is getting clear that at micrometer scale SB is still the dominant behavior [68–72]. Testing of submicron- and nanometer-sized samples is highly desirable but such tests are extremely challenging because of the experimental difficulties in performing neat experiments, that is, fabricating extremely small, nanosized specimens, which are desired to be free of geometrical imperfections and surface contaminations [68–74, 79–82], and testing them quantitatively. Recent literature hinted at homogeneous plastic flow [69–77, 79, 82], but remarkable artifacts arising from geometrical tapering and surface modification due to the use of FIB have also not been clarified. In addition, the driving force and microscopic basic concepts of such transitions in deformation modes remain rather obscure and elusive.

Here, we demonstrate the successful fabrication of MG nanopillars with diameters down to sub-100 nm, and, importantly, free of taper and with clean surfaces. We performed quantitative *in situ* compression tests on these taper-free pillars inside a TEM, and show striking new results that have not been observed previously [141]. Micropillars were cut by FIB from a bulkily brittle Cu-based MG $Cu_{47}Ti_{33}Zr_{11}Ni_6Sn_2Si_1$ [79]. In this study, the side surfaces of the pillars are finally polished with ion beams oriented perpendicular to the pillars in a quasi-parallel milling mode to remove tapering and to avoid surface redeposition. Taper-free MG pillars with sizes ranging from 640 to ~ 70 nm were successfully fabricated. TEM observations confirmed the uniform amorphousness of the pillars and cleanness at the surface. The length to diameter ratio L/D of the pillars is controlled at 3.0 with a deviation smaller than 0.5 in practice. *In situ* compression was performed using a recently developed Hysitron picoindenter equipped on a JEOL 2010F TEM, which is capable of high-resolution measurements of load and displacement ($\sim 0.3\,\mu\text{N}$ in load, ~ 1 nm in displacement) and has rapid instrument response and data acquisition rates [79]. The compression was performed under displacement control at a nominal strain rate of $\sim 10^{-2}\,\text{s}^{-1}$.

Taper-free pillars with relatively large diameters always show catastrophic SB following elastic deformation similar to that observed through testing slightly tapered pillars [78–81]. One example is shown in Figure 30.27a,b, where the pillar with 640 nm diameter shows an initial small local banding to accommodate the imperfect tip-punch contact, and subsequently two major SBs are triggered simultaneously at a larger event. They run fast, leaving behind the platen moving at the programmed displacement rate.

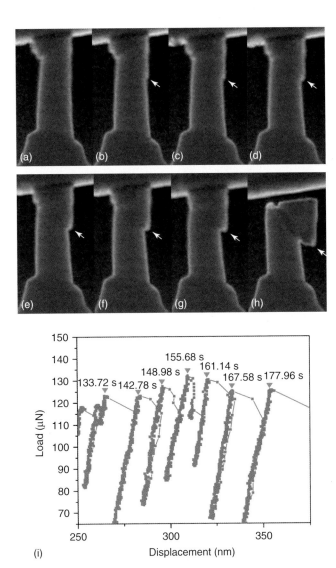

Figure 30.26 (a) Video snaps taken from the *in situ* TEM compression showing repeated interruption and reinitiation of an unconstrained shear band in a 240 nm tip diameter pillar compressed at a programmed relatively low displacement rate of 2 nm s^{-1}. The unconstrained shear band, which is free at both ends, supporting 100% the applied load, and growing intermittently, is associated with seven noticeable load drop events marked in plot (i). Images (b–h) highlight evolution of the shear band after each banding event marked in the load–displacement evolution curve in (i), where the exact time when each event happens is also annotated.

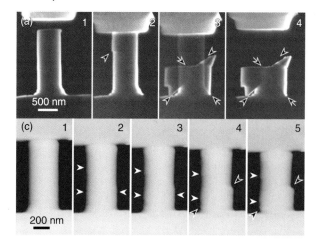 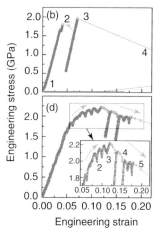

Figure 30.27 Grabbed dark-field TEM video frames showing the deformation of taper-free pillars of (a) 640 nm and (c) 365 nm diameter, respectively, under displacement controlled compression: the numbering of the video frames in (a) and (c) corresponds to the numbers marked in the stress–response curves in (b) and (d) respectively.

Tests of pillars with the diameter down to ∼300 nm regime show interesting new phenomena. For example, the deformation of a 365 nm taper-free pillar in Figure 30.27c shows several features. First, it experienced large engineering strain up to 18%; second, the stress evolution is intermittent; third, the stress evolution is initially "hardening" up to 12.5% strain followed by softening as indicated in Figure 30.27d and zoomed-in in the inset. The initial "hardening" is the result of a general "global fattening" effect, which increases the effective load-bearing-area. The "hardenable" deformation, however, is not uniform, with jerky-type stress drops observable. But no transient shear processes corresponding to the jerky events are observed in the structure evolution; instead, irregular local bumps are gradually developed at the surface. The subsequent softening is due to the development of a nascent SB (indicated by open arrows in (c4) and (c5)) at the edge of a local bump that is developed before the actual "hardening" stage. It should be noted that the initial "hardening" is an intrinsic behavior. It is fundamentally different from the "hardening" observed in tapered pillars, where the stress gradient induced by the tapering promotes the formation of local SBs, that is, the load increase is due to diameter increase associated with the tapered geometry. Also, the transient stress drops at the softening stages observed in Figure 30.27d are associated with the arrest and reinitiation of the single SB, fundamentally different from that observed in bulk specimen because of the interaction of many SBs. It can be interpreted in terms of elastically stored energy averaged to the shear plane, which is proportional to the volume to area ratio (or D for constant L/D ratio) and thus not enough to drive the SB to propagate automatically at this size regime [79].

While further decreasing the size to 120 nm level, the deformation becomes apparently fully homogeneous. With the term *"homogeneous,"* we mean that the deformation is smooth and banding-free (without bandlike features), but not that the deformation is uniform throughout the pillar. In fact, the deformation is not completely uniform because of the tip-punch friction and the constraint at the base. Initially, a homogeneous bulging/swelling occurs at the top, with increasing displacement. The bulge does not extend downward, instead plastic flow starts in a virgin area far ahead of the tip where the side surfaces continuously bow out on compression, producing a barrel-like shape leaving two seemingly "necking" regions at both ends. Importantly, it produces a monotonic "hardening" as a result of a continuous upsetting effect and lack of SBs.

Despite the apparent "hardening," the stress evolution is obviously still intermittent with transient load drops present in Figure 30.28c. The intermittency is presumably the result of very tiny local transitions distributed throughout the pillar, which would not have been detected without ingenious quantitative tests and fast enough machine response. They do not organize into an SB or cause observable bandlike local deformation. By this mechanism, very large engineering strains can be obtained without any indication of damage.

The phenomenon of barreling is often a message of good plasticity. A 3-D finite element analysis (FEA) was performed and the result shows that, by treating the material as an isotropic elastic–plastic body characteristic of homogeneous plastic flow, the deformation of the MG nanopillars can be satisfactorily predicted (Figure 30.28b). The local deformation at the tip of a pillar can be interpreted by either friction or the rounded tip, or their combination as a result of developed local large shear stress shown in Figure 30.28(b4). The continuous barreling and fattening is more likely due to the constraint at the base and especially the shear stress concentration. This is further confirmed by simulating a specimen with a flat top by neglecting the friction in Figure 30.28(b5). Barreling was previously observed only in compression of some BMG-based composites containing nanocrystallites, where the barreling was accommodated, however, by well-developed, profuse SBs whose nucleation, distribution, and propagation are influenced by the embedded nanocrystallites or the platen [142].

HRTEM confirmed the atomically flat surface of the as-milled pillars free of surface redeposition (Figure 30.28d) and the band-free deformation characteristics. It also revealed nanoscale surface modulations associated with local deformation domains of a few nanometer size at either the "bowing out" region or the "necking" region, as shown in Figure 30.28(d2) and (d3), respectively. The intermittency becomes clearly less observable in compression of a thinner pillar with 100 nm diameter in Figure 30.29a, and eventually disappears in Figure 30.29b for the 70 nm pillar, leading to a fully smooth stress response, despite the fact that an initial slight misalignment results in reduced barreling effect in the 100 nm pillar, while a geometric imperfection leads to bending/buckling of the 70 nm pillar at a later stage, indicating extreme sensitivity of the test to external perturbation at this size scale.

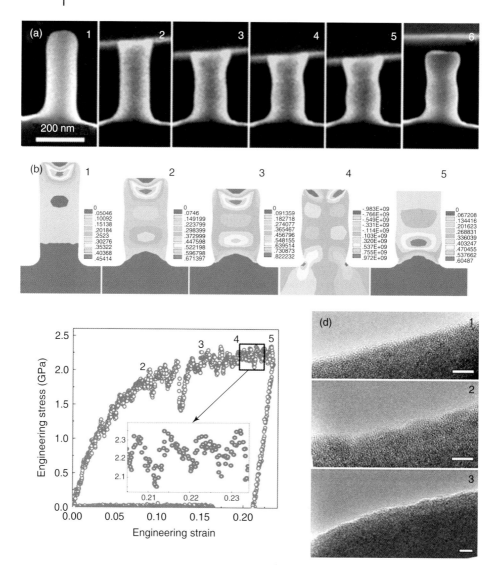

Figure 30.28 Banding-free deformation of a 125 nm diameter pillar: (a) video frames grabbed at points 1–6 marked in the stress-strain plot in (c); (b) 3-D FEM of the compression by treating the MG as an isotopic, elastic ideal–plastic solid, taking into account the rounded tip and the tip-punch friction, with (b1–3) being von Mises-strain at different displacements and (b4) shear strain at the same displacement as in (b3), (b5) compression of a pillar with flat top neglecting friction. (d) HRTEM showing atomically flat and clean surface of the as-milled pillar (d1) and nanoscale inhomogeneity at the "necking" (d2) and "swelling" (d3) regions. Scale bar in (d) represents 5 nm.

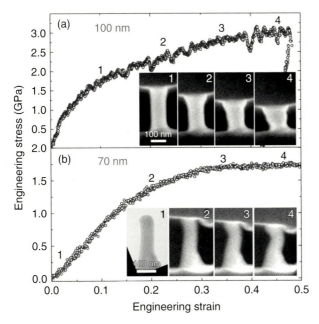

Figure 30.29 Stress and microstructural evolutions of nanopillars with diameters of (a) 100 nm and (b) 70 nm, respectively: the 100 nm pillar still shows noticeable transient stress-drops, and the slightly developed misalignment in loading reduced the barreling effect; the stress evolution in the 70 nm pillar is completely smooth without intermittency; however, a slight geometric imperfection (curved shape) induces bending in the later stage of deformation and thus much lower engineering stress.

The detailed microscopic processes of individual STZs toward the formation of macroscopic SBs have not been visualized directly yet. However, the present results provide strong support that, microscopically, a larger volume on plastic yielding has a higher probability to develop larger flow defects (or percolation of flow defects), which are associated with larger jerky events and SB in micrometer-sized pillars. On decreasing the size of the pillars, smaller local shear processes (flow defects) become critical and they lead to less inhomogeneous deformation. The deformation is still intermittent and may develop to SBs at ∼300 nm, but becomes banding-free although still intermittent when the system approaches the 100 nm scale. It ultimately changes to fully smooth and continuous plastic flow at the sub-100 nm scale. Such a picture of flow defect distribution is reminiscent of flaws/cracks (preexisting and stress induced) commonly observed in many alloys and ceramic systems, whose behaviors can be described statistically, for example, by Weibull statistics (WS), depending on the statistical size distribution of the defects [143]. Active simulation studies on microscopic processes and dynamics of local transitions in MGs indeed suggested inhomogeneity of local mechanical response in MGs and influence of system size (2-D or 3-D) on the size of local processes, although no general scaling laws have been reached [129]. We may tentatively assume that the size of the critical flow defects a in MGs statistically follows the same scaling law with the volume V, as normally assumed in Weibull materials, which is described as

$$a \propto V^m \tag{30.3}$$

with m being a constant [143, 144].

To statistically get such a scaling law, intrinsically it requires that smaller flow defects have a much higher relative density (number per unit volume) in the material. Such dependence can be represented by a size distribution function describing the relative density in terms of the size of the flow defects. A function of a simple form

$$g(a) = g_0 \left(\frac{a}{a_0}\right)^{-r} \tag{30.4}$$

has been shown to be intrinsically consistent with Eq. (30.3), where a is the size of a flow defect, g_0, a_0 are material-related constants, and r is related to m through $m = 2(r-1)$ [144]. The tendency represented by Eq. (30.4) is plotted in Figure 30.30a. Assuming a_c being the size of the smallest critical flow defects that directly contribute to the global plasticity, for evenly distributed flow defects, the total density (number) of critical flow defects is expressed as

$$\rho(a_c) = \int_{a_c}^{\infty} g(a) da = \frac{a_c}{(r-1)} g(a_c) \tag{30.5}$$

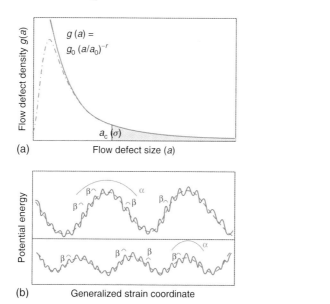

Figure 30.30 Plot of function g(a) defining flow defect density versus size of the defects (a), and a schematic illustration of PELs (b) for a relatively big (top) and small (bottom) system.

and is represented by the shaded area in Figure 30.30a. Here, a_c is determined by Eq. (30.3) and scales with the volume of the specimen. Noticing that, with the size approaching zero, $g(a)$ goes unreasonably to infinity, one expects that the actual density distribution of flow defects smaller than individual STZs (e.g., <1 nm) may deviate from Eq. (30.4) and goes with a tendency represented, for example, by the dashed curve in Figure 30.30a. However, such a deviation may not influence the analysis based on Eq. (30.4), since defects smaller than the critical value a_c do not remarkably influence the phenomena.

The total volume fraction of critical flow defects is expressed as

$$f(a_c) = \int_{a_c}^{\infty} ca^k g(a) da = \frac{ca^{k+1}}{r-(k+1)} g(a_c) \qquad (30.6)$$

c and k being shape-related factors, assuming the ellipsoidal shape of the flow defects that is normally assumed for STZs, k has a value between 3 and 2 that are values for spheres and discs, respectively. Again, assuming that $r = m/2 + 1$ is large compared to $(k+1)$ (in Weibull materials, m is ranging from 10 to 100 for metals and ceramics, and 20–70 for MGs), utilizing Eq. (30.3) and keeping in mind the constant aspect ratio for all the pillars, one obtains approximately

$$f(a_c) \propto a_c^{-r} \propto V^{-m^2/2} = D^{-3m^2/2} \qquad (30.7)$$

where D is diameter. It shows that despite the larger size a_c that can be developed in a bigger specimen, the total volume fraction of the flow defects is rather low. This is consistent with the experimental observation that in larger pillars, deformation is localized in local SBs, which account for a very small fraction of the total volume. In a smaller volume, large flow defects have less chance to be developed and smaller local flow defects are dominating the deformation. This is confirmed by the smoother stress evolution with vanishing stress-drops and less-observable local shear in the structure evolution at smaller size regimes. According to Eq. (30.4), these smaller flow defects will have a higher density. These smaller local transitions, however, account for a larger volume fraction of a specimen and are well distributed in consistence with less localization (Eq. (30.7)). It should be pointed out that in bulk specimens, the scaling of flaw sizes with specimen volume may statistically lead to a size-dependent fracture stress based on the Griffith theory within the framework of continuum mechanics and the weakest link assumption [143], where the flaws are generally treated as cracks. However, the flow defects in the small-sized MGs are approaching nanoscale where the Griffith theory may not hold [79]. In addition, these flow defects are local (shear) transformations without opening and dissimilar with cracks. A flow defect is a cluster of STZs and its strength is directly controlled by the nucleation stress of STZs, equal to the yield stress rather than controlled by its size. Therefore, a significant size-dependent strength based on WS is not expected [145].

It is helpful to understand the microscopic processes underlying the size effects also from a perspective of the potential energy landscape (PEL), which brings hope that the transient local events can be related to the discontinuous transitions and relaxations to different energy minima in the PEL [146]. Recently, Johnson and Samwer et al. [147, 148] have extended Argon's concept of STZ [4] and merged it into the PEL perspective, and proposed a cooperative shear model (CSM) to understand the deformation mechanisms and rheological properties of MGs. According to the CSM, activation of isolated STZs locally confined within the elastic matrix could be associated with the faster β relaxation, which is reversible, while percolation of these transitions leading to an irreversible collapse of the confining matrix and breakdown of elasticity are associated with the slower α process. The β transitions are identified in PEL as stochastically activated (intrabasin) hopping events across "subbasins" within an inherent "megabasin" shown in Figure 30.30b, and the α transitions as (interbasin) hopping events extending across different landscape megabasins. Consequently, the effect of MG sample size on its mechanical response could be correlated to inherent relaxation processes and properties of the PEL. From the PEL perspective, in a smaller volume a large number of hopping events occur with a high frequency, each releasing a small amount of energy, showing a relatively smooth transition in the PEL, and vice versa in larger specimens (Figure 30.30b). In a bulk specimen, excessive energy accumulation may lead to unbalanced relaxation events. When the whole system is unable to find any basin point to rest in the PEL, a shear catastrophe occurs. The experimental observations in this work clearly reveal that the irreversible α events associated with the load drops in Figures 30.27–30.29 are size dependent. With the decrease in the size of pillars, the amplitude of α events decreases but their frequency increases, which are related to the size and density of isolated STZs. In contrast, the amplitude of β transitions is independent of sample size, determined by their local relaxation character. While MG samples reach a size of tens of nanometer, the α alpha hopping events merge with the β transitions, leading to a smooth plastic flow in deformation.

Recent experimental measurement utilizing mechanical loss spectroscopy [149] has shown that by reducing the thickness of an amorphous thin film PdCuSi below 30 nm, the loss of the secondary relaxation is vanishing, which was considered as merging of the primary relaxation and the secondary relaxation because of size effects and associated with less elastic confinement at the surface of the film. Finally, the merging of alpha and beta transition, as well as larger volume fraction involved in deformation, have similarity to that in homogeneous flow of MGs close to glass-transition temperature T_g [150]. More detailed experiments in this direction may provide useful hints on the microscopic processes and elementary transitions within small-sized MGs.

30.6
Conclusions

This contribution highlights recent advances in TEM, in particular concentrating on *in situ* TEM experiments. In particular, we have reviewed recent possibilities of *in situ* TEM indentation, tension, and compression on metallic systems.

The appearance of discrete yield events during nanoindentation of metallic systems depends on the ability of dislocations to propagate into the crystal, and is therefore substantially affected by solute pinning. Under load control, the characteristic yield excursions commonly observed in pure metals are strongly attenuated by solute pinning, leading to a more continuous loading response. Under displacement control, pure metals mostly exhibit full load relaxation during discrete yielding, but in alloys, solutes impede dislocation motion and thereby prevent the load from relaxing completely on reaching a plastic instability. Yield events are resolved more clearly under displacement control, particularly in the presence of solute drag, since displacement-controlled indentation does not require collective dislocation motion to the extent required by load-controlled indentation in order to resolve a yield event. This perception is confirmed by *in situ* TEM displacement-controlled indentations, which show that many dislocations are nucleated *before* the initial macroscopic yield point and that the macroscopic yield event is associated with the rearrangements of the dislocations. *In situ* TEM straining of amorphous metals permit an evaluation of the thickness of the LLL formed because of heat evolution after SB development. The experimental evaluation confirms that the thickness of a LLL present at the last moment of fracture substantially exceeds the generally accepted thickness of an SB. The probability of witnessing a crystalline phase near the fracture surface of amorphous metals is mainly dependent on the size of the shear displacement. Tension and compression tests of MG pillars having diameters ranging above 1 μm, show predominant inhomogeneous and intermittent plastic flow characterized by SB events. The deformation is SB nucleation controlled in larger MG pillars under compression, but becomes propagation controlled in smaller MG pillars under compression.

It is concluded that recent advances in *in situ* TEM can provide new insights about the nucleation and propagation of defects in crystalline and noncrystalline metallic materials that cannot be attained by other means.

Acknowledgments

The author is grateful for the support of and collaboration with Julia Greer (CalTech, USA) in the field of size-dependent material phenomena, and Wouter Soer, ChangQiang Chen, Dave Matthews, Vasek Ocelik, YuTao Pei, Alexey Kuzmin, and Paul Bronsveld for their contributions to the experimental results and discussions on the analyses; and Andy M. Minor, LBL Berkeley USA, with Steven Shan, S.A. Syed Asif, Oden L. Warren, all from Hysitron, USA, and Eric Stach at Purdue Un. USA., for discussions on the *in situ* mechanical testing. Financial support from the Netherlands Science Foundation (Division Physics, FOM-Utrecht, The Netherlands) and from M2i (Materials Innovation Institute) are gratefully acknowledged.

References

1. Nabarro, F.R.N. (1967) *Theory of Crystal Dislocations*, Oxford University Press, Oxford.
2. Hirth, J.P. and Lothe, J. (1968) *Theory of Dislocations*, McGraw-Hill, New York.
3. De Hosson, J.T.M., Kanert, O., and Sleeswyk, A.W. (1983) in *Dislocations in Solids* (ed. F.R.N. Nabarro), North-Holland, Amsterdam, pp. 441–534.
4. Argon, A. (1978) *Acta Metall.*, **27**, 47–58.
5. Spaepen, F. (1977) *Acta Metall.*, **25**, 407–415.
6. Eckert, J., Das, J., Pauly, S., and Duhamel, C. (2007) *J. Mater. Res.*, **22**, 285–301.
7. Gao, M.C., Hackenberg, R.E., and Shiflet, G. (2001) *Mater. Trans.*, **42**, 1741–1747.
8. Hajlaoui, K., Yavari, A.R., Doisneau, B., LeMoulec, A., Botta, W.J., Vaughan, G., Greer, A.L., Inoue, A., Zhang, W., and Kvick, A. (2006) *Scr. Mater.*, **54**, 1829–1834.
9. Hajlaoui, K., Doisneau, B., Yavari, A.R., Botta, W.J., Zhang, W., Vaughan, G., Kvick, A., Inoue, A., and Greer, A.L. (2007) *Mater. Sci. Eng. A*, **449–451**, 105–110.
10. Hufnagel, T.C., Fan, C., Ott, R.T., Li, J., and Brennan, S. (2002) *Intermetallics*, **10**, 1163–1166.
11. Zhang, Y. and Greer, A.L. (2006) *Appl. Phys. Lett.*, **89**, 071907.
12. Wall, M.A. and Dahmen, U. (1997) *Microsc. Microanal.*, **3**, 593.
13. Wall, M.A. and Dahmen, U. (1998) *Microsc. Res. Technol.*, **42**, 248.
14. Stach, E.A., Freeman, T., Minor, A.M., Owen, D.K., Cumings, J., Wall, M.A., Chraska, T., Hull, R., Morris, J.W., Zettl, A., and Dahmen, U. Jr. (2001) *Microsc. Microanal.*, **7**, 507.
15. Minor, A.M., Morris, J.W., and Stach, E.A. Jr. (2001) *Appl. Phys. Lett.*, **79**, 1625.
16. Minor, A.M., Lilleodden, E.T., Stach, E.A., and Morris, J.W. Jr. (2002) *J. Electron Mater.*, **31**, 958.
17. Minor, A.M., Lilleodden, E.T., Stach, E.A., and Morris, J.W. Jr. (2004) *J. Mater. Res.*, **19**, 176.
18. Doherty, R.D., Hughes, D.A., Humphreys, F.J., Jonas, J.J., Juul Jensen, D., Kassner, M.E., King, W.E., McNelley, R., McQueen, H.J., and Rollett, A.D. *Mat.Sci.Eng.*, **A238** (1997), 219.
19. Winning, M., Gottstein, G., and Shvindlerman, L.S. (2001) *Mater. Sci. Eng. A*, **317**, 17.
20. Van Swygenhoven, H., Caro, A., and Farkas, D. (2001) *Mater. Sci. Eng. A*, **309–310**, 440.
21. Van Vliet, K.J., Tsikata, S., and Suresh, S. (2003) *Appl. Phys. Lett.*, **83**, 1441.
22. Shan, Z., Stach, E.A., Wiezorek, J.M.K., Knapp, J.A., Follstaedt, D.M., and Mao, S.X. (2004) *Science*, **305**, 654.
23. Jin, M., Minor, A.M., Stach, E.A., and Morris, J.W. Jr. (2004) *Acta Mater.*, **52**, 5381.
24. Larsson, P.L., Giannakopoulos, A.E., Söderlund, E., Rowcliffe, D.J., and Vestergaard, R. (1996) *Int. J. Solids Struct.*, **33**, 221.
25. Soer, W.A., De Hosson, J.T.M., Minor, A.M., Stach, E.A., and Morris, J.W. Jr. (2004) *Mater. Res. Soc.*, **795**, U9.3.1.
26. Soer, W.A., De Hosson, J.T.M., Minor, A.M., Morris, J.W., and Stach, E.A. Jr. (2004) *Acta Mater.*, **52**, 5783.
27. De Hosson, J.T.M., Soer, W.A., Minor, A.M., Shan, Z., Stach, E.A., Syed Asif, S.A., and Warren, O.L. (2006) *J. Mater. Sci.*, **41**, 7704.
28. Warren, O.L., Downs, S.A., and Wyrobek, T.J. (2004) *Z. Metallk.*, **95**, 287.
29. Minor, A.M. (2002) PhD thesis, University of California, Berkeley.
30. Mondolfo, L.F. (1979) *Aluminum Alloys: Structure and Properties*, Butterworth, London, p. 313.
31. Dahmen, U. and Westmacott, K.H. (1988) *Scr. Metall.*, **22**, 1673.
32. De Hosson, J.T.M., Soer, W.A., Minor, A.M., Shan, Z., Syed Asif, S.A., and Warren, O.L. (2006) *Microsc. Microanal.*, **12** (S02), 890.

33. Gane, N. and Bowden, F.P. (1968) *J. Appl. Phys.*, **39**, 1432.
34. Kramer, D.E., Yoder, K.B., and Gerberich, W.W. (2001) *Philos. Mag. A*, **81**, 2033.
35. Gouldstone, A., Koh, H.J., Zeng, K.Y., Giannakopoulos, A.E., and Suresh, S. (2000) *Acta Mater.*, **48**, 2277.
36. Gerberich, W.W., Venkataraman, S.K., Huang, H., Harvey, S.E., and Kohlstedt, D.L. (1995) *Acta Metall. Mater.*, **43**, 1569.
37. Gerberich, W.W., Nelson, J.C., Lilleodden, E.T., Anderson, P., and Wyrobek, J.T. (1996) *Acta Mater.*, **44**, 3585.
38. Bahr, D.F., Kramer, D.E., and Gerberich, W.W. (1998) *Acta Mater.*, **46**, 3605.
39. Minor, A.M., Shan, Z., Stach, E.A., Syed Asif, S.A., Cyrankowski, E., Wyrobek, T., and Warren, O.L. (2006) *Nat. Mater.*, **5**, 697.
40. Soer, W.A., De Hosson, J.T.M., Minor, A.M., Shan, Z., Syed Asif, S.A., and Warren, O.L. (2007) *Appl. Phys. Lett.*, **90**, 181924.
41. Johnson, K.L. (1985) *Contact Mechanics*, Cambridge University Press, Cambridge.
42. Larsson, P.L., Giannakopoulos, A.E., Söderlund, E., Rowcliffe, F.J., and Vestergaard, R. (1996) *Int. J. Solids Struct.*, **33**, 221.
43. Soer, W.A. and De Hosson, J.Th.M. (2008) in *Electron Microscopy at High Resolution* (ed. F. Banhart), World Scientific, p. 115.
44. Sickafus, K. and Sass, S.L. (1984) *Scr. Metall.*, **18**, 165.
45. Lin, C.H. and Sass, S.L. (1988) *Scr. Metall.*, **22**, 735.
46. Rittner, J.D. and Seidman, D.N. (1997) *Acta Mater.*, **45**, 3191.
47. Lamelas, F.J., Tang, M.-T., Evans-Lutterodt, K., Fuoss, P.H., and Brown, W.L. (1992) *Phys. Rev.*, **B 46**, 15570.
48. Dahmen, U., Hetherington, C.J.D., O'eefe, M.A., Westmacott, K.H., Mills, M.J., Daw, M.S., and Vitek, V. (1990) *Philos. Mag. Lett.*, **62**, 327.
49. Paciornik, S., Kilaas, R., Turner, J., and Dahmen, U. (1996) *Ultramicroscopy*, **62**, 15.
50. Pénisson, J.M., Lançon, F., and Dahmen, U. (1999) *Mater. Sci. Forum*, **294–296**, 27.
51. Merkle, K.L. and Thompson, L.J. (1999) *Phys. Rev. Lett.*, **83**, 556.
52. Medlin, D.L., Foiles, S.M., and Cohen, D. (2001) *Acta Mater.*, **49**, 3689.
53. Medlin, D.L., Cohen, D., and Pond, R.C. (2003) *Philos. Mag. Lett.*, **83**, 223.
54. Westmacott, K.H., Hinderberger, S., and Dahmen, U. (2001) *Philos. Mag.*, **A 81**, 1547.
55. Song, S.G., Vetrano, J.S., and Bruemmer, S.M. (1997) *Mater. Sci. Eng.*, **A 232**, 23.
56. Frank, F.C. (1950) *A Symposium on the Plastic Deformation of Crystalline Solids*, Office of Naval Research, Washington, DC, p. 151.
57. Read, W.T. (1953) *Dislocations in Crystals*, McGraw-Hill, New York.
58. Sutton, A.P. and Balluffi, R.W. (1995) *Interfaces in Crystalline Solids*, Clarendon Press, Oxford.
59. Greer, A.L. and Ma, E. (2007) *MRS Bull.*, **32**, 611.
60. Schuh, C.A., Hufnagel, T.C., and Ramamurty, U. (2007) *Acta Mater.*, **55**, 4067.
61. Falk, M.L. and Langer, J.S. (1998) *Phys. Rev. E*, **57**, 7192.
62. Donohue, A. et al. (2007) *Appl. Phys. Lett.*, **91**, 241509.
63. Hofmann, D.C. et al. (2008) *Nature*, **451**, 1085.
64. Hays, C.C., Kim, C.P., and Johnson, W.L. (2000) *Phys. Rev. Lett.*, **84**, 2901.
65. Delogu, F. (2009) *Phys. Rev. B*, 79.
66. Bailey, N.P. et al. (2007) *Phys. Rev. Lett.*, **98**, 095501.
67. Hentschel, H.G.E. et al. (2010) *Phys. Rev. Lett.*, 104.
68. Lee, C.J., Huang, J.C., and Nieh, T.G. (2007) *Appl. Phys. Lett.*, 91.
69. Volkert, C.A., Donohue, A., and Spaepen, F. (2008) *J. Appl. Phys.*, 103.
70. Chen, C.Q., Pei, Y.T., and De Hosson, J.T.M. (2010) *Acta Mater.*, **58**, 189.
71. Schuster, B.E. et al. (2008) *Acta Mater.*, **56**, 5091.

72. Dubach, A. et al. (2009) *Scr. Mater.*, **60**, 567.
73. Shan, Z.W. et al. (2008) *Phys. Rev. B*, **77**.
74. Guo, H. et al. (2007) *Nat. Mater.*, **6**, 735.
75. Jang, D.C. and Greer, J.R. (2010) *Nat. Mater.*, **9**, 215.
76. Luo, J.H. (2010) *Phys. Rev. Lett.*, **104**, 215503.
77. Ye, J.C. et al. (2010) *Nat. Mater.*, **9**, 619.
78. Volkert, C.A., Donohue, A., and Spaepen, F. (2008) *J. Appl. Phys.*, 103.
79. Chen, C.Q., Pei, Y.T., and De Hosson, J.T.M. (2010) *Acta Mater.*, **58**, 189.
80. Schuster, B.E. et al. (2008) *Acta Mater.*, **56**, 5091.
81. Dubach, A. et al. (2009) *Scr. Mater.*, **60**, 567.
82. Guo, H. et al. (2007) *Nat. Mater.*, **6**, 735.
83. Lee, J.Y., Han, K.H., Park, J.M., Chattopadhyay, K., Kim, W.T., and Kim, D.H. (2006) *Acta Mater.*, **54**, 5271–5279.
84. Lu, J., Ravichandran, G., and Johnson, W.L. (2003) *Acta Mater.*, **51**, 3429–3443.
85. Sergueeva, A.V., Mara, N.A., Kuntz, J.D., Branagan, D.J., and Mukherjee, A.K. (2004) *Mater. Sci. Eng. A*, **383**, 219–223.
86. Rizzi, P. and Battezzati, L. (2004) *J. Non-Cryst. Solids*, **344**, 94–100.
87. Spaepen, F. (2006) *Nat. Mater.*, **5**, 7–8.
88. Zhang, Y., Stelmashenko, N.A., Barber, Z.H., Wang, W.H., Lewandowski, J.J., and Greer, A.L. (2007) *J. Mater. Res.*, **22**, 419–427.
89. Bengus, V.Z., Tabachnikova, E.D., Miskuf, J., Csach, K., Ocelik, V., Johnson, W.L., and Molokanov, V.V. (2000) *J. Mater. Sci.*, **35**, 4449–4457.
90. Argon, A. and Salama, M. (1976) *Mater. Sci. Eng.*, **23**, 219–230.
91. Zhang, Z.F., Eckert, J., and Schultz, L. (2003) *Acta Mater.*, **51**, 1167–1179.
92. Jiang, W.H., Fan, G.J., Liu, F.X., Wang, G.Y., Choo, H., and Liaw, P.K. (2006) in *Bulk Metallic Glasses* (eds P.K. Liaw and R.A. Buchanan), TMS, Warrendale, pp. 3–10.
93. Yang, B., Morrison, M.L., Liaw, P.K., Buchanan, R.A., Wang, G., Liu, C.T., and Denda, M. (2005) *Appl. Phys. Lett.*, **86**, 141904–141907.
94. Lewandowski, J.J. and Greer, A.L. (2006) *Nat. Mater.*, **5**, 15–18.
95. Park, E.S., Lim, H.K., Kim, W.T., and Kim, D.H. (2002) *J. Non-Cryst. Solids*, **298**, 15–22.
96. Inoue, A., Zhang, T. and Masumoto, T. (1990) *Mater. Trans. JIM*, **31**, 177.
97. Lin, X.H., Johnson, W.L., and Rhim, W.K. (1997) *Mater. Trans. JIM*, **38**, 473.
98. Matthews, D.T.A., Ocelík, V., Bronsveld, P.M., and De Hosson, J.T.M. *Acta Mater.*, (2008) **56**, 1762–1773.
99. Kulawansa, D.M., Dickinson, J.T., Langford, S.C., and Watanabe, Y. (1993) *J. Mater. Res.*, **8**, 2543–2553.
100. Russ, J.C. and Dehoff, R.T. (2000) *Practical Stereology*, 2nd edn, Kluwer Academic/Plenum Publishers, Dordrecht, The Netherlands, pp. 56–57.
101. Argon, A.S. and Shi, L.T. (1983) *Acta Metall.*, **31**, 499–507.
102. Bengus, V., Tabachnikova, E.D., Shumilin, S.E., Golovin, Y.I., Makarov, M.V., Shibkov, A.A., Miskuf, J., Csach, K., and Ocelik, V. (1993) *Int. J. Rapid Solidific.*, **8**, 21–31.
103. Bengus, V., Tabachnikova, E., Csach, K., Miskuf, J., and Ocelik, V. (1996) *Scr. Mater.*, **35**, 781–784.
104. Wang, W.H., Dong, C., and Shek, C.H. (2004) *Mater. Sci. Eng. R*, **44**, 45–89.
105. Xing, D., Sun, J., Shen, J., Wang, G., and Yan, M. (2004) *J. Alloys. Compd.*, **375**, 239–242.
106. Lu, K. and Wang, J.T. (1991) *Mater. Sci. Eng. A*, **133**, 500–503.
107. Hall, E.O. (1951) *Proc. Phys. Soc. Lond. Sect. B*, **64**, 747–753.
108. Petch, N.J. (1953) *J. Iron Steel Inst.*, **174**, 25–28.
109. Agena, A.S.M. (2009) *J. Mater. Process. Technol.*, **209**, 856–863.
110. Greer, J.R., Oliver, W.C., and Nix, W.D. (2005) *Acta Mater.*, **53**, 1821–1830.
111. Greer, J.R. and De Hosson, J.Th.M. (2011) *Progress. Mater. Sci.*, **56**, 654.
112. Lowry, M.B., Kiener, D., LeBlanc, M.M., Chisholm, C., and Florando,

J.N. (2010) Morris JWJ *Acta Mater.*, **58**, 5160–5167.

113. Ye, J., Mishra, R.K., Sachdev, A.K., and Minor, A.M. (2011) *Scripta Materialia.*, **64** (3), 292–295.

114. Withey, E.A., Ye, J., Minor, A.M., Kuramoto, S., Chrzan, D.C., and Morris, J.W. Jr. (2010) *Exp. Mech.*, **50**, 37–45.

115. Park, E.S., Lim, H.K., Kim, W.T., and Kim, D.H. (2002) *J. Non-Cryst. Solids*, **298**, 15–22.

116. Inoue, A., Zhang, T., and Masumoto, T. (1990) *Mater. Trans. JIM*, **31**, 177–183.

117. Lin, X.H., Johnson, W.L., and Rhim, W.K. (1997) *Mater. Trans. JIM*, **38**, 473–477.

118. Jang, D. and Greer, J.R. (2010) *Nat. Mater.*, **9**, 215–219.

119. Chen, C.Q., Pei, Y.T., and De Hosson, J.T.M. (2010) *Acta Mater.*, **58**, 189–200.

120. De Hosson, J.T.M. (2009) *Microscopy Research and Technique*, **72** (3), 250–260.

121. Sharma, P., Zhang, W., Amiya, K., Kimura, H., and Inoue, A. (2005) *J. Nanosci. Nanotechnol.*, **5**, 416–420.

122. Ma, E. (2003) *Nat. Mater.*, **2**, 7–8.

123. Dao, M., Lu, L., Asaro, R.J., De Hosson, J.T.M., and Ma, E. (2007) *Acta Mater.*, **55**, 4041–4065.

124. Ma, E. (2004) *Science*, **305**, 623–624.

125. Wang, Y.M., Li, J., and Hamza, A.V. Jr. (2007) *Proc. Natl. Acad. Sci. U.S.A*, **104**, 11155–11160.

126. Liu, Y.H., Wang, G., Wang, R.J., Zhao, D.Q., Pan, M.X., and Wang, W.H. (2007) *Science*, **315**, 1385–1388.

127. Dubach, A., Raghavan, R., Loffler, J.F., Michler, J., and Ramamurty, U. (2009) *Scr. Mater.*, **60**, 567–570.

128. Lu, J., Ravichandran, G., and Johnson, W.L. (2003) *Acta Materialia.*, **51**, 3429–3443.

129. Bailey, N.P. *et al.* (2007) *Phys. Rev. Lett.*, **98**, 095501.

130. Zheng, Q., Cheng, S., Strader, J.H., Ma, E., and Xu, J. (2007) *Scr. Mater.*, **56**, 161–164.

131. Zheng, Q., Cheng, S., Strader, J.H., Ma, E., Xu, J. Critical size and strength of the best bulk metallic glass former in the Mg-Cu-Gd ternary system. (2007) *Scripta Mater*, **56**, 161–164.

132. Senkov, O.N. and Scott, J.M. (2004) *Scr. Mater.*, **50**, 449–452.

133. Yang, B., Morrison, M.L., Liaw, P.K., Buchanan, R.A., Wang, G., and Liu, C.T. (2005) *Appl. Phys. Lett.*, **86**, 141904–141907.

134. Packard, C.E. and Schuh, C.A. (2007) *Acta Mater.*, **55**, 5348–5358.

135. Ye, J.C., Lu, J., Yang, Y., and Liaw, P.K. (2010) *Intermetallics*, **18**, 385–393.

136. Maass, R., Van Petegem, S., Grolimund, D., Van Swygenhoven, H., Kiener, D., and Dehm, G. (2008) *Appl. Phys. Lett.*, **92**, 071905.

137. Hofmann, D.C. *et al.* (2008) *Nature*, **451**, 1085.

138. Shan, Z.W. *et al.* (2008) *Phys. Rev. B*, **77**.

139. Jang, D.C. and Greer, J.R. (2010) *Nat. Mater.*, **9**, 215.

140. Luo, J.H. *et al.* (2010) *Phys. Rev. Lett.*, **104**, 215503.

141. Chen, C.Q., Pei, Y.T., Kuzmin, O., Zhang, Z.F., Ma, E., and De Hosson, J.Th.M. (2011) *Phys.Rev. B.*, **83**, 180201.

142. Das, J. *et al.* (2005) *Phys. Rev. Lett.*, **94**, 205501.

143. Weibull, W. (1951) *J. Appl. Mech.*, **18**, 293.

144. Danzer, R. (2006) *J. Eur. Ceram. Soc.*, **26**, 3043.

145. Jang, D., Gross, C.T. and Greer, J.R. (2010) *Int. J. Plast.*, doi: 10.1016/j.ijplas.2010.09.010

146. Stillinger, F.H. (1995) *Science*, **267**, 1935.

147. Harmon, J.S. *et al.* (2007) *Phys. Rev. Lett.*, **99**, 135502.

148. Demetriou, M.D. *et al.* (2006) *Phys. Rev. Lett.*, **97**, 065502.

149. Bedorf, D. and Samwer, K. (2010) *J. Non-Cryst. Solids*, **356**, 340.

150. Rault, J. (2000) *J. Non-Cryst. Solids*, **271**, 177.

31
Semiconductors and Semiconducting Devices
Hugo Bender

31.1
Introduction

The first transistor was built in polycrystalline Ge by J. Bardeen and W. Brattain in the Bell Labs in December 1947 [1]. It was the start of the fast technological evolution that resulted in electronics all around us in present day life. Although the first devices were based on germanium, silicon soon took over the role of leading semiconductor material. A major reason was the high quality and ease to grow silicon dioxide on silicon wafers. First integrated circuits were produced in 1958 by J. Killy on germanium and half a year later on silicon substrates by R. Noyce. Since then, the evolution of semiconductor technology has been governed by Moore's law, proposed in 1965: the number of transistors on a chip doubles every year [2]. Fulfillment of this prediction has become possible by scaling, that is, a reduction of the 3D-dimensions of the components [3]. It resulted in increased speed and reduced cost per functionality. The smallest lateral dimensions were on the order of 10 µm in 1971 (Intel 4004, 2300 transistors), whereas the latest generations of processors use a 32 nm technology (300–700 million transistors in recent 32 and 45 nm microprocessors), and R&D activities are concentrating on sub-22 nm processes.

Together with shrinkage of the dimensions, the complexity of the processing steps increased and the number of materials involved steeply increased. The first devices consisted of materials based on Si, O, N, Al, and B or P dopants, whereas nowadays a much wider range of materials are involved containing also elements as, for example, Cu, W, Ta, Ti, Co, Ni, Pt, Hf, Zr, Ge, C, voids, and so on. In the search for suitable dielectrics with a high dielectric constant, a wide range of materials based on rare earth elements, Sr and Ba are considered. With the step beyond the 45/32 nm devices, the classical planar Metal Oxide Semiconductor (MOS) device construction (Figure 31.1) will be abandoned and totally new 3D device concepts will be introduced based on nanowire structures [4], for example, multigate field effect transistors (MugFETs), bulk or silicon-on-insulator (SOI) finger field effect transistors (FinFETs) (Figure 31.2), gate-all-around transistors, or tunnel-field effect transistors (TFETs) (Figure 31.3).

Handbook of Nanoscopy, First Edition. Edited by Gustaaf Van Tendeloo, Dirk Van Dyck, and Stephen J. Pennycook.
© 2012 Wiley-VCH Verlag GmbH & Co. KGaA. Published 2012 by Wiley-VCH Verlag GmbH & Co. KGaA.

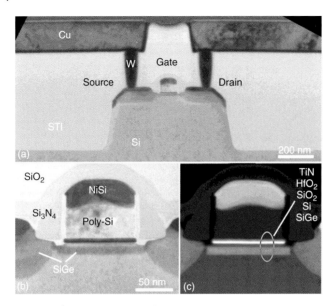

Figure 31.1 Cross-sectional TEM images (a,b) of an advanced planar PMOS transistor with epitaxial $Si_{0.75}Ge_{0.25}$ in source and drain, and $Si_{0.55}Ge_{0.45}/Si$ in the channel. The different SiGe compositions are most obvious on the HAADF-STEM image (c). The high-k/metal gate stack consists of an interfacial SiO_2/HfO_2/poly-Si/NiSi (b,c). Shallow trench isolation (STI) is used between the devices. (Structures: L. Witters; analysis: J. Geypen – imec.)

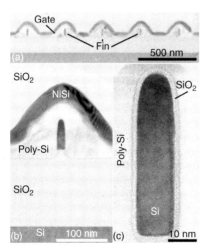

Figure 31.2 31.2 Multigate FinFET structure on a silicon-on-insulator (SOI) substrate. The five single-crystalline Si Fins connect source to drain (not visible), while the silicided polycrystalline Si gate line runs across the Fins. The width of the Fins is on the order of 10 nm (c). In this example, a SiO_2 dielectric is used. (Structures: N. Collaert; analysis: O. Richard – imec.)

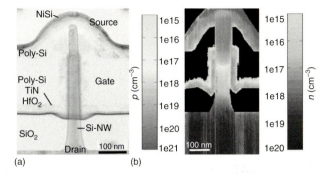

Figure 31.3 Silicon nanowire (NW) tunnel-FET (TFET) device. The drain is n+ doped, while the source is p+ doped. The high-k/metal gate is all around the nanowire. It consists of a HfO$_2$/TiN/poly-Si stack. The diameter of the NW in (a) is 50 nm. (b) Carrier distribution in a 100 nm NW device measured with scanning spreading resistance microscopy (SSRM) [5]. (Structures: A. Vandooren; TEM: J. Geypen; SSRM: A. Schulze – imec). (Reproduced from Ref. [5] by H. Bender with permission from A. Schulze.)

The evolutions into the nanometer scale and 3D architectures continuously pose challenges to the materials characterization techniques for research and development, process control in the production lines, and failure analysis [6–9]. In a nanodevice, the role of each atom in terms of its position, nature, chemical binding, and electrical activity becomes crucial to understand and control the device characteristics. It implies that spatial resolution (preferably in 3D), accuracy, and sensitivity are essential. Almost all microscopic techniques discussed in part 1 of this work find somehow applications in the field of semiconductors. Over the years, the requirements posed by nanoelectronics have been a strong stimulus for the developments of increased resolution scanning electron microscopic (SEM) and transmission electron microscopic (TEM) systems and for the introduction of totally new methods as, for example, a whole series of probe techniques, focused ion beam (FIB) and He-microscopy.

The selection of the physical analysis method will be governed by the requirements of resolution, sensitivity, quantification accuracy, and speed on one hand and the properties of interest on the other: 1D/2D/3D metrology, composition, chemical binding, low-level impurities, and crystallographic or electrical characteristics. Different probe techniques, SEM, and analytical TEM/STEM (scanning transmission electron microscopy) are widely applied for nanoscopy of semiconductor materials and devices. 3D-techniques such as electron tomography, depth sectioning in aberration-corrected STEM, scanning confocal TEM (SCEM), and atom probe tomography (APT) are recently further developed and will gain increasing importance in the near future. Site-specific and artifact-free sample preparation is a crucial step for most nanoscopic techniques. Advanced cleaving systems, FIB, and low-energy ion milling are indispensable preparation methods. In a production environment, nondestructive analysis is an additional requirement, which can be fulfilled by top view imaging (e.g., critical dimension-scanning electron microscopy,

CD-SEM), or site-specific lift-out with FIB with only local contamination on the wafer so that further processing remains possible.

Table 31.1 summarizes some typical analysis questions that can arise and nanoscopic/microscopic techniques that can contribute to the answer. This chapter focuses on Si-based device technology and on techniques that allow nanoscopic analysis. Applications for typical characterization needs at different processing steps of nanodevices are discussed.

31.2
Nanoscopic Applications on Silicon-Based Semiconductor Devices

31.2.1
Metrology

Thickness control of blanket layers or in relatively large areas can be done by a variety of techniques as, for example, spectroscopic elllipsometry or X-ray reflectometry. Measurement of critical dimensions (CDs) in top view or cross section of structured samples is required after many different processing steps, for example, after resist patterning or etch steps. Line shape metrology is possible on special test structures with scatterometry, but in general, for the analysis of nanopatterned structures, microscopic techniques with high spatial resolution are needed. Dedicated CD-SEM and critical dimension atomic force microscopic CD-AFM systems allow highly automated mapping of dimensions at various positions distributed over the wafers (typically 300 mm diameter) [10]. Also, automated dual-beam FIB/SEM cross-sectional analysis is possible. The resolution needed for the metrology of, for example, gate oxides, or etch recess in the substrate demands for TEM analysis also as an in-line measurement tool. Fast, automated, and reproducible specimen preparation is then crucial in order to obtain short turn-around time [11, 12]. Automated TEM specimen loading and imaging (TEM/STEM) is under development. For more detailed nonroutine analysis, in- and off-line SEM and FIB/SEM and off-line TEM/STEM are the major metrology tools. For imaging of electron-beam-sensitive materials, the scanning He ion microscope is a promising new tool.

31.2.2
Epitaxy

For advanced devices, the selective epitaxial growth of doped Si, $Si_{1-x}Ge_x$ (x = 20–70 at %) or $Si_{1-x}C_x$ (x = 1–1.5%) on patterned substrates is an important process to control the doping levels in the surface layer, to form raised source and drain regions, to introduce strain in the silicon channel embedded between the epitaxially grown source/drain regions, or to increase the carrier mobility in the $Si_{1-x}Ge_x$ PMOS channel. Owing to the misfit between the layer and substrate, dislocations can be introduced when the layers relax. Defect control can be done by interference contrast optical microscopy analyzing the cross-hatch patterns formed

Table 31.1 Analysis needs and commonly applied nanoscopic/microscopic methods.

Property to be analyzed	Position in the device	Possible nano/microscopic techniques
1/2/3D metrology	Layer thickness: gate dielectric, barrier layers, and so on Etch depth and profile, line width, and so on Step coverage and filling of high aspect ratio structures	Tilted and cross-sectional scanning electron microscopy (SEM), focused ion beam (FIB/SEM), critical dimension-SEM (CD-SEM), critical dimension atomic force microscopy (CD-AFM), He ion microscopy, transmission electron microscopy (TEM), scanning transmission electron microscopy (STEM), electron tomography
Roughness	Substrate and epitaxial surface Etch bottom and sidewall roughness Interfacial roughness	AFM, SEM, TEM/STEM
Lattice defects	As-grown in substrate Implantation damage Misfit dislocations at epitaxial layer interfaces Threading dislocations in epitaxial layers	X-ray transmission topography, preferential etching and optical microscopy, interference optical microscopy, SEM, TEM
Strain	Strained Si channel due to recessed SiGe or SiC source/drain (S/D) or nitride liner Strained SiGe channel Nanowires	Convergent beam electron diffraction (CBED), nanobeam electron diffraction (NBD), geometrical phase analysis (GPA), holography, µRaman, tip-enhanced Raman spectroscopy (TERS)
Crystallography	Crystal phase: silicides, barriers, dielectrics, and so on	(Nano) electron diffraction, electron backscatter diffraction (EBSD), high-resolution electron microscopy (HREM), high-resolution scanning transmission electron microscopy (HR-STEM)
Grain size	Crystalline layers: silicides, metalization, bond wires, and so on	SEM, FIB, EBSD, dark field (DF)-TEM, DF-STEM
Composition	All layers, in particular: SiGe, silicides, barriers, dielectrics	Energy dispersive spectroscopy (EDS), electron energy loss spectroscopy (EELS), energy-filtered TEM (EFTEM), high-angle annular dark field STEM (HAADF-STEM), atom probe tomography (APT), Auger microscopy

(continued overleaf)

Table 31.1 (Continued)

Property to be analyzed	Position in the device	Possible nano/microscopic techniques
Chemical bonding	Low-k and high-k dielectrics	Electron loss near edge structure (ELNES), EFTEM, EELS, micro-X-ray photoelectron spectroscopy (micro-XPS), scanning transmission X-ray microscopy–near-edge X-ray absorption fine structure spectroscopy (STXM-NEXAFS)
Residues	Particles after cleaning or chemical mechanical polishing (CMP) Etch residue on sidewalls	Light scattering, SEM, AFM, Auger microscopy, nano-secondary ion mass spectroscopy (nano-SIMS), TEM, EDS, EELS, STXM-NEXAFS
Dopant and carrier distribution	Channel and polygate	Scanning spreading resistance microscopy (SSRM), scanning capacitance microscopy (SCM), holography, atom probe tomography
Leakage paths	Gate dielectrics	Conductive AFM (C-AFM), tunneling AFM (TUNA)
Electrical failures	Metalization, gate	E-beam prober, voltage contrast in SEM and FIB, electron-beam-induced current (EBIC), photoemission microscopy, infrared optical beam-induced resistance change microscopy (IR-OBIRC), optical beam-induced current (OBIC), X-ray tomography, scanning acoustic microscopy, TEM/STEM

by the dislocation slip lines in wider areas. A more detailed investigation of the defect formation in nanodevices is possible with cross-sectional or plan-view TEM analysis.

Epitaxial Ge on silicon (Figure 31.4a) is studied to increase the mobility in PMOS devices [13, 14] or as potential substrates for selective epitaxial growth of III–V materials [15, 16] (Figure 31.4b,c). It will allow the introduction of high-mobility materials (InP, InGaAs) in the channel of NMOS devices [17] or to combine optoelectronics with silicon technology.

31.2.3
Strain

Strain has for a long time been considered as a negative property in devices because it can lead to the generation of extended lattice defects that can cross the electrical junctions and result in device leakage. Particularly risky steps are the thermal growth of isolation oxides and silicidation steps. The mechanism

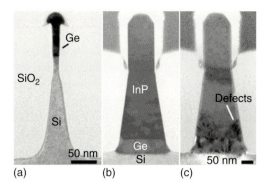

Figure 31.4 Epitaxial Ge selectively grown on a Si bulk FinFET (a) [14]: the Ge at the interface is only 10 nm wide, the hat will be removed by chemical mechanical polishing (CMP) in the next processing step. (b,c) Examples of the selective epitaxial growth of InP on Si in STI trenches [16]. Ge is grown as an intermediate layer. Misfit dislocations are present at the Si/Ge interface. For narrow trenches, the defects in the InP are concentrated near the interface with the Ge (best visible on the two-beam TEM image (c)). (Structures: G. Wang; TEM: P. Favia and O. Richard – imec). ((b) Reproduced from Ref. [16] by H. Bender with permission from G. Wang.)

of line-edge-induced defect generation has extensively been studied by TEM and micro-Raman spectroscopy [18]. More recently, the beneficial effects of strain on the carrier mobility have been explored [19] and are nowadays introduced in the devices. Compressive strain leads to electron mobility enhancement in the PMOS channel, while tensile strain is desired for hole mobility enhancement in the NMOS channel. The lateral resolution of classical micro-Raman spectroscopy is only 0.5 µm. Tip-enhanced Raman spectroscopy (TERS) is an AFM-based method using specially shaped metalized tips with which a resolution below the optical diffraction limit can be obtained. Further development for quantitative strain measurements on silicon devices is still needed. Resolutions below 250 nm [20] and 65 nm [21] have been reported.

TEM-based techniques allow much better lateral resolution and can be applied on cross-sectional specimens so that a 2D distribution of the strain can be derived. Convergent beam electron diffraction (CBED) has been studied extensively for the strain analysis in device structures [22, 23]. The shift of the high order Laue zone (HOLZ) lines in these patterns is sensitive to the local strain. Dedicated fitting software has been developed to interpret the measurements. Drawback of CBED is the requirement to tilt the specimen away from the [110] cross-sectional zone axis along which the device structures are generally aligned, toward [230] (11.3°) or [340] (8.1°). In combination with the need to use quite thick TEM specimens (~300 nm), this leads to an important projection overlap on the edges of narrow structures. The interpretation of CBED measurement is strongly complicated by the splitting of the HOLZ lines due to relaxation of the stress in the TEM lamellae [24, 25]. Although the effect can be modeled well for blanket layers [24], this is no longer a feasible option for device structures where the strain relaxation becomes a 3D effect [26].

Nanobeam electron diffraction (NBD) in microprobe STEM mode uses a 5–10 nm diameter parallel electron beam [27–32]. The shifts of the diffraction spots are directly related to the local lattice parameters so that the strain in the plane of the specimen can be analyzed. The method can be applied along any zone axis of the Si substrate. High-resolution transmission electron microscopy (HR-TEM) imaging combined with geometrical phase analysis (GPA) [33, 34] also maps the in-plane strain distribution. It requires a uniform phase contrast image over the full area of view, that is, very thin specimens with constant thickness. Dark field off-axis [35–37] or in-line [38] holography has been recently developed for strain measurements. The holographic methods have the advantage of a large field of view and, as also GPA, directly give a 2D view of the strain distribution.

A comparison of the properties of the TEM-based strain measurement techniques is given in Table 31.2. CBED is the most sensitive method and is the only technique that yields information on the out-of-plane strain component by the analysis of the higher order diffraction information. It is therefore, however, also more sensitive to relaxation effects in the TEM specimen than the other methods. A comparison of the strain measured by NBD and GPA (Figure 31.5a,b) and NBD and holography (Figure 31.5c,d) for similar devices shows a very good correspondence between the different methods and finite element simulations (FEM). The GPA and holography linescans are averaged over a wider area of the images and are therefore less noisy than the NBD, which is measured over a single line.

Table 31.2 Comparison of different TEM-based techniques for strain measurement.

	Sensitivity (%)	Spatial resolution (nm)	Zone	Field of view (nm)	Reference
CBED	0.02	7–10	<340>: 8.1° off <110> <230>: 11.3° off <110>	500 linescan or mapping	[23]
NBD	0.1	2.7–10	Any low-index zone axis	500 linescan or mapping	[27]
HREM-GPA	0.1–0.3	2–3	Any low-index zone axis	<200 × 200 image	[33, 35]
Dark field off-axis holography	0.1–0.2	4	Two-beam condition: few degrees off zone axis	>250 × 1000 image	[35]
Dark field in-line holography	0.01	1	Two-beam condition: few degrees off zone axis	>250 × 1000 image	[38]

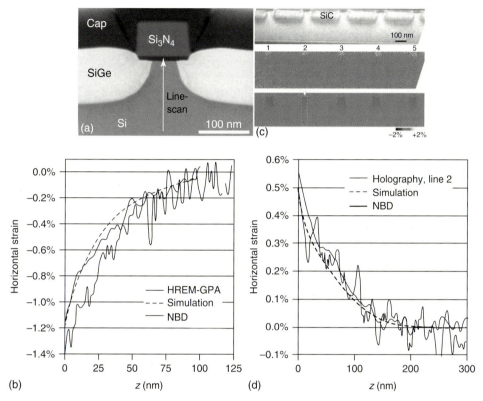

Figure 31.5 HAADF-STEM image of a dummy transistor structure with recessed $Si_{0.80}Ge_{0.20}$ grown in the source and drain and 50 nm wide channel (a). The horizontal strain measured with NBD along the arrow is shown in (b) and compared with HREM-GPA strain analysis and finite element simulation (FEM). (c) Dark field hologram with the (220) diffracted beam, corresponding strain map, and FEM simulation for devices with $Si_{0.99}C_{0.01}$ and 65 nm channel width. The experimental and simulated strain profile of the second device is compared with the NBD measurement (d). (HREM-GPA [34] and holography [36] F. Hüe, CMES, Toulouse; NBD: P. Favia – imec). (c) and GPA, holography, and simulation data curves on (b) and (d) (reproduced from Refs. [34] and [36] by H. Bender with permission from F. Hüe.)

31.2.4
Doping

n or p doping of the source/drain and gate is typically introduced by ion implantation. To optimize the device characteristics of nanodevices, additional extension and halo implants are applied to tune the dopant distribution at the gate edges. New doping methods are explored for the 3D devices [39]. Depending on the conditions (ion, energy, dose, and anneal), ion implantation can result in amorphization of the silicon, formation of end-of-range defects in {113} planes, extended dislocation loops, or twinning. Transmission electron microscopy is extensively applied to study these effects.

With the scaling of the devices, the doping levels increase and steeper dopant profiles are needed while the junction depths decrease. The depth resolution required for the 1D-analysis of the dopant profiles by secondary ion mass spectroscopy (SIMS) has become challenging [40], while for the 2D analysis at the gate edges or in 3D nanodevices it lacks the lateral resolution. Laser-assisted APT [41, 42] has been recently introduced in the semiconductor field. It allows a 3D analysis of the atom distribution in the tip of a needle-shaped sample prepared by FIB in a way similar to that of a TEM specimen. APT has a better depth resolution than SIMS [43] and can be applied for the 3D mapping of dopant distributions, for example, at grain boundaries [44] and at dislocation loops formed after annealing [45]. Application for the analysis of the dopant distribution in 3D FinFET devices has recently been shown [46] (Figure 31.6a,b).

For the device performance, the electrically active dopant distribution is more crucial than the chemical dopant profiling. For 1D profiling in large areas, several techniques can be applied to measure the active dopants [47, 48]. For 2D or 3D profiling of the active dopants, cross-sectional imaging techniques are needed. SEM contrast can be related to active doping levels but is strongly dependent on the surface condition of the sample (cleaved or FIB prepared), which limits the sensitivity and accuracy of the quantification [49–51].

Dopant-selective etching on TEM specimens is hard to control in a reproducible way and does not have the resolution required for nanodevices. Combined with

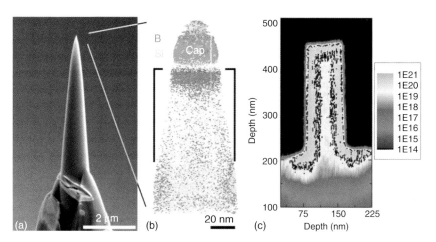

Figure 31.6 Atom probe tip prepared by FIB (a) and atom probe tomographic (APT) measurement of the B distribution in a 80 nm wide FinFET (b). The box in (b) indicates the approximate dimension of the FinFET, which is wider at the top than the diameter of the atom probe sample. Indiffusion of B is observed at the top of the Fin. (c) Scanning spreading resistance microscopic (SSRM) image of the carrier distribution (cm^{-3}) in a similar FinFET device. (APT: A. Kambham, SSRM: J. Mody, imec – [46]). (Reproduced from Ref. [46] by H. Bender with permission from J. Mody.)

AFM, selective etching allows a fast qualitative evaluation, but quantification is also limited by the etching process.

The phase image in electron holography is very sensitive to the potential in the TEM specimen and has a high lateral resolution for dopant profiling [52–55]. Unfortunately, the signal is also very sensitive to the TEM specimen quality, for example, roughness, thickness variations, amorphization of the outer silicon layers, and implantation of Ga by the FIB milling. Thickness variations in the specimen can be minimized by dedicated backside FIB preparation. The formation of a dead layer of >175 nm near both specimen surfaces has been shown for standard 30 kV FIB prepared specimens [54]. It complicates the quantitative analysis of the dopant concentration. The effect of implanted Ga and amorphization can be minimized by lowering the Ga ion beam energy, additional low-energy Ar milling of the specimens, and low-temperature annealing after the FIB preparation [55]. Still, surface charges modify the electrical potential and limit the sensitivity of the technique. By combining holographic and tomographic measurements, the bulk and modified surface regions can be separated [54] in specimens thicker than the critical thickness, that is, thicker than 350 nm for 30 kV Ga FIB preparation.

Different scanning probe techniques can be applied for carrier profiling [56]. The most powerful technique is scanning spreading resistance microscopy (SSRM) [57, 58], which has a good concentration sensitivity (standard deviation below 1%), allows accurate junction delineation (within a few nanometers), and has a very good dopant gradient sensitivity (1–2 nm/decade). It has the ability to quantify quite easily the results by conversion of the measured spreading resistance to resistivity and carrier concentration. Using dedicated staircase calibration structures, concentration quantification accuracy of 20–30% can be obtained. Examples of application of SSRM on nanowire TFET and FinFET devices are given in Figures 31.3b and 31.6c.

31.2.5
Gate Structures

The scaling of the devices reduced the gate dielectric thickness to the order of 1–2 nm in the 65 nm technology. Further reduction leads to unacceptable gate leakage due to the tunneling current through the gate oxide. To overcome this problem, materials with high dielectric constant k are introduced [59, 60]. It allows increasing the physical thickness while reducing the equivalent oxide thickness (EOT). A wide set of materials has been considered in single- or double-layer stacks for application in gates or memory cells.

Although nondestructive control of the layer thickness in large areas is possible by, for example, spectroscopic ellipsometry, X-ray reflectometry, or angular resolved X-ray photoelectron spectroscopy, the analysis of the gate stack (thickness and composition) in small devices is possible only by TEM-based techniques. The accuracy of the metrology [61] is challenged by possible interfacial roughness, overwhelming contrast of the heavy materials that may shadow the roughness in projection over the specimen thickness, and possible interdiffusion. Furthermore, many of these dielectrics are unstable during the electron beam observation, possibly leading to crystallization, growth of the interfacial oxide, damage, and void formation during STEM EELS/EDS (electron energy loss spectroscopic/energy dispersive spectroscopic) analysis.

Instead of the doped poly-Si used in the gates of older technologies, metal gates are introduced. They typically consist of TiN/TaN/poly-Si. Combined with strained SiGe in the channel, this leads to a complicated stack of many layers in the nanometer thickness range (Figures 31.1b,c and 31.7a). High-angle annular dark field scanning transmission electron microscopy (HAADF-STEM) with EELS/EDS analysis is the most appropriate approach for the compositional study of the stack in nanodevices. Deconvolution of the energy loss near edge structure (ELNES) spectral information allows, except for the compositional profiling, also the deduction of chemical profiles and hence the study of possible reaction and interdiffusion of the layers [62] (Figure 31.7b,c). Essential for such analysis is the measurement of zero-loss and core-loss spectra simultaneously, which becomes possible with the latest generations of EELS spectrometers.

Depth sectioning with aberration-corrected STEM allows a 3D reconstruction of the Hf distribution in SiO_2 [63]. Simulation shows that for crystalline materials this is only reliably possible for specimens tilted slightly away from the zone axis [64]. Further experimental investigation of this imaging mode is needed in the future.

Leakage paths and breakdown of dielectric layers can be localized by conductive atomic force microscopy (C-AFM) and directly correlated to topographical AFM measurements. The method is applied to thin SiO_2 layers [65, 66] and different high-k stacks [67, 68]. Lower currents can be measured in the related tunneling AFM (TUNA) mode [69].

31.2.6
Silicides

Silicides are used to lower the contact resistance between source/drain or gate and the metal contacts. A low resistance and thermally stable phase that forms at a relatively low temperature is required. Over the years, technology switched from $TiSi_2$ to $CoSi_2$ and at present to NiSi [70]. Adding a small percentage of Pt to the NiSi is shown to improve the layer stability. NiSi is also investigated as replacement for the poly-Si in fully silicided (Fusi) gate structures [71].

The structural characterization of the phase evolution as a function of anneal conditions is extensively studied by a variety of microscopic techniques. As the layers are mostly thin and polycrystalline and as a wide range of phases exists, applying combined TEM imaging and different analytic modes is needed for full characterization of the material. With low-loss spectra acquired in a monochromated STEM, different Ni-silicide phases can be distinguished (Figure 31.8). The ELNES information can also be used for this purpose [72].

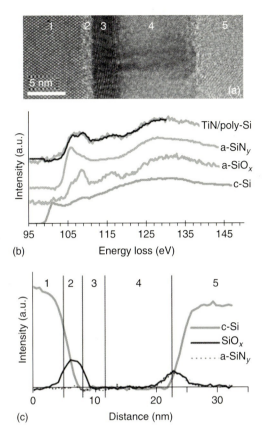

Figure 31.7 Cross-sectional high-resolution TEM image (a) through a high-k/metal gate stack: Si substrate (1), SiO$_x$ (2), HfO$_2$ (3), TiN (4), poly-Si (5). (b) Experimental spectra of the Si L2,3 edges of different Si-phases (gray) are used for multiple linear least squares (MLLS) fit to the edge spectrum acquired at the interface (black). The chemical profiles (c) obtained from the MLLS fit show the presence of SiO$_x$N$_y$ at the TiN/poly-Si interface. (M. MacKenzie, University of Glasgow – [62]). (Reproduced from Ref. [62] by H. Bender with permission from M. MacKenzie.)

31.2.7
Metalization

Cu is commonly used nowadays in the metal lines and interlevel vias, while in the contacts to the transistors either Cu or W is applied. Different kinds of barrier layers are considered between Cu and the dielectrics (e.g., Ta/TaN, Ti/TiN). They are intended to avoid Cu diffusion, improve adhesion, and facilitate nucleation of the Cu seed layers. Hence, good step coverage and continuity of the barrier are required. With the diameter of the vias shrinked (down to 40 nm, Figure 31.9), the allowed thickness of the barriers is reduced to only a few nanometers. The control of void formation in Cu is possible by FIB/SEM, while the detailed analysis of

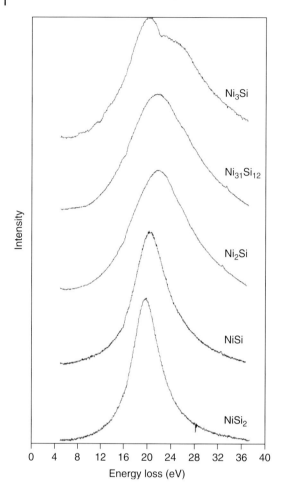

Figure 31.8 Low-loss spectra for different Ni-silicides acquired with a monochromated probe-corrected STEM. Shape and position of the plasmon peak allow distinguishing the phases. (Measurements by D. Klenov, FEI.) (Reproduced from Ref. [72] by H. Bender with permission from E. Verleysen.)

the barrier (and smaller voids in the Cu) will require TEM/STEM investigation. As the vias are cylindrical and on the order of or less than the TEM specimen thickness, projection overlap of the different materials can generally not be avoided (Figure 31.9b). Combined with the roughness on the sidewalls (because of etching or the barrier deposition) and the dark contrast of the high-Z barrier materials on the TEM images (or bright on HAADF-STEM), the investigation of pinholes in the barrier is not straightforward. On the other hand, lack of contrast complicates the characterization of Co barriers or Co-based passivation layers. Both on TEM and STEM images Co cannot be distinguished from Cu. Energy-filtered transmission

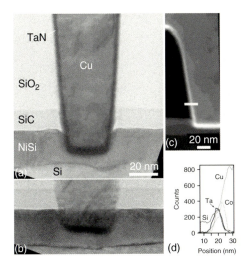

Figure 31.9 TEM image of a 40 nm diameter contact plug filled with TaN/Co barrier and Cu (a). Owing to the small diameter, projection overlap of the silicide with the contact at the bottom of the plug (b) is hard to avoid and complicates the analysis of the layer stack in the plug. (c) HAADF-STEM image and corresponding EDS line profile over the barrier stack (d). (Structures: S. Demuynck; TEM: J. Geypen – imec.)

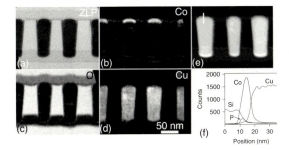

Figure 31.10 Analysis of a CoWxPy passivation layer on 40 nm wide Cu lines: the layer cannot be distinguished from the Cu in the zero-loss TEM (a) or HAADF-STEM (e) images because of lack of contrast. However, the Co clearly shows up in the EFTEM map (b) and on the EDS linescans (f – W not shown). The Ta barrier is thicker on the bottom of the Cu lines and is best observed on the HAADF-STEM image (e). (Structures: M. Pantouvaki; TEM: J. Geypen – imec.)

electron microscopy (EFTEM) (Figure 31.10a–d) or STEM/EDS (Figures 31.9c,d and 31.10e,f) analysis are needed to reveal the presence of the Co layers.

The capacitance increases with shrinkage of the distance between the interconnect lines, which leads to an unacceptable increase of the signal transmission delay and of the dynamic power consumption. Therefore, the permittivity of the dielectric has to shrink. Compared to $k = 3.9$ for SiO_2, this is possible with different kinds of $Si_xC_yO_zH$-based materials. Values below 2 can be reached only by introducing porosity in the material, with airgaps ($k = 1$) as the ultimate low-k material. Most

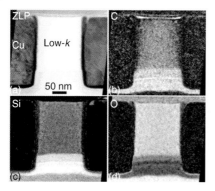

Figure 31.11 (a–d) EFTEM imaging of a low-*k* dielectric between Cu lines [73, 74]. C depletion is observed at the sidewalls of the dielectric because of the trench etch treatment. (Structures: F. Iacopi; TEM: O. Richard – imec.)

low-*k* dielectrics are electron beam sensitive and easily shrink during SEM or TEM investigation. Poor adhesion to the barrier can also lead to delamination during observation. As these problems also occur during SEM or STEM observation at 30 kV or below, little improvement has to be expected with this respect from the new generations of Cs-corrected TEMs operating with high resolution at low voltages (60 kV).

The low-*k* etch process can modify the sidewalls of the material in the vias and lead to an increase of the *k*-value, for example, by oxidation of the material. Such a layer is hard to distinguish on TEM/STEM images due to lack of contrast and a gradient composition change, whereas EFTEM mapping of the C and O distribution can clearly reveal the modified layer (Figure 31.11) [73, 74]. EDS/EELS analysis is also applied to study the carbon depletion and polymer residue layers deposited during the etch [73].

The *k*-value of the dielectric can be calculated based on electrical measurements combined with the metrology of the lines as determined, for example, by cross-sectional TEM. Direct local measurement of the 2D distribution of the *k*-value is studied by valence band EELS [75]. The method is also investigated for other dielectric systems [76, 77]. The newest generations of probe-corrected and monochromated TEM systems are promising for further development of these applications [78].

31.2.8
Three-Dimensional Structures

The structural analysis of nanometer-scaled devices by cross-sectional SEM in a site-specific manner is hard to perform. Even the best cleaving tools do not have the necessary accuracy to cleave through specifically selected nanometer structures. High-resolution cross-sectional SEM is therefore generally only possible on dense test structures. For site-specific analysis, a dual-beam FIB/SEM slice and view approach is usually required. The evolution of the 3D structure can be followed stepwise, and by combining the set of images, a tomographic reconstruction is possible. Owing to the ion beam damage and the smoothness of the milled faces, the SEM resolution and contrast are usually worse than on cleaved faces. Contrast from still buried features can also disturb the interpretation.

The nanodevices have dimensions less than the typical thickness of a TEM specimen, so that projection overlap of different materials cannot be avoided and complicates the interpretation of the TEM/STEM images. The problem is particularly important for 3D devices as, for example, FinFETs (Figure 31.12) and vertical nanowire TFETs (Figure 31.3a), but projection overlaps also occur because of sidewall roughness of long structures. The small dimensions of the devices lead to very stringent requirements for the site-specific specimen preparation. In the top view of the SEM image, the specimen position can often not be recognized accurately enough. Even with a regular control of the progress of the specimen thinning with the SEM image in a dual-beam FIB/SEM, some luck is needed to center a 40 nm contact plug inside the TEM lamellae so that no silicide before/behind the plug is present in the thickness of the specimen (Figure 31.9a,d). Using ultrathin specimens has the disadvantage that the overview of the full structure is lost. Electron tomography [79–81] can overcome these problems to a large extent: slicing through the reconstructed volume (in any direction) circumvents the overlap problem, and slightly larger specimen thickness that facilitates the specimen preparation can be allowed. As the devices contain (poly)-crystalline materials, HAADF-STEM is the most appropriate mode for the tomographic analysis of semiconductor structures so that diffraction effects can be avoided (or minimized). The presence of heavy elements (Hf-based dielectrics, W contacts, and TaN barriers) results in strong contrasts in the images. Using plan parallel specimens, missing wedge artifacts in the reconstructed volume cannot be avoided. They reduce the resolution along the viewing direction. In the frontal

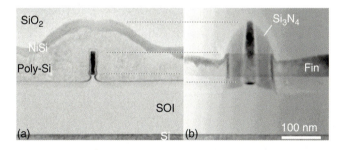

Figure 31.12 FinFET structure viewed along the Fin (a) and along the gate (b) showing overlap of the different materials in the viewing direction. The Fin is about 20 nm wide and silicided where it is not covered by the gate line, which is ~30 nm wide. The lines correlate the different features in the two views. (TEM: O. Richard – imec.)

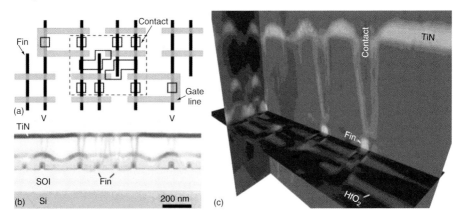

Figure 31.13 Tomographic analysis of a six-transitor FinFET SRAM cell [82]. The full cell has an area of 0.186 µm² (dashed box on the top view layout, (a)). The Fins are 20 nm; the gate lines are 60 nm wide. The contacts are covered with a TiN contrast layer, no Cu is deposited yet. The full cell requires a TEM specimen with thickness >375 nm and hence overlap of all features of the cell in the image (b). The 3D tomographic view shows slices through the top row of contacts and through the HfO$_2$ dielectric on the SiO$_2$ in between the Fins. (Structures: A. Veloso; analysis: O. Richard – imec.)

slices, 2 nm resolution can be obtained for layers that show sufficient contrast [82]. To overcome the missing wedge problem, cone-shaped specimens can be used [83, 84]. Use of a 360° rotation sample holder results in equal spatial resolution in all directions [84].

Figure 31.13 illustrates the use of electron tomography to analyze a full six-transitor FinFET static random access memory (SRAM) cell [82]. The probed volume contains two rows of contacts covered with a Ti/TiN barrier, two wing-shaped contacts, two poly-Si gate lines with HfO$_2$ dielectric, and several single-crystalline Fin lines. A projected TEM image of the ~400 nm thick specimen is hard to interpret (Figure 31.13b), while in the slices of the reconstructed volume the structure can be investigated in detail (Figure 31.13c).

31.2.9
Bonding and Packaging

Different functionalities can be combined by heterogeneous integration of different integrated circuit (IC) technologies in a single package [85, 86]. Systems in a package (SIP) with wire-bonded stacked dies allow only low number of interconnects. For medium- or high-density interconnects other routes are explored, that is, respectively, 3D wafer level packaging (3D-WLP) or 3D stacked ICs (3D-SICs [86]). Both technological options have in common that thin silicon substrates (<50–100 µm) are used with metal interconnects crossing the substrate from one die to the next. For the 3D-SIC approach, the through-silicon-vias (TSVs) have high aspect ratios (5 : 1 to 10 : 1). The characterization of the TSVs focuses on

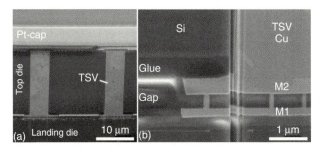

Figure 31.14 FIB/SEM analysis of the through silicon vias (TSV) of a 25 μm thick device (top die) bonded on a landing die with two-layer Cu metalization (a). A zoom of the bond is shown in (b). (Structures: C. Torregiani; FIB: C. Drijbooms – imec.)

the dimensions and the quality of the metal filling, and for stacked dies, also the quality of the bonding of the TSV to the metalization of the landing die. The structural characterization of such structures is challenging because of the large sizes and high aspect ratios. At the same time, an overview of the whole TSV is needed, that is, low-magnification imaging, as well as a detailed analysis of, for example, small voids in the metal filling or the uniformity of the nanometer-thick metal barrier, that is, high-resolution imaging. Milling and imaging in a dual-beam FIB/SEM system is the most appropriate approach to analyze the structures [87]. Owing to the large dimensions long analysis times are required (Figure 31.14). Cleaving the samples close to the structure of interest allows reducing the milling time considerably. Unfortunately, cleaving is not possible for stacked dies because of the risk for damaging the bond. For high-resolution investigation of the barrier uniformity, TEM/STEM is needed. Dedicated specimen preparation is necessary to get uniform thickness over the full height of the structures [87].

31.2.10
Failure Analysis and Debugging

Failure analysis [88] involves the localization of the failure position and identification of the origin of the failure. Many microscopic methods can be applied for the fault localization, for example, voltage contrast in SEM or FIB, electron-beam-induced current (EBIC) in SEM or photoemission microscopy. Several methods that are based on laser-induced signal injection have been developed, for example, infrared optical beam-induced current (OBIC) or infrared optical beam-induced resistance change (IR-OBIRCH) [89].

Owing to the large number (>10) of metal levels on the frontside of the devices or to the use of flip-chip packages where the device is not accessible from the frontside anymore, failure analysis often needs to be done from the backside. It requires backside polishing of the substrate and techniques based on infrared as the incident or detected light. Physical characterization of the localized failure may be done with cross-sectional FIB/SEM or after site-specific FIB specimen preparation with TEM. The latter is in dense nanostructures, often complicated because the accuracy of the position in the device as obtained by the fault localization methods might be insufficient.

Rerouting or cutting of metal interconnects can be done with gas-assisted FIB [90]. Owing to the narrow linewidths and the high number of metal levels, dedicated FIB systems that combine high stage accuracy and stability, navigation software based on the device layouts, infrared imaging, and a range of gas injectors for preferential etching or local material deposition are required. For flip-chip packages modification procedures from the backside of the device are developed. After backside polishing, the silicon substrate is locally etched till the device region with XeF_2-assisted milling [91].

31.2.11
New Materials

Further scaling of the devices below 22 nm will involve the introduction of totally new materials as carbon nanotubes or graphene as well as new device concepts [92]. Microscopy on carbon-based materials also becomes important with the introduction of polymer-based electronic devices and the combination of biomaterials with nanoelectronic structures. For the characterization of these C-based structures, the recent development of aberration-corrected TEM systems that can operate at low voltages (60 kV) with high lateral resolution is important to avoid knock-on damage. Development of damage-free sample preparation methods for these materials, in particular when they are combined with hard materials is an important challenge [84, 93].

31.3
Conclusions

Nanoscopic analysis is essential for the characterization of nanoelectronic devices. The further shrinkage of the dimensions and the introduction of new materials and device concepts demand for ever better resolution and stability. Three-dimensional imaging becomes a crucial property of the nanoscopic methods and needs further development in the future.

Acknowledgments

It is my great pleasure to acknowledge my colleagues of the Structural Analysis team for their outstanding TEM- and FIB-analysis: Olivier Richard, Paola Favia, Patricia Van Marcke, Chris Drijbooms, Eveline Verleysen; for SSRM and APT: Pierre Eyben, Jay Mody, Andreas Schulze, Sebastian Koelling, Ajay Kambham, Matthieu Gilbert, Wilfried Vandervorst; as well as the other members of the Materials and Components Analysis group. The imec processing groups are acknowledged for their continuous flow of challenging nanodevice analysis requests.

References

1. Huff, H.R. (2003) From the lab to the fab: transistors to integrated circuits. Characterization and Metrology for ULSI Technology: 2003 International Conference, AIP Conference Proceedings, Vol. 683 (eds D.G. Seiler, A.C. Diebold, T.J. Shaffner, R. Mc Donald, S. Zollner, R.P. Khosla, and E.M. Secula), pp. 3–39.
2. Moore, G.E. (1965) Cramming more components onto integrated circuits. *Electronics*, **38**, 114–118.
3. International Technology Roadmap for Semiconductors (ITRS) (2010), http://www.itrs.net/Links/2010ITRS/Home2010.htm (accessed 2 January 2011).
4. Colinge, J.-P. (2007) From gate-all-around to nanowire MOSFETS. Proceedings International Semiconductor Conference 2007, CAS 2007, Vol. 1, pp. 11–17.
5. Schulze, A., Hantschel, T., Eyben, P., Verhulst, A.S., Rooyackers, R., Vandooren, A., Mody, J., Nazir, A., Leonelli, D., and Vandervorst, W. (2011) Observation of diameter dependent carrier distribution in nanowire-based transistors, in *Nanotechnology* **22**, 185701-1–185701-7.
6. International Technology Roadmap for Semiconductors (2009) Metrology, 2009 Edition, http://www.itrs.net/Links/2009ITRS/2009Chapters_2009Tables/2009_Metrology.pdf (accessed 19 August 2010).
7. Characterization and metrology for ULSI technology. *AIP Conf. Proc.*, (1998 **449**; (2000) **550**; (2003) **683**; (2005) **780**; (b) Frontiers of metrology and characterization for nanoelectronics. *AIP Conf. Proc.*, (2007) **931**; (2009) **1173**.
8. Schröder, D.K. (2008) Nano characterization of materials. *Int. J. High Speed Electron. Syst.*, **18**, 861–878.
9. Rai, R.S. and Subramanian, S. (2009) Role of transmission electron microscopy in the semiconductor industry for process development and failure analysis. *Progr. Cryst. Growth Char. Mater.*, **55**, 63–97.
10. Marchman, H.M., Lorrusso, G., Adel, M., and Yedur, S. (2007) in *Microlithography, Science and Technology*, 2nd edn (ed. K. Suzuki), CRC Press, Boca Raton, FL, pp. 701–798.
11. Mello, D., DeSouza, Z., Giarrizzo, F., Gagliano, C., and Franco, G. (2007) Advanced strategy for in-line process monitoring using FIB and TEM. *Nucl. Instr. Meth. Phys. Res. B*, **257**, 805–809.
12. Delaye, V., Andrieu, F., Aussenac, F., Faynot, O., and Truche, R. (2008) In-line transmission electron microscopy for micro and nanotechnologies research and development. *Microelectron. Eng.*, **85**, 1157–1161.
13. Loo, R., Wang, G., Souriau, L., Lin, J.C., Takeuchi, S., Brammertz, G., and Caymax, M. (2010) High quality Ge virtual substrates on Si wafers with standard STI patterning. *J. Electrochem. Soc.*, **157**, H13–H21.
14. Wang, G., Rosseel, E., Loo, R., Favia, P., Bender, H., Caymax, M., Heyns, M.M., and Vandervorst, W. (2010) High quality Ge epitaxial layers in narrow channels on Si (001) substrates. *Appl. Phys. Lett.*, **96**, 111903-1–111903-3.
15. Brammertz, G., Mols, Y., Degroote, S., Leys, M., Van Steenbergen, J., Borghs, G., and Caymax, M. (2006) Selective epitaxial growth of GaAs on Ge by MOCVD. *J. Cryst. Growth*, **297**, 204–210.
16. Wang, G., Nguyen, N.D., Leys, M.R., Loo, R., Brammertz, G., Richard, O., Bender, H., Dekoster, J., Meuris, M., Heyns, M.M., and Caymax, M. (2010) Selective epitaxial growth of InP in STI trenches on Off-axis Si (001) Substrates. *ECS Trans.*, **27**, 959–964.
17. Lin, D., Waldron, N., Brammertz, G., Martens, K., Wang, W.-E., Sioncke, S., Delabie, A., Bender, H., Conard, T., Tseng, W.H., Lin, J.C., Temst, K., Vantomme, A., Mitard, J., Caymax, M., Meuris, M., Heyns, M., and Hoffmann, T. (2010) Exploring the ALD Al2O3/In0.53Ga0.47As and Al2O3/Ge interface properties: a common gate stack approach for advanced III-V/Ge CMOS. *ECS Trans.*, **28**, 173–183.

18. De Wolf, I., Vanhellemont, J., Romano-Rodriguez, A., Norström, H., and Maes, H.E. (1992) Micro-Raman study of stress distribution in local isolation structures and correlation with transmission electron microscopy. *J. Appl. Phys.*, **71**, 898–906.
19. Ghani, T., Armstrong, M., Auth, C., Bost, M., Charvat, P., Glass, G., Hoffmann, T., Johnson, K., Kenyon, C., Klaus, J., McIntyre, B., Mistry, K., Murthy, A., Sandford, J., Silberstein, M., Sivakumar, S., Smith, P., Zawadzki, K., Thompson, S., and Bohr, M. (2003) A 90 nm high volume manufacturing logic technology featuring novel 45 nm gate length strained silicon CMOS transistors. International Electron Devices Meetiing Technical Digest, IEEE, pp. 978–980.
20. Zhu, L., Georgi, C., Hecker, M., Rinderknecht, J., Mai, A., Ritz, Y., and Zschech, E. (2007) Nano-Raman spectroscopy with metallized atomic force microscopy tips on strained silicon structures. *J. Appl. Phys.*, **101**, 104305-1–104305-6.
21. Kolanek, K., Hermann, P., Dudek, P.T., Gotszalk, T., Chumakov, D., Weisheit, M., Hecker, M., and Zschech, E. (2010) Local anodic oxidation by atomic force microscopy for nano-Raman strain measurements on silicon-germanium thin films. *Thin Solid Films*, **518**, 3267–3272.
22. Senez, V., Armigliato, A., De Wolf, I., Carnevale, G., Balboni, R., Frabboni, S., and Benedetti, A. (2003) Strain determination in silicon microstructures by combined convergent beam electron diffraction, process simulation, and micro-Raman spectroscopy. *J. Appl. Phys.*, **94**, 5574–5583.
23. Armigliato, A., Balboni, R., and Frabboni, S. (2005) Improving spatial resolution of convergent beam electron diffraction strain mapping in silicon microstructures. *Appl. Phys. Lett.*, **86**, 063508-1–063508-3.
24. Houdellier, F., Roucau, C., Clément, L., Rouvière, J.L., and Casanove, M.J. (2006) Quantitative analysis of HOLZ line splitting in CBED patterns of epitaxially strained layers. *Ultramicroscopy*, **106**, 951–959.
25. Benedetti, A., Bender, H., and Torregiani, C. (2007) On the splitting of high order laue zone lines in CBED analysis of stress in silicon. *J. Electrochem. Soc.*, **154**, H217–H224.
26. Benedetti, A. and Bender, H. (2007) in *Microscopy of Semiconducting Materials 2007*, Springer Proceedings in Physics, Vol. **120** (eds A.G. Cullis and P.A. Midgley), Springer-Verlag, pp. 411–414.
27. Usuda, K., Numata, T., and Takagi, S. (2005) Strain evaluation of strained-Si layers on SiGe by the nano-beam electron diffraction (NBD) method. *Mater. Sci. Semicond. Process.*, **8**, 155–159.
28. Liu, J.P., Li, K., Pandey, S.M., Benistant, F.L., See, A., Zhou, M.S., Hsia, L.C., Schampers, R., and Klenov, D.O. (2008) Strain relaxation in transistor channels with embedded epitaxial silicon germanium source/drain. *Appl. Phys. Lett.*, **93**, 221912-1–221912-3.
29. Armigliato, A., Frabboni, S., and Gazzadi, G.C. (2008) Electron diffraction with ten nanometer beam size for strain analysis of nanodevices. *Appl. Phys. Lett.*, **93**, 161906-1–161906-3.
30. Béché, A., Rouvière, J.L., Clément, L., and Hartmann, J.M. (2009) Improved precision in strain measurement using nanobeam electron diffraction. *Appl. Phys. Lett.*, **95**, 123114-1–123114-3.
31. Sourty, E., Stanley, J., and Freitag, B. (2009) Using STEM with quasi-parallel illumination and an automated peak-finding routine for strain analysis at the nanometre scale. IEEE Proceedings 16th International Symposium on Physical and Failure Analysis, pp. 479–483.
32. Favia, P., Bargallo-Gonzales, M., Simoen, E., Verheyen, P., Klenov, D., and Bender, H. (2011) Nano-beam diffraction: technique evaluation and strain measurement on CMOS devices. *J. Electrochem. Soc.*, **158**, H438–H446.
33. Hÿtch, M.J., Putaux, J.-L., and Pénisson, J.-M. (2003) Measurement of the displacement field of dislocations to 0.03Å by electron microscopy. *Nature*, **423**, 270–273.
34. Hüe, F., Hÿtch, M., Bender, H., Houdellier, F., and Claverie, A. (2008) Direct mapping of strain

in a strained-silicon transistor by high-resolution electron microscopy. *Phys. Rev. Lett.*, **100**, 156602.
35. Hÿtch, M., Houdellier, F. Hüe, F., and Snoeck, E. (2008) Nanoscale holographic interferometry for strain measurements in electronic devices. *Nature*, **453**, 1086–1089.
36. Hüe, F., Hÿtch, M., Houdellier, F., Bender, H., and Claverie, A. (2009) Strain mapping of tensile strained silicon transistors with embedded Si1-yCy source and drain by dark-field holography. *Appl. Phys. Lett.*, **95**, 073103-1–073103-3.
37. Cooper, D., Béché, A., Hartmann, J.M., Carron, V., and Rouvière, J.L. (2010) Strain mapping for the semiconductor industry by dark-field electron holography and nanobeam electron diffraction with nm resolution. *Semicond. Sci. Technol.*, **25**, 095012-1–095012-11.
38. Koch, C.T., Ozdöl, V.B., and van Aken, P.A. (2010) An efficient, simple, and precise way to map strain with nanometer resolution in semiconductor devices. *Appl. Phys. Lett.*, **96**, 091901-1–091901-3.
39. Mody, J., Duffy, R., Eyben, P., Goossens, J., Moussa, A., Polspoel, W., Berghmans, B., van Dal, M.J.H., Pawlak, B.J., Kaiser, M., Weemaes, R.G.R., and Vandervorst, W. (2010) Experimental studies of dose retention and activation in fin field-effect-transistor-based structures. *J. Vac. Sci. Technol. B*, **28**, C1H5–C1H13.
40. Vandervorst, W. (2008) Semiconductor profiling with sub-nm resolution: challenges and solutions. *Appl. Surf. Sci.*, **255**, 805–812.
41. Kellogg, G.L. and Tsong, T.T. (1980) Pulsed-laser atom-probe field-ion microscopy. *J. Appl. Phys.*, **51**, 1184–1193.
42. Kelly, T.F. and Miller, M.K. (2007) Invited review article: atom probe tomography. *Rev. Sci. Instrum.*, **78**, 031101-1–031101-20.
43. Koelling, S., Gilbert, M., Goossens, J., Hikavyy, A., Richard, O., and Vandervorst, W. (2009) High depth resolution analysis of Si/SiGe multilayers with the atom probe. *Appl. Phys. Lett.*, **95**, 144106-1–144106-3.
44. Thompson, K., Booske, J.H., Larson, D.J., and Kelly, T.F. (2005) Three-dimensional atom mapping of dopants in Si nanostructures. *Appl. Phys. Lett.*, **87**, 052108-1–052108-3.
45. Thompson, K., Flaitz, P.L., Ronsheim, P., Larson, D.J., and Kelly, T.F. (2007) Imaging of arsenic cottrell atmospheres around silicon defects by three-dimensional atom probe tomography. *Science*, **317**, 1370–1374.
46. Mody, J., Kambham, A.K., Zschätzsch, G., Schatzer, P., Chiarella, T., Collaert, N., Witters, L., Jurczak, M., Horiguchi, N., Gilbert, M., Eyben, P., Kölling, S., Schulze, A., Hoffmann, T.-Y., and Vandervorst, W. (2010) Dopant and carrier profiling in FinFET-based devices with sub-nanometer resolution. Symposium on VLSI Technology 2010, pp. 195–196.
47. Clarysse, T., Vanhaeren, D., Hoflijk, I., and Vandervorst, W. (2004) Characterization of electrically active dopant profiles with the spreading resistance probe. *Mater. Sci. Eng. R*, **47**, 123–206.
48. Clarysse, T., Dortu, F., Vanhaeren, D., Hoflijk, I., Geenen, L., Janssens, T., Loo, R., Vandervorst, W., Pawlak, B.J., Ouzeaud, V., Defranoux, C., Faifer, V.N., and Current, M.I. (2004) Accurate electrical activation characterization of CMOS ultra-shallow profiles. *Mater. Sci. Eng. B*, **114–115**, 166–173.
49. Perovic, D.D., Castell, M.R., Howie, A., Lavoie, C., Tiedje, T., and Cole, J.S.W. (1995) Field-emission SEM imaging of compositional and doping layer semiconductor superlattices. *Ultramicroscopy*, **58**, 104–113.
50. Sealy, C.P., Castell, M.R., and Wilshaw, P.R. (2000) Mechanism for secondary electron dopant contrast in the SEM. *J. Electron Microsc.*, **49**, 311–321.
51. Kazemian, P., Twitchett, A.C., Humphreys, C.J., and Rodenburg, C. (2006) Site-specific dopant profiling in a scanning electron microscope using focused ion beam prepared specimens. *Appl. Phys. Lett.*, **88**, 212110-1–212110-3.
52. Rau, W.D., Schwander, P., Baumann, F.H., Höppner, W., and Ourmazd, A. (1999) Two-dimensional mapping of the electrostatic potential in transistors by

electron holography. *Phys. Rev. Lett.*, **82**, 2614–2617.
53. Twitchett, A.C., Dunin-Borkowski, R.E., Hallifax, R.J., Broom, R.F., and Midgley, P.A. (2005) Off-axis electron holography of unbiased and reverse-biased focused ion beam milled Si p-n junctions. *Microsc. Microanal.*, **11**, 66–78.
54. Twitchett-Harrison, A.C., Yates, T.J.V., Dunin-Borkowski, R.E., and Midgley, P.A. (2008) Quantitative electron holographic tomography for the 3D characterisation of semiconductor device structures. *Ultramicroscopy*, **108**, 1401–1407.
55. Cooper, D., Rivallin, P., Hartmann, J.-M., Chabli, A., and Dunin-Borkowski, R.E. (2009) Extending the detection limit of dopants for focused ion beam prepared semiconductor specimens examined by off-axis electron holography. *J. Appl. Phys.*, **106**, 064506-1–064506-6.
56. De Wolf, P., Stephenson, R., Trenkler, T., Clarysse, T., Hantschel, T., and Vandervorst, W. (2000) Status and review of two-dimensional carrier and dopant profiling using scanning probe microscopy. *J. Vac. Sci. Technol. B*, **18**, 361–368.
57. De Wolf, P., Clarysse, T., Vandervorst, W., Snauwaert, J., and Hellemans, L. (1996) One- and two-dimensional carrier profiling in semiconductors by nanospreading resistance profiling. *J. Vac. Sci. Technol. B*, **14**, 380–385.
58. Eyben, P., Alvarez, D., Jurczak, M., Rooyackers, R., De Keersgieter, A., Augendre, E., and Vandervorst, W. (2004) Analysis of the two-dimensional-dopant profile in a 90 nm complementary metal-oxide-semiconductor technology using scanning spreading resistance microscopy. *J. Vac. Sci. Technol. B*, **22**, 364–368.
59. Wilk, G.D., Wallace, R.M., and Anthony, J.M. (2001) High-k gate dielectrics: current status and materials properties considerations. *J. Appl. Phys.*, **89**, 5243–5275.
60. Kittl, J., Opsomer, K., Popovici, M., Menou, N., Kaczer, B., Wang, X., Adelmann, C., Pawlak, M., Tomida, K., Rothschild, A., Govoreanu, B., Degraeve, R., Schaekers, M., Zahid, M., Delabie, A., Meersschaut, J., Polspoel, W., Clima, S., Pourtois, G., Knaepen, W., Detavernier, C., Afanasiev, V., Blomberg, T., Pierreux, D., Swerts, J., Fischer, P., Maes, J., Manger, D., Vandervorst, W., Conard, T., Franquet, A., Favia, P., Bender, H., Brijs, B., Van Elshocht, S., Jurczak, M., Van Houdt, J., and Wouters, D. (2009) High-k dielectrics for future generation memory devices. *Microelectron. Eng.*, **86**, 1789–1795.
61. Diebold, A.C., Foran, B., Kisielowski, C., Muller, D.A., Pennycook, S.J., Principe, E., and Stemmer, S. (2003) Thin dielectric film thickness determination by advanced transmission electron microscopy. *Microsc. Microanal.*, **9**, 493–508.
62. MacKenzie, M., Craven, A.J., McComb, D.W., and De Gendt, S. (2006) Interfacial reactions in a HfO2/TiN/poly-Si gate stack. *Appl. Phys. Lett.*, **88**, 192112-1–192112-3.
63. van Benthem, K., Lupini, A.R., Kimb, M., Baik, H.S., Doh, S.J., Lee, J.-H., Oxley, M.P., Findlay, S.D., Allen, L.J., Luck, J.T., and Pennycook, S.J. (2005) Three-dimensional imaging of individual hafnium atoms inside a semiconductor device. *Appl. Phys. Lett.*, **87**, 034104-1–034104-3.
64. Xin, H.L., Intaraprasonk, V., and Muller, D.A. (2008) Depth sectioning of individual dopant atoms with aberration-corrected scanning transmission electron microscopy. *Appl. Phys. Lett.*, **92**, 013125-1–013125-3.
65. Porti, M., Nafr, M., Aymerich, X., Olbrich, A., and Ebersberger, B. (2002) Electrical characterization of stressed and broken down SiO2 films at a nanometer scale using a conductive atomic force microscope. *J. Appl. Phys.*, **91**, 2071–2079.
66. Fiorenza, P., Polspoel, W., and Vandervorst, W. (2006) Conductive atomic force microscopy studies of thin SiO2 layer degradation. *Appl. Phys. Lett.*, **88**, 222104-1–222104-3.
67. Kyuno, K., Kita, K., and Toriumi, A. (2005) Evolution of leakage paths in HfO2/SiO2 stacked gate dielectrics:

a stable direct observation by ultrahigh vacuum conducting atomic force microscopy. *Appl. Phys. Lett.*, **86**, 063510-1–063510-3.
68. Kremmer, S., Wurmbauer, H., Teichert, C., Tallarida, G., Spiga, S., Wiemer, C., and Fanciulli, M. (2005) Nanoscale morphological and electrical homogeneity of HfO2 and ZrO2 thin films studied by conducting atomic-force microscopy. *J. Appl. Phys.*, **97**, 074315-1–074315-7.
69. Paskaleva, A., Yanev, V., Rommel, M., Lemberger, M., and Bauer, A.J. (2008) Improved insight in charge trapping of high-k ZrO2 /SiO2 stacks by use of tunneling atomic force microscopy. *J. Appl. Phys.*, **104**, 024108-1–024108-7.
70. Lauwers, A., Steegen, A., de Potter, M., Lindsay, R., Satta, A., Bender, H., and Maex, K. (2001) Materials aspects, electrical performance, and scalability of Ni-silicide towards sub-0.13μm technologies. *J. Vac. Sci. Technol. B*, **19**, 2026–2037.
71. Kittl, J.A., O'Sullivan, B.J., Kaushik, V.S., Lauwers, A., Pawlak, M.A., Hoffmann, T., Demeurisse, C., Vrancken, C., Veloso, A., Absil, P., and Biesemans, S. (2007) Work function of Ni3Si2 on HfSixOy and SiO2 and its implication for Ni fully silicided gate applications. *Appl Phys Lett.*, **90**, 032103-1–032103-3.
72. Verleysen, E., Bender, H., Richard, O., Schryvers, D., and Vandervorst, W. (2010) Characterization of nickel silicides using EELS-based methods. *J. Microsc.*, **240**, 75–82.
73. Richard, O., Iacopi, F., Bender, H., and Beyer, G. (2007) Sidewall damage in silica-based low-k material induced by different patterning plasma processes studied by energy filtered and analytical scanning TEM. *Microelectron. Eng.*, **84**, 143–149.
74. Iacopi, F., Richard, O., Van Aelst, J., Mannaert, G., Talanov, V.V., Scherz, A., Schwartz, A.R., Bender, H., Travaly, Y., Brongersma, S.H., Antonelli, G.A., Moinpour, M., and Beyer, G. (2006) Understanding integration damage to low-k films: mechanisms and dielectric behaviour at 100kHz and 4GHz. Proceedings of the 2006 International Interconnect Technology Conference, IITC, (IEEE), pp. 12–14.
75. Shimada, M., Otsuka, Y., Harada, T., Tsutsumida, A., Inukai, K., Hashimota, H., and Ogwa, S. (2005) 2-dimensional distribution of dielectric constants in patterned low-k structures by a nm-probe STEM/Valence EELS (V-EELS) technique. Proceedings of the 2005 International Interconnect Technology Conference, IITC, (IEEE), pp. 88–90.
76. Pokrant, S., Cheynet, M., Jullian, S., and Pantel, R. (2005) Chemical analysis of nanometric dielectric layers using spatially resolved VEELS. *Ultramicroscopy*, **104**, 233–243.
77. Potapov, P.L., Engelmann, H.-J., Zschech, E., and Stöger-Pollach, M. (2009) Measuring the dielectric constant of materials from valence EELS. *Micron*, **40**, 262–268.
78. Erni, R., Lazar, S., and Browning, N.D. (2008) Prospects for analyzing the electronic properties in nanoscale systems by VEELS. *Ultramicroscopy*, **108**, 270–276.
79. Midgley, P.A. and Weyland, M. (2003) 3D electron microscopy in the physical sciences: the development of Z-contrast and EFTEM tomography. *Ultramicroscopy*, **96**, 413–431.
80. Kübel, C., Voigt, A., Schoenmakers, R., Otten, M., Su, D., Lee, T.-C., Carlsson, A., and Bradley, J. (2005) Recent advances in electron tomography: TEM and HAADF-STEM tomography for materials science and semiconductor applications. *Microsc. Microanal.*, **11**, 378–400.
81. Bender, H., Richard, O., Kalio, A., and Sourty, E. (2007) 3D-analysis of semiconductor structures by electron tomography. *Microelectron. Eng.*, **84**, 2707–2713.
82. Richard, O., Demuynck, S., Veloso, A., Van Marcke, P., and Bender, H. (2010) Characterization of a FinFET 6T-SRAM cell by tomography. *J. Phys.: Conf. Ser.*, **209**, 012047.
83. Cherns, P.D., Lorut, F., Dupré, C., Tachi, K., Cooper, D., Chabli, A., and Ernst, T. (2010) Electron tomography of Gate-All-Around nanowire transistors. *J. Phys.: Conf. Ser.*, **209**, 012046.

84. Ke, X., Bals, S., Cott, D., Hantschel, T., Bender, H., and Van Tendeloo, G. (2010) Three-dimensional analysis of carbon nanotube networks in interconnects by electron tomography without missing wedge artifacts. *Microsc. Microanal.*, **16**, 210–217.
85. Beyne, E. (2006) The rise of 3rd dimension for system integration. Proceedings of the 2006 International Interconnect Technology Conference, IITC, (IEEE), pp. 1–5.
86. Van Olmen, J., Mercha, A., Katti, G., Huyghebaert, C., Van Aelst, J., Seppala, E., Chao, Z., Armini, S., Vaes, J., Teixeira Cotrin, R.,, Van Cauwenberghe, M., Verdonck, P., Verhemeldonck, P., Jourdain, A., Ruythooren, W., de Potter de ten Broeck, M., Opdebeeck, A., Chiarella, T., Parvais, B., Debusschere, I., Hoffmann, T.-Y., Dehaene, W., Stucchi, M., Rakowski, M., Soussan, P., Cartuyvels, R., Beyne, E., Biesemans, S., and Swinnen, B. (2008) 3D stacked IC demonstration using a trough silicon via first approach. Technical Digest International Electron Devices Meeting, IEDM 2008, pp. 603–606.
87. Bender, H., Drijbooms, C., Van Marcke, P., Geypen, J., Philipsen, H.G.G., and Radisic, A. (2010) Structural characterization of through silicon vias (TSV). *J. Mater. Sci.*, DOI: 10.1007/s10853-010-5144-6.
88. The Electronic Device Failure Analysis Society Desk Reference Committee (eds) (2004) *Microelectronic Failure Analysis Desk Reference*, 5th edn, ASM International, Material Park, OH.
89. Boit, C. (2005) New physical techniques for IC functional analysis of on-chip devices and interconnects. *Appl. Surf. Sci.*, **252**, 18–23.
90. Hooghan, K.N., Wills, K.S., Rodriguez, P.A., and O'Connell, S. (1999) *Proceedings of the 25th International Symposium for Testing and Failure Analysis (ISTFA 99)*, ASM International, Materials Park, OH, pp. 247–254.
91. Schlangen, R., Leihkauf, R., Kerst, U., Lundquist, R., Egger, P., and Boit, C. (2009) Physical analysis, trimming and editing of nanoscale IC function with backside FIB processing. *Microelec. Rel.*, **49**, 1158–1164.
92. Guisinger, N.P. and Arnold, M.S. (2010) Beyond silicon: carbon-based nanotechnology. *MRS Bull.*, **35**, 273–330.
93. Ke, X., Bals, S., Romo Negreira, A., Hantschel, T., Bender, H., and Van Tendeloo, G. (2009) TEM sample preparation by FIB for carbon nanotube interconnects. *Ultramicroscopy*, **109**, 1353–1359.

32
Complex Oxide Materials

Maria Varela, Timothy J. Pennycook, Jaume Gazquez, Albina Y. Borisevich, Sokrates T. Pantelides, and Stephen J. Pennycook

32.1
Introduction

Complex oxides wide range of physical properties that make them a most fascinating class of materials: from high Tc superconductivity (HTCS) [1–4] to colossal magnetoresistance (CMR) [5–7], from ionic conductivity [8–11] to ferroelectricity [12–17] among many other areas. Furthermore, thanks to their robustness, oxides exhibit a high potential for applications in electronics, sensors, spintronics, batteries, or catalysis just to name a few [18–22]. However, the physics underlying many of these behaviors is far from understood. This is, for example, the case with HTCS: 25 years after its discovery, the mechanism responsible for superconductivity above liquid nitrogen temperature in cuprates still remains elusive. The physical properties of these and all other complex oxides depend on the crystal structure and particularly on the oxygen-metal bonding characteristics [23]. These bonds determine the electronic density of states (DOS) and, therefore, the macroscopic properties. Hence, the coordination of the elements bonded to the O atoms is critical to the properties of the system. Complex oxides comprise a very wide range of materials (most of the Earth's crust is made of oxides). Here, we focus on metal oxides, especially transition-metal oxides with the perovskite structure that, within a single class, display the aforementioned physical properties.

In some perovskites, a metal from the 3d series is surrounded by six O atoms in octahedral coordination. This octahedral coordination lifts the degeneracy of the d orbitals. The d_{z2} and d_{x2-y2} (also known as the e_g set) are shifted away from the d_{xz}, d_{xy}, and d_{yz} (the t_{2g} set) [23]. The energy gap separating these two sets is called *crystal field splitting*. The electronic properties of these materials can therefore be studied by means of spectroscopy by exciting electrons from the 2p levels into those 3d bands, that is, studying the fine structure of the $L_{2,3}$ absorption edges [24–35]. Complementary information can be obtained from the O K edge, resulting from excitations of O 1s electrons into available empty states containing O 2p character. These states are strongly hybridized with the transition-metal 3d bands. Electron energy loss spectroscopy (EELS) in the electron microscope is a most suitable technique to carry out such measurements [35–38]. Thanks to the success of aberration correction in the scanning transmission electron microscope (STEM), such experiments are possible with atomic resolution. In this chapter, a number of applications of STEM-EELS techniques to complex oxide materials, mostly those with the perovskite structure, are reviewed. Examples including structural, chemical, and electronic property mapping are presented, along with discussions regarding interpretation, sources of artifacts, ultimate sensitivity levels, and so on. First, we review the state of the art in imaging and especially spectrum imaging in oxides, including the case of sensitivity to isolated impurities and light atoms (oxygen) and then we will move on to a few applications of perovskite systems: manganites, cuprates, cobaltites, and titanates are discussed. Applications of thin films and ionic conductors are reviewed as well, along with the sensitivity of EELS to quantities such as the spin state of magnetic atoms.

32.2
Aberration-Corrected Spectrum Imaging in the STEM

While atomic-resolution annular-dark-field (ADF) imaging and EELS imaging in the STEM was first reported almost two decades ago [39–42], spherical aberration correction has directly impacted the study of complex oxide materials [43–48]. Poor signal-to-noise ratios historically associated with ADF imaging and the lack of atomic resolution in electron energy loss (EEL) images have been overcome with the installation of successful aberration correctors that can now compensate up to fifth-order aberrations in commercial columns. Multiple benefits follow from the better resolution, enhanced contrast, and higher peak intensity that ensue from highly focused electron probes. After the first results hinting at an evolution into the sub-Ångström regime [42, 49], 2004 was a historic year when direct sub-Ångström images of a [112] Si lattice [50] were reported along with the first ever direct detection of single atoms both through imaging and spectroscopy [43, 51–53], as shown in Figure 32.1. Ever since, STEM-EELS has evolved at a very fast pace, from those first reports to element-sensitive 2D images of crystal lattices such as those of manganites reported by Bosman *et al.* [54] (immediately followed by Kimoto *et al.* and Muller *et al.* [55, 56]).

Such atomic-resolution EEL spectrum imaging is becoming a very popular tool that allows simultaneous measurements of the chemical identity of different atomic columns. In addition to composition, electronic properties can be deduced from the EELS fine structure, which directly ensues from the unoccupied DOS [57, 58]. A spectrum image (SI) [58–60] is acquired by raster scanning the electron beam over a predefined area in the sample while acquiring EEL spectra. The three-dimensional data cube built this way consists of two spatial dimensions, corresponding to the x,y coordinates of every pixel in the grid, and in the third dimension, the actual EEL spectra collected in each of these spatial coordinates. Maps with elemental sensitivity can be produced after proper background subtraction and integration of the signal remaining below the relevant edges. Also, for every pixel other

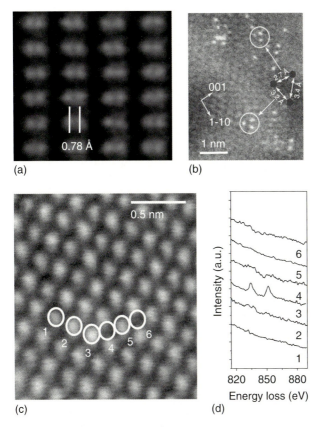

Figure 32.1 (a) ADF image of Si [112] recorded with an aberration-corrected STEM. "Dumbbells" just 78 pm apart are resolved. (Adapted from Ref. [50].) (b) Z-contrast STEM image of Pt on the surface of γ-Al$_2$O$_3$ close to [110] orientation. Two Pt$_3$ trimer structures are circled; inset – measured interatomic distances. A hint of alumina lattice is also visible. (Adapted from Ref. [52].) (c) HAADF image of a single La atom in a CaTiO$_3$ matrix, acquired in the VG Microscopes HB501UX. (d) Electron energy loss spectra obtained from individual columns of the calcium titanate crystal. Spectrum 3 reveals the presence of a single La atom within the calcium column circled 3. The signal is substantially reduced from neighboring columns. (Adapted from Ref. [43]).

signals such as the one scattered to the ADF detector or X-ray-dispersive spectra can be collected. This way, a quite complex data set is built. The data quality depends on a number of factors, such as those related to the specimen (thickness, cleanliness, surface amorphous layers, etc.), instabilities (temperature, spatial drift, electromagnetic fields, and mechanical vibrations), and sample damage induced by the beam or the system optics (e.g., spectrometer collection efficiency, beam current, etc.) itself. Most of these factors amount, in practical terms, to achieving a compromise between parameters such as pixel acquisition times, image size (sampling rate), spatial resolution (beam current), and/or the acceleration voltage.

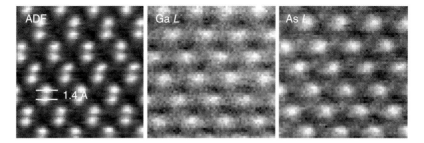

Figure 32.2 Spectroscopic imaging of GaAs in the ⟨110⟩ projection comparing the ADF image to the Ga- and As L edge spectroscopic images, obtained on a Nion Ultra-STEM with a fifth-order aberration corrector operating at 100 kV. Images are 64 × 64 pixels, with a collection time 0.02 s/pixel and a beam current of approximately 100 pA, after noise reduction by principal component analysis (PCA) [61]. (Adapted from Ref. [62]).

Nevertheless, this technique allows the production of elemental images with atomic resolution, as shown in Figure 32.2 for a GaAs specimen.

The capability of producing atomic-resolution images in spectroscopy mode is most useful when it comes to studying low-dimensionality systems. Such investigations benefit in a straightforward way from real-space probes. Examples include, but are not limited to, thin films, interfaces, nanoparticles, and so on. In these systems, average macroscopic diffraction techniques may not provide enough information to understand the detailed structure/chemistry. Real-space techniques such as STEM-EELS are essential to ascertain local properties, which may dominate the properties of the system as a whole. STEM-EELS is a very useful tool to quantify parameters such as local chemical fluctuations [56, 63, 64] or interdiffusion or give an upper estimate of local interface abruptness or map interface termination, as shown in Figure 32.3.

Atomic-resolution spectrum imaging is capable of providing atomic lattice images even when low concentrations of dopants are present. Figure 32.4 shows an image along with several chemical maps of a $Ca_{0.95}La_{0.05}TiO_3/CaTiO_3$ bilayer. Even when the amount of La dopants is relatively small (5% nominal concentration), their distribution can be clearly appreciated on the spectroscopic image.

EELS is also a very powerful technique when it comes to light atom detection, and indeed it is a very useful tool to image O atoms in oxides. Light atoms are difficult to image because of their low scattering powers. Only recently were individual B and N atoms directly imaged by ADF techniques [66]. In perovskites, O atoms are easily masked by adjacent heavy atom columns, but with aberration correction were imaged first by transmission electron microscopy (TEM) and then STEM [67–70]. The signal-to-noise ratio is high enough that their positions can be quantified not only from spectroscopic images but also from Z-contrast and even bright-field (BF) images. The next section deals with explaining these possibilities in detail.

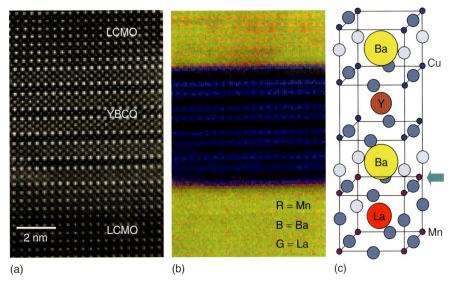

Figure 32.3 (a) ADF signal collected simultaneously with a spectrum image in an LCMO/YBCO/LCMO trilayer. (b) False color image where three atomic-resolution elemental images have been overlayed: a Mn $L_{2,3}$ image in red, a Ba $M_{4,5}$ image in blue, and a La $M_{4,5}$ image in green, for the same area as the Z-contrast image in (a). The EELS data were acquired on a Nion UltraSTEM and processed with PCA in order to remove random noise. (c) Sketch of the observed interface structure. An arrow marks the interface MnO_2 plane, facing a BaO plane from the superconductor. (Adapted from Ref. [65].)

32.3
Imaging of Oxygen Lattice Distortions in Perovskites and Oxide Thin Films and Interfaces

Perovskite materials are rarely perfectly cubic structures but tend to exhibit distortions, mainly in the O sublattice, resulting in tetragonal, orthorhombic, or even more complex structures. The shape of the O octahedra around the metal cations, which controls the atomic bond characteristics, affects the hybridization of the O 2p and the metal 3d or 4d bands and therefore has a profound impact on the macroscopic physical properties of these systems. Therefore, it is very useful to be able to map and quantify small displacements of the O sublattice, which, as explained above, can be now achieved through Z-contrast, BF, or EELS imaging. As an example, Figure 32.5 shows a set of spectroscopic images of $LaMnO_3$, a Jahn-Teller distorted perovskite [71], imaged down the pseudocubic [110] direction acquired at 60 kV in a fifth-order corrected Nion UltraSTEM. After principal component analysis (PCA) [61] was used to remove random noise from the data and the background was subtracted using a power law, the lattices of Mn, La, and O atoms can be imaged by integrating the intensity under the respective edges of interest. All these elemental maps show atomic-resolution images of their respective elements.

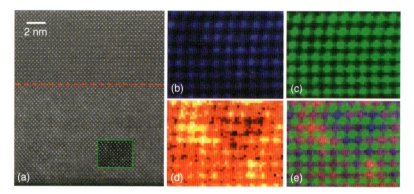

Figure 32.4 (a) Z-contrast image of a $Ca_xLa_{1-x}TiO_3/CaTiO_3$ bilayer doped with $x = 0.05$ La impurities. A red dotted line marks the position of the interface. The inset shows the region where a spectrum image was acquired, 50 × 37 pixels in size, with a current of approximately 100 pA and an exposure time of 0.1 s per pixel. The simultaneously acquired ADF signal is overlaid in the inset. Some spatial drift is observed. (b) Ca $L_{2,3}$ map, (c) Ti $L_{2,3}$ map, and (d) La $M_{4,5}$ map, produced after noise removal using principal component analysis and background subtraction using a power law fit. The Ca and Ti maps were produced by integrating a 20 eV wide window, while for the La map a 30 eV wide window was used. (e) RGB overlay of (b–d), with the La map shown in red, the Ti map in green, and the Ca map in blue. (Specimen courtesy of M. Biegalski and H. Christen from Oak Ridge National Laboratory. Adapted from Ref. [65].)

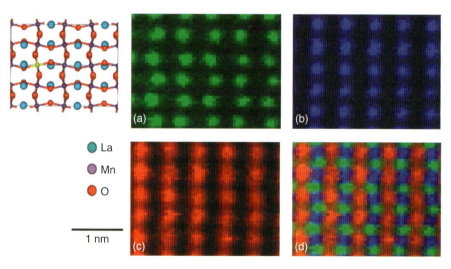

Figure 32.5 (a) Oxygen K edge image of $LaMnO_3$ down the pseudocubic ⟨110⟩ axis, acquired in the Nion UltraSTEM at 60 kV. (b) Simultaneously acquired Mn $L_{2,3}$ image. (c) La $M_{4,5}$ image. (d) RGB overlay of the images in (a–c). The images have been corrected for spatial drift, and principal component analysis has been used to remove random noise from the EEL spectra. (Adapted from Ref. [62].)

32.3 Imaging of Oxygen Lattice Distortions in Perovskites and Oxide Thin Films and Interfaces

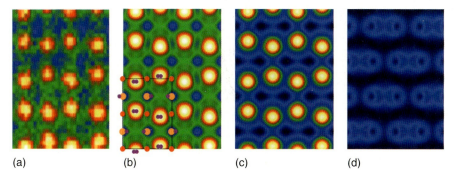

(a) (b) (c) (d)

Figure 32.6 Integrated oxygen K-shell EELS signal from LaMnO$_3$ in the [010] zone axis orientation (pseudocubic ⟨110⟩ axis). (a) Experimental image acquired on the Nion UltraSTEM operating at 60 kV. (b) Simulated image with projected structure inset. (c) Contribution to the total image from the isolated O columns. (d) Contribution to the total image from the O atoms on the La/O columns. (Reproduced from Ref. [77].)

In the O K edge image, the O sublattice can clearly be observed, along with the ripple associated with the distorted O octahedra.

For quantitative interpretation of these images, simulations are needed since elastic scattering cannot be neglected in many specimens [72–76], especially those prepared by methods such as ion milling, which rarely are extremely thin and clean. As an example, the O map shown in Figure 32.5 suffers from clear effects of elastic scattering. Figure 32.6 includes again this map along with simulated images of the O K edge intensity for LMO in the [010] axis (the [110] axis in pseudocubic orientation). Now, there are two types of projected O sites, pure O columns, and LaO columns. Because of the strong scattering when the probe is over the LaO column, the O signal is greatly depleted, in agreement with the experiment.

Nonetheless, from these maps, distortions in the O octahedral sublattice can be quantified, as depicted in Figure 32.5. The position of the different columns can be quantified by several methods. One of the simplest approaches consists of calculating the center of mass for every column. After the coordinates for every column have been determined, differences in coordinates between nearest neighbors, such as the y coordinate, can be calculated and plotted, giving rise to checkerboard patterns such as the one in Figure 32.7.

Mapping such small displacements, be it by EELS imaging or direct analysis of high-resolution TEM or STEM images, can be an important ingredient in the quest to control the physical properties of low-dimensional systems such as nanoparticles or interfaces. An example can be found in multiferroic systems. The multifaceted magnetic, electrical, and structural functionalities of these perovskite ABO$_3$ materials are underpinned by the subtle distortions of the crystallographic lattice from the prototype cubic phase [78]. These distortions include relative displacements of the cations from the centers of the BO$_6$ oxygen octahedra (i.e., ferroelectric displacements), deformations of the oxygen octahedra (e.g., Jahn-Teller distortions), and collective tilts of the octahedral network. Epitaxial films are widely

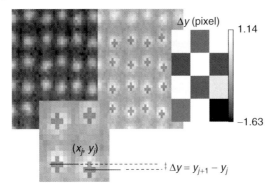

Figure 32.7 Quantification of the position of the different O columns along with Δy coordinate map, for the O K image shown in Figures 32.5 and 32.6 (rotated 90 degrees). The authors are grateful to Andrew Lupini for the column location script.

used to create well-defined strain states; however, a change in symmetry or the octahedral tilt system across such an interface can have an effect on interface and film properties that cannot be generated by the strain alone [79]. Structural changes at epitaxial interfaces can be localized and thus require the capability to map lattice spacings and octahedral distortions at a unit cell scale. Moreover, to investigate the possibility of the coupling of different distortions at and away from the interfaces, it is desirable to map different distortions simultaneously and independently. Aberration-corrected STEM is the perfect tool for such studies, with sufficient resolution and precision to locally quantify changes in structure (similar to the approach developed for high-resolution TEM by Jia et al. [80–82]), as well as the capability to use multiple detectors simultaneously.

These capabilities proved to be instrumental in a study of epitaxial $BiFeO_3$ – $La_{0.7}Sr_{0.3}MnO_3$ (BFO–LSMO) interfaces [83]. In BFO, the octahedral tilts are related to the Fe–O–Fe angles and high spin/low spin (HS/LS) states of Fe atoms, and hence directly control the metal-insulator transitions and magnetic properties [84]. Shown in Figure 32.8a,b are the simultaneously acquired high-angle annular-dark-field (HAADF) and BF STEM images of the STO/5 nm LSMO/5 nm BFO heterostructure (STO is outside of the field of view). The profile in Figure 32.8c (obtained by averaging the 2D map of lattice spacings (not shown) along the interface) shows the behavior of the pseudocubic lattice parameter c (*normal* to the interface) calculated from atomic column positions in (a); the increase from LSMO to BFO can be easily seen. The profile also shows a local increase in the pseudocubic c parameter in the first three to four layers of BFO adjacent to the interface. Note that this increase is localized near the interface and hence cannot be accounted for by the usual volume-conserving Poisson distortion due to epitaxial strain, which would act throughout the entire thickness of the film.

To complement the analysis of the lattice spacings, the octahedral tilts in the vicinity of the interface were also explored, using the image in Figure 32.8b to

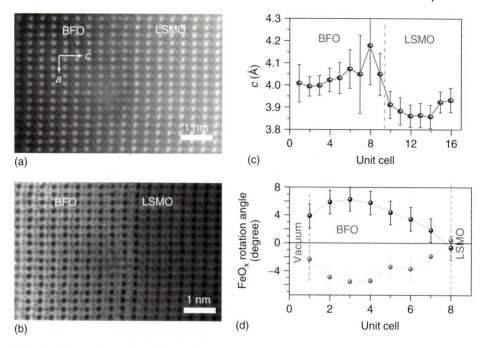

Figure 32.8 (a) Simultaneously acquired STEM HAADF and (b) BF images of the STO/5 nm LSMO/50 nm BFO thin film; (c) line profile of the out-of-plane lattice spacings obtained from (a) and averaged along the interface (d) line profiles of positive (black) and negative (gray) octahedral tilts obtained from (b) and averaged along the interface. (Adapted from Ref. [83].)

determine oxygen positions. The resulting map showed a checkerboard pattern of positive and negative tilt angles characteristic of BFO structure. Figure 32.8d shows the twin profiles of the positive and negative tilts as a function of the distance from the interface. The positive and negative branches appear to have the same magnitude and behavior, and both branches decline on approach to the BFO–LSMO interface. Thus, octahedral tilts are suppressed in the near-interface region, which coincides with the region of anomalous lattice parameter in the profile in Figure 32.8c.

Changes in chemical composition and local dielectric properties of the interface were evaluated using EELS spectrum imaging in the 10–350 eV range (Figure 32.9a). Fourier-log deconvolved spectra of the BFO, LSMO, and STO averaged over regions away from the interface were used as a basis for multiple least squares fits; these spectra are shown in Figure 32.9b as green, blue, and red curves, respectively. The fit was performed in two different energy ranges corresponding to different interactions: core-loss range (35–125 eV, reflecting chemical composition) and plasmonic range (5–35 eV, containing information on dielectric properties [58, 85]). Figure 32.9c,d shows the fitting results for the core-loss range. The fit coefficient map (Figure 32.9c) shows clear contrast associated with

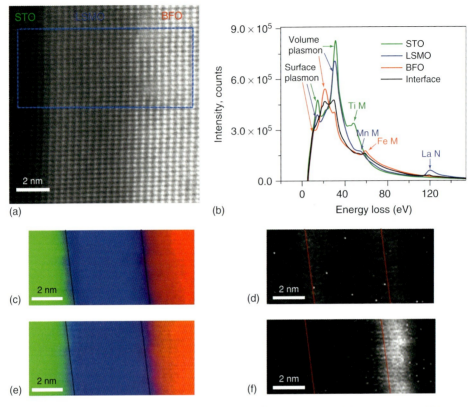

Figure 32.9 Low loss EELS imaging of the SrTiO$_3$/(La, Sr)MnO$_3$/BiFeO$_3$ interface. (a) Imaging area; (b) representative spectra (after Fourier-log deconvolution) of the three components showing distinctive signatures and features; (c–f) results of the multiple least square fit of the image using BFO, LSMO, and STO spectra from (b) in two energy ranges: core-loss (35–125 eV): (c) fit coefficient map, (d) χ^2 map; and plasmonic (5–35 eV): (e) fit coefficient map, and (f) χ^2 map. (Adapted from Ref. [83].)

individual phases, that is, provides chemical identification of the components. It is in good agreement with the simultaneously acquired ADF image, and the map of the χ^2 (Figure 32.9d) shows no anomalies. At the same time, in the plasmonic energy range the χ^2 map (Figure 32.9f) shows an anomalous band, indicating that in this energy range the EELS spectra of the interface region cannot be interpreted as a linear superposition of constituents, suggesting the presence of a layer with anomalous dielectric properties. The band is located over the first ∼2 nm (five unit cells) of the BFO, and closely coincides with the width of the strained layer.

The comparison of EELS (anomalous layer) and structural (tilt angle, lattice expansion) data suggests that the origins of the observed behavior lie in the coupling of the antiferrodistortive order parameter (i.e., octahedral tilts) across the interface. The increase of the Fe–O–Fe angle (toward 180° due to locally suppressed tilts at the interface and forced higher symmetry is expected to increase the bandwidth of occupied and unoccupied bands, W_O and W_{UO}, respectively, reducing the band gap in the BFO; density functional theory (DFT) calculations support this conjecture [79, 83]. These studies demonstrate the new paradigm of an interface phase transition mediated by the antiferrodistortive coupling across the interface, complementing the established strain, polarization, and charge-driven behavior.

32.4
Atomic-Resolution Effects in the Fine Structure–Further Insights into Oxide Interface Properties

As already discussed, in transition-metal oxide perovskites the fine structures of the EELS edges are directly related to the features of the transition-metal 3d bands and also the O 2p bands, since these bands tend to be strongly hybridized near the Fermi level. The O 2p bands can be probed by exploring the O K edge around 530 eV, which is the product of exciting 1s electrons into available 2p-like states. The transition-metal 3d bands can be studied by analyzing the $L_{2,3}$ absorption edges, which again, according to dipole selection rules, are the result of exciting 2p electrons into 3d-like states. In perovskites, where the transition metal is surrounded by six O atoms in octahedral coordination, the O 2p bands are strongly hybridized with the transition-metal 3d bands, whereby the features in those edges typically correlate with each other. A good example can be found in the manganite $La_xCa_{1-x}MnO_3$ (LCMO) system. La is a trivalent cation, whereas Ca is divalent. Hence, the solid solution of both elements into the manganite sublattice results in a Mn^{+3}/Mn^{+4} mixed valence system [6, 23]. Systematic changes take place both in the O K edge and in the Mn $L_{2,3}$ edge as holes are introduced in the Mn 3d e_g band by Ca doping, as can be observed in Figure 32.10. The $L_{2,3}$ intensity ratio decreases, while the pre-peak feature intensity at the O K edge increases. Also, this feature moves away from the next peak in the O K fine structure: the peak separation, ΔE, clearly increases for higher Mn oxidation state (Figure 32.10a) [35].

The spectra in Figure 32.10 have been acquired with illumination of a wide area of the sample, thus giving a spatially averaged signal. Further detail can be observed when employing atomic size electron beams in specimens where inequivalent O species are present, as is the case in LMO. The O octahedra in this compound are Jahn-Teller distorted, leading to the presence of two different O species: the equatorial and the apical O atoms. Theoretical simulations using the $Z+1$ approximation [38, 86] (shown in Figure 32.11) suggest that the O K edge of these distinct oxygen atoms should show some differences around the pre-peak region.

While the differences in the near-edge fine structure between the equatorial (O2 on Figure 32.11) and apical (O1 in Figure 32.11) oxygen atoms are small, they can be detected experimentally. As a matter of fact, ΔE maps can be extracted from EELS linescans or spectrum images. For the image of LMO in Figure 32.5, the ΔE map is shown in Figure 32.12. Spatial variations in the ΔE parameter that correlate with the atomic lattice can be appreciated. These measurements hint toward the

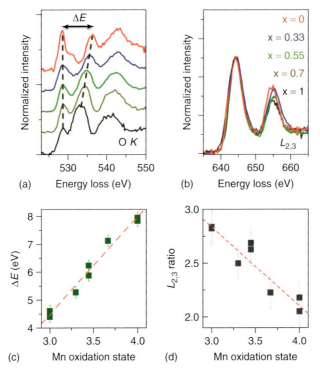

Figure 32.10 O K (a) and Mn $L_{2,3}$ (b) edges for a set of $La_xCa_{1-x}MnO_3$ crystals oriented down the pseudocubic ⟨100⟩ axis for x values of 1 (black), 0.7 (brown), 0.55 (green), 0.33 (blue), and 0 (red). The O K edges have been displaced vertically for clarity, while the Mn L_3 edges have been normalized and aligned for direct visual comparison. (c) Peak separation parameter, ΔE, and $L_{2,3}$ intensity ratio (d) versus nominal Mn oxidation state for the set of $La_xCa_{1-x}MnO_3$ crystals. (Adapted from Ref. [35].)

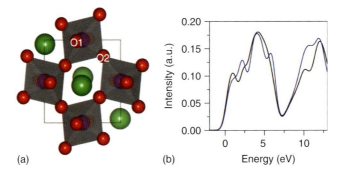

Figure 32.11 Sketch of the LMO structure, extracted from Ref. [71], with the O1 and O2 positions highlighted. (b) O K edge simulations for O1 (black) and O2 (blue) using density functional theory with the $Z+1$ approximation and 1 eV broadening. (Adapted from Ref. [35].)

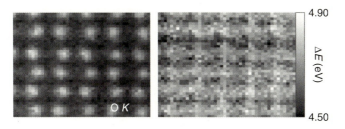

Figure 32.12 O K edge image for LMO down the ⟨110⟩ pseudocubic direction (left, from Figure 32.5), along with the peak separation ΔE map showing a contrast due to the dependence of the fine structure on the atomic lattice (right). Data acquired at 60 kV on the aberration-corrected Nion UltraSTEM. (Adapted from Ref. [65].)

fact that inequivalent O species can be distinguished spectroscopically from each other. Further work is in progress to advance this front.

These atomic-resolution effects tend to be reduced in specimens with thicknesses above 10–20 nm because of the averaging effect of elastic scattering, that is, beam broadening. Alternatively, techniques such as spatial (lateral) averaging over many pixels can render these effects insignificant. An example can be found in Figure 32.13, where the linear relation extracted from Figure 32.10c between ΔE parameter and Mn valence has been used to quantify the oxidation state of the transition metal across $LMO_{17u.c.}/STO_{nu.c.}$ superlattices with different STO layer thicknesses [87]. No atomic-resolution effects are observed when the oxidation state quantification is averaged laterally over a few unit cells (or even better over a few nanometers). For superlattices with relatively thick LMO and STO layers, oxidation states of $+3$ ($+4$) are measured for Mn (Ti) within each layer. However, for superlattices with high LMO to STO layer thickness ratio (such as the $LMO_{17u.c.}/STO_{2u.c.}$ shown here), a significantly reduced Ti oxidation state is measured [87] because of the presence of charge leakage associated with an extra interfacial LaO plane. The "charge leakage" has a strong effect on the macroscopic physical properties of the system, rendering these superlattices not only ferromagnetic but also metallic [87–89].

It is worth noting that, if possible, multiple methods should be used and compared to quantify any given quantity such as the oxidation state. In Figure 32.13, an alternative way to quantify the Ti oxidation state consisting of performing a multiple linear least squares fit to the bulk spectra of STO (Ti^{+4}) and LTO (Ti^{+3}) [90] is compared to the ΔE method. Both avenues give results that are very close to each other, except for the interface regions. This local disagreement is due to elastic scattering and the ensuing beam broadening, which causes an averaging of the O K signals of both materials at the interfaces.

The above is just an example, but it shows how powerful aberration-corrected STEM-EELS can be in the study of oxide thin films and interfaces. Thin films are two-dimensional systems in which one of the system characteristic dimensions (thickness) is much shorter than the other two (lateral dimensions). Unexpected

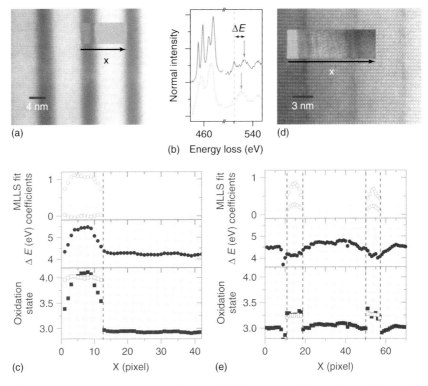

Figure 32.13 (a) Z-contrast image of a $(LMO_{17}/STO_{12})_8$ superlattice. Inset: Color-coded chemical map for a STO/LMO bilayer in the same sample (to scale) – Ti signal in dark gray and Mn in light gray. (b) EELS spectra showing the Ti $L_{2,3}$ and the O K edges for bulk $SrTiO_3$ (solid line) and $LaTiO_3$ (dotted line). (c) Mn and Ti oxidation states measured for the spectrum image shown in the inset. The top panel shows the multiple linear least squares (MLLS) fit coefficients for the LTO spectrum (dotted line) and the STO spectrum (solid line). The middle panel shows the value of the ΔE parameter along the spectrum image. The bottom panel shows the transition-metal oxidation state obtained from the MLLS fit (open squares) and from the ΔE parameter (solid squares). (d) Z-contrast image of the $(LMO_{17}/STO_2)_8$ superlattice. Inset: Color-coded chemical map for an EELS spectrum image (to scale): Ti signal in dark gray and Mn in light gray. (e) MLLS fit coefficients (top), ΔE parameter (middle), and transition-metal oxidation states (bottom) for the $(LMO_{17}/STO_2)_8$ sample. All of the profiles have been obtained by averaging in the direction parallel to the interface across the whole spectrum image. (Adapted from Ref. [87].)

physical behavior can derive from such reduced dimensionality, especially when interfaces between two materials are considered. Epitaxial growth of one material on top of a substrate produces a coherent interface where the properties of the materials show an interplay to some extent. Proximity effects, competition, or effects such as Andreev reflection are widely observed. Furthermore, completely unexpected properties have been reported, for example, the metallicity observed at perovskite interfaces between Mott insulators and band insulators such as LaTiO$_3$ and SrTiO$_3$ (STO) [90, 91]. A similar interface, the LaAlO$_3$/SrTiO$_3$ system, has been reported to display both ferromagnetism and superconductivity. The explanations of such observations are, however, controversial [92–100]. While, in principle, a purely electronic effect was invoked, later reports have highlighted the importance of small amounts of defects such as O vacancies or chemical interdiffusion across the interfaces amounting to doping of the structures [93, 96–98]. Within this context, real-space direct characterization techniques that can provide an atomic-resolution view of these interfaces are very useful tools toward ascertaining the mechanisms underlying the system properties.

32.5
Applications of Ionic Conductors: Studies of Colossal Ionic Conductivity in Oxide Superlattices

Another beautiful example of how aberration-corrected STEM-EELS can unravel the microscopic origin of an extremely interesting physical behavior can be found in the study of ionic conductors. Hydrogen-based fuel cells are a promising alternative to conventional sources of electrical power. They convert chemical energy directly into electricity, emitting only water in the process. Hydrogen production, however, is typically an energy-intensive process. Without a completely clean means of producing H, the efficiency of the fuel cell determines how environmentally friendly it is. Solid-oxide fuel cells (SOFCs), named for their solid oxide electrolytes, are the most efficient currently under development. The electrolytes must conduct O ions from cathode to anode while remaining electrically insulating. Yttria stabilized zirconia (YSZ) $(Y_2O_3)_x(ZrO_2)_{1-x}$ is the most commonly used electrolyte, but requires operating temperatures of at least 800 °C for sufficiently high O conductivity [8–11, 101]. The need for such high temperatures has hampered the application of SOFCs, and a major research effort has been devoted to a search for new materials with enhanced ionic conductivity at lower temperatures.

Although progress has been made toward the goal of higher O ionic conductivities at lower temperatures in bulk materials such as doped cerium oxides, the largest enhancements have been achieved in heterogeneous superlattices. Following work in CaF$_2$–BaF$_2$ multilayers [102, 103], which showed an increase in the F ionic conductivity as the layer thickness was decreased, a variety of thin oxide heterostructures have been fabricated. Enhancements in the YSZ ionic conductivity of one to two orders of magnitude were achieved in the YSZ/Y$_2$O$_3$ multilayers [104] and up to three orders in highly textured YSZ thin films grown on MgO [105]. In these materials, the conductivity increases when the number of interfaces is increased or as the film thicknesses are decreased, indicative of an interfacial conduction pathway.

More recently, a colossal eight orders of magnitude enhancement in the YSZ ionic conductivity was achieved in YSZ/STO multilayers near room temperature [11]. The conductance was found to scale with the number of interfaces, but to be virtually independent of the YSZ layer thickness from 1 to 30 nm, again indicating

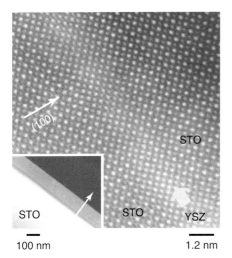 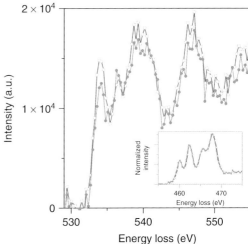

Figure 32.14 (a) Z-contrast STEM image of the STO/YSZ interface of the [YSZ$_{1\,nm}$/STO$_{10\,nm}$]$_9$ superlattice, obtained in the VG Microscopes HB603U microscope. A light gray arrow marks the position of the YSZ layer. (Inset) Low-magnification image obtained in the VG Microscopes HB501UX column. In both cases, a white arrow indicates the growth direction. (b) EEL spectra showing the O K edge obtained from the STO unit cell at the interface plane (gray circles) and 4.5 nm into the STO layer (black squares). (Inset) Ti $L_{2,3}$ edges for the same positions, same color code. (Reproduced from Ref. [11].)

that the majority of the conduction occurs near the interfaces. The contribution of electronic conductivity was determined to be three to four orders of magnitude lower than the ionic conductivity, and experiments also ruled out a protonic contribution to the conductance [106, 107].

Both X-ray and STEM analysis of the YSZ/STO multilayers exhibiting the colossal ionic conductivity showed them to be highly coherent, with the YSZ lattice rotated 45° to that of the STO [11]. Figure 32.14 shows low- and high-magnification Z-contrast images of a YSZ/STO superlattice with 1 nm thick YSZ layers and nine repeats. The layers are continuous and flat over long distances. From the high-magnification image it can be seen that the cation lattices are perfectly coherent across the interface, indicating that the YSZ is strained a large 7%. The YSZ was found to remain strained for YSZ thicknesses up to 30 nm. A superlattice with a nominal YSZ thickness of 62 nm was also grown, but this thicker YSZ layer relaxed, releasing the strain. The conductance of the sample was found to be three orders of magnitude lower than those with strained YSZ, suggesting that strain is vital to colossal ionic conductivity [11]. EELS measurements provide further insight into the nature of the interface. Figure 32.14b displays spectra taken from the interface plane and from the center of the STO layer. No significant change in the Ti fine structure is seen, indicating that a Ti^{+4} configuration is predominant at the interface plane, consistent with the lack of electronic conductivity.

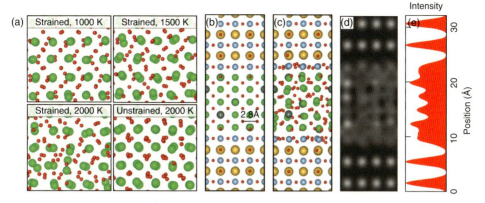

Figure 32.15 (a) Bulk zirconia structures resulting from quantum mechanical simulated annealing at various temperatures. O atoms are shown as small red spheres, and Zr atoms in green. The structures labeled as strained have been expanded 7% in the plane of the page. (b) Model of a strained 1 nm YSZ layer with an ordered O sublattice sandwiched between STO, viewed such that the STO is seen down the ⟨110⟩ direction. Sr atoms are shown in yellow, Ti in blue, and Y in gray. (c) Structure of the YSZ/STO multilayer layer at 360 K as determined by DFT MD calculations. The blurring out of the O EELS intensity caused by the O disorder is illustrated in (c) with a multislice simulation of the O K edge elemental map. The integrated intensity of each column of pixels in the simulated elemental map is shown in (d) in arbitrary units. (Adapted from Refs. [109, 110b].)

In an effort to uncover the origin of the colossal ionic conductivity, the information from the STEM and X-ray experiments was used as a starting point for DFT calculations [108, 109]. Finite temperature quantum molecular dynamics (MD) simulations were performed for both unstrained and 7% strained bulk zirconia. It was found that the large 7% strain completely changes the O sublattice (Figure 32.15a). At temperatures above 1000 K, the O atoms become increasingly disordered, until by 2000 K, the O positions appear completely random. Below 1000 K, the O atoms order into zigzags. However, simulating the full YSZ/STO multilayer with a 1 nm thick 8 mol% yttria YSZ layer, it was found that the disordered O sublattice phase persists in the YSZ down to at least room temperature (Figure 32.15c). The cation lattices remain coherent, but the STO and low-temperature YSZ O sublattices are incommensurate, perturbing the weaker YSZ O sublattice into its disordered phase [109].

By monitoring the mean square displacements of the zirconia O atoms between 1000 and 2500 K, it was clear that the disordered O atoms were far more mobile. From these calculations, it was possible to extract an energy barrier of 0.4 ± 0.1 eV for ionic conductivity in strained bulk zirconia, close to the 0.6 eV barrier obtained experimentally for the colossal ionic conducting YSZ, and far lower than the 1.1 eV unstrained bulk energy barrier. Furthermore, from these theoretical results, it was possible to estimate the ionic conductivity of the multilayer YSZ to be over six orders of magnitude higher than that of unstrained bulk YSZ at 500 K, comparable to the eight orders reported experimentally. It was concluded that the origin of colossal ionic conductivity is a combination of large expansive strain opening up the distances between cations, and the incompatibility O sublattices in YSZ and STO perturbing the YSZ into the disordered O phase at low temperatures.

In light of the theoretical findings, state-of-the-art STEM was employed to search for disorder in the O sublattice in the YSZ layers of coherent YSZ/STO interfaces [110]. In contrast to the sharp features seen in bulk YSZ, EELS spectra show the O K edge fine structure in the YSZ of coherent YSZ/STO multilayers to be blurred out. DFT O K edge fine structure simulations using the $Z+1$ approximation [86] show that this is exactly what one would expect if the O sublattice becomes random. The experimental and simulated fine structures are compared in Figure 32.16.

More impressively, it is actually possible to see O disorder in coherent YSZ/STO multilayers with EELS. We have reached a point in nanoscopy in which we are confident enough about what we should see with EELS elemental mapping that when we cannot resolve something, it can actually be informative. STO O sublattices are now routinely resolved with EELS maps on the UltraSTEM. Ordered bulk YSZ O sublattices can also be resolved with relative ease (Figure 32.17). If YSZ is coherently strained to match the STO in YSZ/STO multilayers, and the YSZ O atoms are ordered, they would have the same 2.76 Å in-plane separation as in STO, when viewed down the ⟨110⟩ STO orientation (Figure 32.15b). The multilayer YSZ O columns should, therefore, be even more easily resolved than in the bulk, if they are ordered. Instead, EELS mapping of the interfaces, Figure 32.18, shows the

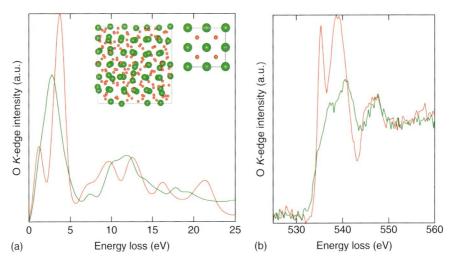

Figure 32.16 (a) EELS O K edge fine structure simulated using DFT and the $Z+1$ approximation for bulk ordered cubic ZrO_2 (red) and the 2000 K strained bulk structure with the extremely disordered O sublattice (green). The energy loss is taken to be zero at the Fermi energy. Both spectra were normalized to the intensities integrated over the first 50 eV and broadened with a 0.5 eV Gaussian. It is seen that the disorder causes the fine structure to blur out. (b) Experimental O K edge fine structure from a thin YSZ layer in a coherent section of a YSZ/STO multilayer (green) and from bulk cubic YSZ (red). (Adapted from Ref. [110].)

32.5 Applications of Ionic Conductors: Studies of Colossal Ionic Conductivity in Oxide Superlattices

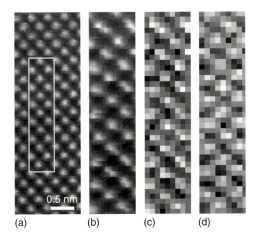

Figure 32.17 (a) High-resolution ADF image of bulk YSZ viewed down the ⟨100⟩ axis. A SI was acquired from the area indicated with the white box simultaneously with the ADF image shown in (b). The Zr $M_{4,5}$ and O K edge integrated intensities of the PCA-processed SI are shown as a function of position in (c) and (d) respectively. (Adapted from Ref. [110b].)

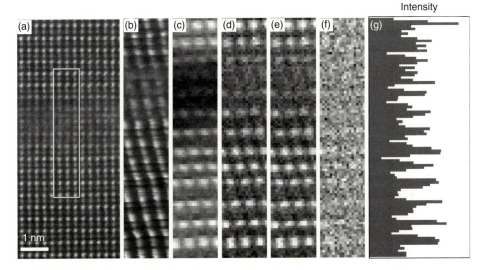

Figure 32.18 (a) High-resolution ADF image of a section of coherent YSZ/STO multilayers viewed down the ⟨110⟩ STO axis. An ADF image recorded simultaneously with an SI taken in the area indicated by the white box is shown in (a). Integrated Ti L and O K edge intensity elemental maps extracted from the PCA-processed SI are shown in (c) and (d) and as a composite map in (e) in which O intensity is shown in white and Ti intensity is shown in black. An O elemental map extracted from the SI without PCA processing is shown in (f), and the integrated intensity of each row of pixels in the raw O map is shown in (g). The intensity is given in arbitrary units, with the background subtracted to improve the contrast. (Adapted from Ref. [110].)

well-resolved O columns of the STO dissolve into just a blur in the YSZ. The ability to actually see the O disorder along with the fine structure evidence for disorder strongly support the conclusion that the origin of colossal ionic conductivity is a combination of strain and O sublattice incompatibility, resulting in a highly mobile disordered YSZ O sublattice. In summary, the disorder in the O sublattice induced near the interfaces by the large epitaxial strain in the system is responsible for the colossal values of the ionic conductivity measured in the system.

32.6
Applications of Cobaltites: Spin-State Mapping with Atomic Resolution

In this section, we present an example of the application of atomic-resolution imaging and EELS to the study of a quantity of direct interest in magnetism, the spin state of magnetic atoms. We focus on the perovskite-type doped lanthanum cobaltite system, $La_xSr_{1-x}CoO_3$ (LSCO), which is relevant for a vast array of materials applications ranging from batteries to catalysts, thermoelectrics, electronics, spintronics, and so on. EEL near-edge structure has been used to measure changes in the electronic structure in bulk LSCO and their evolution as a function of doping x. As in other transition metal perovskites, in cobaltites the Co-3d and the O-2p states lie close to the Fermi level; as a result, the unoccupied part of these states can be investigated by exciting transitions from Co-2p and O-1s levels. The intensity ratio of the L_3 and L_2 lines of the Co $L_{2,3}$ edge (the $L_{2,3}$ ratio) is known to correlate with the oxidation state of Co [111–116]. Figure 32.19a shows a set of Co $L_{2,3}$ edges after background subtraction from a variety of LSCO with increasing values of Sr doping ($x = 0$, 0.15, 0.3, 0.4, 0.5). The spectra in Figure 32.19a have been normalized to the L_3 line intensity to enable direct visual comparison of the data. All spectra exhibit a rather similar profile with the L_3 around 779 eV and the L_2 around 794 eV. Notice that L_2 decreases in intensity with Sr content, although slightly. Sr doping progressively increases the $L_{2,3}$ ratio, as shown in the bottom panel.

The O K edges of these cobaltites exhibit a pre-peak feature and two main peaks (as the manganites and titanates reviewed so far). This near-edge fine structure displays clear changes with Sr content (Figure 32.19b). The undoped $LaCoO_3$ (LCO) compound exhibits a pre-peak around 529 eV that is attributed to the hybridization of the O 2p with Co 3d states. Hence, it is typically referred to as the *hybridization peak*. With the introduction of Sr and the ensuing increasing hole doping, this hybridization peak decreases in intensity, while another spectral contribution increases near the edge onset because of hole states with oxygen character (reaching 527 eV for $x = 0.5$). In the literature, this peak is designated as the hole peak. The features near 535 and 542 eV are related to the bonding between the O 2p with the La 5d (or the Sr 4d) and the Co 4sp bands, respectively [112, 113]. The bottom panel of Figure 32.19b shows the relative intensities of both the hybridization and the hole peaks with Sr content, x. Clear variations that can be used as external calibrations to extract information related to the occupations of these bands and related electronic properties in other systems, such as thin films,

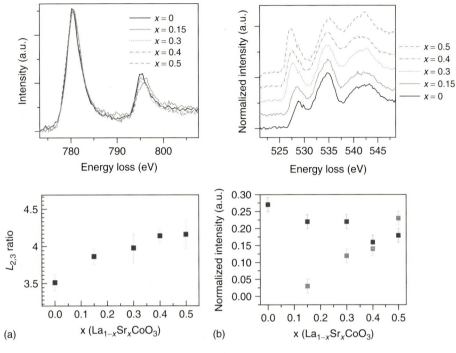

Figure 32.19 (a) Co $L_{2,3}$ edges for a series of LSCO compounds with $x = 0$ (black), $x = 0.13$ (gray), $x = 0.3$ (black dotted line), $x = 0.4$ (dashed-dotted line), and $x = 0.5$ (dashed line). The energy scale has been shifted and the intensity normalized so the L_3 lines match. The bottom panel shows the dependence of the $L_{2,3}$ ratio with the formal oxidation state. (b) O K edges for the series of LSCO compounds (same color code). Spectra have been normalized and displaced vertically for clarity. The bottom panel shows the normalized pre-peak intensity versus Sr content, x, of the hole peak (gray squares) and the hybridization peak (in black). All these spectra have been acquired through illumination of wide sample areas. Data from the aberration-corrected VG Microscopes HB501UX operated at 100 kV. (Samples courtesy of Professor C. Leighton's group at the University of Minnesota. Adapted from Refs. [116, 117].)

are observed. This way, spatially resolved measurements of quantities such as hole doping are possible. Recently, it has been shown that by tracking the depth-wise changes of the complex O K edge pre-peak in $La_{0.5}Sr_{0.5}CoO_{3-\delta}$ (LSCO 0.5) thin films grown on (100) STO, the degradation in magnetization and conductivity in the very thin limit can be understood [117–120]. Depth-dependent inhomogeneities in the La/Sr and O concentrations ensuing from the growth conditions have been shown to underlie the depth dependence of the hole doping in these films (Figure 32.20), and to explain the magnetoelectronic phase separation observed near the substrate interface in these LSCO 0.5 films [117].

In some cases, information about the spin state of Co atoms can be also extracted from the O K edge fine structure. This is an important task, since the rich variety of physical properties shown by the cobaltites is related to the variable spin state

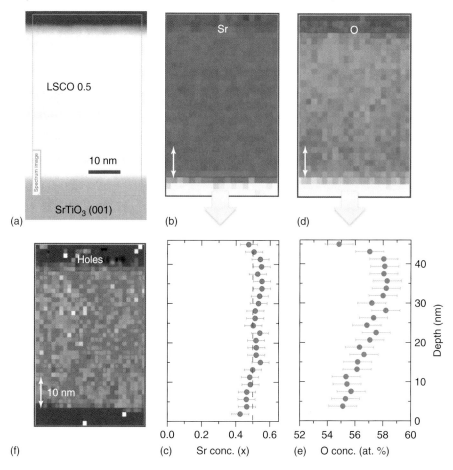

Figure 32.20 (a) Z-contrast STEM image of an LSCO 0.5 film grown on an STO substrate. The rectangle marks the region were a spectrum image was acquired. (b) Sr concentration map, from the Sr $L_{2,3}$ edge, along with depth-dependent laterally averaged profile (c). (d) O elemental map, extracted from the O K edge, along with averaged profile (e). (f) Hole peak map, showing a depleted hole density near the substrate interface. (Adapted from Ref. [117].)

of Co. In these compounds, the spin state of the individual Co ions is determined by the competition between the crystal field splitting and Hund's rule exchange energy in the 3d states [7, 121–134]. Different factors such as temperature, epitaxial strain, or a changing environment of the Co atom might lead to changes in its spin state. Furthermore, it is known that certain bulk Co oxides may present ordered spin states, as determined from neutron scattering, X-ray diffraction, and other techniques that either provide reciprocal-space information or probe large real-space areas [124–126]. Unfortunately, these techniques may not be sensitive to small domains of nanometer dimensions in inhomogeneous systems such as thin films, where chemical inhomogeneities, such as the one shown in Figure 32.20, are likely. Interestingly, the pre-peak intensity in O K EELS spectra has also been proved to be sensitive to the spin state of Co atoms in the undoped parent compound LCO [135]. Atomic-resolution STEM can, therefore, be a very useful tool to spatially map the local spin state of Co in similar compounds with atomic resolution [136] – next, we will examine this possibility.

Unlike other perovskites such as manganites, substituting Sr^{2+} for La^{3+} in LCO results not only in an increase in the average cobalt valence state but also in the formation of oxygen vacancies (for this reason, cobaltites are useful in ionic conductor applications too). At large Sr concentrations, oxygen vacancies are found to interact and form ordered structures with local crystal arrangements similar to the well-known Brownmillerite structure [114, 115, 136]. Furthermore, these ordered phases play an important role in the release of epitaxial strain in LSCO 0.5 thin films. Since this release can take place in an inhomogeneous manner, O vacancy ordered phases have been observed to form nanometer-sized domains with local fluctuations of the O content on this length scale [118, 119]. Figure 32.21a shows an example. The superstructure ensuing from O vacancy ordering is responsible for the contrast observed on the image: every other Co–O plane displays a dimmer contrast in the Z-contrast image; see the magnified view in Figure 32.21b. Some nanometer-size pockets in the films were found to have particularly low oxygen contents. Bulk LSCO 0.5 does not exhibit any Co spin-state ordering, but these heavily deoxygenated nanopockets show a novel spin-state superstructure, with alternating HS and LS Co ions in planes containing O vacancies versus fully oxygenated ones [137].

Spectrum imaging in such nanodomains allows one to study the electronic structure along the different Co–O planes of the superstructure. By examining the fine structure of the Co L and O K edges along different Co–O planes, the effects of the distribution of oxygen vacancies on the local properties of those Co atoms can be addressed. Figure 32.21c,d depicts the Co $L_{2,3}$ and O K edges averaged along the dark (in black) and bright (in gray) Co–O stripes, respectively, marked with matching color arrows on Figure 32.21b. It is noticeable how the Co $L_{2,3}$ ratio remains unchanged when shifting the electron beam from the bright to the dark Co–O stripe, implying that there is no change in Co valence along the different Co–O planes (i.e., no charge ordering). The average Co valence, quantified from the $L_{2,3}$ ratio, is close to Co^{2+}. Co atoms are thus significantly reduced in this nanopocket, as one would expect from an unusually low local O content.

On the other hand, the O K edge shows a very significant decrease in the pre-peak intensity along the dark stripe (Figure 32.21d). Since this pre-edge feature is related to the filling of the hybridized O-2p and Co-3d states, the observed change might indicate a different filling of the Co-3d band and a changing population of the O-2p bands when comparing the dark and bright Co–O stripes. However, the fact that the Co $L_{2,3}$ ratio does not change significantly from plane to plane rules out this scenario. Previous studies showed that changes in the net spin of Co also have a fingerprint in the near-edge fine structure of O K edges. In the case of bulk LCO, where no changes of Co average oxidation state with temperature are expected,

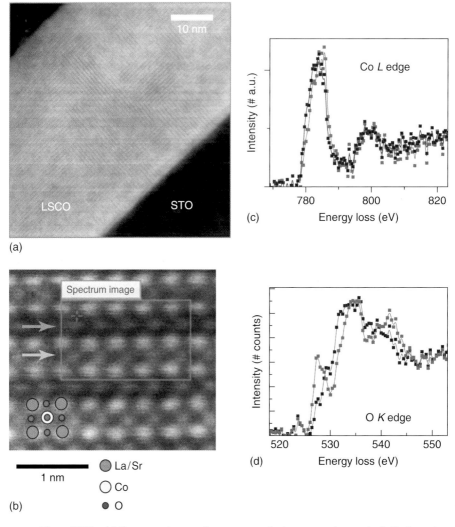

Figure 32.21 (a) Z-contrast image of a $La_{0.5}Sr_{0.5}Co_3O_{3-\delta}$ [001] thin film showing the superstructure and the domain structure. (b) Superstructure at higher magnification. The rectangle highlights the window for the acquired spectrum image. (c,d) Co $L_{2,3}$ edge and O K edges averaged along the dark (black, top arrow) and bright (grey, bottom arrow) Co–O planes of the superstructure, respectively. (Adapted from Ref. [137].)

changes have been observed in the O K edge pre-peak intensity (but not in the Co $L_{2,3}$ ratio) when cooling the compound through the 80 K spin-state transition [116]. In this work, the higher pre-peak intensity was found to correspond to the lowest spin state. Here, a similar behavior is found when probing different atomic planes, pointing to different spin states along the different bright/dark Co–O planes.

DFT calculations have proved useful to interpret these findings. Simulations of the O K edges of the O species along both bright and dark stripes – both the

32.6 Applications of Cobaltites: Spin-State Mapping with Atomic Resolution

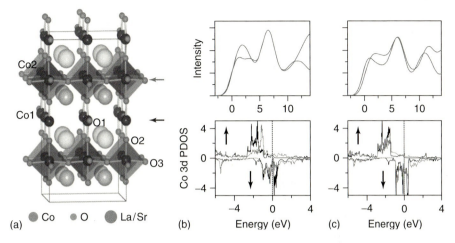

Figure 32.22 (a) Sketch of the $La_{0.5}Sr_{0.5}CoO_{2.25}$ structure, derived from the one reported by Wang and Yin [114] but with different location of oxygen vacancies. (b) Simulated O K edges for the O atoms in the bright (gray) and dark (black) stripes for the $La_{0.5}Sr_{0.5}CoO_{2.25}$ compound before relaxing the DFT calculation. Bottom panel, Co 3d projected densities of states for Co1 (in black) and Co2 (in gray) for the structure in (a). (c) Simulated O K edges for the O atoms in the bright (gray) and dark (black) stripes for the $La_{0.5}Sr_{0.5}CoO_{2.25}$ compound after structure relaxation. Bottom panel, Co 3d projected densities of states for Co1 (in black) and Co2 (in gray) for the relaxed structure. Vertical dashed lines mark the Fermi level position. Black arrows show the spin-up and spin-down channels in (b,c). (Adapted from Ref. [137].)

O1 and the O3 positions – of a heavily deoxygenated $La_{0.5}Sr_{0.5}CoO_{2.25}$ structure (Figure 32.22a), are shown in Figure 32.22b. A noticeably higher pre-peak (arrow) on the O K edge is reproduced on the bright stripe, in good agreement with the experimental data. For this structure, the projected densities of states (PDOS) on both Co species are shown in the bottom panel. The total 3d band occupations for Co1 and Co2 are the same (the sum of the Co 3d orbital occupancies is near 7.3 for both species), consistent with the Co $L_{2,3}$ EELS data. But the PDOS on the Co1 and Co2 positions show noticeably different occupations of the spin-up and spin-down channels for the 3d bands. While the Co2 configuration is consistent with a LS state, Co1 is in a HS state, with three down channels mostly empty. The calculations show that the structural changes associated with the ordering of O vacancies can indeed stabilize a superlattice in the spin state of Co, and that the ensuing changes in the EELS fine structure are consistent with the experimental observation [137].

It is worth noting that such a vacancy superstructure is only stabilized in the films by epitaxial strain, and also that the strain state in these films is not homogeneous, that is, they are in the partially relaxed state. Further theoretical support for this idea comes from allowing the structure in Figure 32.22a to fully relax. Then, a different state is achieved; the PDOS of Co1 and Co2 are no longer consistent with a spin-state ordering (see the bottom panel of Figure 32.22c). Theoretical simulations of the O K edges for the relaxed structure in Figure 32.22c no longer show the large difference in pre-peak structure from Figure 32.21 . These calculations confirm that the modulation of the O K edge fine structure observed in the EELS data arises from a spin-state superlattice, and that the physics of these systems is extremely sensitive to local structural rearrangements and strain.

32.7
Summary

Complex oxides form a very hot field at the forefront of materials physics, since they exhibit some of the most elusive physical behaviors, such as CMR or HTSC. Most of these behaviors ensue from the local crystal chemical and electronic structure of materials. Real-space probes capable of looking at matter with atomic resolution are key to understanding and hence harnessing the properties of these materials. In this chapter, we have shown how imaging and spectroscopy in the aberration-corrected scanning transmission microscope, combined with theoretical calculations, can unravel phenomena such as colossal ionic conductivity, identify spin-state superlattices, or map the minimal lattice distortions of light O atoms near interfaces that drastically change the system's macroscopic properties. The combination of STEM, EELS, and atomic-scale quantum mechanical calculations, therefore, represents a unique and extremely powerful tool for resolving the complex mysteries of oxide materials.

Since this is a fast evolving field, new exciting findings can only be expected with the advent of new developments such as monochromated guns or new advances in magnetic imaging techniques such as electron chiral dichroism at the nanometer scale, just to name a few. There is, however, a need for a better understanding of the physics of diffraction so we can model the propagation of the coherent electron probe and the excitation of EELS fine structure in a unified theory to be able to separate diffraction effects from changes in orbital occupation. With this theory, we would be able to extract the maximum quantitative information on localized electronic structure at interfaces in a large range of novel systems. In this context, complex oxides – in bulk or reduced dimensionality environments – represent a most exciting field with numerous open fronts. We have the tools and we know the issues. It is time to explore!

Acknowledgments

Research supported by the U.S. Department of Energy, Office of Basic Energy Sciences, Materials Sciences and Engineering Division (SJP, AYB, and MV), and by the European Research Council Starting Investigator Award (JG). The authors are grateful to all the collaborators who made this work possible over the years, especially to Rongying Jin, Julia Luck, Weidong Luo, Andrew Lupini, Hyejung Chang, Sergei Kalinin, David Mandrus, Sergio Molina, Mark Oxley, Micah Prange, Jing Tao, Masashi Watanabe, Manish Sharma, Maria A. Torija, Chris Leighton,

Zouhair Sefrioui, Cristina Visani, Vanessa Peña, Javier García-Barriocanal, Flavio Bruno, Carlos Leon, and Jacobo Santamaria. This manuscript has been authored by UT-Battelle, LLC, under Contract No. DE-AC05-00OR22725 with the U.S. Department of Energy. The United States Government retains and the publisher, by accepting the article for publication, acknowledges that the United States Government retains a non-exclusive, paid-up, irrevocable, world-wide license to publish or reproduce the published form of this manuscript, or allow others to do so, for United States Government purposes.

References

1. Bednorz, J.G. and Mueller, K.A. (1986) Possible high Tc superconductivity in the Ba-La-Cu-O system. *Z. Phys. B*, **64**, 189–193.
2. Wu, M.K., Ashburn, J.R., Torng, C.J., Hor, P.H., Meng, R.L., Gao, L., Huang, Z.J., Wang, Y.Q., and Chu, C.W. (1987) Superconductivity at 93K in a new mixed-phase Y-Ba-Cu-O compound system at ambient pressure. *Phys. Rev. Lett.*, **58**, 908–910.
3. Nücker, N., Fink, J., Fuggle, J.C., Durham, P.J., and Temmerman, W.M. (1988) Evidence for holes on oxygen sites in the high-Tc superconductors $La_{2-x}Sr_2CuO_4$ and $YBa_2Cu_3O_{7-y}$. *Phys. Rev. B*, **37**, 5158–5163.
4. Tinkham, M. (1996) *Introduction to superconductivity*, McGraw-Hill, New York.
5. Goodenough, J.B. (1955) Theory of the role of covalence in the perovskite type manganites [La, M(II)]MnO_3. *Phys. Rev.*, **100**, 564–573.
6. Dagotto, E., Hotta, T., and Moreo, A. (2001) Colossal magnetoresistant materials: the key role of phase separation. *Phys. Rep.*, **344**, 1–153.
7. Imada, M., Fujimori, A., and Tokura, Y. (1998) Metal insulator transitions. *Rev. Mod. Phys.*, **70**, 1039–1263.
8. Steele, B.C.H. and Heinzel, A. (2001) Materials for fuel-cell technologies. *Nature*, **414**, 345–352.
9. Goodenough, J.B. (2003) Oxide-ion electrolytes. *Annu. Rev. Mater. Res.*, **33**, 91–128.
10. Ormerod, R.M. (2003) Solid oxide fuel cells. *Chem. Soc. Rev.*, **32**, 17–28.
11. Garcia-Barriocanal, J., Rivera-Calzada, A., Varela, M., Sefrioui, Z., Iborra, E., Leon, C., Pennycook, S.J., and Santamaria, J. (2008) Colossal ionic conductivity at interfaces of epitaxial ZrO_2:Y_2O_3/$SrTiO_3$ heterostructures. *Science*, **321**, 676–680.
12. Ahn, C.H., Rabe, K.M., and Triscone, J.-M. (2004) Ferroelectricity at the nanoscale: local polarization in oxide thin films and heterostructures. *Science*, **303**, 488–491.
13. Dawber, M., Rabe, K.M., and Scott, J.F. (2005) Physics of thin-film ferroelectric oxides. *Rev. Mod. Phys.*, **77**, 1083–1130.
14. Lee, H.N., Christen, H.M., Chisholm, M.F., Rouleau, C.M., and Lowndes, D.H. (2005) Strong polarization enhancement in asymmetric three-component ferroelectric superlattices. *Nature*, **433**, 395–399.
15. Maksymovych, P., Jesse, S., Yu, P., Ramesh, R., Baddorf, A.P., and Kalinin, S.V. (2009) Polarization control of electron tunneling into ferroelectric surfaces. *Science*, **324**, 1421–1425.
16. Garcia, V., Bibes, M., Bocher, L., Valencia, S., Kronast, F., Crassous, A., Moya, X., Enouz-Vedrenne, S., Gloter, A., Imhoff, D., Deranlot, C., Mathur, N.D., Fusil, S., Bouzehouane, K., and Barhelemy, A. (2010) Ferroelectric control of spin polarization. *Science*, **327**, 1106–1110.
17. Bibes, M. and Barthelemy, A. (2011) Ultrathin oxide films and interfaces for electronics and spintronics. *Adv. Phys.*, **60**, 5–84.

18. Bibes, M. and Barthelemy, A. (2007) Oxide spintronics. *IEEE Trans. Electron Devices*, **54**, 1003–1023.
19. Bea, H., Gajek, M., Bibes, M., and Barthelemy, A. (2008) Spintronics with multiferroics. *J. Phys.: Condens. Matter*, **20**, 434221.
20. Reiner, J.W., Walker, F.J., and Ahn, C.H. (2009) Atomically engineered oxide interfaces. *Science*, **323**, 1018–1019.
21. Manhart, J. and Schlom, D.G. (2010) Oxide interfaces – an opportunity for electronics. *Science*, **327**, 1607–1611.
22. Kung, H.H. (1989) *Transition Metal Oxides: Surface Chemistry and Catalysis*, Elsevier.
23. Maekawa, S., Tohyama, T., Barnes, S.E., Ishihara, S., Koshibae, W., and Khaliullin, G. (2004) *Physics of Transition Metal Oxides*, Springer.
24. Sparrow, T., Williams, B., Rao, C., and Thomas, J. (1984) L_3/L_2 white-line intensity ratios in the electron energy-loss spectra of $3d$ transition-metal oxides. *Chem. Phys. Lett.*, **108**, 547–550.
25. Waddington, W.G., Rez, P., Grant, I.P., and Humphreys, C.J. (1986) White lines in the $L_{2,3}$ electron-energy-loss and X-ray absorption spectra of $3d$ transition metals. *Phys. Rev. B*, **34**, 1467–1473.
26. Rask, J.H., Miner, B.A., and Buseck, P.R. (1987) Determination of manganese oxidation states in solids by electron energy-loss spectroscopy. *Ultramicroscopy*, **21**, 321–326.
27. Brydson, R., Sauer, H., Engel, W., Thomas, J.M., Zeitler, E., Kosugi, N., and Kuroda, H. (1989) Electron energy loss and X-ray absorption spectroscopy of rutile and anatase: a test of structural sensitivity. *J. Phys.: Condens. Matter*, **1**, 797–812.
28. Krivanek, O.L. and Paterson, J.H. (1990) ELNES of $3d$ transition-metal oxides I. Variations across the periodic table. *Ultramicroscopy*, **32**, 313–318.
29. Paterson, J.H. and Krivanek, O.L. (1990) ELNES of $3d$ transition-metal oxides II. Variations with oxidation state and crystal structure. *Ultramicroscopy*, **32**, 319–325.
30. Cramer, S.P., de Groot, F.M.F., Ma, Y., Chen, C.T., Sette, F., Kipke, C.A., Eichhorn, D.M., Chan, M.K., Armstrong, W.H., Libby, E., Christou, G., Brooker, S., McKee, V., Mullins, O.C., and Fuggle, J.C. (1991) Ligand field strengths and oxidation states from manganese L-edge spectroscopy. *J. Am. Chem. Soc.*, **113**, 7937–7940.
31. Kurata, H. and Colliex, C. (1993) Electron-energy-loss core-edge structures in manganese oxides. *Phys. Rev. B*, **48**, 2102–2108.
32. Van Aken, P. and Leibscher, B. (2002) Quantification of ferrous/ferric ratios in minerals: new evaluation schemes of Fe L_{23} electron energy-loss near edge spectra. *Phys. Chem. Miner.*, **29**, 188–200.
33. Mitterbauer, C., Kothleitner, G., Grogger, W., Zandbergen, H., Freitag, B., Tiemeijer, P., and Hofer, F. (2003) Electron energy-loss near edge structures of $3d$ transition-metal oxides recorded at high energy resolution. *Ultramicroscopy*, **96**, 469–480.
34. Riedl, T., Gemming, T., Gruner, W., Acker, J., and Wetzig, K. (2007) Determination of manganese valency in $La_{1-x}Sr_xMnO_3$ using ELNES in the (S)TEM. *Micron*, **38**, 224–230.
35. Varela, M., Luo, W., Tao, J., Oxley, M.P., Watanabe, M., Lupini, A.R., Pantelides, S.T., and Pennycook, S.J. (2009) Atomic resolution imaging of oxidation states in manganites. *Phys. Rev. B.*, **79**, 085117.
36. Browning, N.D., Yuan, J., and Brown, L.M. (1993) Determination of local oxygen stoichiometry in $YBa_2Cu_3O_{7-y}$ by electron energy loss spectroscopy in the scanning transmission electron microscope. *Phys. C*, **202**, 12–18.
37. Kurata, H., Lefevre, E., Colliex, C., and Brydson, R. (1993) Electron-energy-loss near-edge structures in the oxygen K-edge spectra of transition-metal oxides. *Phys. Rev. B*, **47**, 13763–13768.
38. Luo, W., Varela, M., Tao, J., Pennycook, S.J., and Pantelides, S.T. (2009) Electronic and crystal-field effects in the fine structure of electron energy-loss spectra of manganites. *Phys. Rev. B.*, **79**, 052405.

39. Crewe, A.V., Wall, J., and Langmore, J. (1970) Visibility of single atoms. *Science*, **168**, 1338–1340.
40. Pennycook, S.J. and Jesson, D.E. (1990) High-resolution incoherent imaging of crystals. *Phys. Rev. Lett.*, **64**, 938–941.
41. Browning, N.D., Chisholm, M.F., and Pennycook, S.J. (1993) Atomic-resolution chemical analysis using a scanning transmission electron microscope. *Nature*, **366**, 143–146.
42. Nellist, P.D. and Pennycook, S.J. (1998) Subangstrom resolution by under-focused incoherent transmission electron microscopy. *Phys. Rev. Lett.*, **81**, 4156–4159.
43. Varela, M., Lupini, A., Christen, H.M., Dellby, N., Krivanek, O.L., Nellist, P.D., and Pennycook, S.J. (2004) Spectroscopic identification of single atoms within a bulk solid. *Phys. Rev. Lett.*, **92**, 095502.
44. Varela, M., Lupini, A.R., van Benthem, K., Borisevich, A.Y., Chisholm, M.F., Shibata, N., Abe, E., and Pennycook, S.J. (2005) Materials characterization in the aberration corrected scanning transmission electron microscope. *Annu. Rev. Mater. Res.*, **35**, 539–569.
45. Klie, R.F., Buban, J.P., Varela, M., Franceschetti, A., Jooss, C., Zhu, Y., Browning, N.D., Pantelides, S.T., and Pennycook, S.J. (2005) Enhanced current transport at grain boundaries in high-Tc superconductors. *Nature*, **435**, 475–478.
46. Pennycook, S.J., Chisholm, M.F., Lupini, A.R., Varela, M., van Benthem, K., Borisevich, A.Y., Oxley, M.P., Luo, W., and Pantelides, S.T. (2008) Materials applicatine of aberration-corrected scanning transmission electron microscopy. *Adv. Imaging Electron Phys.*, **153**, 327–384.
47. Pennycook, S.J., Chisholm, M.F., Lupini, A.R., Varela, M., Borisevich, A.Y., Oxley, M.P., Luo, W.D., van Benthem, K., Oh, S.-H., Sales, D.L., Molina, S.I., Garcia-Barriocanal, J., Leon, C., Santamaria, J., Rashkeev, S.N., and Pantelides, S.T. (2009) Aberration-corrected scanning transmission electron microscopy: from atomic imaging and analysis to solving energy problems. *Philos. Trans. R. Soc. A*, **367**, 3709–3733.
48. Van Aert, S., Verbeeck, J., Erni, R., Bals, S., Luysberg, M., Van Dyck, D., and Van Tendeloo, G. (2009) Quantitative atomic resolution mapping using high-angle annular dark field scanning transmission electron microscopy. *Ultramicroscopy*, **109**, 1236–1244.
49. Batson, P.E., Dellby, N., and Krivanek, O.L. (2002) Sub-Ångström resolution using aberration corrected electron optics. *Nature*, **418**, 617–620.
50. Nellist, P.D., Chisholm, M.F., Dellby, N., Krivanek, O.L., Murfitt, M.F., Szilagyi, Z.S., Lupini, A.R., Borisevich, A.Y., Sides, W.H., and Pennycook, S.J. (2004) Direct sub-Ångström imaging of a crystal lattice. *Science*, **305**, 1741.
51. Voyles, P.M., Muller, D.A., Grazul, J.L., Citrin, P.H., and Grossman, H.-J.L. (2002) Atomic scale imaging of individual dopant atoms and clusters in highly n-type bulk Si. *Nature*, **416**, 826–829.
52. Sohlberg, K.S., Rashkeev, S., Borisevich, A.Y., Pennycook, S.J., and Pantelides, S.T. (2004) Origin of the anomalous Pt-Pt distances in the Pt/alumina catalytic syste. *Chemphyschem*, **5**, 1893–1897.
53. Griffin-Roberts, K., Varela, M., Rashkeev, S., Pantelides, S.T., Pennycook, S.J., and Krishnan, K.M. (2008) Defect-mediated ferromagnetism in insulating Co-doped anatase TiO_2 thin films. *Phys. Rev. B*, **78**, 014409.
54. Bosman, M., Keast, V.J., Garcia-Muñoz, J.L., D'Alfonso, A.J., Findlay, S.D., and Allen, L.J. (2007) Two-dimensional mapping of chemical information at atomic resolution. *Phys. Rev. Lett.*, **99**, 086102.
55. Kimoto, K., Asaka, T., Nagai, T., Saito, M., Matsui, Y., and Ishizuka, K. (2007) Element-selective imaging of atomic columns in a crystal using STEM and EELS. *Nature*, **450**, 702–704.
56. Muller, D.A., Fitting-Kourkoutis, L., Murfitt, M., Song, J.H., Hwang, H.Y., Silcox, J., Dellby, N., and Krivanek, O.L. (2008) Atomic-scale chemical imaging of composition and bonding by aberration-corrected microscopy. *Science*, **319**, 1073–1076.

57. Ashcroft, N.W. and Mermin, N.D. (1976) *Solid State Physics*, Saunders College.
58. Egerton, R.F. (1996) *Electron Energy Loss Spectroscopy in the Electron Microscope*, Plenum Press, New York.
59. Jeanguillaume, C. and Colliex, C. (1989) Spectrum-image: the next step in EELS digital acquisition and processing. *Ultramicroscopy*, **28**, 252–257.
60. Hunt, J.A. and Williams, D.B. (1991) Electron energy-loss spectrum-imaging. *Ultramicroscopy*, **38**, 47–73.
61. Bosman, M., Watanabe, M., Alexander, D.T.L., and Keast, V.J. (2006) Mapping chemical and bonding information using multivariate analysis of electron-energy-loss spectrum images. *Ultramicroscopy*, **106**, 1024–1032.
62. Pennycook, S.J. and Varela, M. (2011) New views of materials through aberration-corrected STEM. *J. Electron Microsc.*, **60**, S213–S223.
63. Varela, M., Pennycook, T., Tian, W., Mandrus, D., Pennycook, S.J., Peña, V., Sefrioui, Z., and Santamaria, J. (2006) Atomic scale characterization of complex oxide interfaces. *J. Mater. Sci*, **41**, 4389–4393.
64. Kourkoutis, L.F., Song, J.H., Hwang, H.Y., and Muller, D.A. (2010) Microscopic origins for stabilizing room-temperature ferromagnetism in ultrathin manganite layers. *Proc. Natl. Acad. Sci.*, **90**, 4731–4749.
65. Pennycook, S.J. and P.D. Nellist (eds) (2011) *Scanning Transmission Electron Microscopy: Imaging and Analysis*, Springer, New York.
66. Krivanek, O.L., Chisholm, M.F., Nicolosi, V., Pennycook, T.J., Corbn, G.J., Dellby, N., Murfitt, M.F., Own, C.S., Szilagyi, Z.S., Oxley, M.P., Pantelides, S.T., and Pennycook, S.J. (2010) Atom-by-atom structural and chemical analysis by annular dark field electron microscopy. *Nature*, **464**, 571–574.
67. Jia, C.L., Lentzen, M., and Urban, K. (2003) Atomic resolution imaging of oxygen in perovskite ceramics. *Science*, **299**, 870.
68. Jia, C.L. and Urban, K. (2004) Atomic resolution measurement of oxygen concentration in oxide materials. *Science*, **303**, 2001–2004.
69. Shibata, N., Chisholm, M.F., Nakamura, A., Pennycook, S.J., Yamamoto, T., and Ikuhara, Y. (2007) Nonstoichiometric dislocation cores in α-alumina. *Science*, **316**, 82–85.
70. Findlay, S.D., Shibata, N., Sawada, H., Okunishi, E., Kondo, Y., Yamamoto, T., and Ikuhara, Y. (2009) Robust atomic resolution imaging of light elements using scanning transmission electron microscopy. *Appl. Phys. Lett.*, **95**, 191913.
71. Rodriguez-Carvajal, J., Hennion, M., Moussa, F., Moudden, A.H., Pinsard, L., and Revcolevschi, A. (1998) Neutron-diffraction study of the Jahn-Teller transition in stoichiometric $LaMnO_3$. *Phys. Rev. B*, **57**, R3189–R3192.
72. Oxley, M.P. and Allen, L.J. (1998) Delocalization of the effective interaction for inner-shell ionization in crystals. *Phys. Rev. B*, **57**, 3273–3282.
73. Allen, L.J., Findlay, S.D., Oxley, M.P., and Rossouw, C.J. (2003) Lattice-resolution contrast from a focused coherent electron probe. Part I. *Ultramicroscopy*, **96**, 47–63.
74. Ankudinov, A.L., Nesvizhskii, A.I., and Rehr, J.J. (2003) Dynamic screening effects in x-ray absorption. *Phys. Rev. B*, **67**, 115120.
75. Findlay, S.D., Allen, L.J., Oxley, M.P., and Rossouw, C.J. (2003) Lattice-resolution contrast from a focused coherent electron probe: part II. *Ultramicroscopy*, **96**, 65–81.
76. Findlay, S.D., Oxley, M.P., Pennycook, S.J., and Allen, L.J. (2005) Modelling imaging based on core-loss spectroscopy in scanning transmission electron microscopy. *Ultramicroscopy*, **104**, 126–140.
77. Oxley, M.P., Chang, H.J., Borisevich, A.Y., Varela, M., and Pennycook, S.J. (2010) Imaging of light atoms in the presence of heavy atomic columns. *Microsc. Microanal.*, **16**, 92.

78. Mitchell, R.G. (2002) *Perovskites – Modern and Ancient*, Almaz Press, Thunder Bay.
79. He, J., Borisevich, A., Kalinin, S.V., Pennycook, S.J., and Pantelides, S.T. (2010) Control of octahedral tilts and magnetic properties of perovskite oxide heterostructures by substrate symmetry. *Phys. Rev. Lett.*, **105**, 227203.
80. Jia, C.L., Nagarajan, V., He, J.Q., Houben, L., Zhao, T., Ramesh, R., Urban, K., and Waser, R. (2007) Unit-cell scale mapping of ferroelectricity and tetragonality in epitaxial ultra-thin ferroelectric films. *Nat. Mater.*, **6**, 64–69.
81. Jia, C.L., Mi, S.B., Urban, K., Vrejoiu, I., Alexe, M., and Hesse, D. (2008) Atomic-scale study of electric dipoles near charged and uncharged domain walls in ferroelectric films. *Nat. Mater.*, **7**, 57–61.
82. Jia, C.L., Mi, S.B., Faley, M., Poppe, U., Schubert, J., and Urban, K. (2009) Oxygen octahedron reconstruction in the $SrTiO_3/LaAlO_3$ heterointerfaces investigated using aberration-corrected ultrahigh-resolution transmission electron microscopy. *Phys. Rev. B*, **79**, 081405.
83. Borisevich, A.Y., Chang, H.J., Huijben, M., Oxley, M.P., Okamoto, S., Niranjan, M.K., Burton, J.D., Tsymbal, E.Y., Chu, Y.H., Yu, P., Ramesh, R., Kalinin, S.V., and Pennycook, S.J. (2010) *Phys. Rev. Lett.*, **105**, 087204.
84. Catalan, G. and Scott, J.F. (2009) Physics and applications of bismuth ferrite. *Adv. Mater.*, **21**, 2463–2485.
85. Garcia de Abajo, F.J. (2010) Optical excitations in electron microscopy. *Rev. Mod. Phys.*, **82**, 209–275.
86. Buczko, R., Duscher, G., Pennycook, S.J., and Pantelides, S.T. (2000) Excitonic effects in core-excitation spectra of semiconductors. *Phys. Rev. Lett.*, **85**, 2168–2171.
87. Garcia-Barriocanal, J., Bruno, F.Y., Rivera-Calzada, A., Sefrioui, Z., Nemes, N.M., Garcia-Hernández, M., Rubio-Zuazo, J., Castro, G.R., Varela, M., Pennycook, S.J., Leon, C., and Santamaría, J. (2010) "Charge leakage" at $LaMnO_3/SrTiO_3$ interfaces. *Adv. Mater.*, **22**, 627–632.
88. Zhai, X., Mohapatra, C.S., Shah, A.B., Zuo, J.-M., and Eckstein, J.N. (2010) *Adv. Mater.*, **22**, 1136–1139.
89. Shah, S.B., Ramasse, Q.M., Zhai, X., Wen, J.G., May, S.J., Petrov, I., Bhattacharya, A., Abbamonte, P., Eckstein, J.N., and Zuo, J.-M. (2010) Probing interfacial electronic structures in atomic layer $LaMnO_3$ and $SrTiO_3$ superlattices. *Adv. Mater.*, **22**, 1156–1160.
90. Ohtomo, A., Muller, D.A., Grazul, J.L., and Hwang, H.Y. (2002) Artificial charge-modulation in atomic-scale perovskite titanate superlattices. *Nature*, **419**, 378–380.
91. Okamoto, S. and Millis, A.J. (2004) Electronic reconstruction at an interface between a Mott insulator and a band insulator. *Nature*, **428**, 630–633.
92. Ohtomo, A. and Hwang, H.Y. (2004) A high-mobility electron gas at the $LaAlO_3/SrTiO_3$ heterointerface. *Nature*, **427**, 423–426.
93. Herranz, G., Basletic, M., Bibes, M., Carretero, C., Tafra, E., Jacquet, E., Bouzehouane, K., Deranlot, C., Hamzic, A., Broto, J.-M., Barthelemy, A., and Fert, A. (2007) High mobility in $LaAlO_3/SrTiO_3$ heterostructures: origin, dimensionality, and perspectives. *Phys. Rev. Lett.*, **98**, 216803.
94. Reyren, N., Thiel, S., Caviglia, A.D., Fitting Kourkoutis, L., Hammerl, G., Richter, C., Schneider, C.W., Kopp, T., Ruetschi, A.-S., Jaccard, D., Gabay, M., Muller, D.A., Triscone, J.-M., and Mannhart, J. (2007) Superconducting interfaces between insulating oxides. *Science*, **317**, 1196–1199.
95. Ariando, Wang, X., Baskaran, G., Liu, Z.Q., Huijben, J., Yi, J.B., Annadi, A., Barman, A.R., Rusydi, A., Dhar, S., Feng, Y.P., Ding, J., Hilgenkamp, H., Venkatesan, T. (2010) Electronic phase separation at the $LaAlO_3/SrTiO_3$ interface. *Nat. Commun.*, **2**, 188.
96. Chambers, S.A., Engelhard, M.H., Shutthanandan, V., Zhu, Z., Droubay, T.C., Qiao, L., Sushko, P.V., Feng, T., Lee, H.D., Gustafsson, T., Garfunkel, E., Shah, A.B., Zuo, J.-M., and

Ramasse, Q.M. (2010) Instability, intermixing and electronic structure of the epitaxial LaAlO$_3$/SrTiO$_3$ heterojunction. *Surf. Sci. Rep.*, **65**, 317–352.

97. Luysberg, M., Heidelmann, M., Houben, L., Boese, M., Heeg, T., Schubert, J., and Roeckerath, M. (2009) Intermixing and charge neutrality at DyScO$_3$/SrTiO$_3$ interfaces. *Acta Mater.*, **57**, 3192–3198.

98. Verbeeck, J., Bals, S., Kratsova, A.N., Lamoen, D., Luysberg, M., Hiujben, M., Rijnders, G., Brinkman, A., Hilgenkamp, H., Blank, D.H.A., and Van Tendeloo, G. (2010) Electronic reconstruction at n-type SrTiO$_3$/LaAlO$_3$ interfaces. *Phys. Rev. B*, **81**, 085113.

99. Sefrioui, Z., Arias, D., Villegas, J.E.M., Peña, V., Saldarriaga, W., Prieto, P., Leon, C., Martinez, J.L., and Santamaría, J. (2003) Ferromagnetic/superconducting proximity effect in LCMO/YBCO superlattices. *Phys. Rev. B*, **67**, 214511.

100. Gonzalez, I., Okamoto, S., Yunoki, S., Moreo, A., and Dagotto, E. (2008) Charge transfer in heterostructures of strongly correlated materials. *J. Phys.: Condens. Matter*, **20**, 264002.

101. Kilner, J.A. (2008) Ionic conductors: feel the strain. *Nat. Mater.*, **7**, 838–839.

102. Maier, J. (2000) Point-defect thermodynamics and size effects. *Solid State Ionics*, **131**, 13–22.

103. Sata, N., Eberman, K., Eberl, K., and Maier, J. (2000) Mesoscopic fast ion conduction in nanometre-scale planar heterostructures. *Nature*, **408**, 946–949.

104. Korte, C., Peters, A., Janek, J., Hesse, D., and Zakharov, N. (2008) Ionic conductivity and activation energy for oxygen ion transport in superlattices – the semicoherent multilayer system YSZ (ZrO$_2$ + 9.5 mol% Y$_2$O$_3$)/Y$_2$O$_3$. *Phys. Chem. Chem. Phys.*, **10**, 4623–4635.

105. Kosacki, I., Rouleau, C.M., Becher, P.F., Bentley, J., and Lowndes, D.H. (2005) Nanoscale effects on the ionic conductivity in highly textured YSZ thin film. *Solid State Ionics*, **176**, 1319–1326.

106. Cavallaro, A., Burriel, M., Roqueta, J., Apostolidis, A., Bernardi, A., Tarancon, A., Srinivasan, R., Cook, S.N., Fraser, H.L., Kilner, J.A., McComb, D.W., and Santiso, J. (2010) Electronic nature of the enhanced conductivity in YSZ-STO multilayers deposited by PLD. *Solid State Ionics*, **181**, 592–601.

107. Garcia-Barriocanal, J., Rivera-Calzada, A., Varela, M., Sefrioui, Z., Iborra, E., Leon, C., Pennycook, S.J., and Santamaria, J. (2009) Response to comment on "colossal ionic conductivity at interfaces of epitaxial ZrO$_2$:Y$_2$O$_3$/SrTiO$_3$ heterostructures". *Science*, **324**, 465.

108. Kohn, W. and Sham, L.J. (1965) Self-consistent equations including exchange and correlation effects. *Phys. Rev.*, **140**, A1133–A1138.

109. Pennycook, T.J., Beck, M.J., Varga, K., Varala, M., Pennycook, S.J., and Pantelides, S.T. (2010) Origin of colossal ionic conductivity in oxide multilayers: interface induced sublattice disorder. *Phys. Rev. Lett.*, **104**, 115901.

110. Pennycook, T.J., Oxley, M.P., Garcia-Barriocanal, J., Bruno, F.Y., Leon, C., Santamaria, J., Pantelides, S.T., Varela, M., and Pennycook, S.J. (2011) Seeing oxygen disorder in YSZ/SrTiO$_3$ colossal ionic conductor heterostructures using EELS. *Eur. Phys. J. Appl. Phys.*, **54**, 33507.

111. Kriener, M., Zobel, C., Reichl, A., Baier, J., Cwik, M., Berggold, K., Kierspel, H., Zabara, O., Freimuth, A., and Lorenz, T. (2004) Structure, magnetization and resistivity of La$_{1-x}$M$_x$CoO$_3$ (M=Ca, Sr and Ba). *Phys. Rev. B*, **69**, 094417.

112. Abbate, M., de Groot, F.M.F., Fuggle, J.C., Fujimori, A., Tokura, Y., Fujishima, Y., Strebel, O., Domke, M., Kaindl, G., van Elp, J., Thole, B.T., Sawatzky, G.A., Sacchi, M., and Tsuda, N. (1991) Soft-X-ray-absorption studies of the location of extra charges induced by substitution in controlled-valence materials. *Phys. Rev. B*, **44**, 5419–5422.

113. Abbate, M., de Groot, F.M.F., Fuggle, J.C., Fujimori, A., Strebel, O., Lopez, F., Domke, M., Kaindl, G., Sawatzky, G.A., Takano, M., Takeda,

Y., Eisaki, H., and Uchida, S. (1992) Controlled-valence properties of $La_{1-x}Sr_xFeO_3$ and $La_{1-x}Sr_xMnO_3$ studied by soft-X-ray absorption spectroscopy. *Phys. Rev. B*, **46**, 4511–4519.

114. Wang, Z.L. and Yin, J.S. (1998) Co valence and crystal structure of $La_{0.5}Sr_{0.5}CoO_{2.25}$. *Philos. Mag. B*, **77**, 49–65.

115. Wang, Z.L., Yin, J.S., and Jiang, Y.D. (2000) EELS analysis of cation valence states and oxygen vacancies in magnetic oxides. *Micron*, **31**, 571–580.

116. Gazquez, J., Luo, W., Oxley, M.P., Orange, M., Torija, M.A., Sharma, M., Leighton, C., Pantelides, S.T., Pennycook, S.J., and Varela, M. (2011) in preparation **23**, 2711–2715.

117. Torija, M.A., Sharma, M., Gazquez, J., Varela, M., He, C., Schmidt, J., Borchers, J.A., Laver, M., El-Khatib, S., and Leighton, C. (2011) Chemically-driven magnetic phase separation at the $SrTiO_3(001)/La_{1-x}Sr_xCoO_3$ interface. *Adv. Mater.*, **23**, 2711–2715.

118. Klenov, D.O., Donner, W., Foran, B., and Stemmer, S. (2003) Impact of stress on oxygen vacancy ordering in epitaxial $La_{0.5}Sr_{0.5}CoO_{3-\delta}$ thin films. *Appl. Phys. Lett.*, **82**, 3427–3429.

119. Torija, M.A., Sharma, M., Fitzsimmons, M.R., Varela, M., and Leighton, C. (2008) Epitaxial $La_{0.5}Sr_{0.5}CoO_3$ thin films: structure, magnetism, and transport. *J. Appl. Phys.*, **104**, 023901.

120. Varela, M., Gazquez, J., Lupini, A.R., Luck, J.T., Torija, M.A., Sharma, M., Leighton, C., Biegalski, M.D., Christen, H.M., Murfitt, M., Dellby, N., Krivanek, O.L., and Pennycook, S.J. (2010) Applications of aberration corrected scanning transmission electron microscopy and electron energy loss spectroscopy to thin oxide films and interfaces. *Int. J. Mater. Res.*, **101**, 21–26.

121. Raccah, P.M. and Goodenough, J.B. (1967) First-order localized-electron collective-electron transition in $LaCoO_3$. *Phys. Rev.*, **155**, 932–943.

122. Senaris-Rodriguez, M.A. and Goodenough, J.B. (1995) $LaCoO_3$ revisited. *J. Solid State Chem.*, **116**, 224–231.

123. Korotin, M.A., Ezhov, S.Y., Solovyev, I.V., Anisimov, V.I., Khomskii, D.I., and Sawatzky, G.A. (1996) Intermediate-spin state and properties of $LaCoO_3$. *Phys. Rev. B*, **54**, 5309–5316.

124. Doumerc, J.P., Coutanceau, M., Demourgues, A., Elkaim, E., Grenier, J.-C., and Pouchard, M.J. (2001) Crystal structure of the thallium strontium cobaltite $TlSr_2CoO_5$ and its relationship to the electronic properties. *Mater. Chem.*, **11**, 78–85.

125. Khomskii, D.I. and Low, U. (2004) Superstructures at low spin-high spin transitions. *Phys. Rev. B*, **69**, 184401.

126. Maignan, A., Caignaert, V., Raveau, B., Khomskii, D., and Sawatzky, G.A. (2004) Thermoelectric power of $HoBaCo_2O_{5.5}$: possible evidence of the spin blockade in cobaltites. *Phys. Rev. Lett.*, **93**, 026401.

127. Phelan, D., Louca, D., Rosenkranz, S., Lee, S.-H., Qiu, Y., Chupas, P.J., Osborn, R., Zheng, H., Mitchell, J.F., Copley, J.R.D., Sarrao, J.L., and Moritomo, Y. (2006) Nanomagnetic droplets and implications to orbital ordering in $La_{1-x}Sr_xCoO_3$. *Phys. Rev. Lett.*, **96**, 027201.

128. Podlesnyak, A., Streule, S., Mesot, J., Medarde, M., Pomjakushina, E., Conder, K., Tanaka, A., Haverkort, M.W., and Khomskii, D.I. (2006) Spin-state transition in $LaCoO_3$: direct neutron spectroscopic evidence of excited magnetic states. *Phys. Rev. Lett.*, **97**, 247208.

129. Khalyavin, D.D., Argyriou, D.N., Amman, U., Yaremchenko, A.A., and Kharton, V.V. (2007) Spin-state ordering and magnetic structures in the cobaltites $YBaCo_2O_{5+\delta}$ ($\delta=0.50$ and 0.44). *Phys. Rev. B*, **75**, 134407.

130. Podlesnyak, A., Russino, M., Alfonsov, A., Vavilova, E., Kataev, V., Buchner, B., Strassle, Th., Pomjakushina, E., Conder, K., and Khomskii, D.I. (2008) spin-state polarons in lightly hole doped $LaCoO_3$. *Phys. Rev. Lett.*, **101**, 247603.

131. He, C., Zheng, H., Mitchell, J.F., Foo, M.L., Cava, R.J., and Leighton, C. (2009) Low temperature Schottky anomalies in the specific heat of $LaCoO_3$: defect-stabilized finite spin states. *Appl. Phys. Lett.*, **94**, 102514.
132. Merz, M., Nagel, P., Pinta, C., Samartsev, A., Lohneysen, H.V., Wissinger, M., Uebe, S., Assmann, A., Fuchs, D., and Schuppler, S. (2010) X-ray absorption and magnetic circular dichroism of $LaCoO_3$, $La_{0.7}Ce_{0.3}CoO_3$ and $La_{0.7}Sr_{0.3}CoO_3$ films: evidence for cobalt-valence-dependent magnetism. *Phys. Rev. B*, **82**, 174416.
133. Herklotz, A., Rata, A.D., Schultz, L., and Dorr, K. (2009) Reversible strain effect on the magnetization of $LaCoO_3$ films. *Phys. Rev. B*, **79**, 092409.
134. Rata, A.D., Herklotz, A., Nenkov, K., Schultz, L., and Dorr, K. (2008) Strain-induced insulator state and giant gauge factor of $La_{0.7}Sr_{0.3}CoO_3$ films. *Phys. Rev. Lett.*, **100**, 076401.
135. Klie, R.F., Zheng, J.C., Zhu, Y., Varcla, M., Wu, J., and Leighton, C. (2007) Direct measurement of the low temperature spin-state transition in $LaCoO_3$. *Phys. Rev. Lett.*, **99**, 047203.
136. Ito, Y., Klie, R.F., Browning, N.D., and Mazanec, T.J. (2002) Atomic resolution analysis of the defect chemistry at domain boundaries in Brownmillerite type strontium cobaltite. *J. Am. Ceram. Soc.*, **85**, 969–976.
137. Gazquez, J., Luo, W., Oxley, M.P., Prange, M., Torija, M.A., Sharma, M., Leighton, C., Pantelides, S.T., Pennycook, S.J., and Varela, M. (2011) Atomic-resolution imaging of spin-stae superlattices in nanopockets within cobaltite thin films. *Nanoletters*, **11**, 973–976.

33
Application of Transmission Electron Microscopy in the Research of Inorganic Photovoltaic Materials
Yanfa Yan

33.1
Introduction

In the past century, the worldwide demand for energy has increased exponentially. The worldwide demand for primary energy is projected to reach 30 TW by 2050. However, the expected fossil energy production in all forms can only supply up to 20 TW of energy. The 10 TW gap can only be closed by producing energy from abundant solar energy. Photovoltaic (PV) technology (or solar cells), which uses semiconductors to convert sunlight to electricity, is a desirable way to produce renewable energy. The PV market has continued to grow rapidly in the past decade and is anticipated to grow even more in the coming decades. Thus, PV research has attracted great attention from both the scientific and public communities.

Current PV technologies can be categorized into two types – p–n junction and carrier injection – according to how photon-generated carriers are separated. The first type typically uses inorganic semiconductors, and the second type usually involves organic, molecular, and nanostructure materials. To date, p–n junction-based PV technology exhibits much higher solar-to-electricity conversion efficiency and better stability. Thus, the majority of the current PV market is p–n junction based, such as wafer Si, thin-film amorphous Si, CdTe, and $Cu(In,Ga)Se_2$ (CIGS) [1–4]. p–n junction-based solar cells usually consists of a front contact, p–n junction, absorbing layer, and back contact. A brief description of the working principle for this type of solar cell is shown in Figure 33.1 [5].

The solar-to-electricity conversion involves three processes: (1) carrier generation (sunlight absorption), (2) carrier separation, and (3) carrier transportation and collection. Subsequently, the performance (efficiency) of a solar cell is determined by three parameters: open-circuit voltage (V_{oc}), short-circuit current density (J_{sc}), and fill factor (FF), which are limited by the performance of the three processes. In process (1), sunlight is absorbed by the semiconductor, and carriers (i.e., electron–hole pairs) are generated. There are two possible energy-loss mechanisms associated with this process: fundamental and practical losses. Photons with energies smaller than the bandgaps of the semiconductors cannot be

Handbook of Nanoscopy, First Edition. Edited by Gustaaf Van Tendeloo, Dirk Van Dyck, and Stephen J. Pennycook.
© 2012 Wiley-VCH Verlag GmbH & Co. KGaA. Published 2012 by Wiley-VCH Verlag GmbH & Co. KGaA.

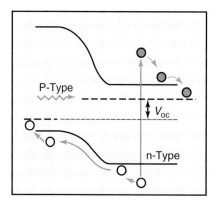

Figure 33.1 Schematic of a p–n junction-based solar cell. The thick dotted lines indicate quasi-fermi levels, and V_{oc} is the open-circuit voltage.

absorbed. For photons with energies larger than the bandgaps of the semiconductors, their generated electrons and holes will thermally relax to the band edges. These are the two fundamental losses for p–n junction-based solar cells and the main causes leading to the Shockley–Queisser limit [6] for solar cells using a single p–n junction. Practical losses occur because part of the sunlight will be reflected, and some semiconductors may not have a sufficient optical absorption coefficient to absorb sunlight before the light travels through the semiconductors.

The fundamental and practical losses associated with process (1) determine the J_{sc} of a solar cell device. The practical loss can be reduced by an antireflection coating and careful light management. In process (2), photon-generated electrons and holes are separated by the built-in field at the p–n junction. When electrons and holes are separated, quasi-fermi levels of the p- and n-type semiconductors will split and the energy difference determines the V_{oc}. Trap-assisted recombination or Shockley-Read-Hall (SRH) recombination [7], which is caused by deep-level defects formed near the built-in field region, reduces the splitting and causes practical loss of the V_{oc}. Thus, the formation of defects such as structural and point defects is very undesirable for solar cells. In process (3), the separated electrons and holes are transported to the front and back contacts to be collected. The energy-band mismatch between interfaces and large electrical resistance in used materials will lead to practical loss of the FF. Therefore, advanced PV technology requires sophisticated interface engineering and careful control of defect formation. Detailed information on extended defects, microstructure, chemistry, and the interface of solar cell materials is critical for understanding the performance of the fabricated devices and providing guidance to improve performance. Transmission electron microscopy (TEM) provides unique methods to obtain such information at the nanometer scale.

33.2
Experimental

In this chapter, TEM samples were prepared by three methods – standard ion milling, cleavage, and focused ion beam – based on the research needs and the material's sensitivity to ion-beam damage. Phase-contrast images, high-resolution transmission electron microscopy (HRTEM) images, electron diffraction patterns, atomic Z-contrast images, electron energy-loss spectroscopy (EELS) spectra, and energy-dispersive X-ray spectroscopy (EDS) spectra were taken in either a Philip CM30 microscope operated at 300 kV or an FEI Tecnai TF20-UT microscope equipped with a Gatan Image Filtering (GIF) system operated at 200 kV. The Z-contrast images were formed by scanning a 1.4 Å probe across a specimen and recording the transmitted high-angle scattering with an annular dark-field (ADF) detector (inner angle ~35 mrad). The image intensity can be described approximately as a convolution between the electron probe and an object function. Thus, the Z-contrast image gives a directly interpretable, atomic-resolution map of the columnar scattering cross section in which the resolution is limited by the size of the electron probe [8, 9], as described in Chapter 23. EELS data were taken with an energy resolution of 1.2 eV.

33.3
Atomic Structure and Electronic Properties of c-Si/a-Si:H Heterointerfaces

Because of the abundance and technology maturity, mono-Si and multicrystalline Si dominate the current PV market. For crystalline Si (c-Si) solar cells, the efficiency is largely limited by surface recombination, which reduces carrier lifetime. Surface passivation of c-Si to enhance carrier lifetime is a critical process for many applications, including solar cells. It is well known that effective surface passivation [10–15] can be achieved by growing a thin layer such as silicon dioxide (SiO_2), silicon nitride (SiN_x), silicon carbide (SiC_x), or hydrogenated amorphous silicon (a-Si:H) on the c-Si surface. Because a-Si:H passivation layers are also effective emitters and back-surface contacts, a silicon heterojunction (SHJ) device structure consisting of a thin layer of a-Si on c-Si was developed to achieve high-efficiency Si solar cells. The excellent surface passivation, junction formation, and contacting capabilities of a-Si:H are demonstrated by Sanyo's heterojunction with intrinsic thin layer (HIT) solar cells, with efficiencies exceeding 21% [14].

Growing an atomically abrupt and flat c-Si/a-Si:H interface is one of the critical issues in achieving a high-performance SHJ solar cell [16]. However, it has been difficult to achieve high-quality c-Si/a-Si:H interfaces for several reasons. First, direct deposition of a-Si:H on a clean c-Si surface can lead to epitaxial or mixed-phase Si growth with a high density of structural defects. Examples are shown in Figure 33.2. The dotted white lines indicate the original surfaces of c-Si. Epitaxial and mixed-phase growth are seen clearly in Figure 33.2. Second, the clean Si surface can easily oxidize before the deposition of a-Si:H, forming

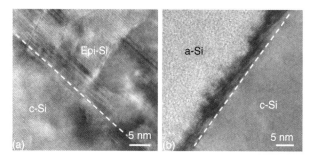

Figure 33.2 (a) Epitaxial and (b) mixed-phase growth on c-Si surfaces.

a very thin layer of SiO_2 at the interface. Insulating interface layers can disrupt carrier transport in SHJ solar cells. Finally, impurities at the interface can cause interface recombination. These problems must be overcome to achieve high-quality c-Si/a-Si:H interfaces and SHJ PV devices.

Typically, high-quality c-Si/a-Si:H interfaces are achieved by direct deposition of a-Si:H on clean c-Si substrates by hot-wire chemical vapor deposition (HWCVD) at low temperatures [16]. It is critical to keep the substrate surface clean and prevent oxidation to obtain high-quality c-Si/a-Si:H interfaces. Scientists at the National Renewable Energy Laboratory (NREL) have discovered a method to overcome these problems through surface pretreatment before the a-Si:H layer is deposited by HWCVD [17]. Before a-Si:H deposition, the c-Si wafer surface was treated by HWCVD of H_2-diluted NH_3 followed by etching with atomic H. The SHJ solar cells were fabricated on polished 350 μm thick, p-type Czochralski-grown (CZ) c-Si wafers with (100) orientation. Before the pretreatment and a-Si:H deposition, the c-Si wafer was heated to about 200 °C. HWCVD decomposition of gases was achieved by heating a 0.5 mm diameter tungsten wire to about 2100 °C with an alternating current [18]. Surface pretreatment was performed by introducing an 80 : 1 mixture of H_2 and NH_3 gas with the W filament on. Following the pretreatment, the surface was subject to etch with hot-wire-decomposed H_2 for well-controlled times. After this two-step pretreatment, a 5 nm intrinsic a-Si:H layer was deposited from pure SiH_4, and a 10 nm n-type a-Si:H layer was deposited from a mixture of SiH_4, PH_3, and H_2 gases. The backside of the cells had an Al-diffused back-surface-field contact made by evaporating Al and annealing for 20 min at 600 °C. The 1 cm^2 cells were finished on the front side with an 80 nm layer of indium-tin-oxide and a fine metal grid.

The pretreatment in an 80:1 mixture of H_2 and NH_3 gas with the W filament on leads to the growth of a thin a-SiN_x:H layer, which prevents oxidation and enables immediate growth of a-Si:H. The performance of the fabricated SHJ cells depends critically on the H etching time. Table 33.1 shows the SHJ solar cell's performance as a function of H etching time. With the two-step treatment, the solar cell performance is improved significantly. Without HW atomic H etching, the cell has a J_{sc} of around 29.9 mA cm^{-2}, a good V_{oc} of 0.55 V, but an FF of only 0.47. With optimized atomic H etching, an identical cell has achieved 13.4%

Table 33.1 Solar cell performance as a function as H etching time.

H_2 etching time (s)	0	20	40	50	60
V_{oc} (V)	0.55	0.54	0.59	0.601	0.55
FF	0.47	0.47	0.59	0.72	0.55
J_{sc} (mA c^{-2})	29.9	31.9	30.5	31.03	36.6
E_{ff} (%)	7.86	8.07	10.8	13.4	11.8

efficiency with a J_{sc} of 31.3 mA cm^{-2}, a V_{oc} of 0.601 V, and an FF of 0.72. To better understand the performance trend, c-Si/a-Si:H interfaces with different H etching time were investigated by HRTEM, atomic-resolution Z-contrast imaging, and EELS. Previously, c-Si/a-SiO$_2$ interfaces were studied extensively using these techniques [19–22]. We directly observed that the atomically abrupt and flat c-Si/a-Si:H interfaces in SHJ solar cells grown after the H-diluted NH$_3$ pretreatment would yield deleterious Si epitaxy under HWCVD conditions. The results correlate with SHJ solar cell performance and therefore provide an understanding of the impact of the different growth strategies.

Without the pretreatment, we observed partial epitaxial growth under these conditions, and this was extremely harmful to SHJ solar cell performance. However, after the HWCVD pretreatment in H$_2$-diluted NH$_3$ (no H$_2$ etch), there was abrupt amorphous growth on the c-Si substrates. With optimal H etching, the abrupt high-quality a-Si/c-Si can be achieved. Figure 33.3 shows HRTEM images of the a-Si/c-Si interface from samples without and with optimal H etching (50 s). The electron beam is parallel to the Si [011] zone axis. The intrinsic and doped a-Si layers are indistinguishable in the images. Figure 33.3a reveals that there is an interlayer with a thickness of about 2 nm between the c-Si and a-Si:H (indicated by the arrow). This layer is amorphous, but has a different contrast from the a-Si:H; EDS and EELS analyses reveal both Si and N but no O and suggest that it is an amorphous SiN$_x$ (a-SiN$_x$) layer with a low N content of $x < 0.05$. This layer is likely formed during the NH$_3$ treatment. Figure 33.3a also reveals that this amorphous layer prevents epitaxial growth and ensures immediate deposition of a-Si:H at 200 °C by HWCVD. Although abrupt growth of a-Si:H normally passivates the

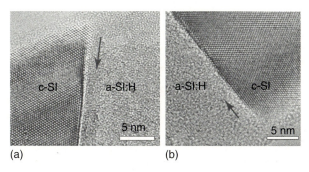

Figure 33.3 HRTEM images of a-Si/c-Si interfaces (a) without H etching and (b) with optimal H etching.

surface and increases carrier lifetime in c-Si [16], we find that the thin a-SiN$_x$ layer is harmful to heterojunction solar cell performance. The 2 nm a-SiN$_x$ layer shown in Figure 33.3a leads to a current voltage (I–V) curve FF of about 0.47, suggesting poor electron transport across the heterojunction. However, if this layer is thinned by controlled atomic H etching, the FF can be significantly improved. Figure 33.3b shows the a-Si/c-Si interface with optimal H etching for 50 s. It reveals a similar a-Si/c-Si interface. However, the interlayer seen in Figure 33.3a is no longer seen here, indicating that the a-SiN$_x$ layer is removed during H etching. If the etching time is longer than the optimal etching time, the a-SiN$_x$ layer can be completely removed and partial epitaxial Si starts to grow, which leads to reduced passivation.

The abruptness of the a-Si/c-Si is a critical parameter that measures the quality of the interface. To obtain quantitative information on the abruptness of the a-Si/c-Si interface, the atomic structures of the a-Si/c-Si interfaces were investigated by the atomic-resolution Z-contrast imaging technique [8, 9]. Figure 33.4a shows an atomic-resolution Z-contrast image of the c-Si/a-Si:H interface from the sample without H etching. On the c-Si side, the elongated bright spots directly represent the two closely spaced (110) Si columns, or Si "dumbbells." The dumbbells are not well resolved because the electron probe of our microscope is slightly larger than the distance between the Si columns. The probe size is estimated to be about

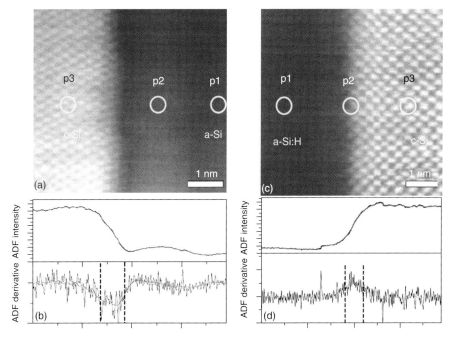

Figure 33.4 (a) Z-contrast image of a-Si/c-Si interface without H etching; (b) the measured intensity profile across the interface; (c) Z-contrast image of a-Si/c-Si interface with H etching for 50 s and (d) the measured intensity profile across the interface.

0.15 nm. The a-SiN$_x$ layer is again clearly seen. We use the method proposed by Diebold et al. [23] to measure the interface roughness using the intensity profile measured from Z-contrast images. Figure 33.4b shows measured intensity profiles and their corresponding derivative. The rough lines are experimental profiles, whereas the smooth lines are fitted profiles. The interface roughness is measured to be <0.45 nm, indicating that the interface between the c-Si and a-SiN$_x$ is abrupt and flat. Figure 33.4c shows an atomic-resolution Z-contrast image of the c-Si/a-Si:H interface from the sample with 50 s H etching. Figure 33.4d shows the measured intensity profiles and their corresponding derivatives. The rough lines are experimental profiles, whereas the smooth lines are fitted profiles. The interface roughness is measured to be <0.40 nm. Both the HRTEM image and Z-contrast image reveal that the c-Si/a-Si:H interface is again atomically abrupt and flat, but the a-SiN$_x$ interlayer is no longer visible. The a-SiN$_x$ layer was thinned down by etching from 2 nm to submonolayer thickness. However, even this thin layer ensures the abrupt growth of a-Si:H on the c-Si wafer, and the c-Si/a-Si:H interface is still atomically abrupt and flat. If the etching time is longer than this optimum, the a-SiN$_x$ layer is completely removed and we observe some epitaxial growth.

To understand the composition and electronic structure changes across the two interfaces, EELS spectra were taken at different points across the interfaces. Figure 33.5 shows the Si-L edge (~100 eV) and N-K edge (~410 eV) EELS spectra taken from three points, indicated as p1, p2, and p3 in Figure 33.4a. Point 1 is inside the a-Si:H layer, point 2 is inside the interlayer, and point 3 is inside the c-Si. Figure 33.5a shows that the intensity of the first peak in the Si-L edge spectra (arrows) is reduced as the electron beam is moved from the c-Si area to the interlayer and the a-Si:H layer. The Si-L edge EELS spectrum represents the transition from the Si 2p band to the conduction band [19, 20, 24]. The intensity reduction of the first peak at p1 and p2 indicates a reduction of density of states around the conduction-band minimum (CBM), which is caused by disorder in the a-SiN$_x$ and a-Si:H. Thus, the low intensity of this peak at p1 indicates a high disorder of the a-Si:H layer. The Si-L edge spectrum obtained from the a-SiN$_x$ layer

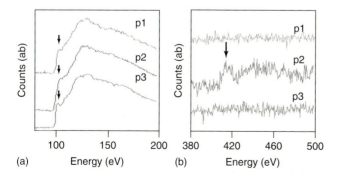

Figure 33.5 (a) Si-L edge and (b) N-K edge EELS spectra taken from three points across the a-Si/c-Si interface without H etching.

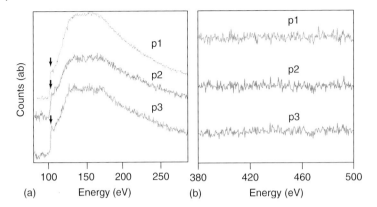

Figure 33.6 (a) Si-L edge and (b) N-K edge EELS spectra taken from three points across the a-Si/c-Si interface with H etching for 50 s.

is very similar to that obtained from the a-Si:H layer, but very different from the Si-L edge of c-Si$_3$N$_4$ [25]. The differences are due to the small concentration of N and the amorphous nature of the SiN$_x$ layer. Figure 33.5b reveals that the N-K edge peak (arrow) is seen only in the interlayer, confirming that the interlayer is SiN$_x$ and the other spots have no N content.

Figure 33.6 shows the Si-L edge and N-K edge EELS spectra taken from three points indicated as p1, p2, and p3 marked in Figure 33.4c. Point 1 is inside the a-Si:H layer, point 2 is at the interface, and point 3 is inside the c-Si. The N-K edge peak is no longer seen in any of the spectra. This confirms that the SiN$_x$ layer has been thinned to below our submonolayer detection limit. Figure 33.6a shows that the intensity of the first peak (arrows) is reduced as the electron beam is moved from the c-Si area to the interface. It is further reduced when the electron beam is moved into the a-Si:H layer. The Si-L edge spectrum taken at point 1 indicates that the a-Si:H layer is highly disordered. Figure 33.6b shows that the SiN$_x$ layer is removed after H etching for 50 s.

In the above, the atomic structure and electronic properties of c-Si/a-Si:H interfaces in SHJ solar cells were investigated by HRTEM, atomic-resolution Z-contrast imaging, EELS, and EDS. The combination of these TEM techniques enables us to demonstrate that with an H$_2$-diluted NH$_3$ surface pretreatment followed by H etching, one can achieve atomically abrupt and flat growth of highly disordered a-Si:H layers on clean c-Si substrates under Si HWCVD growth conditions that would otherwise yield epitaxial growth.

33.4
Interfaces and Defects in CdTe Solar Cells

Owing to its nearly ideal bandgap, high absorption coefficient, and ease of film fabrication, polycrystalline CdTe has been considered as a promising candidate

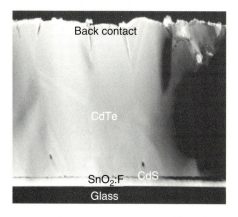

Figure 33.7 Z-contrast image of a CdTe solar cell.

for low-cost, high-efficiency thin-film solar cells. Small-area CdTe/CdS cells with efficiencies of >16.0%, fabricated using various deposition technologies, have been demonstrated [26, 27]. The structure of a typical CdTe solar cell consists of glass/CdS/CdTe/back contact coated with fluorine-doped tin oxide (FTO), as shown in Figure 33.7. The transparent and conducting FTO layer serves as the front contact. The back-contact layer usually contains Cu. Because the CdS window layer does not contribute any photocurrent, the thickness of this layer is limited to around 100 nm. For the record-efficiency CdTe solar cell (16.5% [26]), FTO, which is the conventional front contact, is replaced by a high-quality transparent conducting oxide, Cd_2SnO_4. To allow the use of a thin CdS window layer for improving the blue response, an intrinsic buffer layer, Zn_2SnO_4, is placed between CdS and the front-contact Cd_2SnO_4. This buffer layer effectively prevents the formation of pinholes, which usually form when the CdS window layer becomes too thin. This causes direct contact between the CdTe and the front contact and results in detrimental shorting. The back contact is formed by applying HgTe:CuTe-doped graphite paste, annealed at ~2700 °C for 30 min. Various growth methods, such as chemical-bath deposition (CBD) [28, 29], close-spaced sublimation (CSS) [30, 31], and metal-organic chemical vapor deposition (MOCVD) [32], have been used to grow CdS films. The CdTe/CdS solar cells with CdS films grown by these methods can have different performances. Normally, the CBD CdS films give better cell performance than the CdS films grown by the other techniques.

Typically, polycrystalline solar cells exhibit poor performance compared to their single-crystalline counterparts. The poor performance of polycrystalline materials is commonly attributed to the existence of extended defects such as stacking faults, twins, dislocations, and grain boundaries, which are usually effective recombination centers for photon-generated electrons and holes, as well as scattering centers of free carriers. Therefore, it is important to understand the defect physics in PV materials.

The zincblende (ZB) CdTe films studied in this work were grown by CSS at 450 °C. The substrates are (001) Si wafers. Before the growth of the CdTe films,

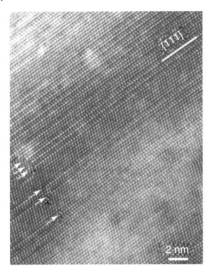

Figure 33.8 HRTEM image from a defective region in CdTe.

80 nm thick CdS films were deposited on the Si wafers. The thickness of the CdTe films is in the range of 2–3 µm. Specimens were prepared for electron microscopy by first mechanical polishing to a ∼100 µm thickness, then dimpling the central portion of the specimens down to ∼5 µm. The samples were subsequently thinned to electron transparency using a 4 kV Ar ion beam at 14° inclination, then cleaned at a lower voltage (1.5 kV). A liquid N_2 cooling stage was used to minimize milling damage. HRTEM images were taken on a Philips CM30 microscope with a Cs = 2.0 mm, operating at 300 kV.

CSS-grown CdTe films are polycrystalline, with an average grain size of 1 µm. X-ray diffraction (XRD) showed that the films are of nominally pure ZB phase. Our conventional TEM examinations revealed a high density of extended defects in most grains. Figure 33.8 shows a typical HRTEM image of these defects, with the electron beam parallel to the [1$\bar{1}$0] zone axis. The extended defects, mainly planar defects, are indicated by white arrows. Detailed analysis reveals that these planar defects are intrinsic and extrinsic stacking faults and lamellar twins. Like lamellar twins, the stacking faults always propagate across the grains because they never end at a partial dislocation inside the grains. The white solid line in the figure indicates a {111} plane of CdTe. It is clearly evident that the planar defects have a habit plane of {111}. Using HRTEM image simulation, we have obtained the atomic structure of these planar defects in CSS-grown ZB CdTe films and found that they are similar to those observed in ZB CdTe films grown by molecular-beam epitaxy [33, 34].

The ZB CdTe structure can be described by the stacking of close-packed double layers of {111} planes in the <111> direction. The bond arrangements of stacking faults and twins had been discussed by Holt in the 1960s [35]. The normal, perfect stacking sequence is ...AaBbCcAaBbCc..., where each letter represents a

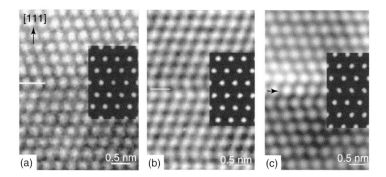

Figure 33.9 HRTEM images of (a) a lamellar twin, (b) an intrinsic stacking fault, and (c) an extrinsic stacking fault observed in CdTe.

stacking plane. The uppercase and lowercase letters indicate the Cd and Te planes, respectively. The letters Aa, Bb, and Cc indicate three possible projected positions of the atoms. Any mistake induced to the perfect stacking sequence will result in a planar defect such as lamellar twins and stacking faults. The stacking sequence of an intrinsic stacking fault is ...AaBbCcAa|CcAaBbCc..., where the symbol | indicates the position where a double layer is removed (in this case, Bb). The stacking sequence of an extrinsic stacking fault is ...AaBbCcAa|Cc|BbCcAaBbCc..., where the two "|" symbols indicate the added double layer (in this case, Cc). The stacking sequence of a lamellar twin is ...AaBbCcAaBbC|cBbAaCcBbAa..., where the symbol "|" indicates the position where the fault starts. Figure 33.9a shows an HRTEM image of a lamellar twin boundary observed in CdTe. The inset in Figure 33.9a is a simulated image using the model with the stacking sequence of lamellar twin. The simulated image agrees very well with the HRTEM image. Figure 33.9b,c shows HRTEM images of an intrinsic and an extrinsic stacking fault, respectively, observed in CdTe. The images are projected along the $[1\bar{1}0]$ zone axis of CdTe. The stacking planes are on the (111) plane and are indicated by the white lines in Figure 33.9a,b and the black arrow in Figure 33.9c. An intrinsic stacking fault is where a stacking layer is removed from the perfect sequence, thus giving a sequence of ...AaBbCcAa|CcAaBbCc... In an extrinsic stacking fault, an additional layer is inserted into the perfect sequence; thus, the stacking sequence of an extrinsic stacking fault is ...AaBbCcAa|Cc|BbCcAaBbCc... The insets in Figure 33.9b,c are the simulated images using these stacking sequence models. They agree very well with the experimental images. Density-functional theory calculations have shown that the formation energies of these stacking faults and twin boundaries are very low, and they do not introduce deep levels in the bandgap of CdTe [36].

We observed many thin layers with a wurtzite (WZ) structure buried in the highly faulted regions of the "pure" ZB CdTe films. Figure 33.10a shows an HRTEM image of such a WZ structure sandwiched by two ZB CdTe regions. The electron beam is parallel to the $[1\bar{1}0]$ zone axis of the ZB structure. The ZB structure stacking sequence is AaBbCcAaBbCc in the $<1\bar{1}0>$ projection, and the WZ structure has a stacking sequence of AaBbAaBb in the $<11\bar{2}0>$ projection [37]. In Figure 33.10a,

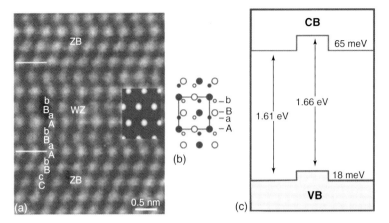

Figure 33.10 (a) HRTEM image of buried WZ CdTe in ZB CdTe. (b) Structure of WZ phase. (c) Calculated band alignment for buried WZ CdTe in ZB CdTe.

the AaBbAaBb stacking feature is clearly seen in the middle region, indicating the existence of a WZ structure. From the Aa and Bb positions, we can also conclude that the WZ layer is oriented such that its [11$\bar{2}$0] zone axis is parallel to the electron beam. Thus, the geometric relationship between the WZ structure and the ZB host structure is [111]ZB||[0001]WZ, [1$\bar{1}$0]ZB||[11$\bar{2}$0]WZ. There is no mismatch found between these two structures. Accordingly, the structural parameters of the WZ phase are calculated to be $a = 0.458$ nm and $c = 0.754$ nm, which are demonstrated by HRTEM image simulations. These values are slightly larger than the first-principles calculated values [38]. Figure 33.10b shows the projection of the WZ structure along the [11$\bar{2}$0] zone axis. The rectangle indicates a unit cell. The inset in Figure 33.10a is the simulated image of the WZ structure using the calculated structural parameters. The excellent match clearly supports the conclusion that the thin layer in the middle of Figure 33.10a is indeed a WZ phase and that our derived structural parameters are correct. Because the buried WZ layers are very narrow and are always in the high-density defect regions, there is no technique currently available to directly determine the effects of these buried WZ layers. We have, therefore, calculated the effects of the buried WZ layer by first-principles electronic structure theory. It is found that the buried WZ structure in the ZB host has type II band alignment, as shown in Figure 33.10c. The valence-band maximum (VBM) of the WZ structure is 18 meV higher in energy than that of the ZB structure, and the CBM of the WZ structure is 65 meV higher than that of the ZB structure. Thus, the hole states will be more localized in the WZ region, whereas the electron states will be more localized in the ZB region. Because the hole effective mass is very large, the hole state localization will be more significant than the electron states. For p-type ZB CdTe, in which holes are the majority carriers, the WZ layers behave like hole trap centers and hence adversely affect its electrical properties. Thus, the buried WZ phase may contribute to the poor electrical quality of the ZB CdTe films.

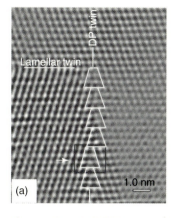 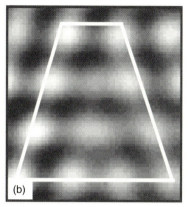

Figure 33.11 (a) HRTEM image of a Σ3(112) twin observed in CdTe. (b) Enlargement of the dislocation indicated by the white arrow in Figure 33.11a.

Beside the lamellar twins and stacking faults, double-positioning (DP) twins, also called Σ3(112) *twins*, were observed [39]. Figure 33.11a shows an HRTEM image of a DP twin boundary along the [1$\bar{1}$0] zone axis obtained in CdTe. It is seen that the DP twin boundary consists of dislocation cores marked by white trapezoids. Figure 33.11b shows an enlarged image of the dislocation core indicated by the white arrow in Figure 33.11a.

On the basis of the guidance from the HRTEM image shown in Figure 33.11b, eight possible structural models for the DP twin boundary can be constructed. These structural models are then relaxed using first-principles density-functional total-energy calculations. We find that there are only two structures that remain stable after the relaxation. Their optimized structures are shown in Figure 33.12a,b. The black balls indicate Cd atoms, whereas the gray balls indicate Te atoms. These two structures, called DP(1) and DP(2), respectively, look similar but have different details. For example, in DP(1) in Figure 33.12a, the Cd atom labeled as "3" has four bonds and Te labeled as "2" has a dangling bond. But in DP(2), the Te atom labeled as "3" has four bonds and Cd atom labeled as "2" has a dangling bond.

Our first-principles density-functional total-energy calculations indicate that both structures could exist [40]. It is not clear which of the two structures should fit the HRTEM image of the DP twin shown in Figure 33.11b. Thus, the two structurally optimized models are used for HRTEM image simulations. The experimental parameters are obtained by through-focus and through-thickness simulations of the HRTEM image of the perfect regions near the boundary. We find that the defocus value is about −65 nm and thickness is about 55 nm. Figure 33.12c,d represents the simulated images using the structural models of Figure 33.12a,b, respectively. For these simulations, the defocus value of −65 nm and thickness of 55 nm are used. It is seen that Figure 33.12c fits with Figure 33.11b much better than Figure 33.12d. Thus, the combination of first-principles density-functional total-energy calculations and HRTEM image simulation demonstrates that the

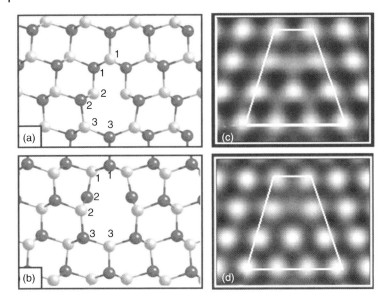

Figure 33.12 Optimized (a) DP(1) and (b) DP(2) structure models. The black balls indicate Cd atoms, and the gray balls indicate Te atoms. Simulated HRTEM image using (c) DP(1) and (d) DP(2) structure models.

DP(1) structure shown in Figure 33.12a is a convincing model for the DP twin boundaries that we have observed experimentally. However, it is noted that the DP(2) structure may also be observed in CdTe. Thus, theoretically, we will consider both structures.

To study the possible effects of the DP twin boundaries on the electronic property of CdTe, we calculate site projected density of states (PDOS) of individual atoms around the boundaries and compare them with the PDOS of a Cd and Te pair in the perfect regions. Figure 33.13a shows the comparison of PDOS of the three Cd atoms on the DP(1) boundary shown in Figure 33.12a. The black solid line marked as Cd + Te is the PDOS of the Cd and Te pair in the perfect region. The PDOS of the Cd atoms marked by numbers 1, 2, and 3 in Figure 33.12a are indicated by Cd1, Cd2, and Cd3, respectively. We see that the Cd3 atom produces a peak in the gap of CdTe. This is likely because the Cd3 atom has a dangling bond. The rest of the Cd atoms are all fourfold and have no dangling bonds; thus, they do not produce deep levels in the gap.

Figure 33.13b shows the PDOS of the three Te atoms on the DP(1) boundary shown in Figure 33.12a and a Cd and Te pair in a perfect region. The black solid line is the PDOS of the Cd and Te pair in the perfect region. The PDOS of the three Te atoms labeled by numbers 1, 2, and 3 in Figure 33.12a are indicated by Te1, Te2, and Te3, respectively. We see that the Te2 atom produces a very large peak in the gap region. This is because the Te2 atom has weak interaction with another Te atom.

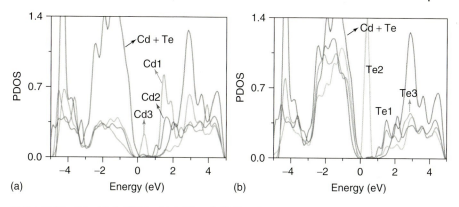

Figure 33.13 Calculated PDOS of (a) Cd and (b) Te atoms around the dislocation core in DP(1) structure.

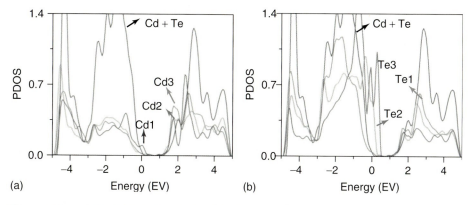

Figure 33.14 Calculated PDOS of (a) Cd and (b) Te atoms around the dislocation core in DP(2) structure.

We also calculated the PDOS of individual atoms around the boundaries in the DP(2) structure. Figure 33.14a shows the comparison of PDOS of the three Cd atoms on the DP(1) boundary shown in Figure 33.12b. The black solid line labeled as Cd + Te is the PDOS of the Cd and Te pair in the perfect region. The PDOS of the Cd atoms labeled by numbers 1, 2, and 3 in Figure 33.12b are indicated by Cd1, Cd2, and Cd3, respectively. We see that the Cd atoms in the DP(2) structure create a much less deep state in the gap than that in the DP(1) structure.

Figure 33.14b shows the PDOS of the three Te atoms on the DP(2) boundary shown in Figure 33.12b and a Cd and Te pair in the perfect region. The black solid line is the PDOS of the Cd and Te pair in the perfect region. The PDOS of the three Te atoms labeled by numbers 1, 2, and 3 in Figure 33.12b are indicated by Te_1, Te_2, and Te_3, respectively. We see that both the Te2 and Te3 atoms produce a large peak in the gap region. Thus, both boundary structures would introduce harmful carrier recombination and adversely affect the cell performance.

33.5
Influences of Oxygen on Interdiffusion at CdS/CdTe Heterojunctions

CdS films grown by CBD and other techniques have two major differences, namely, structural and chemical differences. First, CdS films grown by CBD have a cubic structure [29], whereas the CdS films grown by CSS or MOCVD have a hexagonal structure [27, 32, 41]. However, the energy difference between the cubic and hexagonal CdS bandgap is <0.1 eV. Furthermore, the different structures of CdS do not affect the structure of the deposited CdTe films [42]. Thus, the structural differences are unlikely responsible for the different cell performance. Second, CBD CdS has a high concentration of impurities, mainly O, whereas the CdS grown by other techniques has much less O, unless O is intentionally introduced. It has been reported that the presence of O_2 during the deposition of the CSS CdS improves V_{oc} up to 50 mV [43]. It is also reported that the thermal annealing of CdS in the air atmosphere drastically increases the efficiency. Thus, it is likely that the high O concentration in CBD CdS films is responsible for the better device performance. However, it is still unclear why and how the presence of O in CdS films improves CdTe/CdS solar cell performance.

We find that if the CdS is deposited without the presence of oxygen, the interdiffusion at the CdS–CdTe junction can be substantial, resulting in fully consumed CdS and $CdS_{1-x}Te_x$ regions with high Te concentration. The former have the same effects as pinholes, and the latter have much lower bandgap than CdS. This leads to the reduction in V_{oc} and J_{sc} of the CdS/CdTe solar cell. However, if the CdS contains oxygen, the interdiffusion can be suppressed significantly. This avoids the formation of fully consumed CdS and $CdS_{1-x}Te_x$ regions with high Te concentration. Thus, we propose that O in CdS films achieves its role in improving CdTe/CdS solar cell performance by controlling the interdiffusion at the CdS–CdTe junction.

The CdS films with a thickness of about 150 nm were grown by CBD and CSS with various O pressures on SnO_2-coated Si wafers. The thickness of the SnO_2 layer was about 0.5 μm. The CdTe films were grown by CSS without the presence of O. The CdTe source temperature was 660 °C, and the substrate temperature was 620 °C. The thickness of the CdTe films was about 8 μm. Samples with only CdS deposited were used as thickness references for understanding the consumption and reaction of the CdS at the junction.

Figure 33.15a shows a cross-sectional TEM image of a CdS layer deposited without the presence of oxygen. The thickness of the CdS layer is about 100 nm. Figure 33.15b,c shows cross-sectional TEM images of CdS–CdTe junctions with the CdS films deposited with 0 and 333.3 Pascal O pressure, respectively. The Si, SnO_2, and CdTe layers can be clearly identified and are indicated on both images. In Figure 33.15b, the CdS layer is not seen clearly in some regions. This indicates that during the deposition of CdTe, strong interdiffusion occurred between CdS and CdTe in the sample, with the CdS layer deposited without the presence of O. In some regions, as indicated by the arrow, the CdTe seems to directly contact the SnO_2 layer, showing a complete consumption of CdS near the junction. In some

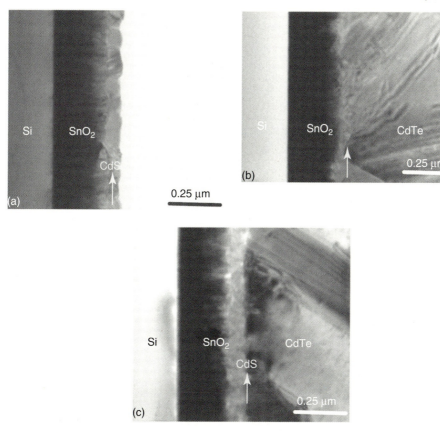

Figure 33.15 Cross-sectional TEM images of (a) as-deposited CdS layer, (b) CdS–CdTe interface for CdS deposited without the presence of O, and (c) CdS–CdTe interface for CdS deposited in the presence of O.

regions, the CdS layer is merely seen but with significantly decreased thickness (compared with the reference sample shown in Figure 33.15a). EDS (not shown here) taken with a small probe size of 20 nm from the thickness-reduced CdS region indicates that the CdS has converted to $CdS_{1-x}Te_x$ alloy. In the CdTe side, $CdTe_{1-x}S_x$ alloys were found near the junction. In Figure 33.15c, however, the CdS layer is seen clearly. The thickness is almost the same as the CdS layer in the reference sample (without a CdTe film). EDS data showed that $CdS_{1-x}Te_x$ alloys can only be found near the junction. As the distance from the junction increases, the concentration of Te decreases rapidly, which indicates that the interdiffusion in the CdS–CdTe junction with CdS layer deposited in the presence of O is much weaker. In other words, the presence of O in the CdS layer suppresses significantly the interdiffusion at the CdS–CdTe junction.

To verify this conclusion, we further annealed the CSS–CdS thin film in an oxygen environment at 400 °C for 10 min to incorporate oxygen into CdS. After the

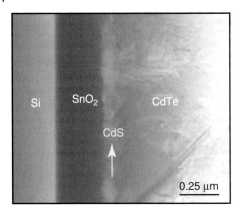

Figure 33.16 Cross-sectional TEM images of CdS/CdTe with CdS deposited without the presence of oxygen, but annealed in oxygen for 10 min. The arrow indicates the CdS layer.

annealing, CdTe was deposited at the same condition. It is reported that oxygen in CdTe produces a positive effect [44]. Figure 33.16 shows a cross-sectional TEM image of the sample for which the CSS–CdS was annealed in oxygen. Unlike in Figure 33.15b, the CdS layer is now seen clearly, confirming that the presence of oxygen in CdS significantly suppresses the interdiffusion at the CdS–CdTe interface. It should be pointed out that CdS layers deposited by CBD contain many other impurities as well. These impurities may also play roles in suppressing interdiffusion at the CBD-CdS/CdTe interface.

We also examined CdS–CdTe junctions, with the CdS layer grown by CBD. CBD CdS films are known to have different structures and contain large amounts of impurities, including a high concentration of O. A cross-sectional TEM image (not shown) of a CdS–CdTe junction with CBD CdS layer revealed the CdS layer clearly. Similar to Figure 33.15c, the thickness of CdS is not changed significantly after the deposition of CdTe. EDS data also showed $CdS_{1-x}Te_x$ alloys near the junction, and the concentration of Te decreases rapidly as the distance from the junction increases. Thus, the CdS–CdTe junction with CBD CdS demonstrates that it is O that suppresses the interdiffusion at CdS–CdTe junctions.

We now discuss the effects of the strong interdiffusion on CdS–CdTe solar cell performance. As stated earlier, strong interdiffusion results in complete consumption of CdS regions and $CdS_{1-x}Te_x$ alloys with high Te concentration. The effects of completely consuming CdS regions are similar to that of pinholes. This leads to reduced V_{oc} and J_{sc} due to the increased shorting of the junction. The $CdS_{1-x}Te_x$ alloys have a lower bandgap energy than CdS, and the bandgap changes with Te concentration. The bandgap of $CdS_{1-x}Te_x$ decreases very quickly as the Te concentration increases. Thus, strong interdiffusion will form a thick lower-bandgap $CdS_{1-x}Te_x$ layer. This layer, which absorbs more photons than CdS, will result in the blue (short wavelength) loss and also decreases the J_{sc}. Thus, when O is present in CdS films, all these problems can be overcome by suppressing interdiffusion. It is noted that for S concentrations less than ~20%, the bandgap of the $CdTe_{1-x}S_x$ alloy decreases as the S concentration increases due to the bowing effect [45]. This enhances the spectral response in long wavelengths and thus increases the photocurrent of the junction [46]. In addition, the $CdTe_{1-x}S_x$ alloy can also decrease the mismatch between the CdS and CdTe, leading to the reduction in interface states. Thus, some degree of interdiffusion is helpful to the cell performance, and only an optimized oxygen concentration in the CdS layer would improve the CdS–CdTe solar cell efficiency.

33.6
Microstructure Evolution of Cu(In,Ga)Se$_2$ Films from Cu Rich to In Rich

CIGS is a promising candidate for high-efficiency solar cells. The efficiency of small laboratory polycrystalline thin-film solar cells based on CIGS has approached >20% [47]. The structure of typical CIGS solar cells is glass/Mo/CIGS/CdS/ZnO, as shown in Figure 33.17. The Mo double layer has a thickness of about 1 μm and is usually deposited by sputtering. The full potential of this absorber material has yet to be realized. The four elements of this multinary polycrystalline film may form different compounds [48–50]. It is known that both the conductivity type and carrier concentration of CIGS depend strongly on the Cu/In and Se/(Cu + In) compositional ratios, which may vary in different growth processes. So far, most of the CIGS absorbers used for high-performance solar cells have been grown by either the "bilayer" or "three-stage" physical vapor deposition processes. Scientists at NREL grow CIGS films using the "three-stage" growth process shown in Figure 33.18. From stages 2 to 3, the composition of the films changes from Cu rich to In rich. During the growth, the measured temperature profile on the sample exhibits a clear drop from stages 2 to 3, indicating phase transitions. This change is important to the final composition and microstructure and, hence, to the electronic properties of the absorber. Thus, it is important to investigate the microstructure evolution of the CIGS films from Cu rich to In rich. For this purpose, we studied the microstructure and composition of the CIGS films obtained by interrupting their

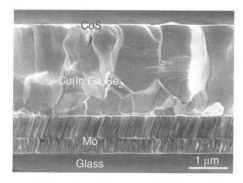

Figure 33.17 Cross-sectional SEM image of a CIGS solar cell.

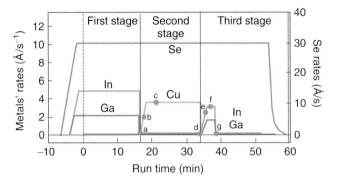

Figure 33.18 Schematic profile of the "three-stage" process.

growth at predetermined points, namely, a–d (Figure 33.18), which correspond to the transition from Cu rich (point a) to In rich (point d).

The CIGS thin films were grown by the sequential physical vapor deposition of metals in the presence of Se in a multisource bell-jar system. A three-stage growth process is used for this growth. Before the deposition of CIGS, a thin layer (∼0.2 μm) of Mo is deposited. From points a to d, the composition of the film changes from Cu rich to In rich. TEM specimens were prepared by first dissolving the glass substrate in hydrofluoric (HF) acid solution. Cross-sectional films were made by sandwiching two samples together, then mechanically thinning them to about 5 μm. The specimens were finally thinned to electron transparency using a 3.5 kV Ar ion beam at 6° inclination. The TEM images were taken with a Philips CM30 microscope operating at 300 kV.

We discuss first the results obtained from sample a. It is reported that Cu_xSe crystals should exist in Cu-rich CIGS samples. Figure 33.19a is a plan-view TEM image of sample a, which clearly shows particles with an average size of 0.2 μm in the film. Figure 33.19b shows the EDS taken from position p1, which is away from the particles. The appearance of Cu, Se, In, and Ga peaks indicates that the large grains are indeed CIGS. Figure 33.19c shows the EDS taken from one of the particles. Only the Cu and Se peaks are detected, which clearly indicates that these particles are Cu_xSe crystals. We have also seen that the Cu_xSe particles are distributed both along the CIGS grain boundaries and on the CIGS grains. Because there are no In and Ga signals detected from the Cu_xSe particles, it is likely that the Cu_xSe particles partially penetrate the grains of the CIGS films or the Cu_xSe particles are embedded in the large CIGS grains.

Figure 33.20 shows the [110] zone-axis electron diffraction patterns taken from the CIGS grains and the Cu_xSe particles. The main diffraction spots (bright spots) in Figure 33.20b are nearly identical to those in Figure 33.20a, indicating that the Cu_xSe particles have very similar structure to that of CIGS. However, satellite diffraction spots, as indicated by the dark gray arrows, are seen in Figure 33.20b. This means that the Cu_xSe particles have an ordered structure. The ordering is identified to be $s = [1/3, -1/3, 0]$ from the diffraction pattern.

33.6 Microstructure Evolution of Cu(In,Ga)Se₂ Films from Cu Rich to In Rich

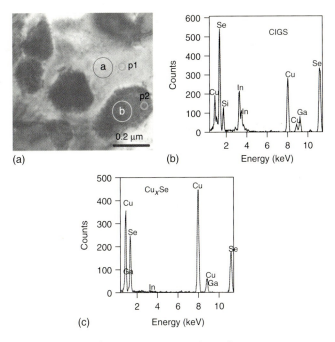

Figure 33.19 (a) Plan-view TEM image of sample a, (b) EDS obtained from position p1, and (c) EDS obtained from position p2.

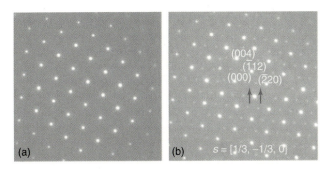

Figure 33.20 [110] zone-axis electron diffraction patterns taken from (a) CIGS and (b) Cu$_x$Se particles.

In sample a, in addition to the Cu$_x$Se crystals, subinterfaces were observed. The subinterfaces are about 0.2 μm below the surface. There are two types of subinterfaces: one composed of dislocations and the other consisting of twin boundaries. Figure 33.21a is a cross-sectional TEM image of sample a, showing the first type of subinterface. As indicated by the arrow, a subinterface consisting of dislocations is seen. The two sides adjacent to the subinterface have the same orientation, and electron diffraction patterns did not show detectable differences in their structures. However, EDS taken from the two sides showed different Cu/In compositional

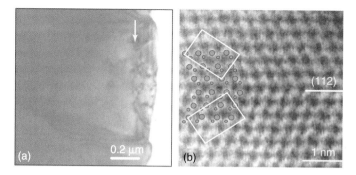

Figure 33.21 (a) Cross-sectional TEM image of sample a, showing a subinterface near the surface. (b) Z-contrast image of a lamellar twin in CIGS.

ratios. We found that the side close to the surface of the film is relatively more Cu rich than the other side of the subinterface. This compositional difference is likely the main cause for the formation of the subinterface. Figure 33.21b shows the Z-contrast image of a lamellar twin in CIGS along the [1$\bar{1}$0] zone axis. The atomic structure of the twin boundary is derived directly from the image and is superimposed on the Z-contrast image.

We now discuss the results obtained from sample b. As in sample a, subinterfaces were also observed in sample b, except at a lower density. EDS taken from the two sides of the subinterfaces in sample b showed similar Cu/In peak ratio changes from the side near the surface to the other side. There are also differences between samples a and b. First, the grain sizes of Cu_xSe crystals in sample b are much smaller than in sample a. The density of Cu_xSe crystals is much lower in sample b because most of the Cu_xSe crystals were consumed by the additional In and Ga as the film transitioned from a to b. Second, we found that domain boundaries started to form in sample b. In sample a, the grains are larger and clean; however, in sample b, subdomain boundaries are present. The subboundaries are conformal and coherent; thus, they cannot be observed easily by scanning electron microscopy (SEM).

In sample d, which is In rich, no Cu_xSe crystals or subinterfaces were observed. The grain sizes are much smaller than in samples a and b. No subboundaries were observed inside the grains. This indicates that the larger grains in samples a and b had decomposed into smaller grains. In this sample, small grains with relatively higher Ga concentration were observed. Figure 33.22a shows a plan-view TEM image of sample d. A dark particle is seen, as indicated by a white arrow. EDS taken from this particle (Figure 33.22b) shows a higher Ga concentration than the EDS taken from position p2 on the CIGS film (Figure 33.22c). EDS data taken from a cross-sectional sample also showed that Ga concentration is inhomogeneous across grains from the surface to the Mo–CIGS interface. The Ga concentration in a region with a thickness of about 0.4 μm and located about 0.2 μm from the surface is lower than in the regions on either side of the layer.

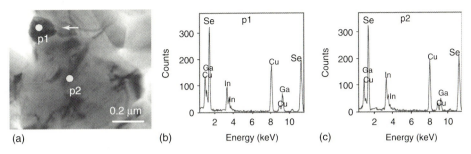

Figure 33.22 (a) Plan-view TEM image of sample d. (b) EDS taken from a small particle marked by p1. (c) EDS taken from position p2 in the CIGS film.

The above observation reveals that in sample a (Cu rich), there is a high density of Cu_xSe crystals on the surface, and these crystals extend between grains deeper into the film. Subinterfaces about 0.2 μm from the surface are observed. In sample b, the density of the Cu_xSe crystals and the subinterfaces is significantly lower than in sample a. Subboundaries are observed inside large grains. In sample d (In rich), we find that there are no Cu_xSe crystals or subinterfaces. In this sample, Ga concentration inhomogeneity across the grains from the surface to the Mo–CIGS interface is observed.

33.7
Microstructure of Surface Layers in Cu(In,Ga)Se₂ Thin Films

In most CIGS thin films, subinterfaces lying about 0.1–0.2 μm below the surface have been observed [51]. Because the surface layer is in the space-charge region, it may have profound effects on the cell performance. Understanding the effects is crucial for further improvements of the cell performance. The CIGS absorber layer is a p-type α-phase, the so-called 112 phase. The surface layer, however, was observed to have less Cu than the bulk of the film [52–54]. A Cu-poor Cu-In-Ca-Se material can be either the Cu-poor α-phase (chalcopyrite structure), the β-phase (ordered defect [Cu vacancy, V_{Cu}] chalcopyrite (ODC) structure), or the γ-phase (ZB structure). A surface layer with different structures will have different electronic properties and charge recombinations and, thus, will have different effects on the junction property and cell performance. It has been assumed that the surface layer in CuInSe₂ (CIS) films has the ODC structure, the β-phase, which is indicated to be n-type. A p-type CIS chalcopyrite-n-type ODC heterojunction model has been proposed to replace the heterojunction between p-type CIS chalcopyrite and n-type CdS [52]. However, so far, direct structural proof of the surface layer is still missing.

We have, therefore, studied the microstructure and the chemical composition of the surface layers in CIGS thin films by convergent-beam electron diffraction (CBED) and EDS. Having high spatial resolution, CBED is an ideal technique for investigating the microstructure of the surface layer. We find that the surface layer and the bulk are structurally similar in our samples, with no ODC structure observed. However, their compositions are slightly different (see below), indicating that they can have different point-defect physics. Our results suggest that the subinterfaces may form electrical homojunctions.

The CIGS thin films used in this study were grown by the sequential physical vapor deposition of metals in the presence of Se in a multisource bell-jar system using a three-stage growth process shown in Figure 33.18. Before the deposition of CIGS, a thin layer (~0.2 μm) of Mo is deposited on soda-lime glass. In the process, a precursor of $(In,Ga)_2Se_3$ is reacted with $Cu + Se$ to produce $Cu(In,Ga)Se_2$ plus Cu_xSe as a secondary phase, followed by the addition of $In + Ga + Se$ to adjust the composition to slightly Cu poor.

The CIGS films were first examined using XRD to check the structure in the overall samples. It revealed that the thin films grown by our three-stage growth process are of the chalcopyrite phase and contain no ODC β-phase. However, it is noted that because the surfaces of the films are usually very rough, the XRD data may not truly represent the structure of the surface, even when the incidence angle is as low as 0.2°. A technique with high spatial resolution is needed to determine the microstructure of the surface region.

TEM investigations revealed that the CIGS films grown by NREL process usually have large grains (1–2 μm) and low density of extended defects, such as dislocations, stacking faults, and twins. As in CIGS films discussed earlier, two types of subinterfaces have also been observed: one is composed of dislocations and the other consists of twin boundaries. In many cases, the subinterface contains dislocations, but occasionally it comprises a planar defect. In some samples, subinterfaces contain a very low density of extended defects, indicating that the defects can be minimized or even eliminated by controlling the growth condition. To examine the local chemical composition, EDS data were taken with a small probe size of 20 nm from both the surface and bulk regions. We found that the Cu/(In,Ga) peak ratio has a small change across the subinterface, that is, the ratio is slightly higher in the bulk region than in the surface layer. This indicates that the surface layer is slightly more Cu poor than the bulk regions. The Ga concentration is not uniform from the surface to bulk regions. To check the concentration of Ga, we have taken EDS spectra by locating a focused electron beam (about 2 nm) at different locations. Figure 33.23 shows EDS data taken from a surface region (a); a subsurface region (b), which is about 0.2 μm below the surface; and a bulk region (c). The Cu/(In,Ga) ratio can be estimated by comparing the first Cu, Ga, and In peak intensities, which were obtained by measuring the total counts of the peaks. It is seen that in the subsurface region, the Ga concentration is significantly lower than that in the surface and bulk regions, as indicated by the arrows. The same trend was obtained from different grains and samples. This is consistent with our secondary-ion mass spectrometry (SIMS) measurements.

Although the α- and β-phases have similar structure and lattice constants, the β-phase has the ODC structure. The ordered structure should give extra diffraction spots in electron diffraction patterns. This allows us to identify the β-phase from

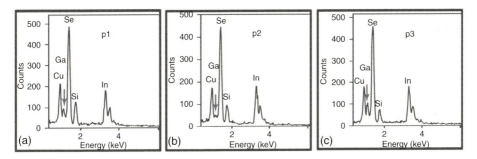

Figure 33.23 EDS taken from a (a) surface region; (b) sub-surface region, which is about 0.2 μm below the surface; and (c) bulk region.

the α-phase. Systematical zone-axis diffraction patterns confirmed that the sample has a chalcopyrite structure. We have obtained CBED patterns along the [1$\bar{1}$0] zone axis from both surface and bulk regions. However, we found that the two CBED patterns are identical. The diffraction spots can only be indexed by the chalcopyrite structure. If the surface layer is the ODC β-phase, there should be extra diffraction spots. We took CBED patterns with long exposure time at various places in the surface layer. However, no extra diffraction spots were observed, indicating that the surface layer is structurally similar to the absorber layer. It should be noted that their lattice constants may be slightly different due to their slightly different compositions. But the difference is beyond the accuracy limit of the electron diffraction patterns.

To support the conclusion, we examined intentionally made very Cu-poor CIGS samples (Cu:(In,Ga):Se, 1 : 3 : 5). In these samples, the ODC β-phase is confirmed by electron diffraction. Figure 33.24 is the CBED pattern taken from an α-phase along the [1$\bar{1}$0] zone axis. This CBED pattern serves as a reference for the chalcopyrite structure without ordered defects (Cu vacancies). Figure 33.24b shows

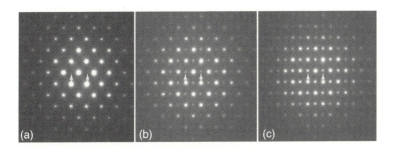

Figure 33.24 CBED pattern taken from (a) an α-chalcopyrite phase, (b) a chalcopyrite phase with ordered defects, and (c) a chalcopyrite phase with high degree of ordered defects. The arrows indicate the location of extra diffraction spots from the ordered defect phase.

a CBED pattern taken from a CIGS sample with composition close to the β-phase. The bright spots are identical to those in Figure 33.24, indicating that the main structure is the chalcopyrite structure. However, additional weak spots, as indicated by arrows in Figure 33.24b, are observed. These extra spots indicate that the phase contains ordered defects. In a more Cu-poor sample, a CBED pattern shown in Figure 33.24c was observed. The extra spots become brighter, indicating a higher degree of ordered defects. The same ordered structure has been found in very Cu-poor Cu-In-Se materials [55–57] and in $CuInS_2$ thin films [58].

On the basis of the above-mentioned study, we conclude that the surface layers have the chalcopyrite α-phase structure, similar to the CIGS bulk region. No ODC structure was observed in either the surface or the bulk region. However, the surface layer is slightly more Cu poor than the CIGS absorber, which may lead to different defect physics in the surface region. Because it has been observed that the junctions in CIGS solar cells are in the subinterface regions, our results suggest that the subinterfaces may form homojunctions.

33.8
Chemical Fluctuation-Induced Nanodomains in Cu(In,Ga)Se$_2$ Films

The efficiency of the CIGS-based device is now over 20%, which is a remarkable performance considering the polycrystalline nature of the film. So far, all high-efficiency devices are made of indium-rich absorbers with a wide deviation of molecularity (Cu:In ratio), typically between 0.95 and 0.82 [59–63]. This composition range of the equilibrium CIS phase diagram lies in the two-phase $\alpha + \beta$ region [64]. Several explanations have been given for the extraordinary performance of CIGS thin films. In general, these explanations fit into two scenarios. The first scenario is the role of grain boundaries. Recent theoretical and experimental studies have both proposed that unlike in other semiconductors, grain boundaries are not very harmful in CIGS thin films [65–68]. The second scenario deals with phase inhomogeneity, which is expected due to the structural mechanism by which the CIGS crystal lattice accommodates large off-stoichiometry [69]. Stanbery has presented an "intra-absorber junction" (IAJ) model [70], which proposes that the CIGS absorbers are in fact two-phase mixtures, and this heterogeneity is fundamental to charge separation in devices made from them. The model predicts that the two phases – the α-like (CIGS) domains and the β-like (copper-poor) domains – segregate on a nanometer-length scale to form an interpenetrating multiply connected network that permits percolation transport of both (electron and hole) photogenerated excess carrier populations in physically distinct paths. Thus, recombination within the absorbers is reduced by their real-space separation.

Here, we show our observation of strong chemical fluctuations at the nanometer scale that result in relatively Cu-poor and Cu-rich nanodomains from high-efficiency devices at the nanometer scale using TEM, EDS, and high-resolution Z-contrast imaging techniques. If we assume that the Cu-poor domains exhibit n-type conductivity [71] (β-like domains) and the Cu-rich domains exhibit p-type conductivity

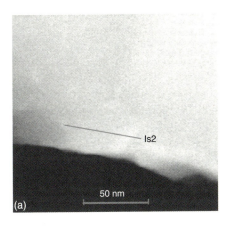
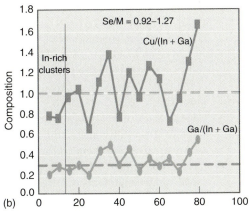

Figure 33.25 (a) Z-contrast image taken from a thick region in a high-efficiency CIGS sample, and (b) profiles of Cu/(In + Ga) and Ga/(In + Ga) ratios.

(α-like domains), then CIGS films are entirely composed of nano-p-n junction networks. Our results provide direct evidence for the phase and chemical inhomogeneity in CIGS thin films and support the IAJ model. Moreover, our results reveal that domains are crystallographically coherent: dislocations and other lattice defects do not have to form at the interface between the domains; rather, the interface will be smeared as a result of gradual interdiffusion, so that carriers generated by photons can be separated and collected rapidly and efficiently into the electron and hole transport paths.

Figure 33.25a shows a low-resolution Z-contrast image of a single CIGS grain at a relatively thick region. The thickness of the sample at this region is estimated to be around 800 Å. The Z-contrast images were formed by scanning a 1.4 Å electron probe across a specimen and recording the transmitted high-angle scattering with an annular detector. The intensity of a Z-contrast image can be described approximately as a convolution between the electron probe and an object function. Thus, the sample regions result in bright contrast and the vacuum results in dark contrast. Although the TEM samples were prepared carefully by gentle ion milling, a very thin layer with slight ion-beam-induced damage is expected to exist on both surfaces. Thus, we carried out quantitative nanoscale EDS chemical analysis at relatively thick regions, so that the effect on the composition from the slightly damaged surface layers is reduced to a minimum. We first carried out EDS data at individual spots and found, surprisingly, that the compositions fluctuate strongly around the average composition, Cu:In:Ga:Se = 23.5 : 19 : 7 : 50.5, determined by electron-probe microanalysis (EPMA). The fluctuations were within 6%, which is significantly larger than the system-detection error bar, that is, normally smaller than 2%. We then carried out a series of individual EDS measurements in a line profile along the black solid line in Figure 33.25a. The line profile contained 15 individual data points that are separated equally by about 5 nm. At each data point, the chemical composition was quantified. In Figure 33.25b, we plot the Cu/(In + Ga) and Ga/(In + Ga) ratios. The two dotted lines indicate the Cu/(In + Ga) and Ga/(In + Ga) ratios for the bulk composition. Both ratios are seen to fluctuate strongly at the nanoscale.

To verify whether these fluctuations were real or not, we carried out area-averaged EDS analysis by scanning the fine electron probe in selected areas, while acquiring EDS data. The averaged composition from a small square of about 50 nm × 50 nm was Cu:In:Ga:Se = 24 : 14 : 8 : 54. The averaged composition from a large square of about 100 nm × 100 nm was Cu:In:Ga:Se = 23 : 18 : 8 : 51, which is very close to the EPMA result. After a large number of analyses, we found that as the averaged area increases, the chemical fluctuation decreases. When the averaged area became close to 100 nm × 100 nm, the averaged compositions became consistent with the real composition and no longer fluctuated significantly. This indicates that the effect on the quantitative chemical compositions from the slightly damaged surface regions is negligible. This is because the EDS analyses were carried out at relatively thick regions, and the majority of the X-rays were generated from regions below the surface. Thus, our results strongly suggest the existence of chemical fluctuations at the nanoscale in CIGS thin films.

The chemical fluctuations may cause the formation of nanodomains. However, such nanodomains cannot be observed in thick regions because of the heavy overlaps in their projections. They can only be seen in very thin regions, where domain overlaps are not an issue. Figure 33.26 shows a Z-contrast image obtained from a relatively thin region, which is near the edge of the vacuum. It clearly reveals domain structures with different contrast. The domain sizes are <10 nm. The intensity variations indicate that these nanodomains may have different chemical compositions or thickness. Nanoscale EDS was used to determine the chemical compositions from the nanodomains, and we found chemical composition fluctuations across the domains; for example, the chemical compositions measured from points p1, p2, and p3 are Cu:In:Ga:Se = 31 : 14 : 7 : 48, 27 : 15 : 9 : 49, and

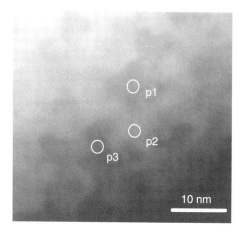

Figure 33.26 Z-contrast image taken from a thin region in a high-efficiency CIGS sample.

30 : 15 : 6 : 49, respectively. The dark domains are relatively Cu rich, whereas the bright domains are relatively Cu poor. It should be pointed out that this region is very thin. The relatively high Cu concentration measured here may be the result of effects of the slightly damaged thin surface layers, which are usually Cu rich. We also found that these damaged surface layers can be removed rapidly without attacking the domain structures by dipping the TEM sample into a NaCN solution.

Because the nanodomains are formed due to chemical fluctuations, they should be crystallographically coherent. Dislocations and other lattice defects do not have to form at the interfaces between the domains; rather, the interfaces will be smeared as a result of gradual interdiffusion. The Cu concentration fluctuation can be considered as scattered Cu vacancies or nanoscale chemical inhomogeneity. The chemical fluctuation introduces interconnected nanodomains, which we think are the result of aggregation of scattered Cu vacancies, as revealed by high-resolution Z-contrast imaging. Such nanodomains are clearly observed in the high-resolution Z-contrast image, as shown in Figure 33.27a. The dotted white line in Figure 33.27a indicates a (112) twin boundary, and the dashed circles show the locations of some of the nanodomains. The nanodomains are found to be distributed randomly and interconnected. Figure 33.27b shows the measured intensity profile from the gray box, revealing again the nanodomain structure. The dotted black line indicates the position of the (112) boundary plane.

We now discuss the possible effects of those chemical-induced nanodomains in CIGS thin films. As seen above, the Cu/(In + Ga) and Ga/(In + Ga) ratios

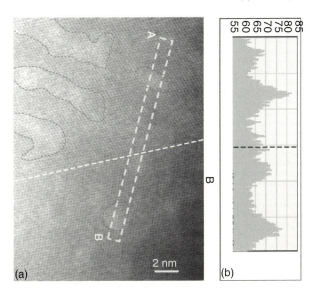

Figure 33.27 (a) High-resolution Z-contrast image showing nanodomains with chemical compositional fluctuation in around a (112) twin boundary in CIGS thin film. (b) Measured intensity line profile from the box shown in Figure 33.27a.

fluctuate across the nanodomains. These two ratio fluctuations may result in different effects. The Cu/(In + Ga) ratio fluctuation may be understood as relatively Cu-poor domains and Cu-rich domains. If we assume that the Cu-poor domains exhibit n-type conductivity (electron-rich β-like domains) and the Cu-rich domains exhibit p-type conductivity (hole-rich α-like domains), then the entire CIGS films are composed of nano-p-n junction networks. Thus, the n-type networks can act as preferential electron pathways and the p-type networks as preferential hole pathways. The carriers generated by photons can be separated effectively, as indicated in Figure 33.28a, and separated carriers can be collected rapidly and efficiently into the electron and hole "freeways," as shown in Figure 33.28b. Because the domains are crystallographically coherent, electron-hole recombination rates are minimized for carriers collected across these IAJs. As a result, the overall recombination can be effectively reduced. The Ga/(In + Ga) ratio fluctuation may be understood as Ga concentration difference across the nanodomains. The Pearson product-moment statistical correlation between the two variables Cu/(In + Ga) and Ga/(In + Ga) is 0.72, yielding 0.3% probability that the apparent correlation between these variables is a consequence of independent random fluctuations. It has been reported that the addition of Ga to CIS raises the CBM, while it does not affect the VBM significantly [72]. Conversely, Cu vacancies will preferentially segregate to form the phase, lowering the VBM while not affecting the CBM significantly. Thus, these nanodomains may have different CBM positions as well as bandgaps, thereby inducing potential fluctuations that may affect solar cell performance [73].

33.9
Conclusions and Future Directions

We have shown that advanced TEM techniques are powerful tools for obtaining critical information on microstructure, defects, interfaces, and chemistry at various scales for solar cell materials. Such information is necessary for understanding the performance of the solar cell materials studied, and it provides guidance for further performance improvement. We have shown results in some key solar cell materials, such as c-Si, and thin-film CdTe and CIGS. For c-Si, our atomic-resolution Z-contrast imaging, EELS, and EDS demonstrated that with a H_2-diluted NH_3 surface pretreatment followed by H etching, we can achieve atomically abrupt and flat growth of highly disordered a-Si:H layers on clean c-Si substrates under Si HWCVD growth conditions that would otherwise yield epitaxial growth. For CdTe, we have shown our results on determining atomic structures of extended defects, such as twins, stacking faults, and dislocation cores. We have further investigated the possible effects of oxygen on the interdiffusion at CdS–CdTe interfaces. For CIGS, we have studied the evolution of microstructures and chemistry of CIGS films at various growth points. We have observed that surface layers have the chalcopyrite α-phase structure, similar to the CIGS bulk region. No ordered defect chalcopyrite structure was observed in either the surface

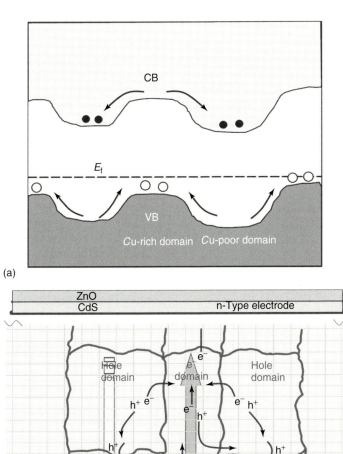

Figure 33.28 (a) Carrier separation between Cu-rich nanodomains and Cu-poor nanodomains, and (b) simplified picture of carrier collection with interconnected nanodomains.

or the bulk region. We have also shown direct evidence of chemical fluctuations at the nanoscale in high-efficiency CIGS thin films, which result in relatively Cu-poor and Cu-rich nanodomains. The nanodomains are crystallographically coherent and contain no lattice defects at the domain walls. These nanodomains may interconnect, forming three-dimensional, intermixed Cu-poor and Cu-rich networks. Such nanostructures may play a role in the high device performance of CIGS thin films.

Besides the above-mentioned information on microstructure, defects, interfaces, and chemistry, other information on carrier recombination, junction formation, and electrical and electronic properties is also very critical for diagnosing the performance of solar cell materials. Currently, this information can be obtained by cathodoluminescence (CL), electron-beam-induced current (EBIC), scanning tunneling spectroscopy (STM), scanning Kelvin probe microscopy (SKPM), and scanning capacitance microscopy (SCM). CL and EBIC are typically performed in SEMs. However, these methods do not have the capability to acquire the structural and chemical information that TEM can provide. Therefore, the direct correlation of electrical and electronic properties with microstructure, defects, and interfaces with high spatial resolution is currently very difficult. A microscopy combining the advantages of SEM, TEM, and scanning probe microscopy (SPM) would make such correlation possible. With today's technical advances made possible through Cs-correction, scanning transmission electron microscopes (STEMs) can exhibit high spatial resolution at low accelerating voltage and with a large space around the specimen, which is necessary for incorporating other techniques. Thus, advanced STEM holds great potential to integrate the mentioned capabilities. In the near future, STEM-CL, STEM-EBIC, and STEM-SPM will be developed and become available for providing information on structural, chemical, electrical, and electronic properties from the same area at high spatial resolution. Such microscopes will play an even more important role in the research of PV materials.

Acknowledgment

The author is grateful to M. M. Al-Jassim, P. Shelton, R. Noufi, Q. Wang, D. Albin, X. Li, T. Wang, H. Branz, M. Page, T. Gessert, C. S. Jiang, K. Jones, S. H. Wei, S. B. Zhang, M. Romero, H. Moutinho, W. Stanbery, and S. Pennycook for providing samples used in this study and/or for their comments and fruitful discussions. The research was supported by the U.S. Department of Energy under Contract No. DE-AC36-08GO28308 to NREL.

References

1. Martinot, E., Chaurey, A., Lew, D., Moreira, J.R., and Wamukonya, N. (2002) *Ann. Rev. Energy Env.*, **7**, 309.
2. Chapin, D.M., Fuller, C.S., and Pearson, G.L. (1954) *J. Appl. Phys.*, **25**, 676.
3. Chopra, K.L., Paulson, P.D., and Dutta, V. (2004) *Prog. Photovoltaics: Res. Appl.*, **12**, 69.
4. Birkmire, R.W. and Eser, E. (1997) *Ann. Rev. Mater. Sci.*, **27**, 625.
5. Nelson, J. (2003) *The Physics of Solar Cells*, Imperial College Press.
6. Shockley, W. and Queisser, H.J. (1961) *J. App. Phys.*, **32**, 510.
7. Shockley, W. and Read, W.T. Jr. (1952) *Phys. Rev.*, **87**, 835.
8. Pennycook, S.J. and Jesson, D.E. (1990) *Phys. Rev. Lett.*, **64**, 938.
9. Pennycook, S.J. and Nellist, P.D. (1998) *J. Microsc.*, **190**, 159.
10. Chen, Z., Pang, S.K., Yasutake, K., and Rohatgi, A. (1993) *J. Appl. Phys.*, **74**, 2856.
11. Aberle, A.G. and Hezel, R. (1997) *Prog. Photovoltaics*, **5**, 29.

12. Tanaka, M., Taguchi, M., Matsuyama, T., Sawada, T., Tsuda, S., Nakano, S., Hanafusa, H., and Kuwano, Y. (1992) *Jpn. J. Appl. Phys.*, **31**, 3518.
13. Taguchi, M., Kawamoto, K., Tsuge, S., Baba, T., Sakata, H., Morizane, M., Uchihashi, K., Nakamura, N., Kiyama, S., and Oota, O. (2000) *Prog. Photovoltaics: Res. Appl.*, **8**, 503.
14. Taguchi, M., Sakata, H., Yoshimine, Y., Maruyama, E., Terakawa, A., and Tanaka, M. (2005) *Proceedings 31st IEEE Photovoltaic Specialists Conference*, IEEE, New York, p. 866.
15. Martin, I., Vetter, M., Orpella, A., Puigdollers, J., Cuevas, A., and Alcubilla, R. (2001) *Appl. Phys. Lett.*, **79**, 2199.
16. Wang, T.H., Iwaniczko, E., Page, M.R., Levi, D.H., Yan, Y., Branz, H.M., Wang, Q., Yelundur, V., and Rohatgi, A. (2005) *Proceedings 31st IEEE Photovoltaic Specialists Conference*, IEEE, New York, p. 955.
17. Wang, Q., Page, M., Yan, Y., and Wang, T. (2005) *Proceedings 31st IEEE Photovoltaic Specialists Conference*, IEEE, New York, p. 1233.
18. Wang, Q., Page, M., Xu, Y., Iwaniczko, E., Williams, E., and Wang, T. (2003) *Thin Solid Film*, **430**, 208.
19. Batson, P.E. (1993) *Nature (London)*, **366**, 727.
20. Muller, D.A., Sorsch, T., Moccio, S., Baumann, F.H., Evans-Lutterodt, K., and Timp, G. (1999) *Natute (London)*, **399**, 758.
21. Himpsel, F., McFeely, F.R., Taleb-Ibrahimi, A., Yarmoff, J.A., and Hollinger, G. (1988) *Phys. Rev. B*, **38**, 6084.
22. Pennycook, S.J., Duscher, G., Buczko, R., and Pantelides, S.T. (1999) *Microsc. Microanal.*, (Suppl. 2), 122.
23. Diebold, A.C., Foran, B., Kisielowski, C., Muller, D.A., Pennycook, S.J., Principe, E., and Stemmer, S. (2003) *Microsc. Microanal.*, **9**, 493.
24. McKenzie, D.R., Berger, S.D., and Brown, L.M. (1986) *Solid State Commun.*, **59**, 325.
25. Tanaka, T., Mizoguchi, T., Sekine, T., He, H., Kimoto, K., Kibayashi, T., Mo, S.-D., and Ching, W.Y. (2001) *Appl. Phys. Lett.*, **78**, 2134.
26. Wu, X. (2004) *Solar Energy*, **77**, 803.
27. Ohyama, H., Aramoto, T., Kumazawa, S., Higuchi, H., Arita, T., Shibutani, S., Nishio, T., Nakajima, J., Tsuji, M., Hanafusa, A., Hibino, T., Omura, K., and Murozono, M. (1997) *Proceedings 26th IEEE Photovoltaic Specialists Conference*, IEEE, New York, p. 343.
28. Britt, J. and Ferekides, C. (1993) *Appl. Phys. Lett.*, **62**, 2851.
29. Martinez, M.A., Guillen, C., and Herrero, J. (1998) *Appl. Surf. Sci.*, **136**, 8.
30. Halliday, D.P., Eggleston, J.M., and Durose, K. (1998) *Thin Solid Films*, **322**, 314.
31. Ferekides, C.S., Marinskiy, D., Viswanathan, V., Tetali, B., Paleski, V., Selvaraj, P., and Morel, D.L. (2000) *Thin Solid Films*, **361–362**, 520.
32. Berrigan, R.A., Maung, N., Irvine, S.J.C., Cole-Hamilton, D.J., and Ellis, D. (1998) *J. Cryst. Growth*, **195**, 718.
33. Xin, Y., Browning, N.D., Rujirawat, R., Sivananthan, S., Chen, Y.P., Nellist, P.D., and Pennycook, S.J. (1998) *J. Appl. Phys.*, **84**, 4292.
34. Smith, D.J., Tsen, S.-C.Y., Chen, Y.P., Faurie, J.-P., and Sivananthan, S. (1995) *Appl. Phys. Lett.*, **67**, 1591.
35. Holt, D.B. (1962) *J. Phys. Chem. Solid*, **23**, 1353.
36. Yan, Y., Al-Jassim, M.M., and Demuth, T. (2001) *J. Appl. Phys.*, **90**, 3952.
37. Amelinckx, S. and Van Landuyt, J. (1978) in *Diffraction and Imaging Techniques in Materials Science* (eds S. Amelinckx, R. Gevers, and J. Van Landuyt), North-Holland, Amsterdam, p. 107.
38. Wei, S.-H. and Zhang, S.B. (2002) *Phys. Rev. B*, **66**, 155211.
39. Yan, Y., Al-Jassim, M.M., and Jones, K.M. (2003) *J. Appl. Phys.*, **94**, 2976.
40. Yan, Y., Al-Jassim, M.M., and Jones, K.M. (2004) *J. Appl. Phys.*, **96**, 320.
41. Lee, Y.H., Lee, W.J., Kwon, Y.S., Yeom, G.Y., and Yoon, J.K. (1999) *Thin Solid Films*, **341**, 172.
42. Cardona, M., Weinstein, M., and Wolff, G.A. (1965) *Phys. Rev.*, **140**, A633.
43. Ferekides, C.S., Dugan, K., Ceekala, V., Killian, J., Oman, D., Swaminathan, R.,

and Morel, D.L. (1995) *Proceedings 24th IEEE Photovoltaic Specialists Conference*, IEEE, New York, p. 99.

44. Emziane, M., Durose, K., Halliday, D.P., Bosio, A., and Romeo, N. (2005) *Appl. Phys. Lett.*, **87**, 261901.
45. Wei, S.-H., Zhang, S.B., and Zunger, A. (2000) *J. Appl. Phys.*, **87**, 1304.
46. Jensen, D.G., McCandless, B.E., and Birkmire, R.W. (1996) *Mat. Res. Soc. Symp. Proc.*, **426**, 325.
47. Repins, I., Contreras, M.A., Egaas, B., DeHart, C., Scharf, J., Perkins, C.L., To, B., and Noufi, R. (2008) *Prog. Photovoltaics: Res. Appl.*, **16**, 235.
48. Boehnke, U.C. and Kuhn, G. (1987) *J. Mater. Sci.*, **22**, 1635.
49. Nishiwaki, S., Satoh, T., Hayashi, S., Hashimoto, Y., Negami, T., and Wada, T. (1999) *J. Mater. Res.*, **14**, 4514.
50. Park, J.S., Dong, Z., Kim, S., and Perepesko, J.H. (2000) *J. Appl. Phys.*, **87**, 3683.
51. Hasoon, F.S., Yan, Y., Jones, K.M., Althani, H., Alleman, J., Al-Jassim, M.M., and Noufi, R. (2000) *Proceedings 28th IEEE Photovoltaic Specialists Conference*, New York, p. 513.
52. Schmid, D., Ruckh, M., Grunwald, F., and Schock, H.W. (1993) *J. Appl. Phys.*, **73**, 2902.
53. Negami, T., Kohara, N., Nishitani, M., and Wada, T. (1994) *Jpn. J. Appl. Phys.*, **33**, L1251.
54. Walter, T., Herberholz, R., Muller, C., and Schock, H.W. (1996) *J. Appl. Phys.*, **80**, 4411.
55. Tseng, B.H. and Wert, C.A. (1989) *J. Appl. Phys.*, **65**, 2254.
56. Xiao, H.Z., Yang, L.-C., and Rockett, A. (1994) *J. Appl. Phys.*, **76**, 1503.
57. Hornung, M., Benz, K.W., Margulis, L., Schmid, D., and Schock, H.W. (1995) *J. Cryst. Growth*, **154**, 315.
58. Su, D.S., Neumann, W., Hunger, R., Schubert-Bischoff, P., Gieysig, M., Lewerenz, H.J., Scheer, R., and Zeitler, E. (1998) *Appl. Phys. Lett.*, **73**, 785.
59. Ramanathan, K., Contreras, M., Perkins, C., Asher, S., Hasoon, F., Keane, J., Young, D., Romero, M.J., Metzger, W., Noufi, R., Ward, J., and Duda, A. (2003) *Prog. Photovoltaics*, **11**, 225.
60. Contreras, M.A., Egaas, B., Ramanathan, K., Hiltner, J., Swartzlander, A., Hasoon, F., and Noufi, R. (1999) *Prog. Photovoltaics*, **7**, 311.
61. Green, M.A., Emery, K., King, D.L., Igari, S., and Warta, W. (2001) *Prog. Photovoltaics*, **9**, 49.
62. Negami, T., Hashimoto, Y., and Nishiwaki, S. (2001) *Sol. Energy Mater. Sol. Cells*, **67**, 331.
63. Kessler, J., Bodegard, M., Hedstrom, J., and Stolt, L. (2001) *Sol. Energy Mater. Sol. Cells*, **67**, 67.
64. Godecke, T., Haalboon, T., and Ernst, F. (2000) *Z. Metallkd.*, **91**, 622.
65. Persson, C. and Zunger, A. (2003) *Phys. Rev. Lett.*, **91**, 266401.
66. Jiang, C.-S., Noufi, R., Ramanathan, K., AbuShama, J.A., Moutinho, H.R., and Al-Jassim, M.M. (2004) *Appl. Phys. Lett.*, **85**, 2625.
67. Jiang, C.-S., Noufi, R., AbuShama, J.A., Ramanathan, K., Moutinho, H.R., Pankow, J., and Al-Jassim, M.M. (2004) *Appl. Phys. Lett.*, **84**, 3477.
68. Yan, Y., Jiang, C.-S., Noufi, R., Wei, S.-H., Moutinho, H.R., and Al-Jassim, M.M. (2007) *Phys. Rev. Lett.*, **99**, 235504.
69. Zhang, S.B., Wei, S.-H., and Zunger, A. (1998) *Phys. Rev. B*, **57**, 9642.
70. Stanbery, B.J. (2005) *Proceedings 31st IEEE Photovoltaics Specialists Conference*, IEEE, New York, p. 355.
71. Contreras, M.A., Wiesner, H., Matson, R., Tuttle, J., Ramanathan, K., and Noufi, R. (1996) *Mat. Res. Soc. Symp. Proc.*, **426**, 243, and references therein.
72. Wei, S.-H., Zhang, S.B., and Zunger, A. (1998) *Appl. Phys. Lett.*, **72**, 1399.
73. Rau, U. and Werner, J.H. (2004) *Appl. Phys. Lett.*, **84**, 3735.

34
Polymers
Joachim Loos

34.1
Foreword

Some time ago, when the idea for this new book with the challenging theme "Nanoscopy" became solid, I was invited by the editors to write a chapter on polymer applications; and I accepted this honorable request with great pleasure. However, it took me a long time to define for myself how to compose such a chapter. When looking at polymers, they are everywhere around us: the case, the keyboard, and a lot more parts of the notebook computer I use to write this chapter are made of polymers. The tables, the wall coating, the picture frames, and the bottles and cups in the University cafeteria, where I spend most of the time for composing this chapter, are all made of polymers. I simply could continue counting for pages applications of polymers, and it was this huge quantity of technologically successful polymer systems we are dealing with in our daily life that made me struggle on which class of polymer systems I should focus on in this chapter. Should I start discussing commodity plastics such as polyolefins, their smart blends and (nano-) composites, and their endless structural and functional applications? It could be a lovely story starting with the very first transmission electron microscopic (TEM) images of solution-grown polyethylene single crystals [1] and ending with the state-of-the-art tailor-made reactor blends, block copolymers, and nanocomposites designed for ultraprecise processing (micromolding) and high-performance applications such as electromagnetic interference (EMI)-shielding, water cleaning membranes, gas diffusion barriers, or self-reinforced ultrastrong but easily recyclable composites [2], to name but a few examples. Similar success stories, I believe, can be easily composed for engineering plastics such as polyamide, polyester, or polycarbonade too.

However, during the past years, a substantial part of my research activity was and is still focused on the overall theme "Nanoscale Organization of Functional Polymer Systems", which comprises understanding and controlling of organization or assembly of functional polymer nanostructures. I try to tune the morphology by physical methods at various length scales from (sub-) nanometer (intra- and intermolecular organization) to hundreds of nanometers (e.g., phase separation and crystal superstructures) toward advanced performance of the corresponding

Handbook of Nanoscopy, First Edition. Edited by Gustaaf Van Tendeloo, Dirk Van Dyck, and Stephen J. Pennycook.
© 2012 Wiley-VCH Verlag GmbH & Co. KGaA. Published 2012 by Wiley-VCH Verlag GmbH & Co. KGaA.

systems and devices. Particularly, I am working on systems with applications in the research and development area of organic electronics with specific applications for printable solar cells (PSCs). Moreover, I am busy with developing methodologies based on high-resolution microscopy to analyze the nanoscale organization and functionality of such systems. Therefore, it was straightforward for me to focus in this chapter on advanced microscopic analysis of PSCs.

34.2
A Brief Introduction on Printable Solar Cells

Solar radiation is a renewable energy source with practically unlimited access. The amount of solar energy reaching the surface of the planet is so vast that in one year, it is about twice as much as will ever be obtained from all of the Earth's nonrenewable resources of coal, oil, natural gas, and mined uranium combined [3]. Solar energy can be utilized in three major ways: by means of concentrated solar power (CSP) technologies aimed at collecting solar heat, concentrating it, and then converting it into electricity; by artificial mimicking of the photosynthesis process as happens everyday in most plants on earth; and by direct conversion of sunlight into electricity in a photovoltaic (PV) process (eventually with the use of solar concentrators too). At present, large expectations are set for PVs to become a significant energy-supplying technology by the end of this century [4].

Beside conventional silicon-based solar cells and promising technological developments of, for example, dye-sensitized and CdTe-based solar cells,inrecent years, an alternative type of solar cells has been intensively studied, viz. thin-film devices printable from purely organic or hybrid solutions/dispersions applying semiconducting polymers for light absorption and charge transport [5–7]. Such printable solar cells (PSCs) have a distinct advantage over inorganic counterparts, that is, their fast and low-cost manufacturing process. Ultimately, they can be fabricated by processing polymers, eventually together with other organic and/or inorganic materials, in solution/dispersion and depositing them by printing or coating in a roll-to-roll manner like newspapers. Thanks to the speed and ease of this manufacturing process, the energy payback time of PSCs may, according to some estimates, be limited to about a year only [8]. Additional advantages include lightweight and flexibility of organic materials, enabling fast and easy applications on, for example, curved surfaces and thus freedom of design.

Most research on PSCs relates to increasing efficiency in converting light to electricity, and as the current record, an efficiency of 9.8% has recently been certified for a laboratory-scale device (end-2011 [9]).

Polymer solar cells (PSCs) are still in the research and development phase; however, the first commercial products were recently introduced in the market. To bring them closer to the stage of practical efficient devices, several issues should still be addressed, including further improvements of their efficiency and stability. These, in turn, are determined to a large extent by the morphological organization of the photoactive layer, that is, the layer in which light is absorbed and converted into electrical charges.

34.3
Morphology Requirements of Photoactive Layers in PSCs

In solar cells, excitons (exciton = bound electron–hole pair) are created when light is absorbed. Organic semiconductors typically possess an exciton binding energy that exceeds kT roughly by more than an order of magnitude [10]. As a consequence, excitons do not directly split into free charges in organic semiconductors, and an additional mechanism is required to achieve this. A successful way to dissociate excitons formed in organic semiconductors into free charges is to use a combination of two materials: an electron donor (the material with low ionization potential) and electron acceptor (the one with large electron affinity). At the donor–acceptor interface, an exciton can dissociate into free charges by rapid electron transfer from the donor to acceptor [11, 12]. The typical exciton diffusion length in most organic semiconductors is, however, limited to 5–20 nm [13–16]. Consequently, acceptor–donor interfaces have to be within this diffusion range for efficient exciton dissociation into free charges within the typical layer thickness of the photoactive layer of 100–200 nm. Currently, the most successful route to maximize the acceptor–donor interface is the so-called bulk heterojunction (BHJ), in which acceptor and donor are intermixed with each other in a controlled manner (Figure 34.1) [17–19].

Beside nanoscale phase segregation between the donor and acceptor components dictated by limited exciton diffusion length, high charge mobility [20] and charge transport via continuous and preferably short percolation pathways to the corresponding electrodes has to be guaranteed [21]. The efficiency of a BHJ PSC thus largely depends on the local nanoscale organization of the photoactive layer in all three dimensions. The key requirements for efficient PSCs, including those

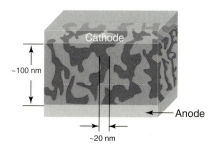

Figure 34.1 Schematic 3D representation of a bulk heterojunction (electron donor and acceptor constituents in different grayscales) with top and bottom electrodes. (Reprinted with permission from Ref. [7]. Copyright (2007) American Chemical Society.)

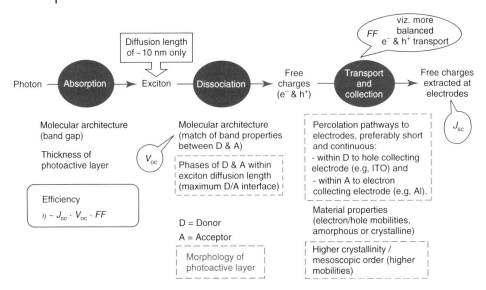

Figure 34.2 The key factors determining the power conversion efficiency (η) of bulk-heterojunction PSCs, together with the parameters of solar cell device performance: short-circuit current density J_{sc}, open circuit voltage V_{oc}, and fill factor FF. All three basic processes: light absorption, characterized by efficiency η_A), exciton dissociation (η_{ed}) and transport, and collection of charges (η_{cc}), should be efficient in order to get high performance PSCs. The efficiency-determining factors are listed under each step; those dealing with photoactive layer morphology are shown in bold.

dealing with photoactive layer morphology, are listed in Figure 34.2, together with the parameters of device performance (J_{sc}, FF, V_{oc}, η).

34.4
Our Characterization Toolbox

The thin-film nature of the active layer with typical thickness of about 100–200 nm and the request for local morphological information are causes that have made high-resolution microscopy techniques the main investigation tool for morphology characterization. TEM, scanning electron microscopy (SEM) – also in combination with preparative and imaging focused ion beam (FIB) and scanning probe microscopy (SPM), in particular atomic force microscopy (AFM), have proven their versatility for detailed characterization of the morphology of the active layer. The main difference between TEM on one hand SPM and SEM on the other hand is that TEM provides mainly morphological information of the lateral organization of thin-film samples by acquisition of projections through the whole film (in transmission), whereas SPM and SEM probe the topography or phase demixing at the surface of such thin-film samples.

Beside conventional transmission electron microscopy (CTEM) and AFM imaging modes for revealing morphological features of the samples under investigation, I will demonstrate how advanced microscopy methodologies can be applied for better understanding of the critical parameters determining the nanoscale organization and performance of functional polymer systems. In more detail, we have applied electron tomography (ET) to gain volume information; we have introduced scanning transmission electron microscopy (STEM), electron energy loss spectroscopy (EELS), and energy-filtered transmission electron microscopy (EFTEM) to create contrast between purely organic components in the BHJ photoactive layer of PSCs; we have shown how conductive atomic force microscopy (C-AFM) visualizes demixed acceptor and donor materials with 10-nm resolution and allows for local I–V spectroscopy; and we have optimized FIB preparation of specimens from whole devices for further investigations by the above-mentioned techniques.

34.5
How It All Started: First Morphology Studies

Intensive morphological studies have been performed on polymer/fullerene systems, in which [6,6]-phenyl-C61-butyric acid methyl ester (PCBM) is applied (for its chemical structure, see Figure 34.3) [7, 22, 23]. PCBM is by far the most widely used electron acceptor, and the most successful polymer solar cells have been obtained by mixing it with the donor polymers such as poly(2-methoxy-5-(3′,7′-dimethyloctyloxy)-1,4-phenylene-vinylene) (MDMO-PPV) [24, 25] and other PPV derivatives, with poly(3-alkylthiophene)s such as regioregular poly(3-hexylthiophene) (P3HT) [21, 26–29] or less studied combination with polyfluorenes [30–33] and other amorphous semiconducting polymers such as poly[2,6-(4,4-bis-(2-ethylhexyl)-4H-cyclopenta[2,1-b;3,4-b′]dithiophene)-alt-4,7-(2,1,3-benzothiadiazole)] (PCPDTBT) [34–37].

The influence of morphology on performance was first observed in the MDMO-PPV–PCBM system; a strong increase in power conversion efficiency was obtained by changing the solvent from toluene (0.9% efficiency) to chlorobenzene (2.5% efficiency) [24]. The better performance of the latter was found to be due to smaller (more favorable) scale of phase segregation (Figure 34.4), viz. smaller PCBM-rich domains in the MDMO-PPV-rich matrix, formed during spin-coating as a result of higher solubility of PCBM in chlorobenzene [24, 38–40].

For a long time, the system MDMO-PPV–PCBM was the standard for morphological investigations of high-performance PSCs. In our group, we have investigated in detail the phase separation caused by different processing routes [41–43] and the sample geometries used for standard investigation of the photoactive layer by TEM and AFM [44]. In general, we have followed several preparation routes to tune the morphology of the BHJ photoactive layer (Figure 34.5). More recently, we have introduced STEM [45, 46], EFTEM [47], and C-AFM [48, 49] to reveal fine morphological details of the photoactive layer of best performing devices, which we will discuss in more detail.

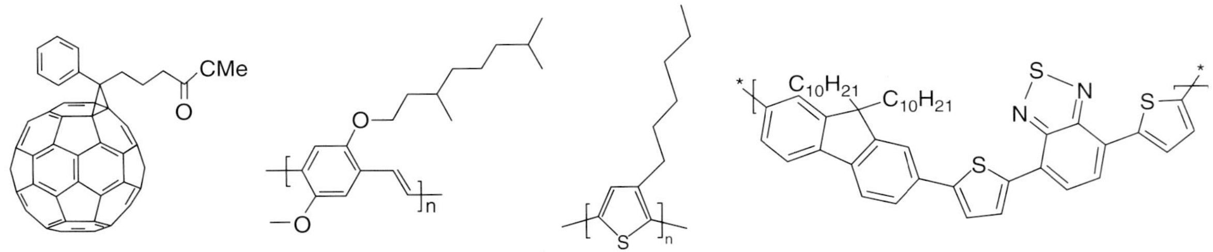

PCBM MDMO-PPV P3HT PFTBT

Figure 34.3 Chemical structures of some most common electron donor and electron acceptor materials: methanofullerene derivative [6,6]-phenyl-C61-butyric acid methyl ester (PCBM), poly(2-methoxy-5-(3',7'-dimethyloctyloxy)-1,4-phenylene-vinylene) (MDMO-PPV), poly(3-hexylthiophene) (P3HT), and poly(9,9-didecanefluorene-alt-bis-thienylene-benzothiadiazole) (PFTBT).

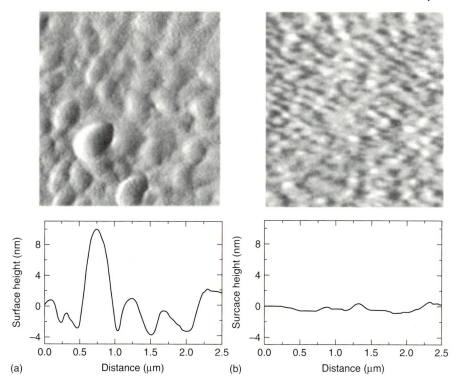

Figure 34.4 AFM images showing the surface morphology of MDMO-PPV–PCBM (1 : 4 by weight) blend films with a thickness of approximately 100 nm and the corresponding cross sections. (a) Film spin coated from a toluene solution. (b) Film spin coated from a chlorobenzene solution. The images show the first derivative of the actual surface heights. (Reprinted with permission from Ref. [24]. Copyright (2001) American Institute of Physics.)

34.6
Contrast Creation in Purely Carbon-Based BHJ Photoactive Layers

34.6.1
Scanning Transmission Electron Microscopy

In a modern TEM, the camera length can be varied from tens of millimeters to several meters, and hence, the minimum collection angle at the annular dark-field (ADF) detector can be varied from about 200 to 2 mrad and less, respectively. Therefore the properties of electrons hitting the detector changes and thus the obtained image contrast can easily be optimized by adjusting the camera length. For more information on STEM, the interested reader should refer to Ref. [48].

The scattering elastic cross section at large angles varies roughly as $Z^{3/2}$. However, looking more carefully at the equation that describes the scattered to

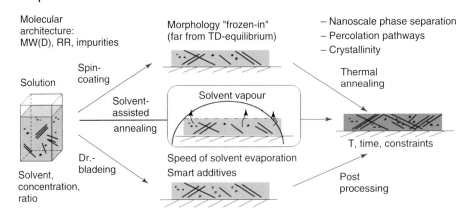

Figure 34.5 Controlled morphology development in P3HT–PCBM photoactive layers depending on the film preparation method. MW(D),molecular weight and distribution; RR,regioregularity; TD,thermodynamic; T,temperature.

incident intensity ratio or the dark-field intensity [51],

$$I/I_0 = 1 - e^{-N\sigma\rho t} \sim N\sigma\rho t; \quad \text{for} \quad N\sigma\rho t \ll 1, \tag{34.1}$$

where $N = N_0/A$ is the Avogadro's constant divided by the atomic weight, σ the scattering cross section, ρ the density, and t the thickness of the specimen, it is evident that variations in density may result in contrast between two different polymer materials.

Beside CTEM images, Figure 34.6 shows ADF-STEM images acquired on the same specimen but with two different camera lengths and collection angle-ranges at the ADF detector. Because of its higher density, the PCBM appears bright and the surrounding MDMO-PPV matrix is dark when imaging in dark-field conditions.

For short camera lengths (i.e., large collection angles) good contrast between the PCBM domains and the matrix exist (Figure 34.6c). Contrast and appearance between the 2 components is somewhat similar to the bright-field (BF) CTEM image acquired at low acceleration voltage (Figure 34.6b); however, the PCBM domains show distinct contrast with the MDMO-PPV matrix. Applying a longer camera length (i.e., lower collection angles) but otherwise similar imaging conditions, the contrast between the PCBM domains and the MDMO-PPV matrix substantially increases (Figure 34.6d,e) because additional diffraction contrast of the PCBM nanocrystals contributes to the image formation. Most of the domains are connected with each other and form aggregates. Only a few PCBM domains are isolated, and both the PCBM as well as the MDMO-PPV form interpenetrating networks throughout the whole photoactive layer. Beside the dominant PCBM domains, tiny PCBM nanobridges are seen that cross the MDMO-PPV matrix and connect the domains with each other. These nanobridges create additional interface in the photoactive film for effective exciton dissipation and a fine dispersed network for charge transport to the respective electrode. These unique morphological features adequately explain the high efficiency of corresponding devices.

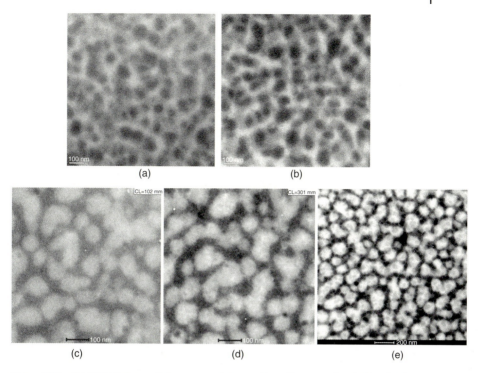

Figure 34.6 BF CTEM images of the PCBM–MDMO-PPV photoactive layer acquired at (a) 300 kV and (b) 80 kV acceleration voltages. HAADF-STEM images of the same sample acquired with camera lengths of (c) 100 mm and (d) 300 mm, that is, collection angles ranging between 73 and 290 mrad and 24 and 120 mrad, respectively. (e) Contrast optimized HAADF-STEM image showing the interconnected network of the PCBM domains. (Reprinted with permission from Ref. [46]. Copyright (2009) Cambridge University Press.)

34.6.2
Energy-Filtered Transmission Electron Microscopy (EFTEM)

Another route to create contrast between purely carbon-based materials is the application of EFTEM. Recently, with the development of modern imaging filters, EFTEM has emerged as a powerful tool for materials analysis [52, 53]. Using electrons with an energy loss characteristic of an atomic core level, quantitative 2D elemental distribution maps can be obtained in a fast, parallel manner, with nanometer resolution and high chemical accuracy [54]. Some examples of successful application of EFTEM in analysis of polymer systems can be found, for example, in the work of Varlot *et al.* [55, 56].

We have applied EFTEM for the morphological determination of a blend of MDMO-PPV with poly[oxa-1,4-phenylene-1,2-(1-cyanovinylene)-2-methoxy-5-(3,7-dimethyloctyloxy)-1,4-phenylene-1,2-(2-cyanovinylene)-1,4-phenylene] (PCNEPV), as it is used for all-polymer solar cells.

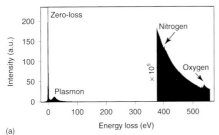 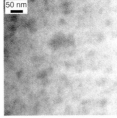 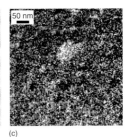

Figure 34.7 (a) Characteristic electron energy spectrum (EEL) spectrum of the MDMO-PPV–PCNEPV blend. (b) Zero-loss filtered TEM; and (c) the corresponding N–K elemental map of the same area; the latter visualizes the nitrogen distribution within the MDMO-PPV–PCNEPV film sample. (Reprinted with permission from Ref. [47]. Copyright (2005) Wiley-VCH.)

Looking to the chemical structure of the two components of the blend under investigation, the PCNEPV component has two nitrogen atoms in each repeat unit of its backbone. Thus, it is possible to apply EFTEM to detect the chemical composition of the materials and to visualize their distribution within thin films.

Figure 34.7a presents a characteristic electron energy loss spectrum of the MDMO-PPV–PCNEPV blend sample. The zero-loss peak and the broad plasmon region are shown at low energy loss. At energy losses of 401 and 532 eV, the presence of nitrogen and oxygen, respectively, can be detected. The high energy loss part of the spectrum is enlarged by a factor of ∼10.000. The nitrogen signal is very weak, and with exposure time, the intensity of the signal diminishes. However, in the present case, we were able to use the electron energy loss signal of nitrogen for the successful detection of its distribution within the sample by applying EFTEM.

By using EFTEM, an electron spectroscopic image (ESI) may be acquired at, or beyond, an ionization-edge absorption energy corresponding to a specific elemental species. Such a postedge image contains information regarding the spatial distribution and concentration of the chosen element, with an additional underlying background contribution corresponding to a variety of inelastic events occurring at lower energy losses. Therefore, to obtain an elemental distribution map from a core-loss image, the spectral background contribution must be taken into consideration. For nitrogen mapping, the three-window technique has been applied [57], that is, measurements in the pre- and postedge region of the nitrogen peak are done for the calculation of the signal's background with the power-law fit. The signal-to-noise ratio is rather low, however, distinct regions showing higher nitrogen intensity can be identified (Figure 34.7b,c). Especially, the central bright domain, which has a size of the order of 50 nm, can be easily correlated with the central dark-gray domain of the corresponding BF TEM image of Figure 34.6a. Keeping in mind that the nitrogen concentration in PCNEPV is only in the order of 5.3% by weight, the result of the elemental mapping is very satisfying. Combining the presented results, the correlation of nitrogen-rich domains with the presence of PCNEPV phases within the MDMO-PPV matrix can be executed.

34.6.3
Conductive Atomic Force Microscopy (C-AFM)

Finally, a very useful analytical tool to create contrast between polymer components in a BHJ layer is SPM, and in particular, AFM equipped with a conductive probe, called C-AFM [58, 59]. Because AFM uses the interaction force between the probe and sample surface as feedback signal, both topography and conductivity of the sample can be mapped independently. Theoretically, the resolution of C-AFM is as small as the tip-sample contact area, which can be <20 nm.

C-AFM is widely used for the characterization of electrical properties of organic semiconductors. For example, single crystals of sexithiophene have been studied [60], in which the $I-V$ characteristics of the samples were measured. Several electrical parameters such as grain resistivity and tip-sample barrier height were determined from these data. In another study, the hole transport in thin films of poly[2-methoxy-5-(2'-ethyl-hexyloxy)-1,4-phenylenevinylene] (MEH-PPV) was investigated and the spatial current distribution and $I-V$ characteristics of the samples were discussed [61].

In our group, recently, the first study on the spatial distribution of electrical properties of realistic BHJ MDMO-PPV–PCNEPV has been performed by applying C-AFM with lateral resolution better than 20 nm [50]. Measurements of the electrical current distribution over the sample surface were performed with an Au-coated tip. In such an experiment, the tip plays the role of the back electrode but having a much more localized contact area. A voltage was applied to the tip, and the indium tin oxide (ITO) front electrode was grounded (Figure 34.8). For C-AFM measurements, the tip was kept in contact with the sample surface while the current through the tip was measured. C-AFM measurements of the same sample area were done several times and resulted in completely reproducible data. Subsequent analysis of the surface showed almost no destruction of the sample surface; only minor changes were detected from time to time (Figure 34.9).

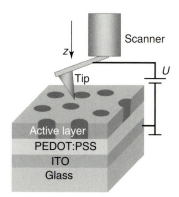

Figure 34.8 Scheme of the sample structure with segregated phases and the conductive AFM experimental setup including scanner and conductive tip.

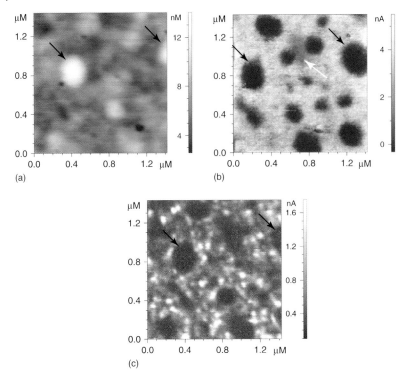

Figure 34.9 C-AFM images of the same area of an MDMO-PPV–PCNEPV active layer: (a) topography and (b) current distribution image with a positive bias at $U_{tip} = +8$ V. The white arrow in (b) indicates a domain with reduced current. (c) Current distribution image with a negative bias at $U_{tip} = -8$ V; black arrows indicate the same domains in all images for easy identification. (Reprinted with permission from Ref. [48]. Copyright 2006 Elsevier.)

It is reported that the electrical contrast measured by C-AFM at the surface of samples depends on the sign of the voltage applied [62]. As shown in Figure 34.9c, the C-AFM measurements at negative bias on the tip showed drastic changes of the contrast in the current images compared to positive bias (Figure 34.9b). PCNEPV domains showed only little current; however, MDMO-PPV showed a heterogeneous spatial current distribution. These electrical heterogeneities indicate small grains with a typical size of 20–50 nm, which differs by the value of current. A similar structure was observed on MEH-PPV films [61]. In the case of MEH-PPV, the authors attribute these substructures to a special and very local organization of the film that could be caused by local crystallization of stereoregular parts of the MEH-PPV molecules or by impurities incorporated during the synthesis.

In addition to topography and current-sensing analysis, we have performed current imaging spectroscopy. The procedure of such measurements is similar to the so-called force volume technique [63], which implies measurements of the force–distance curve at each point of a scan in order to get complete information about the lateral distribution of mechanical properties at the surface. Here, we extend this method to measurements of electrical properties of the sample [64]. Current–voltage ($I-V$, for constant distance; always in contact) dependencies were collected at each point of a scan.

Figure 34.10 shows the typical $I-V$ behavior of the MDMO-PPV–PCNEPV system and the corresponding C-AFM current distribution images for various biases. Three different types of $I-V$ characteristics can be attained in the sample depending on the location of the measurement. For domains of the electron acceptor compound PCNEPV, the current is always low, and the general contrast of the current distribution images depends only on the bias applied. In case of the electron donor matrix compound MDMO-PPV, two different $I-V$ characteristics can be obtained, showing a behavior that is almost the same for positive bias but varies significantly for negative bias. The current distribution images of Figure 34.10 demonstrate this behavior and provide additional information about lateral sizes of the MDMO-PPV heterogeneities. The point resolution of the images is about 15 nm. Some heterogeneities can be recognized with sizes as small as about 20 nm. The C-AFM results obtained on pure MDMO-PPV film are identical to those obtained on the matrix in the heterogeneous film MDMO-PPV/PCNEPV. The log I– log V plot of data obtained shows quadratic dependence of the current I on the voltage V measured on MDMO-PPV. This implies space-charge limited current that is in agreement with $I-V$ measurements on complete devices [65, 66].

34.7
Nanoscale Volume Information: Electron Tomography of PSCs

The performance of PSCs strongly depends on the 3D organization of the compounds within the photoactive layer (Figure 34.1). Donor and acceptor materials should form co-continuous networks with nanoscale phase separation to effectively dissociate excitons into free electrons and holes, and to guarantee fast charge carrier transport from any place in the active layer to the electrodes.

Attempts to gain information on the 3D organization of polymer solar cells have been described by applying, for example, techniques for cross-sectional preparation of TEM or SEM samples. Besides application of FIB to section whole devices that are prepared on glass substrates [67], conventional cross-sectional ultramicrotoming of the photoactive layer is used [39, 68]. Looking from the side perpendicular to the photoactive layer's lateral plane, the organization of segregated phases of the BHJ can be analyzed in thickness direction of the film. Similar to the top view, the cross-sectional side view provides information on sizes and shapes of phases. For instance, Sariciftci et al. has demonstrated by cross-sectional SEM investigations that in a system with MDMO-PPV as the donor and PCBM as the acceptor, the PCBM-rich phases are always imbedded in the MDMO-PPV matrix independent of their size (Figure 34.11) [39]. Most probably, this feature reduces the electron injection capability from PCBM to the electrode.

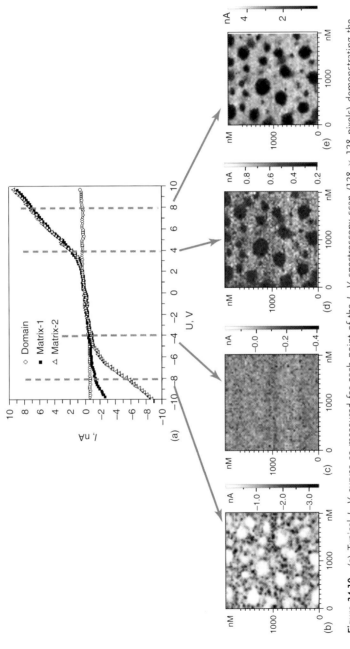

Figure 34.10 (a) Typical $I-V$ curves as measured for each point of the $I-V$ spectroscopy scan (128 × 128 pixels) demonstrating the heterogeneous $I-V$ characteristics of the MDMO-PPV matrix. (b–e) For four biases, the corresponding current distribution images are shown, demonstrating the obvious contrast between PCNEPV and MDMO-PPV for high bias as well as contrast within the MDMO-PPV matrix with heterogeneities as small as few tens of nanometers.

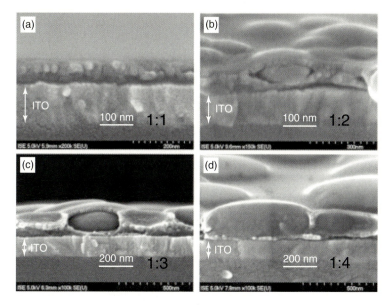

Figure 34.11 SEM side views (cross sections) of MDMO-PPV–PCBM blend films cast from toluene with various weight ratios of MDMO-PPV and PCBM. For ratios 1:4, 1:3, and 1:2 (b–d) the nanoclusters, in the form of discs are surrounded by another phase, called the *skin*, that contains smaller spheres of about 20–30-nm diameter. For the 1:1 film (a), only these smaller spheres are found. (Reprinted with permission from Ref. [39]. Copyright (2004) Wiley-VCH.)

The microscopy technique that provides 3D morphological information with nanometer resolution in all three dimensions is ET, also referred to as *transmission electron microtomography* and *3D TEM* [69–73]. For ET, a series of 2D projections is taken by TEM at different angles by tilting the specimen with respect to the electron beam. The tilt-series thus obtained containing normally more than 100 images of the same specimen spot is then carefully aligned and used to reconstruct a 3D image of the specimen, with nanometer resolution. The outcome of ET can be used voxel by voxel (an abbreviation of "volume pixel") to study in detail the specimen's volume morphological organization.

Although alternative approaches, such as conical tilting [74, 75] and double-axis tilting [76–78], are sometimes employed for ET, in this chapter, we only discuss our results obtained by single-axis tilting. Imaging modes that can be used for ET include BF CTEM [76–80], ADF-CTEM [81, 82] and ADF-STEM [46, 82–85], diffraction [86, 87], and elemental mapping by ESI (as part of EFTEM) [69, 88, 89] and by energy-dispersive X-ray spectroscopy (EDX) [89].

As for any imaging mode, contrast creation is the first step in ET too. Further, tilt-series acquisition, image alignment, reconstruction, and segmentation are consecutive steps that have to be performed with great care; however, imaging and reconstruction artifacts will be introduced, which make interpretation of the obtained data challenging. As an example, I will briefly highlight the influence of the missing wedge on the quality of the obtained volume reconstruction.

To achieve the best resolution and minimize artifacts, it is essential that the tilting range is as large as possible (ideally ±90°). In practice, the tilt range achievable in a TEM is usually restricted by the geometry of the specimen holder, gap between the pole pieces (and of the specimen itself, in case of thin-film samples) and is usually limited to a maximum range of about ±75°, maybe ±80°. The missing information delimitates a wedge-shaped region in Fourier space.

The missing wedge causes two problems. The first problem is that certain features may be distorted or even completely invisible depending on their orientation relative to the tilt axis. For example, if some nanorods lie in the (x,y) plain of the film specimen perpendicular to the tilt axis as shown in Figure 34.12, it is very likely that they will not appear in the 3D reconstruction at all [71, 90]. This results from the fact that the most important information relevant to these nanorods falls inside the missing wedge in the Fourier space. In case such nanorods are not strictly perpendicular to the tilt axis but are very close to this orientation, they may appear in the reconstruction but look vague, that is, they have less strong contrast with the matrix than analogous nanorods having other orientations (e.g., more parallel to the tilt axis or to the projection direction). This may complicate automatic segmentation and analysis of these features.

The second problem caused by the missing wedge is elongation of the reconstructed volume in the direction parallel to the electron beam (along the z-axis in Figure 34.12), which results in distorted features, and means that distance measurements in the z-direction have to be taken with caution. Radermacher has

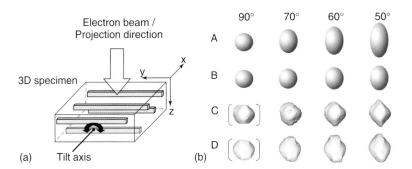

Figure 34.12 (a) Nanorods oriented in relation to the tilt axis as shown here, are very likely to be "invisible" in the 3D reconstruction because of the missing wedge. (b) Illustration of elongation factors in the z-direction as a function of the maximum tilt angle: (A) based on theoretical predictions by Radermacher [91]; (B) determined experimentally by Kawasea et al. [92] by tilting a rod-shaped sample of zirconia/polymer in the full tilting range; and (C,D) gold particles of around 50-nm diameter tilted in the range of ±70°, reconstructed by weighted back projection (WBP) (C) and simultaneous iterative reconstruction technique (SIRT) (D) and segmented afterwards. In (A) and (B), the elongations are visualized on a hypothetical spherical particle. (Reprinted with permission from Ref. [93]. Copyright (2010) Wiley-VCH.)

derived the following expression for this elongation e depending on the size of the missing wedge, viz. maximum tilt angle α_{max} [91]

$$e = \sqrt{\frac{a_{max} + \sin a_{max} \cos a_{max}}{a_{max} - \sin a_{max} \cos a_{max}}}. \qquad (34.2)$$

This implies elongation of 1.3 times in the z-direction for a maximum tilt angle of $70°$, 1.55 times for $\pm 60°$ tilt, and 1.9 times elongation for $\pm 50°$ tilt; these are significant numbers.

Experimental data show, however, that observed elongations can be (considerably) smaller than those predicted by expression (34.2). An elaborate study by Kawasea et al. [92] on the impact of the missing wedge for quantitative structural analysis, carried out on a rod-shaped sample of zirconia fillers in a polymer matrix that could be tilted in the full range of $\pm 90°$, gave the following experimentally determined elongation factors: 1.10 for $\pm 70°$ and 1.23 for $\pm 60°$ tilt. Our own tomography data (acquired in the range of $\pm 70°$) on gold particles also suggest that the experimental elongations in the z-direction are less pronounced than the trend derived by Radermacher.

A consequence of the loss of resolution in the z-direction caused by the missing wedge is that interpretation of the very bottom and very top (say a few first and last nanometers) of the specimen's reconstructed volume can be difficult. In this sense, SPM gives data complementary to that provided by ET.

The effect of the missing wedge can be reduced in one direction by acquiring projection data using two perpendicular tilt axes, the so-called double-axis (or dual-axis) tomography [94–96]. This technique reduces the missing wedge to a "missing pyramid". The resulting reconstructions are more accurate compared to single-axis tomography, but residual artifacts due to the missing angles remain.

34.8
One Example of Electron Tomographic Investigation: P3HT/PCBM

We have applied ET on various BHJ systems including MDMO-PPV–PCBM [7, 97], PF10TBT–PCBM [97], and mainly P3HT–PCBM [79, 97–99], which we will discuss in more detail. Results of the volume reconstruction applied to spin-coated and annealed P3HT–PCBM photoactive layers are shown as snapshots in Figure 34.13a–c. The volume data demonstrate evidently the presence of 3D nanoscale networks of both the P3HT nanowires and PCBM. In the volume data, some indication for inclination of the crystalline P3HT nanowires can be found, which is required for the creation of genuine 3D rather than 2D networks and is beneficial for hole–charge transportation from any place within the photoactive layer to the hole-collecting electrode (Figure 34.13d).

By going slice by slice through the reconstructed volume of the annealed P3HT/PCBM films, the amount of crystalline P3HT nanowires can be quantified in relation to the actual z-position within the photoactive layer (Figure 34.14a–c, exemplarily performed on the thermally annealed P3HT/PCBM layer). As evident

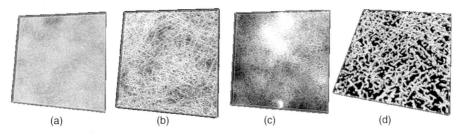

Figure 34.13 Results of TEM tomography; (a–c) snapshots of a reconstructed 3D volume of P3HT/PCBM photoactive layers: (a) as spin-coated, (b) thermally annealed, (c) solvent-assisted annealing for 3 h; the volume dimensions are about 1700 × 1700 × 100 nm, and (d) representation of the 3D network of P3HT nanowires formed in a thermally annealed P3HT–PCBM layer.

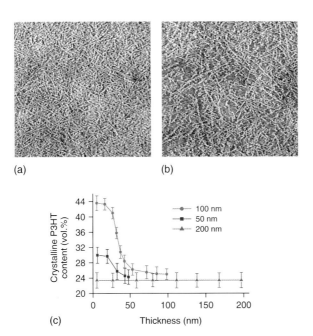

Figure 34.14 Results of electron tomography: quantification of the crystalline P3HT nanowires distribution through the thickness of the thermally annealed P3HT/PCBM photoactive layer. (a,b) Images of the original slices taken of the reconstructed volume of the film, with slice in (a) located close to the bottom of the film and that in (b) close to the top of the film. Dimensions of the slices are 1700 × 1700 nm. (c) The relative area (volume) occupied by P3HT in each slice is determined for all slices through the whole volume of the P3HT–PCBM film and plotted depending on a slice position and thickness of the photoactive layer. (Reprinted with permission from Refs. [79] and [98]. Copyright (2009) American Chemical Society.)

from Figure 34.14c, there is enrichment of crystalline P3HT nanowires in the lower part of the film close to the hole-collecting electrode and, correspondingly, enrichment of PCBM close to the top (electron-collecting) electrode, a situation that is beneficial for collection of free charges.

It should be noted that only crystalline P3HT nanowires are accounted for in this 3D morphology analysis since they have the required contrast with PCBM and get readily visualized in BF TEM. On the basis of the density values of P3HT and PCBM and a 1 : 1 weight ratio of these components in the photoactive layer, P3HT should occupy about 58% of the total volume of the layer. From the plot presented in Figure 34.14c, we can estimate that approximately 35% of the layer volume is actually made up of crystalline P3HT nanowires, which indicates a high crystallinity of P3HT of about 60%.

All the changes in the 3D volume organization of the P3HT–PCBM photoactive layer after annealing, resulting in the formation of nanoscale interpenetrating networks of P3HT nanowires and PCBM nanocrystals with favorable gradients within the thickness of the layer, are paramount for obtaining high-efficiency PSC. Indeed, characterization of the corresponding PSC shows that values of the short-circuit current density (J_{sc}), the open circuit voltage (V_{oc}), and fill factor all increase substantially after optimized thermal or solvent-assisted annealing.

Further, when comparing the volume organization of photoactive layers with different thickness, the basic conclusion is that an optimum morphological organization of a photoactive layer in all three dimensions is more crucial for high-efficiency devices than absorption alone. We have obtained the best device performance using moderately thick (100 nm) P3HT–PCBM photoactive layers characterized by high overall crystallinity of P3HT, namely, more numerous and more perfect crystalline P3HT nanowires forming a genuine 3D network, and by enrichment of P3HT close to the hole-collecting electrode. Thicker films (200-nm thick) absorb more light but show less-favorable morphological organization in photoactive layers, that is, lower crystallinity of P3HT, especially next to the hole-collecting electrode, and as a result produce poorly performing solar cell devices.

What exactly causes vertical segregation of crystalline P3HT is not clear at the moment. However, we infer that the existence and type of composition gradients through the thickness of the active layer are largely determined by the kinetic aspects of film formation because of different solution viscosities, different times taken for the solvent to evaporate, and eventual differences in local solvent concentration. These aspects have a direct impact on how long the (macro)molecules are mobile in the given solution/dispersion on eventual precipitation of components depending on local variations in solvent concentration and thus on formation of nuclei and subsequent growth and distribution of (nano)crystals throughout the active layer. Therefore, molecular characteristics of the components and the chosen processing route strongly determine the nanoscale organization within the photoactive layer.

34.9
Quantification of Volume Data

An alternative to polymer/fullerene solar cells are hybrid solar cells that use a combination of organic and inorganic materials. The concept of hybrid solar cells has been demonstrated by combining semiconducting polymers as donors with different inorganic materials, including CdSe [100, 101], TiO_2 [102, 103], and ZnO [104–106], as acceptors, and recently, hybrid devices with efficiency record (in 2009) were prepared by us [107].

In order to quantify the relevant morphological parameters, an extensive statistical analysis of the 3D datasets provided by ET was performed. The original 3D data of the BHJ photoactive layer were binarized to decide which voxels are ZnO and which are P3HT. The threshold for this binarization has a major impact on the final outcome, and hence error margins were estimated by applying the two extremes for this threshold. We have determined the volume fraction of ZnO for different preparation conditions and the spherical contact distances, defined as the distance from a certain voxel of one material to the nearest voxel of the other material.

Because excitons are mostly generated inside P3HT, we focused on the distance distribution from P3HT to ZnO. Figure 34.15 shows the probability to find P3HT at a certain shortest distance to a ZnO domain for photoactive layers with three different thicknesses. For 100- and 167-nm thick films, most P3HTs lie well within a shortest distance of 10 nm from ZnO. On the other hand, the 57-nm-thick sample displays a large amount of polymer at shortest distances as high as 25 nm from an interface with ZnO. This analysis substantiates that coarser phase separation is present in thinner layers.

The efficiency of the charge carrier generation was then estimated by modeling the exciton diffusion through the P3HT phase. For this, the 3D exciton diffusion equation was solved. As a result, an estimate of the fraction of excitons formed within P3HT reaching the interface with ZnO was obtained. Assuming that excitons efficiently dissociate into free charges at the interface with ZnO [108], the obtained numbers coincide with the efficiency of charge generation.

Beside charge carrier generation, carrier collection is also essential for solar cell operation. Efficient collection relies on continuous pathways for both carriers (Figure 34.15c,d). In view of the large volume excess of polymer in the blend, connectivity of this material will not be a limiting factor. The fraction of ZnO voxels that is interconnected via other ZnO voxels to the top of the investigated slab is quite high, at values well over 90% for all three layers, despite the low ZnO content. The connectivity is smaller for thicker layers, probably because larger distances have to be crossed.

The mere continuity of ZnO phase may not be enough to effectively collect the charges. Within a continuous phase, pathways may exist that do not continue into the direction toward the collecting electrode. Owing to the macroscopic electric field over the active layer of the device, charges may be trapped inside those cul-de-sacs and thus not collected (Figure 34.15d). Therefore, we also determined

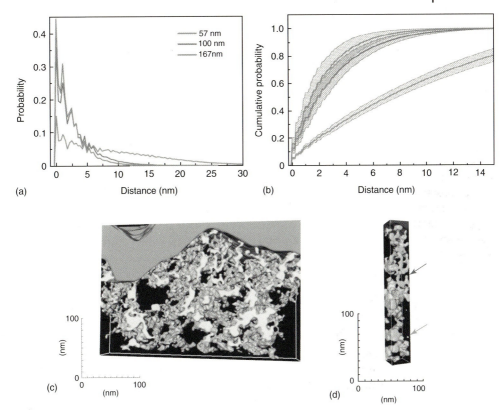

Figure 34.15 Statistical analysis of the 3D morphology: (a) distribution of the probability to find a P3HT voxel at a certain distance from a ZnO domain for mixed P3HT/ZnO films of different thickness. (b) Cumulative probability to have P3HT within a shortest distance to ZnO. The error margins indicated are obtained from the two most extreme thresholds possible for the binarization of the 3D data. (c) Reconstructed volume of a cross-section of the P3HT/ZnO device with light gray ZnO domains in dark P3HT matrix. (d) The part of this volume with the bright arrow indicating an isolated ZnO domain and the dark arrow indicating a ZnO domain connected to the top, but not through a strictly rising path. (Reprinted with permission from Ref. [107]. Copyright (2009) Nature Publishing Group.)

the fraction of ZnO connected to the top through a strictly rising path. The calculated unidirectionally connected fraction of ZnO is still very high (93%) for the 57-nm-thin layer but reduces for thicker layers, reaching 80% for the 167-nm-thick film.

In summary, the performed analysis of the volume data provides us a complete and quantitative morphological view on the nanoscale organization of the BHJ photoactive layer and allows for a straightforward relation with performance of corresponding devices.

34.10
Outlook and Concluding Remarks

Interpretation and reliable quantification of ET data is still challenging and needs further efforts. Already during initial tilt-series acquisition, imaging artifacts might be introduced, and missing wedge and beam sensitivity of organic materials need to be taken into account; not talking about questions arising from the tilt-series alignment, the reconstruction, and the segmentation procedures chosen. Therefore, data interpretation, further processing, and quantification should only be performed when all steps from image acquisition to volume reconstruction are done with care so that introduction of artifacts is minimized.

In this respect, tilt-series acquisition by using ADF-STEM needs further exploration. For example, we have reconstructed the volume of a conductive polymer nanocomposite filled with carbon black (CB) by applying CTEM tomography in slight defocus condition and ADF-STEM tomography [109]. After alignment of the tilt-series, volume reconstruction, and CB volume phase determination, we have measured the average CB volume concentrations and standard deviations for both imaging modes and compared the data with the known CB volume concentration of the nanocomposite. For CTEM tomography data, only defocusing provides enough contrast between the CB particles and the polymer matrix for allocating the CB phase in the tomographic volume reconstruction. The average CB volume concentration calculated from the reconstruction is much larger than the known concentration in the sample. Imaging artifacts and consequent distortion of the particle sizes make the obtained datasets unreliable for further quantification. On the other hand, volume concentration data obtained by ADF-STEM tomography are close to the known CB loading. In both imaging modes, the standard deviation of the CB volume concentration is comparable and thus reflects only the heterogeneity of the sample under investigation at the length scale we have observed.

Therefore, ADF-STEM imaging is able to create good contrast between different phases in polymer systems without the need of staining, and it introduces fewer imaging artifacts because of its incoherent nature so that volume datasets based on this imaging technique are very reliable for further quantification. Another promising aspect of ADF-STEM so far not discussed is that it allows for imaging and ET of micrometer-thick specimens with nanometer resolution when applying the so-called low-convergence angle STEM [110, 111]. At present, in my research group in Glasgow (Kelvin Nanocharacterization Center) we are further developing the above-described methodology and applying it to PSCs too.

A next step toward better understanding of relations between the nanoscale volume organization and the ultimate performance of printable solar cells is advanced control over the complete device architecture. In this respect, sectioning of whole devices by means of FIB seems to be a prospective route, allowing for straightforward preparation of complex multilayered specimens for further investigations [107]. Besides gaining additional information on interfacial organization, investigation of whole devices is the only accurate way to study lifetime-dependent chemical and physical aging. Certainly, volume morphological information can be very helpful to understand the aging mechanisms involved, and chemical volume mapping could tell us something on local oxidation or diffusing electrode material, to name but a few.

In a number of studies, we and others have demonstrated that the performance of PSCs depends critically on the nanoscale organization and functionality of the photoactive layer, the interfaces with and type of the charge collecting electrodes, and the overall device architecture. As demonstrated in this chapter, high-resolution microscopy or nanoscopy as tools to explore the nanoworld have played and continuously play an important role in the research area of printable solar cells. In particular, 3D imaging by means of ET reveals new insights on volume morphology, interface organization, and chemical composition when applying EFTEM. The main advantage of SPM, on the other hand, is the operation mode analyzing simultaneously the local morphology of the photoactive layer and measuring functional (optical and electrical) properties with nanometer resolution at the same spot. Such information provided by "nanoscopes" is mandatory to better understand the operation of printable solar cells at the nanoscale level and to bring such devices to a new level of performance.

Acknowledgment

I would like to use this opportunity to thank Martijn Wienk, Jan Kroon, Sjoerd Veenstra, Volker Schmidt, Thijs Michels, and many others for helpful discussions. Particularly, I want to express my thanks to René Janssen, who introduced me to the PSC theme, Xiaoniu Yang for his incredible hard and knowledgeable work on PSCs, which allowed us to set a mark in the area of morphological analysis of PSC, and Svetlana van Bavel and Erwan Sourty, who developed and applied ET for the first time on PSC systems. Further, I would like to thank the Dutch Polymer Institute (DPI) for support.

References

1. Keller, A. (1959) *Makromol. Chem.*, **34**, 1–28.
2. Peijs, A. (2003) *Mater. Today.* **April** 30–35
3. Hermann, W.A. (2006) *Energy*, **31**, 1685.
4. Dennler, G. and Brabec, C. (2008) in *Organic Photovoltaics* (eds C. Brabec, V. Dyakonov, and U. Scherf), Wiley-VCH Verlag GmbH, Weinheim, pp. 531–566.
5. Gunes, S. *et al.* (2007) *Chem. Rev.*, **107**, 1324.
6. Hoppe, H. and Sariciftci, N.S. (2008) *Adv. Polym. Sci.*, **214**, 1.
7. Yang, X. and Loos, J. (2007) *Macromolecules*, **40**, 1353.
8. Roes, A.L. *et al.* (2009) *Prog. Photovolt. Res. Appl.*, **17**, 372.
9. http://www.heliatek.com.
10. Pope, M. and Swenberg, C.E. (1999) *Electronic Processes in Organic Crystals and Polymers*, Oxford University Press, Oxford.
11. Sariciftci, N.S., Smilowitz, L., Heeger, A.J., and Wudl, F. (1992) *Science*, **I**, 1474.
12. Haugeneder, A., Neges, M., Kallinger, C., Spirkl, W., Lemmer, U., Feldmann, J., Scherf, U., Harth, E., Gügel, A.,

and Müllen, K. (1999) *Phys. Rev. B*, **59**, 15346.
13. Yoshino, K., Hong, Y.X., Muro, K., Kiyomatsu, S., Morita, S., Zakhidov, A.A., Noguchi, T., and Ohnishi, T. (1993) *Jpn. J. Appl. Phys., Part 2*, **32**, L357.
14. Halls, J.J.M., Pichler, K., Friend, R.H., Moratti, S.C., and Holmes, A.B. (1996) *Appl. Phys. Lett.*, **68**, 3120.
15. Savenije, T.J., Warman, J.M., and Goossens, A. (1998) *Chem. Phys. Lett.*, **287**, 148.
16. Kroeze, J.E., Savenije, T.J., Vermeulen, M.J.W., and Warman, J.M. (2003) *J. Phys. Chem. B*, **107**, 7696.
17. Yu, G. and Heeger, A.J. (1995) *J. Appl. Phys.*, **78**, 4510.
18. Halls, J.J.M., Walsh, C.A., Greenham, N.C., Marseglia, E.A., Friend, R.H., Moratti, S.C., and Holmes, A.B. (1995) *Nature*, **376**, 498.
19. Yu, G., Gao, J., Hummelen, J.C., Wudl, F., and Heeger, A.J. (1995) *Science*, **270**, 1789.
20. Sirringhaus, H., Brown, P.J., Friend, R.H., Nielsen, M.M., Bechgaard, K., Langeveld-Voss, B.M.W., Spiering, A.J.H., Janssen, R.A.J., Meijer, E.W., Herwig, P., and de Leeuw, D.M. (1999) *Nature*, **401**, 685.
21. Yang, X., Loos, J., Veenstra, S.C., Verhees, W.J.H., Wienk, M.M., Kroon, J.M., Michels, M.A.J., and Janssen, R.A.J. (2005) *Nano Lett.*, **5**, 579.
22. Hummelen, J.C., Knight, B.W., LePeq, F., Wudl, F., Gao, J., and Wilkins, C.L. (1995) *J. Org. Chem.*, **60**, 532.
23. Hoppe, H. and Sariciftci, N.S. (2006) *J. Mater. Chem.*, **16**, 45.
24. Shaheen, S.E., Brabec, C.J., Sariciftci, N.S., Padinger, F., Fromherz, T., and Hummelen, J.C. (2001) *Appl. Phys. Lett.*, **78**, 841.
25. Mozer, A., Denk, P., Scharber, M., Neugebauer, H., Sariciftci, N.S., Wagner, P., Lutsen, L., and Vanderzande, D. (2004) *J. Phys. Chem. B*, **108**, 5235.
26. Padinger, F., Rittberger, R.S., and Sariciftci, N.S. (2003) *Adv. Funct. Mater.*, **13**, 85.
27. Waldauf, C., Schilinsky, P., Hauch, J., and Brabec, C.J. (2004) *Thin Solid Films*, **451–452**, 503.
28. Al-Ibrahim, M., Ambacher, O., Sensfuss, S., and Gobsch, G. (2005) *Appl. Phys. Lett.*, **86**, 201120.
29. Ma, W., Yang, C., Gong, X., Lee, K., and Heeger, A.J. (2005) *Adv. Funct. Mater.*, **15**, 1617.
30. Svensson, M., Zhang, F., Veenstra, S.C., Verhees, W.J.H., Hummelen, J.C., Kroon, J.M., Inganäs, O., and Andersson, M.R. (2003) *Adv. Mater.*, **15**, 988.
31. Yohannes, T., Zhang, F., Svensson, M., Hummelen, J.C., Andersson, M.R., and Inganäs, O. (2004) *Thin Solid Films*, **449**, 152.
32. Slooff, L.H., Veenstra, S.C., Kroon, J.M., Moet, D.J.D., Sweelssen, J., and Koetse, M.M. (2007) *Appl. Phys. Lett.*, **90**, 143506.
33. Inganäs, O., Zhang, F., and Andersson, M.R. (2009) *Acc. Chem. Res.*, **42**, 1731.
34. Kroon, R., Lenes, M., Hummelen, J.C., Blom, P.W.M., and de Boer, B. (2008) *Polym. Rev.*, **48**, 531.
35. Peet, J., Kim, J.Y., Coates, N.E., Ma, W.L., Moses, D., Heeger, A.J., and Bazan, G.C. (2007) *Nat. Mater.*, **6**, 497.
36. Qin, R., Li, W., Li, C., Du, C., Veit, C., Schleiermacher, H.-F., Andersson, M., Bo, Z., Liu, Z., Inganäs, O., Würfel, U., and Zhang, F. (2009) *J. Am. Chem. Soc.*, **131**, 14612.
37. Chu, T.-Y., Alem, S., Verly, P.G., Wakim, S., Lu, J., Tao, Y., Beaupré, S., Leclerc, M., Bélanger, F., Désilets, D., Rodman, S., Waller, D., and Gaudiana, R. (2009) *Appl. Phys. Lett.*, **95**, 063304.
38. Yang, X., van Duren, J.K.J., Janssen, R.A.J., Michels, M.A.J., and Loos, J. (2004) *Macromolecules*, **37**, 2152.
39. Hoppe, H., Niggemann, M., Winder, C., Kraut, J., Hiesgen, R., Hinsch, A., Meissner, D., and Sariciftci, N.S. (2004) *Adv. Funct. Mater.*, **14**, 1005.
40. Rispens, M.T., Meetsma, A., Rittberger, R., Brabec, C.J., Sariciftci, N.S., and Hummelen, J.C. (2003) *Chem. Commun.*, 2116.
41. van Duren, J.K.J., Loos, J., Morrissey, F., Leewis, C.M., Kivits, K.P.H., van IJzendoorn, L.J., Rispens, M.T.,

Hummelen, J.C., and Janssen, R.A.J. (2002) *Adv. Func. Mat.*, **12**, 665–669.
42. Zhong, H., Yang, X., de With, B., and Loos, J. (2006) *Macromolecules*, **39**, 218–223.
43. Yang, X., van Duren, J.K.J., Rispens, M.T., Hummelen, J.C., Janssen, R.A.J., Michels, M.A.J., and Loos, J. (2004) *Adv. Mater.*, **16** (9–10), 802–806.
44. Yang, X., Alexeev, A., Michels, M.A.J., and Loos, J. (2005) *Macromolecules*, **38**, 4289–4295.
45. Loos, J., Sourty, E., Lu, K., de With, G., and Bavel, S. (2009) *Macromolecules*, **42** (7), 2582–2586.
46. Sourty, E., Bavel, S., Lu, K., Guerra, R., Bar, G., and Loos, J. (2009) *Microsc. Microanal.*, **15**(3), 251–258.
47. Loos, J., Yang, X., Koetse, M.M., Sweelssen, J., Schoo, H.F.M., Veenstra, S.C., Grogger, W., Kothleitner, G., and Hofer, F. (2005) *J. Appl. Polym. Sci.*, **97**, 1001–1007.
48. Alexeev, A., Loos, J., and Koetse, M.M. (2006) *Ultramicroscopy*, **106**, 191–199.
49. Alexeev, A. and Loos, J. (2008) *Org. Electron.*, **9**, 149–154.
50. Reimer, L. and Kohl, H. (2008) *Transmission Electron Microscopy: Physics of Image Formation*, 5th edn, Springer-Verlag, New York, p. 188.
51. Heidenreich, R.D. (1964) *Fundamentals of Transmission Electron Microscopy*, John Wiley & Sons, Inc., New York, p. 31.
52. Krivanek, O.L., Gubbens, A.J., and Dellby, N. (1991) *Microsc., Microanal., Microstruct.*, **2**, 315.
53. Grogger, W., Schaffer, B., Krishnan, K.M., and Hofer, F. (2003) *Ultramicroscopy*, **96**, 481.
54. Kravinek, O.L., Kundmann, M.K., and Kimotot, K. (1995) *J. Microsc.*, **180**, 277.
55. Varlot, K., Martin, J.M., Quet, C., and Kihn, Y. (1997) *Ultramicroscopy*, **68**, 123.
56. Varlot, K., Martin, J.M., Gonbeau, D., and Quet, C. (1999) *Polymer*, **40**, 5691.
57. Jeanguillaume, C., Trebbia, P., and Colliex, C. (1978) *Ultramicroscopy*, **3**, 237.
58. Shafai, C., Thomson, D.J., Simard-Normandin, M., Mattiussi, G., and Scanlon, P. (1994) *J. Appl. Phys. Lett.*, **64**, 342.
59. De Wolf, P., Snauwaert, J., Clarysse, T., Vandervorst, W., and Hellemans, L. (1995) *Appl. Phys. Lett.*, **66**, 1530.
60. Kelley, T.W. and Frisbie, C.D. (2000) *J. Vac. Sci. Technol.*, **B18**, 632.
61. Lin, H.-N., Lin, H.-L., Wang, S.-S., Yu, L.-S., Perng, G.-Y., Chen, S.-A., and Chen, S.-H. (2002) *Appl. Phys. Lett.*, **81**, 2572.
62. Kelley, T.W. and Frisbie, C.D. (2000) *J. Vac. Sci. Technol.*, **B18**, 632.
63. Heinz, W.F. and Hoh, J.H. (1999) *Trends Biotechnol.*, **17**, 143.
64. Eyben, P., Xu, M., Duhayon, N., Clarysse, T., Callewaert, S., and Vandervorst, W. (2002) *J. Vac. Sci. Technol.*, **B20**, 471.
65. Blom, P.W.M., de Jong, M.J.M., and Vleggaar, J.J.M. (1996) *Appl. Phys. Lett.*, **68**, 3308.
66. Tanase, C., Blom, P.W.M., and de Leeuw, D.M. (2004) *Phys. Rev.*, **B70**, 193202.
67. Loos, J., van Duren, J.K.J., Morrissey, F., and Janssen, R.A.J. (2002) *Polymer*, **43**, 7493.
68. Martens, T., D'Haen, J., Munters, T., Beelen, Z., Goris, L., Manca, J., D'Olieslaeger, M., Vanderzande, D., Schepper, L.D., and Andriessen, R. (2003) *Synth. Met.*, **138**, 243.
69. Weyland, M. (2002) *Top. Catal.*, **21**, 175.
70. Weyland, M. and Midgley, P.A. (2004) *Mater. Today*, **7**, 32.
71. Jinnai, H. and Spontak, R.J. (2009) *Polymer*, **50**, 1067.
72. Möbus, G. and Inkson, B.J. (2007) *Mater. Today*, **10**, 18.
73. Cormack, A.M. (1963) *J. Appl. Phys.*, **34**, 2722.
74. Radermacher, M. (1980) PhD thesis, Department of Physics, University of Munich, Munich, Germany.
75. Lanzavecchia, S., Cantele, F., Bellon, P.L., Zampighi, L., Kreman, M., Wright, E., and Zampighi, G.A. (2005) *J. Struct. Biol.*, **149**, 87.
76. Koster, A.J., Ziese, U., Verkleij, A.J., Janssen, A.H., and de Jong, K.P. (2000) *J. Phys. Chem. B*, **104**, 9368.

77. Friedrich, H., Sietsma, J.R.A., de Jongh, P.E., Verkleij, A.J., and de Jong, K.P. (2007) *J. Am. Chem. Soc.*, **129**, 10249.
78. Ziese, U., de Jong, K.P., and Koster, A.J. (2004) *Appl. Catal. A*, **260**, 71.
79. van Bavel, S.S., Sourty, E., de With, G., and Loos, J. (2009) *Nano Lett.*, **9** (2), 507–513.
80. Ikeda, Y., Katoh, A., Shimanuki, J., and Kohjiya, S. (2004) *Macromol. Rapid Commun.*, **25**, 1186.
81. Bals, S., Van Tendeloo, G., and Kisielowski, C. (2006) *Adv. Mater.*, **18**, 892.
82. Friedrich, H., McCartney, M.R., and Buseck, P.R. (2005) *Ultramicroscopy*, **106**, 18.
83. Ziese, U., Kübel, C., Verkleij, A.J., and Koster, A.J. (2002) *J. Struct. Biol.*, **138**, 58.
84. Kübel, C., Voigt, A., Schoenmakers, R., Otten, M., Su, D., Lee, T.C., Carlsson, A., and Bradley, J. (2005) *Microsc. Microanal.*, **11**, 378.
85. Midgley, P.A., Weyland, M., Meurig Thomas, J., and Johnson, B.F.G. (2001) *Chem. Commun.*, **10**, 907.
86. Kolb, U., Gorelik, T., Kübel, C., Otten, M.T., and Hubert, D. (2007) *Ultramicroscopy*, **107**, 507.
87. Kolb, U., Gorelik, T., and Otten, M.T. (2008) *Ultramicroscopy*, **108**, 763.
88. Möbus, G. and Inkson, B.J. (2001) *Appl. Phys. Lett.*, **79**, 1369.
89. Möbus, G., Doole, R.C., and Inkson, B.J. (2003) *Ultramicroscopy*, **96**, 433.
90. Penczek, P.A. and Frank, J. (2006) in *Electron Tomography* (ed. J. Frank), 2nd edn, Springer, pp. 307–330.
91. Radermacher, M. (1988) *J. Electron Microsc. Tech.*, **9**, 359.
92. Kawasea, N., Katoa, M., Nishiokab, H., and Jinnai, H. (2007) *Ultramicroscopy*, **107**, 8.
93. van Bavel, S.S. and Loos, J. (2010) *Adv. Funct. Mater.*, **20**, 3217–3234.
94. Penczek, P., Marko, M., Buttle, K., and Frank, J. (1995) *Ultramicroscopy*, **60**, 393.
95. Arslan, I., Tong, J.R., and Midgley, P.A. (2006) *Ultramicroscopy*, **106**, 994.
96. Mastronarde, D.N. (1997) *J. Struct. Biol.*, **120**, 343.
97. van Bavel, S., Sourty, E., de With, G., Veenstra, S., and Loos, J. (2009) *J. Mater. Chem.*, **19** (30), 5388–5393.
98. Bavel, S., Sourty, E., de With, G., Frolic, K., and Loos, J. (2009) *Macromolecules*, **42** (19), 7396–7403.
99. van Bavel, S.S., Bärenklau, M., de With, G., Hoppe, H., and Loos, J. (2010) *Adv. Funct. Mater.*, **20** (9), 1458–1463.
100. Huynh, W.U., Dittmer, J.J., and Alivisatos, A.P. (2002) *Science*, **295**, 2425.
101. Wang, P., Abrusci, A., Wong, H.M.P., Svensson, M., Andersson, M.R., and Greenham, N.C. (2006) *Nano Lett.*, **6**, 1789.
102. Kwong, C.Y., Djurisic, A.B., Chui, P.C., Cheng, K.W., and Chan, W.K. (2004) *Chem. Phys. Lett.*, **384**, 372.
103. Kuo, C.Y., Tang, W.C., Gau, C., Guo, T.F., and Jeng, D.Z. (2008) *Appl. Phys. Lett.*, **93**, 033307.
104. Beek, W.J.E., Wienk, M.M., and Janssen, R.A.J. (2004) *Adv. Mater.*, **16**, 1009.
105. Beek, W.J.E., Wienk, M.M., and Janssen, R.A.J. (2006) *Adv. Funct. Mater.*, **16**, 1112.
106. Olson, D.C., Shaheen, S.E., Collins, R.T., and Ginley, D.S. (2007) *J. Phys. Chem. C*, **111**, 16640.
107. Oosterhout, S.D., Wienk, M.M., van Bavel, S.S., Thiedmann, R., Koster, L.J.A., Gilot, J., Loos, J., Schmidt, V., and Janssen, R.A.J. (2009) *Nat. Mater.*, **8**, 818.
108. Ravirajan, P., Peiro, A.M., Nazeeruddin, M.K., Graetzel, M., Bradley, D.D.C., Durrant, J.R., and Nelson, J. (2006) *J. Phys. Chem. B*, **110**, 7635.
109. Lu, K., Sourty, E., Guerra, R., Bar, G., and Loos, J. (2010) *Macromolecules*, **43** (3), 1444–1448.
110. Loos, J., Sourty, E., Lu, K., Freitag, B., Tang, D., and Wall, D. (2009) *Nano Lett.*, **9** (4), 1704–1708.
111. Soutry, E., Freitag, B., Wall, D., Tang, D., Lu, K., and Loos, J. (2008) *Microsc. Microanal.*, **14** (2), 1056–1057.

35
Ferroic and Multiferroic Materials
Ekhard Salje

35.1
Multiferroicity

Since 2001 we have witnessed an explosion in the number of publications concerning multiferroic properties of materials. Multiferroicity combines at least two of the three ferroic properties of materials: ferroelasticity, ferroelectricity, and ferromagnetism. Its investigation has a long tradition, with work on boracites in the 1960s and extensive theoretical developments [1] together with a continuous stream of activities on ceramics with perovskite-like structures [2–5]. In addition, it was realized that "ferroelastics" and "martensites" describe the same materials properties that have simply different historic traditions for their naming (so that "ferroelastic" alloys are usually called *"martensites"* and have often, but not always, stepwise phase transitions, whereas ceramics and minerals, which have "martensitic" properties, are usually called "ferroelastics"). Magnetic martensites, for example, are multiferroics and follow the same physical mechanism as the very large number of ferromagnetic and ferroelastic ceramics.

Multiferroic materials may be used to develop improved memory devices [4]. For this purpose, ferroelasticity is of minor importance because reading/writing mechanisms will rely on magnetic or electric fields so that the key for the development of multiferroics is the combination of ferroelectric and ferromagnetic properties. The role of ferroelasticity is nevertheless often key for the performance of such devices: coupling between different ferroic properties can be "strain induced," where both properties couple strongly with some lattice distortion (via, e.g., magnetostriction and electrostriction or piezoelectricity) and, thus, couple with each other. Strain-induced coupling occurs on a mesoscopic scale [6–9], which allows the use of enhanced strains and strain gradients via the formation of microengineered resolution patter, small grain sizes, and surface relaxations. In this chapter, we discuss the advances in microscopy relating to the static properties of multiferroic walls and do not comment on the increasingly important dynamical features of wall movements under external fields.

A concurrent development relates to multiferroic or ferroic properties inside domain walls in situations where no ferroicity may exist in the bulk. This leads the

Handbook of Nanoscopy, First Edition. Edited by Gustaaf Van Tendeloo, Dirk Van Dyck, and Stephen J. Pennycook.
© 2012 Wiley-VCH Verlag GmbH & Co. KGaA. Published 2012 by Wiley-VCH Verlag GmbH & Co. KGaA.

way to memory devices in which the active elements are domain walls. Such walls are thin, and when structured, each active element (pixel) would have geometrical dimensions on a nanometer scale. If devices of this kind could be produced, we would create memory capacities that are several orders of magnitude larger than the best devices in the year 2010. No wonder that a large number of research groups work on this problem worldwide and constitute a field that is best described as "domain boundary engineering" [10, 11].

Now we need to modify the term multiferroic somewhat. Many of the desirable interfacial properties are not just connected with elastic deformation, electric polarization, and magnetic moment but require in addition that information can be read by electric circuits. Such reading is best ensured by strong contrasts in electrical conductivity, so high electric conductivity of interfaces is another desired quality that goes together with its multiferroic properties [12]. We will see that this property can interfere positively or negatively with the other ferroic features of a device.

Ferroelasticity is, in its own right, less important for most multiferroic device applications. The role of the ferroelastic spontaneous strain is often to provide a coupling mechanism between other ferroic or conductivity-related properties. In the latter case, this is typically done in systems that are close to a Mott-type metal/insulator (MI) transition, where the strain plays the same role as the macroscopic deformation that transforms a material from an insulator into a metal. The difference is that the domain wall strain is a strain gradient and is localized around interfaces rather than as a uniform deformation. In this context, the idea of coelastic material is important: ferroelastic materials provide a spontaneous strain together with mobile twin boundaries, and hence a ferroelastic hysteresis. Coelastic materials, on the other hand, do not show a hysteresis while they still possess a spontaneous strain. The reason is that domain boundaries, such as twin boundaries, either do not exist or are so strongly pinned that external forces are not able to move them. Coelastic materials are, thus, similar to polar (but not ferroelectric) materials or so-called hard ferromagnets. Their role in multiferroic domain structures is hence that of a passive "messenger" that provides coupling between ferroic properties rather than an active element of a nonlinear response of the material to external forces.

An instructive example are IV–VI compounds, which have provided a basis for semimetals and small-gap semiconductors with high-temperature phases in various crystal structures transforming into other structural phases on lowering the temperature. The prototype has the cubic rocksalt structure with space group Fm3m. On structurally transforming by lowering temperature, SnTe and GeTe, for example, become polar with space group R3m, and often ferroelectricity has been evoked. In addition, the physical properties of members of this family of compounds depend on the value of their excess carrier and defect concentration. Several ternary compounds, such as $Pb_{1-x}Sn_xTe$ and $Ge_{1-x}Sn_xTe$, have also received considerable attention because changing the carrier concentration was seen as way not only to tailor the size of the band gaps but also to switch the parity of the bands across the gap. Much recent research on these materials was carried out because of a perceived opportunity to create novel tunable multifunctional devices on the nanoscale. Thus, magnetically doped SnTe was perceived to have the potential of being multiferroic with three features – ferroelasticity, ferroelectricity, and ferromagnetism – and interesting electronic properties. A detailed analysis shows that this is not the case; clearly, there are possible pitfalls when multiferroic properties are wrongly anticipated. We are now discussing this case in some more detail.

Group theory shows that the phase transition Fm3m – R3m occurs by a two-step symmetry reduction. The first is the reduction of the cubic phase symmetry to that of a rhombohedral phase (R-3m) with an elastic strain along the cube diagonal. The irreducible representation of this elastic order parameter is T_{2g} [8a], and this resulting phase has an inversion center. The allowed strain components are xy, xz, and yz. The second symmetry reduction removes the inversion center by a relative displacement of the two sublattices whose irreducible representation is T_{1u} [8a]. The excess Gibbs free energy can be written [8] as a sum of the Landau potential of the primary order parameter, the elastic strain energy, the coupling energy between the primary order parameter and the spontaneous strain, and the Ginzburg energy, which describes the domain boundary energy:

$$G(p_i, e_{jk}) = L(P_i) + f_{el}(e_{jk}) + f_{coupling}(P_i, e_{jk}) + f_{Ginzburg}(\nabla P_i)$$

The basic functions of T_{1u} are x, y, and z. The relevant form of the Landau free energy $L(P_i)$ for the rhombohedral (m3m) symmetry is then [8a]

$$L(P) = 1/2\alpha \left(P_1^2 + P_2^2 + P_{31}^{24} + P_2^4 + P_3^4\right)$$

$$+1/4B'\left(P_1^2 P_2^2 + P_2^2 P_3^2 + P_1^2 P_2^2\right) + 1/6C\left(P_1 P_2 P_3\right)^2$$

$$+1/6C'\left(P_1^2 + P_2^2 + P_3^2\right)^3$$

$$+1/6C''\left(P_1^2 + P_2^2 + P_3^2\right)\left(P_1^4 + P_2^4 + P_3^4\right)$$

Within one domain, for example, the (111) domain, one component of P_i is nonzero, while all others are zero. Thus the monodomain Landau free energy has the reduced form

$$L(P) = 1/2\alpha P^2 + 1/4BP^4 + 1/6CP^6$$

with coefficients $B = B(B, B')$ and $C = C(C, C', C'')$. This is the simplest Landau free energy for a continuous phase transition when $B>0$ or $B = 0$ and $C>0$ and

$$\alpha = A\Theta s [\coth(\Theta_s/T) - \coth(\Theta_s/T_c)]$$

where T_c is the transition temperature in a continuous phase transition and Θ_s is the characteristic temperature for the low-temperature quantum saturation [8f].

For the [111] domain, the spontaneous strain is symmetry constrained with $e_4 = e_{xy} + e_{xz} + e_{yz}$. The elastic energy is for a single domain crystal $f_{el}(e_4) = 1/2\, C_{44} e_4^2$, and the coupling energy is $f_{coupling}(P_i, e_{ij}) = 1/2 P^2 e_4$. If we were to relax the free energy with respect to this strain, $dG/de_4 = 0$, we find the same functional form as $L(P)$ with B replaced by a renormalized parameter B^*. Thus, as typical when a secondary order parameter is present, the nature of the transition remains described by the Landau free-energy form of the primary order parameter but with renormalized coefficients. One now has to consider the interaction between the relevant parameters. First, P indicates ferroelectricity only if P is reversible under an external electric field. This is impossible, however, because the sample is highly conducting so that the local field strength is very small and will not exceed the coercive field strength. The sample is, thus, polar because $P \neq 0$ but not ferroelectric because any field is essentially short ircuited. Second, the spontaneous strain e_4 is also finite because of the coupling with P, but one has to check whether e_4 can be inverted by external stress. Experimental work has shown that this is not the case in reasonably pure material. Only when the sample is doped with Cr can one obtain moving domain boundaries.

This example is generic for much research on multiferroics: while the order parameters allow ferroelasticity and ferroelectricity to exist and mix with conductivity, this mixing annuls all multiferroicity. High conductivity destroys the local electric fields so that no ferroelectric domain switching occurs. Extremely small strain e_4 is also insufficient to nucleate twin boundaries so that no moving domain walls exist. SnTe is hence neither multiferroic nor even ferroic at all. It is a weakly coelastic material, which can be made ferroelastic (but not ferroelectric) by including nucleation centers for the twin walls. This demonstrates clearly that the requirement for multiferroicity is twofold, namely, the order parameter must allow for the appearance of ferroic properties and these properties need to be switchable under an appropriate external field. A hypothetical way to establish multiferroicity in SnTe would be to generate domain walls and restrict the conductivity of such domain walls. We have shown below that this was achieved in WO_3.

35.2
Ferroic Domain Patterns and Their Microscopical Observation

We have shown that ferroicity is intrinsically linked with the appearance of domain walls that separate domains with different values of the ferroic order parameter. External forces can then stabilize one of these domains and destabilize the other. The domain boundary between these domains will shift and give rise to the desired ferroic property (we are not commenting here on high-field features where bulk switching can occur without the gradual movement of domain walls). This means that domain boundaries are at the core of research on ferroic materials, and much work has been spent to make such boundaries visible. An essential experimental technique for research in multiferroic/conducting domain boundaries is, thus, microscopy. Traditional optical and transmission electron microscopy (TEM) focused essentially on the orientation of domain boundaries (Figures 35.1 and 35.2) and their capability to form domain patterns. This approach dominated the previous four decades, their application together with the various forms of atomic force microscopy (AFM), piezoforce microscopy (PFM), and so on, have led to significant breakthroughs in the analysis of the internal structure of domain walls and the dynamics of their movement. In addition, magnetic domain

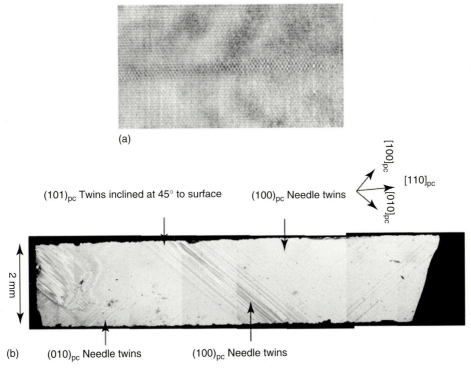

Figure 35.1 (a) Transmission electron microscopic image of NdGaO3 (Pbnm) near a {101} twin boundary in the middle of the figure. The unit cell is $a = 0.5426$ nm, $b = 0.5502$ nm, $c = 0.7706$ nm. The thickness of the interface is $2w \sim 6$ unit layers or ~ 2.3 nm. (Photograph courtesy G. Van Tendeloo, Antwerp.) (b) Needle domains in ferroelastic LaAlO3.

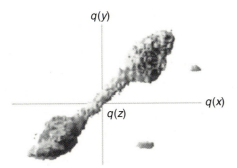

Figure 35.2 Diffuse diffraction of a multitude of twin boundaries in WO3. The "dog-bone" structure consists of two maxima at the two ends surrounding the Bragg peaks of the adjacent domains. The weak middle intensity at the origin relates to the thickness of the individual domain walls. (Reproduced from [8c].)

structures are increasingly been investigated by Lorentz microscopy, which is outside the scope of this chapter.

Under equilibrium conditions, patterns are the result of energy minimization (with appropriate boundary conditions). Simple rules, such as the compatibility relation [8a, 13] have been developed and widely applied. In fact, microscopic images are often used to identify ferroic and multiferroic behavior because crystals that undergo phase transformations are likely to show characteristic microstructures such as twin structures as a typical fingerprint for ferroelastic behavior [8, 14]. The interaction between twin walls leads to the formation of corner domains, needle domains, and comb patterns. These possibilities are illustrated in Figure 35.3. Complex ferroic patterns occur on length scales that bridge the atomic scale and the macroscopic scale, a fact that leads to the definition of the term "*mesoscopic structures*" as being in between these two classic regimes. The patterns are also hierarchical because of the general principle that any pattern can be decomposed into simpler patterns until the most elementary unit, namely the twin wall, is found. Twin domains and their boundaries are the dominant microstructures in many materials and have been shown to be important, theoretically and experimentally, for the interpretation of the thermodynamic behavior of materials. The application of twin walls to control desirable behavior of materials depends on a proper understanding of the fundamental unit of these hierarchical mesostructures, the twin wall. In the context of the derivation of Landau theory [8], the energy associated with spatial variations of the order parameter lies in its spatial gradient energy and leads, in most cases, to profiles of the functional form "$\tanh(x/w)$," where x is the direction perpendicular to the wall and w sets the relevant length scale and stands for half the thickness of the interface.

The result that structural gradients of twin walls extend over several interatomic distances with a length scale w leads to characteristic images of twin walls. A high-resolution TEM image of a twin wall in $NdGaO_3$ is shown in Figure 35.1, where the imaging condition was optimized for atoms inside the twin wall, whereas atoms outside the wall are slightly out of focus by inclination of their lattice plane from the plane of diffraction. The "thickness" of the twin wall can now be estimated by simply counting the number of unit cells in the wall. The resulting wall thickness of about 2 nm compares well with results from diffraction experiments. The wall thickness increases when the transition point is approached. Careful analysis of the diffuse diffraction of wall-related signals in $LaAlO_3$ showed that the wall thickness increases according to the predictions of Landau Ginzburg theory for a second-order phase transition [17]. In first-order martensitic transitions, the effect is smaller, although the increase follows the temperature evolution of the correlation length, which still leads to significant increases near the transformation point in compounds such as NiTi and NiTiFe [18].

Although the functional form of the wall profile ($\tanh x/w$) is essentially independent of temperature, one finds characteristic temperature dependence of the wall thickness w within the framework of Landau theory [8]. The experimental determination of $w(T)$ was first attempted using electron microscopy [19]. Electron paramagnetic resonance (EPR) spectroscopy has also been used to probe the

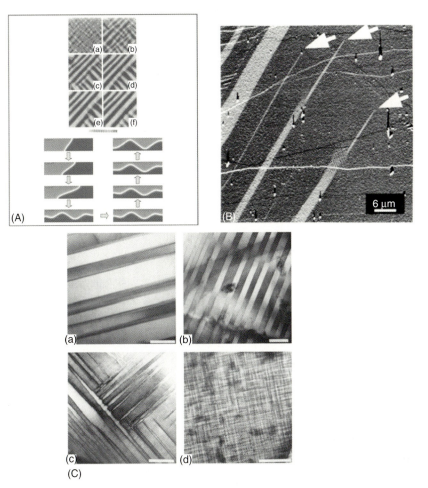

Figure 35.3 (A) Computer simulation of the development of domain patterns in order/disorder systems (top) starting from a tweed pattern to tartan pattern and junction pattern, which develop into a stripe pattern with orthogonal needle domains. The bending of a crystal slab (bottom) rotates a 45° domain wall into an energetically unfavorable direction. This domain wall will "wiggle" with segments in the optimal 45° direction. The surface relaxation repels the wall so that no connection between the surface and the wall is possible. The arrow indicates the increase (left) and decrease (right) of the bending force. Note that the process is irreversible, the stress-free pattern on the left and the right differ. (B) AFM amplitude data of ferroelastic W domains on a (100) surface of $Pb_3(PO_4)_2$. The W domains end as needlelike structures – indicated by arrows. No change in the local surface deformation in the vicinity of a needle tip could be detected. (Reproduced from [15].) (C) Sequence of domain structures in superconducting $YBa_2Cu_3O_7$ doped with Co: (a,b) simple lamellar twins (0 and 1% Co replacing Cu, respectively), (c) two perpendicular twin orientations with junctions and needles (2.5%), and (d) tweed microstructure (2.8%). The scale bar is 0.1 mm. (Reproduced from [16b].

Table 35.1 Values of the twin wall half-width w at room temperature for some materials showing ferroelastic domains, measured by X-ray diffraction methods.

Material	Twin wall half-width (w) (nm)	Reference
$Pb_3(PO_4)_2$	7.2	[23]
$Pb_3(P_{0.77} As_{0.23} O_4)_2$	2.1	[24]
$YBa_2Cu_3O_{7-\delta}$	0.7	[25]
$(Na, K)AlSi_3O_8$	1.3	[22]
WO_3	1.6	[8c]
$LaAlO_3$	2	[17]

structure and thickness of twin walls [20]. X-ray and neutron diffraction methods have proved to be probably the most precise tool together with AFM to study twinning microstructures and their associated domain boundaries, provided care is taken to optimize the collection and processing of the diffraction signals [8, 17]. The characteristic diffraction pattern from a single twin wall, or a set of parallel twin walls, is a weak streak between two twin-related diffraction peaks, as shown in Figure 35.2. Comparing the relative intensities of the bulk and wall diffraction signals allows the twin wall width to be estimated (Table 35.1). These half-widths are typically of the order of 2–5 unit cells, so that there are up to 10 unit cells within the domain wall. Similar results were obtained in a Monte Carlo simulation study of a [001] 90° rotation twin wall in $CaTiO_3$ with $w = 0.57$ nm [21]. By studying the evolution of the twin wall signal with temperature, the temperature dependence of w has been studied in a disordered (Na, K) feldspar [22] and the ferroelastic perovskite $LaAlO_3$ [17]. In both these systems, the expected Landau-type behavior was confirmed.

35.3
The Internal Structure of Domain Walls

Domain boundaries show reduced chemical bonding compared with the bulk material. As a consequence, one expects domain boundaries to react more strongly to external fields than the bulk so that elastic, magnetic, or electric susceptibilities are larger in the boundary than their equivalent bulk values. Such "softening" of the response to fields can hardly be measured macroscopically because the volume proportion of interfaces is relatively small compared with that of the bulk. Exceptions are relaxor materials where the relaxor regions themselves have wall properties and show indeed strong finite size effects and large susceptibilities [26]. Twin boundaries can reach several parts per million of the sample volume so that signals from internal ferroic properties of the boundaries may, under extreme circumstances, compete with the macroscopic properties of the bulk. The advantage of the localized wall properties is their high information density: they are contained

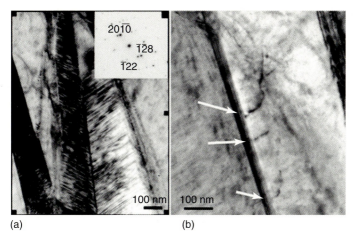

Figure 35.4 (a) Martensite plate showing a twin midrib with the corresponding "selected area diffraction pattern" as inset revealing the respective spot splitting. (b) Dislocation lines arriving at the twin midrib of the martensite plate. (Reproduced from [27].)

within very thin sheets of twin walls and can be addressed spatially with very high resolution. This means that the storage density of information encrypted in twin walls is extremely high so that wall-related devices could theoretically outperform bulk devices by several orders of magnitude.

The microscopy for such walls is less advanced than the investigation of domain pattern as described before. In particular, the imaging of the walls requires that such walls are static in order to obtain high enough spatial resolution. A typical example is the twinning in $Cu_{74.08}Al_{23.13}Be_{2.79}$ [27], which shows that topological pinning is itself a feature of the walls: in this case, the wall structure allows for the nucleation of dislocations that are then pinning the walls (Figure 35.4). Other pinning mechanisms are extrinsic, while intrinsic Peierls pinning was found to be less important in multiferroic materials with rather wide domain walls [28].

The discovery of ferroic properties of interfaces is often related to computer simulations for materials design and the theoretical exploration of extreme physical properties in solids. Such research expands from inorganic materials to biological samples. We first discuss an example in solid state physics, where sizable spontaneous polarization was predicted in {100} twin walls of $CaTiO_3$, a definitely nonpolar material. Theoretical simulations [29] of these walls show an extremely rich texture of the local polarization at and close to the walls. Local distortions include a strong antiferroelectric component, and local nonzero contributions perpendicular to the wall plane, which do not contribute to the macroscopic net dipole moment. Individual Ti displacements of 2 pm off the octahedron center give rise to a net polarization corresponding to a displacement of 0.6 pm in the direction of the bisector of the twin angle. The effect is intrinsically coupled with the appearance of twin boundaries in the matrix, which was already previously identified as locality of oxygen vacancies in $CaTiO_3$ [18a].

While indirect evidence for the polar behavior of twin walls has been reported before [30], as well as in antiphase boundaries (APBs) [31] and grain boundaries [32], the results for $CaTiO_3$ were very instructive, as these results were the first clear indication of twin wall polarity and the underlying structural mechanism for the coupling between strain and dipole moments. $CaTiO_3$ is orthorhombic in its low-temperature form (space group Pnma) and is purely ferroelastic. No ferroelectric features have ever been recorded. The TiO_6 octahedron, on the other hand, is well known for its tendency to form polar groups, where the Ti position is off centered with respect to the geometrical center of the surrounding oxygen atoms (Figure 35.5). Such polar structures exist in compounds such as $BaTiO_3$, $PbTiO_3$, and others. The known competition with octahedral rotation [33] in the tetragonal and orthorhombic phases of $CaTiO_3$ suppresses the off-centering (Figure 35.6). It is, however, restored when the rotation angle vanishes or when the density of the material decreases. Both conditions are met inside the twin wall, and it is thus not entirely unexpected that twin walls should show dipolar moments. What were unknown are the predicted size of the polarization and the texture of the polarization field.

In addition to polar properties of the walls, an increase in the density of oxygen vacancies was also predicted. An oxygen vacancy gains energy when shifted from the bulk of the material into the twin wall [29a]. While this effect is expected from the fact that twin walls in the geological context are known to be decorated by defects, we understand from these calculations that the geometrical requirement for the accommodation of defects may appear negligible, namely about 1% increase of a lattice spacing in $CaTiO_3$. Such small changes are typical for twin boundaries and other interfaces so that the observation that dopants are concentrated in interfaces is not unexpected. These localized dopants, on the other hand, can then be used systematically to modify the properties of the walls, for example, their conductivity or polarity. Doping with magnetic ions may then lead to magnetic properties of the walls, while the same dopants would not necessarily enter the bulk.

The widening of the unit cell at the interface could also lead to a reduction of the local elastic response. This does not mean that the position of a twin wall can be shifted by external forces (which it can), but the compressibility of the wall itself is larger than the equivalent compressibility of the bulk. Although such an effect has been observed in computer simulations [29c], it appears that the effect is much smaller than could be expected by the simple density dependence of the elastic moduli in the wall. In fact, the reduction of the relevant elastic modulus as expected by the increase of the distance between nearest neighbors is partly compensated by the decrease of the distance of the next nearest neighbors so that the relaxation of the structure compensates to a large extent the elastic softening due to the swelling of the interface.

Twin walls can attract defects that lead to the possibility of doping twin boundaries selectively, that is, to introduce defects into the boundaries but not in the bulk. This possibility was first used to change the conductivity in WO_3 in 1998 [12], with the introduction of Na and oxygen vacancies in twin walls. The chemical composition

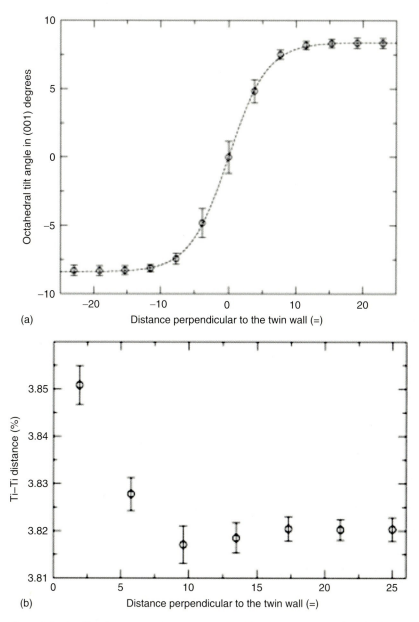

Figure 35.5 Profile of the twin wall in CaTiO$_3$. The primary order parameter is the rotation of the TiO$_6$ octahedra. The twin boundary shows an inversion of the rotation angle (Figure 35.1a), where the dotted line indicates the predictions from Landau–Ginzburg theory. The secondary order parameter is the widening of the unit cell, which is measured by the distance between two adjacent Ti positions. Figure 35.1b shows the increase of the Ti-Ti distance in the wall by about 1%, which is sufficient to induce off-centering of the Ti atom from the middle of the octahedra and also an increase of the mobility of defects. (Reproduced from [21a].)

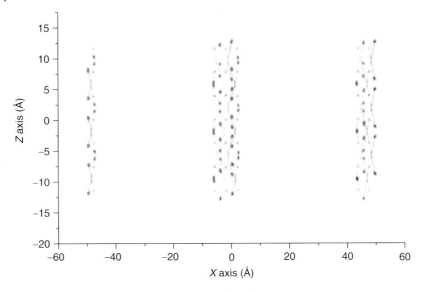

Figure 35.6 Patterns of the off-centering of Ti from the midpoint of the oxygen octahedra in CaTiO$_3$. The graph shows the displacement patterns in the direction to the twin walls [29].

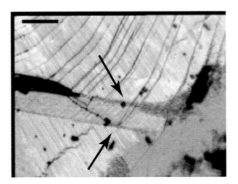

Figure 35.7 Superconducting twin walls (arrows) in WO$_3$ close to the crystal surface. The scale bar in the top left corner is 50 μm.

of the walls was very slightly modified (e.g., from WO$_3$ to WO$_{2.95}$), which induced a MI transition and, at low temperature, led to the appearance of superconductivity in twin walls. The fact that the dopants follow the trajectories of the twin walls means that nanopatterning of the superconducting structure is possible via the initial patterning of the twin boundaries and subsequent doping (Figures 35.7–35.9).

Twinning in tungsten oxide is particularly easy because twin walls in various directions stem from a multitude of structural phase transitions [35] mainly related to shape changes of the WO$_6$ octahedra and their rotations within an octahedral

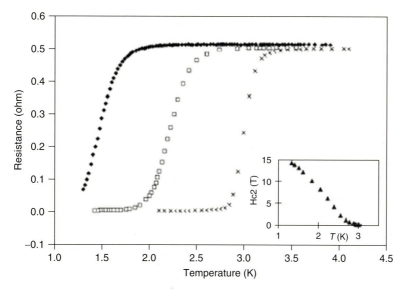

Figure 35.8 Resistance of the superconducting twin wall in WO$_3$. The onset of superconductivity is 3 K; the critical field Hc2 increases to 15 T at low temperatures [12].

network. The facility with which oxygen is subsequently released under reducing conditions is less related to the chemical bonding of oxygen but rather to the low energy required to transfer the valence state of localized surplus electrons on the W^{6+} sites to W^{5+}. This tendency to form W^{5+} states near surfaces and interfaces was directly confirmed by X-ray photoelectron spectroscopy (XPS) experiments [36] and indirectly by scanning tunneling microscope (STM) imaging [37]. W^{5+} states in the bulk are not localized, however, and form bipolarons in the low temperature phase [38]. Consequently, WO$_3$ is a well-known electrochromic, solar cell, and catalytic material [39]. Superconducting twin walls in WO$_3$ are chemically slightly reduced by inserting Na or removing O from the walls. The chemically modified walls (the changes are minor and analytically hard to detect) are then superconducting, with a critical field Hc2 above 15 T and a superconducting transition temperature T_c near 3 K. The surrounding matrix remains insulating so that this arrangement of superconducting twin boundaries with the formation of needle domains and domain junctions is potentially the key for engineering arrays of Josephson junctions and high-sensitivity magnetic scanners. In addition, it has been suggested that surface layers, presumably similar to the interfacial structures in WO$_3$, may display superconductivity at temperatures up to $T_c = 91$ K (Na doping) and 120 K (H doping) [40]. These would constitute extreme values of T_c that have not been reproduced independently, while the lower value in domain boundaries (3 K in Ref. [12]) has been directly observed by transport measurement and subsequently reproduced.

Kim et al. [34] investigated the conductivity of the similar WO$_3$ crystals under reducing conditions. Interfaces were again the loci of high electronic conductivity

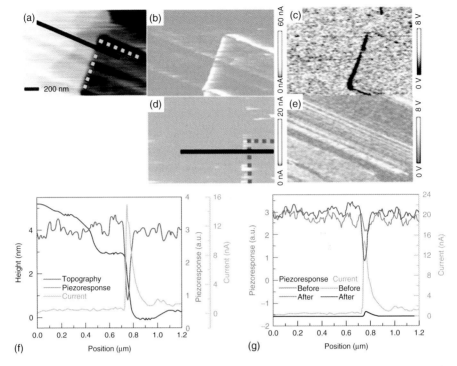

Figure 35.9 Topography (a), current (b), and piezoelectric response PFM (c) images of a freshly reduced WO_{3-x} single crystal. Current (d) and PFM (e) images of the same area of the WO_{3-x} single crystal measured after several weeks. Topography, current, and piezoresponse acquired along the black line in (a–c), respectively. Topography, Current and piezoresponse profiles before and after an elapsed time of several weeks are shown in (f,g) [34].

(Figure 35.9). Piezoelectricity collapsed inside the highly conducting twin boundaries; the thickness of the boundaries was below the resolution limit of the surface microscopy so that only an upper limit of the thickness of some 10 nm could be given. As shown in Figure 35.9, the sample oxidized in air after a few weeks and all conductivity effects disappeared.

However, reduction could be made permanent when not oxygen vacancies or H but Na atoms were used as reducing agents.

If naturally occurring interfaces are not required for some applications, multilayer systems are equally appropriate for the generation of strong structural gradients [41]. In particular, piezoelectric materials, which convert mechanical to electrical energy and vice versa, are typically characterized by the intimate coexistence of two phases across a morphotropic phase boundary where most of the structural strain is located. Electrically switching one to the other domain yields large electromechanical coupling coefficients. Using a combination of epitaxial growth techniques, Seidel *et al.* [42] have demonstrated the formation of a morphotropic phase boundary through epitaxial constraint in piezoelectric bismuth

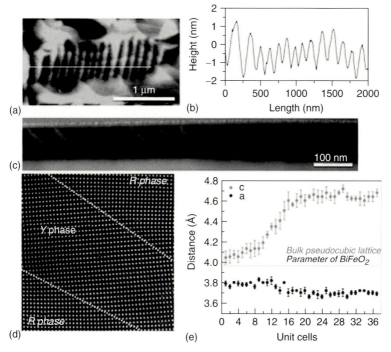

Figure 35.10 (a) High-resolution AFM image of a mixed-phase region. (b) Corresponding line trace at white line in (a) demonstrates ~2–3 nm height changes going from the T (bright) to R (dark) phase. (c) Low-resolution cross-sectional TEM image of a mixed-phase region in an 85 nm thick BFO/LAO film. Here, the light areas correspond to the T phase and the dark areas to the R phase. (d) High-resolution TEM image of the boundaries between R and T regions, indicated by dashed lines. A smooth transition between phases is observed; no dislocations or defects are found at the interface. (e) Corresponding in-plane (a, black) and out-of-plane (c, grey) lattice parameters (mean TSD) demonstrate nearly a 13% change in the out-of-plane lattice parameter in just under 10 unit cells.

ferrite (BiFeO$_3$) films. The role of the intrawall properties is now played by the inner sandwich layer, here the T phase as constrained by the bulk of the R phase (Figure 35.10). Electric-field-dependent studies showed that a tetragonal-like phase could be reversibly converted into a rhombohedral-like phase, accompanied by measurable displacements of the surface. The geometrical surface structures and the local interfaces were carefully characterized using AFM and TEM methods.

35.4
Domain Structures Related to Amorphization

Even in cases of ferroic materials with high defect densities one finds that the local domain walls are fairly straight with rather long Larkin length. The issue arises whether bend, high-density interfaces exist between materials with large density differences but the same chemical composition. An interesting solution to this problem constitutes a new class of coelastic materials in which the domain structure is induced by radiation damage. A generic example is the mineral zircon, which has been discussed as a material to incorporate and passivate nuclear waste. A large number of other materials, such as pyrochlore, SiC, monacite, and titanite, show similar geometrical features. If zircon is mixed with radiogenic atoms such as U or Th, or if the material is irradiated with heavy ions (e.g., Pb), it decays locally by ballistic collisions into a glass state [43]. These glass beads have a diameter of 5 nm [44] and percolate when the number density of the beads reaches high concentrations [45]. The surrounding of the glass beads remain structurally intact and forms the original crystalline host lattice. The density of the host lattice reduces slightly because of the accumulation of point defects, while the total density of the glass is much lower (up to 20% reduced). The interesting interfacial properties arise from the fact that the glass is very different from thermal glass and shows an extreme microstructure: the inner part of the glass beads has a very low density, while the interface has an extremely high density – in fact higher than the surrounding matrix (Figure 35.11).

The rim region is characterized by polymers that do not exist under equilibrium conditions. Careful nuclear magnetic resonance (NMR) studies [46] have shown that the Si-O species in the amorphized areas are not equivalent to those in thermal glass. The ^{29}Si chemical shift for silicon in fourfold coordination with oxygen depends on the nature of the Si second neighbor. The clearest and largest effect is whether this second neighbor is another Si atom, that is, whether the oxygen is bridging or nonbridging. The exact extent of the average polymerization depends on a nonunique interpretation of the relationship between the local structure and the NMR shift. The amorphous phase in a sample with a dose of $1.83 \times 10^{18} \alpha/g$ has an average shift of 289.5 ppm. In alkali silicate glasses, for which there is the largest amount of data on amorphous silicates, this would fall clearly in the middle of the range indicative of two Si second neighbors (Q^2). On this scale, also, the

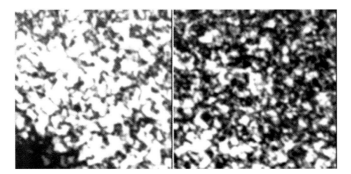

Figure 35.11 Microstructures of zircon under radiation damage as imaged by transmission electron microscopy. The left half of the figure is taken close to the first percolation point, the right half of the figure at the second percolation point. Note the complex branching of the white crystalline matrix and the black amorphized regions. The two parts of the figure are dual: exchanging white and black exchanges left and right.

average number of Si second neighbors in with $15.93 \times 10^{18} \alpha/g$, chemical shift 296.1 ppm, is three (Q^3). However, the nature of the network modifying cation also has an effect on the chemical shift. This may mean that in the amorphous zircon phase the scale is shifted and the observed change in shift would be appropriate for the complete range of Si tetrahedral environments from Q^0 to Q^4. Within this interpretation, the average polymerizations would lie between Q^1 and Q^2 in the first sample and between Q^2 and Q^3 in the second sample. Whatever the subtleties of the structural interpretation, both these NMR peaks cover nearly the whole range of resonance positions observed in crystalline and glassy silicates for Si Qn species and average polymerization increases significantly with the radiation dose.

We can now ask what the consequences of the polymerization of the interface entail. The elastic moduli in zircon are particularly sensitive to amorphization [47]. If the interfacial material could be seen as a simple mixture of crystalline and amorphous material, one would expect that the effective elastic moduli of the partially damaged material fall within the conventional Hashin–Shtrikman bounds of the effective medium approximation. This is not the case, however. The experimental values lay significantly above the calculated values. If, on the other hand, the polymerized interface is taken as a stiffer medium than either phase, we can approximate the elastic moduli by scaling the probability p finding an interface in the sample as a simple mixing behavior ($p = x(1-x)$, where x is the amount of amorphized material). The resulting elastic moduli [46] are much harder than initially expected and follow well the measured data, showing that the behavior of the macroscopic sample is much influenced by the appearance if interfaces rather than a simple volume mixture.

The second issue relates to the doping of samples with impurities or hydrogen. The measurement of the partition coefficient $K = 0.3$ of OH in damaged zircon found a narrow distribution with a predominance of the hydrous species in the crystalline matrix rather than in the amorphized regions [48] (Figure 35.12). This result is highly significant for two reasons. First, the narrow distribution of the data points shows that the partition has reached some thermodynamic equilibrium or steady state for all samples. Recrystallization of zircon after radiation damage is an exceedingly slow process so that one may have expected that at least some samples would show large deviations of K from its equilibrium value. For example, $K = 1$ would signify a random distribution of OH in the two states, which would occur for the uniform distribution of OH in a virgin sample and subsequent random radiation damage. One could envisage this to be the initial scenario. The fact that a reasonably well-defined state of equilibrium partition is observed in all samples also indicates that the hydrogen or proton is mobile enough (over very long timescales) to move from the initial distribution into the equilibrium partitioning, or diffuses from external sources into the crystals. This requires hydrogen to move between the crystalline regions and the amorphized regions through an interface as discussed before. This observation can be contrasted with the behavior of dehydration of titanite in which hydrogen moves predominantly along the interfaces [49]. The second important observation is that OH is highly enriched in the crystalline matrix and depleted in the amorphized regions. A possible explanation for these

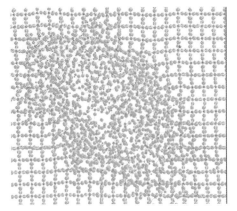

Figure 35.12 Atomic arrangement in the middle of a glass bead as simulated in Ref. [50]. The density is very low in the middle of the damaged region, while the rim between the amorphized region and the crystalline matrix is very high.

experimental observations is that hydrogen acts catalytically to enhance the rate of recrystallization of the amorphized material starting from the interface. The enhanced probability of finding hydrogen in the crystalline matrix would mean that the recrystallization process after the damage event is enhanced by the presence of hydrogen. This idea also concurs with the observation that the spectra of OH in the crystalline parts correspond to highly defective material rather than the virgin crystal. If this interpretation is correct, we may conclude that hydrogen enriches initially in the interface (as in titanite) and then reacts catalytically to break the polymerization in order to enhance the recrystalization of zircon. In this case, we find that doping is again selective in the interface and that such dopants act as catalysts. Each dopant carries an electric dipole moment, which may be switchable under sufficiently high fields.

35.5
Dynamical Properties of Domain Boundaries

Ferroic properties are directly linked (for not too high field strengths) with the mobility of domain walls. We have encountered several cases of extreme coelastic behavior, namely, SnTe, where no domain walls exist; $Cu_{74.08}Al_{23.13}Be_{2.79}$, where the twin walls are pinned by dislocations (Figure 35.4); and amorphized regions, where the domain itself is coelastic with an extremely large volume strain. In these cases, microscopy can show domain walls that do not move. It is much harder to design microscopic experiments in which the movement is observed directly. Some attempts were made to use high-speed imaging, as usually employed for the observation of a flying bullet and other high-speed objects. These attempts were not very successful because the crucial issue of the atomistic mechanism and the

35.5 Dynamical Properties of Domain Boundaries

deformation of the propagating front are issues that need to be investigated on a length scale well below the resolution of the wavelength of light. As a result of such work, the traditional view is that interfaces can move with external fields in a momentum-driven dynamics with little internal deformation even for large fields (reviewed in Ref. [8]). For small field strengths this picture was shown to be wrong, however. Careful measurements under small thermal and elastic driving forces have revealed jerky front propagation and avalanche formation. This phenomenon is well known in shape-memory alloys and magnetic materials, in which the movement of interfaces leads to acoustic emission (AE) or magnetic induction in Barkhausen – type avalanches. Such jerky propagations of one interface releases a multitude of other interfaces so that ultimately an avalanche of propagating phase fronts is observed [51]. Theoretically, such avalanches are expected to obey power law distributions and can be considered to be at or near a point of self-organized criticality [52]. While this idea is appealing for its simplicity, it is hard to imagine that the randomness of the various pinning centers in ferroics will not extend to extremely low values where pinning can occur only at very low temperatures. Most experiments seem to indicate that pinning, depinning, AE, and Barkhausen dynamics is athermal, which means that it is not thermally activated. A key experiment for a ferroelastic material was recently performed [53] where the transition in a $Cu_{67.64}Zn_{16.71}Al_{15.65}$ shape-memory alloy was investigated calorimetrically whereby the thermal driving force was as small as possible for the experiment to be performed. The transition was scanned at rates of some $mK\,h^{-1}$ so that each avalanche could be observed as an individual peak in the latent heat. The resulting DTA curve is shown in Figure 35.13. It consists of two components: the jerks (Figure 35.9) and a continuous background.

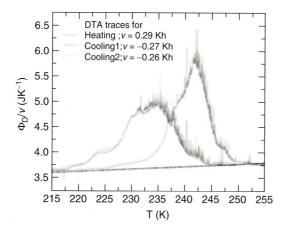

Figure 35.13 DTA traces for cooling and heating experiments. The heating rate was $0.29\,K\,h^{-1}$, the cooling rates were 0.27 and $0.26\,K\,h^{-1}$. Note the coexistence of smooth front propagation and thermal spikes (jerks) even at very low thermal driving forces.

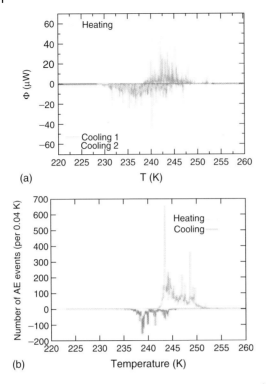

Figure 35.14 Spikes in the calorimetric measurement after removing the smooth baseline (a) and acoustic emission (AE) signals (b) of the same sample. The sign of the peaks has been inverted for clarity between heating and cooling experiments [53,58].

The entropy of the transition is not affected by the jerks and is the same on heating and cooling. Besides, for the strongest avalanches no memory effect was observed for the individual jerks. The statistical analysis of the jerks is the same as that of AE spectra (Figure 35.14) and follows a power law of the energy of the jerks: $P(E) \sim E^{-\varepsilon}$ with an exponent close to $-\varepsilon \sim -2$. This observation shows that the AE exponents are identical with or close to the energy exponents and not the size exponents (Figure 35.15).

A direct observation of the jerks in elastic measurements depends on the smallness of the applied forces. The transition in $Cu_{74.08}Al_{23.13}Be_{2.79}$ was investigated in a very careful dynamical mechanical analyzer (DMA) experiment. The frequency of the three-point bending excitation was chosen 0.1 Hz, the applied forces were extremely small (50 mN maximum amplitude), and the heating/cooling rate was 0.14 K h^{-1}. The mechanical loss is $\sim \tan(\delta)$ and shows spiky avalanche behavior (Figure 35.16) similar to those in Figure 35.9. Statistical analysis of the jerks leads again to a power law. The exponent (-1.3) is significantly smaller than the energy exponent of the colorimetric measurement even though the uncertainty of the fitted

35.5 Dynamical Properties of Domain Boundaries

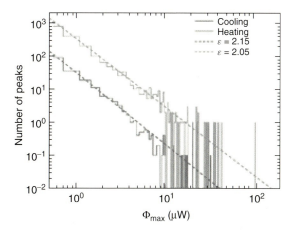

Figure 35.15 Statistical analysis of the heating and cooling curves of the DTA traces in Figure 35.14. Data corresponding to cooling experiments have been shifted one decade downwards in order to clarify the picture [53].

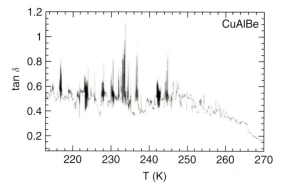

Figure 35.16 Phase lag $\tan(\delta)$ of a Cu74.08 Al23.13 Be2.79 single crystal recorded in three-point bending mode at 0.1 Hz and a heating rate of 0.15 K/h [54].

exponent is very large (Figure 35.17). As an upper bound the exponent of -1.6 was estimated in Ref. [54]. As the elastic response is measured under oscillatory stress, one may expect that the exponent is related to field binning and to the amplitude exponent τ, which was calculated to be in the order of -1.5 in MFT and near -1.3 in simulations [55].

The movement of interfaces between ferroic variants depends theoretically on the dimension of the interface. Planar interfaces have the dimension 2 ($D = 2$), while the tip of a moving needle domain is a line in three dimensions and represents the case $D = 1$. An optical experiment to observe the statistical nature of front propagation was performed by Harrison and Salje [56, 57], in which a single needle domain was pushed by external stress through the bulk of $LaAlO_3$ under

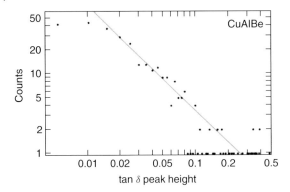

Figure 35.17 Log-log plot of the peak statistics in Figure 35.16. The line represents the best fit to a power law $P(E) \sim E^{-\varepsilon}$ with $\varepsilon = 1.3$ and an upper bound of 1.6 [54].

extremely small forces. Under the optical microscope the superposition of jerky movements and continuous front propagation was observed. In Figure 35.18, the trajectory of a needle domain in LaAlO$_3$ is shown. The advancing or retracting needle domain is pinned by defects that are mostly located at the advancing edge of the hull-shaped domain. Pinning is then described as the local fixation of a line in three-dimensional space, $D = 1, d = 3$. It is not trivial that pinning should occur in this scenario at all: the Larkin length of the edge in elastic systems is assumed to be large, and very strong pinning centers are required in order to obtain the pinning of the advancing needle.

In this context, the recent contribution of Proville [59] is relevant. He showed that in cases where the Larkin length is larger than the system size, one can still expect avalanches with a finite depinning force. This observation calls into question the traditional way of how the Larkin length is simulated in computer experiments: the elasticity of the interface is simply represented by interatomic springs between atoms in the wall. This is not the sole obvious mechanism in ferroic systems, however. The two major forces acting on the interface – in addition to the wall energy itself – are the "anisotropy energy" (which is minimal for walls with an orientation where the compatibility relation is satisfied) and the wall bending energy. Any rotation of a wall segment requires significant energy $\sim \cos \varphi$ (in a local approximation), where φ is the angle between the stress-free equilibrium direction and the actual direction of the wall segment. The second important energy is the wall "bending energy," which resists any curvature of the wall and is particularly large in case of thick walls. These energies will act against the roughening of walls and ensures that twin walls remain globally planar even under doping conditions, which would normally lead to rough walls in magnetic systems. Such walls would meander to take advantage of as many defects as feasible to obtain a maximum pinning force. We expect such walls to exist whenever the elastic forces are sufficiently small (as in magnetic materials with small magnetostriction), while walls in ferroelastic materials tend to be straight.

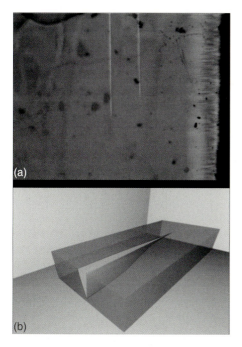

Figure 35.18 Optical microscopic image of needle domains in LaAlO$_3$ (a). Two large needles propagate while their tip is measured optically, a multitude of stationary needles emanate from the surface, which is perpendicular to the base of the big needle domains. (b) The image shows the shape of a needle domain inside the ferroelastic matrix. The advancing front ($D = 1$) is pinned by a small number of defects and followed under the optical microscope.

35.6 Conclusion

A recent breakthrough was the observation of 6pm displacements in twin walls of CaTiO$_3$ [60]. Domain patterns have been extensively studied in ferroic and multiferroic materials. Their origins are rather well understood, and their static properties have been subject to many theoretical investigations. Open questions are multiple, however. First, one wants to detect the secondary order parameter inside domain walls. The current resolution of TEM and AFM is not sufficient to detect – with acceptable degree of certainty – such secondary structural gradients. While structural displacements on the sub-Angstrom scale may be visible in future using improved facilities, it is doubtful that Lorentz microscopy for the analysis of magnetic structures will reach similar resolution. The collapses of dipolar moments inside domain walls are still best observed by PFM, while the spatial resolution remains rather crude.

The second area where progress is needed but where the experimental facilities remain insufficient is the field of the dynamics of domain boundary movement.

The example of LaAlO$_3$ is probably the only high-resolution work that is based on extremely slow increase and decrease of external forces, which makes the optical observation of the moving wall possible. Nevertheless, one would like to understand the kinetic/dynamic behavior on a much smaller length scale, which is excluded using optical microscopy. However, strain-induced movements in TEM or AFM (in whatever guise) have not lead in the past to a detailed avalanche analysis or a high-resolution image of a freely propagating wall. This is obviously due to the experimental difficulty that moving domain walls are hard to image.

Although much progress has been made, even more can be expected in future. The theoretical understanding of multiferroics has progressed rapidly, and indirect measurements of domain wall characteristics and their dynamics has progressed well. Optical microscopy has probably done as much as is possible so that the next step needs to be improved high-resolution work. If this article can stimulate such work it has satisfied its author.

References

1. (a) Ascher, E., Rieder, H., Schmid, H., and Stossel, H. (1966) Some properties of ferromagnetoelectric nickel-iodine boracite NI3B7O13I. *J. Appl. Phys.*, **37**, 1404–1405; (b) Houchmanzadeh, B., Lajzerowicz, J., and Salje, E. (1992) Interfaces and ripple states in ferroelastic crystals- a simple model. *Phase Transitions*, **38**, 77–87; (c) Kinge, S., Crego-Calama, M., and Reinhoudt, D.N. (2008) Self-assembling nanoparticles at surfaces and interfaces. *ChemPhysChem*, **1**, 20–42.

2. (a) Shima, H., Kawae, T., Morimoto, A., Matsuda, M., Suzuki, M., Tadokoro, T., Naganuma, H., Iijima, T., Nakajima, T., and Okamura, S. (2009) Optical properties of BiFeO3-system multiferroic thin films Japanese. *J. Appl. Phys.*, **48**, 09KB01; (b) Singh, A., Patel, J.P., and Pandey, D. (2009) High temperature ferroic phase transitions and evidence of paraelectric cubic phase in the multiferroic 0.8BiFeO(3)-0.2BaTiO(3). *Appl. Phys. Lett.*, **95**, 142909.

3. Fiebig, M. (2005) Revival of the magnetoelectric effect. *J. Appl. Phys. D*, **38**, R123–R152.

4. Lubk, A., Gemming, S., and Spaldin, N.A. (2009) First-principles study of ferroelectric domain walls in multiferroic bismuth ferrite. *Phys. Rev. B*, **80**, 104110.

5. (a) Stenberg, M.P.V. and de Sousa, R. (2009) Model for twin electromagnons and magnetically induced oscillatory polarization in multiferroic RMnO3. *Phys. Rev B*, **80**, 094419; (b) Wang, J., Neaton, J.B., Zheng, H., Nagarajan, V., Ogale, S.B., Liu, B., Viehland, D., Vaithyanathan, V., Schlom, D.G., Waghmare, U.V., Spaldin, N.A., Rabe, K.M., Wuttig, M., and Ramesh, R. (2003) Epitaxial BiFeO3 multiferroic thin film heterostructures. *Science*, **299**, 1719–1722; (c) Eerenstein, W., Mathur, N.D., and Scott, J.F. (2006) Multiferroic and magnetoelectric materials. *Nature*, **442**, 759–765; (d) Lottermoser, T., Meier, D., Pisarev, R.V., and Fiebig, M. (2009) Giant coupling of second-harmonic generation to a multiferroic polarization. *Phys. Rev. B*, **80**, 100101; (e) Spaldin, N.A. and Fiebig, M. (2005) The renaissance of magnetoelectric multiferroics. *Science*, **309**, 391–392; (f) Hur, N., Park, S., Sharma, P.A., Ahn, J.S., Guha, S., and Cheong, S.W. (2004) Electric polarization reversal and memory in a multiferroic material induced by magnetic fields. *Nature*, **429**, 392–395; (g) Ramesh, R. and Spaldin, N.A. (2007) Multiferroics: progress and prospects in thin films. *Nat. Mater.*, **6**, 21–29; (g) Neaton, J.B., Ederer, C., Waghmare, U.V., Spaldin, N.A., and Rabe, K.M.

(2005) First-principles study of spontaneous polarization in multiferroic BiFeO3. *Phys. Rev. B*, **71**, 014113.

6. (a) Marais, S., Heine, V., Nex, C., and Salje, E. (1991) Phenomena due to strain coupling. *Phys. Rev. Lett.*, **66**, 2480–2483; (b) Salje, E.K.H. (1992) Applications of Landau theory for the analysis of phase transitions in minerals. *Phys. Rep.: Rev. Sect. Phys. Lett.*, **215**, 49–99.

7. (a) Saxena, A., Castan, T., Planes, A., Porta, M., Kishi, Y., Lograsso, T.A., Viehland, D., Wuttig, M., and De Graef, M. (2004) Origin of magnetic and magnetoelastic tweedlike precursor modulations in ferroic materials. *Phys. Rev. Lett.*, **92**, 197203; (b) Mori, K. and Wuttig, M. (2002) Magnetoelectric coupling in terfenol-D/polyvinylidenedifluoride composites. *Appl. Phys. Lett.*, **81**, 100–101; (e) Catalan, G., Noheda, B., McAneney, J., Sinnamon, L.J., and Gregg, J.M. (2005) Strain gradients in epitaxial ferroelectrics. *Phys. Rev. B*, **72**, 020102.

8. (a) Salje, E.K.H. (1993) *Phase Transitions in Ferroelastic and Co-elastic Crystals*, Cambridge University Press, Cambridge; (b) Bratkovsky, A.M., Salje, E.K.H., Marais, S.C., and Heine, V. (1995) Theory and computer simulation of tweed structure. *Phase Transitions*, **55**, 79–126; (c) Locherer, K.R., Chrosch, J., and Salje, E.K.H. (1998) Diffuse X-ray scattering in WO3. *Phase Transitions*, **67**, 51–63; (d) Bratkovsky, A.M., Salje, E.K.H., and Heine, V. (1994) Overview of the origin of tweed structure. *Phase Transitions*, **52**, 77–83; (e) Schranz, W., Sondergeld, P., Kityk, A.V., and Salje, E.K.H. (2009) Dynamic elastic response of KMn1−xCaxF3: elastic softening and domain freezing. *Phys. Rev. B*, **80**, 094110; (f) Salje, E.K.H., Wruc B., and Thomas, H. (1991) *Z. Phys. B -Condens. Matter*, **82**, 399.

9. (a) Rossetti, G.A., Khachaturyan, A.G., Akcay, G., and Ni, Y. (2008) Ferroelectric solid solutions with morphotropic boundaries: vanishing polarization anisotropy, adaptive, polar glass, and two-phase states. *J. Appl. Phys.*, **103**, 114113; (b) Salje, E.K.H. and Ishibashi, Y. (1996) Mesoscopic structures in ferroelastic crystals: needle twins and right-angled domains. *J. Phys. Condens. Matter*, **8**, 8477–8495; (c) Ni, Y. and Khachaturyan, A.G. (2007) Phase field approach for strain-induced magnetoelectric effect in multiferroic composites. *J. Appl. Phys.*, **102**, 113506; (d) Salje, E.K.H., Buckley, A., Van Tendeloo, G., Ishibashi, Y., and Nord, G.L. (1998) Needle twins and right-angled twins in minerals: comparison between experiment and theory. *Am. Mineral.*, **83**, 811–822; (e) Salje, E., Kuscholke, B., and Wruck, B. (1985) Domain wall formation in minerals: theory of twin boundary shapes in feldspars. *Phys. Chem. Miner.*, **12**, 132–140; (f) Rossetti, G.A. and Khachaturyan, A.G. (2007) Inherent nanoscale structural instabilities near morphotropic boundaries in ferroelectric solid solutions. *Appl. Phys. Lett.*, **91**, 072909.

10. (a) Salje, E.K.H. (2010) Multiferroic domain boundaries as active memory devices: trajectories towards domain boundary engineering. *ChemPhysChem*, **11**, 940–950; (b) Salje, E.K.H. (2008) A pre-martensitic elastic anomaly in nanomaterials: elasticity of surface and interface layers *J. Phys. Condens. Matter*, **20**, 485003.

11. (a) Salje, E. and Zhang, H.L. (2009) Domain boundary engineering. *Phase Transitions*, **82**, 452–469; (b) Liu, W.K., Karpov, E.G., Zhang, S., and Park, H.S. (2004) An introduction to computational nanomechanics and materials. *Comput. Meth. Appl. Mech. Eng.*, **193**, 1529–1578; (c) Brosseau, C. and Beroual, A. (2003) Computational electromagnetics and the rational design of new dielectric heterostructures. *Prog. Mat. Sci.*, **48**, 373–456.

12. (a) Aird, A. and Salje, E.K.H. (1998) Sheet superconductivity in twin walls: experimental evidence of WO3−x. *J. Phys.: Condens. Matter*, **10**, L377–L380; (b) Aird, A. and Salje, E.K.H. (2000) Enhanced reactivity of domain walls in WO3 with sodium. *Eur. Phys. J. B*, **15**, 205–210; (c) Aird, A., Domeneghetti, M.C., Mazzi F.,

and Salje, E.K.H. (1998) Sheet superconductivity in WO3−x: crystal structure of the tetragonal matrix. *J. Phys.: Condens. Matter*, **10**, L569–L574.
13. (a) Sapriel, J. (1975) Domain-wall orientations in ferroelastics. *Phys. Rev. B*, **12**, 5128–5140; (b) Ball, J.M. and James, R.D. (1987) Fine phase mixtures as minimizers of energy. *Arch. Rat. Mech. Anal.*, **100**, 13–52.
14. (a) Emelyanov, A.Y., Pertsev, N.A., and Salje, E.K.H. (2001) Effect of finite domain-wall width on the domain structures of epitaxial ferroelectric and ferroelastic thin films. *J. Appl. Phys.*, **89**, 1355–1366; (b) Salje, E.K.H. (2000) Ferroelasticity. *Contemp. Phys.*, **41**, 79–91; (c) Nagarajan, V., Roytburd, A., and Stanishevsky, A. (2003) Dynamics of ferroelastic domains in ferroelectric thin films. *Nat. Mater.*, **2**, 43–47; (d) Koukhar, V.G., Pertsev, N.A., and Waser, R. (2001) Thermodynamic theory of epitaxial ferroelectric thin films with dense domain structures. *Phy. Rev. B*, **64**, 214103; (e) Lookman, T., Shenoy, S.R., Rasmussen, K.O. et al. (2003) Ferroelastic dynamics and strain compatibility. *Phys. Rev. B*, **67**, 024114; (f) Harrison, R.J., Redfern, S.A.T., and Salje, E.K.H. (2004) Dynamical excitation and anelastic relaxation of ferroelastic domain walls in LaAlO3. *Phys. Rev. B*, **69**, 144101.
15. Bosbachy, D., Putnis, A., Bismayer, U., and Guttler, B. (1997) An AFM study on ferroelastic domains in lead phosphate, Pb3(PO4)2. *J. Phys.: Condens. Matter*, **9**, 8397–8405.
16. (a) Schmahl, W.W., Putnis, A., Salje, E., Freeman, P., GraemeBarber, A., Jones, S., Singh, K.K., Edwards, P.P., Loram, J., and Mirza, K. (1989) Twin formation and structural modulations in orthorhombic and tetragonal YBa2(Cu1-xCox)3O7 -- delta. *Philos. Mag. Lett.*, **60**, 241–248; (b) Salje, E.K.H., Hayward, S.A., and Lee, W.T. (2005) Ferroelastic phase transitions: structure and micro- structure. *Acta. Crystallogr. A*, **61**, 3–18.
17. Chrosch, J. and Salje, E.K.H. (1999) Temperature dependence of the domain wall width in LaAlO3. *J. Appl. Phys.*, **85**, 722–727.
18. Salje, E.K.H., Zhang, H., Planes, A., and Moya, X. (2008) Martensitic transformation B2-R in Ni-Ti-Fe: experimental determination of the Landau potential and quantum saturation of the order parameter. *J. Phys.: Condens. Matter*, **20**, 275216.
19. (a) Yamaoto, N., Yagi, K., and Honjo, G. (1977) Electron-microscopic studies of ferroelectric ferroelastic Gd2(MoO4)3. *Phys. Status Solidi A*, **42**, 257–265; (b) Ayroles, R., Torres, J., Aubree, J., Roucau, C., and Tanaka, M. (1979) Electron-microscope observation of structure domains in the ferroelastic phase of lead phosphate, Pb3(PO4)2. *Appl. Phys. Lett.*, **34**, 4–6; (c) Viehland, D., Xu, Z., and Huang, W.H. (1995) Structure- property relationship in strontium barium niobate 1. Needle-like nanopolar domains and the metastably-locked incommensurate structure. *Philos. Mag. A*, **71**, 205–217.
20. Zapart, W. (2003) Domain walls structure as studied by EPR. *Ferroelectrics*, **291**, 225–240.
21. (a) Calleja, M., Dove, M.T., and Salje, E.K.H. (2003) Trapping of oxygen vacancies on twin walls of CaTiO3: a computer simulation study. *J. Phys.: Condens. Matter*, **15**, 2301–2307; (b) Goncalves-Ferreira, L., Redfern, S.A.T., Artacho, E., and Salje, E.K.H. (2010) Trapping of oxygen vacancies in the twin walls of perovskite. *Phys. Rev. B*, **81**, 024109.
22. Hayward, S.A. and Salje, E.K.H. (1996) Displacive phase transition in anorthoclase: the plateau effect and the effect of T1-T2 ordering on the transition temperature. *Am. Mineral.*, **81**, 1332–1336.
23. Wruck, B., Salje, E.K.H., and Zhang, M. (1994) On the thickness of ferroelastic twin walls in lead phosphate Pb3 (PO4)2 -- an X-ray -- diffraction study. *Phase Transitions*, **48**, 135–148.
24. Bismayer, U., Mathes, D., Bosbach, D., Putnis, A., van Tendeloo, G., Novak, J., and Salje, E.K.H. (2000) Ferroelastic orientation states and domain walls in lead phosphate type crystals. *Mineral. Magazine*, **64**, 233–239.

25. Chrosch, J. and Salje, E.K.H. (1994) Thins domain -- walls in YBa2Cu3O7-delta and their rocking curves an X-ray -- diffraction study. *Physica C*, **225**, 111–116.
26. (a) Bursill, L.A., Qian, H., Peng, J.L., and Fan, X.D. (1995) Observation and analysis of nanodomain textures in the dielectric relaxor lead magnesium niobate. *Physica B*, **216**, 1–23; (b) Burton, B.P., Cockayne, E., and Waghmare, U.V. (2005) Correlations between nanoscale chemical and polar order in relaxor ferroelectrics and the lengthscale for polar nanoregions. *Phys. Rev.*, **72**, 064113; (c) Tinte, S., Burton, B.P., Cockayne, E., and Waghmare, U.V. (2006) Origin of the relaxor state in Pb(BxB'('))(1-x))O-3 perovskites. *Phys. Rev. Lett.*, **97**, 137601.
27. Salje, E.K.H., Zhang, H., Idrissi, H., Schryvers, D., Carpenter, M.A., Moya, X., and Planes, A. (2009) Mechanical resonance of the austenite/martensite interface and the pinning of the martensitic microstructures by dislocations in Cu74.08Al23.13Be2.79. *Phys. Rev. B*, **80**, 134114.
28. Lee, W.T., Salje, E.K.H., Goncalves-Ferreira, L., Daraktchiev, M., and Bismayer, U. (2006) Intrinsic activation energy for twin-wall motion in the ferroelastic perovskite CaTiO$_3$. *Phys. Rev. B*, **73**, 214110.
29. (a) Goncalves-Ferreira, L., Redfern, S.A.T., Artacho, E., Salje, E.K.H., and Lee, W.T. (2010) Trapping of oxygen vacancies in the twin walls of perovskite. *Phys. Rev. B*, **81**, 024109; (b) Goncalves-Ferreira, L., Redfern, S.A.T., Artacho, E., and Salje, E.K.H. (2008) Ferrielectric twin walls in CaTiO3. *Phys. Rev. Lett.*, **101**, 097602; (c) Goncalves-Ferreira, L., Redfern, S.A.T., Atacho, E., and Salje, E.K.H. (2009) The intrinsic elasticity of twin walls: ferrielectric twin walls in ferroelastic CaTiO3. *Appl. Phys. Lett.*, **94**, 081903.
30. Zubko, P., Catalan, G., Welche, P.R.L., Buckley, A., and Scott, J.F. (2007) Straingradient- induced polarization in SrTiO3 single crystals. *Phys. Rev. Lett.*, **99**, 167601.
31. Tagantsev, A.K., Courtens, E., and Arzel, L. (2001) Prediction of a low-temperature ferroelectric instability in antiphase domain boundaries of strontium titanate. *Phys. Rev. B*, **64**, 224107.
32. Petzelt, J., Ostapchuk, T., Gregora, I., Rychetsky, I., Hoffmann-Eifert, S., Pronin, A.V., Yuzyuk, Y., Gorshunov, B.P., Kamba, S., Bovtun, V., Pokorny, J., Savinov, M., Porokhonskyy, V., Rafaja, D., Vanek, P., Almeida, A., Chaves, M.R., Volkov, A.A., Dressel, M., and Waser, R. (2001) Dielectric, infrared, and Raman response of undoped SrTiO$_3$ ceramics: evidence of polar grain boundaries. *Phys. Rev. B*, **64**, 184111.
33. Zhong, W. and Vanderbilt, D. (1995) Competing structural instabilities in cubic perovskites. *Phys. Rev. Lett.*, **74**, 2587.
34. Kim, Y., Alexe, M., and Salje, E.K.H. (2010) Nanoscale properties of thin twin walls and surface layers in piezoelectric WO3-x. *Appl. Phys. Lett.*, **96**, 032904.
35. (a) Loopstra, B.O. and Boldrini, P. (1966) Neutron diffraction investigation of WO$_3$. *Acta Crystallogr.*, **21**, 158–162; (b) Salje, E.K.H., Rehmann, S., Pobell, F., Morris, D., Knight, K.S., Herrmanndörfer, T., and Dove, M.T. (1997) Crystal structure and paramagnetic behaviour of epsilon-WO3-x. *J. Phys.: Condens. Matter*, **9**, 6563–6577; (c) Diehl, R., Brandt, G., and Salje, E. (1978) Crystal structure of triclinic WO3. *Acta Crystallogr. B*, **34**, 1105–1111; (d) Salje, E. (1977) Orthorhombic phase of WO3. *Acta Crystallogr. B*, **33**, 574–577; (e) Salje, E. and Viswanathan, K. (1975) Physical properties and phase transitions in WO$_3$. *Acta Crystallogr. A*, **31**, 356–359.
36. (a) Salje, E., Carley, A.F., and Roberts, M.W. (1979) Effect of reduction and temperature on the core levels of tungsten and molybdenum in WO$_3$ and $W_xMo_{1-x}O_3$–photoelectron spectroscopic study. *J. Solid State Chem.*, **29**, 237–251; (b) Al-Kandari, H., Al-Kharafi, E., Al-Awadi, N., El-Dusouqui, O.M., and Katrib, A. (2006) Surface electronic structure-catalytic activity relationship of partially reduced WO$_3$ bulk or deposited on TiO$_2$. *J. Electron Spectrosc. Relat. Phenom.*, **151**, 128–134; (c) Leftheriotis, G.,

Papaefthimiou, S., Yianoulis, P., and Siokou, A. (2001) Effect of the tungsten oxidation states in the thermal coloration and bleaching of amorphous WO3 films. *Thin Solid Films*, **384**, 298–299.

37. Jones, F.H., Rawlings, K., Foord, J.S., Egdell, R.G., Pethica, J.B., Wanklyn, B.M.R., Parker, S.C., and Olive, P.M. (1996) An STM study of surface structures on $WO_3(001)$. *Surf. Sci.*, **359**, 107–121.

38. (a) Salje, E.K.H. (1994) Polarons and bipolarons in tungsten-oxide WO_{3-x}. *Eur. J. Solid State Inorg. Chem.*, **31**, 805–821; (b) Schirmer, O.F. and Salje, E. (1980) Conducting bipolarons in low-temperature crystalline WO3-x solid. *State Commun.*, **13**, 333–336; (c) Schirmer, O.F. and Salje, E. (1980) W5+ polaron in crystalline low-temperature WO_3 electron-spin -- resonance and optical absorption. *J. Phys.: Condens. Matter*, **13**, 1067–1072; (d) Salje, E. and Guttler, B. (1984) Anderson transition and intermediate polaron formation in WO_{3-x} transport-properties and optical-absorption. *Philos. Mag. B: Phys. Condens. Matter Stat. Mech. Electron. Opt. Magn. Prop.*, **50**, 607–620.

39. (a) Kim, C.Y., Huh, S.H., and Riu, D.H. (2009) Formation of nano-rod sodium tungstate film by sodium ion diffusion. *Mater. Chem. Phys.*, **116**, 527–532; (b) Zhang, Y.Z., Yuan, J.G., Le, J., Song, L., and Hu, X. (2009) Structural and electrochromic properties of tungsten oxide prepared by surfactant-assisted process. *Sol. Energy Mater. Sol. Cells*, **93**, 1338–1344; (c) Chen, J.P. and Yang, R.T. (1992) Role of WO3 in mixed V_2O_5-WO_3/TiO_2 catalysts for selective catalytic reduction of nitric-oxide with ammonia. *Appl. Catal. A: Gen.*, **80**, 135–148; (d) Katada, N., Hatanaka, T., Ota, M., Yamada, K., Okumura, K., and Niwa, M. (2009) Biodiesel production using heteropoly acid-derived solid acid catalyst H4PNbW11O40/WO3-Nb2O5. *Appl. Catal. A: Gen.*, **363**, 164–168.

40. Reich, S., Leitus, G., Popovitz-Biro, R., Goldbourt, A., and Vega, S. (2009) A possible 2D H (x) WO3 superconductor with a T (c) of 120 K. *J. Supercond. Novel Magn.*, **22**, 343–346, and reference given there.

41. Ramesh, R. and Spaldin, N.A. (2007) Multiferroics: progress and prospects in thin films. *Nat. Mater.*, **6**, 21–29.

42. (a) Seidel, J., Martin, L.W., He, Q., Zhan, Q., Chu, Y.H., Rother, A., Hawkridge, M.E., Maksymovych, P., Yu, P., Gajek, M., Balke, N., Kalinin, S.V., Gemming, S., Wang, F., Catalan, G., Scott, J.F., Spaldin, N.A., Orenstein, J., and Ramesh, R., (2009) Conduction at domain walls in oxide multiferroics. *Nat. Mater.*, **8**, 229–234; (b) Zeches, R.J., Rossell, M.D., Zhang, J.X., Hatt, A.J., He, Q., Yang, C.-H., Kumar, A., Wang, C.H., Melville, A., Adamo, C., Sheng, G., Chu, Y.-H., Ihlefeld, J.F., Erni, R., Ederer, C., Gopalan, V., Chen, L.Q., Schlom, D.G., Spaldin, N.A., Martin, L.W., and Rames, R. (2009) A strain-driven morphotropic phase boundary in BiFeO3. *Science*, **326**, 977–980.

43. (a) Weber, W.J., Ewing, R.C., Catlow, C.R.A., de la Rubia, T.D., Hobbs, L.W., Inoshita, C., Matzke, H., Motta, A.T., Nastasi, M., Salje, E.K.H., Vance, E.R., and Zinkle, S.J. (1998) *J. Mater. Res.*, **13**, 1434–1484; (b) Meldrum, A., Boatner, L.A., Weber, W.J., and Ewing, R.C. (1998) *Geochim. Cosmochim. Acta*, **62**, 2509–2520; (c) Meldrum, A., Zinkle, S.J., Boatner, L.A., and Ewing, R.C. (1998) *Nature*, **395**, 56–58; (d) Rios, S., Salje, E.K.H., Zhang, M., and Ewing, R.C. (2000) *J. Phys.: Condens. Matter*, **12**, 2401–2412.

44. (a) Salje, E.K.H., Chrosch, J., and Ewing, R.C. (1999) *Am. Mineral.*, **84**, 1107–1116; (b) Capitani, G.C., Leroux, H., Doukhan, J.C., Rios, S., Zhang, M., and Salje, E.K.H. (2000) *Phys. Chem. Miner.*, **27**, 545–556.

45. Trachenko, K., Dove, M.T., and Salje, E.K.H. (2002) *Phys. Rev. B*, **65**, 180102.

46. (a) Farnan, I., Cho, H., and Weber, W.J. (2007) *Nature*, **445**, 190–193; (b) Farnan, I. and Salje, E.K.H. (2001) *J. Appl. Phys.*, **89**, 2084–2090; (c) Maekawa, H., Maekawa, K., Aamura, K., and Yokokawa, T. (1991) *J. Non-Cryst. Solids*, **127**, 53–64.

47. Ozkan, H. (1976) *J. Appl. Phys.*, **47**, 4772–4779.
48. (a) Salje, E.K.H. (2007) An empirical scaling model for averaging elastic properties including interfacial effects. *Am. Mineral.*, **92**, 429–432; (b) Salje, E.K.H. (2006) Elastic softening of zircon by radiation damage, *Appl. Phys. Lett.*, **89**, 131902.
49. (a) Salje, E.K.H. and Zhang, M. (2006) *J. Phys.: Condens. Matter*, **18**, L277–L281; (b) Zhang, M. and Salje, E.K.H. (2001) *J. Phys.: Condens. Matter*, **13**, 3057–3071.
50. (a) Rios, S. and Salje, E.K.H. (2004) *Appl. Phys. Lett.*, **84**, 2061–2063; (b) Trachenko, K.O., Dove, M.T., and Salje, E.K.H. (2001) *J. Phys.: Condens. Matter*, **13**, 1947–1959.
51. Vives, E., Ortin, J., and Manosa, L. et al. (1994) Distribution of avalanches in martensitic transformations. *Phys. Rev. Lett.*, **72**, 1694.
52. Kuntz, M.C. and Sethna, J.P. (2000) Noise in disordered systems: the power spectrum and dynamic exponents in avalanche model. *Phys. Rev. B*, **62**, 11699.
53. Gallardo, M.C., Manchado, J., Romero, F.J. et al. (2010) Avalanche criticality in the martensitic transition of Cu67.64Zn16.71Al15.65 shape-memory alloy: a calorimetric and acoustic emission study. *Phys. Rev. B*, **81**, 174102.
54. Salje, E.K.H., Koppensteiner, L., Reinecker, M. et al. (2009) Jerky elasticity: avalanches and the martensitic transition in Cu74.08Al23.13Be2.79 shape-memory alloy. *Appl. Phys. Lett.*, **95**, 231908.
55. Rosso, A., Le Doussal, P., and Wiese, K.J. (2009) Avalanche-size distribution at the depinning transition: a numerical test of the theory. *Phys. Rev. B*, **80**, 144204.
56. Harrison, R.J, and Salje, E.K.H. (2010) The noise of the needle: Avalanches of a single progressing needle domain in $LaAlO_3$. *Appl. Phys. Lett.*, **97**, 021907.
57. Harrison, R.J, and Salje, E.K.H. (2011) Ferroic switching, avalanches , and the Larkin length: needle domains in $LaAlO_3$. *Appl. Phys. Lett.*, **99**, 151915.
58. Romero, F.J, Manchado, J., and Martin-Ollala, J.M.. et al. (2011) Dynamic heat flux experimetns in Cu(67.64)Zn(16.71)Al(15.65): Separating btheb times scales of fast and ultra-slow kinetic processes in martensitic transformations. *Appl. Phys. Lett.*, **99**, 011906.
59. Proville, L. (2009) Depinning of a discrete elastic string from a random array of weak pinning points with finite dimensions. *J. Stat. Phys.*, **137**, 717.
60. Van Aert, S., Turner, S., Delville, R., Schryvers, D., Tendeloo, G.V, and Salje, E.K.H. (2011) Direct Observation of Ferrielectricity at Ferroelastic Domain Boundaries in $CaTiO_3$ by Electron Microscopy. *Adv. Mater.* doi: 10.1002/adma.201103717.

36
Three-Dimensional Imaging of Biomaterials with Electron Tomography

Montserrat Bárcena, Roman I. Koning, and Abraham J. Koster

36.1
Introduction

36.1.1
Electron Microscopy of Biological Samples

Electron microscopy (EM) is a powerful tool to obtain images of biological structures. Its unique capability to record images with nanometer scale resolution makes it possible to visualize molecular arrangements in cells and tissue.

From a physical point of view, biological structures such as cells can be regarded as volumes several hundreds of cubic microns in size. Within these volumes, millions of molecules form interactions with each other in highly dynamic ways, resulting in interacting networks of molecular structures with wildly varying dimensions. The major constituents of biological specimens are elements with a relatively low atomic number.

One has to realize that the low pressure that is present in an electron microscope column is not compatible with the atmospheric pressure and aqueous surroundings under which biological structures thrive. Therefore, an artificial step to fix, or immobilize, the molecular constituents before imaging with the electron microscope is inescapable. This fixation can be done using chemicals and stains, or by using physical means using rapid cryoimmobilization.

It is also important to realize that, because biological tissue and cells are relatively large objects for imaging with transmission electron microscopy (TEM), the specimen has to be thinned for observation. In most cases, the thinning of a specimen is done by cutting sections using an ultramicrotome. In practice, sections with a thickness ranging from 50 to 500 nm are generally imaged with a TEM.

For a biological specimen, the resolution limiting factor of TEM is the radiation sensitivity. When the specimen is exposed to too much electron dose, the specimen will change because of the electron irradiation and the structural integrity of the specimen will be impaired. The allowable electron dose greatly depends on the type of sample and type of fixation. For a cryoimmobilized frozen-hydrated specimen, the allowable dose is seriously limited by radiation damage [1]. For a chemically

Handbook of Nanoscopy, First Edition. Edited by Gustaaf Van Tendeloo, Dirk Van Dyck, and Stephen J. Pennycook.
© 2012 Wiley-VCH Verlag GmbH & Co. KGaA. Published 2012 by Wiley-VCH Verlag GmbH & Co. KGaA.

fixed and stained specimen, the allowable dose is much higher, but the resolution is limited by the preparation method.

36.1.2
Historical Remarks on Biological Electron Microscopy

Right from the conception of the first electron microscope, it was realized that EM could be a powerful tool to image biological material and cells. One of the very first images obtained by Ruska with TEM was a biological structure, a virus [2]. These early electron microscopic images were a starting point for the development of suitable specimen preparation protocols to protect biological material from the harsh environment of vacuum and to the merciless bombardment with electrons.

Since the late 1940s, a wide range of specimen preparation protocols were developed and optimized to preserve biological structures as much as possible [3]. One very successful approach was to chemically immobilize the molecular constituents by using aldehydes for cross-linking and to stain the specimen by adding metal salt solutions with high binding affinity to membranes and proteins. These chemical fixation and staining protocols have the effect that the specimen becomes sturdier and that the images show an increased contrast. The increased contrast is due to the larger difference in the scattering angle between the added presence of high atomic number material (e.g., Os, Pb, Ur) that line out cellular membranes, proteins, and nucleic acids and the inherent low atomic number composition of the specimen embedding resin.

Since the time that the basic steps for fixation and staining were worked out, an overwhelming number of images of cellular structures and organelles have been documented. The TEM images provided insights into cellular architecture and molecular structural arrangements that have been crucial to understand many aspects of cell biology. Although electron microscopic imaging of these chemically fixed specimens was very successful and effective, for many biological questions the preservation of the biological structure appeared still not to be optimal. Especially, questions that required resolvable details on a molecular scale were difficult to answer since these classic specimen preparation protocols can perturb the molecular arrangements.

During the 1970s and 1980s, alternative approaches based on physical fixation by rapid freezing of the biological structures were pioneered by several groups [4, 5]. These cryoimmobilization approaches allowed the preservation of the macromolecular arrangements to subnanometer resolution [6]. In order to actually apply these cryo methods, a variety of dedicated tools was developed. For instance, equipment for rapid cryoimmobilization of biological specimens and dedicated electron microscopes suitable to image specimens cooled down to liquid nitrogen temperature. Today, an array of specialized pieces of cryo-EM-related equipment is available. Examples are high-pressure freezers, plunge freezers, liquid-nitrogen-cooled specimen holders, various types of TEMs capable of imaging specimens cooled to liquid nitrogen temperature, highly coherent field emission guns, and highly sensitive digital image detectors.

An image taken with a TEM is essentially a two-dimensional projection of the specimen in the direction of the illuminating electron beam. Therefore, in a TEM image, structural arrangements at different heights within the specimen will overlap in projection and cannot be separated in one single image. Additionally, as a result of this overlap, the image contrast will appear to be very low. A remedy to this shortcoming is electron tomography (ET) to obtain the 3D information.

With ET, projections are acquired of the specimen from different directions, which are later computationally merged to obtain a reconstruction of the biological specimen. The basic concept of ET is that a 3D reconstruction can be obtained from a set of 2D projections [7]. The approach was put forward in 1968 in three independent articles [8–10]. At the time, the technical challenges to carry out tomography were substantial, and it took 30 years before the technique could be applied in practice. A related three-dimensional imaging approach, which is suitable for revealing the average structure of multiple copies of isolated macromolecular assemblies (particles), is single-particle analysis. With this method, thousands of images are used to compute an average 3D structure [11].

During the 1980s, pioneering research that would later enable the practical acquisition of electron tomograms was carried out in several research groups. In the Wadsworth Center (Albany, New York), particularly Joachim Frank, Bruce McEwen, and Carmen Mannella were on the forefront of developments and applications. The group of Wolfgang Baumeister at the Max Planck Institute for Biochemistry (Martinsried, Germany) realized ground-breaking developments related to ET on highly radiation sensitive frozen-hydrated biological specimens. The group of David Agard and John Sedat at the University of California at San Francisco, California, pioneered many developments related to digital image acquisition with slow-scan charge-coupled device (CCD) cameras.

During the 1990s, fundamental technological developments were pursued related to more suitable instrumentation, incorporation of tools required for automation (e.g., automatic focusing [12, 13]), and to novel approaches in reconstruction and application of statistical image analysis techniques. In 1992, the first automated data collection system for ET was realized and optimized [14–17]. During the same period, the computational power of desktop computing increased rapidly and the required computational methods for ET were optimized. In the beginning of 2000, many of the technological improvements were developed to such a large extent that the technique became more widespread within the academic community [18–20].

Today, ET has become a mature tool for both cell biological and structural biological applications. Currently, software suites for automated acquisition of 3D data sets for both single-particle analysis and ET are available for most types of TEM [21–25]. The applicability of ET is illustrated by the rapidly increasing number of publications using this technique. One of the most challenging steps remains the computational efforts required after the data sets are collected to compute, interpret, and visualize the 3D structural arrangements of biological specimens.

It is to be expected that the ongoing ET method developments will be integrated and combined with other imaging approaches and analytical methods. Techniques such as X-ray microscopy [26, 27], light microscopy [28–31], and mass spectroscopy

[32] will be combined with EM, enabling the identification and imaging of biological structures with a resolution between one and a few hundred nanometers.

36.2
Biological Tomographic Techniques

36.2.1
Basic Principle

The common principle behind all the most widely used 3D-EM techniques is the combination of projection images of the specimen from different directions to reconstruct its three-dimensional structure. It was almost a century ago that Johan Radon laid the mathematical ground supporting the possibility of reconstructing of an object from its projections [7], a general principle that, ever since, has found applications in many other fields such as astronomy or medicine (e.g., computerized axial tomography) [33].

The differences between the various 3D-EM techniques mainly consist of the different approaches to obtain the required set of projection images. For many small biological objects, such as macromolecular complexes or icosahedral viruses, a particular function is closely linked to the maintenance of a specific structure or conformation. Therefore, these specimens occur in multiple copies that can be considered identical, at least to a certain resolution level. When isolated and observed under the EM, each copy can provide a different view of the same structure. The averaging and combination of thousands of these individual images into a 3D-reconstruction is the basis for single-particle 3D-EM methods (Figure 36.1a) [11]. A single copy of a symmetric specimen provides additional projection views of the repetitive structural unit, a fact that is exploited in the reconstruction of, for example, icosahedral [34] or biological helices [35].

As one zooms out to larger biological objects, the plasticity of the structures increases. At the scale of a cell organelle, for example, it is the general architecture and not the precise structure that is essential for a proper function. Every occurrence of a certain specimen becomes unique. The different views needed to retrieve the 3D structure have to be obtained then from the same object, by tilting the holder over a large angular range. This 3D-EM approach is known as *electron tomography* (Figure 36.1b).

Owing to the power of averaging in overcoming radiation damage limitations, single-particle methods can provide a better resolution than ET (subnanometer vs nanometer range for biomaterials). However, they are restricted to a limited type of samples, which need to be extracted and purified from their cellular context before EM imaging. In contrast to single-particle techniques, ET does not rely on averaging, and this makes it applicable not only to unique objects but also to many typical single-particle specimens. Large macromolecular complexes can be analyzed *in situ* by ET, without further purification steps. In this way, individual 3D

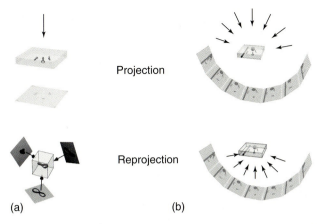

Figure 36.1 3D imaging approaches with electron microscopy. (a) 3D reconstruction based on the single-particle approach, suitable for the computation of the average 3D structure from multiple copies of essentially identical biomolecules. A micrograph of an untilted specimen contains many projections of randomly oriented biomolecules. The average 3D reconstruction can be computed once the correct angular orientation of the multiple biomolecules is determined. (b) 3D reconstruction based on electron tomography. A series of images is recorded of a specimen tilted to a range of angles. From the corresponding images the 3D reconstruction can be computed; in this case, a 3D reconstruction for each structure in the field of view is obtained.

structures, albeit at lower resolution, are obtained for every instance of the complex (Figure 36.1b).

The two main 3D-EM methodologies described here are highly complementary. Single-particle reconstructions can be used to identify the structure of interest in the tomographic reconstruction, in an approach known as *template matching* [36, 37]. Additionally, the individual 3D reconstructions of identical copies can be treated by "3D-single particle" techniques, benefiting from averaging to increased resolution [38, 39]. Single particle approaches can provide quasi-atomic resolution, whereas ET extends our understanding of macromolecular machines by revealing them in their 3D cellular context.

36.2.2
Basic Practice

In ET, the specimen is gradually tilted along an axis perpendicular to the electron beam and images are collected at every small angular increment. The inaccuracies in the *XYZ* mechanical movement of the specimen holder result in displacements that are corrected for through the optical settings of the TEM. The specimen is electron-optically centered back (*XY* shifts) by the image shift coils and refocused with the objective lens to compensate for possible *Z*-shifts. Multiple software packages that control the camera, the optics of the microscope, and the tilting of the specimen make this data collection fully automatic.

Ideally, the set of images should cover the full angular range around the specimen. This poses difficulties in practice, since at high tilt angles the grid bars and the holder block the field of view. This limited angular coverage gives raise to the problem known as the *"missing wedge"* (see Section 36.2.5). In the final reconstruction, the effect of this limited angular range is a loss of resolution and elongation in z, the direction along the specimen depth. These undesired effects can be partially alleviated by collecting a second tilt series around a tilt-axis perpendicular to the first one. This approach is known as *dual-tilt axis electron tomography* [40, 41]. Rodlike specimens can be prepared by focused ion beam (FIB) technology, and special holders that support them are starting to be used in material science, allowing for 360° rotation around the specimen [42]. In life science applications, cell sections have slab geometry. The effective thickness of a slab geometry specimen increases by a factor of $1/\cos(\alpha)$, where α is the tilt angle. Eventually, the effective thickness that the electron beam has to penetrate becomes too large for TEM imaging. Therefore, in practice, the angular range is most often limited to $\pm 60-70°$, where the effective thickness increases by a factor of 2–3, respectively.

After collecting a single-axis or dual-axis tilt series, the images of each data set have to be precisely aligned into a common coordinate system before reconstruction. The direction of the tilt axis as well as minor XY shifts that will have occurred during data collection need to be accurately determined. This can be accomplished by a variety of computational methods such as cross-correlation, common lines, and feature-based alignment [43]. A powerful alternative to these approaches is to use dense colloidal gold particles that can be included during sample preparation as reference markers for the alignment [44]. This fiducial-based alignment easily allows modeling and correcting for additional effects such as magnification changes and image rotations, which are induced by the adjustment of the optical settings along the tilt series, and even for distortions in the sample during data collection [45].

Once the tilt series is aligned, the 3D structure of the specimen can be computationally reconstructed. Different reconstruction algorithms are available, the most widely used being weighted back-projection (WBP) [46]. Iterative algorithms such as algebraic reconstruction technique (ART) and simultaneous iterative reconstruction technique (SIRT) [47, 48], once almost computationally prohibitive for ET, are becoming more popular in the last years.

The 3D reconstruction or tomogram usually marks the beginning of the most time-consuming phase in the tomographic workflow: the analysis of the results. Different computational tools can aid in this analysis. The most straightforward is the visualization of different digital slices through the tomogram. These *in silico* sections are usually a few nanometer thick and can be generated from every possible orientation, thus providing a powerful instrument to gain insight into the 3D structure of the specimen. Next, 3D graphic models can be generated, generally by segmenting different regions in colored surface renderings that emphasize different details of the tomogram, according to the interpretation of the researcher. Different types of quantification may be required as well (sizes, distances, etc.), depending on the biological question at hand. Finally, and as mentioned before, it can be the case that the tomograms contain repetitive structural elements (e.g., ribosomes in the cellular cytoplasm) that can be boxed out into independent subtomograms, for individual analysis and averaging.

36.2.3
Sample Preparation

The tomographic workflow described above started with the data collection at the TEM. In practice, however, there is a prior step that needs to be carefully considered. This step is the preparation of the sample for TEM imaging. As discussed in the Section 36.1, biological samples pose special challenges that need to be tackled at the level of sample preparation. Here, there is always a trade-off between sample preservation on the one hand and contrast and signal to noise ratio on the other. Which one of these aspects should be emphasized determines the choice of the preparative route, a choice that is ultimately driven by the biological question at hand.

From the point of view of sample preservation, cryoimmobilization methods are clearly superior because they maintain the natural aqueous environment of the sample. Immobilization is achieved by freezing water in conditions where ice crystals are not formed (water vitrification), or the crystals are so small that they do not disturb the biological structures at the required resolution level. The simplest way to accomplish this is the method known as *plunge freezing* (Figure 36.2a). A droplet of sample is applied to an EM grid, which is held in a home-made or commercial vitrification system. The excess liquid is blotted with a filter paper and, immediately after, the grid is plunged by a guillotinelike system into a liquid cryogen (e.g., ethane or propane). Samples prepared in this manner can be directly visualized at the TEM. Plunge freezing provides cooling rates that allow vitrification of pure water to a depth of ~1 μm [49], and it is therefore perfectly suited for aqueous suspensions of small specimens such as macromolecular complexes, virus particles (Figure 36.3a), isolated cellular fractions, or even whole small cells (e.g. bacteria). The classic room temperature alternative for these types of small specimens is negative staining [50]: after incubating the sample on the grid and blotting, a drop of a heavy-metal salt solution is applied to the grid, which is then blotted and dried. The heavy metal forms a mold around the particle that strongly scatters electrons, giving rise to images in which the sample appears light against a dark background (Figure 36.3a). Negative staining is relatively simple and fast but prone to artifacts stemming from partial staining or deformation of the sample during dehydration. Still, it is often used before cryo-EM to get preliminary information about the specimen or check the quality of the sample.

Plunge freezing can also be applied to large cells directly grown onto EM grids. This is appropriate when the regions of interests are thin areas that can be found in the periphery of some cells. However, most areas in eukaryotic cells are several micrometers thick. Plunge freezing, with its limited vitrification depth, is not sufficient to avoid ice crystal formation in this thickness range. The alternative

Vitrification (thin samples) HPF and sectioning (thick samples)

(a) (b)

Figure 36.2 Cryoimmobilization methods in biological electron microscopy. (a) Samples that are thinner than 1 μm can be cryoimmobilized by rapid plunging of the electron microscopy grid into liquid ethane, that is, cooled down by liquid nitrogen. A wide variety of home-built and commercial plunger systems can be used, ranging from rather straightforward manual devices to automated systems that allow the control of the environment before plunging. Left, our home-built automatic system, including a humidity and temperature-controlled environmental chamber. Right, a scheme of the vitrification process is depicted. First, the sample is put onto the EM grid. After blotting while suspended with tweezers, the grid is plunged in liquid ethane, after which it is ready for imaging by cryo-EM. (b) Samples that are thicker than 1 μm have to be cryoimmobilized using a high-pressure freezer (HPF) (left) in which the pressure is increased to ∼2000 bars immediately before rapid cooling of the sample. After cryoimmobilization, the sample can be cryosectioned using a cryoultramicrotome (upper right). An alternative is to replace the frozen water by a plastic polymer using freeze substitution (FS) (not depicted) with subsequent sectioning using an ultramicrotome. The bottom right figure schematically illustrates high-pressure freezing and the sectioning.

becomes then high-pressure freezing (HPF) [52]: rapid freezing preceded by a sudden increase in pressure up to ∼2000 bars (Figure 36.2b). The high pressure acts as a physical cryoprotectant, allowing vitrification of samples up to 200–300 μm [53, 54]. Next, the sample needs to be thinned until it is suitable for TEM imaging. This is generally accomplished by cutting vitreous sections of the vitrified sample with a cryoultramicrotome [55, 56]. Sectioning vitreous samples is a challenging technique and quite prone to cutting artifacts (compression and crevasses) that become more severe for sections thicker than ∼100 nm [57]. A possible alternative is thinning the sample by FIB milling, which only recently has started to be explored for cryo samples of biological materials [58, 59]. In FIB milling, the sample is eroded by a focused beam of ions (usually Ga^{2+}) until the desired thickness is reached. The main drawback is that most of the sample is lost, but the resulting FIB-milled sample is devoid of sectioning artifacts.

Vitreous samples prepared by any of the described cryopreparative routes can be directly imaged at the TEM but should be kept frozen all along the process in order to avoid water crystallization. The samples are stored in liquid nitrogen until observation and transferred to special cryoholders that maintain the sample near liquid nitrogen temperature during the TEM session. From the point of view of TEM imaging, cryo-EM poses special challenges, which will be examined in detail in the next sections. The difficulty of these samples stems from their extreme sensitivity to the electron beam. In order not to destroy the specimen, only a limited amount of electrons can be used for imaging. This, together with the fact that the sample is composed of weakly scattering atoms, results in low-contrast

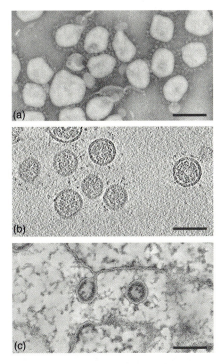

Figure 36.3 Visualization of coronavirus particles using three different preparation techniques. (a) Coronaviruses contain their RNA genomic material within a lipid envelope from which protein spikes stem. In this TEM image of coronavirus (mouse hepatitis virus) particles, prepared by negative staining, it can be seen that the stain forms a mold around the particle so the content inside the envelope can not be visualized. Additionally, the virions appear deformed from their originally spherical shape. (b) After cryoimmobilization and cryo-EM in an aqueous environment, this spherical shape is preserved, which is apparent from this 9 nm thick digital slice of a cryo-tomogram from plunge-frozen coronavirions (mouse hepatitis virus). Owing to the low-dose imaging requirement this image is noisy, but contains molecular resolution details since the contrast stems directly from the viral components (RNA, proteins, and lipids). (c) Plastic embedded and sectioned preparations allow the study of virus morphogenesis inside cells. A 9 nm thick slice of a tomogram taken on a section of a coronavirus-infected (SARS virus) cell after cryoimmobilization and freeze substitution shows newly synthesized virus particles, budding into intracellular compartments. The stains have infiltrated the sample and contrasted the different components. All scale bars are 100 nm. ((b) adapted with permission from Ref. [1]; (c) adapted from Ref. [51]).

and noisy cryo-EM images (Figure 36.3b). Therefore, although the subnanometer resolution of the structures is preserved, it cannot be directly retrieved from the image.

Often, fundamental questions in cell biology, such as the general cellular topology and architecture, do not always require the molecular resolution preservation that cryo-EM provides. In these cases, it can be advantageous to use alternative methods that, with suboptimal preservation, will produce images with higher signal and contrast. The classic path for large samples is chemical fixation: infiltration with cross-linking agents (typically, glutaraldehyde or paraformaldehyde) that fix the cellular components into place. These samples are then dehydrated and embedded into a plastic resin that, after polymerization, results in a hardened block that can be sectioned at room temperature by normal ultramicrotomy. At different steps along this process, depending on the specific protocol, salts of heavy metals (e.g., Os, Ur, Pb) are used as stains that react with different biochemical components, adding contrast to the sample.

Classic chemical fixation is still the method of choice for large size samples (e.g., whole small organisms) but, for smaller scale samples, it has been gradually replaced during the last decade by cryoimmobilization by HPF followed by freeze substitution (FS) [5, 60]. During FS, the cellular water is exchanged at low temperature ($\sim -90\,°C$) by an organic solvent that contains chemical fixatives and a mixture of staining agents. The temperature is then gradually raised and the sample is embedded in resin, ready for normal sectioning. Cross-linking reactions are fast (milliseconds), but in conventional methods, the cross-linking agents have first to diffuse through the sample to reach their target structures and, therefore, cellular processes and structures can be altered before they are fixed. In comparison, HPF directly immobilizes the sample in milliseconds. The dehydration step and the introduction of cross-linking agents during FS occur at a temperature at which the cell still retains its physiological structure. Later on, when the temperature is raised and the cross-linkers become active, they are already at their site of action. This is believed to be crucial for the better ultrastructural preservation that is often apparent in HPF-FS samples [61].

When visualizing cells or tissues prepared in this way, it is important to realize that one is actually imaging the stain that reacted with the cellular components, and not these components themselves (Figure 36.3c). The stain adds contrast to the images and makes the samples far more resistant to the electron beam but limits the information about the specimen that the sample contains. Nevertheless these samples have been shown to provide useful information in the 3–7 nm resolution range, which is sufficient for many biological problems. When the question to be addressed requires 3D-reconstructions of large volumes, plastic sections have two additional favorable characteristics that make them preferable in comparison with cryo samples. First, because of their resistance to the electron beam, montaged tomograms that extend the analyzed region in XY can be easily collected. Second, and in contrast with cryoultramicrotomy, thick sections (generally 200–500 nm) are easily obtained, which helps to extend the volume in the Z dimension. The depth in the reconstructed volume can be further increased by doing tomography in consecutive serial sections, an approach that is known as *serial tomography*.

36.2.4
Thickness and Instrumentation

Considering that ET can retrieve the third dimension from the samples that are examined under the TEM, a natural tendency for many biological problems involving large specimens would be to try to examine them as a whole or at least in sections as thick sections as possible. However, this might not be the best course of action in all cases. The reason is that thickness and resolution are closely intertwined parameters that negatively affect each other. As we will see in this and the following sections, the best compromise will actually depend on the instrumentation available, the type of sample, and the information that needs to be retrieved from the sample.

From the point of view of instrumentation, a crucial factor that conditions the thickness range that can be studied is the acceleration voltage of the microscope. The penetration power of the electrons increases with the acceleration voltage, thus allowing the study of thicker samples. The gain in penetration with the voltage is not linear, and thus, at 300 kV the penetration increases by a factor of 2 relative to 100 kV, whereas the additional gain by using 1.2 MV would be only of 1.5 [62]. TEMs operating at 200–300 kV are often chosen as a suitable compromise between performance and cost for ET applications. Typically, in this voltage range, samples a few hundreds of nanometers in thickness can be analyzed.

When the electrons go through the sample, most electrons will be unaffected (unscattered), whereas others will be either elastically or inelastically scattered. Elastically scattered electrons experience a change in direction but not in energy, while inelastically scattered electrons lose energy because of interactions with the specimen core-shell electrons. Some of the elastically scattered electrons are intercepted by the objective aperture, a removal that is the basis for the scattering contrast mechanism. The inelastically scattered electrons, with low-scattering angles, are not removed by the objective aperture and create a low-resolution inelastic background in the images. Furthermore, given the fact that these electrons have different energies, the chromatic aberration of the TEM would result in a blurring of the image. Since the fraction of inelastically scattered electrons significantly increases with the thickness of the sample (Figure 36.4), they can become a serious problem in ET, which most of the times deals with relatively thick specimens. In this regard, an energy filter becomes a key piece of instrumentation. Energy filters allow the separation and filtering of electrons with different energies [63]. In materials sciences, energy filters are widely used for the analysis of the chemical composition of the sample [64]. In ET of biological samples, they are mainly utilized in a mode that is often referred to as "*zero-loss*" imaging. In this mode, the filter is set in such a way that it selects only those electrons that did not suffer any energy loss, and thus, all the inelastically scattered electrons are removed and do not contribute to the final image.

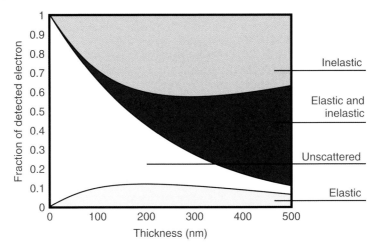

Figure 36.4 The fate of electrons as a function of sample thickness. When an electron beam goes through a specimen, three types of events can take place. Unscattered electrons will transfer through the sample without interaction. For thin specimens this will happen for most electrons, while for thicker specimens most electrons will interact with the sample (scattering). Elastically scattered electrons experience no energy loss. These electrons add to the high-resolution image information. Inelastically scattered electrons experience energy loss and do not add to the high-resolution information content. The graph shows, for instance, that for a 100 nm thick specimen, 50% of the electrons will travel through the specimen unscattered, 10% will elastically scatter, and 30% will be inelastically scattered. For a 400 nm thick specimen these numbers will be quite different. About 10% will be unscattered, 10% elastically scattered, 40% inelastically scattered, and 40% will have experienced a mixture of inelastic and elastic scattering. By using zero-loss energy filtering, the contribution of electrons that experienced inelastic scattering, or a mixture of inelastic and elastic scattering, will be removed by the filter and will not be recorded. Consequently, zero-loss filtered images of thick specimens will show details with a higher contrast and higher resolution. The computations used for this graph assume that an objective lens aperture is not inserted and that the vitrified specimen is visualized in a 300 kV TEM. (Adapted from Ref. [62]).

36.2.5
Thickness, Sampling, and Resolution

The sampling theory establishes limits to the achievable resolution in ET. For single-axis tomography, and in the ideal case of a spherical object with full angular coverage, the final resolution, d, in the direction perpendicular to the tilt axis (Y) and Z (the depth dimension) would be $d = \pi D/N$, where D is the diameter of the particle and N is the number of collected projection images [65] (Figure 36.5). In the direction parallel to the tilt axis (X), the resolution would in principle be limited only by the instrument and the imaging conditions. Therefore, even in this ideal case, because of the geometry of sample collection, the resolution in tomography is not necessarily identical in all directions.

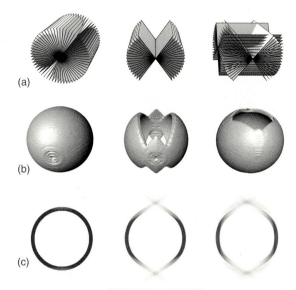

Figure 36.5 Effect of angular sampling and missing wedge on the resolution in tomograms. The effect of the angular sampling and the missing wedge in tomography shown in Fourier space slices (a), surface renderings of a reconstructed thin-shelled sphere (b), and YZ slices through the corresponding tomogram (c). The left column shows the situation when there is full angular sampling along one tilt axis (the X-axis). The top image shows that Fourier space is sampled equally in the YZ plane. The surface rendering and a central slice of the tomogram indicate no anisotropic resolution in the reconstructed sphere in those directions. The middle column shows the situation when there is limited tilt angle (from −60 to 60°) along a single axis. The result will be a missing wedge in Fourier space that can be observed in the surface rendering and as diminishing density at the top and bottom of a reconstructed sphere (b) and an elongation in the YZ slice. The right column shows the situation when two tomograms are combined from two orthogonal tilt axes (dual-axis tomography) with limited rotation between −60 and 60°. In Fourier space, the missing wedge is reduced to a missing pyramid, and the anisotropy in the resolution along the Y-axis is reduced.

It is worth mentioning that, in practice, tomography of biological samples is far from this ideal situation. The samples have slab geometry rather than spherical and a full angular coverage is not possible. This has several consequences. First, instead of the diameter the effective thickness of the sample should be considered, which is a function of the tilt angle. Second, the limited angular range translates into missing information about the object. This can be visualized in Fourier space as a wedge-shaped region that completely lacks information about the object (Figure 36.5). The incompleteness of the data is often referred to as the *"missing wedge"* problem. The consequence of the missing wedge in the final reconstruction is a deterioration of the resolution in Z by a multiplicative elongation factor – for the precise mathematical formulations the reader is referred to [46, 66]. The resolution is anisotropic in all three directions. By examining the situation in Fourier space, the advantages of dual-axis tomography become apparent (Figure 36.5). A second tilt series, around a tilt axis perpendicular to the first, reduces the missing information region from a wedge to a pyramid, and equals the resolution in X and Y. The resolution is still inferior in the Z direction, but the overall quality of the reconstruction significantly increases.

The formula $d = \pi D/N$, even though a simplification, can be used as a rough guideline for a particular experiment. It conveys, in a simple way, the key close relationship between thickness and resolution. For a given number of images, a thinner sample would yield better resolution than a thicker specimen. In other words, for thicker samples more images are necessary to achieve the same resolution. Therefore, the question arises whether it is actually possible to increase the number of images as much as needed to achieve any desired resolution. For stained plastic sections, which are relatively resistant to the electron beam, increasing the number of images is not a major problem. However, as discussed before, the resolution is limited by the preparation method to 3–7 nm. Consequently, eventually more projection images would not bring any practical advantage. For cryo samples, the number of images is limited by the total allowed dose that can be used for data collection and, thus, cannot be increased indefinitely. The role and consequences of radiation damage in cryo-EM is of paramount importance and will be examined in more detail in the next section.

36.2.6
Radiation Damage and Electron Dose

The electron beam constitutes a highly ionizing form of radiation with severe consequences for biological samples. Inelastic scattering events, by depositing energy into the sample, are the main source of beam damage. This leads to a series of destructive effects that include heating of the sample, water radiolysis, the breakage of the covalent bonds that are the backbone of biological material, and the formation of free radicals. In cryo-EM, the radiolytic processes produce small molecules that cannot escape the vitreous layer, accumulate, and eventually lead to the characteristic bubbling that is typically observed in these samples at doses as low as $\sim$10 000–20 000 electrons/nm^2. Bubbling is an indication of excessive beam damage; however, it is important to note that the destruction of the sample starts from the moment the sample is irradiated.

Radiation damage creates a paradoxical situation for cryo-EM. The sample is preserved in its fine details but this high-resolution structure is quickly lost when trying to image it. Unavoidably, the total dose used for imaging has to be limited and this results in images where the high-resolution information, even if properly preserved, is concealed by noise and cannot be directly retrieved. In cryo-ET, this small tolerance of the specimen to the electron beam sets a physical limit to the directly attainable resolution, a limit that has been estimated to be around 2 nm [6]. In single-particle applications, the problem is surmounted by averaging thousands of particles to improve the signal to noise ratio. In a similar manner, subtomogram averaging, whenever applicable, can be used to increase the resolution.

For single-particle applications, the dose that the sample receives is usually kept around 500–1000 electrons/nm^2. In this dose range, the fine details of the sample are preserved, as directly shown by the multiple subnanometer resolution reconstructions that appeared during the last decade. This dose range poses practical problems for cryo-ET. In cryo-ET, the total dose has to be fractionated over a large number of images. At the same time, the dose per image has to be high enough to ensure that the signal to noise ratio still allows proper alignment of the tilt series. This minimum dose per image is usually in the range of 25–200 electrons/nm^2, depending on the detector and the magnification. For a typical cryo tilt series, which can contain more than 100 images, this easily amounts to a total accumulated dose of 5000–15 000 electrons/nm^2. Therefore, in cryo-ET, emphasis is not placed in preserving the finest structural details of the specimen but in choosing a good practical compromise between radiation damage and the quality of the final tomogram.

At present, the resolution achieved in practice by cryotomography is in the 5–10 nm range. The variations are primarily related to the thickness of the sample. When the thickness increases, a larger number of images are required for the same resolution level (see previous section). However, collecting more images is not possible because the total dose has to be kept within similar limits, regardless of the sample thickness. Moreover, the loss in resolution is not linear because, as the thickness increases, the elastically scattered electrons that are selected for zero-loss imaging become a minor fraction of the transmitted electrons (Figure 36.4). Therefore, to ensure enough counting statistics in the images, the dose per pixel needs to be increased. The only options then are either lowering the magnification or collecting fewer images with a higher dose per image, but both alternatives further reduce the resolution. Thick cryospecimens (e.g., whole small cells, in the 0.5–1 μm range) can be reconstructed but the resolution will be seriously hampered. This low resolution is sufficient for biological problems involving large-scale features. However, when solving fine details is a key point for the question at hand, selecting the thinnest possible areas or thinning the sample (e.g., by cryosectioning) becomes essential for cryo-ET.

The situation for tomography of plastic-embedded sections is in sharp contrast with the strict dose requirements of cryo-ET. Actually, these samples are extensively preirradiated before data collection with doses up to $\sim 10^7$ electrons per square nanometer; thus, several orders of magnitude higher than what would be allowed for cryo-EM. The use of these high doses is related with the shrinkage of plastic sections [67]. Owing to mass losses in the organic embedding material produced by the electron beam, plastic sections extensively shrink on irradiation. In a first phase, in which the sections collapse to about 50–60% of their original thickness, the shrinkage is exponential. Preirradiation is used to go beyond this initial phase in order to minimize the changes in the specimen during the collection of the tilt series.

36.2.7
Contrast Formation and Imaging Conditions

The differences in elemental composition between room temperature and cryo samples are also responsible for the marked dissimilarities in the contrast mechanisms and imaging conditions for these two types of specimens.

For room temperature samples, heavy metals are included to selectively stain specific biological structures. The scattering cross sections of these atoms (e.g., Pb, Ur, Os, W) are significantly larger than those of the elements of the organic embedding material (in sections) or the biological structure itself (in negatively stained samples). As a result, a larger proportion of the electrons interacting with the heavy atoms are elastically scattered, intercepted by the objective aperture and removed from the final image. This mechanism, known as *scattering contrast*, is the primary source of contrast for this type of samples [68].

Cryo samples, on the other hand, do not generate significant scattering contrast. The biological material is composed of weakly scattering atoms (e.g., C, H, O, N), and is only marginally denser than the surrounding water. Therefore, cryo-EM has to rely on a different mechanism for image formation known as *phase contrast*. Phase contrast is the result of a phase shift of the diffracted electron wave relative to the nondiffracted wave. The shift is modulated by the specimen and the optical system, particularly, the defocus settings. The optical transfer of information from the specimen to the recorded image can be modeled quite accurately in reciprocal space by the contrast transfer function (CTF) [68]. In cryo-EM, the objective lens is strongly underfocused (often, by several microns) in order to create phase contrast. In these conditions, the CTF is a rapidly oscillating function, that is, the contrast is not linearly transferred for every frequency. The oscillatory behavior implies that, after the first zero of the CTF, there will be the first of a series of frequency bands where the contrast is inverted. Therefore, without any additional measure, the resolution in cryo-EM is limited to the frequency of that first zero-crossing point of the CTF. In single-particle applications, numerous methods have been developed that allow one to accurately determine the CTF from the experimental images and to numerically correct for it [69]. In cryo-ET, the CTF determination and correction faces some additional complications. First, because of dose fractionation, each image in the tilt series has an extremely poor signal to noise ratio, which hampers CTF estimation from the single images. Second, in a tilted image, the defocus is not unique but changes with the distance to the tilt axis. In spite of these difficulties, several methods to correct the CTF in cryo-ET have appeared in the last years [70, 71].

36.3
Examples of Electron Tomography Biomaterials

Biological materials that are visualized using ET vary widely, as do the ways these specimens are prepared. As discussed before, they can be chemically fixed or cryoimmobilized, heavy-metal salts can be used for staining, or the contrast can be generated by the sample itself, samples are sectioned or are thin enough by nature to be visualized directly with the electron microscope. Depending on the biological question to be answered also the size of the reconstructed volume and the attained resolution will vary. In the next part several practical examples are described, covering a variety of samples and preparation techniques, starting off with large stained and sectioned cells and working our way down to cryoimmobilized and purified biomolecules.

36.3.1
Whole Cells

An important landmark in ET on large volumes was carried out by the group of Marsh *et al.* [72, 73]. In their publications, methods to reconstruct whole eukaryotic cells were described and realized in practice. Here, as an example, we show their reconstruction of cells from the islets of Langerhans. The pancreas is a gland that produces several important hormones and pancreatic juice that contain digestive enzymes, which break down carbohydrates, proteins, and fats. Inside the pancreas, there are groups of cells, called the *islets of Langerhans* (Figure 36.6a), which produce the hormones insulin and glucagon, which regulate glucose levels

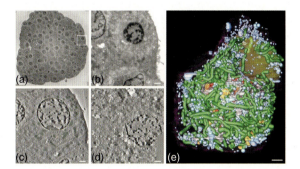

Figure 36.6 Whole-cell tomography on beta cells from islet of Langerhans. (a) Cross section of whole intact islet of Langerhans from adult mouse, preserved for EM by high-pressure freezing, freeze substitution, and plastic embedding. The 2D survey was acquired by combining 144 images at 4700× magnification. (Reproduced from Mastronarde [22].) (b) A beta cell from the Langerhans islet located in (a). (c) One of 1153 XY-slices from the serial sectioned tomogram with drawn contours, taken at a magnification of 3900× (pixel size = 6.0 nm). The color coding scheme is as described for (e). (d) View along the X–Z axis showing 27 serial section tomograms of the surface-rendered 3D model data segmented following serial section ET reconstruction. (e) 3D model from whole cell reveals key compartments involved in insulin production and release by beta cells. Rendering was generated using a segmentation approach that allowed for an efficient and accurate markup of organelles. Shown are the plasma membrane (purple), nucleus (yellow), main Golgi ribbon (gray), trans-most Golgi cisterna (red), penultimate trans-Golgi cisterna (gold), mitochondria (green), multivesicular bodies (orange), and immature granules (light blue). All scale bars are 1 µm. (Reproduced and adapted from Ref. [73], with permission from Elsevier).

in the blood. Electron tomography on these islets of Langerhans provided a fundamental structure–function relationship between the cellular organelles and the release of insulin.

The islets of Langerhans comprise many cells and are roughly 0.1 mm in diameter. To have good preservation of such large structures, the islets were prepared by HPF, followed by FS, sectioning of the plastic-embedded material into 300–400 nm thick sections and staining using heavy-metal salts. Two dimensional imaging of large areas that were needed for the identification of target cells was performed by stitching over 100 images at low magnification. Once a target cell was identified (Figure 36.6b–e) ET was performed at low magnification ($\sim$4000$\times$), and a pixel size of 5.2 nm on several tens of consecutive serial sections. Data collection and reconstruction of these large volumes took between two and three weeks for one cell. Segmentation of the cellular organelles including insulin granules was done manually, which allowed for statistical analysis of segmented volumes. By using this methodology mainly the outline of lipid membranous compartments and vesicles were visualized, and global information about the cell organization was obtained. The serial sectioning tomography generated a cellular volume of roughly $\sim$650 μm^3 with a resolution of $\sim$20 nm.

Serial section ET of whole cells is not performed routinely but can provide valuable data by combining both the large volumes of a whole cell with EM resolution data, visualizing the boundaries and volumes of subcellular structures [74, 75]. The reconstruction of large volumes by the group of Marsh stimulated novel developments [72, 76], especially aimed at stitching of large datasets and development of tools for visualization and annotation of large-scale data.

36.3.2
Interior of the Cell

36.3.2.1 Viral Infection
Many questions in molecular cell biology are confined to the intracellular space inside the cell. To visualize these processes, morphological information at a specific location in the cell is needed, while a larger cellular overview becomes less important. In these cases, ET on single sections can provide the required information.

One example that falls in this category would be the study of virus-induced membrane modifications. A virus is a small piece of genomic material that is surrounded by a layer of proteins (and sometimes also a lipid membrane), which encodes itself and the proteins that are needed for the virus to replicate. Viruses employ all kinds of mechanisms to enter the cell, replicate, and escape to establish further infection.

One of the effects that can be observed in cells after infection with severe acute respiratory syndrome coronavirus (SARS-CoV) is the alteration of the cellular membranes. The virus takes control of some of the host membranes and transforms them into special structures that the virus uses to support viral replication and, possibly, to protect its genome. To investigate these, Vero (monkey kidney) cells were infected with SARS virus in a biosafety level 3 facility. The infected cells were

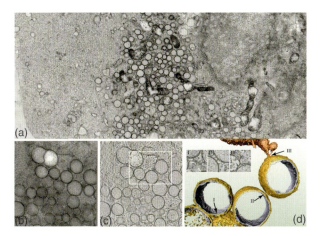

Figure 36.7 Cellular tomography on SARS-infected cell. (a) Electron micrographs of SARS-CoV-infected Vero E6 cells. The cells were cryoimmobilized by vitrification or high-pressure freezing and freeze substituted at 8 h after infection. As infection progressed, double-membrane vesicles were predominantly present in the area around the nucleus (right side of image). (b) A 0° tilt transmission EM image of a 200 nm thick resin-embedded section showing endoplasmaic reticulum and a cluster of double-membrane vesicles. The 10-nm gold particles were layered on top of the sections and were used as fiducial markers during the image alignment for the tomographic reconstruction. (c) Single 1.2 nm thick tomographic slice from the tomogram that was computed from a dual-axis tilt series of the section shown in (b). (d) 3D surface-rendered model showing interconnected double membranes (outer membrane: gold; inner membrane: silver blue) and their connection to an endoplasmatic reticulum stack (bronze). Arrows (I–III) point to three outer membrane continuities. Insets show these connections in corresponding tomographic slices. (Image reproduced and adapted with permission from Ref. [51]).

then prefixed using paraformaldehyde, and cell monolayers were cryoimmobilized by plunge freezing. The samples were then freeze substituted, plastic embedded, stained, and sectioned and ET was performed on 200 nm thick sections and imaged with a pixel size of 1.2 nm providing detailed information at the level of membranous vesicles and lipid membrane structures [51].

An overview of a section of the infected cells showed the presence of SARS-induced membrane alterations located around the nucleus (Figure 36.7a,b). Tomographic imaging showed that these vesicles defined a reticulovesicular network of modified endoplasmic reticulum withconvoluted membranes and many interconnected double-membrane vesicles that were 200–300 nm in diameter (Figure 36.7c,d). Additionally, the budding of new virus particles from vesicles was observed. This outcome aided in the elucidation of the early stages of the SARS-CoV infection and life cycle.

36.3.2.2 Cellular Morphology

In the case of SARS-CoV-infected Vero cells the regions of interest were in the perinuclear area, a cellular region too thick to be directly imaged in the electron microscope. Therefore, the cells were sectioned. However, some types of cells are thin enough to allow vitrification using plunge freezing followed by cryo-ET. The thinnest and most usable areas are localized in the cortex of the cells, although these can extend quite substantially, while the areas including the nucleus and neighboring regions are too thick for TEM.

Cryo-ET on the cortex of mouse embryonic fibroblast cells can be directly performed without the use of staining, chemical fixation, or sectioning. One advantage is that the whole thickness of the cell is imaged and structures are not removed or cut by sectioning. Long structures such as microtubules, actin, and intermediate filaments can be followed using two-dimensional imaging over large distances, and complete mitochondria can be imaged in their cellular context [77]. Another advantage is that structural details to the highest level are preserved, and the contrast is generated by direct interaction of the electron with the sample. On the other hand, the images are noisy because of the limited amount of electrons that are used for imaging, in order to prevent radiation damage. Another disadvantage is that the biological processes that can be investigated are limited to the cortex of thin cells.

In low magnification overviews of vitrified fibroblast cells mainly large and dense structures can be observed since the electron dose used for imaging has to be minimized. Mitochondria can easily be discerned, whereas on close inspection, microtubules, glycosomes, and some lipid vesicles membrane structures can be observed (Figure 36.8a). After cryo electron tomographic imaging, reconstruction of parts of the cellular cortex additionally enables identification of ribosomes, actin filaments, and clathrin-coated vesicles (Figure 36.8b,c). Several publications

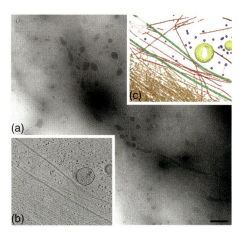

Figure 36.8 Cryo electron tomography on cells. (a) Low magnification cryo electron microscopy overview of a vitrified mouse embryonic fibroblast showing mitochondria, vesicles, microtubules, and filaments. Scale bar is 1 μm. (b) Single slice from a cryo electron tomogram from the cortex of a mouse fibroblast. Compared to (a) structures are shown with an enlargement factor of five. (c) Manually surface-rendered representation of (b). Actin filaments (orange), microtubules (green), intermediate filaments (red), glycosomes (purple), and lipid membrane vesicles (yellow) can be observed.

have specifically described the structure of actin filaments [78], microtubules [79], and ribosomes [80], showing that it is possible to determine structures of macromolecular complexes inside the cell.

36.3.3
Viruses and Macromolecular Complexes

36.3.3.1 Virus Structure
In addition to studying the macromolecular arrangements *in situ*, subcellular components and small biological structures can also be investigated after isolation. Once purified, these samples can be prepared for EM by negative staining or plunge freezing. Viruses are an example of biological specimens of this type. Some viruses are pleomorphic, which means that all the virus particles show structural variations and therefore cannot be averaged. Coronaviruses (Figure 36.3) are an example of pleomorphic viruses: they are not symmetrical, contain a lipid membrane, and show variations in size. To investigate their structure in 3D, cryo-ET is the methodology of choice [1]. This is also the case for many other viruses, such as herpes simplex virus, human immunodeficiency virus (HIV), and influenza [81–83]. Note, however, that many viruses are highly symmetric in structure and can be regarded as identical (excluding their genomic material) and can be investigated by single-particle methods.

An example of a symmetric virus is MS2, which is a small icosahedral bacteriophage (i.e., a virus that infects bacteria) that is well known because its genomic sequence was the first complete RNA sequence ever to be determined [84]. In negative stain and cryo electron microscopic projection imaging, MS2 looks like spherical or sometimes hexagonal particles (Figure 36.9a,b). This represents the icosahedral structure of the capsid, as can also be observed from the atomic structure of the virus capsid as determined by X-ray crystallography [85]. Since all virus capsids are identical, one can combine the projections from cryo-EM into a 3D reconstruction (Figure 36.9c, see also Figure 36.1a). This single-particle 3D icosahedral reconstruction clearly shows the holes in the virus capsid on the outside. Fitting of atomic X-ray structure data into the EM densities additionally showed that on the inside of the capsid small loops of genomic RNA bind on multiple places to the inside of the capsid [86].

MS2 infects bacteria via binding to a bacterial pilus, threadlike appendages that are present on the surface of some bacteria (Figure 36.9d,e). The structure of MS2 decorated pilus is a pleomorphic structure that can be reconstructed via tomographic imaging (Figure 36.9f, see also Figure 36.1b). Although in both cases the MS2 is reconstructed in three dimensions (Figure 36.9c,f), there are significant differences between the biological goals, methodology, and outcome between the two. In the single-particle approach (Figure 36.9c), a few thousand particles were combined into a single reconstruction and icosahedral symmetry was imposed on the data. In contrast, each individual virus particle was reconstructed in the electron tomogram (Figure 36.9f). Therefore, in the single-particle reconstruction the resolution is higher and more isotropic (the projection views are better

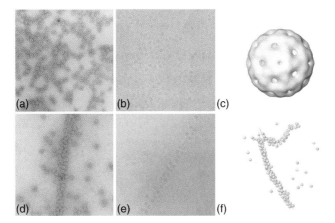

Figure 36.9 Single-particle reconstruction and cryo electron tomography on MS2 bacteriophage particles. Single, isolated, MS2 particles. The samples were prepared by (a) negative staining and (b) cryoimmobilization (plunging). (c) A surface-rendered three-dimensional icosahedral reconstruction of the MS2 particle. The reconstruction was computed using several thousand cryo electron microscopy MS2 particle projections [86]. MS2 particles bound to a bacterial pilus (d) in a negatively stained preparation and (e) after cryoimmobilization (plunging). (f) Surface-rendered three-dimensional representation of one cryo electron tomogram.

distributed). While the resolution in the tomogram is limited by the total dose that can be put on the sample, in the single-particle reconstruction it ultimately depends on the quality and number of the recorded images. The disadvantage is, however, that the asymmetric parts of a virus cannot be resolved using single-particle analysis, whereas that information is present in tomograms.

36.3.3.2 Subtomogram Averaging

While tomography is especially useful to study structures that at a large scale are pleomorphic, at a smaller scale, tomograms can contain substructures that are essentially identical and can be averaged to improve resolution. This approach is often referred to as *subtomogram averaging* [38]. Subtomogram averaging has been exploited, for example, to investigate the structure of the spike proteins that are present on the surface of enveloped viruses [87].

Extensively studied examples are the spikes of human and simian immunodeficiency viruses (HIV/SIV). These are retroviruses, belonging to a family of viruses that build their genome into the genome of the host. They infect immune cells and thereby weaken the immune system, leading to opportunistic infections. HIV is a spherical virus of roughly 120 nm in diameter, whose genome is packed in a conical-shaped protein structure that is inside a spherical lipid vesicle. Tens of copies of a viral spike protein called *Env*, protrude from this lipid envelope. These spike proteins are trimers of heterodimers, consisting of the two glycoproteins gp120 and gp41, and are important since they mediate the attachment of the virus to the host cell to initiate infection.

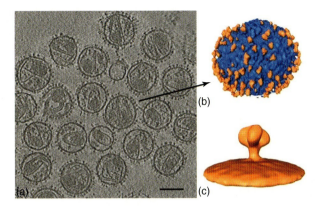

Figure 36.10 Three-dimensional reconstruction of SIV and subtomogram averaging of the spike protein. (a) Slice through a cryo electron tomographic reconstruction showing different SIV particles with spike proteins on the viral surface. Scale bar is 100 nm. (b) Surface rendering of one virus particle from the tomogram with the lipid membrane depicted in blue and the spikes depicted in orange. (c) Surface-rendered representation of subtomogram averaged reconstruction of the spike and part of the lipid surface, showing the threefold appearance of the spike. (Image reproduced and adapted from Ref. [89].)

The 3D structure of these spikes has been investigated using several EM methods, starting with negative stain on purified viruses [88], followed by several publications in which cryo-ET and subtomogram averaging were used on retrovirus spike proteins [87, 89–91], while recently also single-particle processing was used [92]. From these publications, it appeared that in the unprocessed tomograms the surface spikes do not appear to have a specific form (Figure 36.10a,b), while after subtomogram averaging of several thousand spikes, the averaged reconstructions showed molecular details in the trimeric arrangement (Figure 36.10c). Although the published findings are in total agreement on the structural details, they show that subtomogram averaging can increase the resolution and provide better data from multiple identical copies from the same structure.

36.4 Outlook

The strength of ET on biological samples is that it is the only technique that can obtain nanometer resolution 3D information of cellular structures and macromolecular assemblies in their biological context. The goal in structural biology is to unravel structure/function relationships at different size scales and resolution levels and ET will play an increasingly important role in this.

On one hand, attaining the best possible resolution allows docking of atomic structures into EM reconstructions to elucidate the relation between the structure and function of biomolecules. On the other, it is equally important to integrate this information into a larger cellular perspective. While a combination of high resolution and large scale is desirable, in practice there is a trade-off between scale and resolution.

36.4.1
Toward Higher Resolution

For biomaterials, the fundamental limitation to the resolution that can be obtained with TEM is the electron beam sensitivity of the specimen. The potentially subnanometer resolution information that is preserved in cryoimmobilized biological samples can be revealed only partially, since the specimen will be damaged during the imaging process. Many of the technical innovations that were implemented for imaging biological samples are aimed at recording images where the high-resolution information is revealed as good as possible within the constraints of the allowable electron dose.

The performance of the digital image detector on the TEM is a crucial factor that influences the signal to noise ratio in the images recorded. Currently, cooled slow scan CCDs are widely used in the field, with image sizes ranging from 1024 × 1024 pixels to 8096 × 8096 pixels. However, it becomes clear that these CCD-devices will be replaced by a new generation of detectors [93, 94]. These detectors promise to be an order of magnitude more sensitive and faster than the CCD-based detectors and will give a significant push to the resolution obtained in 3D reconstructions obtained by ET or single-particle techniques.

Another technical development that has high potential to push the resolution limits in cryo-EM is the incorporation of phase plates. Phase plates provide an improved contrast and signal to noise ratio in images of radiation sensitive specimens. Their practical implementation has suffered many obstacles, including fabrication, charging, and practical operation, but successful approaches have been reported [95, 96]. Although phase plate devices and techniques are still in their infancy, their potential impact is large and can open the doors to new methods.

Apart from improving EM hardware to acquire images with a higher resolution, significant improvement in resolution is also expected from a more extensive use of image processing tools. Today, postprocessing steps are mainly limited to, for example, median, low-pass filtering, and nonlinear anisotropic diffusion [97, 98]. However, the application of CTF correction in ET has not matured yet and is still under development [70, 71, 89, 99]. The defocus values within individual images of a tomographic tilt series vary because the sample is tilted. Proper correction of these differences is necessary to attain higher resolution tomograms.

Ongoing developments to further improve the resolution are the optimization and application of subtomogram averaging. This way, higher resolution structures of macromolecular machines, can be generated from cryo electron tomograms, albeit only after appropriate 3D classification and averaging of identical structures. Problems that are encountered in subtomogram averaging such as aligning and averaging particles with different missing wedges and classification of the intrinsic structural variability of macromolecules need to be properly addressed [100–103].

36.4.2
Toward Larger Volumes

Technical developments aimed at imaging larger volumes using EM seem to have reached their limits. Current ET software already includes possibilities to automatically record overlapping tomograms and stitch them together to overcome the relative small areas of digital cameras. Data collection of large volumes does not require major technical developments but more practical implementations of data management, image stitching, data storage, visualization and data mining, and improvement of camera speeds.

New systems in scanning electron microscope (SEM) imaging that combine automated ultramicrotomy of plastic-embedded material and repetitive block face scanning within a SEM have been developed [104, 105]. The ultramicrotomy is automatically performed, and the newly developing block faces are imaged and combined *in silico* into a 3D volume. The resolution is limited to the section thickness, several tens of nanometers, but volumes of several microns can be imaged. Next to sectioning, using a diamond knife material can also be removed using a FIB in dual-beam SEM, after which the trimmed face is imaged. This makes the technique also suitable for cryo-cooled samples [106, 107].

The combination of EM with other techniques such as X-ray tomography and light microscopy is being pursued and shows great potential. Using X-ray tomography [108], fully hydrated samples up to 10 µm thick can be imaged with a spatial resolution of 50 nm, enabling 3D imaging of organelles and lipid structures inside cells.

Another great potential technical improvement in covering larger volumes in EM lies in combining light microscopy with EM [29, 31]. In light microscopy, the resolution is 2 orders of magnitude lower than in EM, while the imaged areas are 2 orders of magnitude larger. Acquisitions of large datasets in EM could help bridge this gap between the large volume/areas. On the other hand, nondiffraction limited techniques [109], in which the resolution of light microscopy is improved to the order of 20–70 nm can bridge the resolution gap between light microscopy and EM back to one factor of 10, which will greatly facilitate integration of these two techniques.

References

1. Barcena, M., Oostergetel, G.T., Bartelink, W., Faas, F.G., Verkleij, A., Rottier, P.J., Koster, A.J., and Bosch, B.J. (2009) Cryo-electron tomography of mouse hepatitis virus: insights into the structure of the coronavirion. *Proc. Natl. Acad. Sci. U.S.A.*, **106**, 582–587.
2. Ruska, H., von Borries, B., and Ruska, E. (1940) Ubermikroskopie fur die Virusforschung. *Arch. Gesamte Virusforsch.*, **1**, 155–169.
3. Glauert, A. and Lewis, P. (1998) *Biological Specimen Preparation for Transmission Electron Microscopy*, Princeton University Press, Princeton.
4. Dubochet, J. (2007) in *Cellular Electron Microscopy* (ed. J.R. McIntosh), Elsevier, London, pp. 8–23.
5. McDonald, K. (2007) Cryopreparation methods for electron microscopy of selected model systems. *Methods Cell Biol.*, **79**, 23–56.

6. Henderson, R. (2004) Realizing the potential of electron cryo-microscopy. *Q. Rev. Biophys.*, **37**, 3–13.
7. Radon, J. (1917) Über die Bertimmung von funktionen durch ihre integralwerte längs gewisser mannigfaltigkeiten. *Ber. Verh. König. Sächs. Ges. Wiss. Math. Phys. Klasse*, **69**, 262–277.
8. DeRosier, D.J. and Kug, A. (1968) Reconstruction of three dimensional structures from electron micrographs. *Nature*, **217**, 130–134.
9. Hart, R.G. (1968) Electron microscopy of unstained biological material: the polytropic montage. *Science*, **159**, 1464–1467.
10. Hoppe, W., Langer, R., Knesch, G., and Poppe, C. (1968) Proteinkristallstrukturanalyse mit Elektronenstrahlen. *Z. Naturwiss.*, **55**, 333–336.
11. Frank, J. (2006) *Three-Dimensional Electron Microscopy of Macromolecular Assemblies*, Oxford University Press, New York.
12. Koster, A.J., Vandenbos, A., and Vandermast, K.D. (1987) An autofocus method for A TEM. *Ultramicroscopy*, **21**, 209–221.
13. Koster, A.J., Deruijter, W.J., Vandenbos, A., and Vandermast, K.D. (1989) Autotuning of A TEM Using minimum electron dose. *Ultramicroscopy*, **27**, 251–272.
14. Dierksen, K., Typke, D., Hegerl, R., Koster, A.J., and Baumeister, W. (1992) Towards automatic electron tomography. *Ultramicroscopy*, **40**, 71–87.
15. Koster, A.J., Chen, H., Sedat, J.W., and Agard, D.A. (1992) Automated microscopy for electron tomography. *Ultramicroscopy*, **46**, 207–227.
16. Fung, J.C., Liu, W., de Ruijter, W.J., Chen, H., Abbey, C.K., Sedat, J.W., and Agard, D.A. (1996) Toward fully automated high-resolution electron tomography. *J. Struct. Biol.*, **116**, 181–189.
17. Rath, B.K., Marko, M., Radermacher, M., and Frank, J. (1997) Low-dose automated electron tomography: a recent implementation. *J. Struct. Biol.*, **120**, 210–218.
18. Baumeister, W. (2002) Electron tomography: towards visualizing the molecular organization of the cytoplasm. *Curr. Opin. Struct. Biol.*, **12**, 679–684.
19. McIntosh, R., Nicastro, D., and Mastronarde, D. (2005) New views of cells in 3D: an introduction to electron tomography. *Trends Cell Biol.*, **15**, 43–51.
20. Milne, J.L. and Subramaniam, S. (2009) Cryo-electron tomography of bacteria: progress, challenges and future prospects. *Nat. Rev. Microbiol.*, **7**, 666–675.
21. Bouwer, J.C., Mackey, M.R., Lawrence, A., Deerinck, T.J., Jones, Y.Z., Terada, M., Martone, M.E., Peltier, S., and Ellisman, M.H. (2004) Automated most-probable loss tomography of thick selectively stained biological specimens with quantitative measurement of resolution improvement. *J. Struct. Biol.*, **148**, 297–306.
22. Mastronarde, D.N. (2005) Automated electron microscope tomography using robust prediction of specimen movements. *J. Struct. Biol.*, **152**, 36–51.
23. Nickell, S., Forster, F., Linaroudis, A., Net, W.D., Beck, F., Hegerl, R., Baumeister, W., and Plitzko, J.M. (2005) TOM software toolbox: acquisition and analysis for electron tomography. *J. Struct. Biol.*, **149**, 227–234.
24. Suloway, C., Shi, J., Cheng, A., Pulokas, J., Carragher, B., Potter, C.S., Zheng, S.Q., Agard, D.A., and Jensen, G.J. (2009) Fully automated, sequential tilt-series acquisition with Leginon. *J. Struct. Biol.*, **167**, 11–18.
25. Zheng, S.Q., Sedat, J.W., and Agard, D.A. (2010) Automated data collection for electron microscopic tomography. *Methods Enzymol.*, **481**, 283–315.
26. Le Gros, M.A., McDermott, G., Uchida, M., Knoechel, C.G., and Larabell, C.A. (2009) High-aperture cryogenic light microscopy. *J. Microsc.*, **235**, 1–8.
27. Schneider, G., Guttmann, P., Heim, S., Rehbein, S., Mueller, F., Nagashima, K., Heymann, J.B., Muller, W.G., and McNally, J.G. (2010) Three-dimensional cellular ultrastructure resolved by X-ray microscopy. *Nat. Methods*, **7**, 985–987.

28. Sartori, A., Gatz, R., Beck, F., Rigort, A., Baumeister, W., and Plitzko, J.M. (2007) Correlative microscopy: bridging the gap between fluorescence light microscopy and cryo-electron tomography. *J. Struct. Biol.*, **160**, 135–145.
29. Plitzko, J.M., Rigort, A., and Leis, A. (2009) Correlative cryo-light microscopy and cryo-electron tomography: from cellular territories to molecular landscapes. *Curr. Opin. Biotechnol.*, **20**, 83–89.
30. van Driel, L.F., Valentijn, J.A., Valentijn, K.M., Koning, R.I., and Koster, A.J. (2009) Tools for correlative cryo-fluorescence microscopy and cryo-electron tomography applied to whole mitochondria in human endothelial cells. *Eur. J. Cell Biol.*, **88**, 669–684.
31. Briegel, A., Chen, S., Koster, A.J., Plitzko, J.M., Schwartz, C.L., and Jensen, G.J. (2010) Correlated light and electron cryo-microscopy. *Methods Enzymol.*, **481**, 317–341.
32. Sali, A., Glaeser, R., Earnest, T., and Baumeister, W. (2003) From words to literature in structural proteomics. *Nature*, **422**, 216–225.
33. Cormack, A.M. (1986) Scanning in medicine and other fields. *Cell Biophys.*, **9**, 151–169.
34. Baker, T.S., Olson, N.H., and Fuller, S.D. (1999) Adding the third dimension to virus life cycles: three-dimensional reconstruction of icosahedral viruses from cryo-electron micrographs. *Microbiol. Mol. Biol. Rev.*, **63**, 862–922.
35. Diaz, R., Rice, W.J., and Stokes, D.L. (2010) Fourier-Bessel reconstruction of helical assemblies. *Methods Enzymol.*, **482**, 131–165.
36. Bohm, J., Frangakis, A.S., Hegerl, R., Nickell, S., Typke, D., and Baumeister, W. (2000) Toward detecting and identifying macromolecules in a cellular context: template matching applied to electron tomograms. *Proc. Natl. Acad. Sci. U.S.A.*, **97**, 14245–14250.
37. Frangakis, A.S., Bohm, J., Forster, F., Nickell, S., Nicastro, D., Typke, D., Hegerl, R., and Baumeister, W. (2002) Identification of macromolecular complexes in cryoelectron tomograms of phantom cells. *Proc. Natl. Acad. Sci. U.S.A.*, **99**, 14153–14158.
38. Walz, J., Typke, D., Nitsch, M., Koster, A.J., Hegerl, R., and Baumeister, W. (1997) Electron tomography of single ice-embedded macromolecules: three-dimensional alignment and classification. *J. Struct. Biol.*, **120**, 387–395.
39. Nitsch, M., Walz, J., Typke, D., Klumpp, M., Essen, L.O., and Baumeister, W. (1998) Group II chaperonin in an open conformation examined by electron tomography. *Nat. Struct. Biol.*, **5**, 855–857.
40. Penczek, P., Marko, M., Buttle, K., and Frank, J. (1995) Double-tilt electron tomography. *Ultramicroscopy*, **60**, 393–410.
41. Mastronarde, D.N. (1997) Dual-axis tomography: an approach with alignment methods that preserve resolution. *J. Struct. Biol.*, **120**, 343–352.
42. Kamino, T., Yaguchi, T., Konno, M., Ohnishi, T., and Ishitani, T. (2004) A method for multidirectional TEM observation of a specific site at atomic resolution. *J. Electron Microsc. (Tokyo)*, **53**, 583–588.
43. Brandt, S.S. (2006) in *Electron Tomography: Methods for 3D Visualization of Structures in the Cell* (ed. J. Frank), Springer, New York, pp. 187–215.
44. Lawrence, M.C. (1992) in *Electron Tomography* (ed. J. Frank), Plenum Press, New York, pp. 197–204.
45. Mastronarde, D.N. (2006) in *Electron Tomography: Methods for 3D Visualization of Structures in the Cell* (ed. J. Frank), Springer, New York, pp. 163–185.
46. Radermacher, M. (2006) in *Electron Tomography: Methods for 3D Visualization of Structures in the Cell* (ed. J. Frank), Springer, New York, pp. 245–273.
47. Gordon, R., Bender, R., and Herman, G.T. (1970) Algebraic reconstruction techniques (ART) for three-dimensional electron microscopy and x-ray photography. *J. Theor. Biol.*, **29**, 471–481.
48. Gilbert, P. (1972) Iterative methods for the three-dimensional reconstruction

of an object from projections. *J. Theor. Biol.*, **35**, 105–117.

49. Dubochet, J., Adrian, M., Chang, J.J., Homo, J.C., Lepault, J., McDowall, A.W., and Schultz, P. (1988) Cryo-electron microscopy of vitrified specimens. *Q. Rev. Biophys.*, **21**, 129–228.

50. Brenner, S. and Horne, R.W. (1959) A negative staining method for high resolution electron microscopy of viruses. *Biochim. Biophys. Acta*, **34**, 103–110.

51. Knoops, K., Kikkert, M., Worm, S.H., Zevenhoven-Dobbe, J.C., van der Meer, Y., Koster, A.J., Mommaas, A.M., and Snijder, E.J. (2008) SARS-coronavirus replication is supported by a reticulovesicular network of modified endoplasmic reticulum. *PLoS. Biol.*, **6**, e226.

52. Moor, H. (1987) in *Cryotechniques in Biological Electron Microscopy* (eds R.A. Steinbrecht and K. Zierold), Springer, Berlin, pp. 175–191.

53. Studer, D., Michel, M., Wohlwend, M., Hunziker, E.B., and Buschmann, M.D. (1995) Vitrification of articular cartilage by high-pressure freezing. *J. Microsc.*, **179**, 321–332.

54. Studer, D., Graber, W., Al Amoudi, A., and Eggli, P. (2001) A new approach for cryofixation by high-pressure freezing. *J. Microsc.*, **203**, 285–294.

55. McDowall, A.W., Chang, J.J., Freeman, R., Lepault, J., Walter, C.A., and Dubochet, J. (1983) Electron microscopy of frozen hydrated sections of vitreous ice and vitrified biological samples. *J. Microsc.*, **131**, 1–9.

56. Al Amoudi, A., Chang, J.J., Leforestier, A., McDowall, A., Salamin, L.M., Norlen, L.P., Richter, K., Blanc, N.S., Studer, D., and Dubochet, J. (2004) Cryo-electron microscopy of vitreous sections. *EMBO J.*, **23**, 3583–3588.

57. Al Amoudi, A., Studer, D., and Dubochet, J. (2005) Cutting artefacts and cutting process in vitreous sections for cryo-electron microscopy. *J. Struct. Biol.*, **150**, 109–121.

58. Marko, M., Hsieh, C., Schalek, R., Frank, J., and Mannella, C. (2007) Focused-ion-beam thinning of frozen-hydrated biological specimens for cryo-electron microscopy. *Nat. Methods*, **4**, 215–217.

59. Rigort, A., Bauerlein, F.J., Leis, A., Gruska, M., Hoffmann, C., Laugks, T., Bohm, U., Eibauer, M., Gnaegi, H., Baumeister, W., and Plitzko, J.M. (2010) Micromachining tools and correlative approaches for cellular cryo-electron tomography. *J. Struct. Biol.*, **172**, 169–179.

60. Steinbrecht, R.A. and Müller, M. (1987) in *Cryotechniques in Biological Electron Microscopy* (eds R.A. Steinbrecht and K. Zierold), Springer-Verlag, Berlin, pp. 149–172.

61. Murk, J.L., Posthuma, G., Koster, A.J., Geuze, H.J., Verkleij, A.J., Kleijmeer, M.J., and Humbel, B.M. (2003) Influence of aldehyde fixation on the morphology of endosomes and lysosomes: quantitative analysis and electron tomography. *J. Microsc.*, **212**, 81–90.

62. Koster, A.J., Grimm, R., Typke, D., Hegerl, R., Stoschek, A., Walz, J., and Baumeister, W. (1997) Perspectives of molecular and cellular electron tomography. *J. Struct. Biol.*, **120**, 276–308.

63. Reimer, L. (1995) *Energy-Filtering Transmission Electron Microscopy*, 3rd edn, Springer, Berlin.

64. Egerton, R.F. (1996) *Electron Energy-Loss Spectroscopy in the Electron Microscope*, Plenum Press, New York.

65. Crowther, R.A., Amos, L.A., Finch, J.T., DeRosier, D.J., and Klug, A. (1970) Three dimensional reconstructions of spherical viruses by fourier synthesis from electron micrographs. *Nature*, **226**, 421–425.

66. Radermacher, M. and Hoppe, W. (1980) Properties of 3-D reconstructions from projections by conical tilting compared to single axis tilting. Proceedings of the 7th European Congress on Electron Microscopy, Vol. 1, pp. 132–133.

67. Luther, P.K. (2006) in *Electron Tomography: Methods for 3D Visualization of Structures in the Cell* (ed. J. Frank), Springer, New York, pp. 17–48.

68. Reimer, L. (1993) *Transmission Electron Microscopy*, 3rd edn, Springer, Berlin.

69. Penczek, P.A. (2010) Image restoration in cryo-electron microscopy. *Methods Enzymol.*, **482**, 35–72.
70. Fernandez, J.J., Li, S., and Crowther, R.A. (2006) CTF determination and correction in electron cryotomography. *Ultramicroscopy*, **106**, 587–596.
71. Xiong, Q., Morphew, M.K., Schwartz, C.L., Hoenger, A.H., and Mastronarde, D.N. (2009) CTF determination and correction for low dose tomographic tilt series. *J. Struct. Biol.*, **168**, 378–387.
72. Marsh, B.J. (2007) Reconstructing mammalian membrane architecture by large area cellular tomography. *Methods Cell Biol.*, **79**, 193–220.
73. Noske, A.B., Costin, A.J., Morgan, G.P., and Marsh, B.J. (2008) Expedited approaches to whole cell electron tomography and organelle mark-up in situ in high-pressure frozen pancreatic islets. *J. Struct. Biol.*, **161**, 298–313.
74. Hoog, J.L. and Antony, C. (2007) Whole-cell investigation of microtubule cytoskeleton architecture by electron tomography. *Methods Cell Biol.*, **79**, 145–167.
75. Hoog, J.L., Schwartz, C., Noon, A.T., O'Toole, E.T., Mastronarde, D.N., McIntosh, J.R., and Antony, C. (2007) Organization of interphase microtubules in fission yeast analyzed by electron tomography. *Dev. Cell*, **12**, 349–361.
76. McComb, T., Cairncross, O., Noske, A.B., Wood, D.L., Marsh, B.J., and Ragan, M.A. (2009) Illoura: a software tool for analysis, visualization and semantic querying of cellular and other spatial biological data. *Bioinformatics*, **25**, 1208–1210.
77. Koning, R.I., Zovko, S., Barcena, M., Oostergetel, G.T., Koerten, H.K., Galjart, N., Koster, A.J., and Mieke, M.A. (2008) Cryo electron tomography of vitrified fibroblasts: microtubule plus ends in situ. *J. Struct. Biol.*, **161**, 459–468.
78. Medalia, O., Weber, I., Frangakis, A.S., Nicastro, D., Gerisch, G., and Baumeister, W. (2002) Macromolecular architecture in eukaryotic cells visualized by cryoelectron tomography. *Science*, **298**, 1209–1213.
79. Garvalov, B.K., Zuber, B., Bouchet-Marquis, C., Kudryashev, M., Gruska, M., Beck, M., Leis, A., Frischknecht, F., Bradke, F., Baumeister, W., Dubochet, J., and Cyrklaff, M. (2006) Luminal particles within cellular microtubules. *J. Cell Biol.*, **174**, 759–765.
80. Ortiz, J.O. (2006) Mapping 70S Ribosomes in Intact Cells by Cryoelectron Tomography and Pattern Recognition.
81. Grunewald, K., Desai, P., Winkler, D.C., Heymann, J.B., Belnap, D.M., Baumeister, W., and Steven, A.C. (2003) Three-dimensional structure of herpes simplex virus from cryo-electron tomography. *Science*, **302**, 1396–1398.
82. Briggs, J.A., Grunewald, K., Glass, B., Forster, F., Krausslich, H.G., and Fuller, S.D. (2006) The mechanism of HIV-1 core assembly: insights from three-dimensional reconstructions of authentic virions. *Structure*, **14**, 15–20.
83. Harris, A., Cardone, G., Winkler, D.C., Heymann, J.B., Brecher, M., White, J.M., and Steven, A.C. (2006) Influenza virus pleiomorphy characterized by cryoelectron tomography. *Proc. Natl. Acad. Sci. U.S.A.*, **103**, 19123–19127.
84. Fiers, W., Contreras, R., Duerinck, F., Haegeman, G., Iserentant, D., Merregaert, J., Min, J.W., Molemans, F., Raeymaekers, A., Van den Berghe, A., Volckaert, G., and Ysebaert, M. (1976) Complete nucleotide sequence of bacteriophage MS2 RNA: primary and secondary structure of the replicase gene. *Nature*, **260**, 500–507.
85. Valegard, K., Liljas, L., Fridborg, K., and Unge, T. (1990) The three-dimensional structure of the bacterial virus MS2. *Nature*, **345**, 36–41.
86. Koning, R., van den Worm, S., Plaisier, J.R., van Duin, J., Abrahams, J.P., and Koerten, H. (2003) Visualization by cryo-electron microscopy of genomic RNA that binds to the protein capsid inside bacteriophage MS2. *J. Mol. Biol.*, **332**, 415–422.
87. Forster, F., Medalia, O., Zauberman, N., Baumeister, W., and Fass, D. (2005) Retrovirus envelope protein complex structure in situ studied by

cryo-electron tomography. *Proc. Natl. Acad. Sci. U.S.A.*, **102**, 4729–4734.
88. Zhu, P., Chertova, E., Bess, J. Jr., Lifson, J.D., Arthur, L.O., Liu, J., Taylor, K.A., and Roux, K.H. (2003) Electron tomography analysis of envelope glycoprotein trimers on HIV and simian immunodeficiency virus virions. *Proc. Natl. Acad. Sci. U.S.A.*, **100**, 15812–15817.
89. Zanetti, G., Briggs, J.A., Grunewald, K., Sattentau, Q.J., and Fuller, S.D. (2006) Cryo-electron tomographic structure of an immunodeficiency virus envelope complex in situ. *PLoS. Pathog.*, **2**, e83.
90. Zhu, P., Winkler, H., Chertova, E., Taylor, K.A., and Roux, K.H. (2008) Cryoelectron tomography of HIV-1 envelope spikes: further evidence for tripod-like legs. *PLoS. Pathog.*, **4**, e1000203.
91. Liu, J., Bartesaghi, A., Borgnia, M.J., Sapiro, G., and Subramaniam, S. (2008) Molecular architecture of native HIV-1 gp120 trimers. *Nature*, **455**, 109–113.
92. Wu, S.R., Loving, R., Lindqvist, B., Hebert, H., Koeck, P.J., Sjoberg, M., and Garoff, H. (2010) Single-particle cryoelectron microscopy analysis reveals the HIV-1 spike as a tripod structure. *Proc. Natl. Acad. Sci. U.S.A.*, **107**, 18844–18849.
93. McMullan, G., Chen, S., Henderson, R., and Faruqi, A.R. (2009) Detective quantum efficiency of electron area detectors in electron microscopy. *Ultramicroscopy*, **109**, 1126–1143.
94. McMullan, G., Faruqi, A.R., Henderson, R., Guerrini, N., Turchetta, R., Jacobs, A., and van Hoften, G. (2009) Experimental observation of the improvement in MTF from backthinning a CMOS direct electron detector. *Ultramicroscopy*, **109**, 1144–1147.
95. Danev, R. and Nagayama, K. (2001) Transmission electron microscopy with Zernike phase plate. *Ultramicroscopy*, **88**, 243–252.
96. Danev, R., Kanamaru, S., Marko, M., and Nagayama, K. (2010) Zernike phase contrast cryo-electron tomography. *J. Struct. Biol.*, **171**, 174–181.
97. Frangakis, A.S. and Hegerl, R. (2001) Noise reduction in electron tomographic reconstructions using nonlinear anisotropic diffusion. *J. Struct. Biol.*, **135**, 239–250.
98. Fernandez, J.J. and Li, S. (2003) An improved algorithm for anisotropic nonlinear diffusion for denoising cryo-tomograms. *J. Struct. Biol.*, **144**, 152–161.
99. Winkler, H. and Taylor, K.A. (2003) Focus gradient correction applied to tilt series image data used in electron tomography. *J. Struct. Biol.*, **143**, 24–32.
100. Forster, F., Pruggnaller, S., Seybert, A., and Frangakis, A.S. (2008) Classification of cryo-electron sub-tomograms using constrained correlation. *J. Struct. Biol.*, **161**, 276–286.
101. Bartesaghi, A., Sprechmann, P., Liu, J., Randall, G., Sapiro, G., and Subramaniam, S. (2008) Classification and 3D averaging with missing wedge correction in biological electron tomography. *J. Struct. Biol.*, **162**, 436–450.
102. Scheres, S.H., Melero, R., Valle, M., and Carazo, J.M. (2009) Averaging of electron subtomograms and random conical tilt reconstructions through likelihood optimization. *Structure*, **17**, 1563–1572.
103. Yu, L., Snapp, R.R., Ruiz, T., and Radermacher, M. (2010) Probabilistic principal component analysis with expectation maximization (PPCA-EM) facilitates volume classification and estimates the missing data. *J. Struct. Biol.*, **171**, 18–30.
104. Denk, W. and Horstmann, H. (2004) Serial block-face scanning electron microscopy to reconstruct three-dimensional tissue nanostructure. *PLoS Biol.*, **2**, e329.
105. Knott, G., Marchman, H., Wall, D., and Lich, B. (2008) Serial section scanning electron microscopy of adult brain tissue using focused ion beam milling. *J. Neurosci.*, **28**, 2959–2964.
106. Heymann, J.A., Hayles, M., Gestmann, I., Giannuzzi, L.A., Lich, B., and Subramaniam, S. (2006) Site-specific 3D imaging of cells and tissues with a

dual beam microscope. *J. Struct. Biol.*, **155**, 63–73.

107. Hsieh, C.E., Leith, A., Mannella, C.A., Frank, J., and Marko, M. (2006) Towards high-resolution three-dimensional imaging of native mammalian tissue: electron tomography of frozen-hydrated rat liver sections. *J. Struct. Biol.*, **153**, 1–13.

108. McDermott, G., Le Gros, M.A., Knoechel, C.G., Uchida, M., and Larabell, C.A. (2009) Soft X-ray tomography and cryogenic light microscopy: the cool combination in cellular imaging. *Trends Cell Biol.*, **19**, 587–595.

109. Hell, S.W. (2007) Far-field optical nanoscopy. *Science*, **316**, 1153–1158.

37
Small Organic Molecules and Higher Homologs
Ute Kolb and Tatiana E. Gorelik

37.1
Introduction

Chemicals surround us in our daily life representing a large spectrum of materials ranging from plastics, pharmaceuticals, and petroleum products to pigments and semiconductor electronics. The actual value of a chemical for us is determined by its performance – how well it fulfils the tasks it is designed for. Knowing the *structure* of a material is not only the key to understand its properties but also an essential contribution to develop and enhance its performance.

The *structure* of a material is a complex compilation describing its constitution on different levels. The interplay of different structural levels can cause, modify, or even eliminate a certain physical property. For example, a biopolymer having exactly the same sequence of monomeric subunits as a native material will not exhibit its functionality if the folding of the chain is different or the performance of a semiconductor device may be destroyed if the chip is build using a silicon wafer with inclusion defects.

Structural characterization of materials on different levels is a large area of materials research and is mainly based on microscopic methods. As information on a wide length scale is required, from morphology to molecular and atomic arrangement, a range of microscopic methods has been developed for optimal characterization (see Figure 37.1). The methods can be built based on photons, X-rays, electrons, and different physical means of probing [1–5].

There are two principally different ways to set up an *imaging* experiment: by full illumination of a sample area or by scanning of a confined probing beam across the area. Direct imaging is typically done using parallel illumination and signal readout (optical microscopy (OM) [6], transmission electron microscopy (TEM) [7, 8], holography [9–11], X-ray microscopy (XRM) [12]), while scanning techniques perform a sequential signal readout moving the probe over the sample (scanning electron microscopy (SEM), scanning transmission electron microscopy (STEM), elemental mapping via energy electron loss spectroscopy (EELS) [13] or energy

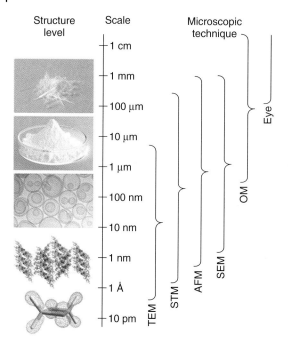

Figure 37.1 Scheme of magnification ranges that can be imaged with different microscopic methods.

dispersive X-ray spectroscopy (EDX) [14]), or the sample (with a piezo precision of the movement) against the probe (scanning probe microscopy – SPM: e.g., atomic force microscopy (AFM), scanning tunneling microscopy (STM)).

Imaging delivers information about the sample in *direct space*, whereas in *diffraction* [15], the experimental data is collected in *reciprocal* space. The two types of data – direct and reciprocal – are in principle related by Fourier transform, thus the structure of a material can be determined in direct space using information collected in reciprocal space after necessary transformations have been applied. Later in the chapter, the structural analysis of materials based on diffraction data is shown in detail.

Diffractive methods are especially efficient when the size of structural details is comparable with the wavelength of the radiation used. Therefore, diffraction is usually chosen to investigate the structure of material at the atomic level – the arrangement of atoms in crystalline lattice, that is, *crystal structure*. Diffraction experiments are built in different ways, therefore, although they all deliver comparable structural information, they can be applied to materials only up to a certain crystal size (Figure 37.2).

The choice of the appropriate method of structural investigation depends mainly on the necessary resolution and the stability of the material throughout the investigation. Important factors are also the achievable contrast or signal-to-noise ratio needed to distinguish structural details and a suitable preparation method.

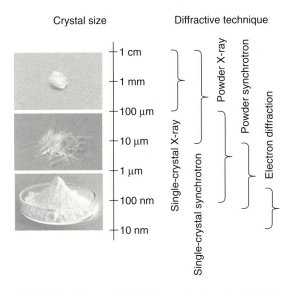

Figure 37.2 Scheme of diffraction methods using different radiations.

The investigation route as well as preparation methods applicable to inorganic and organic material are highly diverse. This chapter mainly focuses on structural investigations of organic compounds, and therefore topics associated with sample preparation, imaging, and diffraction data collection of *soft mater* are described.

All organic molecules can be subdivided into three groups based on their size and function: *small organic molecules, polymers*, and *biopolymers – large biological macromolecular complexes*. The classification of molecules into one of these groups is not strict; some materials such as short oligomers of long polymer chains or larger polyelectrolytes easily bridge the different classes. The materials have in common a tendency to be *soft* in different means: they usually suffer significantly from radiation damage, they may change their structure during investigation or even during sample preparation, they are typically problematic for microscopic methods as they only respond with a weak signal and typically require signal or contrast enhancement.

Small organic molecules represent one of the major parts of chemical industry and touch different aspects of our life: from pharmaceuticals to inks, from electronic components to explosives. The structural investigations on molecular crystals have a particular importance due to the phenomenon called *polymorphism*. Polymorphism is defined as the ability of a molecule to pack in different ways within a crystal. These crystals composed from the same molecule, but having different crystalline structure are then called *polymorphs* [16]. Typically, polymorphs are thermodynamically metastable phases coexisting at given conditions due to kinetical issues. Nevertheless, they can transform one to another under certain circumstances, which can be aging, pressure, humidity, presence of organic solvents, and so on. Structural differences of different polymorphs have enormous effects on the properties of the material; therefore, for example, polymorphic purity is a crucial aspect for pharmaceutical compounds: different crystalline forms can have different solubility and therefore different bioavailability [17]. Organic pigments crystallized as different polymorphs show diverse color and fastness properties [18]. In general, where the organic material is used in its crystalline form, the knowledge of crystal packing is of great importance. Polymorphism also exists for proteins [19], but for these materials, the main interest is the conformation of the molecule itself and the arrangement of the molecules inside a crystal is not a primary concern.

Compared to small organic molecules, the polymer world covers a significantly wider range of products making it impossible to mention all of them. Beside the use of plastics and engineering resins, in a variety of fabricated forms (fibers, films, membranes and filters, mouldings, and extrudates), highly oriented structures often with crystalline domains, polymers blended with nanoparticles [20], and copolymers allow the manufacture of high-performance material.

Properties of polymers strongly depend on the process history that influences mainly the morphology of the material. Polymer morphology describes form, size, as well as distribution and association of structural units (domains) or fillers and additives inside the sample. The ability to tune the properties by small changes in material processing is beneficial for applications, but becomes dangerous for the material characterization, since artifacts can be easily introduced through sample preparation.

The crystallinity of polymers has somewhat different meaning than that of molecular crystals. The structure of a polymer is usually described by a *degree of crystallinity*. The degree of crystallinity ranges from zero for completely irregularly arranged molecules to one for crystalline materials. Usually polymers are built from ordered microcrystalline domains due to regular folding of chains, and completely amorphous material between them. Nevertheless, fully crystalline polymers exist, so polymers may also show polymorphism, for instance, polymorphs were found for poly(1-butene) or polylactides which crystallized epitaxially [21–23].

Owing to complex hierarchical interactions, biopolymers have very sophisticated organization. The monomeric sequence of a biopolymer is called *primary structure*. Folding of the chain and arrangement of individual units are essential for the macromolecular complex functionality. Therefore, the structure of a biopolymer should also describe mutual arrangement of units in space. Efficient and powerful methods were developed to study the structure of the biopolymers allowing complete structural characterization [24, 25]. Although being an exciting topic, the characterization of biological samples has different demands and will not be discussed here in detail. Further information can be derived from many dedicated texts [26–28].

The following sections cover methods suitable for different length scales ranging from OM (also called *light microscopy*), SEM, AFM, STM to TEM in imaging and diffraction mode.

37.2
Optical Microscopy

OM is the most widely spread and common technique to investigate a material. An optical microscope uses a set of glass lenses to magnify the image produced with visible, UV, or IR light. Commonly, visible light is used, and the resolution limit of a classical optical microscope is typically a quarter of a micrometer. Although not showing details at very high resolution, light microscopy can be very useful in the investigation of certain structural features. One of the main advantages of using light microscopy is that the low energy of visible light applied on the sample usually does not modify it – only few materials are sensitive to the visible light. An easy experimental setup often makes possible *in situ* observations of processes.

Bright field light microscopy, based only on amplitude contrast, can visualize features with differences in thickness or transparency in comparison with the support or the embedding matrix (fibers, micelles, and colloids). Figure 37.3 shows the shape change of liquid crystalline (LC)-colloids depending on the temperature. For polymer films or slices, with uniform thickness, contrast can often only be achieved by labeling with optical active markers, staining, or etching. Phase contrast optical techniques enhance contrast between polymers that are transparent but have different optical properties such as refractive index. Reflected light techniques reveal surface structures, and observation of internal textures of thin polymer slices is possible by transmitted light.

Polarized light microscopy is an optical technique that enhances contrast of crystalline materials and allows imaging of individual crystalline domains (Figure 37.4).

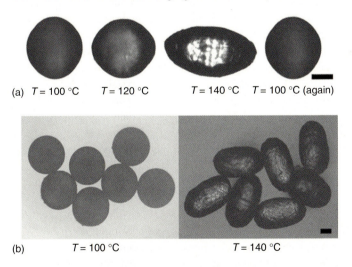

Figure 37.3 Microscopy images of heating experiments with cross-linked LC particles. (a) Shows one particle at different temperatures. It reversibly changes its shape from spherical to cigar like (scale bar = 100 mm). (b) Shows several particles in the liquid crystalline and in the isotropic phase [29].

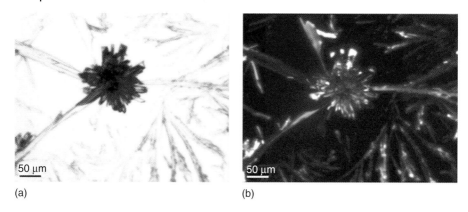

Figure 37.4 Optical microscopy image of an agglomerate of organic crystals: bright-filed image (a) and an image with crossed polarizers (b).

It usually uses a set of two polarizers: the first is built in before the sample, so the material is illuminated with polarized light. The second polarizer is placed after the sample. Crossing the second polarizer regarding to the first blocks directly transmitted light. The crystalline lattice has the ability to rotate the polarization vector, thus crystalline domains will appear bright with different contrast depending on their orientation, whereas the noncrystalline areas (liquid or amorphous material and directly transmitted light – background) will stay dark.

Polarized microscopy has gained particular interest for organic and polymeric materials to visualize *in situ* processes. A combination with the hot-stage evolved into a special method of material research based on melting, eutectic melting, and solid-phase transformation [30]. The thermo microscopy has found vital applications in polymorphs study of molecular crystals including primarily drug relevant materials [31, 32].

Liquid crystals represent another class of materials for which polarized light microscopy is one of the major investigation tools. Just like it was described above for ordinary crystals, polarization microscopy emphasizes domains with the same orientation of the molecules. This makes it easy to investigate different phases of liquid crystals as well as transitions between the phases. Figure 37.5 shows phase transformation of a liquid crystalline material [33]: spherulites of smectic C* liquid crystalline phase of a ferroelectric polymer; the leaves develop "zigzag" domain order after applying an electric field.

During the last decades, there have been a number of successful instrumental implementations to overcome the theoretical resolution limit of light microscopy. The easiest solution of the problem is to use shorter wavelength of light because the resolution limit is proportional to the used wavelength. Optical microscopes with UV sources are routinely applied nowadays for lithographic mask and wafer inspection. Changing the properties of optical lenses in order to improve the resolution was intensively investigated since the introduction of metamaterials with negative refractive index [34, 35]. Ag-coated superlenses allow resolution of

 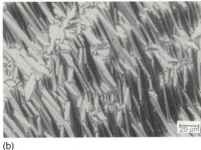

(a) (b)

Figure 37.5 Switching of ferroelectric monomers: overview of spherulite (a) and domains occurring after switching process in the indium tin oxide (ITO) cell (b). Cells for test experiments on liquid crystals built using two glass plates coated with ITO.

one-sixth of the illumination wavelength [36], other metamaterials are investigated expecting a further reduction of this limit [37].

Confocal imaging systems were designed to improve the resolution and allow collection of three-dimensional information. Confocal microscopy, widely used in biological applications, allows imaging of fluorescently marked cells and microorganisms. In a confocal microscope, the incident beam is scanned over a selected region inducing point-by-point fluorescence. The beam is focused within the specimen, so the fluorescent excitation above and below the focus is already reduced. The outcoming signal is then detected through a pin hole, which provides additional attenuation of the out of focus beam. The position of the focus within the sample can be changed, thus recording three-dimensional information.

Confocal microscopy has found a number of applications in investigations of molecular and polymeric systems. Three-dimensional sampling can give spatial information about the dye crystal distribution [38], but is especially interesting for dynamic systems: confocal microscopy can allow imaging of defects in liquid crystals, the interaction of the defects, and can provide as well valuable information about phase transitions [39, 40]. Colloids are one of the target systems for confocal microscopy. The method does not only deliver comprehensive information about the structure of the colloid but also allows tracking of the early stages of instability [41]. In combination with *in situ* experiments, confocal microscopy is especially efficient, and allows, for instance, study of precipitation of organic dye crystals [42].

Another technique probing the specimen in three dimensions and having the resolution superior than that of the confocal imaging is multiphoton microscopy [43, 44]. Incident photons with relatively low energy are focused within the sample. The interference of several photons is sufficient to induce excitation of the dye molecules inside the material and produce a signal which is recorded. Similar to confocal microscopy, the optical sectioning is then achieved by shifting the focus within the sample.

The main advantage of confocal microscopy is the information along the third coordinate. The resolution of confocal microscopy is still restricted by the diffraction limit. It is the use of fluorescent dyes which allows bypassing the diffraction limit [45]. Obviously, the need of fluorescent molecules sets restrictions on the samples to investigate.

A new generation of optical methods was recently developed in order to overcome the diffraction limit for dye-free samples. The technique of Resolution Enhancement with Nanoparticle Random Tracking (RENaRT) [46] uses a thin layer of nanoparticle suspension placed at the top of the sample. The nanoparticles are migrating within the suspension due to Brownian motion eventually crossing different features of the sample below. The brightness of the nanoparticles changes as they move over different parts of the sample. This change in the brightness can then be detected using a standard optical microscope. The resolution limit is determined by the average size of the nanoparticles, which can be as small as 30 nm, thus resulting in a resolution improvement of at least 10 times than that of the classical optical microscope.

All recent developments in OM methods are mainly driven by the demand of the electronic industry for a cheap, high-resolution, and high-throughput inspection method for patterned surfaces. Although organic and polymeric materials are not the primary targets for these developments, they can be studied using the new techniques just as efficient as semiconductor wafers. In the light of increasing importance of organic electronic materials [47], it is likely that the focus of optical developments will shift in the near future.

37.3
Scanning Electron Microscopy–SEM

A comprehensive review on the method can be found in the respective chapter in this book. Here, only a short survey of the method is presented. A SEM uses a focused electron beam to record an image. The beam is scanned across a selected area, the signal at each point is measured and these signals then represent a SEM image. SEMs are primarily used to analyze the texture of surfaces and morphology of materials with the features on the sample from microns down to nanometers. The focused electron beam has a very large depth of focus, the signal is collected from the surface in reflection, and therefore the image has intrinsic three-dimensionality (Figure 37.6). The true shape of the surface can be reconstructed using signals from different detectors placed on different sides of the specimen.

Imaging of the surface topography can be best performed through the detection of low-energy secondary electrons, being more sensitive to the inclination angle of the surface, whereas compositional contrast is better imaged by high-energy backscattered electrons. In order to reduce beam damage and charging, which is an important issue for organic materials, SEM can be performed at low voltage (LVSEM: normally 1–2 kV) [48], or by depositing an ultrathin conductive layer on the surface. In LVSEM, the backscattered electrons are detected as two individual signals, where the sum delivers materials contrast and the difference provides the relief signal for morphological investigation simultaneously [49].

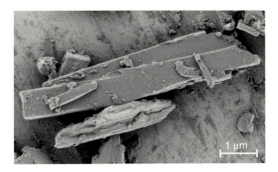

Figure 37.6 A SEM image of β-Cu phthalocyanine crystals; the lower crystal is seen from a side and ripple-like texture is seen originated from twinned domain structure.

In the past years, SEMs with field emission gun (FEG) became popular. FEG sources produce intensive coherent beam that can be focused on a very small area. As a result, a microscope with a FEG source can reach higher magnifications still keeping reasonable contrast compared to conventional SEMs.

Environmental SEMs can study samples in their native environment allowing a gas pressure of up to 10–15 Torr ($\sim$2000 Pa) in the sample chamber. Thus specific questions related to the interaction with fluids of vapors can be studied [50–52].

37.4
Atomic Force and Scanning Tunneling Microscopy (AFM and STM)

Other scanning probe methods – STM and AFM – do not use radiation but use physical properties of the sample. These methods, which are widespread and powerful tools for structural characterization, are described comprehensively in the devoted chapter of this book. Here only some highlights are shortly presented.

Direct imaging of molecules and chemical bonds has been a dream of chemists for many years. Recently, this dream became true: a clear AFM image of a pentacene molecule was recorded [53]. The image shows all carbon atoms of the molecule and all chemical bonds, even bonds to the hydrogen atoms. The ability to image surface molecules directly gives a researcher a unique possibility to not only study termination effect at surfaces but also defects in molecular crystals [54] and stress reaction in crystals [55]. Polymers are often investigated by AFM, which provides direct information about their morphology [56, 57].

37.5
Transmission Electron Microscopy (TEM)

The major advantage of TEM as well as STEM is the ability to characterize a sample over a wide range of magnifications, thus giving information from the morphology of the sample down to its structure on the atomic scale. Diverse samples have been studied by TEM and different experimental setups have been developed including heating and cooling stages. Sample preparation methods have been invented for different samples, so practically there is no material class which cannot be studied in TEM. Although inorganic materials can show fast degradation under the electron beam as well, the beam sensitivity is usually associated with organics. Thus materials such as molecular crystals and polymers are considered most challenging when studied by TEM. The application of TEM to these materials is enlightened on the following pages, starting from the description of the preparation methods, followed by issues of beam sensitivity and possible applications.

37.5.1
TEM: the Instrument

There is a great variety of diverse instruments on the market, but most of them are built using a similar design scheme, first proposed by Knoll and Ruska [58] (Figure 37.7). The source of electrons – also called *electron gun* – is placed at the top of the instrument. A condenser lens system modulates the beam which then will go to the sample; it can adjust the size of the beam at the sample, its convergence defining the intensity of the beam on the sample. After the beam passed through the sample, the image is magnified by the objective lens system, and projected onto a recording media. The most common way to record TEM images in our days is a charge-coupled device (CCD) camera [59], although image plates and photo materials are also used.

Commonly, TEMs are working at high acceleration voltages (80–400 kV). Low-voltage electron microscopy (LVEM at 5 kV) allows increasing of the image contrast significantly, and reduces radiation damage for some materials [60]. With the reduction of the acceleration, voltage increases the effect of magnetic lenses imperfections on the image formation. With the development of aberration correctors, applicable for SEM [61], TEM [62], and STEM [63], the lens aberrations in the low-voltage regime can be significantly reduced, and high-resolution imaging became possible even for highly beam-sensitive material.

Since the interaction of electrons with matter is of coulomb nature and therefore much stronger than for X-rays and neutrons [64], the limiting factor for TEM samples is the *electron beam transparency* depending on the elemental composition of the material. In other words, samples have to be very *thin* [65]. For organic materials composed mainly of light atoms, such as carbon, nitrogen, and oxygen, samples thinner than 200 nm are necessary for TEM and electron diffraction analysis; for low-resolution tomographic reconstructions, cell cuts with a thickness up to 500 nm can still be used [66]. In order to perform high-resolution transmission electron microscopy (HRTEM) or to collect kinematic diffraction data, the thickness should be less than 10 nm [67].

TEM is generally designed and optimized for imaging. The instrumental setup can either deliver high-resolution images or high-contrast images with relatively low resolution. Organic materials which typically do not sustain the radiation level needed for high-resolution imaging are rarely investigated at high magnification.

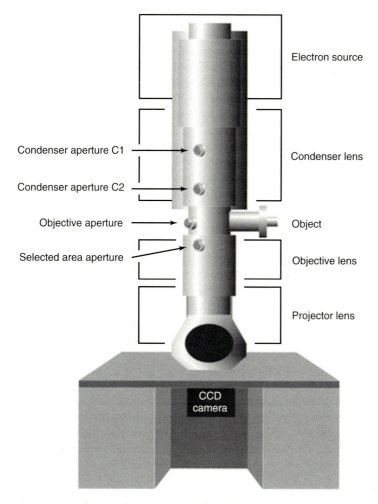

Figure 37.7 Schematical representation of a TEM.

TEM images of organic materials tend to show low contrast, and therefore when possible high-contrast instruments are used to study them.

TEMs have a possibility to collect electron diffraction information together with imaging. A separate section of this chapter is devoted to electron diffraction and the way the diffraction information can be used to analyze a structure.

In TEM, the specimen is simultaneously illuminated to produce an image; STEM mode uses a focused beam and scans it over the desired area. The signal is collected from each sample point using a dedicated detector; all the point information together builds up the STEM image pixel by pixel. The probe size can be set smaller than 1 nm, the current density of the beam can be kept very low, still producing high-contrast images. Scanning of the complete area can be done very fast, providing a possibility to record a high-contrast image using a very low total electron dose; therefore STEM has manifested itself as a preferable imaging mode for beam-sensitive materials.

When a special detector for elemental analysis (EDX or EELS) is attached to the microscope, information about the elemental composition can be collected. Combined with the STEM mode, elemental analysis can be used to create elemental maps of the specimen.

Thus instrumentally TEM allows to cover all questions about materials– morphology, three-dimensional organization, crystal structure, and elemental composition, provided, the sample is prepared in the optimal way.

37.5.2
TEM Sample Preparation

There is a variety of sample preparation techniques for all microscopic methods which have to be selected based on the demands of the particular investigation technique and the material under investigation. Practically, for all microscopic methods, the sample can be directly deposited on an appropriate substrate by applying a droplet, ultrasonic spraying [68], or direct crystallization. Dedicated sample preparation procedures are crucial in order to produce a thin polymer film suitable for TEM investigation. Drop casting – dropping a dilute solution onto a substrate with subsequent drying or annealing – is practically easy but may produce artifacts through drying process [69, 70]. Spin coating – applying a droplet onto a rotating substrate – and dip coating – drawing a substrate slowly out of a solution – result in films with more uniform thickness compared to drop casting. Microtomy and ultramicrotomy allow slicing of a bulk sample with a glass or diamond knife and provide the possibility to investigate polymer samples as processed without the need to dissolve the material. Moreover, preparation of slices from different parts of a material (e.g., surface or bulk) and from particles, too thick for investigation, after being embedded in raisin can be achieved. Vitrification – a very important and fully different approach – is used for investigation of particle structure in solution. The method was originally developed for biological samples in aqueous solution, which are quenched and investigated in the frozen state inside an amorphous water matrix, but is suitable as well for polymer samples in diverse solvents. Preparation methods dedicated to TEM are discussed in detail below. Dependent on the selected microscopic method, the contrast of the features under investigation needs to be enhanced, thus the different possibilities are mentioned directly where these methods are discussed.

37.5.2.1 Direct Deposition
Usually copper grids coated with a thin amorphous carbon film (2–5 nm) are used for sample preparation; alternatively Au, Ni, Mo, Ti, Al, steel, graphite, and nylon grids are available. The carbon film is not only holding the material but also serves heat dissipation and charge transfer. The beam penetrates through material and carbon film. The use of a holey-carbon support allows imaging the material without support contrast, although the holey-carbon films are not recommended for

beam-sensitive material since the reduced heat dissipation gives rise to increased beam damage.

For small molecular crystals, there are three approaches for sample preparation available: crystallization, dispersion, and sublimation. Crystallization from a solution directly onto a grid can produce evenly distributed thin crystals, ideal for TEM investigations. On the other hand, the method has clear limitations: the material has to be soluble in a solvent, which produces nicely defined but small crystals, and neither the solvent nor the material should react with the metal of the grid. Another factor to be considered is the danger to produce a polymorph different from the modification of the bulk phase, which can be avoided if the sample is dispersed rather than dissolved and then sprayed onto the grid. Poorly soluble materials, which are likely to refuse crystallization into crystals suitable for X-ray single-crystal structure analysis, can be nicely prepared using this methodology. A third technique is sublimation directly onto the carbon-coated grid or onto a selected crystalline surface: potassium chloride or mica either precoated with carbon film or not. The crystals grown on a clean surface must then be covered with a thin carbon film; the film is detached from the crystalline block by floatation on water and transferred on a TEM grid. The sublimation method is more likely to provide larger and flatter crystals; however, depending on the crystalline support, different morphology and sometimes different polymorphs can be produced [71].

Molecular crystals may have very strong difference in unit cell parameters causing morphological anisotropy. Often crystals with one long lattice cell parameter and two rather short ones appear as platelets, whereas structures with two relatively long axes tend to appear needle-shaped [72–74]. Crystals with strong anisotropic morphology tend to lie on a support film always in the same way. As a result, a strong preferred orientation of the material is observed. In such situations, observation of certain directions can be problematic due to the missing cone problem. Preferred orientation can be avoided by adjusting crystallization parameters [75]. Sometimes, the employment of epitaxial growth can force the material to change morphology. However, when influencing the crystal morphology, one could easily run into conditions where a new polymorph is produced. Alternatively, the sample can be embedded in an appropriate medium (epoxy resin, ice, etc.) and sliced by a microtome. The thin slices are then placed onto a coated TEM grid. The embedding resin can be removed by an appropriately matched solvent.

If a polymeric material, such as, for example, rigid polymeric nanocomposite, can be dispersed in a liquid which does not affect the structure, the suspension may be directly placed onto a TEM grid and subsequently dried either in vacuum or on air. Under the assumption that the structures do not collapse after they are dried and provide high contrast, the morphology of the material can then be investigated. Usually, this kind of preparation is a first morphology check followed by more sophisticated sample preparation routines.

Relatively short oligomers can be crystallized from dilute solution using a similar methodology to the small molecule crystals [76]. Another possibility is to place a drop of a polymer on the surface of a nonsolvent liquid and to collect the resulting thin film with a TEM grid subsequently. For long-chain polymer sample preparation, usually two strategies are taken into account: cutting thin slices out of bulk using an ultramicrotomy (at cryogenic temperatures for material with low glass transition temperature) or a focused ion beam (FIB) or preparing thin films on a suitable support. Recently, an extended review was published summarizing sample preparation procedures and the artifacts to be taken into account for TEM imaging [77]. Polymer films can be extra ordered during film preparation, either by mechanical sliding over a glass surface (shearing), spin coating, or epitaxial growth on a template.

37.5.2.2 Vitrification

A large number of materials lose their structure on drying: vesicles tend to collapse when the solvent is removed; large protein macromolecular complexes change their structure and disintegrate when dried. To preserve the material in its native state in TEM, a dedicated cryo-TEM method was developed. The method, developed originally for biological samples, is based on rapid cooling of the sample preventing crystallization of the solvent. Instead of crystallizing, the solvent stays amorphous – vitrified – and serves as sample support. While water is the standard solvent used for biological samples, cryo-TEM from other organic solvents is often performed for polymer solutions. The vitrified sample is kept constantly cooled to prevent spontaneous crystallization of the solvent on temperature rise. Cryo-transfer techniques were developed and are now routinely used to transfer the vitrified sample into a TEM. Figure 37.8 shows a scheme of the vitrification process. The quality of the resulting vitrified film can be controlled through a range of parameters such as humidity and temperature of the atmosphere in the vitrification chamber, surface properties of the carbon support film, blotting time, time intervals between the steps, and so on. It must be noted that vitrification does not only preserve the material in the natural state but also reduces temperature-induced radiation damage.

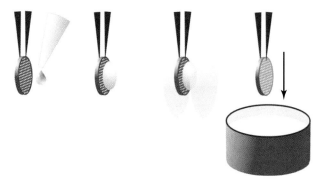

Figure 37.8 Cryo-TEM sample preparation (vitrification) scheme: a drop of the sample-containing suspension is placed onto a holey (commercially available QUANTI-foil grids are often used for their homogeneous holes) carbon TEM grid, the excess of the suspension is blotted away by filter paper, the grid with the homogeneous thin layer of the suspension is quenched into freezing media (liquid ethane, propane, or mixtures).

A cryo-TEM image is then a projection of an object fixed within an amorphous solvent matrix. Since the method is usually used for organic matter, the question of the relative contrast of the object compared to the amorphous matrix is very important. Usually, the images are taken at high defocus, as this enhances contrast of the boundaries. Otherwise, staining methods can be used to improve the contrast.

On the basis of the recent developments in instrumentation and automation, cryo-TEM has evolved in an established approach to study the structure of biocolloids, polymeric nanostructures, and proteins [78]. Cryo methods can be combined with tomographic approach and deliver valuable 3D information of objects in their native media.

37.5.2.3 Staining

Different elemental composition, that is, mass contrast or chemical contrast, between different morphological elements in a sample or between the sample and the support matrix allow imaging in TEM without any further preparation. For instance, the material constituting a vesicles wall may have stronger scattering elements such as silicon, so the whole structure can be easily imaged on the background of the amorphous ice. However, contrast between different structural details of organic materials is often poor because they are mainly composed of the same light elements. Staining is a technique used in microscopy to enhance contrast in the image. Staining is often used in biology; however, it is not limited to biological materials, and can also be used to study the morphology of other materials, for example, the lamellar structures of semicrystalline polymers or domain structures of block copolymers.

Staining artificially increases contrast of particular parts of a sample by depositing high-density metal compound (Figure 37.9). The staining agent usually selectively binds to certain parts of the sample thus enhancing contrast of particular elements of the structure. Osmium tetroxide (OsO_4) – an often used staining agent – reacts with double bonds similar to those present in polyisoprene, but does not react with conjugated double bonds. It is used to stain both biological samples and polymers. The mechanism of staining with RuO_4 is different from that of OsO_4:

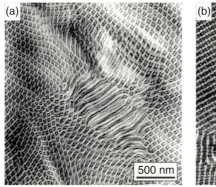 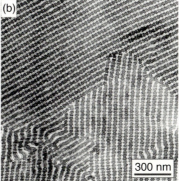

Figure 37.9 Block copolymers stained with (a) OsO_4 and (b) RuO_4 [80].

osmium tetroxide does not react directly with chemical species, but forms clusters instead. It stains most polymers, and the staining degree depends on the diffusion rate. For instance, in semicrystalline polymers, RuO_4 diffuses preferentially into amorphous regions, leaving the contrast of crystalline areas not modified. Both, OsO_4 and RuO_4 are very volatile and extremely toxic; therefore, special rules must be followed when working with them. The recipes of staining with osmium and rhenium tetroxide can be found for instance in [79]. Uranyl acetate, as a 1–2% aqueous solution, is one of the widely used staining agents. In the case of negative staining, the solution is excluded by the polymer or biological material due to surface tension interactions. Thus, a stained TEM grid is homogeneously covered by uranyl acetate, excluding the areas with the specimen. The staining penetrates open irregularities of the particle surface, therefore making visible the shape of the particle.

There is a large list of staining agents used in TEM [81]. Some of them are highly selective, enhancing the contrast of certain parts of a sample; others underline the general morphology of the complete structure. Depending on the task, the operator should decide if a staining agent is needed, and which part of the structure has to be clarified. Staining with uranyl acetate cannot be applied to polyelectrolytes (synthetic and natural) because these materials change their structure on change of salt concentration.

37.5.2.4 Surface Topology: Freeze Fracture and Etching

Freeze-fracture technique is used for biological as well as synthetic polymeric systems. The material is quickly frozen to avoid disruption of the structure due to components (mostly water) crystallization. Then the frozen material is fractioned simply by breaking with a sharp edge. This fracture is irregular and occurs along lines of weakness such as the plasma membrane, surfaces of organelles, or other morphological boundaries. The fracture surface may need additional cleaning at this stage to remove ice crystals formed at the surface before it is shadowed by metal (platinum or gold) deposition at an average angle around 45°. The metal layer is additionally covered by a thin carbon film (in some cases, the carbon film may be deposited onto the clean fraction surface before metal shadowing). The organic material is then removed by etching, and the metal replicas are transferred onto TEM copper grids for subsequent investigation in a TEM. For some polymers, the etching can be problematic and may destroy the replica, so instead of shadowing the material itself, a wax print of the surface can be used. The wax print inverts the surface profile, but reflects the same morphological features. The wax profile is then coated with amorphous carbon and shadowed by metal. The carbon film with the metal pattern is then floated from the wax surface and imaged in TEM.

Surface etching is based on selective erosion resulting in removal of one of the components of the sample. Etching can be done chemically using a specific solvent (typically acids or bases) or physically using ion etching or glow discharge. The etching process modifies the topology of the surface creating a map of a certain

Figure 37.10 Freeze fraction sample preparation: zigzag structures obtained from ferroelectric LC quenched after switching in ITO cell [30].

component of the sample. This surface topology can then be copied by wax and imaged using carbon and metal coating as described above (Figure 37.10).

37.5.3
Beam Sensitivity

Radiation damage is the main factor limiting the resolution of TEM data of organic materials leaving far behind instrumental resolution limits. The electron beam irradiation often leads to structural changes or even to a complete decomposition of the material. Usually, the damage mechanism is a complex process including many intermediate states of different stability. Many publications describe the radiation decay rate in different systems, measured by various means and at various irradiation parameters. At the same time, there was no attempt to study individual damage mechanism within a given system, because the intermediate short-living species are very difficult to detect.

The damage of material occurs due to inelastic scattering, which deposits energy into the material. The deposited energy manifests itself in two ways: one, heating of the specimen (molecular vibrations) or two, excitation of individual molecules which eventually leads to bond cleavage (ionization). Knock-on damage, as it is defined for metals or carbon, is usually not considered for organic materials because its contribution to the damage is negligible compared to the other mechanisms. The heat transfer effects can be significantly reduced by sample cooling, either to liquid nitrogen, of even down to liquid helium temperatures [82]. Cooling reduces molecular motion, keeping the relative position of radicals generated by electron beam and increasing the likelihood for the bond reformation. Radicals created after the primary ionization step usually react further and produce other longer-living radicals, or cross-linked aggregates. These processes are in a dynamic equilibrium and the final damage mechanism may be more dependent on the stability of the intermediates or the final construction, than on the dissociation energy in the primary bond cleavage.

There are a few ways to measure damage effects quantitatively. Electron-beam-induced molecular fragmentation leads to mass loss. The mass of the specimen can be measured before and after the irradiation. Obviously, a relative large amount of material has to be used in order to make these measurements. Depending on the composition of the sample, exponential to linear decay of mass with irradiation was observed. Changes in the molecular structures can be quantitatively measured by spectroscopic methods such as infrared spectroscopy. EELS, which can be measured directly in a TEM, can even deliver information about changes in the element ratio during irradiation. Radiation damage on crystalline material can be quantified by diffraction. Collective damage to single molecules causes deterioration of the periodicity in the crystal structure. The intensities of reflections generally fade as the degree of crystallinity decreases (Figure 37.11). The *critical dose*, defined as a dose at which the peak intensities of diffraction spots decrease to $1/e$ (37%) of the initial value, can be measured from the diffraction intensities fading curve. Dependent on the crystal structure and molecular shape, intensities do not decay uniformly. In comparison to high-resolution reflections, low-indexed reflections decay slower but those connected to structural features such as layers may even gain intensity [83].

No matter which mechanism is responsible for the material decay, the damage level primarily depends on the amount of energy deposited in a given volume and the rate of dissipation away from this volume. The amount of energy received

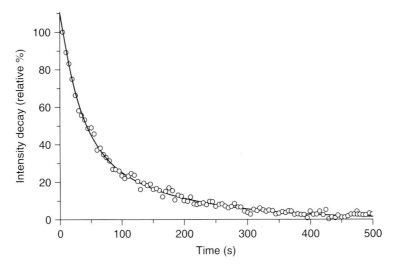

Figure 37.11 Decay of the intensity of electron diffraction reflections of β-Cu phthalocyanine at the radiation dose rate of 3 $e^-/\text{Å}^2$ s.

during irradiation is proportional to the number of incident electrons per unit area. The incident charge density is also called *electron dose q* and is usually measured in (C cm^{-2}) or ($\bar{e}$ Å^{-2}) with a conversion factor given in Eq. (37.1). Different concepts of electron dose are used to quantify the damaging process [84]: *critical or characteristic dose* defined as the point at which the electron diffraction intensities are reduced to $1/e$ (corresponding to 37% reduction of the initial intensity value) of its initial value, and *total end-point dose* at which the crystallinity is completely lost or no further change in intensity is observed.

Thus, copper phthalocyanine (Pc) crystals, which are completely destroyed at a total end-point electron dose of $q = 0.4$ C cm^{-2} = 250 $\bar{e}$ Å^{-2}, will degrade after 25 s of irradiation with the electron beam conditions set at 10 $\bar{e}$ Å^{-2} s^{-1} (experimental conditions are usually described *by electron dose rate* measured in ($\bar{e}$ Å^{-2} s^{-1})).

$$q\left[C \cdot cm^{-2}\right] = \frac{10^{16}}{1.6022 \times 10^{-19}} \left[\bar{e} \cdot A^{-2}\right] = 625 \left[\bar{e} \cdot A^{-2}\right] \tag{37.1}$$

The energy dissipation rate can be difficult to estimate but there are some basic rules. Typically, organic crystals appear more stable under the electron beam if the crystal is partially illuminated, so the irradiation-induced heat can be transported to the adjacent parts of the crystal. Crystal thickness has a similar effect – thicker crystals are more stable. A large crystal area contact with the supporting film is of crucial importance: crystals with minimal surface contact degrade more rapidly under irradiation. Techniques for increasing the crystal stability is the use of an additional carbon layer on top of the specimen [85, 86], this increases charge transfer and heat dissipation and therefore the stability of the material, or insertion of an objective aperture in the vicinity of the sample [87].

Sometimes, the critical electron dose is correlated with the melting or degradation temperature of the material. The melting is associated with the flexibility and mobility of a molecule, while irradiation damage is generally connected with radical creation and subsequent cross-linking. Significant deviations from the melting temperature/characteristic dose correlation are observed for conjugated systems. Typically, extended aromatic moieties have higher stability, likely connected with the stability of the ionization/dissociation products. Close derivatives of the same molecule can show strikingly different irradiation stability as for instance Pc molecule being much less stable than its halogenated derivatives [88].

Material degradation is also dependent on the energy of the incident electron beam determined by the microscope setup. The electron beam energy is one of the main parameters defining the interaction cross section; therefore it is natural to expect different damage behavior for different electron beam energies. Depending on the exact damage mechanism for a selected material, the damage can be reduced by changing the beam energy. Thus, there are two general ways to improve the stability of a specimen: reducing the *electron dose* in terms of reducing the number of electrons irradiating a certain area and adjusting the *microscope parameters* in order to influence scattering (accelerating voltage) or heat transfer (specimen cooling) processes.

37.5.4
Morphological Questions

Real materials rarely appear fully disordered due to very well-defined interaction between molecules constituting them. Often, intermolecular forces lead to self-organization of the molecules into certain structures. Although these aggregates are not strictly periodic on the large scale, the way they are built together is very well defined. The applications of the self-organized structures are spread from solar cells to drug delivery. Understanding of different levels of organization – chemical composition, molecular architecture (the way in which the monomer building blocks are bound together in the polymer chain), and supramolecular architecture (the organization of the chains relative to each other) – is the key point in further development.

The simplest polymers are isolated linear chains, which tend to form random coils. More complex polymers include branched polymers, block copolymers, polymer networks, cyclic polymers, and tree-like structures (such as dendrimers). Block copolymers can form spherical and cylindrical micelles, vesicles, spheres with face-centered cubic (fcc) and body-centered cubic (bcc) packing, hexagonally packed cylinders, minimal surfaces (gyroid, F surface, and P surface), simple lamellae, and modulated and perforated lamellae [89]. Applications of TEM to study some of these basic morphologies are summarized in the following sections.

37.5.4.1 Vesicles and Micelles

Surfactants tend to self-assemble into closed aggregates. These can be either bilayered and enclose some of the aqueous phase – vesicles, or filled with the hydrophobic part – micelles (Figure 37.12). Both vesicles and micelles are often spherical but can form other shapes. Micelles are important in many surfactant applications for their capacity to solve water-insoluble compounds. Vesicles are model systems for biological cells and can be used for entrapping active compounds inside the cavity. Besides, vesicles and micelles are used as microreactors for performing chemical reactions and preparing solid nanoparticles of varied shapes.

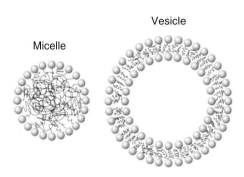

Figure 37.12 Schematic representation of a micelle and a vesicle, both spherical shape.

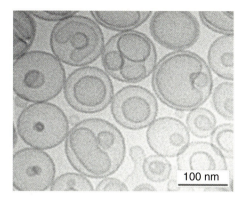

Figure 37.13 Cryo-TEM image of polyethylene glycol (PEG)-liposome vesicles.

One of the most widely used and successful applications of cryo-TEM is imaging of nanostructures created in lipid-surfactant systems in their native media. Vesicles are probably most extensively studied self-assembled systems due to high demands from the industrial side – the vesicles are often used in pharmacy, cosmetics, and personal care products. Vitrified TEM samples can deliver exact information about the size distribution and morphology of vesicles (Figure 37.13).

Phospholipid vesicles morphology and their interaction with membrane associated proteins is a key question in understanding the transport process in living organisms. Application of cryo-TEM techniques allow resolving of double wall membrane and formation of gap-junction-like structures between adjacent vesicles in the presence of membrane protein [90].

Filled vesicles can also be studied by cryo-TEM revealing not only the overall morphology and the structure of the vesicle wall but also the structure of the interior part. Recently liquid-crystal-filled polymeric vesicles were imaged by cryo-TEM [91]. The study showed that two forms are possible. The liquid crystalline phase can exist as a smectic colloid with straight nearly unbend smectic layers. These colloids are especially interesting as nanoscaled anisotropic liquid crystalline phase. The other type of molecular packing is shown in Figure 37.14 where the smectic layers follow the surface forming a concentric spherical structure.

Micelles are typically smaller than vesicles. The evolution of micelles in order to reach the equilibrium state involves the transformation of their structure and size and requires a considerable time. Studies by cryo-TEM can provide a possible scenario of the equilibration process [92]: the polystyrene-poly(2-vinylpyridine) (PSPVP) micellar size distribution, which is relatively narrow after dissolution, becomes broader with time. The micellar structure changes with time toward a spherical form with narrow interfacial zone between the core and corona regions.

Micelles can pack together in 2D and 3D assemblies. The packing preferences strongly depend on the composition of the micelles and the relative concentration of the substances, thus fcc and bcc symmetry as well as 2D hexagonal packing can be favored by some systems [93, 94] (Figure 37.15).

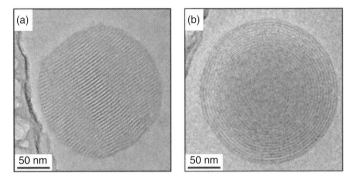

Figure 37.14 Cryo-TEM images of liquid crystal filled vesicles showing different microphase separation. The Si inside the backbone provides the contrast to the pure organic liquid crystalline side chains. a) layered structure; b) concentric structure.

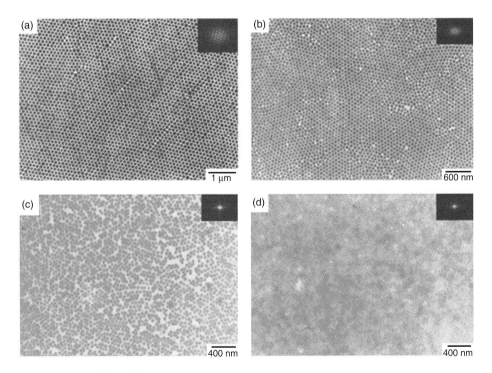

Figure 37.15 TEM images of copolymer micelle monolayer films formed in *p*-xylene at a concentration of 2.0 mg/ml by copolymer 1 (a), copolymer 2 (b), copolymer 3 (c), and copolymer 4 (d). Inset at right top of each figure is the (fast) Fourier transform (FFT) of the obtained particle arrays.

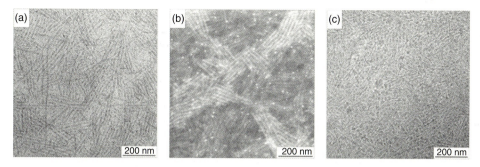

Figure 37.16 Images of oligo(*p*-benzamide)-*b*-poly(ethylene glycol) fibers prepared from chloroform: (a) drop casting TEM image, (b) drop casting STEM, and (c) cryo-TEM [97].

37.5.4.2 Fibers

Macromolecular engineering using different polymers as building blocks hands a chemist unlimited flexibility in creating new systems. Such block copolymers allow designing a huge number of self-organized structures with different architecture. Fibers among them are probably one of the simplest and nevertheless rich in their diversity.

Fibers can be build using a rigidly stack rods linked to a coil polymer, which would curl around the construction thus stabilizing it [95]. Oligo(*p*-benzamide)s (OPBAs) molecules, for example, can be used as rod motifs. The molecular packing of OPBA molecules is strongly driven by intermolecular H-bonds. Therefore, the molecular packing in shorter OPBA oligomers in the crystal structure is of high importance to get the insights into the molecular arrangements in the fibers [96] (Figure 37.16).

37.5.4.3 Electron Tomography

The TEM collects data in *transmission*, thus only one 2D projection of a sample can be recorded at a time. The easiest way 3D information can be collected in a TEM is the tilting of a sample through data collection (Figure 37.17). A set of different 2D projections, obtained in such a way is geometrically connected and includes the complete 3D information about the sample. The method of reconstruction of the 3D scattering potential from 2D TEM imaged is nowadays well elaborated and is known as *electron tomography* [98]. Electron tomography can be successfully applied to organic [99] as well as purely biological material [66]. TEM tomography is of high interest especially for materials with increasing complex structural features such as multicomponent polymer systems, for example, copolymers showing phase separation or composite material. Electron tomography can also be done using STEM images, which is especially beneficial for thick samples [100, 101].

Tomographic reconstruction allows the determination of three-dimensional symmetry of an object. Clathrin is a protein which plays a major role in the formation of coated vesicles mediating vesicle traffic in cells. The molecule forms a polyhedral cage which encapsulates a vesicle, therefore being a challenging

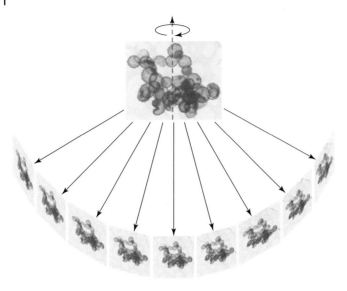

Figure 37.17 Scheme of electron tomography experimental setup: a sample is tilted around a goniometer axis producing a set of projected images. These are used to create a 3D reconstruction of the object.

candidate for cryo-electron tomography study [102]. Vitrified samples were studied in a TEM. Different symmetry of cages was observed with pentagonal, hexagonal, and occasionally heptagonal facets. Analysis of the 3D electron density maps allowed drawing conclusions about the interaction between the protein cage and the vesicle (Figure 37.18).

37.5.5
Structural Questions

Packing of molecules inside material is the most needed information for understanding properties of the material. Large organic molecules due to their complexity have problems packing in a regular way. Nevertheless, the packing may have a regular character. Features of molecular packing can be directly imaged by TEM.

37.5.5.1 Polymer Chain Packing
If polymeric chains provide strong contrast, their packing can directly be imaged in TEM. In case the contrast is too low, the material can be stained. Energy-filtered images provide a clearer signal, making the chains directly visible. Thus, poly(butadiene) and poly(isobutylene), stabilized by a saturated poly(butadiene) copolymer serving as a surfactant, could be imaged by energy-filtering TEM [103].

Another application of TEM to study packing of polymeric systems is imaging of cholesteric textures in diffraction contrast [104] (Figure 37.19). The diffraction contrast of the dark lines in bright field is due to an intermolecular distance of

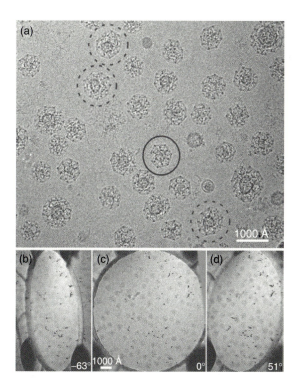

Figure 37.18 (a) Cryo-electron microscopy image of clathrin-coated vesicles embedded in vitreous ice. The particle marked by a continuous circle is probably the top view of a D6 barrel. From the projection image alone, it is difficult to determine whether the coat contains a vesicle. The particles marked by broken circles are larger and contain vesicles obviously offset from the centers of the coats. (b–d) Three representative raw images of a tilt series used to calculate cryo-tomograms of vitrified clathrin-coated vesicles. The tilt axis is approximately along the vertical direction.

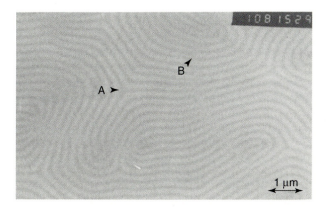

Figure 37.19 Fingerprint texture in cholesteric liquid crystals.

4.7 Å oriented along the cholesteric axis. Detailed investigation of layer junctions observed in TEM allowed studying the structural defects within the liquid crystalline state. Different models of layers splitting were proposed based on the TEM images.

37.5.5.2 High-Resolution TEM

High-resolution images imply not only higher magnification but also special type of contrast formation mechanism – phase contrast. With the increase of magnification, the relative amount of illumination per area unit increases. Reducing the illumination level is restricted by the quality of the recording media and can influence the contrast of the final image as well as the signal-to-noise ratio. Therefore, the possibility to obtain a high-resolution image of a crystal is primarily defined by the beam stability of the material. Once the material allows high-resolution imaging, the contrast, and thus the quality, of the image is also defined by the composition of the material: the presence of relatively heavy elements increases the contrast and therefore the quality of high-resolution images.

The short-range order and disorder within conducting polymers and especially polyaniline (PANI) remains a subject of intense discussion. The conformation of the polymeric chain, the location of the anions, and the chain-to-chain and chain-to-anion packing are all aspects of the short-range order, which may determine the electric (electron and proton) conductivity. Therefore, the packing of PANIs was studied by high-resolution TEM [105]. Two kinds of grains with varying degrees of crystallinity were observed: rolled sheets (20–100 nm in diameter) with an amorphous surface layer and crystalline grains (20–50 nm) exhibiting well-characterized 0.59 nm interreticular spacing.

Careful analysis of lattice planes can give precious information about the material structure. Although true high-resolution imaging of polymers at atomic resolution is a challenging task, several papers were published recently demonstrating the use of HRTEM [106, 107].

MPDI (poly(*m*-phenylene isophthalamide)) is a commercially available aromatic polyamide, known as *Nomex*. The MPDI single strains were shown to crystallize into regularly twisted fibers [108]. High-resolution imaging allowed determination of the twisting geometry and showed that the single twisted strains are combined into helical structures. The helical packing created dense constructions, nevertheless having a free channel in the middle, filled with solvent.

High-resolution TEM of molecular crystals has the same general problems as for polymers – mainly associated with the radiation stability of the material. Pcs are relatively stable conjugated molecules and they potentially can provide high contrast due to the presence of strong metal scatterers allowing large interplanar distances of Pcs being imaged in high-resolution mode. Figure 37.20 provides a high-resolution image of a β-CoPc crystal exhibiting these Pc columns.

Halogenated Pc molecules are usually relatively stable under the electron beam, therefore attractive candidates for HRTEM imaging. Kobayashi *et al.* [109] imaged Zn Pc molecules in thin films of different polymorphs in the beginning of the

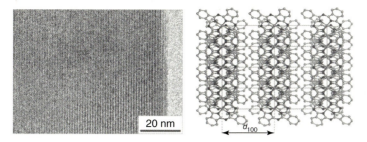

Figure 37.20 High-resolution TEM image of β-CoPc crystal showing lattice planes separated by 12.60 Å (d_{100}) corresponding to columns of Pc molecules.

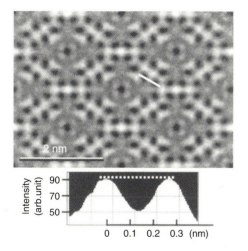

Figure 37.21 Contrast profile in Cl16CuPc where the chlorine atoms equally substituted at the 1-, 2-, 3-, and 4-positions [110].

1980s. Modern microscopes can provide much more clear images [110] allowing distinguishing between the halogen atoms (Figure 37.21).

37.5.5.3 Electron Diffraction

If the material is periodic, diffractive methods can be of high benefit deriving structural information. Electron diffraction patterns demand significantly lower electron dose thus allowing to access structural features not visible even by low-dose HRTEM. In contrast to small organic molecules, polymers seldom deliver prominent electron diffraction patterns.

Sometimes, the structure of monomeric model compounds is studied beforehand in order to understand the diffraction patterns obtained for the corresponding polymer. For example, the process of polymerization of diacetylene 1,6-di(N-carbazolyl)-2,4 hexadyine (DCHD) induced by the electron beam irradiation could be monitored in the diffraction pattern of [001] zone. During polymerization, the angle between (200) and (010) increases continuously indicating the transition from monomeric DCHD (94°) to the polymer (108°) [111]. For ferroelectric polymers, a combination of electron diffraction data with packing energy minimization allowed the complete description of the microphase separation and packing of side chains [30].

37.5.6
Crystallographic Questions

The ultimate aim of a structural study is to find electrostatic potential distribution, which is usually associated with atomic coordinates and their relative positions within the material. If the materials periodicity is suitable to provide separated and defined Bragg diffraction in the reciprocal space, this data can be used to determine the atomic structure. In classical crystallography, only positions and intensities of reflections are used for analysis. The reflections positions describe the periodicity of the material; the intensities describe the distribution of the electrostatic potential within the periodically repeated unit cell.

Having "reciprocal" information about the crystal (periodicity of the reciprocal lattice and the intensities of the spots) in principle, it is possible to extract the information about the direct-space electrostatic potential distribution within the direct-space crystalline lattice. To a certain degree, the reciprocal lattice is a Fourier transformation of the direct lattice. Therefore, one could imagine an *inverse* Fourier transformation of the diffraction data would deliver direct-space information. Unfortunately, the reflection intensities measured are real values, with the complex part missing, so this task cannot be solved strictly mathematically. Nevertheless, there are routines, which are covered later in this chapter, able to *calculate* direct-space data from reciprocal data.

37.5.6.1 Lattice Cell Parameters Determination

Determination of the unit cell parameters is the first step in the structural analysis. In order to find the three linearly independent basis vectors, three-dimensional diffraction data should be collected. Usually, a set of diffraction patterns is collected with known mutual orientation, although it is also possible to reconstruct the unit cell from a set of randomly oriented patterns [112].

Practically, there are two approaches to the zonal data collection: use of a sample holder with double-tilt possibility or rotation-tilt combination. The double-tilt holder allows a free rotation around the goniometer axis (α – primary tilt axis), and perpendicular to it. All crystallographic zones appearing within the available wedge can be collected. When a rotation-tilt holder is used, before the tilting, the crystal has to be oriented with a low-index crystallographic axis along the goniometer axis. One of these tilting schemes is normally selected on the basis of the hardware available in the laboratory, or on the taste of the operator. A hybrid double-tilt rotational holder is also available (GATAN) providing an additional degree of freedom.

In a typical electron diffraction experiment, a starting low-index electron diffraction pattern is selected. Then the crystal is being tilted around one of its crystallographic axes and other low-index crystallographic zones are collected for further processing. These are normally diffraction patterns with a relatively high density of reflections and strong symmetry imprint. Together with the diffraction patterns, the reciprocal angular relationship is stored.

From an electron diffraction tilt series, all unit cell parameters can be deduced. The procedure includes basic 3D geometry. Knowing the relative tilt ranges between the zones, a Vainshtein plot [113] can be constructed, which is a projection of the reciprocal lattice. Software packages are also available to image the three-dimensional network (TRICE [114], numerous homemade software packages). The metric of the unit cell usually is the basis for the crystal system determination. The true symmetry and space groups are additionally confirmed by evaluation of the intensity distribution in the recorded diffraction patterns.

Lattice parameter determination through a zonal electron diffraction tilt series (using rotation-tilt sample holder) is demonstrated for a 9,9′-bianthracene-10-carbonitrile (CNBA) crystal [79]. The diffraction tilt series were collected through a tilt around the monoclinic axis b^* (Figure 37.22). The starting zone has an index of [100] with the monoclinic axis β^* vertical and the axis c^* horizontal. From this pattern, the reciprocal lattice parameters b^*, c^*, and α^* can be measured directly. For the determination of other lattice parameters, the crystal was tilted around the β^* axis until centrosymmetrical patterns were obtained. In this way, zones [501], [401], [301], [201], and [101] were recorded. A Vainshtein plot is essentially a combination of (h0l) lines extracted from each zone plotted according to the angular distances between the zones. Thus, the first zone [100] is plotted as a row of spots along the c^* axis. When all available reflections are sketched, a two-dimensional net of reflections can be drawn through them. This net is the [010] zone, not directly observed but reconstructed. The missing lattice parameters a^* and β^* can be measured directly from the Vainshtein plot. In all diffraction patterns, (h0l) appear orthogonal to the tilt axis β^*, thus the plane of the Vainshtein plot is orthogonal to β^*. Since α^* is within the Vainshtein plot, the angle γ^* ought to be $90°$, as it is the angle between the tilt axis β^* and α^*. Therefore, the complete lattice cell metric can be calculated.

The symmetry of the structure can also be deduced from the analysis of the diffraction patterns. All diffraction patterns collected in the CNBA tilt series have mm symmetry. This means that the structure in the projection has either 2, m, or $2/m$ symmetry. This allows an assignment of the tilt axis as unique monoclinic axis β^*. Along this axis, each odd reflection is missing satisfying the reflection conditions $0k0 : k = 2n$ due to a 2_1 screw axis. Besides, from the Vainshtein plot (Figure 37.22), it is evident that full rows of reflections are extincted (marked as hollow spots in the Vainshtein plot), the observed reflection following the rule h0l : l = $2n$, due to a c-type glide plane. Therefore, the space group of the structure can be unambiguously determined as $P2_1/c$.

Knowing the cell metric and the space group, all available diffraction patterns can be indexed and intensities of reflections can be integrated. The approach to the reflection integration is more or less standardized, and CRISP/ELD (Calidris [115]) is the most common software package used for extracting intensities of diffraction

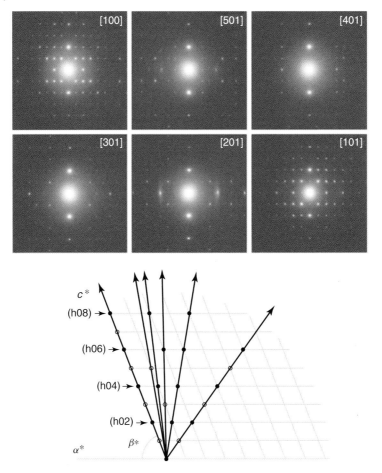

Figure 37.22 Electron diffraction tilt series of CNBA collected around the monoclinic axis b*: electron diffraction patterns, Vainshtein plot showing (010) slice of the reciprocal space.

spots from zonal patterns in order to produce a standard crystallographic data set (h k l Int). There is no strict definition for the determination of intensity error sigma, most general approaches being using a square root of the intensity value or 10% of the determined intensities [116].

Merging of the data originating from different zones is a serious problem: often zones are collected from different crystals, which may have had different thickness, and therefore a scaling factor would be necessary to combine the data from different patterns together. The diffraction patterns may have different degree of structural damage due to electron beam irradiation. Usually, there are some common reflections present in different zones, thus based on their intensities, a scaling factor can be estimated. It is a known effect that the intensities along a row of reflections may appear different for the same row present in different zone patterns. This effect is associated with multiple scattering paths that are realized differently when another set of reflections is excited. Therefore, even this intuitive approach is basically wrong, but nevertheless most commonly used. Yet, a comprehensive data set allowing a structure solution can be extracted from a manual tilt series.

37.5.6.2 Structural Analysis Based on Electron Diffraction Data

Electron diffraction not only can deliver information to be combined with other data but can also provide a complete *ab initio* structure solution. Generally, electron diffraction data (meaning the collected indexed reflection intensities) suffers from two problems: the quantity problem, that is, low amount of data leading to low completeness and the quality problem caused by perturbations in the intensity values.

The quantity problem is mostly connected with the strategy of data collection: the total tilt range of the goniometer scanned, and how fine the data is sampled within the accessible range. Owing to geometrical restrictions, a sample cannot be tilted all around 180°, as then the TEM grid itself would block the beam. Further restrictions can be introduced by sample holder and TEM configuration. Modern microscopes optimized for high-resolution imaging may have available tilt range as low as $\pm 12°$. However, nowadays, a tilt range of at least $\pm 30°$ is achieved, and specially designed high-tilt holders allow recording data within up to $\pm 80°$ wedge. The other point, connected with the amount of data collected, is how the data is distributed within the available range. Manually, collected electron diffraction tilt series usually consist of a few relatively low-index zones, which cover a high number of symmetrically dependent reflections, higher index zones being difficult to recognize, align, and index. Therefore, although a high total tilt range may have been scanned, only a few data comes out. Nevertheless, these problems can be principally optimized from the experimental part.

The quality of the data collected can be reduced by different reasons: perturbation of the intensities is usually associated with dynamical scattering effects. Although dynamical effects modulating the electron intensities are a more serious issue for stronger scatters such as inorganic material, it may also affect intensities for organics. Besides, the patterns may have a different degree of structural decomposition or even features of another packing produced at the early stages of irradiation. Additionally, a slight misorientation of a zonal pattern may change intensities severely.

Despite all mentioned problems, electron diffraction is a very valuable and sometimes the only source of information able to give insights into the structure of nanomaterial.

37.5.6.3 Combination of Electron Diffraction with Different Methods to Facilitate Structure Analysis

Electron diffraction information is often combined with other methods to support structure solution procedure. As banal it sounds, often combined with simple *chemical considerations*, electron diffraction information can give a reasonable model for a structure solution. A structure model can additionally be supported by energy minimization, geometrical constrains, or combination with other diffraction data such as X-ray powder diffraction.

For less complicated molecular packing, for instance for π-stacked layered materials, the structure can be solved when the stacking direction is known. Electron diffraction can easily find and index the strongest reflection corresponding to the stacking direction, thus supporting the structure solution. These questions are often met for low-symmetry layered systems. Cross-like molecular packing can also be resolved from visual inspection of electron diffraction intensities [117]. Planar molecules of ζ phase of P.R. 53 : 2 form dimers coordinating calcium ions. The angles these molecules form to each other could be measured from electron diffraction patterns, and so the structural model was build (Figure 37.23). Although a structural model of ζ-P.R. 53 : 2 could be constructed from electron diffraction data, the refinement procedure could not be performed on electron diffraction intensities, but was made using X-ray powder diffraction profile.

X-ray powder diffraction is often used together with electron diffraction data. The two methods can be applied to the same material providing information of different type: X-ray powder pattern describes the *bulk* material using one-dimensional averaged data and electron diffraction provides single-crystal data from a few individual crystals. Reflections with the same interplanar distances can be identified in X-ray powder and electron diffraction patterns.

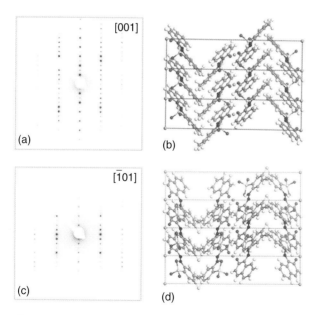

Figure 37.23 (a–d) Selected electron diffraction patterns and corresponding views of the structure of ζ phase of P.R. 53 : 2.

Even when no tilt series data and thus clear unit cell parameters are available, angles between reflections measured from electron diffraction patterns can facilitate unit cell parameter determination from X-ray powder data. Unit cell parameters determined from electron diffraction tilt series directly allow indexing of X-ray powder profiles, thus supporting structure determination. Besides, such profile indexing may identify impurities and additional phases in mixed systems. These questions become particularly important for polymorphic materials [118] where due to peak overlap, indexing of X-ray powder profile fails.

Combination of powder X-ray and electron diffraction is probably the most widely spread data complementarity. Generally, three cases can be classified: separate electron diffraction patterns support indexing of X-ray powder profile, unit cell parameters obtained from electron diffraction tilt series allow indexing the powder profile, and electron diffraction structural model is refined using X-ray powder data. Numerous examples can be found in literature demonstrating these cases.

Also *two-dimensional X-ray data* can be efficiently combined with electron diffraction [119]. Usually, within this combination, a structural model is obtained from electron diffraction data, and the refinement of the structure is done using X-ray intensities. Extraction of initial structure model is one of the most difficult and most important stages of structure analysis of semicrystalline polymer. X-ray diffraction pattern of such a sample consists of limited number of broad and diffuse reflections. On the other hand, the electron diffraction may give large number of sharp reflections if the crystalline domains of the appropriate size can be obtained. Thus, electron diffraction data set of polyoxymethylene was collected through tilting experiments [120]. Low-completeness data set was collected, nevertheless allowing construction of a reasonable structural model. This model was then refined using X-ray data.

Energy minimization of structural model is one of the most widely used methods to support electron diffraction structure analysis of organic materials. It can be applied both to small organic molecules [121] and polymers [122–125]. There is no principal restriction for the energy calculation method used. The only demand is a reliable force field describing both the contact between the atoms within the molecule and the molecular packing.

Practically, after the initial structure model is created, the energy minimization steps are always followed by quantitative comparison of the simulated and experimental electron diffraction data. Once a good agreement, in terms of diffraction data, is reached and a reasonable packing energy could be achieved, the structure is considered solved. This approach was often employed for small organic molecules, being almost a standard electron diffraction structure solution route until recent times [126–128]. Recently, commercial software was developed automating the procedure [129].

The iterative "energy minimization – electron diffraction match" method can be efficiently applied to polymers. This combination of electron diffraction data and energy minimization techniques was successfully used to build a structural model of poly(3-hexylthiophene) (P3HT) [130, 131]. Prominent electron diffraction patterns were collected corresponding to the specific directions of the structure – view onto the molecules, along the side chain direction and perpendicular to the side chain direction. From these data, a rough model was constructed, which then was refined iteratively using packing energy minimization and qualitative comparison with electron diffraction data.

Use of *geometrical considerations* to support electron diffraction data is in a way similar to energy minimization approach, with considering only the energetically "best" conformations of the molecule. Since, there are only a limited number of these conformations, they all can be tried out until simulated electron diffraction data matches the experimental patterns. The best matching model corresponds then to the structure solution. Most perspective systems for this structure solution path should have a limited number of clearly distinguishable conformations. Historically, this was the first electron diffraction involved method used for structure solution from organic material [132]. It can also be applied to polymers. A structural model of high-temperature polymorph of chitosan was constructed based on the likely molecular conformation fitting unit cell found through electron diffraction tilt experiments. Three torsion angles described the flexibility of the structure, so the values for these angles were refined against available electron diffraction data. The structure crystallizes in $P2_12_12_1$ space group with cell parameters $a = 8.07$ Å, $b = 8.44$ Å, and c (chain axis) $= 10.34$ Å. Forty seven reflection intensities were collected. For a noncentrosymmetric structure, this data would be way too few to achieve a structure solution using direct methods, but together with geometrical constrains produced a reliable structural model [133].

Rather exotic combinations may also occur – so, electron diffraction data can be supported by *spectroscopic measurements*. Thus, solid-state structure of symmetric trans-alkenes was studied using a combination of electron diffraction data and infrared spectroscopy [134].

Once a high-coverage data set is available with intensities approaching kinematical scattering conditions, a structure solution can be obtained from electron diffraction data *solely*. The structure solution is then performed using one of the programs used for single-crystal X-ray diffraction data with scattering factors for electrons such as SIR, SHELX, MICE.

One of the oldest and striking examples of direct method structure determination from only electron diffraction data is the structure solution of diketopiperazine [108]. The data was collected using an electron diffraction camera and textured polycrystalline sample. The structure crystallizes in $P2_1/a$ space group, with the lattice parameters of $a = 5.20$ Å, $b = 11.45$ Å, $c = 3.97$ Å, $\beta = 81.9°$. Recently, the structure determination was revisited: modern automated structure solution routines allow determination of all atoms in the structure including hydrogens [135].

A structure can be found relatively easily if the unit cell parameter in one of the directions is very short – that is, the molecular packing in this direction is known due to geometrical restrains. Although these systems are not true 2D crystals – they have periodicity in three directions, from the point of structure solution they can be considered as 2D case. One-zone data recorded into this short direction would allow resolving the complete structure. This method is relatively seldom used, since it demands special procedures for sample preparation. Typically, long crystallographic axes represent slowly growing directions in a crystal. Therefore, the crystal is usually *thin* in the direction of the long axis. Crystals tend to lay on the support rather than stand, therefore most of the flat crystals would tend to be oriented with their long axis vertically. In order to produce crystals oriented with their short axis vertically, sublimation or epitaxial growth methods specially adapted to each system should be used. Once the sample preparation succeeded and the zonal data was recorded at sufficient resolution, the structure solution routine is straight. Crystal structures of two polymorphs of red pigment perylenetetracarboxylic dianhydride (PTCDA) were solved using this approach [136].

These examples basically state that if enough data is collected and special care is taken about the reflection intensities, it must be possible to solve a structure using direct methods just like it is done in single-crystal X-ray structure analysis. Therefore, basically two problems of electron diffraction data should be solved or at least optimized: completeness of the data and dynamical effects. The completeness can be increased by collecting all available diffraction data rather than only low-index zones, and dynamical effects can be reduced by avoiding all low-index zones. A combination of these two approaches is elaborated into a new method of electron diffraction data collection – automated diffraction tomography (ADT) [137, 138].

37.5.6.4 Off-Zone Data Collection: Automated Diffraction Tomography (ADT)

Instead of tilting a crystal around a low-index crystallographic axis with relatively large gaps between the zones, applying ADT, the crystal is tilted in small fixed tilt steps around an arbitrary axis. This allows fine sampling of a large reciprocal volume containing the diffraction information. The patterns are arbitrary slices through the reciprocal space and usually do not show any high symmetry, only by chance are low-index zones recorded. Additional use of electron precession unit [139] allows more precise integration of the reflection intensities.

ADT data can be collected with any TEM, with no restrictions on the sample holder – only one degree of freedom – primary tilt possibility is necessary. *Ab initio* structure analysis of various inorganic materials was recently demonstrated using manually collected ADT data [140–142].

Automated acquisition module for ADT data collection, helpful but not compulsory, was developed in collaboration with FEI for Tecnai class TEMs. In the module, during the ADT data acquisition, the crystal is imaged in STEM mode. Simple calculations show that for the beam settings used in ADT acquisition module, due to the scanning of the beam instead of large area illumination, the electron dose received by a sample during STEM image acquisition is at least 2 orders of magnitude lower than during TEM imaging or electron diffraction recording. A suitable crystal is selected in STEM mode at the starting tilt position. Then diffraction patterns are recorded at different tilt positions sequentially. Optionally, crystal tracking during the tilt can be used. The tracking procedure cross-correlate STEM images taken at neighboring tilt steps and takes account for the shift of the crystal.

Figure 37.24 ADT diffraction patterns of a CNBA single crystal. The tilt axis position in the diffraction patterns is marked by a dotted line.

Data processing for unit cell parameter determination and reconstruction of the three-dimensional volume, including background correction, correction of diffraction pattern shift, refinement of the tilt axis position was performed by bespoke Matlab routines produced by authors and by recently developed ADT3D software processing sequential electron diffraction data.

Here, the application of ADT data for *ab initio* structure solution is demonstrated for CNBA. This material exhibits nonlinear optical (NLO) activity in solution but, due to the centrosymmetric space group it adapts, the NLO activity vanishes in the crystalline state. Initially, the structure was solved from manually collected electron diffraction data using maximum entropy methods [143]. Subsequently, the structure was confirmed by single-crystal X-ray analysis [143].

Selected ADT diffraction patterns for CNBA are shown in Figure 37.24. Projections of the collected reciprocal volume along the main directions are shown in Figure 37.25. There is a principal difference between single zones and the visualization of three-dimensional ADT. Single-electron diffraction zones comprise "cuts" through the reciprocal space showing only the reflections on this cut (Figure 37.25b), whereas 3D ADT data is presented here as projections of the reciprocal space along rows orthogonal to the view direction, producing an overlapping of reflections obscuring small intensities or extinctions by larger reflections (Figure 37.25a). In the projection (100), a row of extinctions along the vertical direction can be observed due to a *c*-type glide plane placed perpendicular to the *b* axis.

The unit cell parameters were determined automatically by cluster analysis of the difference vector space: $a = 14.75$ Å, $b = 9.48$ Å, $c = 15.36$ Å, $\alpha = 89.4°$, $\beta = 112.6°$, $\gamma = 90.5°$, corresponding to a lattice of $a = 14.70$ Å, $b = 9.47$ Å, $c = 15.42$ Å, $\beta = 112°$.

ADT data was collected in a tilt range of 120° (±60°) delivering a coverage of two-third of the complete reciprocal volume. Within the resolution range of 0.8 Å, 9415 reflections were collected. After merging symmetry equivalent pairs, 3519 independent reflections were produced covering 87% of the reflections possible within the achieved resolution sphere. The R_{sym} of the data set was 21.77%. Using the collected reflections in SIR2008 package [144], the space group could be

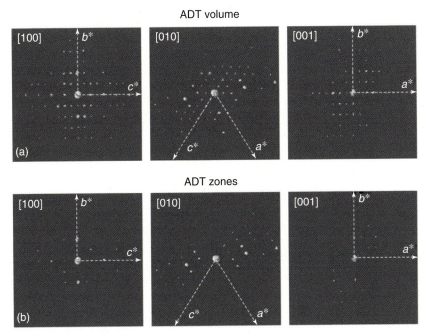

Figure 37.25 (a) Views of reciprocal volume reconstructed from ADT data for CNBA single crystal along the main crystallographic directions of the reciprocal space. (b) Main crystallographic zones clipped from the ADT volume. Zonal extinctions due to c-glide plane are seen both in zones and projections, while serial extinctions of a 2_1 screw axes can only be visualized in the zonal cuts.

determined automatically to be $P2_1/c$ leading to a density of $d = 1.21$ g cm^{-3} with four molecules in the unit cell.

The structure was solved *ab initio* using direct methods implemented in SIR2008. The isotropic thermal factor determined from the Wilson plot U was 0.013 Å^2. The best solution delivered the R factor of 24.43%. The structure solution as it came out of SIR2008 is shown in Figure 37.26. No ghost atoms were detected in the structure and the first 30 positions of the solution could be assigned to the 30 non-hydrogen atoms of the molecule. In comparison to the previously conducted single-crystal X-ray analysis of CNBA, the average deviation of the atom positions was 0.186 Å, with the maximal deviation of 0.524 Å. These deviations are within the range found for other structure solutions performed with ADT [145–147].

CNBA molecule has a rather rigid conformation. The only degree of conformational freedom is the torsion angle between the two anthracene moieties. The anthracene fragments were found basically in a flat geometry, but additional refinement of the geometry was clearly necessary. Structure refinement was performed in SHELXS [148, 149]. It turned out to be stable without any restraints but did not deliver a significant improvement in molecular geometry. In order to improve the geometry, soft restraints on C–C bond lengths of 1.4 Å with standard deviation 0.05

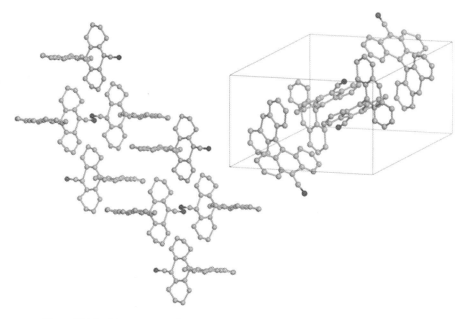

Figure 37.26 SIR structure solution of CNBA from ADT data: unit cell and packing of the molecules in the crystal.

Å and later strong constraints keeping the anthracene moieties flat were applied. The refinement delivered as expected a flat geometry with a torsion angle between the anthracene moieties of 93.8°. The R value increased to 39%. As a last step and in order to verify the results, geometry optimization was performed using forcefield calculations with the recently available COMPASS force field as implemented in Material Studio package (Accelrys). The torsion angle was found to be 94.0°.

In principle, the correct refinement procedure should be performed based on dynamical calculations. Unfortunately, the only existing program package for dynamical refinement – MSLS – is restricted to zonal diffraction data as it is collected by traditional approach. Understanding and description of the reduced but still present dynamical effects in nonzonal diffraction patterns are crucial for refinement and structure validation based on ADT data. The adaption of dynamical refinement procedures to ADT data is under development.

37.6
Summary

Microscopic methods generally provide a possibility of direct visualization of structural features which play a major role in material characterization. For organic materials – both polymeric and monomeric – it delivers structural information on a broad dimension scale revealing features associated with morphology, supramolecular organization, and the molecular packing in a crystalline structure. Light microscopy has an advantage of material stability during the investigation, but is restricted in resolution. Newly developed light microscopic techniques allow overcoming the resolution limit, but usually have strict demands on the sample and cannot be applied to a large diversity of material systems. SEM is an ultimate method to investigate surface textures. AFM and STM allow imaging of surfaces at higher magnification – down to single molecules.

TEM is an ultimate instrument allowing structural investigation of architectures starting from few microns down to the atomic scale. Acquisition of TEM data requires high electron dose to the samples, which is a critical issue for organic materials. Nevertheless, special methods have been developed to reduce the illumination level during the investigation and protect the sample. TEM imaging has been extensively used for morphological and structural studies of organic materials. Besides, in the past years, the application of electron diffraction gains more and more interest as a possibility to perform a fully quantitative structure analysis on nanocrystalline materials.

References

1. Rochow, T.G. and P.A., Tucker (eds) (1994) *Introduction to Microscopy by Means of Light, Electrons, X-rays, or Acoustics*, 2nd edn, Springer, New York.
2. Heath, J.P. (ed.) (2005) *Dictionary of Microscopy*, John Wiley & Sons, Ltd, West Sussex.
3. Hawkes, P.W. and J.C.H., Spence (eds) (2006) *Science of Microscopy*, vol. I and II, Springer, New York.
4. Magonov, S. and Yerina, N.A. (2005) in *Visualization of Nanostructures with Atomic Force Microscopy in Handbook of Microscopy for Nanotechnology* (eds N. Yao and Z.L. Wang), Kluwer Academic Press, New York, pp. 113–156.
5. Egerton, R.F. (ed.) (2005) *Physical Principles of Microscopy*, Springer, New York.
6. Lipson, S.G., Lipson, A., and Lipson, H. (eds) (2010) *Optical Physics*, 4th edn, Cambridge University Press, Cambridge.
7. Spence, J.E.H. (ed.) (2003) *High-Resolution Electron Microscopy*, 3rd edn, Oxford University Press, Oxford.
8. Reimer, L. (ed.) (1997) *Transmission Electron Microscopy: Physics of Image Formation and Microanalysis*, 4th edn, Springer-Verlag, New York.
9. Lichte, H. (1986) Electron holography approaching atomic resolution. *Ultramicroscopy*, **20** (3), 293–304.
10. Tonomura, A. (1987) Applications of electron holography. *Rev. Mod. Phys.*, **59** (3), 639–669.
11. Gabor, D. (1948) A new microscopic principle. *Nature*, **161** (4098), 777–778.
12. Schmahl, G., Rudolph, D., Niemann, B., Guttmann, P., Thieme, J., and Schneider, G. (1996) X-ray microscopy. *Naturwissenschaften*, **83** (2), 61–70.
13. Ahn, C.C. (ed.) (2004) *Transmission Electron Energy Loss Spectrometry in Materials Science and the EELS Atlas*, 2nd edn, John Wiley & Sons, Inc., Hoboken, NJ.
14. Bell, D.C. and Garratt-Reed, A.J. (eds) (2003) *Energy Dispersive X-Ray Analysis in the Electron Microscope*, Garland Science.
15. Hirsch, P.B. (ed.) (1999) *Topics in Electron Diffraction and Microscopy of Materials*, CRC Press, Boca Raton.
16. Bernstein, J. (ed.) (2002) *Polymorphism in Molecular Crystals*, Clarendon Press, Oxford.
17. Snider, D.A., Addicks, W., and Owens, W. (2004) Polymorphism in generic drug product development. *Adv. Drug Delivery Rev.*, **56**, 391–395.

18. Schmidt, M.U. (2003) Crystal engineering and polymorphism of organic pigments. *Adv. Colour Sci. Technol.*, **6**, 59–61.
19. Legrand, L., Ries-Kautt, M., and Robert, M.-C. (2002) Two polymorphs of lysozyme nitrate: temperature dependence of their solubility. *Acta Crystallogr.*, **D 58**, 1564–1567.
20. Simon, G.P. (ed.) (2003) *Polymer Characterization Techniques and Their Application to Blends*, Oxford University Press, Oxford.
21. Kopp, S., Wittmann, J.C., and Lotz, B. (1994) Epitaxial crystallization and crystalline polymorphism of poly(1-butene): forms III and II. *Polymer*, **53**, 908–915.
22. Di Lorenzo, M.L., Righetti, M.C., and Wunderlich, B. (2009) Influence of crystal polymorphism on the three-phase structure and on the thermal properties of isotactic Poly(1-butene). *Macromolecules*, **42**, 9312–9320.
23. Cartier, L., Okihara, T., Ikada, Y., Tsuji, H., Puiggali, J., and Lotz, B. (2000) Epitaxial crystallization and crystalline polymorphism of polylactides. *Polymer*, **41**, 8909–8919.
24. Kessel, A. and Ben-Tal, N. (eds) (2011) *Introduction to Proteins: Structure, Function, and Motion*, CRC Press, Boca Raton, FL.
25. Whitford, D. (ed.) (2009) *Proteins: Structure and Function*, John Wiley & Sons, Ltd, West Sussex.
26. Bozzola, J.J. and Russell, L.D. (eds) (1999) *Electron Microscopy: Principles and Techniques for Biologists*, 2nd edn, Jones & Bartlett, Sudbury, MA.
27. Hayat, M.A. (ed.) (2000) *Principles and Techniques of Electron Microscopy: Biological Applications*, 4th edn, Cambridge University Press, Cambridge.
28. Jensen, G.J. (ed.) (2010) *Methods in Enzymology*, vols. 481–483, Elsevier Inc., Cambridge.
29. Ohm, C., Serra, C., and Zentel, R. (2009) A continuous flow synthesis of micrometer-sized actuators from liquid crystalline elastomers. *Adv. Mater.*, **21**, 4859–4862.
30. Kuhnert-Brandstätter, M. (1982) Thermomicroscopy of organic compounds, in *Wilson and Wilson's Comprehensive Analytical Chemistry* (ed. G. Svehla), Vol. XVI, Elsevier, Amsterdam, pp. 329–498.
31. Griesser, U. and Stowell, J.G. (2003) Solid-state analysis and polymorphism, in *Pharmaceutical Analysis* (eds D.C. Lee and M.L. Webb), Continuum International Publishing Group, Ltd, Oxford, pp. 240–294.
32. Kuhnert-Brandstätter, M. (ed.) (1971) *Thermomicroscopy in the Analysis of Pharmaceuticals*, Pergamon Press, Oxford.
33. Loos-Wildenauer, M. (1997) Electron microscopic structural investigations on ferroelectric liquid crystalline polymers and their monomeric model compounds Electron microscopic structural investigations on ferroelectric liquid crystalline polymers and their monomeric model compounds (in German). PhD Thesis, Johannes Gutenberg-University, Mainz.
34. Melville, D.O.S., Blaikie, R.J., and Wolf, C.R. (2004) Submicron imaging with a planar silver lens. *Appl. Phys. Lett.*, **84**, 4403–4405.
35. Blaikie1, R.J. and Melville, D.O.S. (2005) Imaging through planar silver lenses in the optical near field. *J. Opt. A: Pure Appl. Opt.*, **7**, S176–S183.
36. Fang, N., Lee, H., Sun, C., and Zhang, X. (2005) Sub-diffraction-limited optical imaging with a silver superlens. *Science*, **308**, 534–537.
37. Boltasseva, A. and Shalaev, V.M. (2008) Fabrication of optical negative-index metamaterials: recent advances and outlook. *Metamaterials*, **2**, 1–17.
38. Simpson, J., Nien, C.J., Flynn, K., Jester, B., Cherqui, S., and Jester, J. (2011) Quantitative in vivo and ex vivo confocal microscopy analysis of corneal cystine crystals in the $Ctns^{-/-}$ knockout mouse. *Mol. Vis.*, **17**, 2212–2220.
39. Lavrentovich, O.D. (2003) Fluorescence confocal polarizing microscopy: three-dimensional imaging of the director. *Pramana*, **61** (2), 373–384.

40. Dierking, I. (2010) Recent developments in polymer stabilised liquid crystals. *Polym. Chem.*, **1**, 1153–1159.
41. Prasad, V., Semwogerere, D., and Weeks, E.R. (2007) Confocal microscopy of colloids. *J. Phys.: Condens. Matter.*, **19**, 1–25.
42. Ibanez, A., Maximov, S., Guiu, A., Chaillout, C., and Baldeck, P.L. (1998) Controlled nanocrystallization of organic molecules in sol-gel glasses. *Adv. Mater.*, **10**, 1540–1543.
43. Denk, W., Strickler, J.H., and Webb, W.W. (1990) Two-photon laser scanning fluorescence microscopy. *Science*, **248**, 73–76.
44. Helmchen, F. and Denk, W. (2005) Deep tissue two-photon microscopy. *Nat. Methods*, **2**, 932–940.
45. Hell, S.W. (2009) Microscopy and its focal switch. *Nat. Methods*, **6**, 24–32.
46. Mikliaev, Y. and Asselborn, S. (2009) Method for obtaining a high resolution image. US Patent N 20090116024.
47. Inzelt, G. (2008) in *Conducting Polymers A New Era in Electrochemistry*, Series: Monographs in Electrochemistry (ed. F. Scholz), Springer, Heidelberg.
48. Butler, J.H., Joy, D.C., Bradley, G.F., and Krause, S.J. (1995) Low-voltage scanning electron microscopy of polymers. *Polymer*, **36**, 1781–1790.
49. Martin, D.C., Wu, J., Shaw, C.M., King, Z., Spanninga, S.A., Richardson-Burns, S.M. *et al.* (2010) The morphology of poly(3,4-ethylenedioxythiophene). *Polym. Rev.*, **50** (3), 340–384.
50. He, C.C., Gu, Z.Y., and Tian, J.Y. (2003) Studies on acrylic acid–grafted polyester fabrics by electron beam preirradiation method. II. Novel intelligent immersion-resistant and moisture-permeable fabrics. *J. Appl. Polym. Sci.*, **89**, 3939–3943.
51. Wei, Q.F., Wang, X.Q., Mather, R.R., and Fotheringham, A.F. (2004) New approaches to characterisation of textile materials using environmental scanning electron microscope. *Fibres Text. East. Eur.*, **12**, 79–83.
52. Jenkins, L.M. and Donald, A.M. (1999) Contact Angle Measurements on Fibers in the Environmental Scanning Electron Microscope. *Langmuir*, **15**, 7829–7835.
53. Gross, L., Mohn, F., Moll, N., Liljeroth, P., and Meyer, G. (2009) The chemical structure of a molecule resolved by atomic force microscopy. *Science*, **325**, 1110–1114.
54. Ward, M.D. (2001) Bulk crystals to surfaces: combining X-ray diffraction and atomic force microscopy to probe the structure and formation of crystal interfaces. *Chem. Rev.*, **101**, 1697–1725.
55. Kaupp, G., Schmeyers, J., and Hangen, U.D. (2002) Anisotropic molecular movements in organic crystals by mechanical stress. *J. Phys. Org. Chem.*, **15**, 307–313.
56. Magonov, S.N. and Reneker, D.H. (1997) Characterization of polymer surfaces with atomic force microscopy. *Annu. Rev. Mater. Sci.*, **27**, 175–222.
57. Guillard1, H., Sixou, P., and Tottreau, O. (2002) Atomic force microscopy study of the polymer growth in a polymer stabilized liquid crystal. *Polym. Adv. Technol.*, **13**, 491–497.
58. Knoll, M. and Ruska, E. (1932) Das Elektronenmikroskop. *Z. Phys.*, **78**, 318–339.
59. Mooney, P.E., Pan, M., van Hoften, G., and Haas, F. (2004) Quantitative evaluation of fiber-optically coupled CCD cameras for use in cryo-microscopy. *Microsc. Microanal.*, **10** (Suppl. 02), 168–169.
60. Drummy, L.F., Yang, J., and Martin, D.C. (2004) Low voltage electron microscopy of polymer and organic molecular thin films. *Ultramicroscopy*, **99** (4), 247–256.
61. Haider, M., Braunshausen, G., and Schwan, E. (1995) Correction of the spherical-aberration of a 200-KV TEM by means of a hexapole-corrector. *Optik*, **99**, 167–179.
62. Zach, J. and Haider, M. (1995) Aberration correction in a low voltage SEM by a multipole corrector. *Nucl. Instrum. Method*, **363A**, 316–325.
63. Krivanek, O., Dellby, N., and Lupini, A.R. (1999) Towards sub-Å electron beams. *Ultramicroscopy*, **78**, 1–11.

64. Shmueli, U. (ed.) (2001) *International Tables for Crystallography*, Reciprocal Space, Vol. B, Kluwer Academic Publishers, London.
65. Williams, D.B. and C.B., Carter (eds) (1996) *Transmission Electron Microscopy*, Plenum Press, New York.
66. Milne, J.L.S. and Subramaniam, S. (2009) Cryo-electron tomography of bacteria:progress, challenges and future prospects. *Nat. Rev. Microbiol.*, **7**, 666–675.
67. Spence, J.C.H. (ed.) (2003) *High-Resolution Electron Microscopy*, Oxford University Press, New York.
68. Mugnaioli, E., Gorelik, T., and Kolb, U. (2009) "Ab initio" structure solution from electron diffraction data obtained by a combination of automated diffraction tomography and precession technique. *Ultramicroscopy*, **109**, 758–765.
69. Kübel, C., Mio, M.J., Moore, J.S., and Martin, D.C. (2002) Molecular packing and morphology of oligo-(m-phenylene ethynylene) foldamers. *J. Am. Chem. Soc.*, **124** (29), 8605–8610.
70. Wilson, P.M. and Martin, D.C. (1992) Dislocation mediated lattice bending in 1,6-Di(N-Carbazolyl) 2,4-Hexadiyne. *J. Mater. Res.*, **7** (11), 3150–3158.
71. Fryer, J.R. and Smith, D.J. (1982) High resolution electron microscopy of molecular crystals I, quaterrylene, $C_{40}H_{20}$. *Proc. R. Soc. Lond., A Math. Phys. Sci.*, **381**, 225–240.
72. Bravais, A. (ed.) (1913) *Etudes Crystallographiques*, Academie de Sciences, Paris.
73. Donnay, J.D.H. and Harker, D. (1937) A new law of crystal morphology extending the law of Bravais. *Am. Miner.*, **22**, 446–467.
74. Swarbrick, J. (ed.) (2007) *Encyclopedia of Pharmaceutical Technology*, vol. **4**, Informa Healthcare.
75. Dorset, D.L. (ed.) (1995) *Structural Electron Crystallography*, Plenum, New York.
76. Martin, D.C., Chen, J., Yang, J., Drummy, L.F., and Kübel, C. (2005) High resolution electron microscopy of ordered polymers and organic molecular crystals: recent developments and future possibilities. *J. Polym. Sci., B Polym. Phys.*, **43**, 1749–1778.
77. Tivol, W.F., Briegel, A., and Jensen, G.J. (2008) An improved cryogen for plunge freezing. *Microsc. Microanal.*, **14** (5), 375–379.
78. Frederik, P.M. and Sommerdijk, N. (2005) Spatial and temporal resolution in cryo-electron microscopy – a scope for nano-chemistry. *Curr. Opin. Colloid. Interface Sci.*, **10**, 245–249.
79. Michler, G.H. (ed.) (2008) *Transmission Electron Microscopy on Polymers*, Springer-Verlag, Berlin, Heidelberg.
80. Krappe, U. (1995) "Supramolecular helices" and "mesoscopic knitting motives" PhD thesis, Johannes Gutenberg-University, Mainz.
81. Dykstra, M.J. and Reuss, L.E. (1995) *Biological Electron Microscopy: Theory, Techniques, and Troubleshooting*, Plenum Press.
82. Fryer, J.R., McConnell, C.H., Zemlin, F., and Dorset, D.L. (1992) Effect of temperature on radiation damage to aromatic organic molecules. *Ultramicroscopy*, **40**, 163–169.
83. Kolb, U., Gorelik, T.E., Mugnaioli, E., and Stewart, A. (2010) Structural Characterization of Organics Using Manual and Automated Electron Diffraction. *Polym. Rev.*, **50**, 385.
84. Kumar, S. and Adams, W.W. (1990) Electron beam damage in high temperature polymers. *Polymer*, **31**, 15–19.
85. Fryer, J.R. and Holland, F. (1983) The reduction of radiation damage in the electron microscope. *Ultramicroscopy*, **11**, 67–70.
86. Fryer, J.R. (1984) Radiation damage in organic crystalline films. *Ultramicroscopy*, **14**, 227–236.
87. Rossmenn, M.G. and Arnold, E. (2001) *International Tables for Crystallography*, Crystallography of Biological Macromolecules, Vol. **F**, Kluwer Academic Publishers, London.
88. Koshino, M., Kurata, H., and Isoda, S. (2007) Stability due to peripheral halogenation in phthalocyanine complexes. *Microsc. Microanal.*, **13**, 96–107.
89. Bucknall, D.G. and Anderson, H.L. (2003) Polymers get organized. *Science*, **302** (5652), 1904–1905.

90. Stoilova-McPhie, S. and Parmenter, C. (2007) Cryo-electron microscopy of phospholipid vesicles with bound coagulation factorVIII. *Microsc. Anal.*, **21** (6), 9–11.
91. Vennes, M., Zentel, R., Rössle, M., Stepputat, M., and Kolb, U. (2005) Smectic liquid-crystalline colloids by miniemulsion techniques. *Adv. Mater.*, **17**, 2123–2127.
92. Esselink, F.J., Dormidontova, E., and Hadziioannou, G. (1998) Evolution of block copolymer micellar size and structure evidenced with cryo electron microscopy. *Macromolecules*, **31**, 2925–2932.
93. Park, S.-Y., Sul, W.-H., and Chang, Y.-J. (2007) A study on the selectivity of toluene/ethanol mixtures on the micellar and ordered structures of poly(styrene-b-4-vinylpyridine) using small-angle X-ray scattering, generalized indirect fourier transform, and transmission electron microscopy. *Macromolecules*, **40**, 3757–3764.
94. Tu, Y., Grahamb, M.J., Van Horn, R.M., Chen, E., Fan, X., Chen, X., Zhou, Q., Wan, X., Harris, F.W., and Cheng, S.Z.D. (2009) Controlled organization of self-assembled rod-coil block copolymer micelles. *Polymer*, **50**, 5170–5174.
95. Schleuss, T.W., Abbel, R., Gross, M., Schollmeyer, D., Frey, H., Maskos, M., Berger, R., and Kilbinger, A.F.M. (2006) Hockey puck-micelles from oligo(p-benzamide)-b-PEG rod-coil block copolymers. *Angew. Chem.*, **45** (18), 2969–2975.
96. Gorelik, T., Matveeva, G., Kolb, U., Schleuß, T., Kilbinger, A.F.M., van de Streek, J., Bohled, A., and Brunklausd, G. (2010) H-bonding schemes of di- and tri-p-benzamides assessed by a combination of electron diffraction, X-ray powder diffraction and solid-state NMR. *CrystEngComm*, **12**, 1824–1832.
97. König, H.M., Gorelik, T., Kolb, U., and Kilbinger, A.F.M. (2007) Supramolecular PEG-co-Oligo(p-benzamide)s Prepared on a Peptide Synthesizer. *J. Am. Chem. Soc.*, **129**, 704–708.
98. Weyland, M. and Midgley, P.A. (2004) Electron tomography. *Mater. Today*, **7**, 32–40.
99. Kohjiya, S., Katoh, A., Shimanuki, J., Hasegawa, T., and Ikeda, Y. (2005) Three-dimensional nano-structure of in situ silica in natural rubber as revealed by 3D-TEM/electron tomography. *Polymer*, **46**, 4440–4446.
100. Loos, J., Sourty, E., Lu, K., Freitag, B., Tang, D., and Wall, D. (2009) Electron tomography on micrometer thick specimens with nanometer resolution. *Nano Lett.*, **9**, 1704–1708.
101. Aoyama, K., Takagi, T., and Miyazawa, A. (2008) Stem tomography for thick biological specimens. *Ultramicroscopy*, **109**, 70–80.
102. Cheng, Y., Boll, W., Kirchhausen, T., Harrison, S.C., and Walz1, T. (2007) Cryo-electron tomography of clathrin-coated vesicles: structural implications for coat assembly. *J. Mol. Biol.*, **365**, 892–899.
103. Gomez, E.D., Ruegg, M.L., Minor, A.M., Kisielowski, C., Downing, K.H., Glaeser, R.M., and Balsara, N.P. (2008) Interfacial concentration profiles of rubbery polyolefin lamellae determined by quantitative electron microscopy. *Macromolecules*, **41**, 156–162.
104. Pierron, J., Boudet, A., Sopena, P., Mitov, M., and Sixou, P. (1995) Cholesteric textures observed by transmission electron microscopy in diffraction contrast. *Liq. Cryst.*, **19** (2), 257–267.
105. Mazerolles, L., Folch, S., and Colomban, P. (1999) Study of polyanilines by high-resolution electron microscopy. *Macromolecules*, **32**, 8504–8508.
106. Tosaka, M., Kamijo, T., Tsuji, M., Kohjiya, S., Ogawa, T., Isoda, S., and Kobayashi, T. (2000) High-resolution transmission electron microscopy of crystal transformation in solution-grown lamellae of isotactic polybutene-1. *Macromolecules*, **33**, 9666–9672.
107. Brinkmann, M. and Rannou, P. (2009) Molecular weight dependence of chain packing and semicrystalline structure in oriented films of regioregular

poly(3-hexylthiophene) revealed by high-resolution transmission electron microscopy. *Macromolecules*, **42**, 1125–1130.
108. Kübel, C., Lawrence, D.P., and Martin, D.C. (2001) Super-helically twisted strands of poly(m-phenylene isophthalamide) (MPDI). *Macromolecules*, **34**, 9053–9058.
109. Kobayashi, T., Fujiyoshi, Y., Iwatsu, F., and Uyeda, N. (1981) High-resolution tem images of zinc phthalocyanine polymorphs in thin films. *Acta Crystallogr.*, **A 37**, 692–697.
110. Kobayashi, T., Ogawa, T., Moriguchi, S., Suga, T., Yoshida, K., Kurata, H., and Isoda, S. (2003) Inhomogeneous substitution of polyhalogenated copper phthalocyanine studied by high-resolution imaging and electron crystallography. *J. Electron Microsc. (Tokyo)*, **52** (1), 85–90.
111. Liao, J. and Martin, D.C. (1999) Dynamic transmission electron microscopy of the [1,6-di(N-carbazolyl)-2,4 hexadyine] diacetylene monomer-polymer phase transition. *Philos. Mag.*, **79** (6), 1507.
112. Jiang, L., Georgievaa, D., and Abrahamsa, J.P. (2011) EDIFF: a program for automated unit-cell determination and indexing of electron diffraction data. *J. Appl. Cryst.*, **44**, 1132–1136.
113. Vainshtein, B.K. (1964) *Structure Analysis by Electron Diffraction*, Pergamon Press, New York.
114. Zou, X., Hovmöller, A., and Hovmöller, S. (2004) TRICE – A program for reconstructing 3D reciprocal space and determining unit-cell parameters. *Ultramicroscopy*, **98**, 187–193.
115. Zou, X., Sukharev, Y., and Hovmöller, S. (1993) Quantitative electron diffraction – new features in the program system ELD. *Ultramicroscopy*, **52**, 436–444.
116. Dorset, D. and Gilmore, C. (2000) Prospects for kinematical least-squares refinement in polymer electron crystallography. *Acta Crystallogr.*, **A 56**, 62–67.
117. Gorelik, T., Schmidt, M.U., Brüning, J., Bekö, S., and Kolb, U. (2009) Using electron diffraction to solve the crystal structure of a laked azo pigment. *Cryst. Growth Des.*, **9** (9), 3898–3903.
118. Bernstein, J. (ed.) (2002) *Polymorphism in Molecular Crystals*, Clarendon Press, Oxford.
119. Casas, M.T., Gestí, S., and Puiggalí, J. (2005) Structural data on regular poly(ester amide)s derived from even diols, glycine, and terephthalic acid. *Cryst. Growth Des.*, **5** (3), 1099–1107.
120. Tashiro, K., Kamae, T., Asanaga, H., and Oikawa, T. (2004) Structural analysis of polyoxymethylene whisker single crystal by the electron diffraction method. *Macromolecules*, **37**, 826–830.
121. Dorset, D.L., McCourt, M.P., Gao, E., and Voigt-Martin, I.G. (1998) Electron crystallography of small organic molecules: criteria for data collection and strategies for structure solution. *J. Appl. Crystallogr.*, **31**, 544–553.
122. Ritcey, A.M., Brisson, J., and Prud'homme, R.E. (1992) Electron diffraction of optically active polyesters. *Macromolecules*, **25**, 2705–2712.
123. Almontassir, A., Gestí, S., Franco, L., and Puiggalí, J. (2004) Molecular packing of polyesters derived from 1,4-Butanediol and even aliphatic dicarboxylic acids. *Macromolecules*, **37**, 5300–5309.
124. Paredes, N., Casas, M.T., and Puiggali, J. (2000) Packing of sequential poly(ester amide)s derived from diols, dicarboxylic acids, and amino acids. *Macromolecules*, **33**, 9090–9097.
125. Lieser, G., Oda, M., Miteva, T., Meisel, A., Nothofer, H.-G., Scherf, U., and Neher, D. (2000) Ordering, graphoepitaxial orientation, and conformation of a polyfluorene derivative of the "Hairy-Rod" type on an oriented substrate of polyimide. *Macromolecules*, **33**, 4490–4495.
126. Voigt-Martin, I.G., Zhang, Z.X., Kolb, U., and Gilmore, C. (1997) The use of maximum entropy statistics combined with simulation methods to determine the structure of 4-dimethylamino-3-cyanobiphenyl. *Ultramicroscopy*, **68**, 43–59.
127. Yu, R.C., Yakimansky, A.V., Kothe, H., Voigt-Martin, I.G., Schollmeyer,

D., Jansen, J., Zandbergen, H., and Tenkovtsev, A.V. (2000) Strategies for structure solution and refinement of small organic molecules from electron diffraction data and limitations of the simulation approach. *Acta Crystallogr., A*, **56**, 436–450.

128. Voigt-Martin, I.G., Yan, D.H., Yakimansky, A., Schollmeyer, D., Gilmore, C.J., and Bricogne, G. (1995) Structure determination by electron crystallography using both maximum-entropy and simulation approaches. *Acta Crystallogr., A*, **51**, 849–868.

129. Putz, H., Schoen, J.C., and Jansen, M. (1999) Combined method for 'ab initio' structure solution from powder diffraction data. *J. Appl. Crystallogr.*, **32**, 864–870.

130. Brinkmann, M. (2007) Directional epitaxial crystallization and tentative crystal structure of poly(9,9-di-n-octyl-2,7-fluorene). *Macromolecules*, **40**, 7532–7541.

131. Kayunkid, N., Uttiya, S., and Brinkmann, M. (2010) Structural model of regioregular poly(3-hexylthiophene) obtained by electron diffraction analysis. *Macromolecules*, **43**, 4961–4967.

132. Karpov, V.L. (1941) Electron diffraction from crystals of retene and β-methylanthracene. *Zh. Fiz. Khim.*, **15**, 577–591.

133. Mazeau, K., William, J., Winter, T., and Chanzy, H. (1994) Molecular and crystal structure of a high-temperature polymorph of chitosan from electron diffraction data. *Macromolecules*, **27**, 7606–7612.

134. Dorset, D.L. and Snyder, R.G. (1999) Effect of chain unsaturation on polymethylene lamellar packing: solid-state structure of a symmetric trans-Alkene. *Macromolecules*, **32**, 8139–8144.

135. Dorset, D.L. (2010) Direct methods and refinement in electron and X-ray crystallography – diketopiperazine revisited. *Z. Kristallogr.*, **225** (2–3), 86–93.

136. Ogawa, T., Kuwamoto, K., Isoda, S., Kobayashi, T., and Karl, N. (1999) 3,4:9,10-Perylenetetracarboxylic dianhydride (PTCDA) by electron crystallography. *Acta Crystallogr., B* **55**, 123–130.

137. Kolb, U., Gorelik, T., Kübel, C., Otten, M.T., and Hubert, D. (2007) Towards automated diffraction tomography: Part I – Data acquisition. *Ultramicroscopy*, **107**, 507–513.

138. Kolb, U., Gorelik, T., and Otten, M.T. (2008) Towards automated diffraction tomography. Part II – Cell parameter determination. *Ultramicroscopy*, **108**, 763–772.

139. Own, C.S. (2005) System design and verification of the precession electron diffraction technique. PhD thesis, Northwestern University, Evanston, IL, http://www.numis.northwestern.edu/Research/Current/precessions. Accessed 2007.

140. Palatinus, L., Klementová, M., Dřínek, V., Jarošová, M., and Petříček, V. (2011) An incommensurately modulated structure of eta '-phase of Cu(3+x)Si determined by quantitative electron diffraction tomography. *Inorg. Chem.*, **50**, 3743–3751.

141. Gorelik, T.E., Stewart, A.A., and Kolb, U. (2011) Structure solution with automated electron diffraction tomography data: different instrumental approaches. *J. Microsc.* **244**, 325–331.

142. Gemmi, M., Campostrini, I., Demartin, F., Gorelik, T., and Gramaccioli, C.M. (2012) The structure of the new mineral sarrabusite, $Pb_5CuCl_4(SeO_3)_4$ solved by manual electron diffraction tomography. *Acta Crystallogr., B*.

143. Voigt-Martin, I.G., Yan, D.H., Wortmann, R., and Elich, K. (1995) The use of simulation methods to obtain the structure and conformation of 10-cyano-9,9'-bianthryl by electron diffraction and high-resolution imaging. *Ultramicroscopy*, **57**, 29–43.

144. Burla, M.C., Caliandro, R., Camalli, M., Carrozzini, B., Cascarano, G.L., De Carlo, L., Giacovazzo, C., Polidori, G., Siliqi, D., and Spagna, R. (2007) IL MILIONE: a suite of computer programs for crystal structure solution of proteins. *J. Appl. Crystallogr.*, **40**, 609–613.

145. Rozhdestvenskaya, I., Mugnaioli, E., Czank, M., Depmeier, W., Kolb,

U., Reinholdt, A., and Weirich, T. (2010) The structure of charoite, (K,Sr,Ba,Mn)15–16(Ca,Na)32 [(Si70(O,OH)180)](OH,F)4.0 * nH$_2$O, solved by conventional and automated electron diffraction. *Mineral. Mag.*, **74**, 159–177.

146. Birkel, C.S., Mugnaioli, E., Gorelik, T., Kolb, U., Panthöfer, M., and Tremel, W. (2010) Solution synthesis of a new thermoelectric Zn1+xSb nanophase and its structure determination using automated electron diffraction. *J. Appl. Comput. Sci.*, **132**, 9881.

147. Kolb, U., Gorelik, T., and Mugnaioli, E. (2009) Automated diffraction tomography combined with electron precession: a new tool for ab initio nanostructure analysis. *Mater. Res. Soc. Symp. Proc.*, Vol. 1184, pp. 1184-GG01-05.

148. Sheldrick, G.M. (1997) *SHELXL97. Program for the Refinement of Crystals Structures*, University of Göttingen, Germany.

149. Sheldrick, G.M. (2008) A short history of shelx. *Acta Crystallogr., A*, **64**, 112–122.